Thermodynamics of Aqueous Systems with Industrial Applications

Thermodynamics of Aqueous Systems with Industrial Applications

Stephen A. Newman, EDITOR
Foster Wheeler Energy Corporation

ASSOCIATE EDITORS
Herbert E. Barner
Max Klein
Stanley I. Sandler

Based on a symposium cosponsored by the American Institute of Chemical Engineers, the National Bureau of Standards, and the National Science Foundation at Airlie House Conference Site, Warrenton, Virginia, October 22–25, 1979.

ACS SYMPOSIUM SERIES 133

AMERICAN CHEMICAL SOCIETY
WASHINGTON, D. C. 1980

Library of Congress CIP Data

Thermodynamics of aqueous systems, with industrial applications.
(ACS symposium series; 133 ISSN 0097-6156)

Includes bibliographies and index.

1. Thermodynamics—Congresses. 2. Chemistry, Technical—Congresses.
I. Newman, Stephen A., 1938- . II. American Institute of Chemical Engineers. III. United States. National Bureau of Standards. IV. United States. National Science Foundation. V. Series: American Chemical Society. ACS symposium series; 133.

QD504.T47 660.2'969 80-16044
ISBN 0-8412-0569-8 ASCMC 8 133 1–771 1980

PRINTED IN THE UNITED STATES OF AMERICA

ACS Symposium Series

M. Joan Comstock, *Series Editor*

FOREWORD

The ACS SYMPOSIUM SERIES was founded in 1974 to provide a medium for publishing symposia quickly in book form. The format of the Series parallels that of the continuing ADVANCES IN CHEMISTRY SERIES except that in order to save time the papers are not typeset but are reproduced as they are submitted by the authors in camera-ready form. Papers are reviewed under the supervision of the Editors with the assistance of the Series Advisory Board and are selected to maintain the integrity of the symposia; however, verbatim reproductions of previously published papers are not accepted. Both reviews and reports of research are acceptable since symposia may embrace both types of presentation.

CONTENTS

THERMODYNAMICS OF SYNTHESIS GAS AND RELATED SYSTEMS

PROPERTIES OF AQUEOUS SOLUTIONS—THEORY, EXPERIMENT, AND PREDICTION

PREFACE

This volume contains most of the papers presented at a conference on The Thermodynamics of Aqueous Systems with Industrial Applications, held October 22–25, 1979 at Airlie House, Warrenton, Virginia. The conference, cosponsored by the American Institute of Chemical Engineers, the National Bureau of Standards, and the National Science Foundation, was organized by the following members of the AIChE Subcommittee on Thermodynamics (Research Committee): Stephen A. Newman, Herbert E. Barner, Stanley S. Grossel, Michael G. Kesler, Max Klein, and Stanley I. Sandler.

The conference was subdivided into four sessions, and chapters within this text are arranged according to these categories. Papers included in the first section (Thermodynamics of Electrolytes for Pollution Control) provide the reader with insights into the practical aspects of pollution control, as well as an overall appreciation of applied electrolyte phase equilibria. Other chapters include detailed descriptions of thermodynamic models that recently have been developed to describe important industrial pollution control processes with emphasis on acid gas absorption/sour water stripping and flue gas desulfurization.

The increasing importance of coal gasification, liquefaction, and shale oil processing has focused attention on a new class of thermodynamic problems. In the second section (Thermodynamics of Synthesis Gas and Related Systems) new data and calculation methods appropriate to water-containing synthesis gas systems are emphasized. As background to this work, chapters describing industrially important processes and cooperative research organizations dealing with emerging synthesis gas technology have been included.

Essentially all of the engineering thermodynamic correlations used in pollution control models and synthesis gas phase equilibria, chemical equilibria, and enthalpy calculation schemes have their foundations in fundamental theory. Experimental data, in addition to being directly useful to designers, allows the correlation developer to assess the validity and suitability of his model. Included within the third section (Properties of Aqueous Solutions—Theory, Experiment, and Prediction) are chapters providing both comprehensive reviews and detailed descriptions of specific areas of concern in the theory and properties of aqueous solutions.

The increasing interest in hydrometallurgical processing stems from an increased awareness of the necessity of environmental protection and pollution. The final section (Hydrometallurgy, Oceanography, and Geology) includes chapters highlighting theory and applications of electrolyte thermodynamics to hydrometallurgy and related concepts important in oceanography and geology.

Each session was initiated by state-of-the-art reviews, summarizing both the theoretical and applied aspects of the subject; these were followed by invited technical papers. A panel discussion among speakers/coauthors, including audience participation, concluded each session. This volume contains only the technical papers of the conference; the questions and answers for the papers will be printed as a separate publication.

One of the major goals of our conference was to bring together outstanding investigators in diverse areas to discuss topics of mutual interest. Elaboration on this point is made in the Conference Overview. The constituency of our audience (40% academic, 40% industrial, and 20% government) was indeed varied and consisted of both theoretical and applied workers. Our speakers spoke on topics ranging from the most recent concepts of electrolyte phase equilibria theory to the principles and applications of coal processing. Throughout the presentations it was apparent that a common thread linked each theme and that both the practical and theoretical aspects of the subject were important, each providing the other with sustenance.

We are confident that in addition to fostering the presentation of many superb technical papers, which we anticipate will be used as reference material for many years to come, the conference provided a means for many investigators to become acquainted with colleagues working in related fields. Hopefully, these contacts will be maintained and flourish.

We were fortunate in having several guest speakers and we acknowledge their important contributions to our conference. Specifically, we thank Larry Resen (AIChE), Emanuel Horowitz (NBS), Davis Hubbard (NSF), Donald Ehreth (EPA), and Bernard Lee (IGT). Additionally, we are grateful to J. Charles Forman and Joel Henry of AIChE and Marshall Lih of NSF for their part in allowing the conference to grow from a concept into a reality. A special thanks goes to Marie Kennedy of AIChE who played an important part in cheerfully executing many of the day-to-day chores associated with planning a large meeting, including handling the registration duties at Airlie House.

We enjoyed planning the meeting and participating in its evolution and look forward to developing other such rewarding activities.

STEPHEN A. NEWMAN

Foster Wheeler Energy Corporation
110 South Orange Avenue
Livingston, NJ 07039
February 14, 1980

1

Conference Overview

STEPHEN A. NEWMAN

Foster Wheeler Energy Corporation, Livingston, NJ 07039

The theme of the symposium has relevance to workers engaged in a wide variety of activities; it is anticipated that the concepts described in the specific papers will be useful in areas superficially different from each other, yet in a fundamental manner, indeed very similar. The invited conference papers have been organized into reasonably related subject matter, yet run the gamut from the very practical to the very theoretical with the intent of providing the theorist an appreciation of the practical problems facing the technologist, and the technologist, an awareness of the theoretical tools he can use to solve his problems.

Several applied areas of concern, all involved with water, served as focal points for the meeting. Two of these, pollution control and coal processing, have current social significance and are subjects that everyone can readily appreciate. Practical insights into the fields were given in the industrially oriented review papers, and the role of thermodynamics in contributing to the solution of the many specific problems encountered in carrying out an engineering design were brought out in other papers. As support to the applied thermodynamic techniques described in the application-oriented sessions, one session was devoted to theory. It is this fundamental work that enables the thermodynamic practitioner, be he either academic or industrial, to develop in a rational matter his correlations which are absorbed into the design techniques applied by process engineers. A very fine example was provided by the extensive use of Professor Pitzer's electrolyte activity coefficient theory within several acid gas phase equilibrium models.

Another important group of papers provided insight into several very successful cooperative industrial programs, those sponsored by the Gas Research Institute, the American Petroleum Institute, the Gas Processors Association and the American Institute of Chemical Engineers. The designation infratechnology has been applied to those aspects of technical investigation and knowledge that permeate a given industry, yet are not specific

0-8412-0569-8/80/47-133-001$05.00/0

enough to individual constituents to justify extensive investigation on their own. Frequently, the problems are so complex and diffuse that it would be essentially useless for a single company to embark on an investigation that would take many man-years of investment with very little monetary rate-of-return. Certain segments of the chemical engineering industry have found a solution to such problems by consolidating interests in cooperative research organizations such as those sponsored by API, GPA, GRI and AIChE. The growth of these industrial cooperative groups is encouraging; besides conserving resources and eliminating duplication of effort, these groups provide a forum for exchange of ideas and non-proprietary technology. The involvement of academia in these organizations, usually as research contractors, allows university researchers an intimate insight into real-world problems and helps close the breach between academia and industry that has become so prevalent in recent years. All of these factors substantially increase output per unit input and by its very definition promote an increase in productivity. One very important element missing from the scene until recently has been government. Government's involvement has obviously been increasing and will become a welcome addition by providing additional cohesiveness as well as financial backing and technical expertise to fledgling groups.

As an introduction to the technical aspects of the conference, the results of some studies conducted by the writer on two relevant subjects are presented below. The first commentary is concerned with the design of sour-water strippers and the effects of thermodynamic data on these designs; the second commentary is concerned with the calculation of enthalpies of steam-containing mixtures, essential to the design of coal processing and related plants.

Sour-water Thermodynamic Data

For the past several years there has been considerable interest in the phase equilibria of sour-water systems. This interest largely stems from the need to design sour-water strippers so that their liquid effluents attain the very low levels mandated by current environmental regulations. The EPA liquid emission level goals, based on ecological effects, are ten PPM (mass) for ammonia and two PPM (mass) for hydrogen sulfide. The guidelines for tower design established many years ago (for example, 0.06 kgs stripping steam per liter of hot feed) are outmoded and are not exact enough in an era when sophisticated data and calculation methods can be applied to the problem once considered so complicated that only empirical rules were accepted practice, for lack of anything better.

A number of models have been developed to describe the chemistry occurring in sour-water stripping and absorption.

The many simultaneous chemical reactions whose solution results in the establishment of the phase equilibria of the components are listed in Figure 1. This system is in several respects considerably more complicated than many hydrocarbon and chemical systems in that it is necessary to contend with both chemical and phase equilibria in establishing the design. Each of the cited reactions and Henry's law constants, besides being temperature dependent, are also functions of the type and concentration of acid-gas components in solution and, in some cases, the ionic strength of the solution. In a practical sense, the most important reason for considering these electrolyte reactions is that only those portions of ammonia, hydrogen sulfide, and carbon dioxide that remain in molecular, rather than ionic form, can be stripped from solution. The object of the calculations, therefore, is the determination of the ionic and molecular distribution of species as a function of temperature,pressure,and composition.

The designer can employ a computer program embodying the calculations to carry out his design, or he can use charts such as those shown in Figure 2 as a basis for his stripping or absorption calculations. The two charts shown can be used either to calculate the ammonia and hydrogen sulfide concentrations in the liquid-phase from a knowledge of their partial pressures in the gas phase; or, alternately, the partial pressures can be directly read from the charts given the solution concentration of the dissolved gases. Calculating the liquid-phase composition from that of the gas requires a trial-and-error procedure. A liquid-phase molar ratio of the acid gases is assumed, and from the known partial pressures, a liquid-phase composition is read from the graphs and a calculated liquid-phase molar ratio is obtained. This calculation is repeated two or three times until the calculated liquid-phase molar ratio equals the assumed value, and at this point the equilibrium liquid-phase composition has been determined.

For most sour water stripper design work, a computer is used to perform the calculations. Several of the proposed sour-water modules were incorporated into a tower program and a series of designs on a typical sour-water stripper have been undertaken.

Comparisons of data methods with each other such as those shown in Figure 3 for the Wilson (1) and Mason-Kao (2) methods demonstrate that the Wilson method generates the smaller ammonia and hydrogen sulfide partial pressures. In practical terms, Wilson's predictions imply a greater amount of stripping steam to achieve a desired level of acid gas removal when compared to the Mason-Kac predictions. Unfortunately, such comparisons are only semi-quantitative from a designer's viewpoint and it is nenessary to employ the full strength of a rigorous tower computer simulation to demonstrate the significance that the data differences imply in terms of equipment specification and utility comsumption. Some insight is provided in the comparisons depicted in Figures 4 and 5.

SOUR-WATER FEED

Qc

Q_R

PPM NH_3, H_2S, CO_2 = ?

SUMMARY OF CHEMICAL EQUILIBRIA INVOLVED IN CALCULATING $NH_3-H_2S-CO_2-H_2O$ VLE

$$CO_2 + H_2O \overset{K_1}{\rightleftharpoons} HCO_3^- + H^+$$

$$HCO_3^- \overset{K_2}{\rightleftharpoons} CO_3^= + H^+$$

$$NH_3 + H^+ \overset{K_3}{\rightleftharpoons} NH_4^+$$

$$NH_3 + HCO_3^- \overset{K_4}{\rightleftharpoons} H_2NCO_2^- + H_2O$$

$$H_2S \overset{K_5}{\rightleftharpoons} HS^- + H^+$$

$$HS^- \overset{K_6}{\rightleftharpoons} S^= + H^+$$

$$H_2O \overset{K_7}{\rightleftharpoons} H^+ + OH^-$$

$\sum$ OF ALL ELECTRONIC CHARGES $\equiv 0$

$\bar{P}_{NH_3} = H_{NH_3}\ NH_3$; $\bar{P}_{H_2S} = H_{H_2S}\ H_2S$

$\bar{P}_{CO_2} = H_{CO_2}\ CO_2$

IPC Science and Technology

Figure 1. Sour-water absorber/stripper VLE (4)

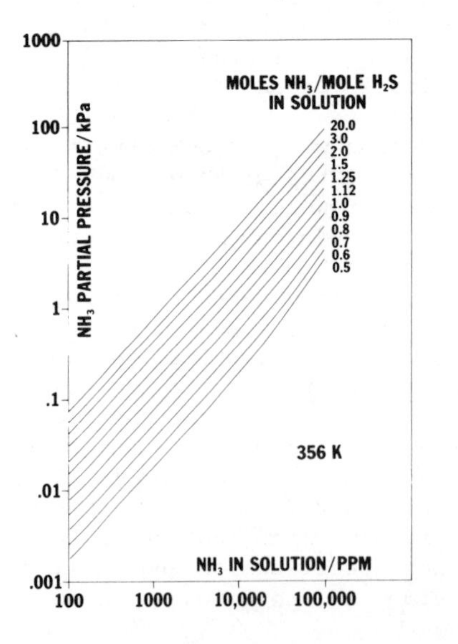

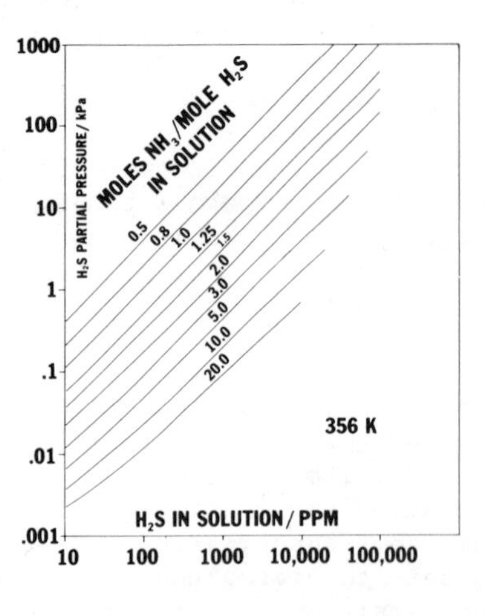

Figure 2. Typical design data charts for NH_3–H_2S–H_2O VLE

The PPM ammonia remaining in the bottoms product stream of the stripper tower are shown in Figure 4 as a function of the kilograms of steam injected into the tower. The number of theoretical stages used is shown as a cross-parameter. What is observed is that in a practical sense, the Wilson and Mason-Kao methods yield essentially the same ammonia purity in the stripped water product, whereas very substantial differences are obtained when the classical van Krevelen correlation is applied to design the wastewater stripper.

Using the amount of hydrogen sulfide remaining in the tower bottoms as the criterion, Figure 5 shows similar design comparisons. Large differences between the methods are observed; however, the Mason-Kao predictions and van Krevelen results are very similar. The Wilson method generates a more conservative design in that to achieve a designated level of hydrogen sulfide purity a designer would specify a greater number of trays, or alternately, provide for more stripping steam.

Enthalpies of Steam-containing Mixtures

For those engaged in designing plants for processing syngas streams, consisting of mixtures of hydrogen, carbon monoxide, carbon dioxide, nitrogen and methane, there is a constant need to calculate the enthalpies of mixtures of these gases with steam. These data are required in the design of reactors, heat exchangers and separation equipment. Unfortunately, despite the prevalence of such systems in design work, until very recently, there have been no experimental calorimetric data with which to assess the validity of proposed design correlations. With the increased interest in substitute gas processes in which enthalpies for mixtures containing as much as 50 percent water, at high pressures, are required, efforts to obtain experimental enthalpy data for these systems have been initiated, and several measurement programs are underway.

Some data comparisons were made of several predictive methods with steam mixture enthalpy data obtained by Professor Wormald (3). To provide a basis of comparison, Figure 6 illustrates how three methods currently in vogue in the thermodynamics world perform in predicting the enthalpy departures from ideality of methane. The predictions of the Lee-Kesler equation-of-state seem to best replicate the data, with a maximum error of 1.2 kJ/kg.

A portion of the enthalpy-temperature diagram for steam is presented in Figure 7. Frequently, in performing heat balance calculations for syngas processing, it is necessary to develop enthalpies for steam in mixtures at conditions that do not conform to pure component conditions. Design engineers frequently develop their steam enthalpies from the pure component data by either using saturation values or by extrapolating into the pure component dome region from the linear portion of the supercritical isobars. Some advocates of this latter procedure

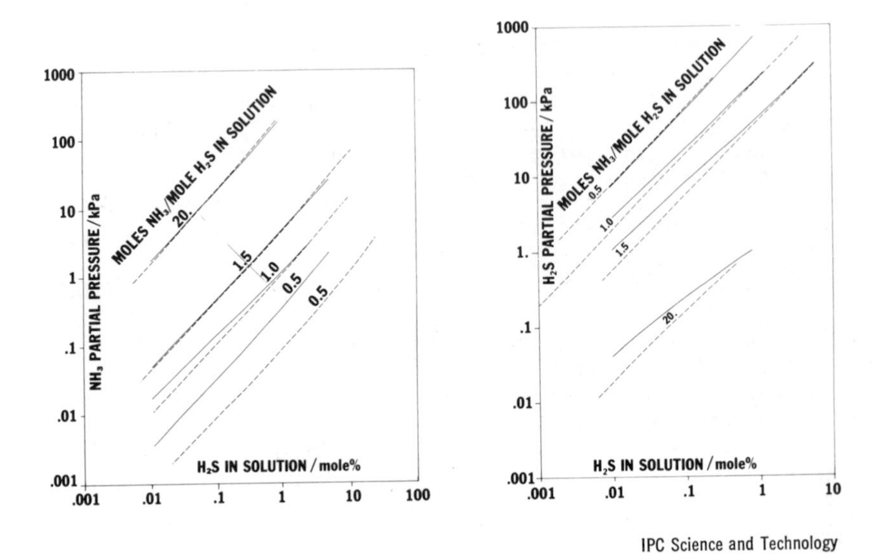

IPC Science and Technology

Figure 3. Comparison of methods for predicting NH_3 and H_2S partial pressures in NH_3–H_2S–H_2O system (methods compared at 356 K; (——) Mason–Kao; (– – –) Wilson) (4)

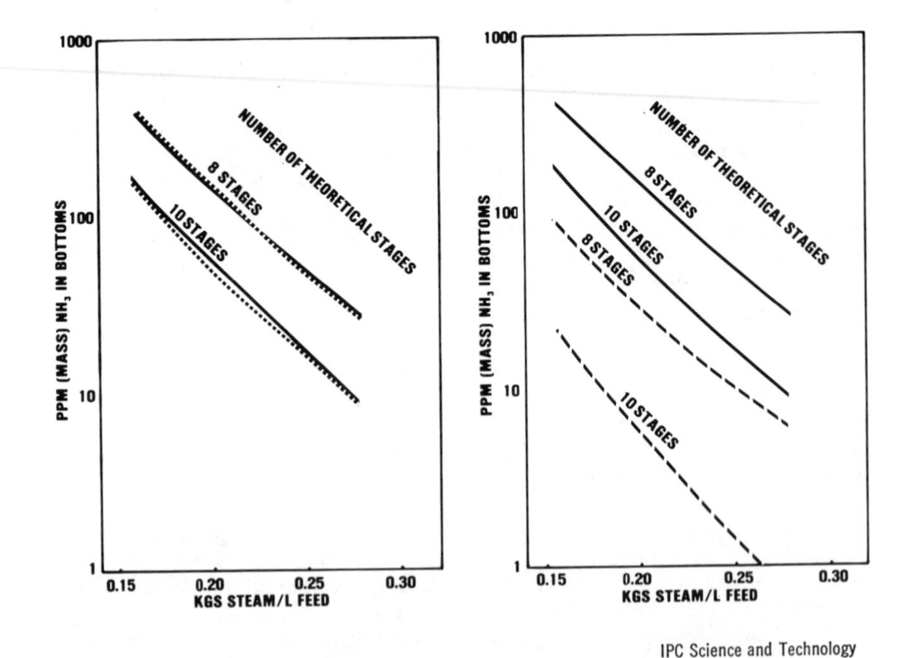

IPC Science and Technology

Figure 4. Predicted PPM NH_3 in tower bottoms ((——) Wilson; (– – –) Mason–Kao; (— — —) van Krevelen) (4)

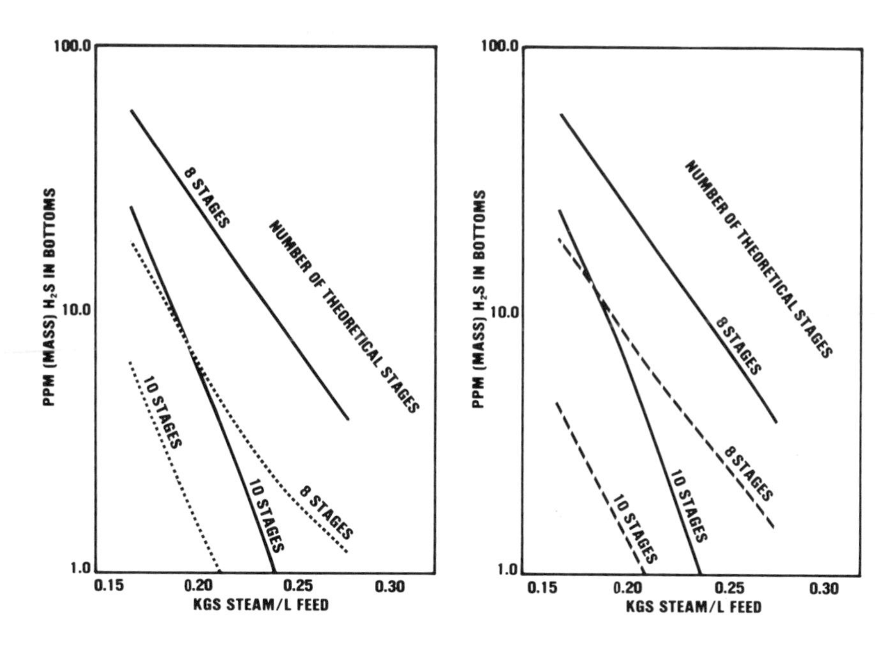

Figure 5. Predicted PPM H_2S in tower bottoms ((——) Wilson; (– – –) Mason–Kao; (— — —) van Krevelen)

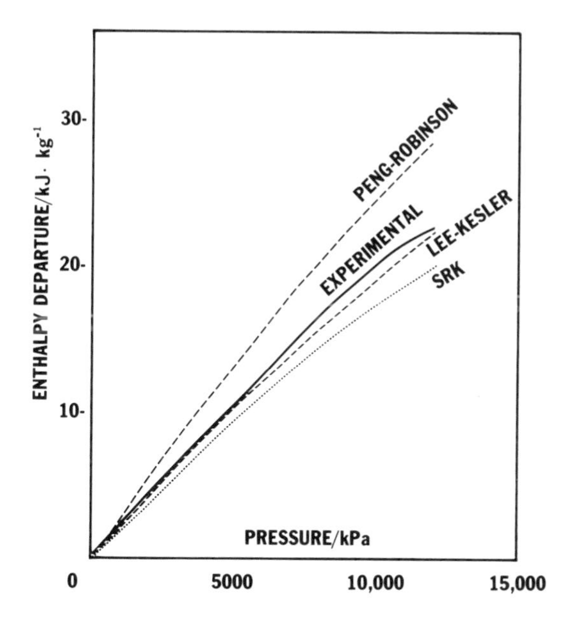

Figure 6. Comparisons of predicted and experimental enthalpy departures for pure methane at 598 K

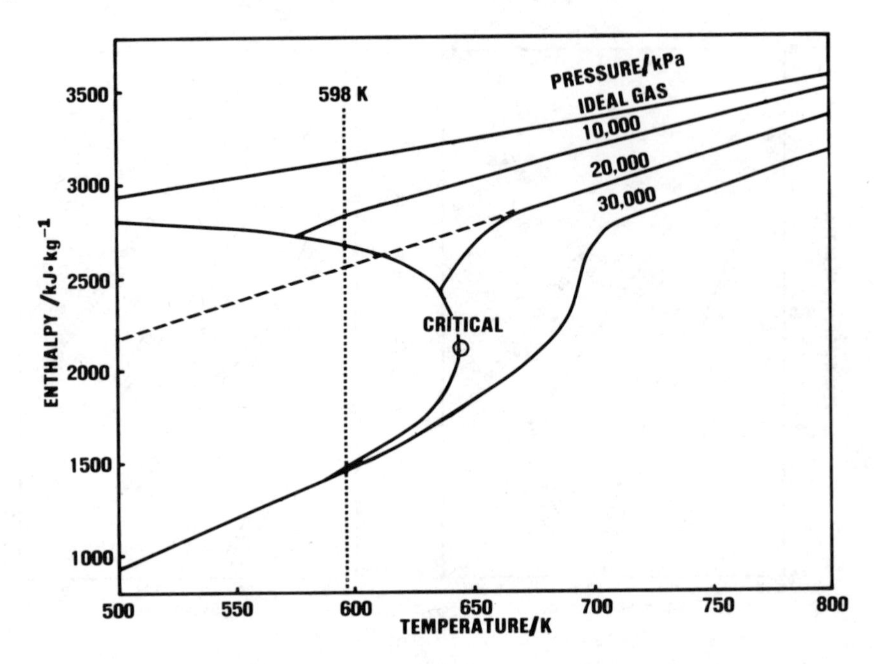

Figure 7. Enthalpy diagram for steam

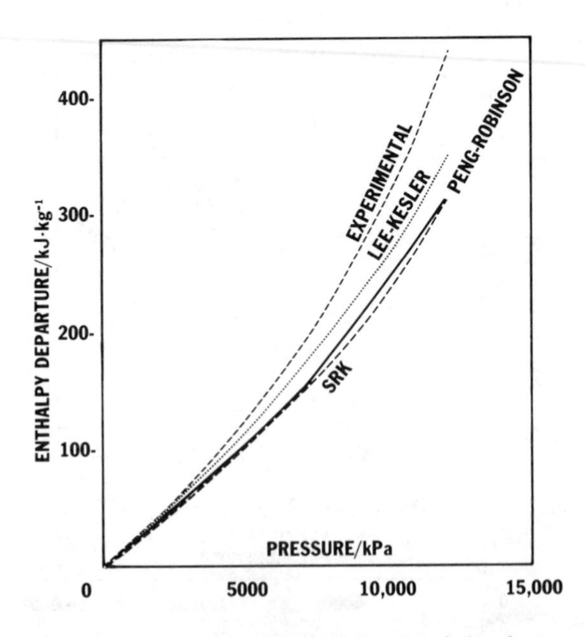

Figure 8. Comparisons of predicted and experimental enthalpy departures for pure steam at 598 K

use enthalpy values at the partial pressure of steam, others, at the total system pressure. In the enthalpy diagram, the temperature at which the only published steam mixture calorimetric data were obtained is indicated. The pressure range of the data spans from ideal gas to saturation pressure, about 12,000 kPa.

Figure 8 depicts how the three popular equation-of-state methods cited previously perform on pure steam. From a theoretical viewpoint, none of the methods has the foundation to handle mixtures of polar/non-polar components. Although the agreement with experimental data is not very satisfactory for any of the methods, the Lee-Kesler equation-of-state does best. It was also found that by slightly adjusting the acentric factor of water, improvement in the representation of the enthalpy of steam can be obtained by this method at 598 K, the conditions of the experimental mixture data, and at other temperatures as well.

Figure 9 provides a comparison of the predictions of empirical methods with Wormald's data for a 50/50 mole percent mixture of steam and methane. As can be seen, the frequently used artifices of calculating mixture enthalpies by blending the pure component enthalpies at either total or partial pressures are very inaccurate. Likewise, the assumption of ideal gas enthalpy for the real gas mixture, equivalent to a zero enthalpy departure on the diagram, is an equally poor method.

In Figure 10 are shown comparisons of the equation of state methods with the experimental data. The Lee-Kesler methods represent the data the best. Again, if the water acentric factor determined to best represent the pure steam enthalpy data is applied to the mixtures, further improvement is noted for the predictions by the Lee-Kesler method. Use of interaction constants within the Lee-Kesler, or other models, would undoubtedly provide even better representation of the data.

Although comparisons for the steam-methane system have been presented, similar trends were noted for the other binary systems previously published by Wormald, namely mixtures of steam with nitrogen, carbon dioxide, n-hexane, and benzene.

Professor Wormald, in a paper presented at this conference, has provided additional steam mixture enthalpy data and some correlation work he has done on the data using an association model.

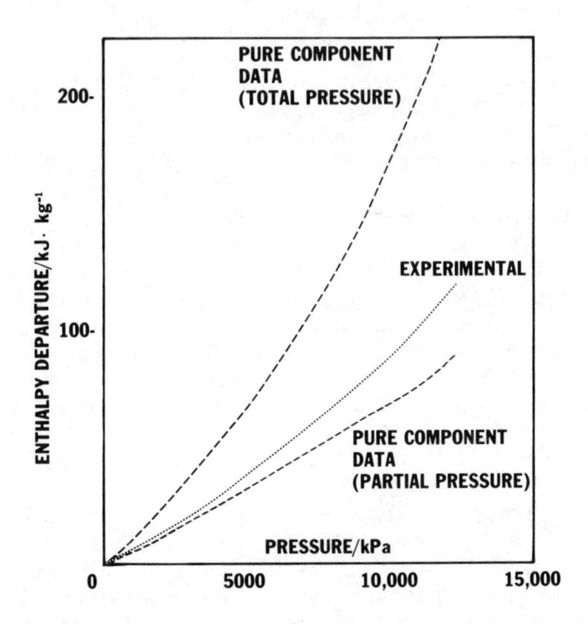

Figure 9. Comparisons of predicted and experimental enthalpy departures for an equimolar steam–methane mixture at 598 K (equation-of-state methods)

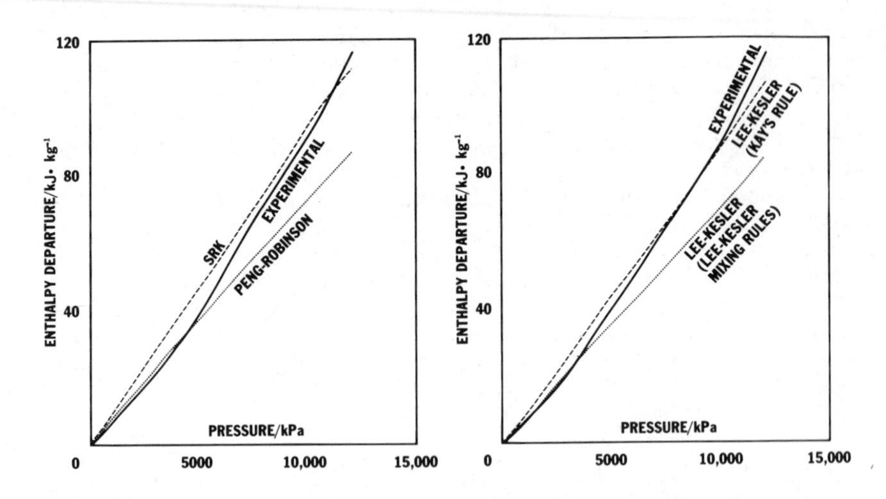

Figure 10. Comparisons of predicted and experimental enthalpy departures for an equimolar steam–methane mixture at 598 K (empirical methods)

Literature Cited

1. Wilson, G.M., "A New Correlation of NH_3, CO_2 and H_2S Volatility Data from Aqueous Sour Water Systems." Final Report to API Committee on Refinery Environmental Control under EPA Grant No. R804364010, Thermochemical Institute, BYU, Provo, Utah, February 9, 1978.

2. Mason, D., "Vapor-Liquid Equilibria in the NH_3-CO_2-H_2S-H_2O System", Project 8979, "Preparation of a Coal Conversion Systems Technical Data Book," Quarterly Reports by Institute of Gas Technology for U.S.Department of Energy Contract EX-76-C-01-2286, February and May, 1978.

3. Wormald, C.J., "Thermodynamic Properties of Some Mixtures Containing Steam." Paper presented at National Physical Laboratory Conference on Chemical Thermodynamic Data on Fluids and Fluid Mixtures, Teddington, Middlesex, U.K., September 11-12, 1978. Conference Proceedings published by IPC Science and Technology Press.

4. Newman, S.A., "Novel Applications of Phase Equilibria Methods to Process Design Problems." Paper presented at National Physical Laboratory Conference on Chemical Thermodynamic Data on Fluids and Fluid Mixtures, Teddington, Middlesex, U.K., September 11-12, 1978. Conference Proceedings published by IPC Science and Technology Press.

5. Lee, B.-I. and Kesler, M.G., AIChE Journal, 1975, 12(5), 510.

6. Peng, D.-Y. and Robinson, D.B., Ind. Eng. Chem., Fundam, 1976, 15, 59

7. Soave, G., Chem. Eng. Science, 1972, 27, 1197.

8. van Krevelen, D.W., Hoftijzer, P.J. and Huntjens, F.J., Recueil des Travaux Chimiques des Pays-Bas, 1949, 68, 191.

RECEIVED January 17, 1980.

THERMODYNAMICS OF ELECTROLYTES FOR POLLUTION CONTROL

2

A Survey of Some Industrial Waste Treatment Processes

W. T. ATKINS, W. H. SEWARD, and H. J. TAKACH

Mittelhauser Corporation, Downers Grove, IL 60515

This paper was prepared to give a brief overview of some of the technologies used for industrial pollution control. Among the many areas of possible discussion under this broad a topic, this paper focuses on sulfur control technologies due to the potentially major cost impact of these technologies on the industrial processes with which they are associated. The process areas covered are acid gas removal, sulfur recovery, sulfur dioxide removal, and wastewater treating. In the first three process areas, alternative process types are described and guidelines for process selection are briefly reviewed. Because of the operating difficulties encountered with utility sulfur dioxide removal processes, information on industrial installation of these processes is given. For wastewater treating, a state-of-the-art industrial wastewater treatment system is discussed along with some major items to consider in process selection.

Acid Gas Removal

Acid gas removal is the removal of sulfur compounds and CO_2 (acid gas) from process gas streams. The following sections describe available process alternatives, design options, and guidelines for selection among alternatives.

Process Alternatives. Acid gas removal processes have been extensively surveyed in the published literature (1,8,3). Figure 1 shows how acid-gas-bearing process gases can be generally treated in industrial processes. The sulfur compounds and CO_2 may be absorbed in a liquid medium, such as amines, alkali salts (NaOH, K_2CO_3), physical solvents (methanol, propylene carbonate), or water (3). The absorbed acid gases are released by reduction of pressure and/or by application of heat. Alternatively, the H_2S and CO_2 may chemically combine with the absorbent (as in NaOH scrubbing) to form salts which are removed in a liquid treatment unit. This requires continual and expensive makeup of sodium to the system.

0-8412-0569-8/80/47-133-015$08.25/0

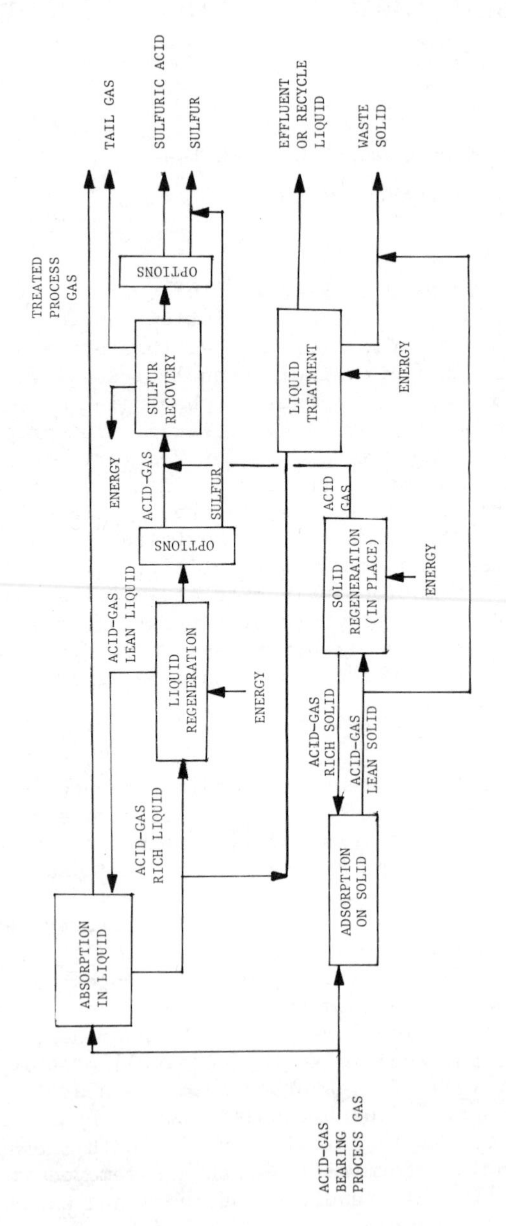

Figure 1. Schematic of acid gas removal process alternatives

Released acid gases in the same form as captured, i.e., H_2S, COS, CO_2, etc. are converted in the sulfur recovery unit to forms in which they may be exported from the industrial facility. Sulfur recovery is discussed later in the paper. Some liquid absorption processes produce two separate acid gas streams (selective AGR). One stream, containing the majority of sulfur compounds, is sent to the sulfur recovery unit, while the other is vented to atmosphere, environmental regulations permitting.

Alkanolamines, generally referred to as amines, are organic compounds of the form $H_n\text{-N-(ROH)}_{3-n}$ (3); the hydroxyl group generally provides for the compounds' solubility in water, while the HN group provides the alkalinity in water solutions to cause the absorption of acid gases. Amine processes used commercially are shown in Table I. These compounds are chemical solvents; they combine chemically with H_2S, CO_2, and other sulfur compounds. They are customarily regenerated by the application of heat.

Alkali salt solutions are aqueous solutions of sodium or potassium salts. They, too, are chemical solvents, reacting with H_2S and CO_2 as follows (e.g., for K_2CO_3):

$$K_2CO_3 + CO_2 + H_2O \rightarrow 2KHCO_3$$

$$K_2CO_3 + H_2S \rightarrow KHS + KHCO_3$$

This solution is regenerated by pressure letdown and steam stripping of the solution (2).

The physical solvents shown in Table I operate by dissolving the acid gases in the absorbing medium at elevated pressures and low temperatures. Regeneration of the solvent is principally by reduction of pressure, although heating is often necessary in high-efficiency applications, where H_2S is to be removed to a few ppmv (3).

Mixed solvents are combinations of physical and chemical solvents which increase the flexibility of treating (1). The chemical solvent allows for treatment of lower-pressure streams while the physical solvent allows for bulk removal of the acid gas.

Absorption-oxidation processes oxidize absorbed H_2S directly to elemental sulfur in solution (1). The principal example in current industrial use is the Stretford process (3). The chemistry of the process can be represented by the following idealized equations (ADA represents anthraquinone disulfonic acid):

$$Na_2CO_3 + H_2S \rightarrow NaHS + NaHCO_3$$

$$4\ NaVO_3 + 2\ NaHS + H_2O \rightarrow Na_2V_4O_9 + 2S + 4\ NaOH$$

$$Na_2V_4O_9 + 2\ NaOH + H_2O + 2\ ADA \rightarrow 4\ NaVO_3 + 2\ ADA\ (reduced)$$

$$2\ ADA\ (reduced) + O_2 \rightarrow 2\ ADA + H_2O$$

Table I. Acid Gas Removal Processes

Principal Absorbent	Process Name	Typical Absorber Temperature, °F	Absorber Pressure, psi	Components Removed		
				H_2S	Organic Sulfur	CO_2
AMINE PROCESSES						
MEA	Monoethanolamine	80-110	Atm-1000; not highly sensitive	Yes, down to 1 ppm level	COS, CS_2, some mercaptan	Yes, down to ppm level
DEA	Diethanolamine	100-130	Atm-1000; not highly sensitive	Yes, down to 4 ppm	COS, CS_2 absorbed reversibly (some mercaptans)	Yes, down to ppm level
DEA	Societe National des Petrols d'Aquitaine SNPA-DEA	90-120	500-1100; high acid-gas pressure desirable (<50 psi)	Yes, down to 4 ppm	Some COS removed	Yes, <1%
MDEA	Methyldiethanolamine	80-120	Not highly pressure sensitive	Yes, down to 4 ppmv (but not selectively)	Mercaptans not removed	Yes, normally 25-75% removal
DIPA	Shell ADIP	90-130	Not highly sensitive	Yes, <10 ppmv	Partial removal of COS (some mercaptans)	Yes, normally 30-90% removal
DGA	Fluor Econamine diglycolamine	90-130	Not highly sensitive	Yes, <4 ppmv	COS, CS_2 degrades solution	Yes, below that normally provided by MEA
ALKALI SALT PROCESSES						
K_2CO_3	Hot Pot, Potassium Carbonate	Amb-240	100-1000	Yes, to 4 ppmv (CO_2 must be present also)	COS partially hydrolyzed to CO_2 + H_2S and removed	Yes, to 1.5% - single stage <0.8% - two stage
K_2CO_3 (activated)	Benfield and HiPure	Amb-280; nominally 230	100-2000	Yes, to 4 ppmv Benfield, <1 ppm HiPure	Can remove 100% COS, 75% CS_2, 70-80% mercaptans	Yes, to 500 ppmv Benfield, <50 ppm HiPure
K_2CO_3 (activated)	Catacarb	Amb-280; nominal bulk temp 230	Nominally 100-1000	Yes, <4 ppmv	COS, CS_2	Yes, <0.05%
K_2CO_3 + AsO_3	Giammarco Vetrocoke CO_2	120-220	200-1000	Removed in Vetrocoke H_2S process	Minor amounts only	Yes, to 0.05%
K+weak organic acid	Alkazid "M"	70-100	Atm-1000	Yes, to 5 ppmv	Unknown	Yes
K+weak organic acid	Alkazid DIK	70-100	Atm-1000	Yes, to 5 ppmv (but non-selective)	No	Yes
NaOH	Caustic Scrubing	60-100	100-250	Yes, to <1 ppm	?	Yes

H_2S/CO_2 Selectivity	Relative Solvent Cost and Process Complexity	Commercial Status	References
Not selective	Low-cost solvent medium complexity	Commercial - used on N.G. and refinery gases (excluding COS)	2,3,5
Not selective	Slightly higher solvent cost than than MEA but less complex	Commercial - used on gases which contain COS	1,2,3,5
Not selective	High-pressure absorption may require intermediate flashing	Commercial - developed for high-pressure natural gas treatment	1,2,3,5
Partially selective	Much higher solvent cost than MEA	Commercial - limited application to date	1,2,3,5
Partially selective	Much higher solvent cost than MEA	Commercial - used for refinery gases which include COS	1,2,3,5
Not selective	Higher solvent cost than DIPA, as complex as MEA	Commercial	1,2,3,5
Can be selective	Low-cost solvent; simple operation for bulk CO_2	Commercial for CO_2 $+H_2S$ high-pressure bulk removal	1,2,3,5
Selectivity increases complexity and/or residual impurities	Staged operation and cooling for high purity	Commercial high acid-gas pressure; bulk removal favored	1,2,3,4,5
Selectivity increases complexity and/or residual impurities	Staged operation and cooling for high purity	Commercial high acid-gas pressure; bulk removal favored	1,2,3,5
Designed for CO_2 removal after H_2S is removed	Simple operation	Commercial overseas - used to remove CO_2 after separate H_2S removal	1,2,5
Not selective	As complex as amine systems	Commercial overseas	2,5
Less selective than K_3PO_4 (tripotassium phosphate)	As complex as amine systems	Commercial overseas	2,5
Can be selective	Simple operation, very high solvent cost	Commercial - used for fuel cleanup	19

Table I. Continued

Principal Absorbent	Process Name	Typical Absorber Temperature, °F	Absorber Pressure, psi	Components Removed: H_2S	Components Removed: Organic Sulfur	Components Removed: CO_2
PHYSICAL SOLVENTS						
Methanol	Lurgi Rectisol	Down to -100, normally -40 to -70	300-2000	Yes, down to 0.02 ppmv	Yes, COS <0.1 ppmv	Yes, down to 1 ppmv, normally to 0.1%
N-methyl 2-pyrrolidone	Lurgi Purisol	80-105	∿1000	Yes, <4 ppmv attainable	Yes, but process not designed for such	Yes, <0.1%
dimethylether polyethylene glycol	Allied Chemical Selexol	20-100	300-1000	Yes, <1 ppmv	Yes, <1 ppmv	Yes, normally to 1%
propylene carbonate	Fluor Solvent	20-80	400-2000	Yes, <4 ppmv	Yes	Yes, <1%
MIXED SOLVENTS						
Sulfolane plus diisopropanolamine	Shell Sulfinol	110	Atm-1000; higher pressures favored	Yes, <4 ppmv	COS, mercaptans, total sulfur <16 ppmv	Yes, 50 ppmv possible normal 0.3%
MEOH + MEA or DEA	Lurgi Amisol	95	Higher pressures favored (440 psia operation reported)	Yes, 0.1 ppmv	Yes, COS to 0.1 ppmv	Yes, below 5 ppmv possible
ABSORPTION-OXIDATION PROCESS						
Na_2CO_3 + anthraquinone disulfonic acid and sodium meta vanadate	Stretford	70-110	Atm-300	Yes, <10 ppmv	Light mercaptan removed but COS/CS_2 not affected	HCN-essentially complete removal
Na_2CO_3 + plus napthoquinone sulfonic acid	Takahax	Ambient	Atmospheric	Yes	Unknown	HCN
Na_2CO_3 + AsO_3	Giammarco-Vetrocoke H_2S	Amb-300	Atmospheric	Yes, <0.5 ppmv	Unknown	HCN
SOLID-BED PROCESSES						
FeO	Appleby-Frodingham Hot Ferric Oxide	600-750	Normally low	Yes, 98% removal by two stages, >99% in four stages	COS	
Zno	Appleby-Frodingham Hot Ferric Oxide	600-850	atm-1000	Yes, to below measurement threshold	Some COS	
Molecular Sieves	Appleby-Frodingham Hot Ferric Oxide	∿100	0-1000	Yes	Produced in sieves	CO_2
Activated Carbon	Appleby-Frodingham Hot Ferric Oxide	<140	?	Yes	Yes, in two-stage unit	

H_2S/CO_2 Selectivity	Relative Solvent Cost and Process Complexity	Commercial Status	References
Can be selective in staged operation	Low solvent cost but extremely complex process	Commercial - on coal and oil gasification processes	1,3,4,5
More selective than MeOH	Selectivity or high purity require high complexity	Commercial - high acid-gas pressure, bulk removal favored	5
Can be selective, with stages and recycle	High solvent cost, moderate complexity for bulk removal	Commercial - designed for high acid-gas pressures, bulk removal	1,2,3,4,5
Can be selective (multistage)	Flash regeneration when only bulk removal required	Commercial - designed for high acid-gas pressures, bulk removal	1,3,5
Not very selective	High solvent cost, high complexity	Commercial - high acid-gas pressure favored	1,3,5
Not operated selectively	Low solvent cost, less complex than Rectisol	Commercial - for gas from partial oxidation of oil	1,3
Completely selective but CO_2 decreases absorber efficiency		Commercial - designed for H_2S removal from coal gas	1,3,5
Selective		Commercial - designed for H_2S removal from coke-oven gas	3
Selective		Commercial overseas	3,5
Selective		Commercialized for coke oven gas (obsolete)	1,3
Selective		Commercial	21
Can be selective		Commercial for dehydration, H_2S, CO_2 removal from natural gas	1
Selective		Commercial	3

Water is often used as a medium for removing particulates from process gas streams (3). In the process, significant quantities of CO_2 and H_2S may be removed particularly if NH_3 is present in the process gas. Because water is not generally used commercially for the express purpose of acid gas removal, it will not be discussed further.

Liquid membrane processes for removing H_2S from process gases are potentially attractive because they may require less energy than conventional techniques. Research is now going on to develop these technologies, but they have not yet achieved commercial application.

Molecular sieves are crystalline metal aluminosilicates (1). Openings in their crystal structure permit passage of many gas constituents while preferentially adsorbing large, polar, or unsaturated compounds. Acid gas compounds may be adsorbed by certain types of molecular sieves. When used for H_2S removal, the sieve is regenerated by a thermal swing cycle (1), being heated to release the H_2S for downstream sulfur recovery.

Iron oxide is one of the oldest media used for H_2S removal from process gas. Its use is limited by the necessity of periodically replacing the solid adsorbent (3), as the adsorbent cannot be regenerated completely and gradually loses its effectiveness. Hot-gas purification processes using iron oxide include the Appleby-Frodingham Process and the METC Fixed-Bed Process (1). In these processes the iron oxide removes the H_2S at elevated temperatures (700-1000°F) and is also regenerated at these temperatures. None of these or other hot-gas purification processes is commercial today.

Activated carbon, used commercially to remove small quantities of H_2S from synthesis gas (3), is useful as a follow-on unit to a bulk-removal process. Operation is cyclic, with periodic regeneration of the carbon by steam or inert gas.

Process Selection. Table I presents general performance data to guide process selection of acid gas removal facilities. Several surveys have recently appeared in the literature (4,5). The information in Table I and these surveys should be used only as a rough screening tool for very preliminary studies. Specific applications should be discussed with the licensors or builders of such systems.

The selection of an AGR removal process should be guided by the following guidelines:

Solvent type should be compatible with the processing pressure, selectivity desired, and potential contaminants (organic sulfur, NH_3, HCN, hydrocarbons) expected in the feed gas.

Selectivity - solvents generally have different affinities for H_2S and CO_2. Some processes can remove essentially all of the H_2S and many of the other sulfur compounds while leaving most of the CO_2 in the treated gas, potentially reducing the cost of the downstream sulfur recovery facilities (4). However, such selective removal processes may be more costly than nonselective processes. In petroleum refineries, the CO_2 content of the gases to be treated is usually low enough that nonselective amines may be used without unduly diluting the Claus plant feed with CO_2; this approach would not work in a high-BTU coal gasification plant where the ratio of CO_2 to H_2S is much higher.

Extent of removal required by downstream processing will also dictate the choice of process. Complete removal of H_2S to prevent poisoning of downstream catalysts may require both a liquid absorption step and a solid-bed trace sulfur removal step, such as zinc oxide.

Processing pressure - generally, the carrying capacity of physical solvents increases with absorption pressure much more rapidly than that of chemical solvents (4). Therefore, AGR at high-pressures may make physical solvents look attractive, while low-pressure operation favors chemical solvents.

Economics is the dominant factor in process selection by industrial users. In weighing these economics, capital investments for alternative processes must be obtained on a consistent basis, and operating costs must be calculated using realistic unit costs for such items as electricity and process steam. Perhaps most importantly, the potential markets for the ultimate products from the plant must be accurately assessed.

Commercial status is another major criterion. Industry generally tends to be suspicious of processes which have not been proved in commercial application. Therefore, processes with multiple installations in the same service as the currently proposed application, of the same scale or larger, will generally be preferred, even if at a somewhat higher projected cost, than processes which do not meet these criteria.

Sulfur Recovery

Sulfur recovery processes convert acid gases containing H_2S and other sulfur compounds to elemental sulfur and sulfuric acid. Table II is a summary of many of the available sulfur recovery processes.

Process Alternatives.

Process alternatives for sulfur recovery are shown schematically in Figure 2. The choice of either elemental sulfur or sulfuric acid will depend on economics and markets related to each plant location. Elemental sulfur may be produced by gas-phase oxidation (the Claus process) or liquid-phase oxidation (e.g., the Stretford process). Stretford units were described in Section 1 and are well discussed in the literature (1,2,5). Claus sulfur recovery efficiency is usually less than required by current air emission standards. Therefore, some form of tail-gas treating is required. Sulfuric acid may be produced by the well-known contact process (6). This process is licensed by a number of firms, each of which has its own

Table II. Sulfur Recovery Processes

Process	U.S. Licensors	Required Feed Sulfur Content	Process Performance
Claus Process	Parsons, Shell, Amoco Production Co., through its licensees such as Ford, Bacon & Davis, Ortloff, Olsen, and many others	Normally 15-20% H_2S or more, with product sulfur-burning, can handle H_2S contents down to 5%	92-97% removal with 3 catalytic stages. COS CS_2 not converted without provisions in design
Tail-Gas Treating Processes			
Low-Efficiency Processes			
Sulfreen	American Lurgi	Claus tail gas. Sulfur compounds 1-3%	Raises overall sulfur recovery to 99%. 2000-3000 ppmv sulfur in treated gas. No COS/CS_2 conversion
SNPA/Haldor-Topsoe	Haldor-Topsoe	Claus tail gas	About 500 ppmv sulfur in treated gas. All sulfur compounds are handled
Cold-Bed Adsorption	Amoco Production Company	Claus tail gas	Raises overall sulfur recovery to 99-99.3%, <1500 ppmv sulfur in treated gas. No COS/CS_2 conversion
IFP	IFP	Claus tail gas or other H_2S-containing feeds, up to 20%	<1500 ppmv sulfur in treated gas. No COS/CS_2 conversion
Beavon Mark II	Parsons	H_2S content of <5% or Claus tail gas	Up to 99% overall sulfur recovery. Some COS and CS_2 conversion
High-Efficiency Processes			
Beavon Mark I	Parsons	Claus tail gas	99.8% overall sulfur recovery, <300 ppmv sulfur compounds in treated gas, generally <10 ppmv H_2S. Some COS/CS_2 conversion
SCOT	Shell Development Co.	Claus tail gas	99.8% overall sulfur recovery if COS/CS_2 content is not excessive. <300 ppmv sulfur compounds in treated gas. H_2S is about 150-200 ppmv
Trencor	Trentham Corp.	Claus tail gas	Same as SCOT
Wellman-Lord	Davy Powergas	Claus tail gas	<200 ppmv sulfur compounds in treated gas. All sulfur compounds are handled
Contact Sulfuric Acid Process	Monsanto, Parsons, Davy Powergas, others	Can accept elemental sulfur, or H_2S and SO_2-bearing streams down to about 5% sulfur content	A double contact/double absorption plant can recover up to 99.8% of the sulfur fed to it. All sulfur compounds handled

Sulfur Removal Mechanism	Commercial Status	References
Claus reaction: $2H_2S+SO_2 \rightarrow 3S+2H_2O$ one thermal stage, multiple catalytic stages. Several commercial catalysts are available	Fully commercial in natural gas and refinery applications. Feed impurities in gases from synfuel plants can create problems	2,3,7,10,16
Extended Claus reaction below sulfur dewpoint over alumina catalyst	11 commercial units now operating	3,9,10
Independent conversion by catalytic oxidation of sulfur compounds to SO_3, followed by absorption to produce 94% to H_2SO_4	1 commercial unit at Lacq, France	9,10
Extended Claus reaction below sulfur dewpoint over Claus catalyst at 260°F	1 commercial unit at East Crossfield, Alberta, Canada	9
Extended Claus reaction in liquid phase, with polyethylene glycol as liquid carrier for catalyst	Over 20 commercial units now in operation	9,10
Independent catalytic conversion of sulfur compounds to H_2S, followed by successive oxidation and Claus reaction over proprietary Parsons/Union Oil Catalyst	1 commercial unit operating in Germany on Claus tail gas	13
Independent catalytic conversion of sulfur compounds to H_2S, followed by direct oxidation of H_2S to sulfur in Stretford unit	19 commercial units now operating or under construction	3,8,10,13
Recycle process with catalytic conversion of sulfur compounds to H_2S followed by absorption with di-isopropanol amine. H_2S and CO_2 from amine regenerator recycled to Claus plant	Several commercial units now in operation	3,8,10
Same as SCOT, except amine used is MDEA	1 commercial unit has been constructed; 99.8% recovery not yet demonstrated	11
Recycle process with oxidation of all sulfur species to SO_2, followed by absorption in sodium sulfite solution. Evaporation of solution releases SO_2 which is recycled to Claus plant. Up to 10% of entering sulfur may be purged as sulfate solution	18 commercial installations on boiler flue gas; 7 on Claus tail gas	8,10,12
Thermal oxidation of sulfur compounds to SO_2, followed by catalytic oxidation to SO_3. Absorption of SO_3 by concentrated sulfuric acid creates net product acid	Many commercial units in operation on sulfur feed, or smelter gases containing SO_2. Process not yet commercially demonstrated for 5% H_2S feed	6,14,15

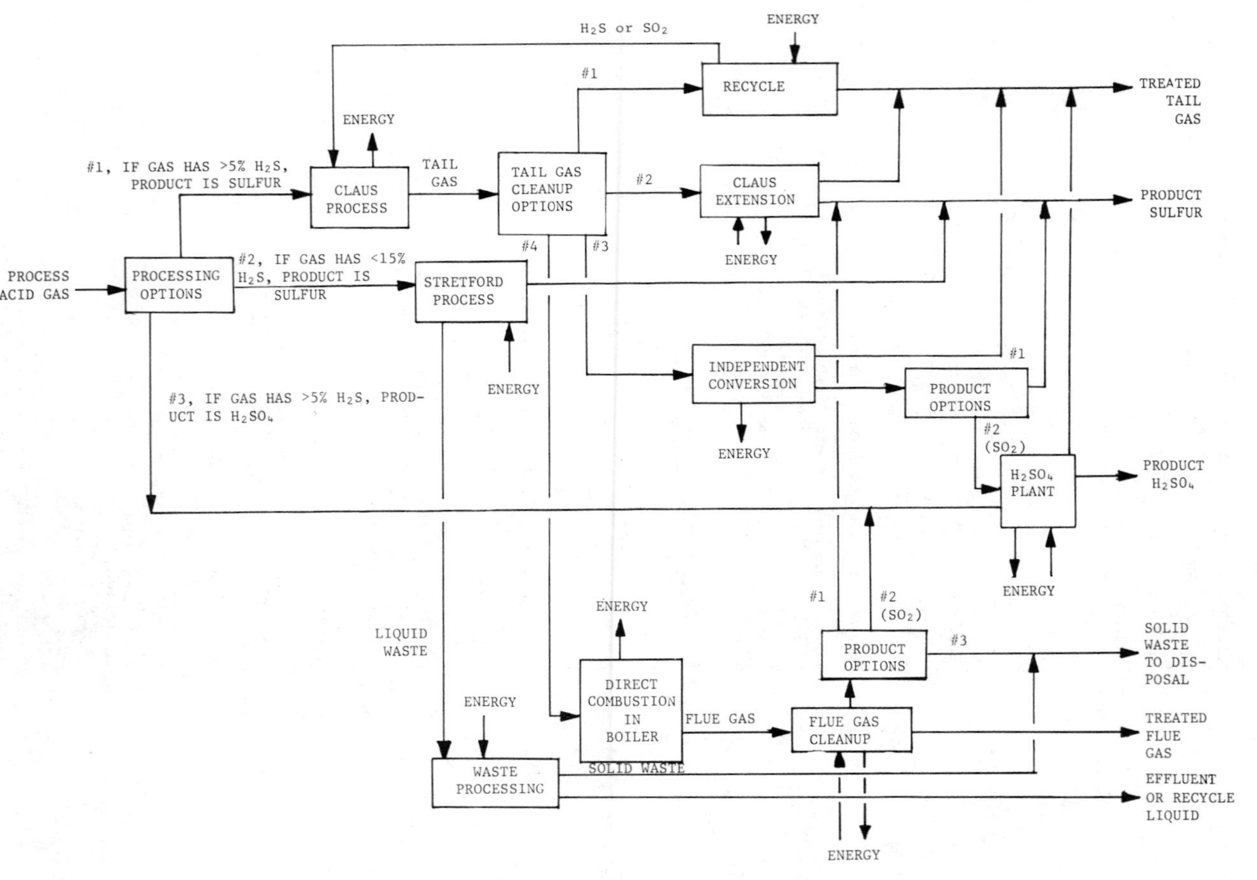

Figure 2. Schematic of sulfur recovery process alternatives

variations on the basic process. The remaining paragraphs describe some of the currently available sulfur recovery processes.

The main processing steps in the Claus process are thermal conversion, sulfur condensation, reheat, and catalytic conversion. In the thermal conversion step, one-third of the feed H_2S is oxidized by air:

$$H_2S + \frac{3}{2} O_2 \rightarrow SO_2 + H_2O \text{ (exothermic)}$$

The desired reaction temperature is in the 2000-2400° range. At this temperature, the direct formation of sulfur vapor occurs:

$$2H_2S + SO_2 \rightarrow \frac{3}{2} S_2 \text{ (vapor)} + 2H_2O \text{ (endothermic)}.$$

This thermal conversion is generally between 50 and 80% of the total reactant sulfur, being lower for feeds more dilute in H_2S. Many other side reactions occur, as discussed in the literature.

The sulfur vapor produced in the thermal conversion step is partially condensed at about 350°F by raising steam. After condensation of liquid sulfur, the vapor is reheated to approximately 400-450°F (7) before entering the catalytic reactors. There are several methods of reheating the vapor, including hot gas exchange, in-line burners, hot gas bypass, or indirect steam reheat. Each has advantages and disadvantages; generally indirect steam reheat is most expensive but yields higher recoveries.

The reheated vapor is passed through a fixed bed of catalyst where the Claus reaction takes place:

$$2H_2S + SO_2 \rightarrow \frac{3}{X} S_x + 2H_2O \text{ (exothermic)}$$

After each catalytic conversion, the condensation/reheat/conversion sequence is repeated. The number of cycles is dictated by the tradeoff between desired recovery and economics. Usually, a fourth catalytic stage is not justified.

Tail gas cleanup is required because a well-designed Claus plant with three catalytic stages and fresh catalyst will recover only 95-97% of its feed sulfur (8), which is not generally sufficient to meet current emission standards. In addition, feed impurities and catalyst aging will reduce overall recovery in some plants to about 92% just before catalyst changeout. Therefore, tail-gas cleanup is required. Tail-gas treating processes are generally classified as follows:

1. Extensions of the Claus reaction
2. Recycle processes
3. Independent conversion

Extended Claus processes maintain the 2:1 ratio of HS to SO_2 in the tail gas. The reaction is extended at low temperatures, below the dewpoint of sulfur. Some processes of this type are the Sulfreen, IFP Clauspol, and CBA processes (9).

Recycle processes convert the sulfur species in the tail gas to a single compound which is captured, concentrated, and returned to the Claus plant. Examples are the SCOT (10), Trencor (11), and Wellman-Lord (8,12) processes. The first two recover and recycle H_2S while the Wellman-Lord process recovers and recycles SO_2.

Independent conversion processes may not employ the Claus reaction for sulfur production and do not recycle the captured sulfur compounds to the Claus plant. Examples are the Beavon Mark I Process (Hydrogenation + Stretford) (13), the Beavon Mark II Process (Hydrogenation + Claus) (13), and the SNPA/Haldor-Topsöe Process (Catalytic Oxidation to H_2SO_4) (9,10).

Sulfuric acid production is one alternative to the manufacture of elemental sulfur from acid gas streams. If a market for the product acid can be found, sulfuric acid may be economically attractive relative to elemental sulfur (14,15).

The feed stream containing H_2S is incinerated with air to oxidize all sulfur compounds to SO_2. The heat of combustion is partially recovered by raiding medium-pressure steam (400-600 psig) in a waste heat boiler. The gas is washed and demisted, then dried with concentrated H_2SO_4. The gas is then compressed, preheated, and enters a fixed-bed catalytic reactor. In the reactor SO_2 is converted to SO_3:

$$SO_2 + \frac{1}{2} O_2 \xrightarrow{V_2O_5} SO_3$$

In successive steps, the reactor outlet gas is cooled, enters the next catalytic stage, is again cooled, and flows through an absorber where the SO_3 is absorbed by concentrated H_2SO_4. The absorber overhead is reheated and goes through another 2-stage reaction sequence to final absorption. The absorbed SO_3 is diluted with water to make product strength acid (usually 93 or 98% by weight) which is then exported.

Generally, one of two main types of mist eliminator is used for removal of SO_3 and acid mist (6). The Brink "HV" type, or equivalent is used to remove particles of three microns and larger in size, while for removal of finer particles, the Brink "HE" type, or equal, is used.

Process Selection. Rough guidelines for sulfur recovery process selection are given in Table III. These guidelines are intended only as a rough screening tool for preliminary studies. Individual process licensors may be able to "tailor" these processes to applications outside the range of these guidelines.

Process selection should be made through a "sulfur-management" approach. Sulfur management is the determination of

Table III. Guidelines for Selection—Sulfur Recovery Processes

	H_2S Content in Acid Gas, Vol %				
	>60%	25-60	10-25	5-10	<5
Claus Process					
Straight-through	x				
Split-flow		x	x		
Preconcentration					x
Preheat			x		
Sulfur burning				x	
Stretford				x	x
Tail Gas Processes					
Extended Claus Processes	x	x	x	x	x
Recycle Processes	x	x	x	x	x
Independent Conversion Processes	x	x	x	x	x
Sulfuric Acid	x	x	x	x	
Integrated Processing	x	x	x	x	x

sulfur processing steps in an industrial plant, such that all plant product and effluent streams are environmentally and economically acceptable (18). Such an approach involves many sequential decisions, and development of a decision-tree will aid in making these decisions. The decision-tree approach identifies major studies which must be performed to provide input for the process decisions, and provides for the proper sequencing of these studies (17).

The selection of a sulfur recovery process should mainly be guided by economics and commercial status. Here are some other possible guidelines:

Claus Processing of Low-H_2S Gases - If the H_2S content of the acid gas is below about 60%, either split-flow, preheat, preconcentration or sulfur burning must be used for Claus processing. In a typical split-flow Claus plant (16), two-thirds of the feed is bypassed around the reaction furnace while the remaining one-third is burned with nearly stoichiometric air, thus raising the flame temperature. This method can be used in conjunction with preheat to treat acid gases of down to about 10 volume percent H_2S. Alternatively, the flame temperature in the furnace may be increased by preheat (7,16) of either the acid gas or the air to the furnace or both. Either steam or indirect fuel firing is often used. Preconcentration of the H_2S through scrubbing with a selective amine could be used to permit use of a straight-through Claus plant instead of a split-flow process, with possible rejection of hydrocarbons and impurities picked up by a physical solvent process in the acid-gas removal step, an added benefit. With H_2S contents of between 5 and 10% in the acid gas, sulfur burning has been used (7,16). Some of the product sulfur is recycled and burned in the reaction furnace, raising the flame temperature to desired levels. Another option is available in plants having an oxygen plant (7). Pure oxygen can be used instead of air for the combustion step, avoiding dilution of the combustion products with nitrogen.

Impure Feeds - Impure feeds cause Claus and Stretford plant problems (3,16). The most troublesome impurities are NH_3, HCN, and hydrocarbons. These can cause catalyst fouling, plugging, chemical losses, and unstable operations. Remedies include catalytic conversion of impurities, use of special burners (16), and the use of different AGR processes to reduce the amounts of these compounds in acid gases.

Tail Gas Cleanup Process Efficiency - Required process efficiency depends on applicable emission regulations. Low-efficiency processes result in up to 99.0-99.5% overall sulfur recovery when combined with the Claus plant and include the Sulfreen, SNPA/Haldor-Topsoe, CBA, IFP, and Beavon Mark II processes. High-efficiency tail-gas treating processes can achieve overall sulfur recoveries of 99.8% and above under ideal conditions. These include the Beavon Mark I, SCOT, Trencor, and Wellman-Lord processes.

Sulfuric Acid Options - To meet current standards, sulfuric acid plants must generally be designed as double-contact, double-absorption plants (6), or they must use a tail-gas scrubbing step to generate an alternate product, e.g., ammonium sulfate. Firm markets for alternative products should be carefully investigated and established before a commitment is made to implement such an alternative.

Integrated Processing - In a large industrial complex, a number of separate sulfur removal and conversion activities will be performed. These will include acid-gas removal, flue-gas cleanup, sourwater stripping, and sulfur recovery. It may be economically attractive to integrate two or more of these processing activities. For example, integration of flue gas cleanup with sulfur recovery involves combining the treating of SO_2-containing boiler flue gas and H_2S-containing acid gas in a single unit. This concept has been applied to refinery processing, and coal gasification systems as well (17). The overall performance of an integrated system should be excellent. The treated flue gas should contain no more than 150-200 ppmv of sulfur, while the overall recovery of SO_2 and H_2S in the sulfur recovery or acid plant should be about 99.7%.

Sulfur Dioxide Removal

Sulfur dioxide removal processes can be used to treat flue gas from industrial boilers, heaters, or other process gases where sulfur compounds are oxidized. These processes have generally been proven in utility applications. More recently, several industrial SO_2 removal installations have been completed.

Process Alternatives. Sulfur dioxide removal processes can be categorized as throwaway or recovery. Throwaway processes produce a liquid or solid waste that requires disposal. Recovery processes convert the sulfur dioxide to elemental sulfur or sulfuric acid. Throwaway processes have been used in most utility applications, but there could be greater incentives for using the recovery processes in industry.

Throwaway processes generally remove sulfur dioxide by absorption into a lime or limestone slurry or a clear solution. Figures 3-5 show general diagrams for these processes (22,23,24,25).

The lime and limestone processes, as indicated in Figure 3, produce a sludge consisting mainly of calcium sulfite and calcium sulfate by the following reactions (limestone):

$$SO_2 + CaCO_3 + \frac{1}{2} H_2O \rightarrow CaSO_3 \cdot \frac{1}{2} H_2O + CO_2$$

$$SO_2 + CaCO_3 + 2 H_2O + \frac{1}{2} O_2 \rightarrow CaSO_4 \cdot 2H_2O + CO_2$$

The poor physical and structural properties of this sludge make utilization impractical and disposal costly (26).

Clear solution processes used are either once through or regeneration (double alkali). Once-through processes, as shown in Figure 4, generally use caustic or soda ash in the following reactions. Ammonia could also be used under special circumstances where there is a ready market for ammonium sulfate.

$$NaOH + SO_2 \rightarrow NaHSO_3$$

$$NaOH + SO_2 + \frac{1}{2} O_2 \rightarrow Na_2SO_4 + H_2O$$

$$Na_2CO_3 + SO_2 \rightarrow Na_2SO_3 + CO_2$$

$$Na_2SO_3 + \frac{1}{2} O_2 \rightarrow Na_2SO_4$$

Sometimes the solution used is spent byproducts, like caustic, from nearby process units. Generally, the absorbent is purchased and the solution is made up on site. Once-through processes are used when there is a large process wastewater stream available to dilute the dissolved solids concentration to an acceptable level.

Regenerable processes, as shown in Figure 5, utilize solutions of sodium sulfite or dilute sulfuric acid (Chiyoda Process) to absorb the sulfur dioxide by the following reactions:

$$Na_2SO_3 + SO_2 + H_2O \rightarrow 2NaHSO_3$$

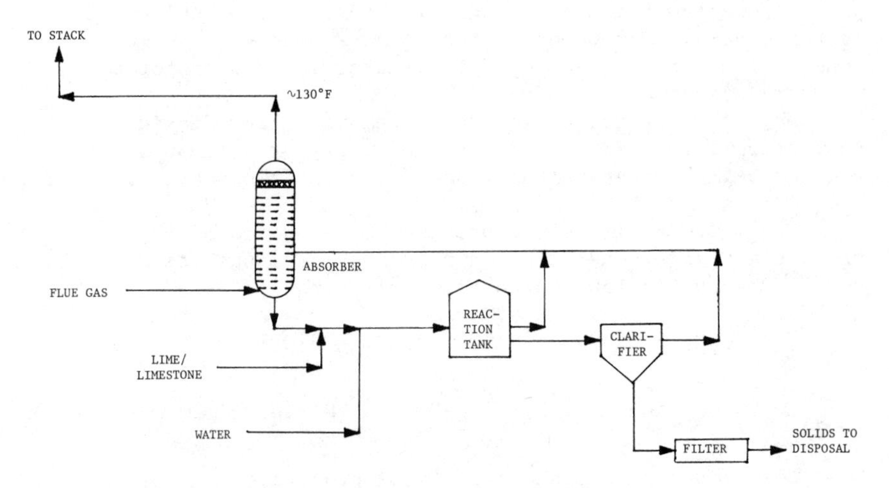

Figure 3. Lime/limestone process

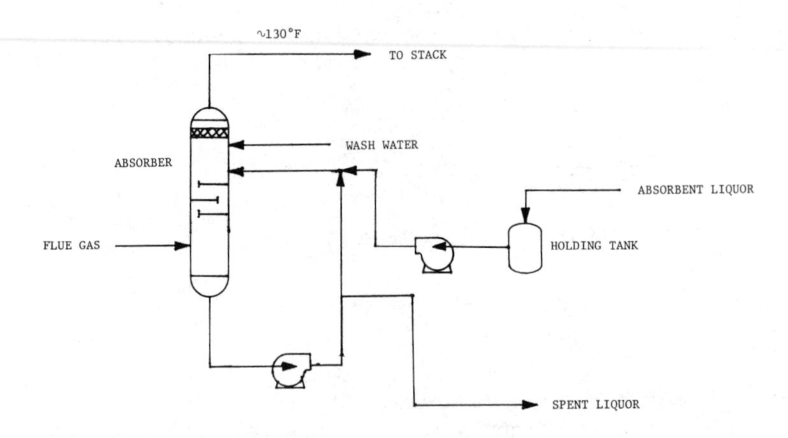

Figure 4. Clear solution once-through process

$$H_2O + SO_2 \rightarrow H_2SO_3$$

$$H_2SO_3 + \frac{1}{2} O_2 \rightarrow H_2SO_4$$

The spent solution is regenerated by mixing it with lime or limestone where the following reactions occur:

$$NaHSO_3 + Ca(OH)_2 \rightarrow Na_2SO_3 + CaSO_3 \cdot \frac{1}{2} H_2O + \frac{3}{2} H_2O$$

$$Na_2SO_3 + Ca(OH)_2 + \frac{1}{2} H_2O \rightarrow 2NaOH + 2CaSO_3 \cdot \frac{1}{2} H_2O$$

$$H_2SO_4 + CaCO_3 \rightarrow CaSO_4 + H_2O + CO_2$$

The calcium sulfite or sulfate solids are allowed to settle from the solution. The regenerated solution is returned to the absorber. The solids are concentrated to around 70%. Because these solids are not a mixture of the sulfite and sulfate, their properties are far superior to lime or limestone process sludge (unless oxidation is used) and disposal should be easier.

Recovery processes generally involve one of the following three concepts:

- Solution absorption followed by release of the absorbed SO_2 (by heat). An example is the Wellman-Lord process shown in Figure 6. The concentrated SO_2 stream is sent to a separate unit for conversion to sulfuric acid or elemental sulfur.
- Solution absorption followed by conversion in solution to elemental sulfur, as in the Citrate process shown in Figure 7.
- Reaction with a solid acceptor followed by regeneration to produce a concentrated stream of SO_2. This is done in the UOP/SFGD process shown in Figure 8. Here the regeneration step is done chemically using a hydrogen-containing reducing gas. The concentrated SO_2 stream can be sent to a separate unit for conversion into sulfuric acid or elemental sulfur.

If sulfuric acid is produced, the SO_2 containing gas does not require fuel gas or reducing gas, except for providing process heat. If elemental sulfur is to be produced, it is generally done by the Claus reaction previously discussed. In that case, two-thirds of the SO_2 is catalytically hydrogenated to SO_2:

$$SO_2 + 3H_2 \rightarrow H_2S + 2H_2O$$

Hydrogen sulfide could be produced by reacting steam, methane, and elemental sulfur.

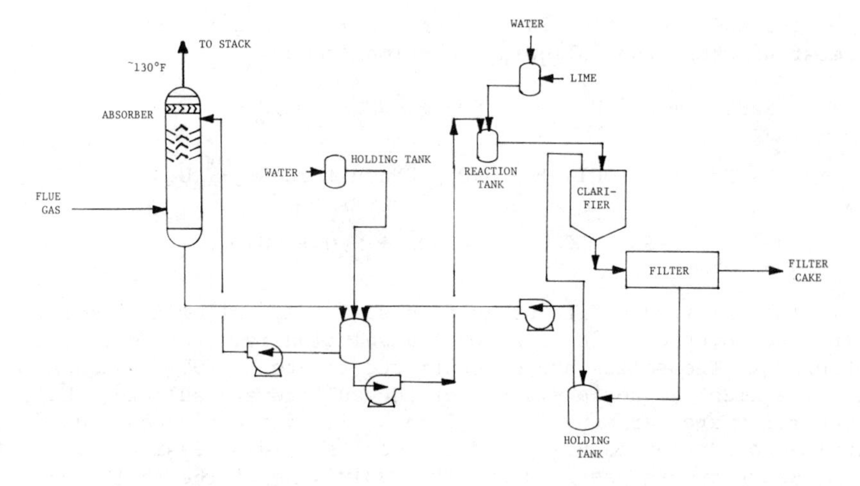

Figure 5. Clear solution regenerable process (double alkali)

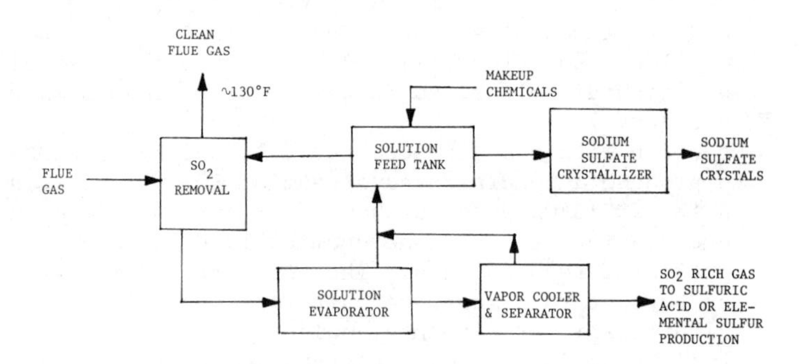

Figure 6. Wellman–Lord process

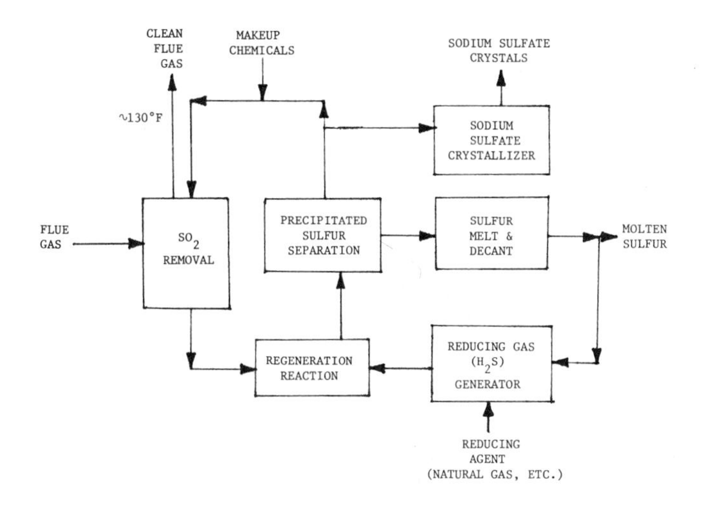

Figure 7. Citrate process

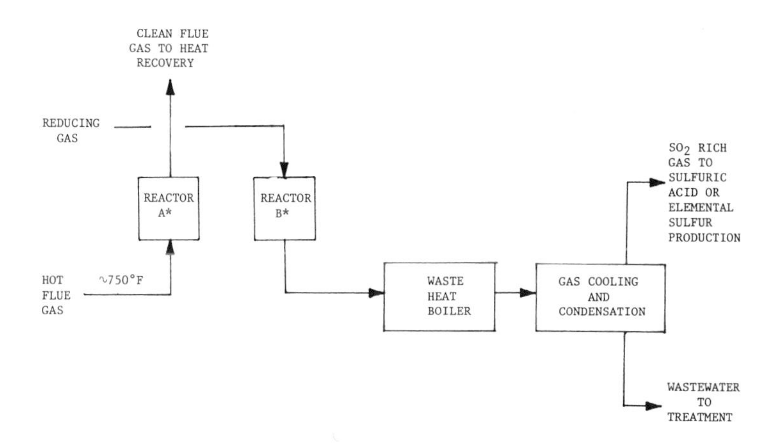

Figure 8. UOP/SFGD process

The generation of the required reducing gas is very expensive because natural gas or low sulfur oil are used. Both of these fuels are in short supply and do not offer long-term solutions to the problem. However, in certain industrial processes, like petroleum refineries, a reducing gas could be readily available. Also, if a Claus sulfur recovery plant existed on-site, the concentrated SO_2 stream could be sent to the Claus plant where it would mix with the H_2S containing gas streams. Final adjustment of the $H_2S:SO_2$ ratio would be necessary. If the overall sulfur balance were favorable, the need for a reducing gas could be avoided. Either of these options could make the use of a recovery process economically attractive for industrial applications.

Process Selection. The following guidelines are appropriate to the selection of an SO_2 removal process.

Absorbent availability could have significant impact upon process costs. Most spent alkali streams could be used depending on the other impurities present. Corrosion or precipitation tests should be considered for these cases. Any other readily available alkali, like sodium carbonate or sodium hychloride, should be considered. Slurry solutions of lime or limestone should be avoided because of past operating problems.

Land availability could dictate the choice between throw-away and recovery processes. A considerable quantity of land is required for disposal of waste solids.

Other process wastewater streams would provide the needed dilution for a once-through clear solution process. The total dissolved solids content of the combined waste should be determined to ensure that all discharge requirements can be met.

The availability of a reducing gas (or light hydrocarbons for its production) could make the recovery processes that produce elemental sulfur economically attractive. The impacts of utilizing this gas on the overall plant energy balance should be considered in the economic evaluation.

The presence of other sulfur recovery facilities that could be integrated with the SO_2 removal process should be considered. If there is an existing Claus process nearby, the SO_2 rich gas could be combined with the H_2S feed. If the ratio of H_2S to SO_2 is about 2:1, a minor adjustment could be made and the gas sent to the catalytic reactor.

The reheat of the clean flue gas can consume a significant quantity of energy. This energy must be supplied by steam or a low sulfur fuel (28). The wet scrubbing processes cool the flue gas to around 120°F when treating it. However, some processes like the UOP/SFGD can remove the SO_2 at elevated temperatures. These processes not only save reheat energy, but allow for additional cooling below the dewpoint without fear of sulfuric acid corrosion.

Process Installations. Because of past operating problems with utility SO_2 removal processes, information on the type and number of industrial SO_2 removal process installations is of interest here. As of April 1979, 36 industrial SO_2 removal processes were in operation and an additional 21 installations were in some stage of planning, design, or construction. These 57 installations represent a total of around 8.5 million SCFM capacity with 163 systems operating on 305 different boilers. Table IV summarizes the parameters of interest for the operating SO_2 removal processes (23).

Where most utility installations are the lime or limestone processes, it can be seen from Table IV that a very small percentage of industrial installations are of this type. Most of these installations are the once-through sodium carbonate, sodium hydroxide, and double alkali processes. Where the utility installations have been plagued with corrosion, erosion, scaling and fouling problems, the industrial installations, have to date performed much better. A number of systems showed a process reliability of greater than 85%.

Wastewater Treatment

The successful design of an industrial wastewater treatment system is a complex decision and often integrated into site and process specific considerations. The quantity of wastewater can be established from an overall water balance. The quality is determined from the process design. The wastewater treatment unit operations will vary as a function of discharge requirements, reuse considerations, and economic reviews.

Typical Processing Steps. The range of treatment processes is also quite variable. A generalized flowsheet of the unit operations that represents the state-of-the-art in the treatment of industrial wastewater from the petroleum/petrochemical industry is shown in Figure 9 (29,32). The unit operations included in this system are as follows:

- Immiscible Liquids and Solids Separation (33,35)
- Wastewater Stripping (32,29)
- Equalization and Neutralization (29,35)
- Biological Treatment (29,30,35)
- Powdered Activated Carbon Treatment (29,31,33,34)

The separation of immiscible liquids and solids is performed by gravity separation in API Separators (or equivalent) or even the wastewater stripper feed drum. The purpose is to remove as much of the nondissolved organics from the wastewater as possible prior to biological treatment. Settleable solids are also removed by air flotation after stripping to further reduce the load on the biological treatment unit.

The wastewater is stripped to remove dissolved gases, like H_2S, NH_3, and CO_2. Over 90% of the H_2S and NH_3 present must be

Table IV. Sulfur Dioxide Process Installations

SO_2 Process	Location	Process Vendor	Fuel Type/ w %/Sulfur
Ammonia Scrubbing	Inland Container Corp. New Johnsville, TN	Neptune Airpol, Inc.	Wood/Spent Liquor <3.0% Sulfur
	Minn-Dak Farmer's Co-op Wahpeton, N.D.	Koch Engineering	Lignite 1.% Sulfur
Caustic Scrubbing	Nekaasa Papers, Inc. Ashdown, AR	Neptune Airpol, Inc.	Coal 1%-1.5% Sulfur
Caustic Waste Stream	Canton Textiles Canton, GA	FMC Environmental Equipment	Coal 0.8% Sulfur
	Georgia-Pacific Paper Co. Crossett, AR	Neptune Airpol,Inc.	Bark/Coal/Oil 1.5%-2.0% Sulfur
	Great Southern Paper Co. Cedar Springs, GA	Neptune Airpol, Inc.	Bark/Coal/Oil 1.0%-2.0% Sulfur
Citrate Process	St. Joe Zinc Co. Monaca, PA	Bureau of Mines	Coal 2.5%-4.5% Sulfur
Double Alkali	C.A.M.(Carbide-Amoco- Monsanto Houston, TX	Not yet selected	Not yet selected
Double Alkali	Dupont, Inc. Athens, GA	Not yet selected	Coal 1.5% Sulfur
Double Alkali (Concen- trated)	Arco/Polymers, Inc. Monaca, PA	FMC Environmental Equipment	Coal 3.0% Sulfur
	Caterpillar Tractor Co. E. Peoria, IL	FMC Environmental Equipment	Coal 3.2% Sulfur
	Caterpillar Tractor Co. Mapleton, IL	FMC Environmental Equipment	Coal 3.2% Sulfur
	Caterpillar Tractor Co. Massville, IL	FMC Environmental Equipment	Coal 3.2% Sulfur
	Firestone Tire & Rubber Pottstown, PA	FMC Environmental Equipment	Coal 2.5-3.0% Sulfur
	Gressom AFB Bunder Hill, IN	Neptune Airpol, Inc.	Coal 3.0%-3.5% Sulfur
	Santa Fe Energy Corp. Bakersfield, CA	FMC Environmental Equipment	Oil 1.5% Sulfur
Double Alkali (Dilute)	Caterpillar Tractor Co. Joliet, IL	Zurn Industries	Coal 3.2% Sulfur
	Caterpillar Tractor Co. Morton, IL	Zurn Industries	Coal 3.2% Sulfur
	General Motors Corp. Parma, OH	G.M. Environmental	Coal 1.2% Sulfur
Dry Lime Scrubbing	Celanese Corp. Cumberland, MD	Wheelabator Ferge Rockwell Industries	Coal 1.0-2.0% Sulfur
	Stratmore Paper Co. Waronoco, MA	Mikropech Corp.	Coal/Oil 0.75-3.0% Sulfur
Lime Scrubbing	Carborendum Abrasives Buffalo, N.Y.	Carborundum Envir. Sys. LTD	Coal 2.2% Sulfur
	Pfizer, Inc. E. St. Louis, IL	In-House Design	Coal 3.5% Sulfur
Limestone Scrubbing	Richenbacker AFB Columbus, OH	Research-Cottrell/ Bakco	Coal 3.6% Sulfur
Process Not Yet Selected	Shell Oil Co. Bakersfield, CA	Not Yet Selected	Oil 1.1% Sulfur
	Shell Oil Co. Taft, CA	Not Yet Selected	Oil 1.1% Sulfur
Sodium Carbonate Scrubbing	Chevron U.S.A., Inc. Bakersfield, CA	Koch Eng.	Oil 1.1% Sulfur
	Chevron U.S.A., Inc. Bakersfield, CA	Koch Eng.	Oil 1.1% Sulfur

Fluegas Rate, SCFM	Startup Date	Status	Cost	
			Capital	Operating
154,000	5/79	Construction	$350,000	-
164,000	6/77	Operational	$300,000	-
211,000	2/76	Operational	$250,000	$207,000
25,000	6/74	Operational	$138,000	$34,000
220,000	7/75	Operational	$275,000	-
420,000	6/75	Operational	$1,800,000	-
142,000	4/79	Construction	$12,700,000	-
1,200,000	6/84	Planned - Considering SO_2 Control		
280,000	12/85	Planned - Considering SO_2 Control	-	-
305,000	6/80	Construction	$11,600,000	$2,400,000
210,000	4/78	Operational	-	-
131,000	3/79	Operational	-	-
140,000	10/75	Operational	-	-
8,070 (13,000)	1/75	Operational	$163,000	$ 60,000
32,000	11/79	Construction	$1,610,000	-
70,000	5/79	Construction	$1,500,000	-
67,000	9/74	Operational	-	-
38,000	1/78	Operational	-	-
92,000	6/75	Operational	$3,200,000	$644,000
50,700	12/79	Planned - Contract Awarded	-	-
22,000	5/79	Construction	$1,400,000	$162,000
30,000	6/80	Planned Contract Award	-	-
40,000	9/78	Operational	$1,800,000	$500,000
55,000	3/76	Operational	$2,200,000	$207,000
99,000	N/A	Planned Consideration SO_2 Control	-	-
25,000	N/A	Planned Consideration SO_2 Control	-	-
248,000	7/78	Operational	$2,800,000	$920,000
146,000	7/79	Construction	-	-

Table IV. Continued

SO_2 Process	Location	Process Vendor	Fuel Type/ w %/Sulfur
Sodium Carbonate Scrubbing (Contd.)	FMC (Soda Ash Plant) Green River, WY	FMC Environmental Equipment	Coal 1% Sulfur
	Getty Oil Co. Bakersfield, CA	FMC Environmental Equipment	Oil 1.1% Sulfur
	Getty Oil Co. Bakersfield, CA	In-House Design	Oil
	Kerr-McGee Chem. Corp. Trona, CA	Combustion Equip. Assoc.	Coke/Coal/Oil 0.5-5.0% Sulfur
	Mead Paperboard Co. Stevenson, AL	Neptune Airpol, Inc.	Oil 1.5-3% Sulfur
	Mobil Oil Col Buttonwillow, CA	Heater Technology	Oil 1.1% Sulfur
	Phillip Morris, Inc. Chesterfield, VA	Flaht, Inc.	Coal 1.4% Sulfur Design
	Reichhold Chemicals, Inc. Pensacola, FL	Neptune Airpol, Inc.	Wood and Oil 2% Sulfur
	Tenneco Oil Co. Green River, WY	Vendor Not Selected	Coal 1.5% Sulfur (maximum)
	Texaco, Inc. San Ardo, CA	Duncan Co.	Oil 1.7% Sulfur
	TexasGulf Granger, WY	Swenco, Inc.	Coal 0.75% Sulfur
Sodium Hydroxide Scrubbing	Alyeska Pipeline So.Co. Valdez, AK	FMC Environmental Equipment	Oil 0.03-0.1% Sulfur
	Belridge Oil Co. McKittrick, CA	C-E Natco	Crude Oil 1.1% Sulfur
	Belridge Oil Co. McKittrick, CA	Thermatics, Inc.	Crude Oil 1.1% Sulfur
	Double Barrel Oil Co. Bakersfield, CA	C-E Natco	Oil 1.1% Sulfur
	General Motors Corp. St. Louis, MO	A. D. Little	Coal 3.2% Sulfur
	General Motors Corp. Dayton, OH	Entelecter, Inc.	Coal 0.7-2.0% Sulfur
	General Motors Corp. Tonowande, NY	FMC Environmental Equipment	Coal 1.2% Sulfur
	General Motors Corp. Pontiac, MI	G.M. Environmental	Coal 0.84% Sulfur
	Getty Oil Co. Orcutt, CA	In-House Design	Oil 4.0% Sulfur
	ITT Rayonier, Inc. Fernandina Beach, FL	Neptune Airpol, Inc.	Bark, Oil 2.0-2.5% Sulfur
	Mobil Oil Co. San Ardo, CA	In-House Design	Oil 2.0-2.25% Sulfur
	Northern Ohio Sugar Co. Fremont, OH	Great Western Sugar	Coal 1% Sulfur
	Sun Production Co. Fellows, CA	C-E Natco	Oil 1.4% Sulfur
	Sun Production Co. Oildak, CA	C-E Natco	Oil 1.2% Sulfur
	Texaco, Inc. San Ardo, CA	Ceilcote	Oil 1.7% Sulfur
Sulf-X Score Process	Western Correctional Inst. Pittsburgh, PA	Pittsburgh Env. & Engrg. Systems	Coal 3.5% Sulfur

Fluegas Rate, SCFM	Startup Date	Status	Cost	
			Capital	Operating
446,000	5/76	Operational	$10,000,000	-
72,000	6/77	Operational	$400,000	
891,000	12/78	Operational	$5,400,000	$5,220,000
490,000	6/78	Operational	$6,000,000	-
100,000	6/75	Operational	$173,000	$840,000
80,500	4/79	Construction	$500,000	-
39,000	6/79	Construction	-	-
80,000	6/75	Operational	$270,000	-
140,000	1/82	Planned-Requesting/Evaluating Bids	-	-
99,000	3/79	Construction	-	-
140,000	9/76	Operational	$250,000	-
50,000	6/77		-	-
12,000	1/79	Operational	$106,000	$79,000
12,000	7/78	Operational	$172,000	$60,500
12,000	6/78	Operational	-	-
64,000	6/72	Operational	$773,000	$172,000
36,000	9/74	Operational	$668,000	-
92,000	6/75	Operational	$2,200,000	-
107,300	4/76	Operational	$600,000	-
5,000	6/77	Operational	-	-
176,000	6/75	Operational	$500,000	-
175,000	6/74	Operational	$2,900,000	$1,290,000
40,000	10/75	Operational	-	-
6,000	5/79	Construction	$100,000	$35,000
6,000	4/79	Construction	$100,000	$35,000
347,000	11/73	Operational	-	-
10,000	1/80	Planned-Contract Awarded	-	-

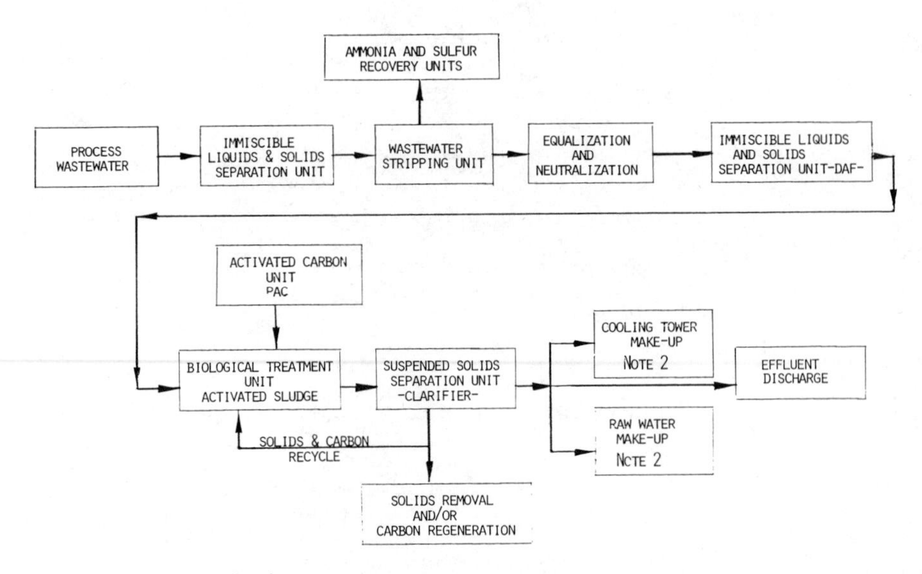

Figure 9. General water treating schematic flowsheet

removed in order to produce an acceptable effluent without the need for added process units for biological nitrification and denitrification. With increasingly strict wastewater discharge standards, much more attention is required in sour water stripping. In the past the feed was heated to a stripper bottom temperature of around 200-250°F and around 1.0 to 2.0 pounds of steam were added per gallon of feed. Around 99% of the H_2S and 90 to 95% of the ammonia was removed. These removals, and better, can be achieved if there are few electrolytes other than NH_3 and H_2S present in the water. If other electrolytes are present, like CO_2, HCN, organic acids, they can chemically fix more of the NH_3 and greatly reduce stripping efficiency. This requires adding caustic to allow release of the NH_3. Also, acid could be added to displace any H_2S that is similarly fixed.

Wastewater equalization is necessary to reduce biological system upsets from large variations of influent compositions or other physical parameters such as temperature. Equalization is accomplished by providing sufficient residence time and often mixing energy. Neutralization is often necessary to protect both the treatment system biological activity as well as the materials of construction. Accidental spills of significant impact upon wastewater pH occur relatively frequently in the hydrocarbon industry.

Biological treatment of industrial wastewater is a proven means of removing soluble, biodegradable organic contaminants from the process effluent. Depending upon the type of components and concentration, physical-chemical treatment processes may also be considered. The design details of these systems are beyond the survey scope of this document but include such steps as adsorption, extraction, evaporation, ion exchange or membranes unit operations in order to recover or concentrate the organic constituents for further processing.

For this review, the complete mix activated sludge (CMAS) process is presented as the representative biological treatment process. To design a CMAS system, the wastewater influent characteristics, the biological reactor kinetics, and desired effluent quality need to be determined. The influent composition and flow are a function of the industrial process and can be determined from the designer's energy and material balances for a new plant or stream sampling for an existing facility. The biological reactor growth kinetics are determined from bench-scale tests which can be conducted by trained engineers. Effluent quality is set by regulatory requirements on the specific industrial process and possibly amended by site specific permit limitations as mandated by the Federal Clean Water Act and state or regional discharge prohibitions.

An overall treatment concept using the above unit operations is shown in Figure 9. This flow sketch presents some of the process considerations which may be required in industrial wastewater treatment design. The biological reactor is the key to effective organic removals. This figure depicts a one-stage process to treat the wastewater. The concept of staged biological systems using either trickling filters, anaerobic and/or aerobic treatment or combinations of the above for high strength industrial wastes can both enhance organic removals and prove cost effective.

If a significant portion of the organics present are found to be not biologically degradable, then further treatment with activated carbon may be required. Treatment with powdered activated carbon (PAC) has recently been shown to be an accepted means of removing refractory organic constituents from refinery wastewater. PAC is introduced into the activated sludge basin with the biological floc where intimate contact with the wastewater results and organics are adsorbed by the carbon. The carbon, along with the biological solids, is separated from the wastewater in the clarifier and recycled back to the aeration basin.

The waste from this treatment scheme should be suitable for discharge or reuse with a possible need for removal of inorganic dissolved solids if a "zero discharge" system is used.

Process Selection. The selection of the final wastewater treatment system depends on a number of variables. Some of the more significant items are summarized as follows.

The discharge or reuse of the treated water will dictate the final treatment scheme. If the water is to be reused, treatment requirements could be more strict than discharge standards. This is especially true for refractory organics and potentially corrosive dissolved inorganics.

The type and concentration of dissolved compounds can affect the performance of the wastewater stripper and, as previously discussed, could necessitate the need for caustic or acid injection.

The biodegradability of the organics is important in that it determines if activated carbon is required for removal of refractory organics. Also, materials could be present that would reduce the activity of the biological floc.

Land availability for the wastewater treatment system should be determined. There are a number of tradeoffs that can be made in the biological processing step and biological solids disposal to reduce land requirements.

Literature Cited

1. Edwards, M.S. "H_2S-Removal Processes for Low-BTU Coal Gas," Oak Ridge National Laboratory, Oak Ridge, Tennessee, ORNL/TM-6077.

2. Maddox, R.N. "Gas and Liquid Sweetening," Campbell Petroleum Series, 1974.

3. Kohl, A.L. and Riesenfeld, F.C. "Gas Purification," 3rd ed., Gulf Publishing Company, Houston, TX, 1979.

4. Christensen, K.G. and Stupin, W.J. "Merits of Acid-Gas Removal Processes," Hydrocarbon Processing, February 1978.

5. Tennyson, R.N. and Schaaf, R.P. "Guidelines can Help Choose Proper Process for Gas-Treating Plants," The Oil and Gas Journal, January 10, 1977.

6. Rinckhoff, J.B. and Friedman, L.J. "Design Options for Sulfuric Acid Plants," presented at 69th Annual AIChE Meeting, November 30, 1976.

7. Fischer, H. "Sulfur Costs Vary with Process Selection," Hydrocarbon Processing, March 1979.

8. GPA H_2S Removal Panel-4. "Processes Cleanup Tail Gas," The Oil and Gas Journal, August 28, 1978.

9. GPA H_2S Removal Panel-5 (Conclusion). "More Claus Cleanup Processes," The Oil and Gas Journal, September 11, 1978.

10. Crynes, B.L., Ed. "Chemical Reactions as a Means of Separation Sulfur Removal," Marcel Dekker, Inc., New York.

11. Crow, J.H. and Baumann, J.C. "Versatile Process Uses Selective Absorption," Hydrocarbon Processing, October 1974.

12. Laengrich, A.R. and Cameron, W.L. "Tail-Gas Cleanup Addition May Solve Sulfur-Plant Compliance Problem," The Oil and Gas Journal, March 27, 1978.

13. The Ralph M. Parsons Co. "Sulfur Recovery Plants and Tailgas Purification Units by Parsons."

14. Todd, F.A., "Sulfuric Acid Versus Elemental Sulfur as By-Products," C.F. Braun & Co., FE-2240-54, January 1978.

15. Bucy, J.I. and Ransom, T.M. "Potential Markets for Sulfur Dioxide Abatement Products," presented at FGD Symposium, November 8-11, 1977.

16. Chute, A.E. "Tailor Sulfur Plants to Unusual Conditions," Hydrocarbon Processing, April 1977.

17. Chia, W.S.; Todd, F.A.; and Stupin, W.J."Sulfur Recovery in a Coal Gasification Plant," (Preprint). Presented to the Tenth Synthetic Pipeline Gas Symposium, Chicago, IL, October 30-November 1, 1978, C.F. Braun & Co., Alhambra, CA.

18. Seward, W.H. "Process Alternatives for Sulfur Management: Overview Report," C.F. Braun & Co., Alhambra, CA, FE-2240-41, January 1978.

19. Raab, Dr. M. "Caustic Scrubbers can be Designed for Exacting Needs," The Oil and Gas Journal, October 11, 1976.

20. Atkins, W.T. and Takach, H.J. "Problems Associated with Controlling Sulfur Emissions from High-BTU Coal Gasification Plants, Interim Report," C.F. Braun & Co., Alhambra, CA, FE-2240-13.

21. Burklow, B.W. and Coleman, R.L. "Coal Processing: Developments in Natural Gas Desulfurization," CEP, June 1977.

22. Ayer, F.A., Compiler. "Proceedings: Symposium on Flue Gas Desulfurization - Hollywood, Florida, November 1977 (Vol. I). Environmental Protection Agency, Research Triangle Institute, Research Triangle Park, NC, EPA-600/7-78-058a, March 1978.

23. Tuttle, J., Patkar, A.; Kothari, S.; Osterhout, D.; Heffling, M.; and Eckstein, M. "EPA Industrial Boiler FGD Survey: First Quarter 1979," PEDCo Environmental, Inc., Cincinnati, OH, EPA-600/7-79-067b, April 1979.

24. Head, H.N. "EPA Alkali Scrubbing Test Facility: Advanced Program, Third Progress Report," Bechtel Corp., San Francisco, CA, EPA-600/7-77-105, September 1977.

25. VanNess, R.P.; Somers, R.P.; Frank, T.; Lysaght, J.M.; Jashnani, I.L.; Lunt, R.R.; LaMantia, C.R. "Project Manual for Full-Scale Dual-Alkali Demonstration at Louisville Gas and Electric Co. - Preliminary Design and Cost Estimate," Louisville Gas and Electric Co., Louisville, KY, EPA-600/7-78-010, January 1978.

26. Duvel, W.A., Jr.; Atwood, R.A.; Gallagher, W.R.; Knight, R.G.; McLaren, R.J.; "FGD Sludge Disposal Manual: Final Report," Michael Baker, Jr., Inc., Beaner, PA, EPRI FP-977, January 1979.

27. Aul, E.F.; Delleney, R.D.; Brown, G.D.; Page, G.C.; Stuebner, D.O.; "Evaluation of Regenerable Flue Gas Desulfurization Process, Vol. I. Final Report," Radian Corp., Austin, TX, EPRI FP-722, March 1976.

28. Choi, P.S.K.; Bloom, S.G.; Rosenberg, H.S.; DiNovo, S.T. "Desulfurization Systems, Final Report," Battelle Columbus Laboratories, Columbus, OH, EPRI FP-361, February 1977.

29. "Proceedings of the Second Open Forum on Management of Petroleum Refinery Wastewater," Environmental Protection Agency, EPA 600/2-78-058, March 1978.

30. Crame, L.W. "Pilot Studies on Enhancement of the Refinery-Activated Sludge Process," American Petroleum Institute, Report No. 953, October 1977.

31. Culp, R.L. "Handbook of Advanced Wastewater Treatment," 2nd ed., Van Nostrand-Reinhold & Co., New York, NY, 1978.

32. Beychok, M.R. "Aqueous Wastes from Petroleum and Petrochemical Plants," John Wiley & Sons, New York.

33. Metcalf and Eddy, Inc., "Wastewater Engineering Treatment Disposal Reuse," 2nd ed., revised by George Tchobanoglous; McGraw-Hill Book Co., New York.

34. Weber, W.J., Jr. "Physicochemical Processes for Water Quality Control," a Wiley-Interscience Series of Texts and Monographs.

35. Associated Water and Air Resources Engineers, Inc. (Aware, Inc.), Nashville, TN. "Process Design Techniques for Industrial Waste Treatment," Enviro Press, Nashville, TN.

RECEIVED January 31, 1980.

3

Phase Equilibria in Aqueous Electrolyte Solutions

A. E. MATHER[1] and R. D. DESHMUKH

University of Alberta, Edmonton, Alberta, Canada

Many of the undesirable substances present in gaseous or liquid streams form volatile weak electrolytes in aqueous solution. These compounds include ammonia, hydrogen sulfide, carbon dioxide and sulfur dioxide. The design and analysis of separation processes involving aqueous solutions of these materials require accurate representation of the phase equilibria between the solution and the vapor phase. Relatively few studies of these types of systems have been published concerning solutions of weak electrolytes. This paper will review the methods that have been used for such solutions and, as an example, consider the alkanolamine solutions used for the removal of the acid gases (H_2S and CO_2) from gas streams.

In general, the formulation of the problem of vapor-liquid equilibria in these systems is not difficult. One has the mass balances, dissociation equilibria in the solution, the equation of electroneutrality and the expressions for the vapor-liquid equilibrium of each molecular species (equality of activities). The result is a system of non-linear equations which must be solved. The main thermodynamic problem is the relation of the activities of the species to be measurable properties, such as pressure and composition. In order to do this a model is needed and the parameters in the model are usually obtained from experimental data on the mixtures involved. Calculations of this type are well-known in geological systems (1) where the vapor-liquid equilibria are usually neglected.

Solutions of Weak Electrolytes Van Krevelen et al. (2) measured the vapor pressures of aqueous

[1] Current address: Lehrstuhl für Physikalische Chemie II, Ruhr-Universität Bochum, 463o Bochum 1, W.-Germany

0-8412-0569-8/80/47-133-049$05.00/0

solutions of NH_3+CO_2, NH_3+H_2S and the ternary mixture NH_3+CO_2+H_2S, at temperatures between 20 $^{\circ}$ and 60 $^{\circ}$C. The experimental data were obtained in NH_3-rich solutions and did not extend to the dilute region of interest in pollution control. In their theoretical work, Van Krevelen et al. used pseudo-equilibrium constants, defined as follows:

$$K' = K \left[\frac{\gamma_{H_2S}\ \gamma_{NH_3}}{\gamma_{HS^-}\ \gamma_{NH_4^+}} \right] = \frac{m_{HS^-}\ m_{NH_4^+}}{m_{H_2S}\ m_{NH_3}} \qquad (1)$$

From their experimental data, values of K' for the various equilibria were obtained. In some cases the values were almost constant at a given temperature, while in other cases log K' was found to be a linear function of the ionic strength:

$$I = \frac{1}{2} \sum_i z_i^2\ m_i \qquad (2)$$

In their model Van Krevelen et al. neglected the second ionization of H_2S and the concentration of OH^- and H^+ ions. While this method was able to reproduce the experimental data, the model does not lend itself to extrapolation to regions where data were not obtained and it is not useful for dilute solutions as the accuracy of the predictions is poor.

Van Krevelen et al. used molarity instead of molality as the measure of their concentrations. The use of molality here does not alter the essential features of the method.

Dankwerts and McNeil (3) have employed the method of Van Krevelen et al. to predict the partial pressure of carbon dioxide over carbonated alkanolamine solutions. The central feature of this model is the use of pseudo-equilibrium constants and their dependence on ionic strength. The ratio of the pseudo-equilibrium constant at a certain ionic strength to that at zero ionic strength has been termed the "ionic characterization factor". However, ionic strength alone is insufficient to determine the ionic characterization factors. As well the ionic characterization factors are sometimes not a simple linear function of ionic strength.

Lemkowitz et al. (4) used a similar model to that proposed by Van Krevelen et al. They correlated the equilibria in the CO_2+NH_3+Urea+H_2O system. The pseudo-equilibrium constant for urea formation, as well as the vapor pressure of NH_3 and the Henry's constant for CO_2 were treated as parameters and were determined by using the model to predict bubble point pressures.

Kent and Eisenberg (5) also correlated solubility data in the system H_2S+CO_2+alkanolamines+H_2O using pseudo-equilibrium constants based on molarity. Instead of using ionic characterization factors, they accepted published values of all but two pseudo-equilibrium constants and found these by fitting data for MEA and DEA solutions. They were able to obtain excellent fits by this approach and also discovered that the fitted pseudo-equilibrium constants showed an Arrhenius dependence on temperature.

This procedure of lumping all non-idealities into a few adjustable parameters is unsatisfactory for many reasons. Thermodynamic rigor is lost if experimentally determined dissociation constants or vapor pressures are disregarded. Also the parameters determined in this way are accurate only over the range of variables fitted and usually the model cannot be used for extrapolation to other conditions. The attractive feature of these models in the past was their need for little input information and the simple equations could often be solved algebraically.

The first rigorous method for weak electrolyte solutions was that of Edwards et al. (5). Because comparisons with the models of other workers will be made, the thermodynamic framework will be outlined and the assumptions that were made stated. For a single solute which dissociates only in the aqueous solution, the model is based on four principles:

1. Mass balance on the electrolyte in the liquid phase.
2. The ratio of the molecular to the ionic concentrations of the electrolyte is determined by the dissociation constant K. The activity is related to the molality through the activity coefficient γ_i:

$$a_i = \gamma_i m_i \tag{3}$$

   where $\gamma_i \rightarrow 1$ as $\sum m_j \rightarrow 0$. The subscript j refers to all solute species.
3. Bulk electroneutrality of the liquid phase.
4. For the molecular solute, equilibrium between the vapor phase and the liquid phase is given by:

$$\phi_a y_a P = \gamma_a m_a H_a \tag{4}$$

For the solvent, water, the vapor-liquid equilibrium is given by:

$$\phi_w y_w P = a_w P_w^s \phi_w^s \exp\left[\frac{\bar{v}_w (P-P_s)}{RT}\right] \quad (5)$$

Edwards et al. (6) made the assumption that ϕ_a was equal to $\phi_{pure\ a}$ at the same pressure and temperature. Further they used the virial equation, truncated after the second term to estimate $\phi_{pure\ a}$. These assumptions are satisfactory when the total pressure is low or when the mole fraction of the solute in the vapor phase is near unity. For the water, the assumption was made that ϕ_w, ϕ_w^s, a_w and the exponential term were unity. These assumptions are valid when the solution consists mostly of water and the total pressure is low. The activity coefficient of the electrolyte was calculated using the extended Debye-Hückel theory:

$$\ln \gamma_i = \frac{-A_\phi z_i^2 \sqrt{I}}{1 + \sqrt{I}} + 2 \sum_{k \neq w} \beta_{ik}\, m_k \quad (6)$$

In applying this equation to multi-solute systems, the ionic concentrations are of sufficient magnitude that molecule-ion and ion-ion interactions must be considered. Edwards et al. (6) used a method proposed by Bromley (7) for the estimation of the ß parameters. The model was found to be useful for the calculation of multi-solute equilibria in the NH_3+H_2S+H_2O and NH_3+CO_2+H_2O systems. However, because of the assumptions regarding the activity of the water and the use of only two-body interaction parameters, the model is suitable only up to molecular concentrations of about 2 molal. As well the temperature was restricted to the range 0° to 100 °C because of the equations used for the Henry's constants and the dissociation constants. In a later study, Edwards et al. (8) extended the correlation to higher concentrations (up to 10 - 20 molal) and higher temperatures (0° to 170 °C). In this work the activity coefficients of the electrolytes were calculated from an expression due to Pitzer (9):

$$\ln \gamma_i = -A_\phi z_i^2 \left[\frac{\sqrt{I}}{1+1.2\sqrt{I}} + \frac{2}{1.2}\ln(1+1.2\sqrt{I})\right] + 2\sum_{j \neq w} m_j \left\{\beta_{ij}^{(o)} + \frac{\beta_{ij}^{(1)}}{2I}\left[1-(1+2\sqrt{I})\exp(-2\sqrt{I})\right]\right\}$$

$$- \frac{z_i^2}{4I^2} \sum_{j \neq w} \sum_{k \neq w} m_j m_k \beta_{jk}^{(1)} \left[1-(1+2\sqrt{I}+2I)\exp(-2\sqrt{I})\right] \qquad (7)$$

The activity of the water is derived from this expression by use of the Gibbs-Duhem equation. To utilize this equation, the interaction parameters $\beta_{ij}^{(0)}$ and $\beta_{ij}^{(1)}$ must be estimated for moleculemolecule, molecule-ion and ion-ion interactions. Again the method of Bromley was used for this purpose. Fugacity coefficients for the vapor phase were determined by the method of Nakamura et al. (10).

About the same time Beutier and Renon (11) also proposed a similar model for the representation of the equilibria in aqueous solutions of weak electrolytes. The vapor was assumed to be an ideal gas and ϕ_a was set equal to unity. Pitzer's method was used for the estimation of the activity coefficients, but, in contrast to Edwards et al. (8), two ternary parameters in the activity coefficient expression were employed. These were obtained from data on the two-solute systems. It was found that the equilibria in the systems NH_3+H_2S+H_2O, NH_3+CO_2+H_2O and NH_3+SO_2+H_2O could be represented very well up to high concentrations of the ionic species. However, the model was unreliable at high concentrations of undissociated ammonia. Edwards et al. (12) have recently proposed a new expression for the representation of the activity coefficients in the NH_3+H_2O system, over the complete concentration range from pure water to pure NH_3. It appears that this area will assume increasing importance and that one must be able to represent activity coefficients in the region of high concentrations of molecular species as well as in dilute solutions. Cruz and Renon (13) have proposed an expression which combines the equations for electrolytes with the non-random two-liquid (NRTL) model for non-electrolytes in order to represent the complete composition range. In a later publication, Cruz and Renon (14), this model was applied to the acetic acid-water system.

Application to Alkanolamine Solutions Aqueous alkanolamine solutions are widely used for the removal of H_2S and CO_2 from gaseous streams, because they can reduce the concentration of H_2S and CO_2 to low levels, even if the gas stream is at a low total pressure. The most commonly used alkanolamines are monethanolamine (MEA) and diethanolamine. However, diisopropanolamine

(DIPA) and methyldiethanolamine (MDEA) have also been employed. Earlier, Atwood et al. (15) proposed a thermodynamic model for the equilibria in H_2S+alkanolamine+H_2O systems. The central feature of this model is the use of mean ionic activity coefficient. The activity coefficients of all ionic species are assumed to be equal and to be a function only of ionic strengths. Klyamer and Kolesnikova (16) utilized this model for correlation of equilibria in CO_2+alkanolamine+H_2O systems and Klyamer et al. (17) extended it to the H_2S+CO_2+alkanolamine+H_2O system. The model is restricted to low pressures as the fugacity coefficients are assumed unity and it has been found that the predictions are inaccurate in the four-component system since the activity coefficients are not equal when a number of different cations and anions are present.

Deshmukh and Mather (18) have recently presented a model for the equilibria in alkanolamine solutions using the ideas of Edwards et al. (6) for calculation of the activity coefficients. Here only the salient features of this model will be presented. The main reactions occuring in the CO_2+H_2S+alkanolamine+H_2O system are as follows:

Ionization of water

$$H_2O = H^+ + OH^- \tag{8}$$

Dissociation of hydrogen sulfide

$$H_2S = H^+ + HS^- \tag{9}$$

Dissociation of carbon dioxide

$$H_2O + CO_2 = H^+ + HCO_3^- \tag{1o}$$

Dissociation of alkanolamine

$$H_2O + RR'NH = RR'NH_2^+ + OH^- \tag{11}$$

Formation of carbamate

$$RR'NH+CO_2 = RR'NCOO^-+H^+ \tag{12}$$

Dissociation of bisulfide ion

$$HS^- = H^+ + S^= \tag{13}$$

Dissociation of bicarbonate ion

$$HCO_3^- = H^+ + CO_3^= \qquad (14)$$

In these equations RR'NH is the chemical formular of the alkanolamine. Tertiary amines, such as triethanolamine, lack the extra hydrogen atom and do not form carbamates. The chemical reactions are accompanied by the vapor-liquid equilibria of the volatile species: CO_2, H_2S and H_2O. Under the conditions of interest, the vapor pressures of the amines are very small and it can be assumed that the amine is present only in the liquid phase.

The equilibrium relations can be written as follows:

$$K_w = \gamma_{H^+} \gamma_{OH^-} m_{H^+} m_{OH^-} / a_w \qquad (15)$$

$$K_{1c} = \gamma_{H^+} \gamma_{HS^-} m_{H^+} m_{HS^-} / \gamma_{H_2S} m_{H_2S} \qquad (16)$$

$$K_{1y} = \gamma_{H^+} \gamma_{HCO_3^-} m_{H^+} m_{HCO_3^-} / \gamma_{CO_2} m_{CO_2} a_w \qquad (17)$$

$$K_i = \gamma_{RR'NH_2^+} \gamma_{OH^-} m_{RR'NH_2^+} m_{OH^-} / \gamma_{RR'NH} m_{RR'NH} a_w \qquad (18)$$

$$K_A = \gamma_{H^+} \gamma_{RR'NCOO^-} m_{H^+} m_{RR'NCOO^-} / \gamma_{RR'NH} \gamma_{CO_2} m_{RR'NH} m_{CO_2} \qquad (19)$$

$$K_{2c} = \gamma_{H^+} \gamma_{S^=} m_{H^+} m_{S^=} / \gamma_{HS^-} m_{HS^-} \qquad (2o)$$

$$K_{2y} = \gamma_{H^+} \gamma_{CO_3^=} m_{H^+} m_{CO_3^=} / \gamma_{HCO_3^-} m_{HCO_3^-} \qquad (21)$$

$$\phi_{CO_2} y_{CO_2} P = H_{CO_2} \gamma_{CO_2} m_{CO_2} \qquad (22)$$

$$\phi_{H_2S} y_{H_2S} P = H_{H_2S} \gamma_{H_2S} m_{H_2S} \qquad (23)$$

$$\phi_w y_w P = a_w P_w^s \phi_w^s \exp\left[\frac{\bar{v}_w (P - P_w^s)}{RT}\right] \qquad (24)$$

Values of the dissociation constants and Henry's constants were determined from the literature. The fugacity coefficients, ϕ_i, were calculated using the Peng-Robinson (19) equation of state. The activity of the solvent, water, was set equal to its mole fraction. Also the fugacity coefficient of water at its vapor pressure, ϕ_w^s and the Poynting correction were assumed to be unity. The model is hence restricted to relatively dilute solutions, but this restriction can be removed by determining the expression for a_w using the Gibbs-Duhem equation, as shown by Edwards et al. (8). The activity coefficients of the solute species have been determined from the extended Debye-Hückel expression given by Guggenheim (20), Equation (6). This equation was used by Edwards et al. (6). The major problem in applying it to alkanolamine solutions is the estimation of the β's since the procedure of Bromley cannot be used as the input parameters - the ionic entropies or salting-out parameters have not been determined for ethanolammonium or carbamate ions.

The following balance equations for the reacting species can be formed

Electroneutrality

$$m_{H^+} + m_{RR'NH_2^+} = m_{OH^-} + m_{HS^-} + m_{HCO_3^-} + m_{RR'NCOO^-} + 2\, m_{S^=} + 2\, m_{CO_3^=} \qquad (25)$$

Mass balances

$$m_A = m_{RR'NH} + m_{RR'NH_2^+} + m_{RR'NCOO^-} \qquad (26)$$

$$m_A\, \alpha_{CO_2} = m_{CO_2} + m_{HCO_3^-} + m_{CO_3^=} + m_{RR'NCOO^-} \qquad (27)$$

$$m_A\, \alpha_{H_2S} = m_{H_2S} + m_{HS^-} + m_{S^=} \qquad (28)$$

Here α_{CO_2} and α_{H_2S} are the mole ratios in the liquid phase (carbon to nitrogen and sulfur to nitrogen) and are the experimentally measured concentrations.

The mathematical problem is to solve Equations (15) to (28). Twelve species exist: H_2S, CO_2, RR'NH,

HS^-, $S^=$, HCO_3^-, CO_3^-, $RR'NH_2^+$, $RR'NCOO^-$, H^+, OH^- and H_2O. Hence there are twenty-three unknowns (m_i and γ_i for all species except water plus x_w). To solve for the unknowns there are twenty-three independent equations: Seven chemical equilibria, three mass balances, electroneutrality, the use of Equation (6) for the eleven activity coefficients and the phase equilibrium for x_w. The problem is one of solving a system of non-linear algebraic equations. Brown's method (21, 22) was used for this purpose. It is an efficient procedure, based on a partial pivoting technique, and is analogous to Gaussian elimination in linear systems of equations.

The application of this model to alkanolamine solutions is not possible directly since the specific interaction parameters (β's) for alkanolammonium ions and carbamate ions are not available. Also the dissociation constant for the simplest amines (MEA, DEA, TEA) is known only over the range of temperatures between 0° and 50 °C and the equilibrium constant for carbamate formation is known only at 18 °C for MEA and DEA.

In monoethanolamine solutions the unknown interaction parameters and equilibrium constants were determined by fitting the model to data for the three component systems CO_2+MEA+H_2O and H_2S+MEA+H_2O. The agreement of the fitted model with the data was found to be good. The parameters obtained in this way were then used to predict the partial pressures of mixtures of H_2S and CO_2 over aqueous MEA solutions. The predictions were in good agreement with experimental data, except at the higher partial pressures.

This procedure could not be employed for diisopropanolamine (DIPA) solutions since data were available only for one amine concentration at two temperatures. In this case data for mixtures of H_2S+CO_2+DIPA+H_2O were used together with the data for H_2S+DIPA+H_2O and CO_2+DIPA+H_2O to obtain the interaction parameters and equilibrium constants. The results are shown in Figures 1 and 2 to be in good agreement with the experimental data (23). In this case, however, in contrast to the case of MEA, the predictions use parameters evaluated from data for the four component system.

Conclusions The correlation of vapor-liquid equilibria in aqueous solutions of weak electrolytes is important for the separation of undesirable components from gases and liquids. The major problem in such correlations is the estimation of the activity

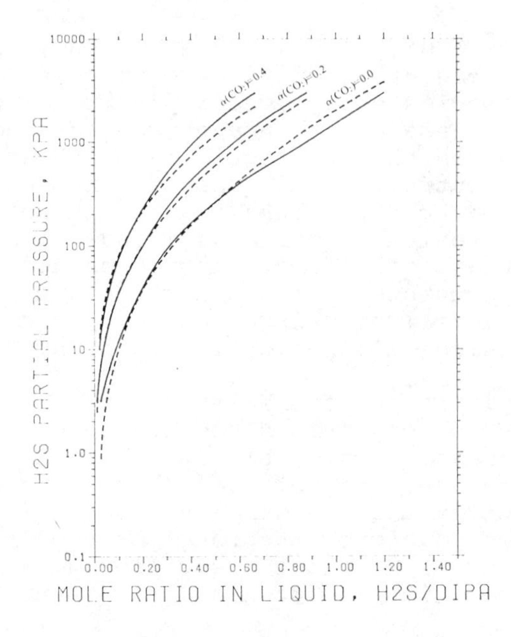

*Figure 1. Effect of CO_2 on the solubility of H_2S in 2.5*N *DIPA solutions at 100°C ((——) experimental (23); (– – –) predicted)*

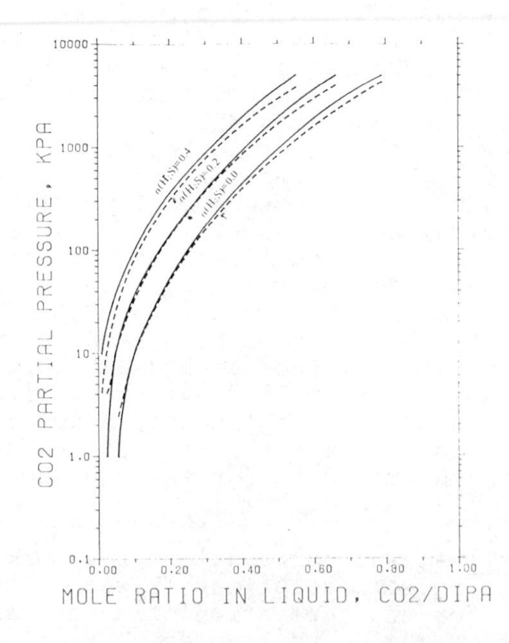

*Figure 2. Effect of H_2S on the solubility of CO_2 in 2.5*N *DIPA solutions at 100°C ((——) experimental (23); (– – –) predicted*

coefficients of the ionic species and although a number of models have been proposed, the determination of the parameters in a new case is not a simple matter. As well dissociation constants and Henry's constants for a species must be available over the temperature range of interest. Both these problems occur in the application of the fundamental thermodynamics to alkanolamine solutions containing H_2S and CO_2. However, by using limited experimental data, the parameters in the model may be obtained and the representation of the equilibria is good over the range of importance in industrial processes.

Nomenclature

a = activity
A_ϕ = Debye-Hückel parameter
H = Henry's constant for molecular solute
I = ionic strength = $0.5 \sum m_i z_i^2$
K = equilibrium constant
m = molality, mole kg^{-1}
P = pressure, Pa
R = gas constant, J mol^{-1} K^{-1}
T = temperature, K
$\bar{v}$ = partial molar volume, cm^3 mol^{-1}
y = vapor phase mole fraction
z_i = ionic charge on species i

Greek letters

α = mole ratio in the liquid phase, mole/mole amine
β, $\beta^{(o)}$, $\beta^{(1)}$ = interaction parameters, kg mol^{-1}
ϕ = vapor phase fugacity coefficient

Superscripts

s = saturation
' = pseudo-equilibrium constant

Subscripts

a = molecular species
A = carbamate equilibria, amine
i, j, k = species or component, amine equilibria
w = water
1c, 2c = carbonic acid equilibria
1y, 2y = hydrogen sulfide equilibria

Literature cited

1. Crerar, D.A. Geochim. Cosmochim. Acta, 1975, 39, 1375.
2. Van Krevelen, D.W.; Hoftijzer, P.J.; Huntjens, F.J. Rec. Trav. Chim., 1949, 68, 191.
3. Dankwerts, P.V.; McNeil, K.M. Trans. Inst. Chem. Eng., 1967, 45, T32.
4. Lemkowitz, S.M.; de Cooker, M.G.R.T.; Van den Berg, P.J. J. Appl. Chem. Biotechnol., 1973, 23, 63.
5. Kent, R.L.; Eisenberg, B. Hydrocarbon Processing, 1976, 55, (2), 87.
6. Edwards, T.J.; Newman, J.; Prausnitz, J.M. A.I.Ch.E.J.,1975, 21, 248.
7. Bromley, L.A. J. Chem. Thermo., 1972, 4, 669.
8. Edwards, T.J.; Maurer, G.; Newman, J.; Prausnitz, J.M. A.I.Ch.E.J., 1978, 24, 966.
9. Pitzer, K.S. J. Phys. Chem., 1973, 77, 268.
1o. Nakamura, R.; Breedveld, G.J.F.; Prausnitz, J.M. Ind. Eng. Chem. Proc. Des. Dev., 1976, 15, 557.
11. Beutier, D.; Renon, H. Ind. Eng. Chem. Proc. Des. Dev., 1978, 17, 22o.
12. Edwards, T.J.; Newman, J.; Prausnitz, J.M. Ind. Eng. Chem. Fundam., 1978, 17, 264.
13. Cruz, J.-L.; Renon, H. A.I.Ch.E.J., 1978, 24, 817.
14. Cruz, J.-L.; Renon, H. Ind. Eng. Chem. Fundam., 1979, 18, 168.
15. Atwood, K.; Arnold, M.R., Kindrick, R.C. Ind. Eng. Chem., 1957, 49, 1439.
16. Klyamer, S.D.; Kolesnikova, T.L. Zhur. Fiz. Khim., 1972, 46, 1o56.
17. Klyamer, S.D.; Kolesnikova, T.L.; Rodin, Yu.A. Gazov. Prom., 1973, 18(2), 44.
18. Deshmukh, R.D.; Mather, A.E. Chem. Eng. Sci. (in press).
19. Peng, D.-Y.; Robinson, D.B. Ind. Eng. Chem. Fundam., 1976, 15, 59.
2o. Guggenheim, E.A. Phil. Mag., 1935, 19, 588.
21. Brown, K.M. SIAM J. Numer. Anal., 1969, 6, 56o.
22. Brown, K.M. in Numerical Solution of Nonlinear Algebraic Equations, ed. by G.D. Byrne and C.A. Hall, Academic Press, 1973, p. 281.
23. Isaacs, E.E.; Otto, F.D.; Mather, A.E. Can. J. Chem. Eng., 1977, 55, 21o.

RECEIVED January 31, 1980.

Two New Activity Coefficient Models for the Vapor-Liquid Equilibrium of Electrolyte Systems

CHAU-CHYUN CHEN, HERBERT I. BRITT, JOSEPH F. BOSTON, and LAWRENCE B. EVANS

Department of Chemical Engineering and Energy Laboratory, Massachusetts Institute of Technology, Cambridge, MA 02139

The use of modern process simulators for the analysis and design of processes involving electrolytes has been greatly limited by the lack of adequate correlations for electrolyte thermodynamics. For most systems of industrial importance, empirical correlations are applicable only to one particular system, over a limited range of conditions. The empirical correlations do not provide a framework for treating new systems or for extending the range of existing data, because the nonidealities have not been accounted for in a general and consistent manner.

As in the nonelectrolyte case, the problem of representing the thermodynamic properties of electrolyte solutions is best regarded as that of finding a suitable expression for the non-ideal part of the chemical potential, or the excess Gibbs energy, as a function of composition, temperature, dielectric constant and any other relevant variables.

Recently, there have been a number of significant developments in the modeling of electrolyte systems. Bromley (1), Meissner and Tester (2), Meissner and Kusik (3), Pitzer and co-workers (4,5,6), and Cruz and Renon (7), presented models for calculating the mean ionic activity coefficients of many types of aqueous electrolytes. In addition, Edwards, et al. (8) proposed a thermodynamic framework to calculate equilibrium vapor-liquid compositions for aqueous solutions of one or more volatile weak electrolytes which involved activity coefficients of ionic species. Most recently, Beutier and Renon (9) and Edwards, et al.(10) used simplified forms of the Pitzer equation to represent ionic activity coefficients.

In this paper, two new models for the activity coefficients of ionic and molecular species in electrolyte systems are presented. The first is an extension of the Pitzer equation and is covered in more detail in Chen, et al. (11). The second is based on the local composition concept and represents work in progress.

0-8412-0569-8/80/47-133-061$07.25/0

Nature of Electrolyte Systems

The thermodynamic properties of a mixture depend on the forces which operate between the species of the mixture. Electrolyte systems are characterized by the presence of both molecular species and ionic species, resulting in three different types of interaction. They are ion-ion interaction, molecule-molecule interaction, and ion-molecule interaction. The forces involved in each interaction are briefly discussed in the following paragraphs.

The ion-ion interaction is characterized by electrostatic forces between ions. These electrostatic forces are inversely proportional to the square of the separation distance and therefore have a much greater range than intermolecular forces which depend on higher powers of the reciprocal distance. Except at short-range, other forces are relatively insignificant compared to the interionic electrostatic force.

Many different types of forces arise from molecule-molecule interaction. They may be electrostatic forces between permanent dipoles, induction forces between a permanent dipole and induced dipoles, or dispersion forces between non-polar molecules, etc. (Prausnitz, (12)). Forces involved in molecule-molecule interaction are known to be short-range in nature.

The forces involved in ion-molecule interaction are also short-range in nature. The dominant forces are electrostatic forces between ions and permanent dipoles. As discussed by Robinson and Stokes (13) regarding aqueous electrolyte systems, it seems likely that the ion-molecule interaction energies of the water molecules in the first layer about a monatomic ion would be large compared with the thermal energy (RT), and the second layer of water molecules will be much less strongly bound to the ion than the first. It is probably only with polyvalent monatomic ions of small size that the interaction energies of the water molecules in the second layer would be comparable to the thermal energy.

The excess Gibbs energy of electrolyte systems can be considered as the sum of two terms, one related to long-range forces between ions and the other to short-range forces between all the species. As discussed by Robinson and Stokes (13), long-range forces dominate in the region of dilute electrolyte concentration and short-range forces dominate in the region of high electrolyte concentration. It is the long-range nature of the electrostatic forces between ions that have no counterpart in nonelectrolyte systems.

The Pitzer Equation

In a series of papers, Pitzer and his co-workers (4,5,6) proposed a very useful semiempirical equation for the unsymmetric excess Gibbs free energy of aqueous electrolyte systems. The basic equation is

$$\frac{G^{ex*}}{n_w RT} = f(I) + \sum_{ij}\sum \lambda_{ij}(I) m_i m_j + \sum_{ijk}\sum\sum \mu_{ijk} m_i m_j m_k \quad (1)$$

The function f(I) expresses the effect of long-range electrostatic forces between ions. It is a function of ionic strength, temperature and solvent properties. The empirical form chosen by Pitzer for f(I) is

$$f(I) = -A_\phi \frac{4I}{1.2} \ln(1+1.2\sqrt{I}) \quad (2)$$

The parameters λ_{ij} are second virial coefficients giving the effect of short-range forces between solutes i and j; the parameters μ_{ijk} are corresponding third virial coefficients for the interaction of three solutes i, j, and k. The second virial coefficients are a function of ionic strength. Dependence of the third virial coefficients on ionic strength is neglected. The λ and μ matrices are taken to be symmetric.

To make the basic Pitzer equation more useful for data correlation of aqueous strong electrolyte systems, Pitzer modified it by defining a new set of more directly observable parameters representing certain combinations of the second and third virial coefficients. The modified Pitzer equation is

$$\frac{G^{ex*}}{n_w RT} = f(I) + \sum_{cc'}\sum m_c m_{c'} (\theta_{cc'} + \sum_a m_a \psi_{cc'a}) + \sum_{aa'}\sum m_a m_{a'} (\theta_{aa'} + \sum_c m_c \psi_{aa'c})$$
$$+ 2\sum_{ca}\sum m_c m_a [B_{ca}(I) + (\sum_c m_c Z_c C_{ca})/\sqrt{Z_c Z_a}] \quad (3)$$

Essentially, the new parameters B and θ are binary ion-ion parameters and C and ψ ternary ion-ion parameters. The ion-ion interaction parameters, B and C, are characteristic of each aqueous single-electrolyte system. The ion-ion difference parameters, θ and ψ , are characteristic of each aqueous mixed-electrolyte system.

Recognizing the ionic strength dependence of the effect of short range forces in binary interactions, Pitzer was able to develop an empirical relation for $B_{ca}(I)$. The expression for systems containing strong electrolytes with one or both ions univalent is

$$B_{ca}(I) = \beta^{(0)}_{ca} + \beta^{(1)}_{ca} [1-(1+2\sqrt{I})\exp(-2\sqrt{I})]/2I \quad (4)$$

Therefore, the adjustable parameters in the modified Pitzer equation are $\beta^{(0)}$, $\beta^{(1)}$, C, θ, and ψ. The modified Pitzer equation has been successfully applied to the available data for many pure electrolytes (Pitzer and Mayorga, (5)) and mixed aqueous electrolytes (Pitzer and Kim, (6)). The fit to the experimental data is within the probable experimental error up to molalities of 6.

However, in most aqueous electrolyte systems of industrial interest, not only strong electrolytes but also weak electrolytes and molecular nonelectrolytes are present. While the modified Pitzer equation appears to be a useful tool for the representation of aqueous strong electrolytes including mixed electrolytes, it cannot be used in the form just presented to represent the important case of systems containing molecular solutes. A unified thermodynamic model for both ionic solutes and molecular solutes is required to model these kinds of systems.

Previous Applications of the Pitzer Equation to Weak Electrolytes

Recently, the Pitzer equation has been applied to model weak electrolyte systems by Beutier and Renon (9) and Edwards, et al. (10). Beutier and Renon used a simplified Pitzer equation for the ion-ion interaction contribution, applied Debye-McAulay's electrostatic theory (Harned and Owen, (14)) for the ion-molecule interaction contribution, and adopted Margules type terms for molecule-molecule interactions between the same molecular solutes. Edwards, et al. applied the Pitzer equation directly, without defining any new terms, for all interactions (ion-ion, ion-molecule, and molecule-molecule) while neglecting all ternary parameters. Bromley's (1) ideas on additivity of interaction parameters of individual ions and correlation between individual ion and partial molar entropy of ions at infinite dilution were adopted in both studies. In addition, they both neglected contributions from interactions among ions of the same sign.

There are drawbacks with the approaches taken by both Beutier and Renon, and Edwards, et al. First, the Pitzer equation is a virial-expansion type equation and semi-empirical in nature. Estimating interaction parameters using Bromley's approach is based on an interpretation of the parameters that is uncertain at best and in any case is not valid at high ionic strength. Second, ternary parameters in the Pitzer equation can be significant for systems of high ionic strength. These parameters should not be neglected in a model covering high ionic strength electrolyte systems. Third, Bronsted's principle of specific ion interaction is the basis for assuming that interactions between ions of the same sign can be neglected. However, Bronsted's principle of specific ion interaction is not always valid, as discussed in Pitzer's paper (4). Fourth, theoretical aspects of the physical chemistry of salting-out effects on non-electrolytes are still in a development stage. While many theories have been proposed in the literature, none is quantitatively satisfactory, including Debye-McAulay's electrostatic theory.

Extension of the Pitzer Equation

In this study the Pitzer equation is also used, but a different, more straightforward approach is adopted in which the drawbacks just discussed do not arise. First, terms are added to the basic virial form of the Pitzer equation to account for molecule-ion and molecule-molecule interactions. Then, following Pitzer, a set of new, more observable parameters are defined that are functions of the virial coefficients. Thus, the Pitzer equation is extended, rather than modified, to account for the presence of molecular solutes. The interpretation of the terms and parameters of the original Pitzer equation is unchanged. The resulting extended Pitzer equation is

$$\frac{G^{ex*}}{n_w RT} = f(I) + \sum_{cc'}\sum m_c m_{c'}(\theta_{cc'} + \sum_a m_a \psi_{cc'a}) + \sum_{aa'}\sum m_a m_{a'}(\theta_{aa'} + \sum_c m_c \psi_{aa'c}) + 2\sum_{ca}\sum m_c m_a [B_{ca}(I) + (\sum_c m_c Z_c C_{ca})/\sqrt{Z_c Z_a}] + \sum_{mm'}\sum m_m m_{m'} \lambda_{mm'} + \sum_m (\sum_a D_{ca,m} m_m m_a - \sum_{c'} \omega_{cc',m} m_m m_{c'}) \tag{5}$$

The parameters $D_{ca,m}$ are binary parameters representing the interactions between salt ca and molecular solute m in an aqueous single salt, single molecular solute system. Binary parameters $\omega_{cc',m}$ and $\omega_{aa',m}$ represent the differences between the interactions of a specific molecular solute with two unlike salts sharing one common anion or cation. Ternary molecule-ion virial coefficients are neglected in this study to simplify the extension.

It is interesting to note that the molecule-ion interaction contribution in equation (5) is consistent with the well-known Setschenow equation. The Setschenow equation is used to represent the salting-out effect of salts on molecular nonelectrolyte solutes, when the solubilities of the latter are small (Gordon, (15)). The Setschenow equation is

$$\ln\gamma^*_m = k_{s,m} m_s \tag{6}$$

where $k_{s,m}$ is the Setschenow constant (a salt-molecule interaction parameter) and m_s is the molality of the salt. The D's are equivalent to the Setschenow constants and the ω's are equivalent to differences between Setschenow constants.

The third virial coefficients for molecule-molecule interactions can be taken as zero for aqueous systems containing molecular solutes at low concentration. The remaining term for the molecule-molecule interaction contribution is equivalent to the unsymmetric two-suffix Margules model.

Summarizing the results of this section, an excess Gibbs free energy model for both ionic and molecular solutes in aqueous electrolyte systems is obtained by extending the Pitzer equation in order to account for the presence of molecular solutes. Model parameters include binary ion-ion interaction and difference parameters, ternary ion-ion interaction and difference parameters, salt-molecule interaction parameters or Setschenow constants, salt-salt difference parameters for molecular solute salting, and unsymmetric Margules parameters for molecule-molecule interactions. Like the Pitzer equation, the model is designed for convenient and accurate representation of aqueous electrolyte systems, including mixtures with any number of molecular and ionic solutes.

Application of the Extended Pitzer Equation

To test the validity of the extended Pitzer equation, correlations of vapor-liquid equilibrium data were carried out for three systems. Since the extended Pitzer equation reduces to the Pitzer equation for aqueous strong electrolyte systems, and is consistent with the Setschenow equation for molecular non-electrolytes in aqueous electrolyte systems, the main interest here is aqueous systems with weak electrolytes or partially dissociated electrolytes. The three systems considered are: the hydrochloric acid aqueous solution at 298.15°K and concentrations up to 18 molal; the NH_3-CO_2 aqueous solution at 293.15°K; and the K_2CO_3-CO_2 aqueous solution of the Hot Carbonate Process. In each case, the chemical equilibrium between all species has been taken into account directly as liquid phase constraints. Significant parameters in the model for each system were identified by a preliminary order of magnitude analysis and adjusted in the vapor-liquid equilibrium data correlation. Detailed discusions and values of physical constants, such as Henry's constants and chemical equilibrium constants, are given in Chen et al. (11).

T-P-x-y data for hydrochloric acid concentration up to 18 molal were obtained from Vega and Vera (16). The following reactions occur in the liquid phase.

$$HCl \rightleftharpoons H^+ + Cl^-$$
$$H_2O \rightleftharpoons H^+ + OH^-$$

The least squares data correlation was carried out on HCl vapor mole fraction and total pressure with $\beta^{(0)}_{HCl}$, $\beta^{(1)}_{HCl}$, C_{HCl},

$D_{HCl,HCl}$, $\lambda_{HCl,HCl}$, and the Henry's constant for hydrogen chloride as adjustable parameters. Figure 1 shows experimental data and correlation results. The average percentage deviation for total pressure is 0.44, and that for HCl vapor fraction is 0.35. The same data was previously correlated with the same objective function by Cruz and Renon (7). Their results were 0.99 percent deviation for total pressure and 0.34 percent deviation for HCl vapor fraction.

The data reported by van Krevelen, et al. (17) at 293.15°K were used for data correlation of the NH_3-CO_2 aqueous solution system. The following reactions occur in the liquid phase.

$$NH_3 + H_2O \rightleftharpoons NH_4^+ + OH^-$$
$$CO_2 + H_2O \rightleftharpoons H^+ + HCO_3^-$$
$$HCO_3^- \rightleftharpoons CO_3^= + H^+$$
$$NH_3 + HCO_3^- \rightleftharpoons NH_2COO^- + H_2O$$
$$H_2O \rightleftharpoons H^+ + OH^-$$

The least squares data correlation was carried out on partial pressures of NH_3 and CO_2 with $\beta^{(0)}_{NH_4HCO_3}$, $\beta^{(0)}_{(NH_4)_2CO_3}$, $\beta^{(0)}_{NH_4NH_2COO}$, $D_{NH_4HCO_3,NH_3}$, $D_{(NH_4)_2CO_3,NH_3}$, and $D_{NH_4NH_2COO,NH_3}$ as adjustable parameters. Experimental data and calculated results are shown in Figure 2. The average percent deviation of calculated versus measured partial pressure is 11% for CO_2 and 3.9% for NH_3. The same system and the same least squares objective function have been studied by Beutier and Renon (9). Their results, on the same basis, were 16% for CO_2 and 5% for NH_3. Edwards, et al. (10) also studied vapor-liquid equilibrium of a NH_3-CO_2 aqueous system at 373.15°K. However, the accuracy of the fit was not reported quantitatively.

The equilibrium data obtained by Tosh and coworkers (18) were used for data correlation of the K_2CO_3-CO_2 aqueous solution system. The data have a temperature range from 343.15°K to 413.15°K and a range from 20 to 40 percent equivalent concentration of potassium carbonate. The following reactions occur in the liquid phase.

$$CO_2 + H_2O \rightleftharpoons HCO_3^- + H^+$$
$$HCO_3^- \rightleftharpoons CO_3^= + H^+$$
$$H_2O \rightleftharpoons OH^- + H^+$$

The least squares data correlation was carried out on partial pressures of CO_2 and H_2O with appropriate weight and with $\beta^{(0)}_{KHCO_3}$, $\beta^{(0)}_{K_2CO_3}$, $\theta_{HCO_3^-,CO_3^=}$, C_{KHCO_3}, $C_{K_2CO_3}$, and $\psi_{K^+,HCO_3^-,CO_3^=}$ as adjustable parameters. The average percent deviation of calculated versus measured partial pressure of CO_2 at 383.15°K is 11.5% and for H_2O is 10.5%.

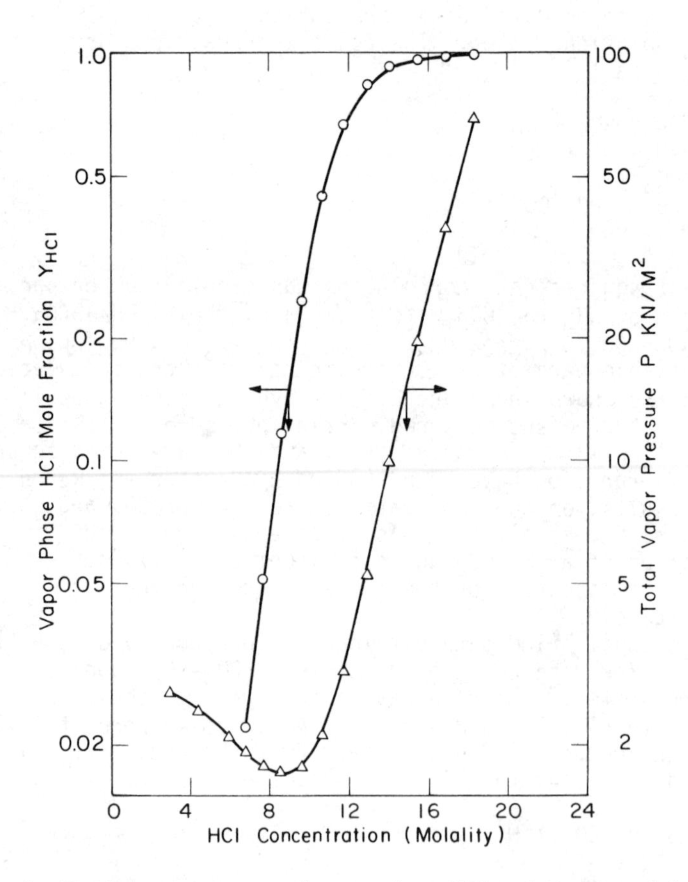

Figure 1. Total vapor pressure and vapor phase HCl mole fraction of the HCl aqueous solution at 298.15 K ((——) calculated; (○, △) data from Ref. 16)

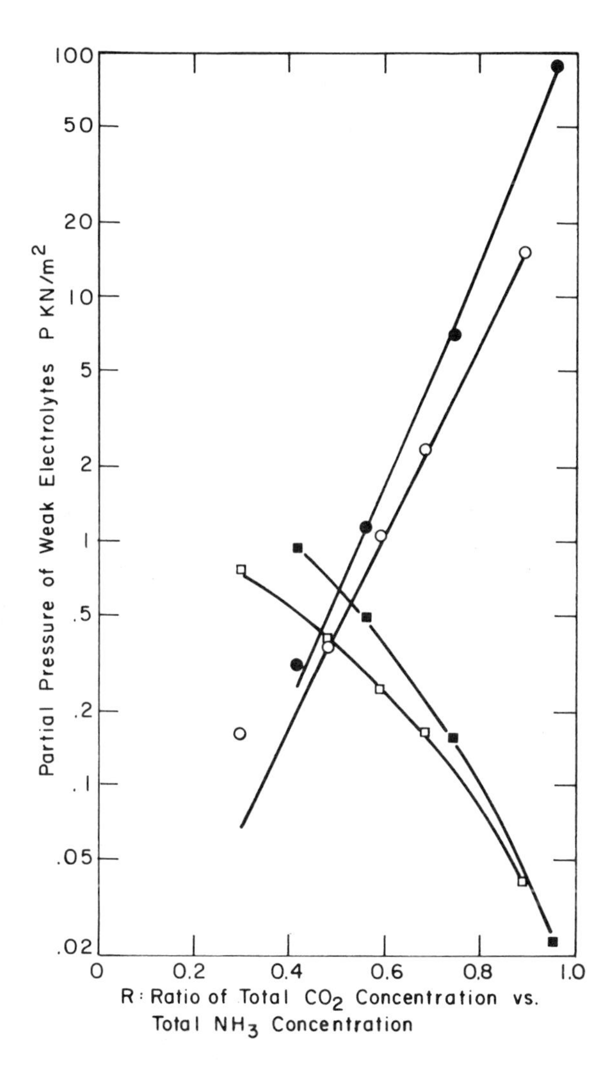

Figure 2. Partial pressures of NH_3 and CO_2 of the NH_3–CO_2 aqueous solution at 293.15 K (experimental data (17):(□) NH_3 (2N NH_3); (■) NH_3 (1N NH_3); (○) CO_2 (2N NH_3); (●) CO_2 (1N NH_3); (——) calculated)

Limitations of the Pitzer Equation

The Pitzer equation just presented is subject to all of the limitations of a virial-expansion type equation. The equation parameters, denoting the short-range interactions between and among solute species, are arbitrary, highly temperature-dependent and are characteristic of the solvent. Binary parameters are expressed as empirical functions of ionic strength. Different empirical functions are proposed for different types of electrolytes. Ternary parameters are required for electrolyte systems at higher ionic strength. Furthermore, for the industrially important class of mixed solvent electrolyte systems, the Pitzer equation is not applicable because its parameters are unknown functions of solvent composition and because the empirical expressions required are available only for water. Therefore, although the Pitzer equation has been shown to be a convenient and accurate representation of aqueous electrolyte systems, a more versatile model is needed to cover a wider variety of electrolyte systems.

The Local Composition Model

Nonelectrolyte systems, which are characterized by short-range forces between molecules, have frequently been studied using the local composition concept. Models such as Wilson (19), NRTL (Renon and Prausnitz, (20)), and UNIQUAC (Abrams and Prausnitz, (21)) have resulted. Such models have proven to be a great advancement over older models based on algebraic expansions of mole fraction, such as the Margules model. In this study the local composition concept is applied to the short range interaction forces occurring in electrolyte systems with the hope that a similar advance over the Pitzer model will result. However, it must be emphasized that the local composition concept is in no sense rigorous. It is used to develop correlating expressions, with adjustable parameters, for experimental data. The purpose in adopting a quasi-theoretical approach is to develop expressions with a small number of parameters, that apply over wide concentration ranges, may be expressed as simple functions of temperature, and may be used to predict the behavior of multicomponent systems. The validity of this approach may be determined only by its empirical success or failure. We believe that the model to be presented in this study is quite successful by these criteria, as demonstrated by the examples shown in this paper and additional work we have done with multicomponent systems involving weak electrolytes.

A fundamental difference between electrolyte systems and nonelectrolyte systems is the presence of long range ion-ion electrostatic forces in electrolyte systems. No attempt was made to develop a long-range contribution model based on the local compostion concept. Instead, the Debye-Huckel formula as proposed by Fowler and Guggenheim (22) was used without modification to represent the unsymmetric excess Gibbs energy contribution arising from the long-range ion-ion electrostatic forces. Nevertheless, the local composition concept is consistent with the Debye-Huckel formula in the sense that the Boltzmann distribution law is assumed in both models. The Debye-Huckel formula is a function of solvent density, dielectric constant, and ionic strength. It is known to correctly account for the ion-ion electrostatic contribution in the limit of infinite dilution. When electrolyte concentration increases, short-range forces start to play a role and finally dominate in the region of high electrolyte concentration (Robinson and Stokes, (13)).

The general approach taken in the present study is as follows. The Debeye-Huckel formula is used to represent long-range ion-ion interactions while the local composition concept is used to represent short range interactions of all kinds. The local composition model is based on two fundamental assumptions; 1) that the local composition of anions around anions is zero, and similarly for cations, which is equivalent to assuming that repulsive forces between ions of like charge are large, 2) that the distribution of anions and cations around solvent molecules is such that the net ionic charge is zero. The latter assumption we refer to as local electroneutrality.

The local compostion model is developed as a symmetric model, based on pure solvent and hypothetical pure completely-dissociated liquid electrolyte. This model is then normalized by infinite dilution activity coefficients in order to obtain an unsymmetric local composition model. Finally the unsymmetric Debye-Huckel and local composition expressions are added to yield the excess Gibbs energy expression proposed in this study.

Development of the Local Composition Model

Among the various models incorporating the local composition concept for short-range interactions, the NRTL equation is adopted in this study. Electrolyte systems are characterized by extraordinarily large heats of mixing. Compared to the heat of mixing, the nonideal entropy of mixing is negligible, which is consistent with the basic assumption behind the NRTL equation. In addition, the NRTL equation is algebraically simple while applicable to mixtures which exhibit phase splitting. No specific volume or area data are required.

In the NRTL model, the local mole fractions x_{ji} and x_{ii} of species j and i, respectively, in the immediate neighborhood of a central molecule of species i are related by

$$x_{ji}/x_{ii} = (x_j/x_i)G_{ji} \tag{7}$$

where

$$G_{ji} = \exp(-\alpha\tau_{ji})$$
$$\tau_{ji} = (g_{ji}-g_{ii})/RT$$

The quantities g_{ji} and g_{ii} are, respectively, energies of interaction between j-i and i-i pairs of species, and are inherently symmetric ($g_{ji}=g_{ij}$). The nonrandomness factor, α, was fixed at a value of 0.2 in this study.

For convenience in representing other local mole fraction ratios, we introduce additional notation as follows:

$$x_{ji}/x_{ki} = (x_j/x_k)G_{ji,ki} \tag{8}$$

where

$$G_{ji,ki} = \exp(-\alpha\tau_{ji,ki})$$
$$\tau_{ji,ki} = (g_{ji}-g_{ki})/RT$$

While the derivation that follows may be generalized to handle all types of electrolyte systems, for the sake of simplicity, the derivation will be based on a single completely-dissociated electrolyte, single solvent system. In a binary mixture of single completely-dissociated electrolyte and single solvent, we assume that there are three types of cells. One type consists of a central solvent molecule with solvent molecules, anions and cations in the immediate neighborhood. The other two types have either an anion or cation as the central species, and an immediate neighborhood consisting of solvent molecules and oppositely-charged ions, but no ions of like charge (i.e., $x_{cc}=x_{aa}=0$). The local mole fractions are related by:

$$\begin{aligned} &x_{cm}+x_{am}+x_{mm}=1 && \text{(central solvent cells)} \\ &x_{mc}+x_{ac} = 1 && \text{(central cation cells)} \\ &x_{ma}+x_{ca} = 1 && \text{(central anion cells)} \end{aligned} \tag{9}$$

Among the three types of cells there are four distinct local mole fraction ratios: x_{cm}/x_{mm}, x_{am}/x_{mm}, x_{mc}/x_{ac}, and x_{ma}/x_{ca}. It is notable that the assumption that $x_{cc}=x_{aa}=0$ is equivalent to the assumption that g_{cc} and g_{aa} are much greater than the other interaction energies.

By combining equations (7), (8) and (9), the following expressions for the local mole fractions in terms of overall mole fractions may be derived:

$$x_{im} = x_i G_{im}/(x_a G_{am} + x_c G_{cm} + x_m G_{mm}) \quad (i = c, a, m)$$
$$x_{ac} = x_a/(x_a + x_m G_{mc,ac}) \tag{10}$$
$$x_{ca} = x_c/(x_c + x_m G_{ma,ca})$$

In order to obtain an expression for the excess Gibbs energy, we first define $g^{(a)}$, $g^{(c)}$, and $g^{(m)}$ as the residual Gibbs energies per mole of cells of central anion, central cation and central solvent molecule, respectively. These Gibbs energies are related to the local mole fractions as follows:

$$g^{(a)} = Z_a(x_{ma}g_{ma} + x_{ca}g_{ca})$$
$$g^{(c)} = Z_c(x_{mc}g_{mc} + x_{ac}g_{ac}) \tag{11}$$
$$g^{(m)} = x_{am}g_{am} + x_{cm}g_{cm} + x_{mm}g_{mm}$$

We then adopt the pure solvent as the reference state for the solvent, and a hypothetical pure completely-dissociated liquid electrolyte as the reference state for the electrolyte. The reference Gibbs energies per mole are then:

$$g_{ref}^{(c)} = Z_c g_{ac}$$
$$g_{ref}^{(a)} = Z_a g_{ca} \tag{12}$$
$$g_{ref}^{(m)} = g_{mm}$$

In both equations (11) and (12) the charge number Z_c and Z_a are introduced to account for the fact that the ratio of the coordination number of central anion cells to that of central cation cells must be equal to the corresponding ratio of charge numbers.

The molar excess Gibbs energy may now be derived by summing the changes in residual Gibbs energy resulting when x_m moles of solvent are transfered from the solvent reference state to their cells in the mixture, and when x_a moles of anions and x_c moles of cations are transfered from the electrolyte reference state to their respective cells in the mixture. The expression is:

$$g^{ex} = x_m(g^{(m)} - g_{ref}^{(m)}) + x_c(g^{(c)} - g_{ref}^{(c)}) + x_a(g^{(a)} - g_{ref}^{(a)}) \tag{13}$$

Substituting equations (11) and (12) into equation (13) we obtain

$$g^{ex}/RT = x_m x_{cm}\tau_{cm} + x_m x_{am}\tau_{am} + x_c x_{mc} Z_c \tau_{mc,ac} + x_a x_{ma} Z_a \tau_{ma,ca} \tag{14}$$

The assumption of local electroneutrality applied to the cells of central solvent molecules may be stated as

$$x_{am}Z_a = x_{cm}Z_c \tag{15}$$

Substituting equation (7) into this relationship leads to the following equality:

$$g_{am} = g_{cm} \tag{16}$$

Since the interaction energies are symmetric, it may be inferred from this result that:

$$\tau_{am} = \tau_{cm} = \tau_{ca,m} \tag{17}$$
$$\tau_{mc,ac} = \tau_{ma,ca} = \tau_{m,ca} \tag{18}$$

The binary parameters $\tau_{ca,m}$ and $\tau_{m,ca}$ then become the only two independent adjustable parameters for a single completely-dissociated electrolyte, single solvent system.

In order to combine equation (14) with the Debye-Huckel formula, which accounts for the long-range force contribution, it is necessary to normalize to the infinite dilution reference state for the ions:

$$g^{ex*}/RT = g^{ex}/RT - x_c \ln\gamma_c^\infty - x_a \ln\gamma_a^\infty \tag{19}$$

After employing equation (14) to obtain $\ln\gamma_c^\infty$ and $\ln\gamma_a^\infty$ and substituting back into equation (19), the final result is:

$$\begin{aligned} g^{ex*}/RT = {} & x_m(x_{cm}+x_{am})\tau_{ca,m} \\ & + x_c x_{mc} Z_c \tau_{m,ca} + x_a x_{ma} Z_a \tau_{m,ca} \\ & - x_c(Z_c \tau_{m,ca} + G_{cm}\tau_{ca,m}) \\ & - x_a(Z_a \tau_{m,ca} + G_{am}\tau_{ca,m}) \end{aligned} \tag{20}$$

The equations for binary systems just presented can be generalized to multicomponent systems consisting of any combination of weak and strong electrolytes, molecular solvents, and molecular solutes.

Discussion of the Local Composition Model

The local composition model makes it possible to study electrolyte thermodynamics over a wide range of compositions. It assumes that the Debye-Huckel formula is adequate to represent the long-range ion-ion electrostatic contribution and the local composition model can account for the short-range interactions among all species. While the validity of the Debye-Huckel formula at high ionic strength is questionable, it is hoped that the short-range contribution will dominate at high ionic strength, so that accounting for the long-range ion-ion electrostatic contribution accurately is not critical.

Systems with weak electrolytes, or partially dissociated electrolytes, can be studied if chemical equilibrium among ionic species and molecular species is considered. Multisolvent systems can be investigated with the knowledge of mixed-solvent dielectric constant and density which are required for the Debye-Huckel formula. The hypothetical pure completely-dissociated liquid electrolyte model has nothing to do with solid salt crystals. However, salt precipitation can also be studied if solubility product constants are known.

It should be noted that the local composition model is not consistent with the commonly accepted solvation theory. According to the solvation theory, ionic species are completely solvated by solvent molecules. In other words, the local mole fraction of solvent molecules around a central ion is unity. This becomes unrealistic when applied to high concentration electrolyte systems since the number of solvent molecules will be insufficient to completely solvate ions. With the local composition model, all ions are, effectively, completely surrounded by solvent molecules in dilute electrolyte systems and only partially surrounded by solvent molecules in high concentration electrolyte systems. Therefore, the local composition model is believed to be closer to the physical reality than the solvation theory.

Application of the Local Composition Model

A wide variety of data for mean ionic activity coefficients, osmotic coefficients, vapor pressure depression, and vapor-liquid equilibrium of binary and ternary electrolyte systems have been correlated successfully by the local composition model. Some results are shown in Table 1 to Table 10 and Figure 3 to Figure 7. In each case, the chemical equilibrium between the species has been ignored. That is, complete dissociation of strong electrolytes has been assumed. This assumption is not required by the local composition model but has been made here in order to simplify the systems treated.

In general, data are fit quite well with the model. For example, with only two binary parameters, the average standard deviation of calculated $\ln\gamma^*$ versus measured $\ln\gamma^*$ of the 50 uni-univalent aqueous single electrolyte systems listed in Table 1 is only 0.009. Although the fit is not as good as the Pitzer equation, which applies only to aqueous electrolyte systems, with two binary parameters and one ternary parameter (Pitzer, (5)), it is quite satisfactory and better than that of Bromley's equation (1).

Data correlation results for single-salt, single-solvent binary systems are shown in Table 1 to Table 6 and Figure 3 to Figure 6. There is an obvious trend between $\tau_{m,ca}$ and standard deviation of calculated $\ln\gamma^*$ versus measured $\ln\gamma^*$. When the absolute value of $\tau_{m,ca}$ increases, standard deviation also increases. This is consistent with the physical meaning of $\tau_{m,ca}$. The larger the absolute value of $\tau_{m,ca}$, the stronger the interaction between cation and anion. In

Table 1. Data and Results of Fit for Aqueous Solutions of uni-univalent electrolyte at 298.15°K - Mean Ionic Activity Coefficient Data

Electrolyte	No. of data* Points	molality	$\tau_{m,ca}$	$\tau_{ca,m}$	$\sigma_{ln\gamma^*}$
$AgNO_3$	23	0.1-6.0	7.295	-3.059	0.012
CsAc	18	0.1-3.5	8.462	-4.500	0.009
CsBr	21	0.1-5.0	8.381	-4.034	0.008
CsCl	23	0.1-6.0	8.368	-4.043	0.009
CsI	17	0.1-3.0	8.280	-3.963	0.007
$CsNO_3$	12	0.1-1.4	8.988	-4.057	0.003
KAc	18	0.1-3.5	8.459	-4.476	0.007
KBr	22	0.1-5.5	7.901	-3.962	0.002
KCl	20	0.1-4.5	7.917	-3.944	0.002
KCNS	21	0.1-5.0	7.319	-3.644	0.001
KF	19	0.1-4.0	8.679	-4.373	0.004
KH malonate	21	0.1-5.0	7.338	-3.462	0.006
KH succinate	20	0.1-4.5	7.982	-3.861	0.002
KH_2PO_4	14	0.1-1.8	8.924	-4.017	0.003
KI	20	0.1-4.5	7.620	-3.892	0.004
KNO_3	18	0.1-3.5	7.642	-3.327	0.008
KOH	23	0.1-6.0	9.733	-4.945	0.019
LiAc	19	0.1-4.0	8.304	-4.278	0.004
LiBr	23	0.1-6.0	10.331	-5.251	0.046
LiCl	23	0.1-6.0	9.900	-5.046	0.036
$LiClO_4$	19	0.1-4.0	9.464	-4.986	0.022
LiI	17	0.1-3.0	9.157	-4.889	0.024
$LiNO_3$	23	0.1-6.0	8.804	-4.562	0.010
LiOH	19	0.1-4.0	8.920	-4.275	0.027
LiTol	20	0.1-4.5	7.396	-3.710	0.012
NaAc	18	0.1-3.5	8.257	-4.356	0.006
NaBr	19	0.1-4.0	8.672	-4.435	0.008
$NaBrO_3$	16	0.1-2.5	7.587	-3.589	0.001
Na butyrate	18	0.1-3.5	7.230	-4.104	0.008
NaCl	23	0.1-6.0	8.715	-4.400	0.014
$NaClO_3$	18	0.1-3.5	7.128	-3.527	0.004
$NaClO_4$	23	0.1-6.0	7.799	-3.937	0.007
NaCNS	19	0.1-4.0	7.770	-4.078	0.009
NaF	10	0.1-1.0	7.517	-3.677	0.001
Na formate	18	0.1-3.5	7.295	-3.776	0.003
NaH_2PO_4	23	0.1-6.0	8.138	-3.711	0.002
NaI	18	0.1-3.5	8.752	-4.535	0.009
Na malonate	21	0.1-5.0	7.527	-3.659	0.003
$NaNO_3$	23	0.1-6.0	7.071	-3.381	0.003
NaOH	23	0.1-6.0	9.225	-4.647	0.026
Na propionate	17	0.1-3.0	8.277	-4.435	0.006
Na succinate	21	0.1-5.0	8.075	-3.968	0.002
NH_4Cl	23	0.1-6.0	7.614	-3.800	0.002

Table 1. Continued

NH_4NO_3	23	0.1-6.0	7.170	-3.295	0.014
RbAc	18	0.1-3.5	8.602	-4.545	0.008
RbBr	21	0.1-5.0	7.920	-3.891	0.004
RbCl	21	0.1-5.0	8.086	-3.983	0.003
RbI	21	0.1-5.0	8.052	-3.949	0.004
$RbNO_3$	20	0.1-4.5	7.648	-3.287	0.013
TlAc	23	0.1-6.0	7.683	-3.618	0.014

*(Robinson and Stokes, (13))

Table 2. Data and Results of Fit for Acids at 298.15°K Assuming Complete Dissociation - Mean Ionic Activity Coefficient Data

Acid	No. of Data* Points	Highest Molality	$\tau_{m,ca}$	$\tau_{ca,m}$	$\sigma_{\ln\gamma^*}$
HBr	17	3.0	9.742	-5.087	0.014
HCl	23	6.0	9.957	-5.106	0.031
$HClO_4$	23	6.0	10.488	-5.328	0.058
HI	17	3.0	9.483	-5.059	0.017
HNO_3	17	3.0	8.327	-4.341	0.008

*(Robinson and Stokes, (13))

Table 3. Data and Results of Fit for Aqueous Solutions of Bi-bivalent Electrolytes at 298.15°K - Mean Ionic Activity Coefficient Data

Electrolyte	No. of Data* Points	Molality	$\tau_{m,ca}$	$\tau_{ca,m}$	$\sigma_{\ln\gamma^*}$
$BeSO_4$	18	0.2-4.0	11.728	-6.905	0.049
$MgSO_4$	16	0.2-3.5	11.623	-6.827	0.047
$MnSO_4$	18	0.2-4.0	11.499	-6.732	0.046
$NiSO_4$	15	0.2-2.5	11.704	-6.826	0.042
$CuSO_4$	11	0.2-1.4	12.128	-7.043	0.043
$ZnSO_4$	17	0.2-3.5	11.693	-6.827	0.046
$CdSO_4$	17	0.2-3.5	11.481	-6.704	0.053
UO_2SO_4	22	0.2-6.0	11.316	-6.646	0.078

*(Robinson and Stokes, (13))

Table 4 Data and Results of Fit for Uni-univalent Electrolyte Effect on the Vapor Pressure of Methanol at 298.05°K

Electrolyte	No. of Data* Points	Highest Molality	$\tau_{m,ca}$	$\tau_{ca,m}$	σ_p
LiCl	9	5.3554	11.783	-5.562	0.034
NaBr	9	1.556	10.717	-5.176	0.002
NaOH	9	5.9413	10.372	-5.633	0.019
NaI	16	4.5200	9.716	-5.186	0.011
KI	9	1.1219	10.765	-5.138	0.002

*(Bixon et al., (25))

Table 5 Temperature Effect on Data and Results of Fit for Aqueous NaCl Solutions - Mean Ionic Activity Coefficient Data

T°K	No. of Data* Points	Molality	$\tau_{m,ca}$	$\tau_{ca,m}$	$\sigma_{\ln\gamma}$*
273.15	28	0.05-6.0	8.831	-4.406	0.018
298.15	28	0.05-6.0	8.744	-4.409	0.014
323.15	28	0.05-6.0	8.629	-4.380	0.011
348.15	28	0.05-6.0	8.510	-4.334	0.008
373.15	28	0.05-6.0	8.420	-4.288	0.005

*smoothed data (Silvester and Pitzer, (23))

Table 6 Temperature Effect on Data and Results of Fit for Aqueous KBr Solutions - Mean Ionic Activity Coefficient Data

T°K	No. of Data* Points	Molality	$\tau_{m,ca}$	$\tau_{ca,m}$	$\sigma_{\ln\gamma}$*
333.15	14	0.1-4.0	7.860	-3.994	0.001
343.15	14	0.1-4.0	7.831	-3.989	0.001
353.15	14	0.1-4.0	7.773	-3.970	0.001
363.15	14	0.1-4.0	7.769	-3.971	0.002
373.15	14	0.1-4.0	7.760	-3.969	0.002

*(Robinson and Stokes, (13))

Table 7. Data and Results of Fit for the Mean Ionic Activity Coefficients of HCL and HBr in Halide Solutions at 298.15°K Acid Concentration = 0.01M (approximated as 0.01m)

Acid	Salt	No. of Data* Point	Highest Salt Molality	$\tau_{MX,HX}(=-\tau_{HX,MX})$	$\sigma_{\ln\gamma^*}$
HCl	KCl	13	3.5	-1.633	0.020
HCl	NaCl	9	3.0	-1.129	0.008
HCl	LiCl	11	4.0	-0.106	0.022
HBr	KBr	12	3.0	-1.464	0.009
HBr	NaBr	11	3.0	-1.113	0.007
HBr	LiBr	15	3.0	-0.307	0.027

*(Harned and Owen, (14))

Table 8. Data and Results of Fit on Solubility of Carbon Dioxide in Aqueous Solutions at 298.15°K

Salt	No. of Data+ Points	Highest Molality	$\tau_{m,ca}$	$\tau_{ca,m}$	$\sigma_{\ln\gamma^*_m}$
NaCl	9	5.732	5.733	-4.115	0.014
KCl	5	3.942	10.414	-6.109	0.008

$\tau_{CO_2,H_2O} = -1.644917+(0.320488D-1)*(T-273.15)$
$\tau_{H_2O,CO_2} = \tau_{CO_2,H_2O}*(2.442172)$
(CO_2-H_2O binary data obtained from Houghton, G., A. M. McLean, and P. D. Ritchie, (26))

+(Yasunishi, A. and F. Yoshida, (24))
+original data were expressed in terms of molarity and the Ostwald coefficient

Table 9. Vapor-Liquid Equilibrium Data Correlation for Methanol-Water-NaBr system at 298.15°K

x_1'	x_2'	Salt Molality	Expt. Data* P**	Expt. Data* y_1	Calc. Value P**	Calc. Value y_1	diff P	diff y
0.148	0.852	1	49.8	0.603	50.4	0.622	+0.6	+0.019
0.148	0.852	2	50.1	0.627	52.7	0.655	+2.6	+0.028
0.148	0.852	4	51.1	0.675	53.2	0.686	+2.1	+0.011
0.148	0.852	7.1	50.4	0.756	47.7	0.693	-2.7	-0.063
0.292	0.708	1	68.0	0.742	68.7	0.766	+0.7	+0.024
0.292	0.708	2	68.9	0.762	70.9	0.783	+2.0	+0.021
0.292	0.708	4	68.6	0.798	69.3	0.792	+0.7	-0.006
0.292	0.708	5.7	65.2	0.820	64.2	0.788	-1.0	-0.032
0.500	0.500	1	85.1	0.850	86.9	0.859	+1.8	+0.009
0.500	0.500	2	84.35	0.860	86.7	0.862	+2.35	+0.002
0.500	0.500	4	80.4	0.884	79.9	0.853	-0.5	-0.031
0.700	0.300	1	99.5	0.920	100.4	0.916	+0.9	-0.004
0.700	0.300	2	95.7	0.926	97.2	0.911	+1.5	-0.015
0.700	0.300	2.8	91.4	0.932	92.7	0.906	+1.3	-0.026
0.900	0.100	1	114.2	0.977	113.6	0.969	-0.6	-0.008
0.900	0.100	1.9	107.3	0.979	107.1	0.966	-0.2	-0.013
						mean dev.	1.35	0.0195

$\tau_{H_2O,NaBr}$ =8.672+0.244*(78.48-D)/(78.48-32.66)
τ_{NaBr,H_2O} =-4.435+0.244*(78.48-D)/(78.48-32.66)
$\tau_{CH_3OH,NaBr}$ =10.717-3.493*(32.66-D)/(78.48-32.66)
τ_{NaBr,CH_3OH} =-5.176-3.493*(32.66-D)/(78.48-32.66)
τ_{H_2O,CH_3OH} =0.2944
τ_{CH_3OH,H_2O} =0.1936
(methanol-water binary data obtained from Gmehling, J. and U. Onken, (27))

d =0.9971+(-0.163939D-2)*(x_1')+(0.1701563D-5)*(x_1')^2
-(0.6285073D-7)*(x_1')^3

D =78.48-(0.4233608)*(x_1')-(0.3307047D-3)*(x_1')^2
-(0.3434429D-7)*(x_1')^3

$x_1' = x_{CH_3OH}/(x_{CH_3OH}+x_{H_2O})$
$x_2' = x_{H_2O}/(x_{CH_3OH}+x_{H_2O})$

*(Ciparis, (28))
**(unit: mmHg)

Table 10. Vapor-Liquid Equilibrium Data Correlation for Methanol-Water-LiCl system at 298.15°K

x_1'	x_2'	Salt Molality	Expt. Data* P**	Expt. Data* y_1	Calc. Value P**	Calc. Value y_1	diff P	diff y
0.152	0.848	1	47.3	0.605	47.7	0.597	+0.4	-0.008
0.298	0.702	1	65.3	0.765	65.2	0.751	-0.1	-0.014
0.470	0.530	1	80.0	0.860	81.0	0.840	+1.0	-0.020
0.700	0.300	1	96.3	0.930	98.6	0.917	+2.3	-0.013
0.958	0.042	1	115.3	0.993	117.4	0.988	+2.1	-0.005
						mean dev.	1.2	0.012

$\tau_{H_2O,LiCl}$ =9.900-0.2239*(78.48-D)/(78.48-32.66)
τ_{LiCl,H_2O} =-5.046-0.2239*(78.48-D)/(78.48-32.66)
$\tau_{CH_3OH,LiCl}$ =11.783-1.853*(32.66-D)/(78.48-32.66)
τ_{LiCl,CH_3OH} =-5.562-1.853*(32.66-D)/(78.48-32.66)
τ_{CH_3OH,H_2O} =0.1936
τ_{H_2O,CH_3OH} =0.2944
(methanol-water binary data obtained from Gmehling, J. and U. Onken, (27))

$$d = 0.9971+(-0.163939D-2)*(x_1')+(0.1701563D-5)*(x_1')^2 -(0.6285073D-7)*(x_1')^3$$

$$D = 78.48-(0.4233608)*(x_1')-(0.3307047D-3)*(x_1')^2 -(0.3434429D-7)*(x_1')^3$$

$$x_1' = x_{CH_3OH}/(x_{CH_3OH}+x_{H_2O})$$
$$x_2' = x_{H_2O}/(x_{CH_3OH}+x_{H_2O})$$

*(Ciparis, (28))
**(unit: mmHg)

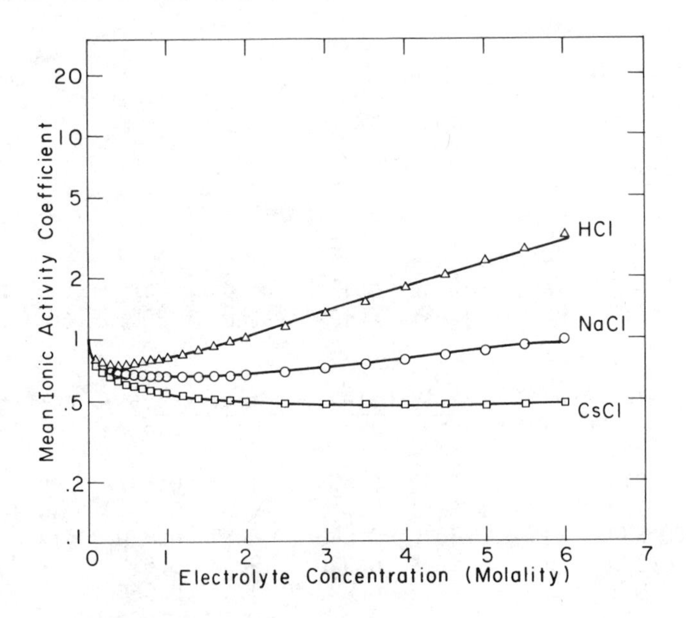

Figure 3. Comparison of the calculated and experimental mean ionic activity coefficients of three uni-univalent electrolytes at 298.15 K: (——) calculated; (△, ○, □) data from Ref. 13

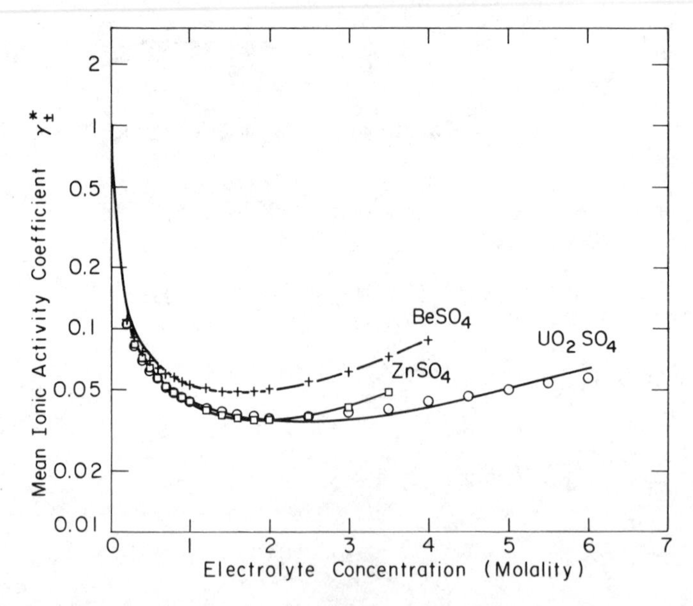

Figure 4. Comparison of the calculated and experimental mean ionic activity coefficients of three bi-bivalent electrolytes at 298.15 K: (——) calculated; (+, □, ○) data from Ref. 13

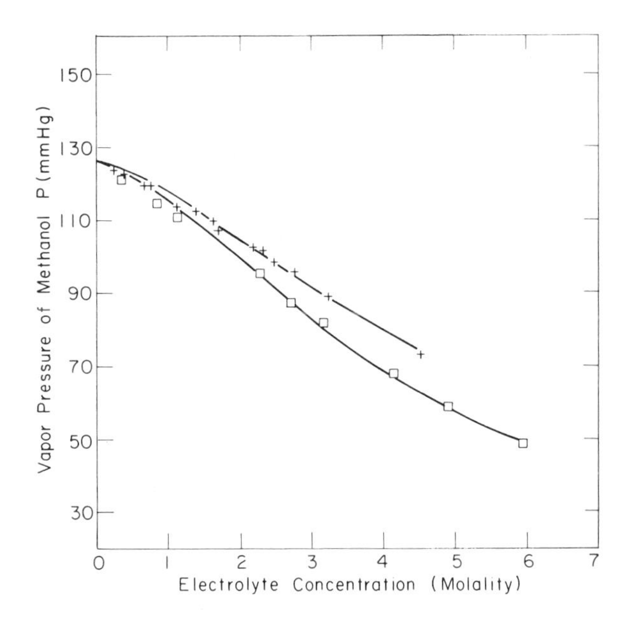

Figure 5. Salt effect on the vapor pressure of methanol at 298.05 K (experimental data (25): (□) NaOH–methanol; (+) NaI–methanol)

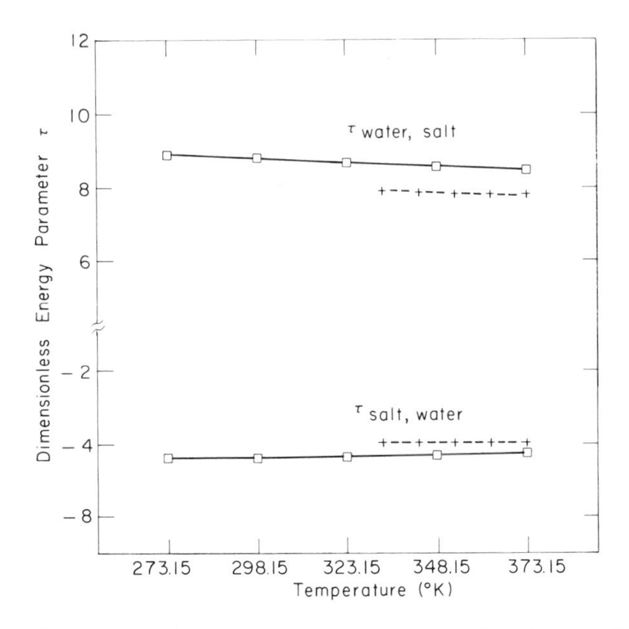

Figure 6. Dimensionless interaction parameters as determined from isothermal fits at various temperatures ((+) KBr–water; (□) NaCl–water)

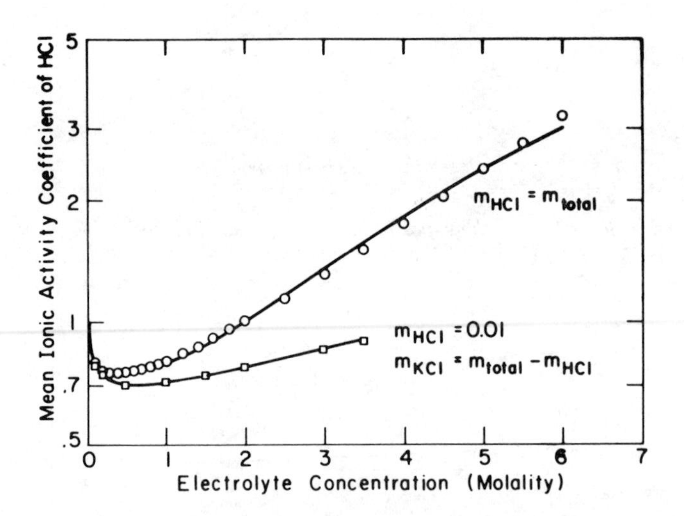

Figure 7. Comparison of the calculated and experimental mean ionic activity coefficient of HCl in the KCl aqueous solution at 298.15 K: (——) calculated; (○, □) data from Ref. 14

other words, the ions tend to associate and lack of fit on data correlation can be related to this ion association. When this occurs, chemical equilibrium of ion association should be taken into consideration. Larger standard deviations were found for several higher valent electrolytes. It also should be noted that the electrolyte concentration in these systems is very high and it seems resaonable that association should occur to some extent.

The interaction parameters are weak, linear functions of temperature, as shown in Table 5, Table 6 and Figure 6. These tables and figure show the results of isothermal fits for activity coefficient data of aqueous NaCl and KBr at various temperatures. The Pitzer equation parameters are, however, strongly dependent on temperature (Silvester and Pitzer, (23)).

There can be many different types of ternary electrolyte systems. The HCl-KCl-H_2O system is an example of a two-electrolyte, one-solvent ternary electrolyte systems. Some data correlation results for the activity coefficients of salts in ternary electrolyte systems of this type are shown in Table 7 and Figure 7. Water-electrolyte binary parameters were obtained from Table 1.

A second type of ternary electrolyte systems is solvent - supercritical molecular solute - salt systems. The concentration of supercritical molecular solutes in these systems is generally very low. Therefore, the salting out effects are essentially effects of the presence of salts on the unsymmetric activity coefficient of molecular solutes at infinite dilution. The interaction parameters for NaCl-CO_2 binary pair and KCl-CO_2 binary pair are shown in Table 8. Water-electrolyte binary parameters were obtained from Table 1. Water-carbon dioxide binary parameters were correlated assuming dissociation of carbon dioxide in water is negligible. It is interesting to note that the Setschenow equation fits only approximately these two systems (Yasunishi and Yoshida, (24)).

Another type of ternary electrolyte system consists of two solvents and one salt, such as methanol-water-NaBr. Vapor-liquid equilibrium of such mixed solvent electrolyte systems has never been studied with a thermodynamic model that takes into account the presence of salts explicitly. However, it should be recognized that the interaction parameters of solvent-salt binary systems are functions of the mixed solvent dielectric constant since the ion-molecular electrostatic interaction energies, g_{ma} and g_{mc}, depend on the reciprocal of the dielectric constant of the solvent (Robinson and Stokes, (13)). Pure component parameters, such as g_{mm} and g_{ca}, are not functions of dielectric constant. Results of data correlation on vapor-liquid equilibrium of methanol-water-NaBr and methanol-water-LiCl at 298.15°K are shown in Tables 9 and 10.

Conclusions

Two activity coefficient models have been developed for vapor-liquid equilibrium of electrolyte systems. The first model is an extension of the Pitzer equation and is applicable to aqueous electrolyte systems containing any number of molecular and ionic solutes. The validity of the model has been shown by data correlation studies on three aqueous electrolyte systems of industrial interest. The second model is based on the local composition concept and is designed to be applicable to all kinds of electrolyte systems. Preliminary data correlation results on many binary and ternary electrolyte systems suggest the validity of the local composition model.

Abstract

The semi-empirical Pitzer equation for modeling equilibrium in aqueous electrolyte systems has been extended in a thermodynamically consistent manner to allow for molecular as well as ionic solutes. Under limiting conditions, the extended model reduces to the well-known Setschenow equation for the salting out effect of molecular solutes. To test the validity of the model, correlations of vapor-liquid equilibrium data were carried out for three systems: the hydrochloric acid aqueous solution at 298.15°K and concentrations up to 18 molal; the NH_3-CO_2 aqueous solution studied by van Krevelen, et al. (17) at 293.15°K; and the K_2CO_3-CO_2 aqueous solution of the Hot Carbonate Process with temperatures from 343.15°K to 413.15°K and concentrations up to 40 weight percent equivalent potassium carbonate. The success of the correlations suggests the validity of the model for aqueous electrolyte systems of industrial interest.

For the industrially important class of mixed solvent, electrolyte systems, the Pitzer equation is not useful because its parameters are unknown functions of solvent composition. A local composition model is developed for these systems which assumes that the excess Gibbs free energy is the sum of two contributions, one resulting from long-range forces between ions and the other from short-range forces between all species. The long-range term has been satisfactorily described by the Debye-Huckel formula and is retained. The short-range contribution is modeled by utilizing the concept of local compositions in a manner similar to Renon and Prausnitz (20) but with additional assumptions appropriate for electrolyte systems. Preliminary results suggest the validity of the model since good fits to experimental data have been obtained for a wide range of binary and ternary systems with only binary parameters.

Acknowledgement

This work was part of the Advanced System for Process Engineering Project (ASPEN) supported by the U.S. Department of Energy under contract number E(49-18)-2295, Task No. 9.

Notation

A_ϕ = Debye-Huckel constant for the osmotic coefficient
B_{ca} = binary ion-ion interaction parameter
C_{ca} = ternary ion-ion interaction parameter
$D_{ca,m}$ = binary salt-molecule interaction parameter or Setschenow constant
D = dielectric constant
G^{ex} = excess Gibbs free energy
I = ionic strength ($=1/2 \sum_i m_i Z_i^2$)
P = pressure
R = gas constant
T = temperature (^oK)
Z = absolute value of ionic charge
d = density
g^{ex} = molar excess Gibbs free energy
m = molality (g-mole/kg of solvent)
n_w = number of kg of the solvent, water
x = liquid phase mole fraction
y = vapor phase mole fraction

Greek Letters

$\beta^{(0)}$, $\beta^{(1)}$ = parameters of the empirical expression of B(I)
γ = activity coefficient
θ = binary ion-ion difference parameter
λ = second virial coefficient of the basic Pitzer equation
μ = third virial coefficient of the basic Pitzer equation
υ = stoichiometric coefficient
σ = standard deviation
τ = binary interaction energy parameter
ψ = ternary ion-ion difference parameter
ω = binary salt-molecule difference parameter

Superscripts

* = unsymmetric convention
∞ = infinite dilution

Subscripts

a, a' = anion
c, c' = cation
ca = salt ca
i, j, k = any solute in the Pitzer equation
any species in the local composition model
m = molecular solute in the Pitzer equation
any molecular species in the local composition model
ref = reference
s = salt

Literature Cited

1. Bromley, L. A., "Thermodynamic Properties of Strong Electrolytes in Aqueous Solutions," AIChE J., 1973, 19, 313.
2. Meissner, H. P. and J. W. Tester, "Activity Coefficients of Strong Electrolytes in Aqueous Solutions," Ind. Eng. Chem. Process Des. Dev., 1972, 11, 128.
3. Meissner, H. P. and C. L. Kusik, "Activity Coefficients of Strong Electrolytes in Multicomponent Aqueous Solutions," AIChE J., 1972, 18, 294.
4. Pitzer, K. S., "Thermodynamics of Electrolytes. I. Theoretical Basis and General Equations," J. Phys. Chem., 1973, 77, 268.
5. Pitzer, K. S. and Guillermo Mayorga, "Thermodynamics of Electrolytes. II. Activity and Osmotic Coefficients for Strong Electrolytes with One or Both Ions Univalent," J. Phys. Chem., 1973, 77, 2300.
6. Pitzer, K. S. and J. J. Kim, "Thermodynamics of Electrolytes. IV. Activity and Osmotic Coefficients for Mixed Electrolytes," J. Am. Chem. Soc., 1974, 96, 5701.
7. Cruz, Jose-Luis and H. Renon, "A New Thermodynamic Representation of Binary Electrolyte Solutions Nonideality in the Whole Range of Concentrations," AIChE J., 1978, 24, 817.
8. Edwards, T. J., John Newman and J. M. Prausnitz, "Thermodynamics of Aqueous Solutions Containing Volatile Weak Electrolytes," AIChE J., 1975, 21, 248.
9. Beutier, D., and H. Renon, "Representation of NH_3-H_2S-H_2O, NH_3-CO_2-H_2O, and NH_3-SO_2-H_2O Vapor-Liquid Equilibrium," Ind. Eng. Chem. Process Des. Dev., 1978, 17, 220.
10. Edwards, T. J., Gerd Maurer, John Newman and J. M. Prausnitz, "Vapor-Liquid Equilibria in Multicomponent Aqueous Solutions of Volatile Weak Electrolytes," AIChE J., 1978, 24, 966.
11. Chen, C., H. I. Britt, J. F. Boston, and L. B. Evans, "Extension and Application of the Pitzer equation for Vapor-Liquid Equilibrium of Aqueous Electrolyte Systems with Molecular Solutes," AIChE J., 1979, 25, 820.
12. Prausnitz, J. M., Molecular Thermodynamics of Fluid-Phase Equilibria, Prentice-Hall, Englewood Cliffs, N. J., 1969.
13. Robinson, R. A. and R. H. Stokes, Electrolyte Solutions, 2nd ed., Butterworths, 1970.
14. Harned, H. S., and B. B. Owen, "The Physical Chemistry of Electrolyte Solutions," ACS Monograph Series No. 137, 3rd Ed. Reinhold, New York, 1958.
15. Gordon, J. E., The Organic Chemistry of Electrolyte Solutions, Wiley, New York, 1975.

16. Vega, R., and J. H. Vera, "Phase Equilibria of Concentrated Aqueous Solutions Containing Volatile Strong Electrolytes," Can. J. Chem. Eng., 1976, 54, 245.
17. van Krevelen, D. W., P. J. Hoftijzer, and F. J. Huntjens, "Composition and Vapor Pressures of Aqueous Solutions of Ammonia, Carbon Dioxide, and Hydrogen Sulfide," Rec. Trav. Chim. Pays-bas, 1949, 68, 191.
18. Tosh, J. S., J. H. Field, H. E. Benson and W. P. Waynes, "Equilibrium Study of the System Potassium Carbonate, Potassium Bicarbonate, Carbon Dioxide and Water," U. S. Bureau of Mines, Report of Investigations, 1959, 5484.
19. Wilson, G. M., "Vapor-Liquid Equilibrium XI. A New Expression for the Excess Free Energy of Mixing," J. Am. Chem. Soc., 1964, 86, 127.
20. Renon, H. and J. M. Prausnitz, "Local Compositions in Thermodynamic Excess Functions for Liquid Mixtures," AIChE J. 1968, 14, 135.
21. Abrams, D. S., and J. M. Prausnitz, "Statistical Thermodynamics of Liquid Mixtures: A New Expression for the Excess Gibbs Energy of Partly or Completely Miscible Systems," AIChE J., 1975, 21, 116.
22. Fowler, R. H., and E. A. Guggenheim, Statistical Thermodynamics, Cambridge Univ. Press, 1949.
23. Silvester, L. F. and K. S. Pitzer, "Thermodynamics of Geothermal Brines, I. Thermodynamic Properties of Vapor-Saturated $NaCl_{(aq)}$ Solutions From 0-300 ^{o}C," LBL-4456, University of California, Berkeley, California, 1976.
24. Yasunishi, A. and F. Yoshida, "Solubility of Carbon Dioxide in Aqueous Electrolyte Solutions," J. of Chem. and Eng. Data, 1979, 24, 11.
25. Bixon, E., R. Guerry and D. Tassios, "Salt Effect on the Vapor Pressure of Pure Solvents: Methanol with Seven Salts at 24.9^{o}C," J. of Chem. and Eng. Data, 1979, 24, 9.
26. Houghton, G., A. M. McLean, and P. D. Ritchie, "Compressibility, fugacity, and Water-Solubility of Carbon Dioxide in the Region 0-36 Atm. and 0-100°C," Chem. Eng. Sci., 1957, 6, 132.
27. Gmehling, J., and U. Onken, Vapor-Liquid Equilibrium Data Collection, Aqueous-Organic Systems, Chemistry Data Series, Vol. 1, Part 1, DECHEMA, 1977.
28. Ciparis, J. N., "Data of Salt Effect in Vapor-Liquid Equilibrium," Lithuanian Agricultural Academy, Kaunas, Lithuania, USSR, 1966.

RECEIVED January 31, 1980.

5

Chemical Equilibria in Flue Gas Scrubbing Slurries

CLAYTON P. KERR

Tennessee Technological University, Cookeville, TN 38501

There are several ways for controlling emissions of sulfur dioxide from coal-fired power plants, but flue gas desulfurization is the most highly developed method with about 20,000 megawatts of generating capacity equipped with flue gas desulfurization equipment and another 20,000 megawatts being constructed or designed. Within 10-15 years, the total electrical generating capacity so equipped will exceed 100,000 megawatts. The subject of this paper is the chemical equilibria of the more prevalent aqueous scrubbing processes for flue gas desulfurization. Because the equilibria of the liquids are highly dependent upon the alkali used and certain design and operating conditions, a few words of discussion about some of the more prevalent processes make the subsequent discussion more easily understood.

Flue gas desulfurization processes can be divided into two broad categories: 1. throw-away processes where the removal reagent is not regenerated and a waste product containing sulfur is created, and 2. regenerative processes where the removal reagent is regenerated and a salable product containing sulfur is created. Both categories of processes can be further subdivided into wet and dry processes.

The throw-away processes with aqueous slurries of lime or limestone as the scrubbing media are the most extensively installed processes. These processes create a waste sludge containing calcium sulfite, calcium sulfate, fly ash, unreacted alkali, and other minor dissolved species in the free water contained in the sludge. Since flue gas contains oxygen, some of the dissolved sulfur dioxide is oxidized, and calcium sulfate is formed.

Several power plants have been equipped with dual alkali processes. These are throw-away processes with two liquid loops. In one common process, the scrubbing liquid is a clear solution of sodium sulfite. The absorption of sulfur dioxide converts the sodium sulfite to sodium bisulfite. In the regeneration loop, an alkali such as lime slurry is added; the sodium sulfite solution is regenerated; and a mixture of calcium sulfite and calcium sulfate is precipitated. The slurry is

0-8412-0569-8/80/47-133-091$05.00/0

filtered, and the clear liquid is returned to the scrubbing loop. This process has the advantages of excellent sulfur dioxide removal and a reduced scaling and plugging tendency because the scrubbing is performed with a clear liquid solution.

Sodium sulfite scrubbing can be configured as a regenerative process if the regeneration is performed thermally in an evaporator system. Thermal regeneration removes gaseous sulfur dioxide with the sodium bisulfite being converted back to sodium sulfite. The sulfur dioxide is then reduced to elemental sulfur or converted to sulfuric acid.

Several plants have been equipped with processes that use an aqueous slurry of magnesium sulfite and magnesium oxide as the scrubbing material. Sulfur dioxide is absorbed, and a mixture of magnesium sulfite and magnesium sulfate is precipitated. A bleed stream from the absorption part of the process is dewatered and dried. The crystals of magnesium sulfite and magnesium sulfate are then calcined; solid magnesium oxide is returned to the scrubbing process, and sulfur dioxide which is released with the calcination is then converted into a salable product such as sulfuric acid.

Coals contain chloride in varying amounts, and the burning of coal releases the chloride as gaseous hydrogen chloride. Unless the hydrogen chloride is removed separately, the alkaline liquid of the scrubber will convert the hydrogen chloride into very soluble chloride salts. Depending upon the chloride content of the coal and the amount of water purged from the process, the chloride content of the liquid can be as high as 20,000 parts per million by weight. The chloride content of the circulation liquid has considerable effect on the equilibria in the liquid. It also is an important factor in the choice of materials of construction. Some processes such as magnesium oxide scrubbing require separate removal of chloride.

For various reasons, it is often desirable to perform equilibrium or quasi-equilibrium calculations in the design or operation of a flue gas desulfurization facility. In this paper, equilibrium reactions, formulas for the calculation of temperature dependent equilibrium constants, and methods of calculation of activity coefficients of ions and ion-pairs are presented. Calculation of charged species is based on modified Debye-Hückel theory. The equilibrium calculations are almost always iterative. A way of directing these calculations by using an optimum seeking method is presented. The method is easily computer programmed, and if the problem is properly formulated, machine running time is modest. The method developed is very general and can be used in a wide variety of applications. Two examples of equilibrium calculations for lime or limestone processes are outlined: calculation of calcium sulfite and calcium sulfate supersaturation ratios and calculation of dissolved alkalinity.

Solution Equilibria in Aqueous Flue Gas Desulfurization Processes

Lowell et al. (1) have prepared a general list of the equilibria present in aqueous flue gas desulfurization processes. The relative importance of the dissolved species will be dependent upon the type of process under consideration. For example, in the magnesium oxide process, magnesium is one of the dominant species while sodium is one of the dominant species in the regenerative sodium process. A general list of equilibria is presented in Table 1. Expressions of the following form have been developed for the temperature dependent equilibrium constants (1).

$$\log K = -\frac{B}{T} - C \log T - DT + E \tag{1}$$

K is the temperature dependent equilibrium constant; T is the absolute temperature in degrees Kelvin; and B, C, D, and E are constants. Numerical values of these constants are presented in Table 2.

Values of the equilibrium constants at 298°K can also be calculated from tabulated thermodynamic properties. The standard Gibbs' free energy of the reaction at 298°K is first calculated, and the equilibrium constant at 298°K is then determined from the equation

$$\Delta G^o = -RT\ln K \tag{2}$$

ΔG^o is the standard Gibbs' free energy for the reaction at 298°K; R is the ideal gas constant; and T is 298°K. Since the actual temperature of most slurries or solutions in flue gas scrubbing applications usually does not exceed 50°C, the value of the equilibrium constant can be determined at some temperature other than 298°K by using the van't Hoff equation

$$K = K_{298} \exp\left(-\frac{\Delta H^o}{R}\left(\frac{1}{T} - \frac{1}{298}\right)\right) \tag{3}$$

ΔH^o is the standard heat of reaction at 298°K.

Calculation of Ion and Ion-Pair Activity Coefficients

For the concentrations of dissolved species encountered in flue gas scrubbing applications, the equilibria must be formulated in terms of activities rather than molalities. The activities, molalities, and activity coefficients are related by

$$a_i = \gamma_i m_i \tag{4}$$

where a_i, γ_i, and m_i are respectively the activity, activity coefficient, and molality of component i.

Table I. Equilibria Present in Flue Gas Scrubbing Slurries for the Lime or Limestone Processes

$H_2O \rightleftharpoons H^+ + OH^-$	(1)	$HSO_3^- \rightleftharpoons H^+ + SO_3^{--}$	(2)
$H_2SO_3 \rightleftharpoons H^+ + HSO_3^-$	(3)	$HSO_4^- \rightleftharpoons H^+ + SO_4^{--}$	(4)
$HCO_3^- \rightleftharpoons H^+ + CO_3^{--}$	(5)	$H_2CO_3 \rightleftharpoons H^+ + HCO_3^-$	(6)
$CaOH^+ \rightleftharpoons Ca^{++} + OH^-$	(7)	$CaSO_3^o \rightleftharpoons Ca^{++} + SO_3^{--}$	(8)
$CaSO_4^o \rightleftharpoons Ca^{++} + SO_4^{--}$	(9)	$CaCO_3^o \rightleftharpoons Ca^{++} + CO_3^{--}$	(10)
$CaHCO_3^+ \rightleftharpoons Ca^{++} + HCO_3^-$	(11)	$MgOH^+ \rightleftharpoons Mg^{++} + OH^-$	(12)
$MgSO_3^o \rightleftharpoons Mg^{++} + SO_3^{--}$	(13)	$MgSO_4^o \rightleftharpoons Mg^{++} + SO_4^{--}$	(14)
$MgCO_3^o \rightleftharpoons Mg^{++} + CO_3^{--}$	(15)	$MgHCO_3^+ \rightleftharpoons Mg^{++} + HCO_3^-$	(16)
$NaOH^o \rightleftharpoons Na^+ + OH^-$	(17)	$NaSO_4^- \rightleftharpoons Na^+ + SO_4^{--}$	(18)
$NaCO_3^- \rightleftharpoons Na^+ + CO_3^{--}$	(19)	$NaHCO_3^o \rightleftharpoons Na^+ + HCO_3^-$	(20)
$CaSO_4 \cdot 2H_2O \rightleftharpoons Ca^{++} + SO_4^{--} + 2H_2O$	(21)		
$CaSO_3 \cdot \frac{1}{2}H_2O \rightleftharpoons Ca^{++} + SO_3^{--} + \frac{1}{2}H_2O$	(22)		
$CO_2(g) + H_2O \rightleftharpoons H_2CO_3$	(23)	$SO_2(g) + H_2O \rightleftharpoons H_2SO_3$	(24)

Table II. Expressions for Equilibrium Constants from Lowell et al. (1)

$$\log K = -B/T - C \log T - DT + E$$

Reaction	B	C	D	E
1	4.4710E 03	0.0000	1.7060E-02	6.0875E 00
2	-6.3384E 02	0.0000	0.0000	-9.3320E 00
3	-8.4367E 02	0.0000	0.0000	-4.7171E 00
4	4.7514E 02	0.0000	1.8222E-02	5.0435E 00
5	2.9024E 03	0.0000	2.3790E-02	6.4980E 00
6	3.4047E 03	0.0000	3.2786E-02	1.4843E 01
7	-2.7300E 02	0.0000	0.0000	-2.2900E 00
8	-5.0480E 02	0.0000	0.0000	-5.0910E 00
9	2.5721E 03	2.3150E 01	0.0000	6.3600E 01
10	-4.7548E 02	0.0000	0.0000	-4.7954E 00
11	-3.0185E 02	0.0000	0.0000	-2.2720E 00
12	-5.1799E 02	0.0000	0.0000	-4.3223E 00
13	-4.3250E 02	0.0000	0.0000	-4.3715E 00
14	-1.0579E 03	0.0000	0.0000	-5.7950E 00
15	-5.0480E 02	0.0000	0.0000	-5.0910E 00
16	-2.3508E 02	0.0000	0.0000	-1.7470E 00
17	0.0000	0.0000	0.0000	5.7000E-01
18	-2.4100E 02	0.0000	0.0000	-1.5290E 00
19	-3.0341E 02	0.0000	0.0000	-2.2852E 00
20	0.0000	0.0000	0.0000	2.5000E-01
21	4.9440E 03	3.7745E 01	0.0000	1.0536E 02
22	0.0000	0.0000	0.0000	-7.0757E 00
23	-1.0150E 03	0.0000	0.0000	-4.8700E 00
24	-1.3700E 03	0.0000	0.0000	-4.5100E 00

Semi-empirical methods based upon Debye-Hückel theory can be used to calculate the activity coefficients. The activity coefficients are dependent upon several variables and among them is the ionic strength of the liquid. The ionic strength of the flue gas scrubbing process is highly dependent upon the type of process and certain parameters such as the chloride content of the coal, the amount of free water purged from the process and the addition of soluble additives to the scrubbing liquid. Some flue gas desulfurization processes such as dual alkali processes or regenerative sodium scrubbing have inherently higher ionic strengths because of the soluble nature of the dissolved species. Processes operating with lime or limestone slurries have lower ionic strengths because the dissolved species are less soluble. Lime or limestone processes typically operate with ionic strengths in the range of 0.1 to 0.2 M. If the chloride content of the coal is high and the process has extensive dewatering equipment, the ionic strength might be as high as 1 M. The addition of magnesium oxide to lime or limestone slurries to promote the removal of sulfur dioxide will also increase ionic strength.

The original Debye-Hückel expression for the calculation of the activity coefficients of ions is

$$\log \gamma_i = -Az_i^2 I^{\frac{1}{2}} \tag{5}$$

γ_i is the activity coefficient of the ith ion; A is a constant dependent upon the solvent and the temperature; z is the charge of the ith ion; and I is the ionic strength.

$$I = 0.5 \Sigma m_i z_i^2 \tag{6}$$

A derivation of this expression is presented by Daniels and Alberty (2). Equation 5 is limited to very dilute solutions, I less than 0.01 M. Equation 5 can be extended to more concentrated solutions by incorporating a term $1 + \beta \varepsilon I^{\frac{1}{2}}$ in the denominator of equation 5.

$$\log \gamma_i = -\frac{Az_i^2 I^{\frac{1}{2}}}{1 + \beta \varepsilon I^{\frac{1}{2}}} \tag{7}$$

β is a constant that is dependent upon temperature and the type of solvent, and ε is the mean distance of closest approach of the ion in solution. Values of ε, expressed in angstroms, have been tabulated by Lowell et al. (1). A somewhat simpler result is obtained if a typical value is chosen for ε, so that $\beta\varepsilon$ is a constant, and a deviation parameter b is incorporated into equation 7. Choosing unity for $\beta\varepsilon$, equation 7 becomes

$$\log \gamma_i = Az_i^2 \{- \frac{I^{1/2}}{1 + I^{1/2}} + b_i I\} \quad (8)$$

Values of the parameters ε and b_i are listed for several ions in Table 3. It has been observed that b_i for many ions is about 0.3 (3). With this simplification, equation 8 can be written as

$$\log \gamma_i = Az_i^2 \{- \frac{I^{1/2}}{1 + I^{1/2}} + 0.3I\} \quad (9)$$

Equations 8 and 9 can be used for values of I up to 1. M. The second term in these equations accounts for the reversal of slope of activity coefficient versus ionic strength from negative to positive as ionic strength increases. Equations 8 and 9 have been widely used in the equilibrium calculations of the lime or limestone processes. With coals of moderate chloride content and for systems without extensive sludge dewatering, the ionic strength is well below 1.0 M, and equations 8 and 9 reasonable.

For applications where the ionic strength is as high as 6 M, the ion activity coefficients can be calculated using expressions developed by Bromley (4). These expressions retain the first term of equation 9 and additional terms are added to improve the fit. The expressions are much more complex than equation 9 and require the molalities of the dissolved species to calculate the ion activity coefficients. If all of the molalities of dissolved species are used to calculate the ion activity coefficients, then the expressions are quite unwieldy. However, for the applications discussed in this paper many of the dissolved species are of low concentration and only the major dissolved species need be considered in the calculation of ion activity coefficients. For lime or limestone applications with a high chloride coal and a tight water balance, calcium chloride is the dominant dissolved specie. For this situation Kerr (5) has presented these expressions for the calculation of ion activity coefficients.

For dual alkali scrubbing, the major dissolved species are sodium chloride, sodium bisulfite, sodium sulfite, and sodium sulfate. If a separate prescrubber is provided for the removal of fly ash and hydrogen chloride before the removal of sulfur dioxide, then sodium chloride will not be present in the sulfur dioxide scrubber.

For the regenerative magnesium oxide scrubbing process, the dominant dissolved species are magnesium sulfite, bisulfite, and sulfate. For this process, chloride and fly ash are removed separately to avoid their accumulation in the main scrubbing loop. For this reason, chloride is not a significant

Table III. Activity Coefficient Parameters (1)

Species	ε	b_i
H^+	6.0	0.4
Na^+	5.0	0.1
Ca^{++}	4.5	0.1
SO_4^{--}	3.0	0.0
Cl^-	4.0	0.0
SO_3^{--}	4.5	0.0
HSO_3^-	4.5	0.0
CO_3^{--}	4.5	0.0
HCO_3^-	4.5	0.0

All other charged species: $\varepsilon = 3.0$
$b_i = 0.3$

specie in the main scrubbing loop.

For the regenerative sodium sulfite scrubbing process, the dominant dissolved species are sodium sulfite, bisulfite, and sulfate. This process is usually configured so that chloride and fly ash are removed separately.

The activity coefficients of uncharged species can be calculated using the expression (6).

$$\log \gamma_i = 0.076I \qquad (10)$$

For most applications, the activity of water may be taken as unity. A better approximation for the activity of water is its mole fraction (7).

Computational Methods

Although one can probably find exceptions, most equilibrium calculations involving flue gas slurries are performed with temperature as a known variable. With temperature known, the numerical values of the appropriate equilibrium constants can be immediately calculated. The remaining unknown variables to be determined are the activities, activity coefficients, molalities, and the gas phase partial pressures. The equations used to determine these variables are formulated from among the equilibrium expressions presented in Table 1, the expressions for the activity coefficients, ionic strength, material balance expressions, and the electroneutrality balance. Although there are occasionally exceptions, the solution sequence generally is an iterative or cyclic sequence.

Successive substitution is the simplest although generally the least effective method for performing these calculations. Values are assumed for one or more of the unknown variables; other variables are then determined from some of the equations; new values of the assumed variables are determined from the remaining equations. The process is repeated until convergence is obtained. For flue gas desulfurization examples, many of the variables are highly constrained, and the calculation sequence can easily move into infeasible regions. The solution sequence frequently oscillates.

Linearization methods such as the Newton-Raphson algorithm might be used. The equations are linearized using a first order Taylor series expansion. Values of the first derivatives are calculated using assumed values of the variables. The resulting system of linearized equations is solved using a matrix inversion technique. The process is then repeated until convergence is obtained. This method is quite tedious to implement. The calculations can also move into the infeasible region. The solution sequence often oscillates.

The use of optimum seeking methods to direct the iterative calculations has been found to be an excellent method of performing equilibrium calculations described in this paper. Convergence is rapid; computer time is minimal; and the computer programming is simple. Variables are divided into two categories: search variables and state variables. Variables manipulated by the optimum seeking method to zero or minimize an artificial objective function are search variables. Variables that are solved explicitly from an equation are called state variables. There are two types of state variables. Iterative state variables are those calculated from certain equations each time the optimum seeking method supplies a new set of search variables. Noniterative state variables are those calculated from certain equations only once: before, after, or in between searches if more than one search is required. Equations have two uses. They are used to solve for state variables and to formulate artificial objective functions. There is a separate objective function for each search. Most of the applications of the type described in this paper require only a single search. Each term in the artificial objective function corresponds to a specific equation in the statement of the problem. The search dimensionality equals the number of terms in the objective function.

An optimum seeking method is a systematic way of manipulating a set of variables to find the values of the variables to maximize or minimize some criterial. Their most popular uses have been economic ones such as profitability or costs or technical criterial such as conversion of raw materials or product recovery. For the applications described in this paper, the optimization criterial are the minimization of the squared deviations from zero of the equations chosen to constitute the objective function. The equations chosen to formulate the objective function can be written as

$$f(\bar{x}) = 0 \tag{11}$$

The minimization criteria y can then be expressed as

$$y = \Sigma f^2 \tag{12}$$

If the equations used in formulating equation 12 are dimensionally dissimilar, then the equations should be written as fractional deviations from zero, f^*. This requires that equation 12 be rewritten as

$$y = \Sigma(f^*)^2 \tag{13}$$

The optimum seeking methods which have been found to be particularly useful are the modified Fibonacci search (search by golden section) for one-dimensional searches and the Hooke-Jeeves search for multi-dimensional searches. Beveridge and Schechter (8) give a complete description of these searches.

Kuester and Mize (9) have listed computer programs written in Fortran IV for both of these searches.

If all of the variables appeared in all of the equations, then the use of optimum seeking methods for the direction of the calculations would be impractical because the search dimensionality would become excessive. However, the opposite is true for these applications. The system of equations is sparse; only a few variables are present in each equation. This requires that only a few variables need to be search variables with the rest being state variables. Search variables must be chosen carefully. Generally, the most constrained variables should be chosen as search variables, and the least constrained variables chosen as state variables. The opposite choice will often drive the highly constrained variables into the infeasible region causing computational difficulties. Also for the applications illustrated in this paper, minor equilibrium species should not be chosen as search variables.

With regard to computer programming, the optimum seeking methods should be written separately and stored as a library program for repetitive use. The main program reads the input data, calls the search routine, performs any noniterative calculations, and handles the output. The calculation of iterative state variables and the calculation of the objective function should be performed within function subprograms.

Calculation of Supersaturation Ratios of Flue Gas Scrubbing Slurries

The scaling tendency of the lime or limestone processes for flue gas desulfurization is highly dependent upon the supersaturation ratios of calcium sulfate and calcium sulfite, particularly calcium sulfate. The supersaturation ratios cannot be measured directly. They are determined by measuring experimentally the molalities of dissolved sulfur dioxide, sulfate, carbon dioxide, chloride, sodium and potassium, calcium, magnesium, and pH. Then by calculation, the appropriate activities are determined, and the supersaturation ratio is determined. Using the method outlined in Section IV, the concentrations of all ions and ion-pairs can be readily determined. The search variables are the molalities of bisulfite, bicarbonate, calcium, magnesium, and sulfate ions. The objective function is defined from the mass balance expressions for dissolved sulfur dioxide, sulfate, carbon dioxide, calcium, and magnesium. This equation is

$$y = (m_{SO_2} - m_3 - m_4 - m_5 - m_{13} - m_{19})^2$$

$$+ (m_{SO_4} - m_6 - m_7 - m_{14} - m_{20} - m_{25})^2$$

$$+ (m_{CO_2} - m_8 - m_9 - m_{10} - m_{15} - m_{16} - m_{21}$$

$$- m_{22} - m_{26} - m_{27})^2$$

$$+ (m_{Ca} - m_{11} - m_{12} - m_{13} - m_{14} - m_{15} - m_{16})^2$$

$$+ (m_{Mg} - m_{17} - m_{18} - m_{19} - m_{20} - m_{21} - m_{22})^2 \quad (14)$$

With the search variables listed above, the calculation of the remaining molalities (iterative state variables) can be easily performed sequentially by using the equilibrium expressions for the reaction shown in Table 1 and the calculated activity coefficients. The equations are used in the order in which they appear in Table 1. The sequence of calculation of molalities is OH^-, SO_3^{--}, H_2SO_3, HSO_4^-, CO_3^{--}, H_2CO_3, $CaOH^+$, $CaSO_3^o$, $CaSO_4^o$, $CaCO_3^o$, $CaHCO_3^+$, $MgOH^+$, $MgSO_3^o$, $MgSO_4^o$, $MgCO_3^o$, $MgHCO_3^+$, $NaOH^o$, $NaSO_4^-$, $NaCO_3^-$, and $NaHCO_3^o$. For the first function evaluation, the ionic strength is calculated using only the known molalities. As the search progresses past the first point, ionic strength is calculated more precisely by using known molalities and the remaining molalities from the previous point. Likewise, on the first cycle of the search, sodium is assumed to be completely ionized. As the search progresses, a more precise value of the molality of sodium ion is obtained by subtracting the sodium ion-pair molalities obtained from the previous point. When y becomes suitably small, the search is terminated and the supersaturation ratios are calculated using two remaining equations of Table 1. The ionic imbalance can also be calculated and used as a guide for assessing the reliability of the measured compositions. A complete numerical example has been outlined by Kerr (5).

Calculation of Dissolved Alkalinity

The absorption of sulfur dioxide into flue gas scrubbing slurries is enhanced considerably because of the reactions of dissolved sulfur dioxide with various dissolved alkaline species. The alkaline species of interest are the following ions and ion-pairs: OH^-, SO_3^{--}, HCO_3^-, CO_3^{--}, $MgSO_3^o$, $MgCO_3^o$,

$MgOH^+$, $MgHCO_3^+$, $CaOH^+$, $CaSO_3^o$, $CaCO_3^o$, and $CaHCO_3^+$. It is often desired to calculate the sum of the above molalities for a specified temperature, carbon dioxide partial pressure, pH, chloride molality, total molality of dissolved magnesium, and supersaturation ratio of calcium sulfite and calcium sulfate.

An examination of Table 1 shows that if the molalities of Mg^{++} and SO_4^{--} are used as search variables, then the activity coefficients and the remaining molalities can be calculated sequentially as iterative state variables. The sequence of molalities is OH^-, Ca^{++}, SO_3^{--}, HSO_3^-, H_2SO_3, HSO_4^-, HCO_3^-, CO_3^{--}, $CaOH^+$, $CaSO_3^o$, $CaSO_4^o$, $CaSO_3^o$, $CaHCO_3^+$, $MgOH^+$, $MgSO_3^o$, $MgSO_4^o$, $MgCO_3^o$, and $MgHCO_3^+$. The sequence in which the equilibrium expressions of Table 1 are used is 1, 21, 22, and 2 through 16. In this problem formulation sodium was not considered.

The artificial objective function was formulated from the electroneutrality balance and a mass balance on the total dissolved magnesium. This result is

$$y = (\Sigma z_i m_i)^2 + (m_{Mg} - m_{17} - m_{18} - m_{19} - m_{20} - m_{21} - m_{22})^2 \quad (15)$$

The starting values of the search variables were the total dissolved magnesium molality for m_{Mg}^{++} and $m_{Mg} - 0.5m_{Cl}$ for $m_{SO_4^{--}}$. The activity coefficients were calculated using equation 8. At the end of the search, the dissolved alkalinity is calculated by summing the appropriate molalities.

The computed results are listed in Table 4. Inspection of this table shows that dissolved alkalinity increases with increasing concentration of dissolved magnesium and decreases with increasing concentration of chloride.

Table IV. Calculated Values of Dissolved Alkalinity

Chloride Molality	Magnesium Molality	Dissolved Alkalinity Molality
0.0857	0.1250	0.001544
0.1142	0.1250	0.001494
0.1427	0.1250	0.001448
0.0857	0.1666	0.001835
0.1142	0.1666	0.001778
0.1427	0.1666	0.001721
0.0857	0.2082	0.002137
0.1142	0.2082	0.002076
0.1427	0.2082	0.002015

These values are based upon a pH of 5.5, temperature of 50°C, carbon dioxide partial pressure of 0.12 atm, calcium sulfite supersaturation ratio of 1.0, and calcium sulfate supersaturation ratio of 1.25.

Abstract

The most significant chemical equilibria present in flue gas scrubbing slurries are outlined. Expressions for temperature dependent equilibrium constants are presented that are suitable for the temperature ranges encountered in scrubbing applications. Expressions for activity coefficients of ions and ion-pairs are presented that are suitable for the ranges of ionic strengths encountered for this type of applications.

A novel method of performing equilibrium calculations based on optimum seeking methods is developed. The method is easily computer programmed, has good convergence properties, and requires only modest amounts of machine time. All of the above points are illustrated with two examples: calculation of gypsum supersaturation ratio and calculation of dissolved alkalinity. Gypsum supersaturation ratio is an important operating parameter for scale control. The second example will show how the presence of magnesium increases dissolved alkalinity in a calcium system and how the presence of chloride decreases dissolved alkalinity.

Nomenclature

Symbol	Definition
A	Debye-Hückel constant
B,C,D,E	Constants in expressions for equilibrium constants
ΔG^o	Standard Gibbs' energy of reaction
ΔH^o	Standard heat of reaction
I	Ionic strength
K	Equilibrium constant
R	Ideal gas constant
T	Absolute temperature
a	Activity
b	Parameter for calculation of activity coefficients
f	Algebraic expression or deviation from zero
f*	Algebraic expression or deviation from zero (dimensionless)
m	Molality
x	Variable
y	Value of an artificial objective function
z	Ion charge
β,ε	Parameters for calculation of activity coefficients
γ	Ion activity coefficient

Component Subscripts	Definition
Ca	Total calcium
CO_2	Total dissolved carbon dioxide
Mg	Total magnesium
SO_2	Total dissolved sulfur dioxide
SO_4	Total dissolved sulfate
1	H^+
2	OH^-
3	HSO_3^-
4	SO_3^{--}
5	$H_2SO_3^o$
6	HSO_4^-
7	SO_4^{--}
8	HCO_3^-
9	CO_3^{--}
10	H_2CO_3
11	$CaOH^+$
12	Ca^{++}
13	$CaSO_3^o$
14	$CaSO_4^o$
15	$CaCO_3^o$
16	$CaHCO_3^+$
17	$MgOH^+$
18	Mg^{++}
19	$MgSO_3^o$
20	$MgSO_4^o$
21	$MgCO_3^o$
22	$MgHCO_3^+$
23	$NaOH^o$
24	Na^+
25	$NaSO_4^-$
26	$NaCO_3^-$
27	$NaHCO_3^o$
28	Cl^-

Literature Cited

1. Lowell, P. S., Ottmers, D. M., Strange, D. M., Schwitzgebel, K., and DeBerry, D. W., "A Theoretical Description of the Limestone Injection-Wet Scrubbing Process," Vol. I, (National Air Pollution Control Administration Contract Number CPA 22-69-138), NTIS Number PB 193-029 (1970).

2. Daniels, F., and Alberty, R. A., "Physical Chemistry," Second Edition, Wiley, New York, 1963.

3. Davies, Cecil W., "Ion Association," Butterworth, Washington, D.C., 1962.

4. Bromley, L. A., AIChE J. (1973) 19, 313.

5. Kerr, C. P., AIChE J. (1976) 22, 403.

6. Harned, H. S., and Owen, B. B., "The Physical Chemistry of Electrolytic Solutions," Third Edition, Reinhold, New York, 1958.

7. Han, S. T., and Bernardin, L. J., Tappi (1958) 41, 540.

8. Beveridge, G. S. C., and Schechter, R. S., "Optimization: Theory and Practice," McGraw-Hill, New York, 1970.

9. Kuester, J. L., and Mize, J. H., "Optimization Techniques with Fortran," McGraw-Hill, New York, 1973.

RECEIVED January 31, 1980.

6

Correlation of Vapor-Liquid Equilibria of Aqueous Condensates from Coal Processing

D. M. MASON and R. KAO

Institute of Gas Technology, 3424 South State St., ITT Center, Chicago, IL 60616

Ammonia is produced along with carbon dioxide and hydrogen sulfide in many coal conversion processes such as hydrogasification for the production of substitute natural gas (HYGAS®, Synthane, etc.), medium- and low-Btu gasification, and liquefaction. When reactor product gases are cooled, an aqueous solution is obtained that may contain substantial amounts of ammonium bicarbonate and hydrosulfide. For example, after letting the solution down to atmospheric pressure, condensates from the HYGAS Process Development Unit ranged up to about 2 molar ammonia concentration accompanied by lesser amounts of bicarbonate and sulfide.(1) Small amounts of hydrogen cyanide and hydrogen chloride are also produced in such processes. In auxiliary processing under more oxidizing conditions, sulfur dioxide may be produced together with ammonia. This is true for the HYGAS Process when the feed is bituminous coal, which must be pretreated by reaction with air under relatively mild conditions to reduce its agglomerating tendency.

Successive steps of hydrogasification processes in which the equilibria with bicarbonate and hydrosulfide are involved include cooling of the raw product gas either indirectly or by quenching with water, de-pressuring of the condensate or quench water with evolution of gases, processing of the condensate or quench water to recover ammonia and hydrogen sulfide, and cleanup of waste water for disposal. Vapor-liquid equilibria of nonelectrolyte gases including hydrogen, methane, carbon monoxide, and carbonyl sulfide are also involved in some of these process steps. Pressures range up to about 90 atmospheres in hydrogasification and perhaps up to 200 atmospheres in liquefaction. Temperatures of interest range up to about 230°C in gasification; the maximum is lower in liquefaction.

Of the problems presented, correlation of the NH_3-CO_2-H_2S-H_2O system is most important. Data that might be used for direct empirical correlation of partial pressures or fugacities with total concentrations of ammonia, carbon dioxide, and hydrogen sulfide in the liquid are available for relatively limited ranges

0-8412-0569-8/80/47-133-107$08.00/0

of temperature, composition, and pressure. Thus, analysis in terms of a model is necessary for extension to areas where data are not available or are of poor accuracy. In the approach taken by Prausnitz and co-workers(2, 3), reliance is placed on ionization constants, Henry's law correlations based on binary mixtures of each gas with water, and ionic activity coefficients based on correlations of Bromley(4) and Pitzer(5). They used ternary vapor-liquid equilibria data for verification only. Beutier and Renon(6) have used a similar model but adjusted their ionic activity coefficient parameters to fit selected ternary data. Our own approach, initiated before we became aware of the Prausnitz work, was to analyze the ternary data by modification of the method of van Krevelen, Hoftijzer, and Huntjens(7) and to extend its range by use of ionization constants, Henry's law correlations, and correlations of activity coefficients as needed. Thus, in many areas we needed the same basic data as in the Prausnitz or Renon approach.

Solubility of Gases in Water

The solubilities of ammonia, carbon dioxide, and hydrogen sulfide were obtained from binary data and expressed in terms of a Henry's constant for infinite dilution and an interaction parameter:

$$\log \gamma = bm$$

Variation of the Henry's constants with temperature are presented in Figures 1, 3, and 5 for the respective gases, and their interaction parameters are presented in Figures 2, 4, and 6. For carbon dioxide, the interaction parameter in Figure 4 is expressed on the mole fraction basis, as in the Krichevsky-Ilenskaya equation. Details of the data reduction and results have been presented elsewhere(8), except for the recent data of Wells(9) on ammonia-water and the data of Mather and Lee(10) and Mather(11) on hydrogen sulfide-water. Wells reduced his data according to the methods of Edwards, Maurer, Newman, and Prausnitz (EMNP)(3). Mather's data were correlated by Neuburg *et al.*(11), according to the Krichevsky-Ilenskaya equation. They calculated apparent molar volumes of hydrogen sulfide (invariant with concentration and thus considered equal to partial molar volumes), from density data of Murphy and Gaines(12). The molar volumes are presented in Table I together with data on carbon dioxide and ammonia from other sources.

The Henry's constant for ammonia at temperatures from 20° to 150°C is admirably represented by the equation of EMNP(3). Haas and Fisher(13) obtained constants from 150° to 300°C from the data of Jones (14) on low concentrations of ammonia.

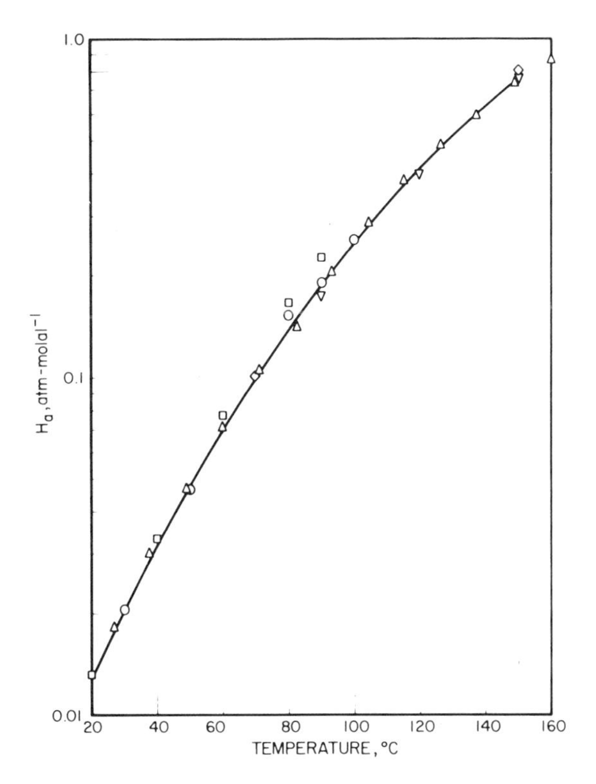

Figure 1. Henry's constant for solubility of ammonia in water: (○) Clifford and Hunter; (△) Macriss; (□) Van Krevelen; (▽) Frohlich; (◇) Wells (9); also see *Ref.* 8

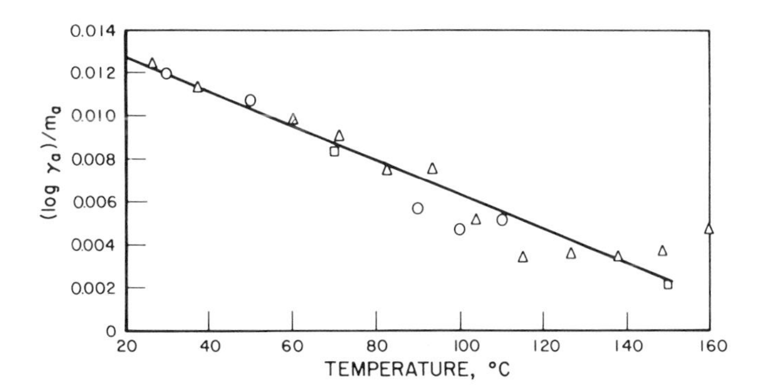

Figure 2. Effect of temperature on the activity coefficient of ammonia in water: (○) Clifford and Hunter; (△) Macriss et al.; (□) Wells (9); also see *Ref.* 8

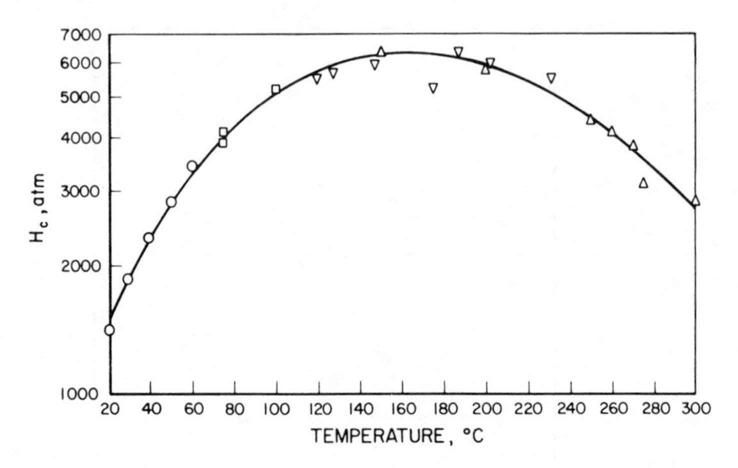

Figure 3. Henry's constant for solubility of carbon dioxide in water: (○) int'l. crit. table; (△) calc. from Takenouchi and Kennedy, and Malinin; (□) calc. from Wiebe and Gaddy, Houghton et al.; (▽) Malinin, low pressure; see *Ref.* 8

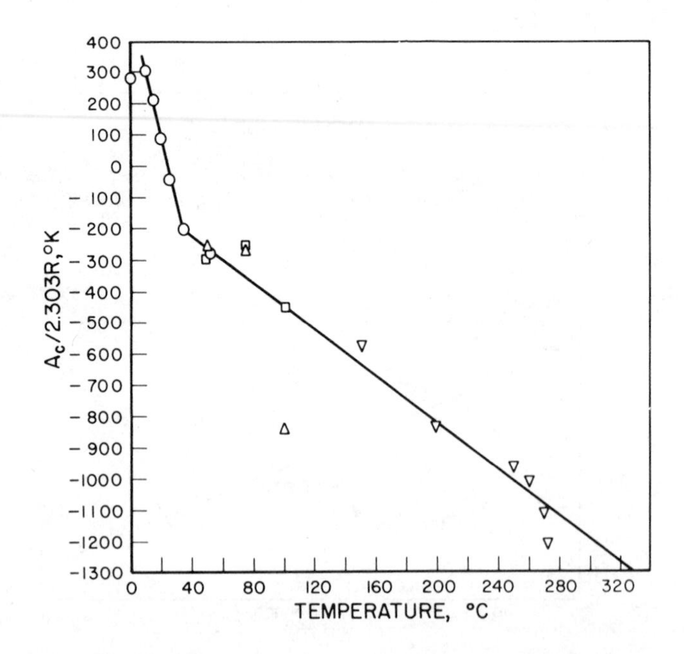

Figure 4. Effect of temperature on Margules constant for carbon dioxide in water: (○) Gibbs and Van Ness; calculated by Malinin: (△) Houghton et al.; (□) Wiebe and Gaddy; (▽) Takenouchi and Kennedy; see *Ref.* 8

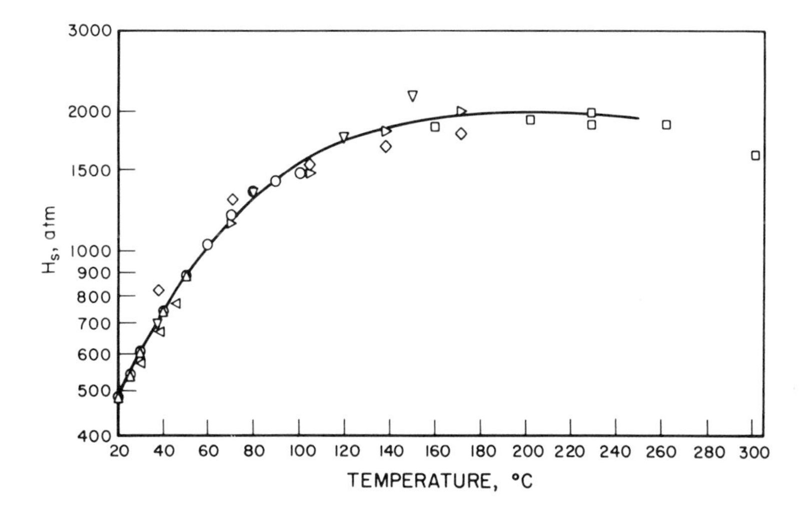

Figure 5. Henry's constant for solubility of hydrogen sulfide in water: (○) intl. crit. table; (△) Clark and Glew; (□) Kosintseva; (▽) Miles and Wilson; (◇) Selleck et al.; (◁) Froning et al.; (▷) Mather (11); also see *Ref.* 8

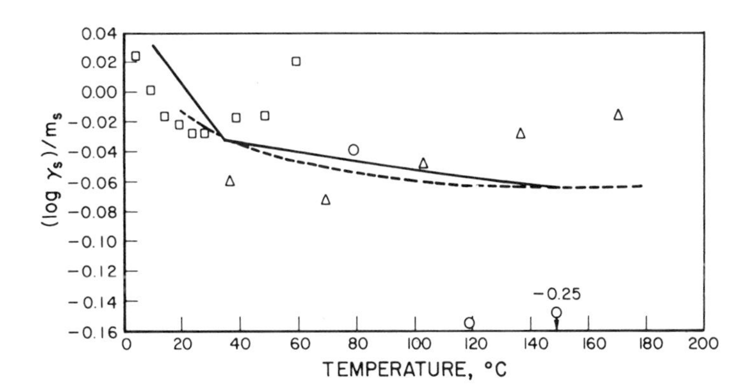

Figure 6. Effect of temperature on the activity coefficients of hydrogen sulfide and carbon dioxide in water: (——) carbon dioxide; (– – –) Mather; hydrogen sulfide: (○) Miles and Wilson; (△) Selleck et al.; (□) Wright and Maass; see *Refs.* 8 *and* 11

Table I. MOLAR VOLUMES OF AMMONIA, CARBON DIOXIDE, AND HYDROGEN SULFIDE IN WATER

	Pressure, atm	Temperature, °C								
		25	50	75	100	150	200	250	300	350
		cm^3/mol								
Ammonia,[a] This Work	--	24.7	25.5	26.2	27.6[e]	28.5[e]	--	--	--	--
Carbon Dioxide (8)										
Parkinson and DeNevers[b]	0-5	37.6	37.7	--	--	--	--	--	--	--
Ellis and McFadden[c]	--	33.2	34.1	35.2	35.7	36.4	36.8	37.2	--	--
Malinin[b]	290	--	--	35 ± 3	32 ± 3	35 ± 4	40 ± 4	55 ± 5	80 ± 6	140 ± 20
Malinin[b]	484	--	--	35 ± 3	32 ± 3	35 ± 4	40 ± 4	55 ± 5	70 ± 6	110 ± 15
Malinin[b]	968	--	--	35 ± 3	32 ± 3	35 ± 4	40 ± 4	55 ± 5	65 ± 6	75 ± 10
Hydrogen Sulfide[d] (11)	1-18	34.7	35.9	37.2[e]	38.4[e]	40.8[e]	--	--	--	--

[a]Partial molar volume at infinite dilution.

[b]Partial molar volume at concentrations yielded by the indicated total pressure of gas and water.

[c]At infinite dilution. The variation with temperature above 25°C was taken to be equal to that of boric acid.

[d]Apparent molar volume. No trend with pressure was observed.

[e]Extrapolated.

For carbon dioxide, we expressed the Henry's constant in atm-kg H_2O/mole as:

$$\log H_c = 3.822 - 7.8665 \times 10^{-4} \exp(T/100) - 0.04145 (T/100)^2 - 17.457 (T/100)^{-2}$$

for the temperature range of 20° to 300^oC. Values from the equation of EMNP agree within 7%.

We adopted the equation of Clarke and Glew(15), shown in Figure 5, for the Henry's constant of hydrogen sulfide in atm-kg H_2O/mole:

$$\log H_s = 102.325 - 4423.11 T^{-1} - 36.6296 \log T + 0.013870 T$$

It fits the data well from 20° to about 260^oC. At temperatures up to 150^oC, the equation of EMNP(3) is in good agreement.

We fitted the interaction parameters with equations as represented in the graphs(8) and given in Table II. Values do not differ significantly from those of the equations of EMNP(3).

Table II. EFFECT OF TEMPERATURE ON THE ACTIVITY COEFFICIENT OF THE GAS IN BINARY MIXTURE WITH WATER

Gas	Equation $\log \gamma = bm$	Temperature Range
NH_3	$b_a = 0.0356 - 0.00008\ T$	$20^o - 150^oC$
CO_2	$b_c = -0.767 + 226.7/T$	$10^o - 35^oC$
CO_2	$b_c = -0.143 + 34.56/T$	$35^o - 325^oC$
H_2S	$b_s = -0.143 + 34.56/T$	$20^o - 180^oC$

Ionization Constants

Equations for the ionization of water(16) and ammonia(17), each incorporating new data extending to 300^oC, have appeared recently (Table III). At temperatures up to 150^oC earlier data selected by Barnes *et al.*(18) and represented by the equation of EMNP(3) agree within 2.5% for water and 6% for ammonia. The first apparent ionization constant of carbon dioxide ($CO_2 + H_2CO_3$) was represented by EMNP in an equation based on values selected by Barnes *et al.*(18); later experimental data by Kryukov agree with the equation within 2%(19). Data on the first ionization constant of hydrogen sulfide vary more widely, as shown in Figure 7. Professor H. L. Barnes(20) has recommended the equation of Naumov(21), applying to the range of 0° to 350^oC (Table III).

Salting Out of Gases

Salting-out Coefficients at 25^oC. The effect of an electrolyte on the solubility or activity of a gas dissolved in an aqueous solution is commonly expressed as a salting out coefficient:

Table III. EFFECT OF TEMPERATURE ON IONIZATION CONSTANTS

$$\log K = \frac{A}{T} + B \ln T + CT + \frac{D}{T^2} + E$$

Gas	A	B	C	D	E	Temperature Range, °C	Reference
H_2O	3.12860×10^4	94.9734	-0.097611	-2.17087×10^6	-606.522	0–300	(16)
NH_3	2.74967×10^4	81.2824	-0.0905795	-1.71772×10^6	-513.761	0–300	(17)
CO_2	-5251.53	-15.97405	--	--	102.2685	0–225	(3)
H_2S	-3539.1	--	-0.02522	--	12.41	0–350	(20)

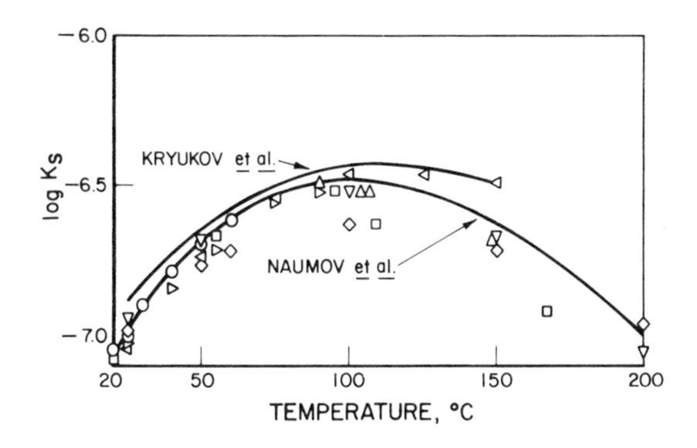

Figure 7. First ionization constant of hydrogen sulfide: (○) Barnes et al.; (△) Ellis and Milestone; (□) Ellis and Giggenbach; (▽) Ellis; (◇) Helgeson; (◁) Tsonoupolos et al.; (▷) Sretenskaya; see *Ref.* 8

$$\log \frac{S_o}{S} = \log \gamma = kC$$

Such coefficients are needed for the effect of ammonium bicarbonate, carbamate, and hydrosulfide salts on the liquid phase activities of ammonia, carbon dioxide, and hydrogen sulfide. They cannot be directly measured because of the chemical reactions of the dissolved molecular components, but must be calculated theoretically or estimated by correlation. Electrostatic theory does not predict negative coefficients, which are characteristic of ammonia with some salts. To us, it appears that scaled particle theory(22) is probably the best method of calculation, but the required parameters (polarizability and ion size) are not available for the salts of interest.

Long and McDevit(23) obtained an expression for prediction of salting out of nonpolar nonelectrolytes based simply on the work required to create the volume occupied by the neutral molecules:

$$k = \overline{V}_i^o(V_s - \overline{V}_s^o)/2.3\ RT\ \beta_o$$

where k is the salting-out coefficient on a molar basis, $\overline{V}_i^o$ is the partial molar volume of the molecular specie at infinite dilution, V_s is the molar volume of the hypothetical liquefied salt, $\overline{V}_s^o$ is the partial molar volume of the salt in water at infinite dilution, and β_o is the compressibility of pure water.

This expression is a limiting law for infinite dilution. Its physical interpretation is that the contraction in total volume that occurs on mixing salt and water can be thought of as a compression of the solvent; this compression makes it more difficult to insert the molecule of nonelectrolyte. This theory has been more successful than those based on electrostatic effects in explaining the variation in salting-out coefficient from salt to salt, including the occurrence of negative values. However, coefficients calculated from this equation or from its alternative expression using experimental compressibility data, do not always agree well with experiment (22). Instead, we use the relationship as the basis of an empirical correlation of the effect of temperature on the salting-out coefficient, as described later.

Van Krevelen and Hoftijzer developed an empirical correlation in which the coefficient, on an ionic strength basis, is considered to be the sum of contributions from the cation, the anion, and the gas(24):

$$\log \frac{S_o}{S} = (x_g + x_{ca} + x_{an})I$$

where x_g, x_{ca}, and x_{an} are salting-out contributions from gas, cation, and anion, respectively, and I is ionic strength on the molar basis. Once a contribution has been determined from one or more combinations, coefficients for unmeasured combinations can be calculated. However, no data were then available for bicarbonate or hydrosulfide salts. More recently, Onda and co-workers(25,26)published a correlation using the same scheme. They measured the salting out of ethylene by mixed $KHCO_3$-K_2CO_3 and $NaHCO_3$-Na_2CO_3 solutions and obtained the contribution of the bicarbonate ion by difference from their correlated values for ethylene in K_2CO_3 and Na_2CO_3. They give a contribution value for the hydrosulfide ion also, but do not reference its source. For a similar correlation, Hikita and co-workers measured the salting out of nitrous oxide by mixtures of sodium carbonate and sodium bicarbonate.(27)

Onda found a standard deviation of 0.052ℓ/mol for his complete set of data, including multicharged ions. These correlations thus provide a coefficient for gases in ammonium bicarbonate that may be in error by more than 0.05 ℓ/mol: Onda's coefficients for gases in ammonium hydrosulfide are of unknown origin and accuracy, and coefficients for ammonium carbamate are not provided. In short, this type of correlation does not provide the needed information.

To establish a data base for correlation and prediction of salting-out coefficients, we have gathered all the data we could find in the literature on salting out of ammonia, carbon dioxide, hydrogen sulfide, and nitrous oxide by ammonium, potassium, and sodium salts of single- and double-charged anions. Because molalities are commonly used for expressing concentrations in vapor-liquid equilibria correlations, we have calculated the coefficients with salt concentration in gram equivalents per kilogram of water and gas solubility in amount per kilogram of water. Except for salts of low solubility such as potassium sulfate, bromate, and iodate, measurements were usually available over a range of concentration. When calculated on the molal basis, the salting-out coefficient for many salts was constant to a few units in the third decimal, but for others it varied by a maximum of about 0.02 over a 2 molal concentration range. For this compilation, we have chosen mostly coefficients obtained at 1 gram equivalent per kilogram of water. Values obtained are presented in Table IV.

Table IV shows that potassium and sodium salts have been studied much more extensively than ammonium salts. However, we found that, for any one gas, differences in the coefficient between potassium and ammonium salts of the same anion and between sodium and ammonium salts of the same anion are nearly constant. For gases other than carbon dioxide, we have used data on bromides, chlorides, nitrates, and sulfates from a single investigator to obtain average differences (Table V). For carbon dioxide, such data were not available, and we used averages of results on chlorides, nitrates, and sulfates from different investigators.

Table IV. EXPERIMENTAL SALTING-OUT COEFFICIENTS[a] AT 25°[b]

	Ammonia		Hydrogen Sulfide		Carbon Dioxide		Nitrous Oxide	
	Coefficient	Ref.	Coefficient	Ref.	Coefficient	Ref.	Coefficient	Ref.
NH_4Br	-0.052	(29)	-0.018	(30)	--	--	0.026	(36)
$NH_4CH_3CO_2$	--	--	-0.061[c]	(30)	--	--	--	--
NH_4Cl	-0.023	(29)	0.001	(30)	0.022,0.026	(34,35)	0.033,0.035	(32,36)
NH_4NO_3	-0.039	(29)	-0.016	(30)	-0.002	(25)	0.009	(36)
$(NH_4)_2SO_4$	0.040	(29)	0.071	(30)	0.040	(34)	0.086	(36)
KBr	0.026,0.002	(28,29)	0.008	(30)	0.044	(32)	0.071,0.063	(32,36)
$KBrO_3$	0.083	(28)	--	--	--	--	--	--
KCH_3CO_2	0.087	(28)	--	--	--	--	--	--
KCN	0.049	(28)	--	--	--	--	--	--
KCNS	0.049	(28)	--	--	--	--	--	--
KCl	0.046,0.027	(28,29)	0.055	(30)	0,059,0.063, 0.066	(32,33,35)	0.082,0.083, 0.074	(32,33,36)
$KClO_3$	0.103[c],0.000	(28,29)	--	--	--	--	--	--
KF	0.149	(28)	--	--	--	--	--	--
$KHCO_2$	0.096	(28)	--	--	--	--	--	--
KI	0.004,-0.027	(28,29)	-0.011	(30)	0.035	(32)	).059	(32)
KIO_3	0.070	(28)	--	--	--	--	--	--
KOH	0.136,0.135	(28,29)	--	--	--	--	).128	(32)
KNO_2	0.049	(28)	--	--	--	--	--	--
KNO_3	0.043,0.032	(28,29)	0.021	(30)	0.024,0.025, 0.026	(32,33,34)	0.056,0.052	(33,36)
K_2CO_3	0.178,0.122	(28,29)	--	--	--	--	--	--
$K_2C_2O_4$	0.097,0.079	(28,29)	--	--	--	--	--	--
K_2CrO_4	0.110	(28)	--	--	--	--	--	--

Table IV, Cont. EXPERIMENTAL SALTING-OUT COEFFICIENTS[a] AT 25°C[b]

	Ammonia		Hydrogen Sulfide		Carbon Dioxide		Nitrous Oxide	
	Coefficient	Ref.	Coefficient	Ref.	Coefficient	Ref.	Coefficient	Ref.
K_2HPO_4	0.113	(28)	--	--	--	--	--	--
K_2SO_3	0.102	(28)	--	--	--	--	--	--
K_2SO_4	0.099,0.086	(28,29)	0.097	(30)	0.109	(34)	0.133	(36)
NaBr	0.020,-0.007	(28,29)	0.018	(30)	--	--	0.095	(36)
NaCl	0.041,0.017	(28,29)	0.063, 0.064	(30,31)	0.097,0.094	(33,36)	0.117,0.106	(33,36)
$NaClO_3$	-0.001	(29)	--		--	--	--	--
$NaClO_4$	--	--	0.037	(31)	--	--	--	--
NaI	-0.012,-0.041	(28,29)	--	—	--	--	--	--
NaOH	0.103, 0.116	(28,29)	--	--	--	--	--	--
$NaNO_3$	-0.002	(29)	0.035	(30)	0.063,0.061	(33,34)	0.075	(36)
Na_2CO_3	0.108	(29)	--	—	--	--	0.194	(27)
Na_2HPO_4	--	--	--	--	--	--	0.183	(36)
Na_2S	0.099	(28)	--	--	--	--	--	—
Na_2SO_4	0.083	(29)	0.131	(30)	0.148,0.146	(33,34)	0.175,0.167	(33,36)

[a] k in the equation log (S°/S) = km where S° and S are the respective solubilities, in water and salt solution, per unit weight of water, and m is g equivalents of salt per kg of water.

[b] Except data of Dawson and McCrea(29), obtained at 20°C.

[c] These values were rejected in the correlation study.

These yielded poorer agreement than was obtained on the other gases.

Table V. EFFECT OF CATION ON SALTING-OUT COEFFICIENT

	Average Difference		Average Absolute Deviation
	$K^+ - NH_4^+$	$Na^+ - NH_4^+$	
	kg H_2O/mol		
Ammonia	0.055	0.041	0.006
Carbon Dioxide	0.045	0.082	0.017
Hydrogen Sulfide	0.036	0.052	0.009
Nitrous Oxide	0.042	0.072	0.004

We have used these average differences between potassium and ammonium salt solutions and between sodium and ammonium salt solutions to calculate average values of salting-out coefficients for the four gases in ammonium salts from determinations on salts with the same anion but different cations. A sample calculation from data on ammonia in bromide solutions is shown in Table VI.

Table VI. SAMPLE CALCULATION

		Coefficient, kg H_2O/mol		
Salt	Investigator	Observed (Table IV)	Cation Difference (Table V)	NH_4^+ Value
NH_4Br	Dawson and McCrea	-0.052	0.000	-0.052
KBr	Abegg and Riesenfeld	0.026	0.055	-0.029
KBr	Dawson and McCrea	0.002	0.055	-0.053
NaBr	Abegg and Riesenfeld	0.020	0.041	-0.021
NaBr	Dawson and McCrea	-0.007	0.041	-0.048
Average				-0.041

Results thus obtained on single-charged anions and on sulfate are presented in Table VII. Other salts with double-charged anions could have been added, but we have not been able to use the resulting values in correlations for prediction of salting out by bicarbonate, hydrosulfide, and carbamate salts. Note that the result for hydroxyl ion is for a hypothetical ammonium hydroxide that behaves as a strong electrolyte. Also, results obtained from potassium and/or sodium salts may differ from actual behavior of ammonium salts because of pairing of ammonium ion with the anion. For later discussion we have also calculated an average difference in salting-out coefficient between ammonia and each other gas.

For correlation of the coefficients we have found two parameters for which values are known for all or most of the single-charged anions in the data base and also for bicarbonate and hydrosulfide. These are the entropy of the ion and the thermochemical radius of the ion as calculated by Yatsimirskii(37,38).

Table VII. SALTING-OUT COEFFICIENTS AND PARAMETERS FOR AMMONIUM SALTS

Anion	NH_3		H_2S			CO_2			N_2O			Anion Entropy cal/ °K mol	Anion Radius, Å
	k, kg H_2O/mol	No. of Det'n.	k, kg H_2O/mol	No. of Det'n.	k(H_2S) -k (NH_3)	k, kg H_2O/mol	No. of Det'n.	k(CO_2) -k (NH_3)	k, kg H_2O/mol	No. of Det'n.	k(N_2O) -k (NH_3)		
Br^-	-0.041	5	-0.027	3	0.014	-0.001	1	0.040	0.025	4	0.066	24.6	1.96
BrO_3^-	0.028	1	--	--	--	--	--	--	--	--	--	43.8	1.91
$CH_3CO_2^-$	0.032	1	-0.061	1	-0.093*	--	--	--	--	--	--	26.0	1.59
CN^-	-0.006	1	--	--	--	--	--	--	--	--	--	33.5	1.82
CNS^-	-0.006	1	--	--	--	--	--	--	--	--	--	41.3	1.95
Cl^-	-0.017	5	0.011	4	0.028	0.018	7	0.035	0.037	7	0.054	18.5	1.81
ClO_3^-	-0.048	2	--	--	--	--	--	--	--	--	--	44.3	2.00
ClO_4^-	--	--	-0.013	1	--	--	--	--	--	--	--	48.5	2.36
F^-	0.094	1	--	--	--	--	--	--	--	--	--	3.0	1.33
HCO_2^-	0.042	1	--	--	--	--	--	--	--	--	--	27.2	1.58
HCO_3^-	--	--	--	--	--	--	--	--	0.142	1	--	28.0	1.63
I^-	-0.047	4	-0.047	1	0.000	-0.010	1	0.037	0.017	1	0.064	31.4	2.20
IO_3^-	0.015	1	--	--	--	--	--	--	--	--	--	33.3	1.82
OH^-	0.074	4	--	--	--	--	--	--	0.086	1	0.012*	2.8	1.40
NO_2^-	-0.006	1	--	--	--	--	--	--	--	--	--	35.2	1.55
NO_3^-	-0.023	5	-0.016	3	0.007	-0.014	6	0.009	0.009	4	0.032	40.3	1.89
SO_4^{-2}	0.039	4	0.071	3	0.031	0.056	3	0.017	0.094	4	0.055	--	--
Avg.	--	--	--	--	0.016	--	--	0.028	--	--	0.054		

* Not included in the average.

Salting-out coefficients for ammonia in ammonium salts are plotted in Figure 8 against entropy of the anion based on $\bar{S}^o$ of hydrogen equal to −5.3 entropy units (cal/mol K). This basis is adopted from Friedman and Krishnan(39), but the entropy values are taken from a more recent compilation(40). With some exceptions, values fall into two groups: one consisting of chloride, bromide, and iodide, and the other consisting mostly of oxygenated anions. Fluoride, hydroxide, and cyanide appear to be intermediate. A similar effect, also showing a difference between the halides and oxygenated anions of about 20 entropy units, was found by Bromley on a plot of activity coefficients of anions(4).

Yatsimirskii's thermochemical radii are obtained from Kapustinskii's empirical formula for the lattice energy of crystalline salts. The lattice energy is the heat evolved when the gaseous cation and anion combine to form the salt:

$$U_o = \Delta H_f \text{ anion (g)} + \Delta H_f \text{ cation (g)} - \Delta H_f \text{ salt (c)}$$

where U_o is the lattice energy in kcal/mol and ΔH_f anion (g), ΔH_f cation (g), and ΔH_f salt (c) are the respective heats of formation of gaseous cation, gaseous anion, and crystalline salt. Kapustinskii's formula for univalent salts is —

$$U_o = \left(\frac{574.4}{r_{an} + r_{ca}}\right)\left(1 - \frac{0.345}{r_{an} + r_{ca}}\right)$$

where r_{an} and r_{ca} are the respective thermochemical radii of anion and cation in Å. Kapustinskii established his formula with the use of Goldschmidt radii of monatomic ions. To extend the scheme to multiatomic anions of unknown radius and heat of formation of the gaseous anion, the formula is applied to two salts of the same anion with cations of differing radii. The two simultaneous equations can then be solved for the radius and heat of formation of the anion.

The correlation of salting-out coefficients with thermochemical radii of anions from Yatsimirskii is presented in Figure 9. The bromate ion again shows a higher than expected salting-out coefficient, which indicates that the coefficient is erroneous. The coefficient for nitrite lies beneath the curve. We have recalculated its radius from Kapustinskii's formula with current thermochemical data on barium, calcium, potassium, and sodium nitrites(41,42) to obtain values ranging from 1.60 to 1.68 Å. An average of these values reduces the disagreement with our correlation, but not sufficiently to bring the deviation into the range, about ±0.02 kg H_2O/mol, found for other salts.

In the entropy correlation, the bicarbonate ion should fall on the oxygenated anion line, and we assume that the hydrosulfide

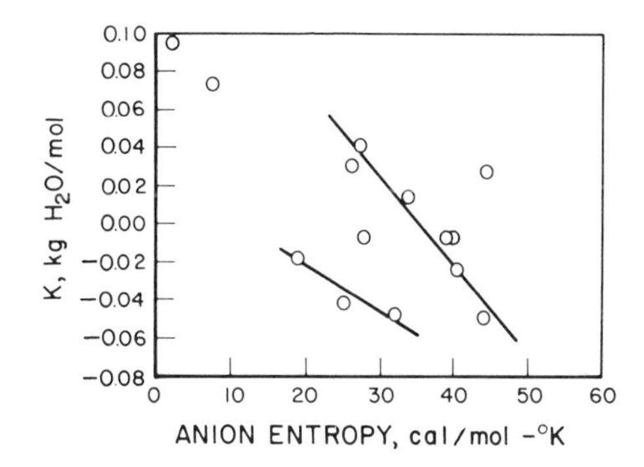

Figure 8. Correlation of salting-out coefficients for ammonia in ammonium salts with entropy of the anion

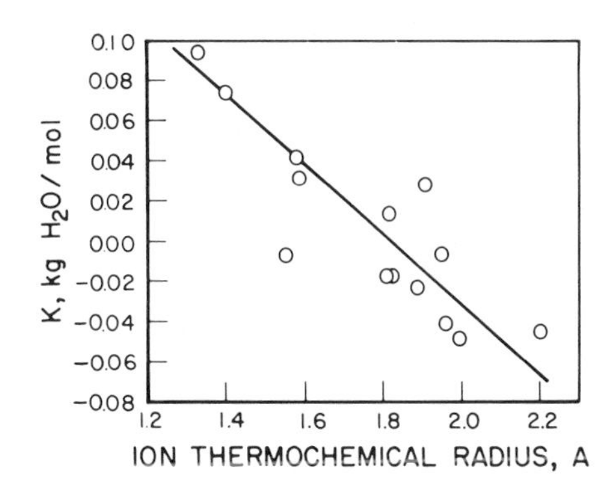

Figure 9. Correlation of salting-out coefficients for ammonia in ammonium salts with radius of the anion

ion falls on the halide line. We then obtain the following coefficients for the salting out of ammonia by ammonium bicarbonate and ammonium hydrosulfide at 25°C:

	Entropy Correlation	Ion Radius Correlation
HCO_3^-	0.038	0.034
HS^-	-0.022	-0.022

The two correlations thus agree very well.

The correlation parameters are not well established for the carbamate ion. The determination of entropy of the aqueous salt is uncertain because of its tendency to decompose to ammonia and bicarbonate ions. Entropy of the anion has been estimated by Wagman and Goldberg of NBS at 22 cal/mol K ($\overline{S}^\circ$ of H^+ = 0.0). (43) This yields a salting-out coefficient for ammonia by ammonium carbamate of 0.036 kg H_2O/mol.

From thermochemical data of Bernard and Borel(44) we obtain 1.69 Å as the thermochemical radius of the carbamate ion (instead of 1.93 Å as apparently miscalculated by Bernard and Borel). Our value gives a salting-out coefficient of 0.024 kg H_2O/mol, in good agreement with the coefficient obtained from the entropy of the anion.

Thus, averages from the two correlations yield tentative values of 0.036, −0.022, and 0.033 kg H_2O/mol for the salting out of ammonia by ammonium bicarbonate, hydrosulfide, and carbamate solutions, respectively, at 25°C. However, for ammonia in ammonium hydrosulfide we prefer to use a value from our correlation of the ternary data of Miles and Wilson(45,46), as discussed later.

The coefficients for the other gases in ammonia salts at 25°C (Table VII), show patterns similar to that of ammonia, but there are insufficient data to obtain independent correlations. Instead, we have averaged the differences between each gas and ammonia in ammonium salts, as shown in Table VII, to obtain for any given salt solution:

$$k_c = k_a + 0.028$$

$$k_s = k_a + 0.016$$

Effect of Temperature on Salting-Out Coefficients. Long and McDevit (23) point out that on differentiation of their equation with respect to temperature, several terms are obtained of which the $d\overline{V}_S^o/dT$ one is dominant, and the predicted temperature coefficient is approximated by the equation:

$$\frac{dk}{dT} = \frac{-\overline{V}_i^o}{2.3RT\,\beta_o} \left(\frac{d\overline{V}_S^o}{dT}\right)$$

The partial molar volume of carbon dioxide and the product $\beta_0 T$ are essentially constant from 0 to about 60°C; this suggests use of the partial molar volume of the salt as the only correlation parameter. The correlation was tested in the form:

$$k_{t_2} - k_{t_1} = c\,(\overline{V}^o_{t_2} - \overline{V}^o_{t_1})$$

where $k_{t_2} - k_{t_1}$ are salting-out coefficients on the molal basis for a given salt at the respective temperatures t_2 and t_1; c is a correlation constant; and $\overline{V}^o_{t_2}$ and $\overline{V}^o_{t_1}$ are partial molar volumes of the salt at infinite dilution at these temperatures. We tested the relationship with data on carbon dioxide in the temperature range of 0 to 40°C. Markham and Kobe(33) report precise data at 0.2°, 25° and 40°C for salting out by a number of salts. Two additional data points with 15° to 35°C temperature intervals were obtained from data of Yasunishi and Yoshida(34). Partial molar volumes of salts of infinite dilution were obtained from a compilation by Millero.(47) Results are presented in Figure 10. (Note that the salts are listed there in the order of increasing difference in partial molar volume.) The correlation is surprisingly good, with the constant c equal to −0.0056 kg H_2O/cm^3.

Ellis and Golding(48) and Malinin and coworkers (49,50) have obtained data on the salting out of carbon dioxide by aqueous sodium chloride solutions at still higher temperatures, as shown in Table VIII. According to these data, the salting-out coefficient has a minimum at about 140°C, whereas the maximum in the partial molar volume of aqueous sodium chloride occurs at about 65°C. Malinin's salting-out coefficients for carbon dioxide in calcium chloride solutions, also reported in Table VIII, show a minimum at about 60°C, whereas the maximum in the partial molar volume of calcium chloride occurs at about 35°C. The salting out of methane by sodium chloride solutions, on the other hand, does show a minimum at about 70°C(51), close to the temperature predicted by the Long-McDevit equation. Probably the behavior of carbon dioxide can be rationalized on the basis of the formation of carbonic acid or of other interaction with water as shown, for example, by equations of state of water-carbon dioxide mixtures (52). In any case, it appears that uncertainty in the variation of the coefficient for carbon dioxide in sodium chloride solution is not more than about 0.01 kg H_2O/mol in the temperature range from 25° to 150°C.

To apply the correlation of Figure 10 for prediction of the salting out of carbon dioxide by ammonium hydrosulfide and bicarbonate solutions we need to correct for the differences of their partial molar volumes from that of sodium chloride. Partial molar volumes were obtained from Ellis and McFadden(53). Volume changes of the hydrosulfide and bicarbonate are equal within 0.2 cm^3/mol at temperatures up to 100°C and differ very little at still higher temperatures: thus, we assume that the changes with temperature of the salting-out coefficients of the two salts are equal up to

Table VIII. EFFECT OF TEMPERATURE ON SALTING-OUT OF CARBON DIOXIDE BY SODIUM CHLORIDE AND CALCIUM CHLORIDE SOLUTIONS

	NaCl Solution					$CaCl_2$ Solution	
	1 molal			0 molal		1 molal	0 molal
Salt Concentration	Markham and Kobe(32)	Ellis and Golding(48)	Malinin (49,50)	Markham and Kobe(33)	Malinin (49,50)	Malinin (49,50)	Malinin (49,50)
Temp., °C	Salting-Out Coefficients, kg H_2O/mol						
25	0.097	--	0.096*	0.102	0.1007	0.171*	0.1851
50	--	0.095	0.089*	--	0.0946	0.168*	0.1768
75	--	0.088	0.088*	--	0.0958	0.1685*	0.180
100	--	0.087	0.087*	--	0.0931	0.174*	0.192
150	--	0.076	0.084*	--	0.0905	0.176*	0.201
200	--	0.089	--	--	--	--	--
250	--	0.128	--	--	--	--	--
300	--	0.176	--	--	--	--	--

* Interpolated or extrapolated at IGT. Note that the coefficients for $CaCl_2$ are on the molal basis, not kg H_2O/g equivalent.

about 150°C. We have smoothed the temperature-change data for salting out of carbon dioxide by sodium chloride and have adjusted them to account for the difference in partial molar volume of the ammonium salts to yield changes in the salting-out coefficient for carbon dioxide in ammonium hydrosulfide and bicarbonate solutions. Temperature effects with ammonium carbamate are tentatively assumed to be equal to those of ammonium bicarbonate. The equations for carbon dioxide in Table IX represent these changes and the values at 25°C adopted as described in the previous section.

Table IX. CORRELATION EQUATIONS FOR SALTING-OUT OF AMMONIA, CARBON DIOXIDE, AND HYDROGEN SULFIDE

Gas	Salt	Coefficient, kg H_2O/mol
CO_2	NH_4HS	$0.1519-6.870 \times 10^{-4}\ T + 0.8318 \times 10^{-6}\ T^2$
CO_2	NH_4HCO_3	$0.1949-6.870 \times 10^{-4}\ T + 0.8318 \times 10^{-6}\ T^2$
CO_2	$NH_4NH_2CO_2$	$0.1919-6.870 \times 10^{-4}\ T + 0.8318 \times 10^{-6}\ T^2$
NH_3	NH_4HS	$0.0802-4.559 \times 10^{-4}\ T + 0.552 \times 10^{-6}\ T^2$
NH_3	NH_4HCO_3	$0.1229-4.559 \times 10^{-4}\ T + 0.552 \times 10^{-6}\ T^2$
NH_3	$NH_4NH_2CO_2$	$0.1199-4.559 \times 10^{-4}\ T + 0.552 \times 10^{-6}\ T^2$
H_2S	NH_4HS	$0.1399-6.870 \times 10^{-4}\ T + 0.8318 \times 10^{-6}\ T^2$
H_2S	NH_4HCO_3	$0.1829-6.870 \times 10^{-4}\ T + 0.8318 \times 10^{-6}\ T^2$
H_2S	$NH_4NH_2CO_2$	$0.1799-6.870 \times 10^{-4}\ T + 0.8318 \times 10^{-6}\ T^2$

Data on the effect of temperature on salting out of ammonia are even less satisfactory than those for carbon dioxide. Perman obtained some data on a potassium sulfate solution at temperatures of 40° to 59°C and on two ammonium chloride solutions at temperatures from 19° to 58°C.(54). His ammonia concentrations were in the range of 5 to 13 molal. His data indicate only small changes in the salting-out coefficient, but the coefficient for ammonium chloride increases with temperature, which is contrary to the effect found with carbon dioxide.

Riesenfeld(55) reported data on ammonia obtained by Konovalov(56) at 60°C for comparison with his own data at 35°C and that of Abegg and Riesenfeld(28) at 25°C. Results from the Abegg and Riedenfeld data at 25°C and the Konovalov data at 60°C are shown in Figure 11, plotted against change in partial molar volume of the salt. The data show much scatter even aside from the points for lithium chloride and the hydroxides. Ignoring this erratic data, we have tentatively adopted the correlation developed for carbon dioxide, modified to take account of the difference in the partial molar volumes of the two gases(Table I): that is, the slope of the correlation line found for carbon dioxide was multiplied by the factor 25/37.6. The resulting correlation is shown as the line on Figure 11.

The partial molar volume data of Table I on ammonia was calculated from density data in Landolt-Bornstein(57). The values of partial molal volume at infinite dilution can be expressed as:

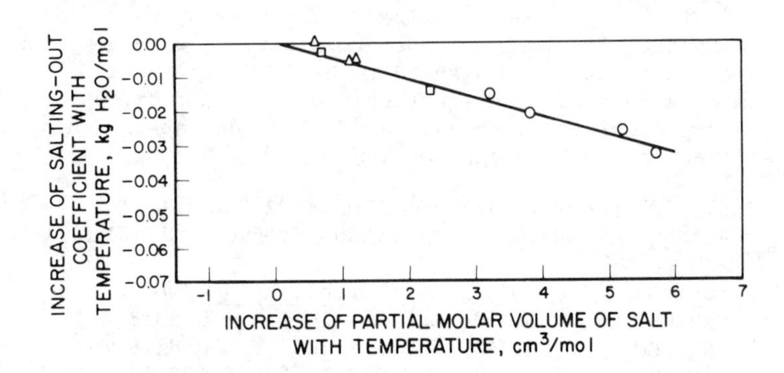

Figure 10. Effect of temperature on salting out of carbon dioxide (coefficient source: (○) (32); salts: KCl, NaCl, KNO_3, $NaNo_3$; $t_1 = 0.2°C$; $t_2 = 25°C$; (△) (32); KCl, NaCl, KNO_3; $t_1 = 25°C$; $t_2 = 40°C$; (□) (33); NH_4Cl, $NaNO_3$; $t_1 = 15°C$; $t_2 = 35°C$)

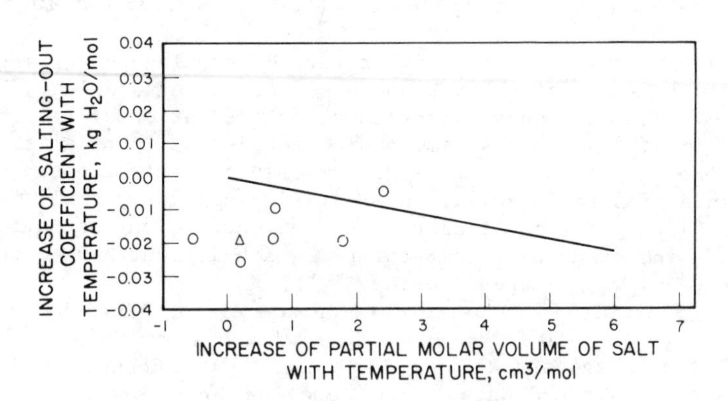

Figure 11. Effect of temperature on salting out of ammonia (coefficient source: (○) at 25°C (28), at 60°C, (55,56); salts: LiCl, KOH, NaOH, KCl, NaCl, KNO_3; $t_1 = 25°C$; $t_2 = 60°C$; (△) (58); KOH; $t_1 = 25°C$; $t_2 = 60°C$; (——) modified carbon dioxide correlation)

$$\bar{V}_a^o = 15.8 + 0.030\ T$$

The partial molar volume of hydrogen sulfide is nearly equal to that of carbon dioxide (Table I), and we tentatively assume that changes in its salting-out coefficients with temperature are the same as those of carbon dioxide.

We believe that the most accurate salting-out coefficient for ammonia in ammonium hydrosulfide solution is the value, −0.012 kg H_2O/mol, that we calculated from Miles and Wilson vapor-liquid equilibrium data at 80°C(45). With the temperature change correlation that we have adopted as discussed above, this yields the value −0.0067 kg H_2O/mol at 25°C. This is in reasonable agreement with the value −0.022 kg H_2O/mol obtained from our correlations. For ammonia in bicarbonate and carbamate at 25°C, we adopt the values from the correlations, namely 0.036 and 0.033 kg H_2O/mol. As discussed previously, for carbon dioxide and hydrogen sulfide at 25°C, we adopt values 0.028 and 0.016 higher, respectively, than those for ammonia. Equations incorporating these values and the changes with temperature discussed above are presented in Table IX.

Salting-Out of Other Gases. Gases other than ammonia, carbon dioxide, and hydrogen sulfide are also commonly present in streams from the processing of coal, and, thus, their solubilities in aqueous condensates of the NH_3-CO_2-H_2S-H_2O system are of concern. In hydrogasification processes such gases include hydrogen, carbon monoxide, methane, carbonyl sulfide, and hydrogen cyanide. Nitrogen and argon are, of course, present in low-Btu gases from gasification with air.

As a data base for correlation of salting-out coefficients of such gases, we have collected data on salting-out by sodium chloride solutions at 25°C; more data are available on this salt for a wide variety of gases than on any other. We found that the salting-out coefficient can be correlated against Henry's constants for solubility in water at infinite dilution and the Lennard-Jones force constant σ (a collision diameter), as shown in Figure 12. The gases, in the order of increasing Henry's constant, are ammonia, hydrogen cyanide, sulfur dioxide, hydrogen sulfide, chlorine, acetylene, carbon dioxide, nitrous oxide, phosphine, ethylene, xenon, krypton, ethane, argon, methane, oxygen, hydrogen, neon, helium, and sulfur hexafluoride. Details of the parameters are discussed elsewhere(59).

The data for the polar gases, ammonia, hydrogen cyanide, and sulfur dioxide, show that the correlation does not apply to such gases but only to non-polar gases. The correlation does allow us to predict salting-out coefficients for carbon monoxide and carbonyl sulfide in sodium chloride solutions. Further, by application of the principles of the correlations we have developed for the effects of cations and anions, we should be able to predict salting-out coefficients for methane, oxygen, nitrogen,

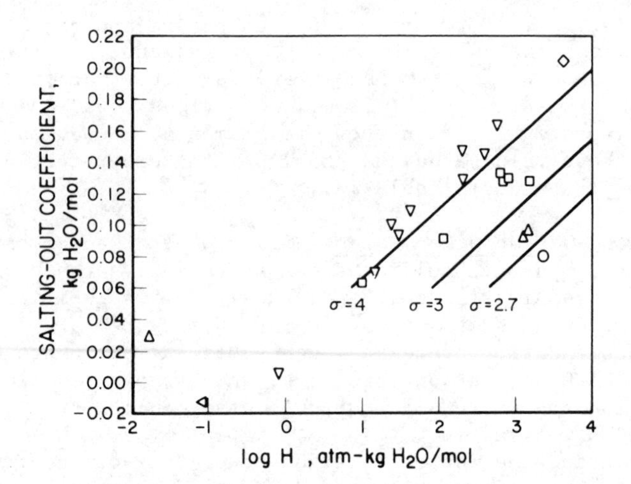

Figure 12. Correlation of salting-out coefficients of various gases in sodium chloride solution at 25°C: force constant σ (Å): (○) 2.6; (△) 2.8–3.2; (□) 3.4–3.9; (▽) 4.1–4.6; (◇) 5.5; (<) unknown

carbon monoxide, and carbonyl sulfide in solutions of the NH_3-CO_2-H_2S-H_2O system.

Analysis of Ternary and Quaternary Data

Equilibrium data for the NH_3-H_2S-H_2O subsystem have been reported by van Krevelen *et al.* at 20°, 40°, and 60°C(7), by Miles and Wilson at 80° and 120°C(46), and by Ginzburg *et al.* at temperatures from 57° to 87°C (obtained at constant total pressure rather than at constant temperature) (60,61) We correlated data having ionic strengths above 0.2 molal in terms of an equilibrium coefficient K_4:

$$K_4 = \frac{(NH_4^+)\ (HS^-)}{m_a\ \gamma_a\ f_s}$$

The effect of concentration of free (molecular) ammonia on the activity of the electrolyte was derived mainly from two 80°C data points of Miles and Wilson having 16 to 17 molal free ammonia concentration. Data points below 0.2 ionic strength were fitted by application of Kielland's estimation of ionic activity coefficients(62). Details are presented elsewhere(45), together with graphs giving partial pressures of ammonia and hydrogen sulfide for temperatures from 80° to 260°F over a range of liquid concentration.

Data reported by van Krevelen et al.(7) on the NH_3-CO_2-H_2O subsystem (including data of Pexton and Badger) and data of Frohlich(63) were analyzed during an early period of our investigation.(8) We plan to revise our correlation with use of Meissner's(64) treatment of ionic activity coefficients, which is better suited than van Krevelen's for application to mixtures of three or more electrolyte components.

Conclusions

Vapor-liquid equilibria in the NH_3-CO_2-H_2S-H_2O system is of importance in the processing of coal for production of gaseous and liquid fuels. Direct vapor-liquid equilibrium data are available for only limited ranges of temperature and composition and must be supplemented by correlation with models utilizing other data. Data on the solubility of each of the component gases in water and data on their ionization constants have been reviewed. Correlations for the salting out of these and other gases by the dissolved electrolytes have been developed. Work on the NH_3-CO_2-H_2O ternary and on the quaternary system is still in progress.

Notation

A = Coefficient expressing the effect of concentration of gas on its activity coefficient in binary solution; $\ln \gamma = A\ (1-x_w^2)/RT$.

b = Coefficient expressing the effect of concentration of gas on its activity coefficient, kg H_2O/mole
c = Coefficient expressing the effect of change of partial molal volume of electrolyte (with temperature) on the salting-out coefficient, kg H_2O/cm^3.
C = Salt concentration, mol/ℓ
f = Fugacity, atm
H = Henry's constant, atm or atm kg H_2O/mol
I = Ionic strength, mol/ℓ.
k = Salting-out coefficient, ℓ/mol or kg H_2O/mol
K = Ionization constant, molal
K_4 = Equilibrium coefficient, mol/kg H_2O-atm
m = Concentration, mol/kg H_2O
$(NH_4^+),(HS^-)$ = Concentration of indicated ion, mol/kg H_2O
r_{an}, r_{ca} = Thermochemical radius of anion and cation, Å
R = Gas constant
S, S_o = Solubility of gas in salt solution and water, respectively.
T = Temperature, °K
U_o = Lattice energy of crystalline salt, k cal/mol
$\overline{V}_i^o$ = Partial molar volume of molecular specie at infinite dilution, cm^3/mol
V_s = Molar volume of hypothetical molten salt, cm^3/mol
x_g, x_{ca}, x_{an} = Contribution of gas, cation, and anion, respectively, to a salting-out coefficient

Greek

β_o = Compressibility of water
γ = Activity coefficient

Subscripts

a = Ammonia
an = Anion
c = Carbon dioxide
ca = Cation
g = Gas
s = Hydrogen sulfide

Acknowledgement

The work reported here was conducted as part of a project, sponsored by the U.S. Department of Energy, on preparation of a Coal Conversion Systems Technical Data Book.

Literature Cited

1. IGT Process Research Division, HYGAS™: 1964 to 1972, Pipeline Gas from Coal – (IGT Hydrogenation Process), R&D Report No. 22 Final Report (FE-381-T9-P3) (Ca.400 pp), Vol. 3, Part VI: "Coal and Char Characterization," prepared by IGT for ERDA. ERDA: Wshington, D.C., July 1975, pp. 66-69.

2. Edwards, T.J.; Newman, J.; Prausnitz, J.M. "Thermodynamics of Aqueous Solutions Containing Volatile Weak Electrolytes," A.I.Ch.E. J. , 1975, 21, 248-59.

3. Edwards, T.J.; Maurer, G.; Newman, J.; Prausnitz, J.M. "Vapor-Liquid Equilibria in Multicomponent Aqueous Solutions of Volatile Weak Electrolytes," A.I.Ch.E. J., 1978, 24, 966-76.

4. Bromley, L.A. "Approximate Individual Ion Values of β (or B) in Extended Debye-Hückel Theory for Uni-univalent Aqueous Solutions At 298.15 K," J.Chem. Thermo., 1972, 4, 669-73.

5. Pitzer, K.S. "Thermodynamics of Electrolytes, I, "J. Phys. Chem., 1973, 77, 268-77.

6. Beutier, D.; Renon, H. "Representation of NH_3-H_2S-H_2O, NH_3-CO_2-H_2O, and NH_3-SO_2-H_2O Vapor-Liquid Equilibria," Ind. Eng. Chem. Process Des. Dev., 1978, 17, 220-30.

7. van Krevelen, D.W.; Hoftijzer, P.J.; Huntjens, F.J. "Composition and Vapour Pressures of Aqueous Solutions of Ammonia, Carbon Dioxide, and Hydrogen Sulphide," Rec. Trav. Chim., Pays-Bas, 1949, 68, 191-216.

8. Institute of Gas Technology. "Preparation of a Coal Conversion Systems Technical Data Book," U.S. Department of Energy Project No. FE-2286-16., Chicago, 1977, pp. I-87.

9. Wells, H. M.S. Thesis, University of California, Berkeley, CA, 1979.

10. Lee, J.I.; Mather, A.E."Solubility of Hydrogen Sulfide in Water," Ber.Bunsenges. physik. Chem., 1977, 81 (10), 1020-23.

11. Neuburg, H.J.; Atherley, J.F.; Walker, L.G. "Girdler-Sulfide Process Physical Properties," AECL-5702. Chalk River Nuclear Laboratories: Chalk River, Ontario, May 1977.

12. Murphy, J.A.; Gaines, G.L., Jr. "Density and Viscosity of Aqueous Hydrogen Sulfide Solutions at Pressures to 20 Atm." J. Chem. Eng. Data, 1974, 19, 359-62.

13. Haas, J.L., Jr.; Fisher, J.R. "Simultaneous Evaluation and Correlation of Thermodynamic Data, "Amer. J. Sci., 1976, 276, 525-45.

14. Jones, M.E. "Ammonia Activity Between Vapor and Liquid Phases at Elevated Temperatures," J. Phys. Chem., 1963, 67, 113-15.

15. Clarke, E.C.W.; Glew, D.N. "Aqueous Nonelectrolyte Solutions. Part VIII. Deuterium and Hydrogen Sulfides Solubilities in Deuterium Oxide and Water," Can. J. Chem., 1971, 49, 691-98.

16. Sweeton, F.H.; Mesmer, R.E.; Baes, C.F.,Jr. "Acidity Measurements at Elevated Temperatures. VII. Dissociation of Water," J. Solution Chem., 1974, 3, 191-214.

17. Hitch, B.F.; Mesmer, R.E. "The Ionization of Aqueous Ammonia to 300°C in KCL Media," J. Solution Chem., 1976, 5, 667-79.

18. Barnes, H.L.; Helgeson, H.C.; Ellis, A.J. "Ionization Constants in Aqueous Solutions," in Clark, S.P., Jr., Ed. "Handbook of Physical Constants"; Memoir 97, Geological Society of America: New York, 1966, pp. 401-13.

19. Kryukov, P.A.; Starostina, L.I.; Tarasenko, S.Ya; Pavlyuk, L.A.; Smolyakok, B.S.; Larionov, E.G. "Ionization Constants of Carbonic Acid, Hydrogen Sulfide, Boric Acid, and Sulfuric Acid at High Temperatures," Mezhdunar. Geokhim., Kongr. (Dokl.) 1st, 1971, 2, 186-98; C.A., 1974, 84 (69193).

20. Barnes, H.L., Personal communication, 1976.

21. Naumov, G.B.; Ryzhenko, B.N.; Khodakovsky, I.L. "Handbook of Thermodynamic Data"; Report No. USGS-WRD-74-001, NTIS: Arlington, Va., January 1974.

22. Masterton, W.L.; Lee, T.P. "Salting Coefficients from Scaled Particle Theory," J. Phys. Chem., 1970, 74, 1776-80.

23. Long, F.A.; McDevit, W.F. "Activity Coefficients of Nonelectrolyte Solutes in Aqueous Salt Solutions," Chem. Rev., 1952, 51, 119-69.

24. van Krevelen, D.W.; Hoftijzer, P.J. "On the Solubility of Gases in Aqueous Solutions," Proc. of the 21st Cong., Intern. de Chimie Industrielle, 1948 (March) 21, 168-73.

25. Onda, K.; Sada, E.; Kobayashi, T.; Kito, S.; Ito, K. "Salting Out Parameters of Gas Solubility in Aqueous Salt Solutions," J. Chem. Eng. Japan, 1970, 3, (1), 18-24.

26. Onda, K.; Sada, E.; Kobayashi, T.: Kita, S.; Ito, K. "Solubility of Gases in Aqueous Solutions of Mixed Salts," J. Chem. Eng. Japan, 1970, 3, (2), 137-42.

27. Hikita, H.; Asai, S.; Ishikawa, H.; Esaka, N. "Solubility of Nitrous Oxide in Sodium Carbonate-Sodium Bicarbonate Solutions at 25°C and 1 atm," J. Chem. Eng. Data, 1974, 19, 89-92.

28. Abegg, R.; Riesenfeld, H. "The Solubility of Ammonia in Salt Solutions as Measured by Its Partial Pressure. I., "Z Physik. Chem., 1902, 40, 84-108.

29. Dawson, H.M.; McCrae, J. "LIV.-Metal-Ammonia Compounds in Aqueous Solution. Part II. The Absorptive Powers of Dilute Solutions of Salts of the Alkali Metals," J. Chem. Soc. Trans., 1901, 79, 493-511.

30. McLauchlan, W.H. "The Influence of Salts on the Solubility of Hydrogen Sulfide, Iodine, and Bromine," Z. Physik. Chem., 1903, 44, 600-33.

31. Gamsjager, H.; Schindler, P. "Solubilities and Activity Coefficients of H_2S in Electrolyte Mixtures," Helv. Chim. Acta,1969, 52, 1395-1402.

32. Geffcken, G. in Linke, W.F., Ed. "Seidell's Solubilities of Inorganic and Metal-Organic Compounds A-Ir," 4th Ed., Vol. I, Van Nostrand: Princeton, 1958, pp. 468-72.

33. Markham, A.E.; Kobe, K.A. "Solubility of Carbon Dioxide and Nitrous Oxide in Salt Solutions," J. Amer. Chem. Soc., 1941, 63, 449-54.

34. Yasunishi, A.; Yoshida, F. "Solubility of Carbon Dioxide in Aqueous Electrolyte Solutions," J. Chem. Eng. Data, 1979, 24, 11-14.

35. Findlay, A.; Shen, B. in Linke, W.F., Ed. "Seidell's Solubilities of Inorganic and Metal-Organic Compounds A-Ir," 4th Ed., Vol. I, Van Nostrand: Princeton, 1958, pp.468-72

36. Manchot, W.; Jahrstorfer, M.; Zepter, H. in Linke, W.F., Ed. "Solubilities of Inorganic and Metal-Organic Compounds K-Z"; Vol. II, 4th Ed., Amer. Chem. Soc.: Washington, D.C., 1965, pp. 797-98.

37. Yatsimirskii, K.B. "Thermochemical Radii and Heats of Hydration of Ions," Izvest. Akad. Nauk. S.S.S.R., Otdel. Khim. Nauk., 1948, 398-405; Chem. Abstr., 1948, 42, 8604.

38. Yatsimirskii, K.B. "Thermochemical Radii of Ions and Heats of Formation of Salts, "Izvest. Akad. Nauk. S.S.S.R., Otdel. Khim. Nauk., 1947, 453-7; Chem. Abstr.,1948, 42, 2168.

39. Friedman, H.L.; Krishnan, C.V. in Franks, F., Ed. "Water. A Comprehensive Treatise. Vol. 3 Aqueous Solutions of Simple Electrolytes"; Plenum: New York, 1973, pp 55-59.

40. Wagman, D.D.; Evans, W.H.; Parker, V.B.; Halow, I.; Bailey, S.M.; Schumm, R.H. "Selected Values of Chemical Thermodynamic Properties U.S. Bureau of Standards Technical Note 270-3. U.S. Gov't. Printi Office: Washington, D.C., 1968.

41. Wagman, D.D.; Evans, W.H.; Parker, V.B.; Schumm, R.H. "Chemical Thermodynamic Properties of Compounds of Sodium, Potassium and Rubidium: An Interim Tabulation of Selected Values," NBSIR 76-1034

42. Wagman, D.D.; Evans, W.H.; Parker, V.B.; Halow, I.; Bailey, S.M.; Schumm, R.H.; Churney, K.L. "Selected Values of Chemical Thermodynamic Properties," U.S. Bureau of Standards Technical Note 270-6 U.S.Gov't Printing Office: Washington, D.C., 1971.

43. Goldberg, N., Personal communication, 1979.

44. Bernard, M.A.; Borel, M.M. "Thermochemical Study of Some Carbamates," Bull. Soc. Chim. Fr., 1968 (6), 2362-6.

45. Institute of Gas Technology. "Preparation of a Coal Conversion Systems Technical Data Book," U.S. Department of Energy Project No. FE-2286-32, Chicago, 1978, pp. I-5 to I-11.

46. American Petroleum Institute. "A New Correlation of NH_3, CO_2 and H_2S Volatility Data From Aqueous Sour Water Systems"; Publication 955, American Petroleum Institute: Washington, D.C., March 1978.

47. Millero, F.J. in Horne, R.A., Ed. "Water and Aqueous Solutions — Structure, Thermodynamics, and Transport Properties"; Wiley Interscience: New York, 1972, pp. 565-95.

48. Ellis, A.J.; Golding, R.M. "The Solubility of Carbon Dioxide Above 100°C in Water and in Sodium Chloride Solutions," Amer. J. Sci., January 1963, 261, 47-60.

49. Malinin, S.D.; Savelyeva, N.I. "Solubility of Carbon Dioxide in Sodium Chloride and Calcium Chloride Solutions at Temperatures of 25, 50, and 75°C and Elevated Carbon Dioxide Pressures," Geochem. Intern., 1972, 9, (3), 410-18.

50. Malinin, S.D.; Kurovskaya, N.A. "Solubility of Carbon Dioxide in Chloride Solutions at Elevated Temperatures and Carbon Dioxide Pressures," Geochem. Intern., 1975, 12, 199-201.

51. Clever, H.L.; Holland, C.J. "Solubility of Argon Gas in Aqueous Alkali Halide Solutions — Temperature Coefficient of the Salting Out Parameter," J. Chem. Eng. Data, 1968, 13, 411-14.

52. de Santis, R.; Breedveld, N.J.F.; Prausnitz, J.M. "Thermodynamic Properties of Aqueous Gas Mixtures at Advanced Pressures," Ind. Eng. Chem. Process Des. Dev., 1974, 13 374-77.

53. Ellis, A.J.; McFadden, I.M. "Partial Molar Volumes of Ions in Hydrothermal Solutions," Geochem. Cosmochim. Acta, 1972, 36, 413-26.

54. Perman, E.P. "XLIX — The Influence of Salts and other Substances on the Vapour Pressure of Aqueous Ammonia Solution," J. Chem. Soc., 1902, 81, 480-89.

55. Riesenfeld, H. "The Solubility of Ammonia in Salt Solutions as Measured by its Partial Pressure, II," Z. Physik. Chem., 1903, 45, 461-64.

56. Konovalov, D. J. Russ. Phys. Chem. Soc., 1898, 30, 367; 1899, 31, 910; 1899, 31, 895; Chem Zentr.,1900, I, 938.

57. Synowietz, C. in "Landolt — Bornstein Numerical Constants and Functions, New Series, Group IV, Macroscopic and Technical Properties of Matter, Vol. 1, Densities of Liquid Systems and their Heat Capacities, Part b"; Springer Verlag: Berlin, 1977, pp. 69-70.

58. Katan, T.; Campa, A.B. "Vapor Pressure of Ammonia in Aqueous Potassium Hydroxide Solutions," J. Chem. Eng. Data, 1963, 8, 574-75.

59. Institute of Gas Technology. "Preparation of a Coal Conversion Systems Technical Data Book," U.S. Department of Energy Project No. FE-2286-52, pp. I-1 to I-20. Chicago, in preparation.

60. Ginzburg, D.M.; Pikulina, N.S.; Litvin, V.P. "The System NH_3-H_2S-H_2O," Zh. Prikl. Khim., 1965, 38, (9), 2117-19.

61. Ginzburg, D.M.; Pikulina, N.S.; Litvin, V.P. "The System NH_3-H_2S-H_2O at 600 mm," Zh. Prikl. Khim., 1966, 39 (10) 2371-73.

62. Kielland, J. "Individual Activity Coefficients of Ions in Aqueous Solutions," J. Amer. Chem. Soc., 1937, 59, 1675-78.

63. Frohlich, G.J. "Vapor-Liquid Equilibria of Aqueous Systems Containing Ammonia and Carbon Dioxide," Polytechnic Institute of Brooklyn, D.Ch. E., 1957, Engineering, Chemical. University Microfilms, Inc., Ann Arbor, Mich.

64. Meissner, H.P. "Prediction of Activity Coefficients of Strong Electrolytes in Aqueous Systems," paper presented at symposium on "Thermodynamics of Aqueous Systems with Industrial Application," Washington, D.C., October 22-25, 1979

RECEIVED January 31, 1980.

7

On the Solubility of Volatile Weak Electrolytes in Aqueous Solutions

G. MAURER

Institut für Technische Thermodynamik und Kältetechnik, Universität Karlsruhe (TH), D-7500 Karlsruhe, Fed. Rep. of Germany

The solubility of gaseous weak electrolytes in aqueous solutions is encountered in many chemical and petrochemical processes. In comparison to vapor-liquid equilibria in non reacting systems the solubility of gaseous weak electrolytes like ammonia, carbondioxide, hydrogen sulfide and sulfur dioxide in water results not only from physical (vapor-liquid) equilibrium but also from chemical equilibrium in the liquid phase. This interaction between physical and chemical equilibria complicates considerably the description of vapor-liquid equilibria in multicomponent aqueous solutions. The development of thermodynamic correlations for those equilibria is also hindered by the limited experimental material available on that subject.

This contribution describes and compares three procedures for representing vapor-liquid equilibria in multicomponent aqueous solutions of volatile weak electrolytes. Starting from the basic thermodynamic relations, the approximations and simplifications applied by van Krevelen, Hoftijzer and Huntjens (1), Beutier and Renon (2) and Edwards, Maurer, Newman and Prausnitz (3) are discussed; the necessary information for using these correlations is compiled. Results calculated with these procedures are discussed and compared with literature data.

Thermodynamics

As shown in figure 1, a volatile weak electrolyte in water at a given temperature and pressure distributes itself between vapor and liquid phase. Phase equilibrium determines the concentration of the weak electrolyte in the gaseous phase at a known concentration of molecular electrolyte in water. But due to

0-8412-0569-8/80/47-133-139$08.50/0

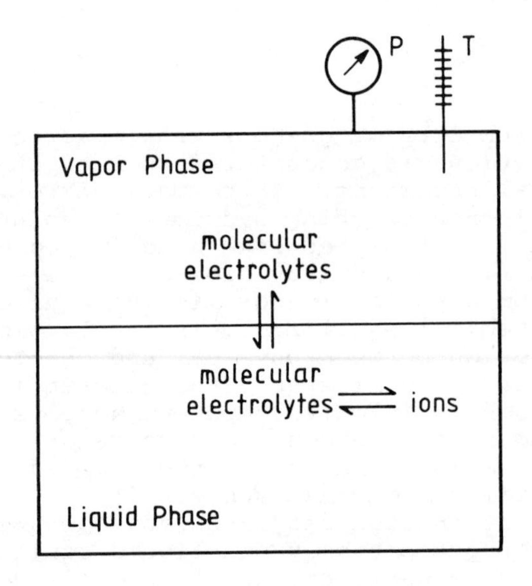

Figure 1. Vapor–liquid equilibrium in an aqueous system of volatile weak electrolytes

chemical reactions this concentration of molecular electrolyte in liquid water may be considerably different from its total concentration. (Total concentration is the amount of weak electrolyte in liquid phase, if there would not be dissociation).

In a multicomponent aqueous system encountering ammonia, carbon dioxide, hydrogen sulfide and sulfur dioxide, the vapor phase contains molecules of only five different species, e.g. NH_3, CO_2, H_2S, SO_2 and H_2O, while in the liquid phase 15 different species are present: besides the molecular species also 10 ionic species, e.g. NH_4^+, HCO_3^-, HS^-, HSO_3^-, OH^-, H^+, NH_2COO^-, $S^=$, $CO_3^=$, $SO_3^=$. For given temperature and total molalities of the weak electrolytes in the liquid phase a system of 20 mostly nonlinear equations has to be solved, in order to find the total pressure of the system and the composition of the vapor phase.

These equations include equilibrium-constants $K_i(T)$ for nine chemical reactions, expressed by activities a_i:

First dissociations of ammonia (reaction 1), carbon dioxide (reaction 2), hydrogen sulfide (reaction 3), sulfur dioxide (reaction 4):

$$NH_3+H_2O \rightleftharpoons NH_4^+ + OH^- \qquad K_1(T) = \frac{a_{NH_4^+}\, a_{OH^-}}{a_{NH_3}\, a_w} \qquad (I)$$

$$CO_2+H_2O \rightleftharpoons HCO_3^- + H^+ \qquad K_2(T) = \frac{a_{HCO_3^-}\, a_{H^+}}{a_{CO_2}\, a_w} \qquad (II)$$

$$H_2S \rightleftharpoons HS^- + H^+ \qquad K_3(T) = \frac{a_{HS^-}\, a_{H^+}}{a_{H_2S}} \qquad (III)$$

$$SO_2+H_2O \rightleftharpoons HSO_3^- + H^+ \qquad K_4(T) = \frac{a_{HSO_3^-}\, a_{H^+}}{a_{SO_2}\, a_w} \qquad (IV)$$

Second dissociations of carbon dioxide (reaction 5), hydrogen sulfide (reaction 6), sulfur dioxide (reaction 7)

$$HCO_3^- \rightleftharpoons CO_3^= + H^+ \qquad K_5(T) = \frac{a_{CO_3^=}\, a_{H^+}}{a_{HCO_3^-}} \qquad (V)$$

$$HS^- \rightleftharpoons S^= + H^+ \qquad K_6(T) = \frac{a_{S^=}\, a_{H^+}}{a_{HS^-}} \qquad (VI)$$

$$HSO_3^- \rightleftharpoons SO_3^= + H^+ \qquad K_7(T) = \frac{a_{SO_3^=}\, a_{H^+}}{a_{HSO_3^-}} \qquad (VII)$$

Carbamate reaction (reaction 8)

$$NH_3 + HCO_3^- \rightleftharpoons NH_2COO^- + H_2O \qquad K_8(T) = \frac{a_{NH_2COO^-}\, a_w}{a_{NH_3}\, a_{HCO_3^-}} \qquad (VIII)$$

Dissociation of water (reaction 9)

$$H_2O \rightleftharpoons H^+ + OH^- \qquad K_9(T) = \frac{a_{H^+}\, a_{OH^-}}{a_w} \qquad (IX)$$

Furthermore, 5 vapor-liquid phase equilibria are involved, e.g. for molecular NH_3, CO_2, H_2S and SO_2 and for water. Applying the concept of Henry's constant H_i for the solution of a gas i in pure water and fugacity-coefficient φ_i for describing the influence of intermolecular forces in the vapor phase, the resulting equations are:

$$a_i\, H_i = P\, y_i\, \varphi_i \qquad (X)-(XIII)$$

$$i = NH_3,\ CO_2,\ H_2S,\ SO_2$$

and

$$a_w\, f_w = P\, y_w\, \varphi_w \qquad (XIV)$$

(P, y_i and f_w designate the total pressure, the mole fraction of component i in vapor phase and the fugacity of pure water). The mass balance in the liquid phase results in four additional equations:

$$m_{tot,NH_3} = m_{NH_3} + m_{NH_4^+} + m_{NH_2COO^-} \qquad (XV)$$

$$m_{tot,CO_2} = m_{CO_2} + m_{HCO_3^-} + m_{NH_2COO^-} + m_{CO_3^=} \qquad (XVI)$$

$$m_{tot,H_2S} = m_{H_2S} + m_{HS^-} + m_{S^=} \qquad (XVII)$$

$$m_{tot,SO_2} = m_{SO_2} + m_{HSO_3^-} + m_{SO_3^=} \qquad (XVIII)$$

(m_i designates the molality of species i in liquid phase). The remaining two equations result from the condition of bulk electroneutrality in the liquid phase and the mole balance in the gaseous phase:

$$m_{H^+} + m_{NH_4^+} = m_{OH^-} + m_{HCO_3^-} + m_{HSO_3^-} + m_{HS^-} + m_{NH_2COO^-} + 2(m_{S^=} + m_{CO_3^=} + m_{SO_3^=}) \qquad (XIX)$$

$$y_{NH_3} + y_{CO_2} + y_{H_2S} + y_{SO_2} + y_w = 1 \qquad (XX)$$

In principle, this system of 20 equations can be solved provided the equilibrium constants, activities, Henry-constants and fugacities are available. While some results for most of these properties are available, there exists no approved method for calculating activities in concentrated aqueous solutions of weak electrolytes; therefore, several approximations were developed.

Method by van Krevelen, Hoftijzer and Huntjens (1) (KHH)

This method applies very restrictive approximations, resulting in a limited applicability. The method cannot be applied to the complete multicomponent system described above; it is suitable only for ammonia rich subsystems of

$NH_3-CO_2-H_2O$: $m_{tot,NH_3} > m_{tot,CO_2}$

$NH_3-H_2S-H_2O$: $m_{tot,NH_3} > m_{tot,H_2S}$

and $NH_3-CO_2-H_2S-H_2O$: $m_{tot,NH_3} > (m_{tot,CO_2} + m_{tot,H_2S})$.

As the model is rather simplified, it was based on experimental results which cover the temperature range from 20 to 60 °C at the following molalities:

$$NH_3-CO_2-H_2O: \; 0.5 \leqslant m_{tot,NH_3} \leqslant 2.0; \; 0.2 \leqslant \frac{m_{tot,CO_2}}{m_{tot,NH_3}} \leqslant 0.67$$

$$NH_3-H_2S-H_2O: \; 0.3 \leqslant m_{tot,NH_3} \leqslant 3.5; \; 0.17 \leqslant \frac{m_{tot,H_2S}}{m_{tot,NH_3}} \leqslant 0.65$$

$NH_3-CO_2-H_2S-H_2O$:

$$0.7 \leqslant m_{tot,NH_3} \leqslant 2.3 \qquad 0.1 \leqslant m_{tot,CO_2} \leqslant 1.4 \qquad 0.44 \leqslant \frac{m_{tot,CO_2} + m_{tot,H_2S}}{m_{tot,NH_3}} \leqslant 0.84.$$

The simplified equations are discussed here only for $NH_3-CO_2-H_2O$; for $NH_3-H_2S-H_2O$ and $NH_3-CO_2-H_2S-H_2O$ they are given in Appendix I. Van Krevelen et al. neglect the dissociation of water (eq. IX), thereby reducing the number of ionic species in liquid phase to NH_4^+, HCO_3^-, $CO_3^=$, NH_2COO^-. For given temperature and total molalities of NH_3 and CO_2 there remain 10 equations to determine all liquid concentrations, the total pressure and the composition of the vapor. Two purely chemical equilibria: The carbamate reaction (reaction 8) and the following combination of reactions 1,5 and 9

$$NH_3 + HCO_3^- \rightleftharpoons NH_4^+ + CO_3^= \text{ (reaction 10) } K_{10} = \frac{a_{NH_4^+}\, a_{CO_3^=}}{a_{NH_3}\, a_{HCO_3^-}} .$$

The equilibrium constant for the carbamate reaction (eq.VIII) was simplified by assuming $a_{H_2O} = 1$ and replacing all other activities by molalities. Numbers for $K_8(T)$ at 20, 40 and 60 °C were determined from experimental results. (Van Krevelen et al. only report discrete numbers or diagrams for some constants. For inter- and extrapolation these numbers were replaced by equations, wherein the dimensions of m_i and T are moles/kg H_2O and Kelvin, respectively.):

$$K_8(T) = \frac{m_{NH_2COO^-}}{m_{NH_3}\, m_{HCO_3^-}} = \exp\left[-5.6 + \frac{2000.6}{T}\right]. \qquad (K\ 1)$$

In the equilibrium constant K_{10} activities were replaced by molalities. Numbers for K_{10} were taken from literature:

$$K_{10}(T) = \frac{m_{NH_4^+}\quad m_{CO_3^=}}{m_{NH_3}\quad m_{HCO_3^-}} = \exp\left[-18.26 + 4780/T\right]. \qquad (K\ 2)$$

The combination of a chemical equilibrium with the phase equilibrium for CO_2: Reactions 1,2 and 9 were combined to

$$NH_3 + CO_2 + H_2O \rightleftharpoons NH_4^+ + HCO_3^- \qquad \text{(reaction 11)}$$

The appropriate equilibrium constant, combined with eq. (XI) depends on temperature and ionic strength:

$$\ln K_{11}^*(T) = \ln \frac{m_{NH_4^+}\, m_{HCO_3^-}}{m_{NH_3}\; P\; y_{CO_2}} \qquad (K\ 3)$$

$$= -26.112 + 7040/T + I\ (-0.564 + 326/T) - 1.05\ \exp(-3.7\ I)$$

(P in mm Hg, $I = \frac{1}{2} \sum_i m_i z_i^2$ = ionic strength,

z_i = charge number of ion i).

Phase equilibria: For NH_3 and H_2O eqs.(X) and (XIV) were simplified:

$$m_{NH_3} = H_{NH_3}^*\ y_{NH_3}\ P \qquad (K\ 4)$$

$$P_w^s = (1 + 0.018 \sum_i m_i)\, P\, y_w \qquad \text{(K 5)}$$

where, when in (K 4) P is in mm Hg,

$$H^*_{NH_3} = [-17.13 + 4350\,(T/K)^{-1}]\ \exp(-0.0576\ m_{NH_3})$$

and P_w^s is the vapor pressure of pure water. The remaining equations are the mass balances for NH_3 and CO_2 (eq. XV and XVI), the condition of bulk electroneutrality (eq. XIX), the mole balance in the vapor phase (eq. XX) and the assumption that - as the model should be applied to ammonia-rich solutions - in the mass balance molecular carbon dioxide can be neglected. A very simple numerical iteration procedure to solve these equations is discussed by van Krevelen et al.(1).

Procedure of Beutier and Renon (2) (BR)

The procedure of Beutier and Renon as well as the later on described method of Edwards, Maurer, Newman and Prausnitz (3) is an extension of an earlier work by Edwards, Newman and Prausnitz (4). Beutier and Renon restrict their procedure to ternary systems NH_3-CO_2-H_2O, NH_3-H_2S-H_2O and NH_3-SO_2-H_2O; but it may be expected that it is also useful for the complete multisolute system built up with these substances. The concentration range should be limited to mole fractions of water $x_w \gtrsim 0.7$; a temperature range from 0 to 100 °C is recommended. Equilibrium constants for chemical reactions 1 to 9 are taken from literature (cf. Appendix II). Henry's constants are assumed to be independent of pressure; numerical values were determined from solubility data of pure gaseous electrolytes in water (cf. Appendix II). The vapor phase is considered to behave like an ideal gas. The fugacity of pure water is replaced by the vapor pressure. For any molecular or ionic species i, except for water, the activity is expressed on the scale of molality m_i

$$a_i = m_i\ \gamma_i\ . \qquad \text{(B 1)}$$

Activity coefficients γ_i are normalized:

$$\lim_{\text{all solutes} \to 0} \gamma_i = 1.0\ . \qquad \text{(B 2)}$$

The activity of water is expressed on the scale of mole fraction x_w:

$$a_w = x_w\ \gamma_w \qquad \text{(B 3)}$$

with the normalization $\lim_{x_w \to 1} \gamma_w = 1.0.$ (B 4)

Activity coefficients of solute species are calculated from

$$\ln \gamma_i = \left[\frac{\partial}{\partial n_i} (G^E/RT) \right]_{P,T,\ n_j = i} \quad \text{(B 5)}$$

where n_i stands for the mole number of species i and G^E for the Gibbs excess energy of the liquid mixture. The Gibbs energy G of the ionic solution is

$$G(T,p,n_i) = G^{id}(T,p,n_i) + G^E(T,p,n_i) \quad \text{(B 6)}$$

with the Gibbs energy of an ideal solution

$$G^{id}(T,p,n_i) = n_w [\mu_w^o(T,p) + RT \ln x_w]$$

$$+ \sum_i n_i [\mu_i^o(T,p) + RT \ln m_i] \quad \text{(B 7)}$$

where μ_k^o represents the chemical potential of k in the reference state. The excess Gibbs energy is assumed to result from three different parts:

$$G^E = G_{ij}^E + G_{ia}^E + G_{aa}^E \quad \text{(B 8)}$$

representing ion-ion interactions - G_{ij}^E -, ion-molecule interactions - G_{ia}^E - and molecule-molecule interactions - G_{aa}^E -. Pitzer's proposal (5) was used for G_{ij}^E:

$$\frac{G_{ij}^E}{0.018\ n_w RT} = f(I) + \sum_k \sum_l \lambda_{kl} m_k m_l + \sum_k \sum_l \sum_h \mu_{klh} m_l m_k m_h \quad \text{(B 9)}$$

(k,l,h represent ionic species only).

$$f(I) = -\frac{A\ I}{0.9} \ln(1+1.2\ \sqrt{I}), \quad \text{(B 10)}$$

A is a Debye-Hückel parameter (cf. Appendix II) and I is the ionic strength. Pitzer found that binary interaction parameter λ_{kl} depends on ionic strength and may conveniently be expressed as:

$$\lambda_{kl} = \beta_{kl}^{(o)} + \beta_{kl}^{(1)} \cdot \frac{1}{2I} [1-(1+2\sqrt{I}) \exp(-2\sqrt{I})] \quad \text{(B 11)}$$

while for given klh μ_{klh} is a constant, representing ternary ion-ion interactions. As coulombic forces between ions of like charge do not allow those ions to approach one another, binary and ternary interactions between those ions are neglected:

$$\beta^{(o)}_{kl} = \beta^{(1)}_{kl} = \mu_{klh} = 0 \qquad \text{(B 12)}$$

whenever k,l,h represent ions of like charge.

$\beta^{(o)}_{kl}$ and $\beta^{(1)}_{kl}$-numbers for unlike charged ionic species k and l were determined from correlations for strong electrolytes.

Contributions to G^E_{ij} from ternary interactions are only important when concentrations of both acid and base are large, then the predominant ions are NH_4^+, HCO_3^-, NH_2COO^-, HS^-, HSO_3^-, $S^=$ and $SO_3^=$. μ_{klh}-numbers are assumed to differ from zero only if all three species k, l and h are members of that group of predominant ions. Four of those ternary parameters were adjusted to experimental results, one was estimated from results for strong electrolytes; the remaining parameters were determined using empirical mixing rules (cf. Appendix II).

Molecule-ion interaction contribution G^E_{ia} is estimated from the work necessary to transfer ions from a solution which does not contain any neutral solute molecule to the real ionic solution. The expression for G^E_{ia} based on the Debye-McAulay electrostatic theory is given in Appendix II.

Molecule-molecule interaction contribution G^E_{aa} is considered to result only from interactions between like neutral solutes. A Margules-type expression is used for G^E_{aa}:

$$\frac{G^E_{aa}}{0.018\ n_w\ RT} = \sum_{\substack{\text{all neutral} \\ \text{solute species}}} (\lambda_{aa}\, m_a^2 + \mu_{aaa}\, m_a^3) \qquad \text{(B 13)}$$

where μ_{aaa} is a pseudoternary parameter, determined from binary parameter λ_{aa}:

$$\mu_{aaa} = -\frac{1}{55.5}\left(\lambda_{aa} + \frac{1}{166.5}\right). \qquad \text{(B 14)}$$

Binary interaction parameters λ_{aa} are assumed to be constants and were determined from solubility data for gaseous species a in water.

Using these equations for the excess Gibbs energy and the wellknown relation

$$\frac{G^E}{n_w RT} = \ln a_w - \ln x_w + 0.018 \sum_{\substack{\text{all solute} \\ \text{species}}} m_i \ln \gamma_i \qquad \text{(B 15)}$$

the activity of water, a_w, is calculated. (For more detailed information on the procedure of Beutier and Renon cf. Appendix II).

Procedure of Edwards, Maurer, Newman and Prausnitz (3) (EMNP)

This method is applicable to the complete multisolute aqueous solution described before. It is estimated that total solute concentrations up to 10 or 20 molal may be handled. The limitation on temperature results mainly from the limited temperature range for which experimental results for equilibrium constants and Henry's constants are available (cf. Appendix II and tables I and II). Although for some constants this range only extends up to 60 °C, it is expected that by an appropriate extrapolation the method may be used also at temperatures up to 170 °C.

To calculate the multicomponent vapor-liquid equilibrium, equilibrium constants for chemical reactions 1-9 are taken from literature; in comparison to the original publication, in the present work different numerical values for the second dissociations of hydrogen sulfide and sulfur dioxide were chosen (cf. Appendix III). Henry's constants are evaluated from single solute solubility data without neglecting Poynting corrections:

$$H_i = H_i(T,p) = H_i(T,p = p_w^s) \exp \frac{\bar{v}_i (P-P_w^s)}{RT}, \qquad \text{(E 1)}$$

where $\bar{v}_i$ stands for the partial molal volume of component i in water at infinite dilution (cf. Appendix III). Fugacity coefficients are calculated from the equation of state by Nakamura, Breedveld and Prausnitz (6). For pure water the fugacity is replaced by

$$f_w = p_w^s \quad \exp \frac{v_w (P-P_w^s)}{RT}, \qquad \text{(E 2)}$$

where v_w stands for the specific volume of pure liquid water. Activities of all liquid components are expressed by Pitzer's equation (eq. B 9). This equation was adopted to describe not only the influence of ion-ion interactions, but of intermolecular forces between all solute species. Furthermore, ternary interactions are neglected. The resulting equation for the activity coefficient of any solute species i is:

Table I. Comparison of calculated partial pressures of NH_3 and CO_2 with experimental results by van Krevelen et al. for NH_3-CO_2-H_2O

$t/^{o}C$	m_{NH_3}	m_{CO_2}	y_{NH_3} P/mm Hg				y_{CO_2} P/mm Hg			
			exp.	KHH	BR	EMNP	exp.	KHH	BR	EMNP
20	1.96	0.68	9.4	8.1	8.0	7.9	0.83	0.95	1.41	0.68
mean deviation / per cent				12	13	15		29	73	15
40	0.50	0.257	4.4	4.4	4.0	4.2	10.1	13.9	16.5	10.6
	1.00	0.513	9.0	7.5	6.8	7.2	16.7	19.3	30.4	15.6
	2.00	1.026	11.5	23.5	12.0	27.0	11.2	57.9	12.0	21.4
mean deviation / per cent				9	14	10		31	91	14
60	0.50	0.257	10.5	11.1	9.7	10.7	46.0	54.3	70.9	45.3
	1.00	0.513	19.0	19.2	16.9	19.1	73.0	76.4	129.3	66.1
	2.00	1.026	35.0	31.8	28.5	32.8	89.0	109.9	240.8	90.0
mean deviation / per cent				7	15	7		25	98	17

Table II. Comparison of calculated partial pressures of NH_3 and CO_2 with experimental results by Pexton and Badger for NH_3-CO_2-H_2O

t/°C	m_{NH_3}	m_{CO_2}	y_{NH_3} P/mm Hg exp.	KHH	BR	EMNP	y_{CO_2} P/mm Hg exp.	KHH	BR	EMNP
20	0.128	0.0865	0.30	0.30	0.28	0.28	2.65	3.65	2.81	2.38
	0.500	0.355	0.69	0.70	0.63	0.62	10.0	12.6	16.2	10.0
	1.000	0.580	1.92	1.88	1.72	1.67	6.6	7.2	11.5	5.9
	2.000	1.374	1.73	1.52	1.45	1.36	30.7	31.4	81.0	27.6
mean deviation / per cent				11	14	16		25	86	18
30	1.000	0.350	8.1	7.5	7.1	7.1	2.0	2.3	2.8	1.8
	1.000	0.765	1.36	1.41	1.29	1.33	65.1	70.9	130.4	55.6
mean deviation / per cent				4	10	9		14	86	12
40	0.128	0.076	1.01	1.13	1.03	1.07	7.3	9.5	8.1	6.5
	0.500	0.251	4.65	4.54	4.12	4.32	11.0	12.9	15.1	9.8
	1.000	0.449	9.4	9.2	8.4	8.8	10.4	11.8	17.3	9.6
	2.000	1.088	11.9	10.5	9.8	10.5	31.1	35.9	79.5	27.9
mean deviation / per cent				7	12	10		38	88	15

$$\ln \gamma_i = -\frac{z_i^2 A}{3}\left[\frac{\sqrt{I}}{1+1.2\sqrt{I}} + \frac{1}{0.6}\ln(1+1.2\sqrt{I})\right] \qquad \text{(E 3)}$$

$$+ 2\sum_j \left\{\beta_{ij}^{(o)} + \frac{\beta_{ij}^{(1)}}{2I}\left[1-(1+2\sqrt{I})\exp(-2\sqrt{I})\right]\right\} m_j$$

$$- \frac{z_i^2}{4I^2}\left[1-(1+2\sqrt{I}+2I)\exp(-2\sqrt{I})\right]\sum_j\sum_k \beta_{jk}^{(1)} m_j m_k$$

where j and k designate all species in liquid phase except water. The acitivity of water is calculated analogously to eq. B 15.

Parameters $\beta_{ii}^{(o)}$- for interactions between like molecules were evaluated from single solute solubility data in water. These parameters proved to depend on temperature (cf. Appendix III). Parameters $\beta_{ij}^{(o)}$ for interactions between different neutral solute species i and j were estimated by:

$$\beta_{ij}^{(o)} = \frac{1}{2}\left[\beta_{ii}^{(o)} + \beta_{jj}^{(o)}\right]. \qquad \text{(E 4)}$$

Ion-ion interaction parameters were set equal to zero whenever i and j are ions with charges of the same sign:

$$\beta_{ij}^{(o)} = \beta_{ij}^{(1)} = 0. \qquad \text{(E 5)}$$

For ionic species i and j with charges of different signs $\beta_{ij}^{(o)}$-numbers were estimated from Pitzer's results for strong electrolytes using a method proposed by Bromley (7); $\beta_{ij}^{(1)}$-numbers were determined from an empirical correlation suggested by Pitzer and Mayorga (8):

$$\beta_{ij}^{(1)} = 0.018 + 3.06\ \beta_{ij}^{(o)}. \qquad \text{(E 6)}$$

Due to very limited experimental data, ion-ion interaction-parameters had to be assumed to be independent of temperature. Ion-molecule interaction parameters $\beta_{ij}^{(o)}$ were estimated from experimental results on salting-out effects, while $\beta_{ij}^{(1)}$ were set equal zero.

In the present work most of the original β-numbers were accepted, only a few were changed: $\beta_{ij}^{(o)}$-numbers for NH_4^+-HS^-, NH_4^+-NH_2COO^-, NH_4^+-HCO_3^- and NH_3-NH_4^+, NH_3-HCO_3^-, NH_3-HS^- as well as $\beta_{ij}^{(1)}$-numbers for

NH_3-NH_4^+, NH_3-HCO_3^- and NH_3-HS^- were fitted to 68 selected ternary data points (partial pressures of weak electrolytes) measured by Otsuka et al. for NH_3-CO_2-H_2O at 40, 60, 80 and 100 °C (9) and Ginzburg et al. for NH_3-H_2S-H_2O at temperatures between about 40 and 90 °C (10). While with the original β-numbers the mean deviations

$$(= 100 \left[\sum_{i=1}^{N} (P_{i,exp} - P_{i,calc})^2 / (P_{i,exp}^2 \; (N-1)) \right]^{1/2})$$

in the partial pressure amount to 26 per cent for ammonia, 42 per cent for carbon dioxide and 15 per cent for hydrogen sulfide, the corresponding numbers using the fitted β-values are 19, 36 and 13 per cent; thus indicating a small improvement. Similarly, 14 selected data points by Boublik et al. (11) for NH_3-SO_2-H_2O at 50 and 90 °C were used to fit $\beta^{(o)}$ and $\beta^{(1)}$ for NH_4^+-HSO_3^-, resulting in a considerable improvement for sulfur dioxide (75 to 28 per cent) and a slight change for the partial pressure of ammonia. All β-numbers - fitted as well as accepted ones - are compiled in Appendix III.

Comparison with Experimental Results

System NH_3-CO_2-H_2O: Experimental VLE data for NH_3-CO_2-H_2O are compared with calculated results in figures 2 and 3 and in tables I to IV. At temperatures between 20 and 60 °C partial pressures of ammonia calculated from different methods agree well with each other; there is also a good agreement with experimental results by van Krevelen et al. (1) (cf. table I) and Pexton and Badger (12) (cf. table II). For molalities $m_{tot,NH_3} \leqslant 2$ and $m_{tot,CO_2} \leqslant 1.4$ the deviations amount to between 5 and 15 per cent. With increasing concentrations the deviations between calculated and measured partial pressures of ammonia increase, as can be seen from the comparison with experimental results by Otsuka et al. (9) (cf. table III). At temperatures between 20 and 60 °C in ammonia-rich solutions with total solute molalities up to about 15 molal, partial pressures of ammonia calculated with different procedures agree within about 10 to 20 per cent, but deviate from the experimental results by Otsuka et al. by up to about 50 per cent. Some data points are reproduced well, some rather badly, indicating a scattering of the experimental results. At higher temperatures the KHH-method tends to overestimate the partial pressure of ammonia; the deviations are still rather small at low concentrations (cf. table IV), but increase rapidly with increasing solute concentration (cf. table III). In that region

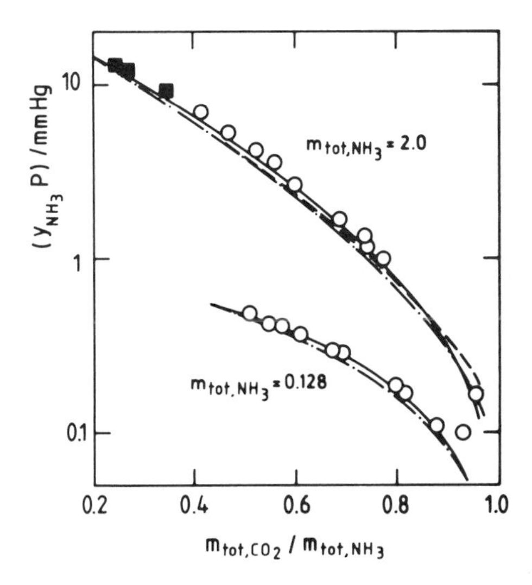

Figure 2. Comparison of calculated and experimental partial pressures of NH_3 at 20°C; (○) Pexton and Badger exp.; (■) van Krevelen et al. exp.; (——) KHH calc.; (– – –) BR calc.; (– · – ·) EMNP calc

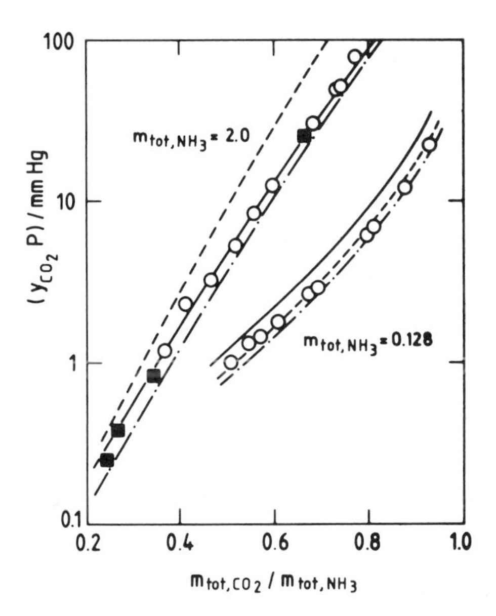

Figure 3. Comparison of calculated and experimental partial pressures of CO_2 at 20°C; (○) Pexton and Badger exp.; (■) van Krevelen et al. exp.; (——) KHH calc.; (– – –) BR calc.; (– · – ·) EMNP calc

Table III. Comparison of calculated partial pressures of NH_3 and CO_2 with experimental results by Otsuka et al. for NH_3-CO_2-H_2O

			y_{NH_3} P/mm Hg				y_{CO_2} P/mm Hg			
t/°C	m_{NH_3}	m_{CO_2}	exp.	KHH	BR	EMNP	exp.	KHH	BR	EMNP
20	1.23	0.804	1.4	1.5	1.4	1.3	16.8	15.8	30.0	13.5
	4.59	0.551	48.4	43.2	40.4	39.8	0.1	0.04	0.04	0.03
	11.33	2.034	120	112	96.5	99.8	0.2	0.06	0.13	0.06
mean deviation/per cent				19	35	25		35	200	39
40	2.17	1.89	3.4	1.5	1.8	1.8	304	810	1907	503
	9.13	1.72	208	209	179	178	1.2	0.6	1.3	0.7
	15.28	3.67	322	345	269	293	1.1	0.6	3.0	1.0
mean deviation/per cent				24	23	20		79	340	61
60	1.99	1.39	22.8	14.1	13.4	16.3	442	499	1171	326
	7.51	2.48	364	211	190	204	14.9	19.3	69.1	20.8
	11.32	3.87	485	309	272	316	20.0	14.7	98.2	18.2
mean deviation/per cent				38	46	37		34	725	36
80	3.15	1.36	109	140	120	139	256	241	555	228
	4.82	0.70	408	543	424	416	17.5	11.4	18.1	19.4
	8.66	0.663	802	1407	952	905	19.0	2.7	3.6	5.4
mean deviation/per cent				44	16	20		50	62	33
100	2.90	1.450	157	217	183	222	1211	1304	2502	1180
	4.29	0.905	540	777	592	596	146	102	165	184
	7.37	1.099	1022	1759	1243	1168	107	43	65	95
mean deviation/per cent				50	15	20		46	52	16

the BR- and EMNP-procedures yield remarkable improvements over the correlation of van Krevelen et al..

In comparison to ammonia, for carbon dioxide calculated partial pressures deviate more from the experimental results. At temperatures up to about 60 °C and total solute molalities up to about 3.5 molal the mean deviation to the experimental results by van Krevelen et al. and Pexton and Badger is about 30 per cent for the KHH-method, 15 per cent for the EMNP-procedure and rises to more than 80 per cent for the procedure of Beutier and Renon. Again at higher temperatures and higher concentrations the deviations are still larger. As is shown in table III, in that region the EMNP-procedure yields the best agreement with the experimental results.

System NH_3-H_2S-H_2O: Van Krevelen et al. reported partial pressures of hydrogen sulfide over aqueous solutions containing up to 3.5 moles NH_3 and up to 2 moles H_2S per kg H_2O at temperatures between 20 to 60 °C. As is shown in table V, the KHH-method gives the best agreement with these experimental results; the deviations are about 2.5 times less than those from the BR- and EMNP-procedures. Ginzburg et al. (10) investigated ammonia-rich NH_3-H_2S-H_2O solutions at 600 and 760 mm Hg in the temperature range between about 60 and 90 °C at $m_{tot,NH_3} \leqslant 7.3$ and $m_{tot,H_2S} \leqslant 3.3$. For 40 data points the mean deviations between measured and calculated partial pressures are given in table VI. (Some data points were neglected, because the measured hydrogen sulfide vapor pressures are obviously wrong.) Van Krevelen's procedure yields the best agreement for NH_3, while the EMNP-method is best for H_2S. Oratovskii et al. (14) reported compositions of saturated vapor over ammonium sulfide solutions at temperatures between 70 and 90 °C and liquid molalities $m_{tot,NH_3} < 7$ and $m_{tot,H_2S} < 3.6$. In table VI the mean deviations between these results and calculated data are compiled. While for the BR- and EMNP-models the mean deviations are nearly independent of temperature, for the van Krevelen procedure they increase with temperature, indicating the difficulties in extrapolating the KHH-method to higher temperatures.

Table VII shows a comparison with experimental data by Leyko and Piatkiewicz (15) at 80 to 110 °C. At high temperatures partial pressures calculated from the BR- and EMNP-methods deviate by up to 20 per cent from the experimental results, whereas van Krevelen's method - extrapolated to 110 °C - yields partial pressures of hydrogen sulfide which are only about 1/4 to 1/5 of the measured values.

Table IV. Comparison of calculated partial pressure of NH_3 with experimental results by Badger and Wilson for NH_3-CO_2-H_2O at 90 °C

m_{NH_3}	m_{CO_2}	y_{NH_3} P/mm Hg exp.	KHH	BR	EMNP
1.090	0.0050	193	200	157	157
0.635	0.0349	103	105	84	86
1.260	0.1147	192	201	159	162
0.380	0.0509	51	56	45	47
1.700	0.3298	205	223	177	185
1.430	0.3647	152	165	132	142
1.200	0.3204	128	136	109	118
mean deviation/per cent			7	16	13

Table V. Mean deviations between calculated partial pressures of H_2S and experimental results by van Krevelen et al. for NH_3-H_2S-H_2O

t/°C	f_{rel,H_2O} /per cent KHH	BR	EMNP
20	11	36	32
40	7	17	15
60	12	24	15

Table VI. Mean deviations between calculated and experimental results for NH_3-H_2S-H_2O

	mean deviation to partial pressures measured by Ginzburg et al. f_{rel}/per cent		
	KHH	BR	EMNP
NH_3	12	18	20
H_2S	27	93	21
	mean deviation to mole fractions is vapor phase measured by Oratovskii et al. f_{rel}/per cent		
	KHH	BR	EMNP
NH_3	20	27	10
H_2S	39	23	28

Table VII. Comparison between calculated and experimental results for the pressure over aqueous solutions containing 6.15 mole NH_3/kg H_2O and 1.9 mole H_2S/kg H_2O (exp.results from Leyko and Piatkiewicz - assuming molality = molarity)

	y_{NH_3} P/atm				y_{H_2S} P/atm				P/atm			
t/°C	exp.	KHH	BR	EMNP	exp.	KHH	BR	EMNP	exp.	KHH	BR	EMNP
80	0.92	0.88	0.73	0.74	0.34	0.23	0.43	0.22	1.60	1.52	1.57	1.37
90	1.27	1.24	1.00	1.00	0.49	0.26	0.67	0.36	2.27	2.11	2.26	1.96
100	1.61	1.71	1.34	1.32	0.73	0.27	1.02	0.56	3.10	2.85	3.22	2.76
110	1.98	2.31	1.78	1.70	1.13	0.25	1.43	0.85	4.26	3.80	4.42	3.80

Table VIII. Comparison of calculated partial pressures of NH_3 and H_2S with data by Wilson and Miles on the system NH_3-H_2S-H_2O at 80 °C.

Molality		y_{NH_3} P/mm Hg				y_{H_2S} P/mm Hg			
NH_3	H_2S	exp.	KHH	BR	EMNP	exp.	KHH	BR	EMNP
0.960	0.971	12.0	-	12.2	11.6	2389	-	2539	2133
1.063	1.452	4.36	-	4.94	4.26	8556	-	9172	9143
2.332	1.151	136.0	156.7	138.6	151.7	319	265	409	264
5.112	1.143	574	618	494	526	94.8	78.0	114	80.9
5.538	5.305	72.6	29.3	133.1	70.9	12140	12201	22123	15638
9.245	7.935	208	175	334	176	9506	2832	40721	22956
10.201	5.983	597	667	619	474	1916	746	10404	2469
22.627	5.561	2285	5651	2570	2416	241	174	1296	196

Table VIII comprehends a comparison with experimental data by Wilson and Miles as cited in (2) for 80 °C. These results are represented best by the EMNP-procedure.

System NH_3-CO_2-H_2S-H_2O: The partial pressure of hydrogen sulfide over this quaternary solution was measured by van Krevelen et al. at 20, 40 and 60 °C at total solute concentrations up to 4 molal. The mean deviations between measured and calculated results are 9 per cent for KHH, 10 per cent for EMNP and 59 per cent for BR. Calculated numbers using EMNP are mostly too small, while results from the method of Beutier and Renon are always too large and results using van Krevelen's procedure scatter around the experimental results.

System NH_3-SO_2-H_2O: For comparison with calculated data only the experimental results of Johnstone (16) and Boublik et al. (11) were used. (Boublik et al. investigated the system NH_3-SO_2-SO_3-H_2O; only some of their results with very low SO_3/SO_2 ratios were used for comparison with calculated data). Experimental results by other authors mostly cover very high solute concentrations in the liquid phase (20 molal and more) and are, therefore, not suitable for comparison with the models discussed here. As van Krevelen's method cannot be used for this system, the comparison is limited to the other procedures. Partial pressures of ammonia calculated from the BR-model are generally too large; the calculated values exceed the experimental results mostly by a factor larger than 5. The EMNP method generally yields partial pressures which are only about half as large as the measured ones. The calculated partial pressures of SO_2 are always too small, for temperatures between 50 and 90 °C the mean deviations amount from 20 to 40 per cent for the EMNP-model and from 40 to 70 per cent for the BR-model.

Conclusion

The comparison between measured and calculated results for vapor-liquid equilibria in aqueous systems of weak electrolytes confirms the applicability of van Krevelen's method for moderate temperatures and concentrations. The comparison also indicates that the procedure of Edwards, Maurer, Newman and Prausnitz yields reliable results also at temperatures around 100 °C; therefore, it may be expected that it is also useful at higher temperatures where experimental material, necessary for checking that procedure, is not available in the literature. The excellent agreement between measured and calculated vapor-liquid equilibrium reported in a publication by Beutier and Renon could not be reproduced in all cases; whereas for example for the system NH_3-CO_2-H_2O results for the method of Beutier and Renon calculated in this work were confirmed by independent calculations (17). Further improvements of the procedures described here seem possible, but they obviously require more and better experimental data for high temperature equilibria constants, Henry's constants and multicomponent vapor liquid equilibria in aqueous solutions of weak electrolytes.

APPENDIX I:

Procedure of van Krevelen et al. for calculating VLE

A. System NH_3-H_2S-H_2O. The dissociation of water (reaction 9) and the second dissociation of H_2S (reaction 6) are neglected; at given temperature and total molalities of NH_3 and H_2S there remain four unknown molalities in the liquid phase (e.g. NH_3, NH_4^+, H_2S and HS^-), the composition of the vapor phase and the total pressure, which are calculated from 8 equations: The dissociation constants of ammonia and hydrogen sulfide (eqs.I and III) together with the phase equilibrium for hydrogen sulfide (eq. XII) are combined resulting in a equilibrium constant K_{12}^*

$$K_{12}^* = (m_{NH_4^+} \; m_{HS^-})/(m_{NH_3} \; y_{H_2S} \; P) \qquad \text{(A I.1)}$$

Numbers for K_{12}^* were derived by van Krevelen et al. from experimental VLE data and can be summarized by the following equations (when p is in mm Hg):

$$\log_{10}K_{12}^{*} = -k_x + 0.089\ m_{tot,H_2S} + k_y \qquad \text{(A I.1a)}$$

$$\ln k_x = -11.09 + 8920/T - 1.653 * 10^6/T^2 \qquad \text{(A I.1b)}$$

$$\text{and } k_y = 0\ . \qquad \text{(A I.1c)}$$

The remaining equations are: phase equilibria for ammonia (eq. K4) and water (eq. K5), mass balances for ammonia (eq. XV) and hydrogen sulfide (eq. XVII), the condition of bulk electroneutrality (eq. XIX), the mole balance in the vapor phase (eq. XX), and the assumption that, for the ammonia-rich systems considered exclusively, in the mass balance for the liquid hydrogen sulfide may be neglected. The system of eight equations can easily be solved:

$$y_{NH_3}\ P = (m_{tot,NH_3} - m_{tot,H_2S})/H_{NH_3}^{*} \qquad \text{(A I.2)}$$

$$y_{H_2S}\ P = m_{tot,H_2S}^{2}/((m_{tot,NH_3} - m_{tot,H_2S})K_{12}^{*}) \qquad \text{(A I.3)}$$

$$y_w\ P = p_w^s/(1 + 0.018(m_{tot,NH_3} + m_{tot,H_2S}) \qquad \text{(A I.4)}$$

<u>B. System NH_3-CO_2-H_2S-H_2O:</u> This quaternary system is dealt combining the equations and assumptions used to describe the ternary systems NH_3-CO_2-H_2O and NH_3-H_2S-H_2O. In the mass balances for the liquid phase molecular CO_2 and H_2S are neglected. At given temperature and total liquid molalities there are eleven unknown properties, e.g. m_{NH_3}, $m_{NH_4^+}$, $m_{NH_2COO^-}$, $m_{CO_3^=}$, m_{HS^-}, the total pressure and the composition of the vapor. The corresponding eleven equations are K1, K2, K3, K4, K5, A I.1, the mass balances (eqs. XV to XVII), the condition of bulk electroneutrality (eq. XIX) and the mole balance for the vapor (eq. XX). Only equation A I.1c is changed by setting

$$k_y = (1.932 - 540/T)\ m_{tot,CO_2}\ .$$

APPENDIX II

On the method of Beutier and Renon used for the calculations of the present work

The temperature dependency of equilibrium constants and of Henry's constants are compiled in tables A II.I and A II.II.

For the calculations described here the vapor pressure of pure water was approximated by

$$\ln(p_w^s/\mathrm{atm}) = 70.435 - \frac{7362.7}{T/K} + 0.006952\,\frac{T}{K} - 9.0\,\ln(T/K) \qquad \text{(A II.1)}$$

The ion-molecule contribution to G^E is:

$$\frac{G_{ia}^E}{0.018 n_w RT} = \left[\frac{D_w}{D_S}\,\frac{1+Y_f/2}{1-Y_f} - \frac{1+Y_i/2}{1-Y_i}\right] \cdot \left[\sum_i L_j\, m_j\right] \qquad \text{(A II.2)}$$

$D_w = 4\pi\varepsilon_o\varepsilon_w$ = absolute dielectric constant of pure water;

ε_o = dielectric constant of water;

ε_w = relative dielectric constant of water.

The following equations for ε_w were used:

$t/^oC < 100.0$

$$\varepsilon_w = 78.54(1-4.579\,t^* + 11.9\,t^{*2} + 28\,t^{*3}); \quad t^* = (t/^oC-25)/1000 \qquad \text{(A II.3a)}$$

$t/^oC \geq 100.0$

$$\varepsilon_w = 5321/(T/K) + 233.76 - 0.92951\,T/K + 1.417\cdot 10^{-3}(T/K)^2 - 8.292\cdot 10^{-7}(T/K)^3 \qquad \text{(A II.3b)}$$

$$D_S = D_w\left[1 + \sum_{\text{neutral solutes}} \alpha_a \frac{m_a}{V_a}\right] = \text{absolute dielectric constant of the real solution without ions} \qquad \text{(A II.4)}$$

α_a = dielectric coefficient of neutral solute a

$$V_a = v_w + \sum_{\text{neutral solutes}} m_a \bar{v}_a = \text{volume of the solution containing 1 kg } H_2O \text{ and neutral solutes} \qquad \text{(A II.5)}$$

v_w = volume of 1 kg liquid water

$\bar{v}_a$ = partial molar volume of solute a

$$Y_i = \frac{4}{3}\pi\,\frac{N_A\cdot 10^{-27}}{V_i} \sum_{\text{ionic species}} r_j^3\, m_j = \text{volume fraction of ionic cavities} \qquad \text{(A II.6)}$$

Table A II.I. Coefficients for equilibrium constants K;
$\ln K = A_1/T + A_2 \ln T + A_3 T + A_4$, T in Kelvin
(A_i: method of Beutier and Renon; A_i^*: method of Edwards, Maurer, Newman and Prausnitz)

constant / reaction	A_1	A_2	A_3	A_4
1	-8451.61	-31.4335	0.0152123	191.97
2	-80063.5	-478.653	0.714984	2767.92
3	-12995.40	-33.5471	0.0	218.5989
4	-3768.0	-20.0	0.0	122.53
5	-6286.89	0.0	-0.050628	12.405
6	-2048.99	15.65	0.0	-114.45
7	1333.4	0.0	0.0	21.274
8	1998.0	0.0	0.0	-5.593
9	-10294.83	0.0	-0.039282	14.01708

constant / reaction	A_1^*	A_2^*	A_3^*	A_4^*	temp. limits of A_i^* °C
1	-3335.7	1.4971	-0.0370566	2.76	0-225
2	-12092.1	-36.7816	0.0	235.482	0-225
3	-12995.40	-33.5471	0.0	218.5989	0-150
4	637.396	0.0	-0.0151337	-1.96211	0-50
5	-12431.7	-35.4819	0.0	220.067	0-225
6	-7211.2	0.0	0.0	-7.489	20-100
7	1333.4	0.0	0.0	-21.274	0-50
8	2895.65	0.0	0.0	-8.5994	20-60
9	-13445.9	-22.4773	0.0	140.932	0-225

Table A II.II. Coefficients for Henry's constant H_i/atm (kg H_2O) mol^{-1};
$\ln H = B_1/T + B_2 \ln T + B_3 T + B_4$

(B_i: method of Beutier and Renon; B_1^*: method of Edwards, Maurer, Newman and Prausnitz)

constant / component	B_1	B_2	B_3	B_4
NH_3	-8621.06	-25.6767	0.035388	160.559
CO_2	-34417.2	-182.28	0.25159	1082.37
H_2S	7056.07	74.6926	-0.14529	-403.658
SO_2	-5160.4	-7.61	0.0	60.538

constant / component	B_1^*	B_2^*	B_3^*	B_4^*	temp. limits of B_i^* °C
NH_3	-157.552	28.1001	-0.049227	-149.006	0-150
CO_2	-6789.04	-11.4519	-0.010454	94.4914	0-250
H_2S	-13236.8	-55.0551	0.0595651	342.595	0-150
SO_2	-5578.8	-8.76152	0.0	68.418	0-100

N_A = Avogadro's number

$$V_i = v_w + \sum_{\text{ionic species}} m_i \bar{v}_i = \text{volume of the solution without neutral solute molecules} \quad \text{(A II.7)}$$

$$Y_f = Y_i \frac{V_i}{V_f} \quad \text{(A II.8)}$$

$$V_f = V_w + \sum_{\text{all solute species}} m_k \bar{v}_k = \text{volume of the solution with neutral and ionic species} \quad \text{(A II.9)}$$

For the present calculations the specific density $\rho_w = 1/v_w$ of liquid water was approximated by:

$$\rho_w/\text{kg dm}^{-3} = 0.12365 + 509.65/(T/K) - 74146.2/(T/K)^2 \quad \text{(A II.10)}$$

L_j includes properties of ionic species j:

$$L_j = \frac{e^2 \, z_j^2}{2r_j \, kT \, D_w} \cdot 10^8 \quad \text{(A II.11)}$$

e = the charge of an electron
k = Boltzmann's constant
z_j = number of charge of ion j.

Numbers for α_a, $\bar{v}_k$ and r_j given by Beutier and Renon are compiled in table A II.III.

Table A II.III. Physical data for dielectric effects

	$\bar{v}_k/\text{dm}^3 \text{ mol}^{-1}$	$\alpha_a/\text{dm}^3 \text{ mol}^{-1}$	$r_j/\text{Å}$
NH_3	0.030	-0.0235	
CO_2	0.037	-0.037	
H_2S	0.034	-0.031	
SO_2	0.045	-0.037	
NH_4^+	0.0134		2.5
HCO_3^-	0.0288		2.7
HS^-	0.0232		2.3
$CO_3^=$	0.0065		4.0
$S^=$	-0.0037		3.3
H^+	-0.0047		3.8
OH^-	0.0005		3.5
NH_2COO^-	0.0459		2.7
HSO_3^-	0.0375		2.7
$SO_3^=$	0.0197		2.8

For the Debye-Hückel parameter A in this work the following equation was used:

$t/^{o}C \leq 100.0$:

$$A = 3.7323 - 1354.21/(T/K) + 176349/(T/K)^2 \qquad \text{(A II.12)}$$

$t/^{o}C > 100.0$:

$$A = 4.2051*10^6 \, x_o^{1/2} \, [\varepsilon_w(T/K)]^{-3/2} \qquad \text{(A II.13)}$$

$$x_o = 0.5771 + 142.81/(T/K) \qquad \text{(A II.14)}$$

Parameters for binary molecule-molecule (λ_{aa}) and ion-ion interactions ($\beta_{ij}^{(o)}$ and $\beta_{ij}^{(1)}$) are given in tables A II.IV and A II.V.

Ternary parameters μ_{klh} are given in table A II.VI. μ_{klh} is set equal zero, whenever at least one of k, l and h is not an ion mentioned in table A II.VI. Furthermore, the following assumptions are made:

$$\mu_{klh} = \frac{1}{2}\left[\mu_{kll} + \mu_{khh}\right] \text{ and } \mu_{kll} = \mu_{kkl} \qquad \text{(A II.15)}$$

<u>Complete expressions for activity coefficients:</u>
<u>Molecular species a</u>

$$\ln \gamma_a = (\ln \gamma_a)_{aa} + (\ln \gamma_a)_{ia}$$

$$(\ln \gamma_a)_{aa} = 2\,\lambda_{aa}\, m_a + 3\,\mu_{aaa}\, m_a^2$$

$$(\ln \gamma_a)_{ia} = \frac{D_w}{D_f}\left[-\frac{D_w}{D_f}\frac{\alpha_a + \overline{v}_a}{V_a} + \frac{\overline{v}_a}{V_a} - \frac{D_s}{D_f}\frac{\overline{v}_a}{V_f}\frac{1.5Y_f}{(1+\frac{Y_f}{2})^2}\right] * \left[\sum_{\substack{\text{ionic}\\ \text{species}}} L_j\, m_j\right] \qquad \text{(A II.16)}$$

$$\text{with } D_f = D_s(1-Y_f)/(1 + 0.5\, Y_f) \qquad \text{(A II.17)}$$

<u>Ionic species i</u>

$$\ln \gamma_i = (\ln \gamma_i)_{ia} + (\ln \gamma_i)_{ij}$$

$$(\ln \gamma_i)_{ia} = L_i\left(\frac{D_w}{D_f} - \frac{D_w}{D_s}\right) + \left[\frac{D_w^{\cdot}}{D_i^2}\, d_i - \frac{D_w}{D_f^2}\, d_f\right]\left[\sum_{\substack{\text{ionic}\\ \text{species}}} m_j L_j\right] \qquad \text{(A II.18)}$$

$$D_i = D_w(1-Y_i)/(1 + 0.5\, Y_i) \qquad \text{(A II.19)}$$

$$d_i = -D_w \frac{1.5}{(1+\frac{Y_i}{2})^2} \frac{1}{V_i} \left[\frac{4}{3}\pi N_A*10^{-27} r_i^3-\bar{v}_i\ Y_i\right] \quad (A\ II.20)$$

$$d_f = -D_s \frac{1.5}{(1+\frac{Y_f}{2})^2} \frac{1}{V_f} \left[\frac{4}{3}\pi N_A*10^{-27} r_i^3-\bar{v}_i\ Y_f\right] \quad (A\ II.21)$$

$$(\ln \gamma_i)_{ij} = \frac{-z_i^2 A}{3}\left[\frac{\sqrt{I}}{1+1.2\sqrt{I}} + \frac{1}{0.6}\ln(1+1.2\sqrt{I})\right] + 2\sum_j \lambda_{ij} m_j$$

$$- \frac{z_i^2}{4I^2}\left[1-(1+2\sqrt{I}+2I)\exp(-2\sqrt{I})\right] \sum_j \sum_k \beta_{jk}^{(1)} m_j m_k$$

$$+ 3 \sum_j \sum_k \mu_{ijk}\ m_j m_k \quad (A\ II.22)$$

(j and k designate ionic species only).

Table A II.IV. Binary molecule-molecule interaction parameters λ_{aa}

a	NH_3	CO_2	H_2S	SO_2
λ_{aa}	0.017	0.010	0.005	-0.05

Table A II.V. Binary ion-ion interaction parameters $\beta^{(o)}_{ij}$ and $\beta^{(1)}_{ij}$

i \ j		HS^-	HCO_3^-	NH_2COO^-	HSO_3^-
H^+	$\beta^{(o)}$	0.18	0.126	0.085	-0.06
	$\beta^{(1)}$	0.32	0.294	0.255	-0.54
NH_4^+	$\beta^{(o)}$	0.055	-0.054	0.0	0.0
	$\beta^{(1)}$	0.193	0.594	0.5	0.45

i \ j		OH^-	$S^=$	$CO_3^=$	$SO_3^=$
H^+	$\beta^{(o)}$	0.04	0.0	0.0	0.12
	$\beta^{(1)}$	0.12	0.0	0.0	1.08
NH_4^+	$\beta^{(o)}$	0.115	0.041	0.041	0.041
	$\beta^{(1)}$	0.345	0.659	0.659	0.659

Table A II.VI. Ternary interaction parameters $\mu_{kkl}*10^3$

k \ l	HS^-	HCO_3^-	NH_2COO^-	HSO_3^-	$S^=$	$SO_3^=$
NH_4^+	-1/6	-10/6	-1	1.6/6	$25\sqrt{2}/9$	$2.2\sqrt{2}/9$

APPENDIX III

On the method of Edwards et al.

The temperature dependency of equilibrium as well as of Henry's constants is given in tables A II.I and A II.II (cf. Appendix II). In comparison with the original publication for the equilibrium constants of the second dissociations of hydrogen sulfide and sulfur dioxide (reactions 6 and 7) numbers derived from Cobble (18) and Arkhipova et al. (19) were used.

Partial molar volumes at infinite dilution were adopted from the correlation of Brelvi and O'Connell (20). (In the pressure range regarded here ($p \ll 100$ atm) Poynting corrections are very small and can be neglected for all electrolytes as well as for water (eqs. E1, E2)).

Most available experimental results for vapor-liquid equilibria of aqueous solutions are low pressure data. As in that region the fugacity coefficients do not deviate remarkably from unity, one may set $\varphi_i = 1.0$ without causing an important error.

The vapor pressure of pure water and the Debye-Hückel parameter A are expressed by eqs. A II.1, A II.12 and A II.13 (cf. Appendix II).

Coefficients $\beta_{ii}^{(o)}$ for interactions between molecular solute species were expressed by

$$\beta_{ii}^{(o)} = E_i + F_i/T \quad \text{(T in Kelvin)}$$

Numbers of E_i and F_i are given in table A III.I.

Table A III.I. Parameters E_i and F_i

	E_i/(kg H_2O) mol^{-1}	F_i/(kg H_2O) K mol^{-1}
NH_3	-0.0260	12.29
CO_2	-0.4922	149.2
H_2S	-0.2106	61.56
SO_2	+0.0275	0

Numbers for ion-ion-interaction parameters $\beta_{ij}^{(o)}$ are given in table A III.II.

Table A III.II. Binary ion-ion parameters $\beta_{ij}^{(o)}$

i \ j	HCO_3^-	NH_2COO^-	HS^-	HSO_3^-
NH_4^+	-0.0435*	0.0505*	0.0638*	-0.0466*
H^+	0.071	0.198	0.194	0.085

i \ j	OH^-	$CO_3^=$	$S^=$	$SO_3^=$
NH_4^+	0.06	-0.062	-0.021	-0.045
H^+	0.208	0.086	0.127	0.103

* = fitted to experimental results

$\beta_{ij}^{(1)}$ for NH_4^+-$CO_3^=$ was set equal zero: $\beta_{ij}^{(1)} = 0.0$

$\beta_{ij}^{(1)}$ for NH_4^+-HSO_3^- was fitted: $\beta_{ij}^{(1)} = 0.0876$;

all other $\beta_{ij}^{(1)}$-values were calculated from

$$\beta_{ij}^{(1)} = 0.018 + 3.06\ \beta_{ij}^{(o)}.$$

All ion-molecule interaction parameters $\beta_{ij}^{(o)}$ which are not zero are given in table A III.III.

Table A III.III. β_{ij} -parameters for interactions between molecular and ionic species

Molecule-ion	$\beta_{ij}^{(o)}$
$NH_3-NH_4^+$	0.0117*
$NH_3-HCO_3^-$	-0.0816*
$NH_3-CO_3^=$	0.068
NH_3-HS^-	-0.0449*
$NH_3-S^=$	0.032
$NH_3-HSO_3^-$	-0.038
$NH_3-SO_3^=$	-0.044
NH_3-H^+	0.015
NH_3-OH^-	$0.227 - 1.47 * 10^{-3}\ T + 2.6 * 10^{-6}\ T^2$
$CO_2-NH_4^+$	$0.037 - 2.38 * 10^{-4}\ T + 3.83 * 10^{-7}\ T^2$
$CO_2-S^=$	0.053
$CO_2-HSO_3^-$	-0.03
$CO_2-SO_3^=$	0.068
$CO_2-NH_2COO^-$	0.017
CO_2-OH^-	$0.26 - 1.62 * 10^{-3}\ T + 2.89 * 10^{-6}\ T^2$
CO_2-H^+	0.033
$H_2S-NH_4^+$	$0.120 - 2.46 * 10^{-4}\ T + 3.99 * 10^{-7}\ T^2$
$H_2S-HCO_3^-$	-0.037
$H_2S-CO_3^=$	0.077
$H_2S-HSO_3^-$	-0.045
$H_2S-SO_3^=$	0.051
$H_2S-NH_2COO^-$	-0.032
H_2S-H^+	0.017
H_2S-OH^-	$0.26 - 1.72 * 10^{-3}\ T + 3.07 * 10^{-6}\ T^2$
$SO_2-NH_4^+$	-0.05
$SO_2-HCO_3^-$	-0.86
$SO_2-CO_3^=$	0.94
SO_2-HS^-	-0.58
$SO_2-S^=$	0.28
$SO_2-NH_2COO^-$	-0.79
SO_2-OH^-	0.08
$NH_3-NH_4^+$ **	-0.020
$NH_3-HCO_3^-$ **	0.4829
NH_3-HS^- **	0.406

* = fitted , ** = $\beta_{ij}^{(1)}$

Literature Cited

1 van Krevelen, D.W.; Hoftijzer, P.J.; Huntjens,F.F.; Rec.Trav.Chim.Pays-bas, 1949, 68, 191.

2 Beutier, D.; Renon, H.; Ind.Eng.Chem.Process Des. Dev, 1978, 17, 220.

3 Edwards, T.J.; Maurer, G.; Newman, J.; Prausnitz, J.M.; AIChE J., 1978, 24, 966.

4 Edwards, T.J.; Newman, J.; Prausnitz, J.M.; AIChE J., 1975, 21, 248.

5 Pitzer, K.S.; J.Phys.Chem., 1973, 77, 268.

6 Nakamura, R.; Breedveld, G.J.F.; Prausnitz, J.M.; Ind.Eng.Chem.Process Des.Dev., 1976, 15, 557.

7 Bromley, L.A.; J.Chem.Thermodynamics, 1972, 4, 669.

8 Pitzer, K.S.; Mayorga, G.; J.Phys.Chem., 1973, 77, 2300.

9 Otsuka, E.; Yoshimura, S.; Yakabe, M.; Inoue, S.; Kogyo Kagaku Zasshi, 1960, 62, 1214.

10 Ginzburg, D.M.; Pikulina, N.S.; Litvin, V.P.; Zh.Prikl.Khim., 1965, 38, 2117.

11 Boublik, T.; Dworak, E.; Hala, E.; Schauer, V.; Coll.Czechoslov.Chem.Commun., 1963, 28, 1791.

12 Pexton, S.; Badger, E.H.M.; J.Soc.Chem.Ind., 1938, 57, 106.

13 Badger, E.H.M.;, Wilson, D.S.; J.Soc.Chem.Ind., 1947, 66, 84.

14 Oratovskii, V.I.; Gamols'ski, A.M.; Klimenko, N.W.; Zh.Prikl.Khim., 1964, 37, 2392.

15 Leyko, J.; Piatkiewicz,J.; Bull.Acad.Polon.Sci., Ser.Sci.Chim., 1964, 12, 445.

16 Johnstone, H.F.; Ind.Eng.Chem.,1935, 27, 587.

17 Müller, G.; Universität Kaiserlautern, Fed.Rep.of Germany, private communication.

18 Cobble, J.W.; J.Am.Chem.Soc., 1964, 88, 5349.

19 Arkipova, G.P.; Flis, I.E.; Mishchenko, K.P.; Russ.J.Appl.Chem., 1964, 37, 2275.

20 Brelvi, S.W.; O'Connell, J.P.; AIChE J., 1972, 18, 1239.

RECEIVED January 31, 1980.

8

Representation of NH_3-H_2S-H_2O, NH_3-SO_2-H_2O, and NH_3-CO_2-H_2O Vapor-Liquid Equilibria

H. RENON

Group Commun Reacteurs et Processus, ENSTA-Ecole Des Mines,
60, Boulevard St. Michel, 75006 Paris, France

A renewed interest in the behavior of volatile electrolyte solutions appeared around 1975. It was raised by the need of better design of industrial processes , especially pollution control processes, elimination of acid gases from natural gas, removal of sulfur from liquid and solid fuels and more recently coal conversion processes.

VAN AKEN et al. (1) and EDWARDS et al. (2) made clear that two sets of fundamental parameters are useful in describing vapor-liquid equilibria of volatile weak electrolytes, (1) the dissociation constant(s) K of acids, bases and water, and (2) the Henry's constants H of undissociated volatile molecules. A thermodynamic model can be built incorporating the definitions of these parameters and appropriate equations for mass balance and electric neutrality. It is complete if deviations to ideality are taken into account. The basic framework developped by EDWARDS, NEWMAN and PRAUSNITZ (2) (table 1) was used by authors who worked on volatile electrolyte systems : the difference among their models are in the choice of parameters and in the representation of deviations to ideality.

Table 1. Thermodynamic Framework of Representation of Vapor-Liquid Equilibria of Weak Electrolytes

Vapor-Liquid Equilibrium
Dissociation Balances
Mass Balances
Electroneutrality
Deviations to Ideality

An application to one binary mixture of a volatile electrolyte and water will illustrate the choice of parameters H and K, an approach is proposed to represent the vapor-liquid equilibrium in the whole range of concentration. Ternary mixtures with one acid and one base lead to the formation of salts and high ionic strengths can be reached. There, it was found useful to take into account

0-8412-0569-8/80/47-133-173$05.00/0

ternary parameters in PITZER'S (3) development and improved results using original BEUTIER' S ideas are presented.

Acetic Acid-Water Mixture. CRUZ (4) chose this example to illustrate his method of representation of vapor-liquid equilibria of volatile weak electrolyte and to show how to obtain simply from experimental vapor-liquid equilibrium data the significant parameters.

He uses the dissociation constant given in the litterature to represent the distribution of acetic acid in dilute solution from his own measurements (mole fraction of acetic acid between 10^{-3} and 10^{-7}). Equation (1) where x_A is the measured apparent mole fraction of acid in the liquid phase gives H by plotting Py_A vs.αx_A

$$Py_A = (\alpha x_A)^2 \frac{H}{K} \frac{1}{M_w} \frac{1}{\phi_A^\infty} \tag{1}$$

the dissociation coefficient α is obtained from equation (2)

$$K = \frac{\alpha^2 x_A}{(1-\alpha)(\alpha x_A + 1)} \frac{1}{M_w} \tag{2}$$

ϕ_A^∞, the fugacity coefficient of acetic in the vapor phase is taken from NOTHNAGEL et al. (5) correlation and the activity coefficients of true species in the mole fraction scale are taken equal to unity.

Another way to obtain HENRY'S constant H of undissociated acid is from high concentration vapor-liquid equilibria where dissociation is negligible. Using NRTL equation for the representation of the data of BROWN and EWALD (6) at high concentration in acetic acid ($10^{-2} < x_A < 1$), he finds the limiting activity coefficient of undissociated acid at 100°C

$$\gamma^\infty = \frac{H}{f^\circ} = 2.29$$

where f° is the reference fugacity (pure acetic acid). Both value of HENRY'S constant are in agreement within one per cent (455.2 or 457.7 torr respectively). Therefore it would be possible to obtain K and H from the analysis of dilute and concentrated vapor-liquid data. The lower limit of concentration for the application of the dilute solution treatment is related to the dissociation of water. Below this limit the ionization constant of water K_W should be taken into account to obtain α from x_A.

CRUZ (7) equation for g^E of binary electrolyte solution which incorporates a DEBYE - HÜCKEL term, a BORN - DEBYE - MAC. AULAY contribution for electric work, and NRTL equation, can be used to represent the vapor-liquid equilibria of volatile electrolyte in the whole range of concentration.

NH_3 - Acid gases - Water Systems. The same type of treatment could be applied to other volatile acids or base like NH_3 - SO_2 - CO_2 and SH_2 but limitations in the determination of constants can arise because on unsufficient data base.

BEUTIER (8) did not use CRUZ model because he started his work earlier and that an extension of PITZER'S semi-empirical description of deviations to ideality as proposed by EDWARDS (2) seemed sufficient to represent NH_3 - CO_2 - H_2O, NH_3 - H_2S - H_2O and NH_3 - SO_2 - H_2O vapor-liquid equilibria.

The ion-ion electrostatic interaction contribution is kept as proposed by PITZER. BEUTIER estimates the ion - undissociated molecules interactions from BORN - DEBYE - MAC. AULAY electric work contribution, he correlates $\beta^{(0)}$ and $\beta^{(1)}$ parameters in PITZER'S treatment with ionic standard entropies following BROMLEY'S (9) approach and finally he fits a very limited (one or two) number of ternary parameters on ternary vapor-liquid equilibrium data.

Because a few errors were found in the original article complete expressions for activity coefficients are given in the appendix. BEUTIER took H, K, λ_{aa} parameter from the earlier EDWARDS et al. (2) work. A new treatment of ternary data is presented using H, $\overline{K}$, λ_{aa} parameters from the newer EDWARDS et al. (10) work. All parameters necessary in the calculation are listed in tables 2. a, b, c. Numerical results and comparison of calculated and experimental partial pressures are given in tables 3. a, b for a few typical data sets. In this work deviations to ideality in the vapor phase are calculated according to NOTHNAGEL et al. (5).

Table 2. a. Temperature Sensitive Parameters, HENRY'S Constants H, Equilibrium Constants K and EDWARDS Constants λ_{aa}

	C_1	C_2	C_3	C_4
H (NH_3)	-149.006	- 157.552	28.1001	-0.049227
H (SH_2)	342.595	-13236.8	-55.0551	0.0595651
H (CO_2)	94.4914	- 6789.04	-11.4519	-0.010454
H (SO_2)	60.538	- 5160.4	- 7.61	0.
K (NH_3) (I)	2.76080	- 3335.71	1.4971	-0.0370566
K (H_2S) (II)	218.5989	-12995.40	-33.5471	0.
K (HS^-) (III)	-114.45	- 2048.99	15.65	0.
K (CO_2) (II)	235.482	-12092.1	-36.7816	0.
K (HCO_3-) (III)	220.067	-12431.7	-35.4819	0.
K (SO_2) (II)	122.53	- 3768.	-20.	0.
K (SO_3H^-) (III)	- 21.274	1333.4	0.	0.
K (NH_3COO^-) (IV)	- 8.6	2900.	0.	0.
K (H_2O) (V)	140.932	13455.9	22.4773	0.
λ_{aa}	- 0.0260	12.29		
λ_{aa}	- 0.2106	61.56		
λ_{aa}	- 0.4922	149.20		
λ_{aa}	- 0.05	0.		

Table 2. b. Parameters for Dielectric Effects

Species	$\bar{v}_a$ or $\bar{v}_j$	α_a	r_j
NH_3	0.0245	-0.019	
H_2S	0.0349	-0.032	
CO_2	0.035	-0.035	
SO_2	0.036	-0.030	
NH_4^+	0.0134		2.5
HS^-	0.018		2.3
$S^=$	-0.0037		3.3
HCO_3^-	0.0288		2.7
$CO_3^=$	0.0065		4.0
HSO_3^-	0.035		2.7
$SO_3^=$	0.0197		2.8
$NH_2\ COO^-$	0.0459		2.7
H^+	-0.0047		3.8
OH^-	0.0005		3.5

Table 2. c. PITZER Ionic Interaction Parameters

		HS^-	$S^=$	HCO_3^-	$CO_3^=$	HSO_3^-	$SO_3^=$	NH_2COO^-	OH^-
NH_4^+	S	0.248	0.70	0.54	0.7	0.45	0.7	0.5	0.46
	Q	0.22	0.059	-0.1	0.059	0.	0.059	0.	0.25
	C^ϕ	0.002	0.0255 *	-0.08	0.	0.0014	0.0045 *	-0.006	0.
H^+	S	0.50	0.	0.42	0.	-0.6	1.2	0.34	0.16
	Q	0.36	0.	0.3	0.	0.1	0.1	0.25	0.25
	C^ϕ	0.	0.	0.	0.	0.	0.	0.	0.

* $\mu_{NH_4^+, S^=, S^=} = 0.004$ $\mu_{NH_4^+, SO_3^=, SO_3^=} = 0.0007$

Table 3. a. Calculated and Experimental Results for NH_3-SH_2-H_2O System

T(C)	Apparent molalities of		Partial Pressures of			
			NH_3		SH_2	
	NH_3	SH_2	EXP	CALCD	EXP	CALCD
			Data of MILES AND WILSON (1975)			
80	0.960	0.971	12.1	10.7	2389.	2074.
80	1.063	1.452	4.4	3.9	8556.	8554.
80	2.332	1.151	136.	138.	319.	270.
80	5.112	1.143	575.	475.	94.8	91.1
80	5.538	5.305	72.6	84.1	12140.	10537.
80	9.245	7.935	209.	221.	9506.	11949.
80	10.201	5.983	597.	588.	1916.	2019.
80	22.627	5.561	2285.	2383.	241.	468.
120	0.0106	0.0005	2.14	3.10	.65	.20
120	0.442	0.181	81.2	86.9	333.	277.
120	0.526	0.109	177.	134.	82.5	80.2
120	0.534	0.712	23.7	22.7	6410.	6425.
120	0.992	0.493	183.	174.	916.	783.
120	1.131	0.998	95.4	97.5	4632.	4015.
120	2.031	0.433	569.	529.	280.	236.
120	10.081	2.114	3179.	3019.	973.	925.

	Data of LEYKO (1959)					
20	2.25+	1.07+	11.	13.	7.5	10.5
20	4.91	2.34	21.	31.	23.	21.
20	9.36	4.46	44.	65.	38.	43.
20	15.66	7.46	80.	110.	62.	64.
30	9.36	4.46	79.	103.	74.	76.
30	15.65	7.45	129.	173.	124.	112.
40	2.25	1.07	27.	31.	31.	33.
40	4.91	2.34	58.	75.	73.	65.
40	9.35	4.46	119.	159.	137.	131.
40	15.65	7.45	215.	265.	238.	192.
50	2.25	1.07	45.	47.	60.	55.
50	4.91	2.34	89.	112.	131.	110.
50	9.35	4.45	190.	237.	235.	220.
45	15.65	7.45	255.	324.	291.	249.
	Data of LEYKO and PIATKEWICZ (1964)					
80	7.340+	2.279+	722.	649.	266.	249.
90	7.340	2.279	996.	878.	387.	393.
100	7.340	2.280	1262.	1162.	570.	612.
110	7.341	2.280	1558.	1510.	889.	939.

+ converted when necessary from concentration (mol/l) in a consistent way with the model using $\underline{c}$ v and calculated concentration of real species.

Table 3. b. Summury of Comparison of Calculated and Experimental Results

SYSTEM	DATA	T(c)	Number of points	r.m.s. deviation in partial pressure of NH_3	ACID
NH_3-H_2S-H_2O	VAN KREVELEN et al(1949)	20	21	-	.25
	"	40	19	-	.09
	"	60	23	-	.13
	MILES and WILSON (1975)	80-120	16	.15	.31
	LEYKO (1959)	20-45	14	.30	.15
	LEYKO and PIATKEWICZ(1964)	80-110	4	.09	.06
NH_3-CO_2-H_2O	VAN KREVELEN (1942)	20	11	-	.12
	"	60	19	.07	.33
	PEXTON and BADGER (1938)	20	30	.18	.23
NH_3-SO_2-H_2O	JOHNSTONE (1935)	35	8+	-	.10
	"	50	9	-	.11
	"	70	8	-	.07
	"	90	8	-	.19

References :

JOHNSTONE H.F. Ind. Eng. Chem. , 27, 587 (1935)
LEYKO J. Bull. Acad. Polon. Sci. Ser. Sci. Chem. , 7, 675 (1959)
LEYKO J.; PIATKEWICZ J, Bull Acad. Polon. Sci. Ser. Sci. Chem. , 12, 445 (1964)
MILES D. H.; WILSON G. M. Center for Thermodynamical Studies of Brigham Young University, Utah, Annual Report (1975)
PEXTON S.; BADGER E.H.M. J. Soc. Chem. Ind. , 57, 106 (1938)
VAN KREVELEN D.W.; HOFTIJZER P.J.; HUNTJENS F.J. Recl. Trav. Chim. Pays-Bas, 68, 191 (1949)

+ only reliable data points are kept, see BEUTIER (8)

Appendix. Formalism of BEUTIER'S Model of Volatile Weak Electrolytes Vapor-Liquid Equilibria

It is shown with the example of a ternary mixture of one Base (B), one Acid (A) with two ionizing steps (A^- and $A^=$) and water (W). See nomenclature at end of Appendix for symbols and additional equations.

Vapor-Liquid Equilibrium (2 equations)

$$y_a \phi_a P = m_a \gamma_a H_a \mathcal{P}_a \quad (3)$$

$$y_w \phi_w P = a_w \pi_w \phi_w(S) \mathcal{P}_w \quad (4)$$

Dissociation Equilibrium (5 equations)

(I) $B + H_2O \rightleftarrows B^+ + OH^-$

(II) $A \rightleftarrows A^- + H^+$

(III) $A^- \rightleftarrows A^= + H^+$

type of dissociation (I), (II) or (III) is reported in table 2. a. for acids and base.

(IV) $B + A^- \rightleftarrows BA^- + H_2O$, for example, formation of carbamate

(V) $H_2O \rightleftarrows H^+ + OH^-$

$$K = \frac{\prod_i \gamma_i m_i}{\prod_j \gamma_j m_j}$$

i, components on the left side of dissociation equilibrium, j , components on the right side of the same. (5)

Mass Balance (2 equations)

$$B = m_B + m_{B^+} + m_{BA^-} \quad (6)$$

$$A = m_A + m_{A^-} + m_{A^=} + m_{BA^-} \quad (7)$$

Electroneutrality

$$m_{B^+} + m_{H^+} = m_{A^-} + 2\, m_{A^=} + m_{OH^-} \quad (8)$$

Given (P, T, A, B), the 10 equations (3) to (8) yield the ten unknown (m_B, m_A, m_{A^-} , $m_{A^=}$, m_{B^+} , m_{AB^-} , m_{H^+} , m_{OH^-} , y_A , y_B) if the expressions of ϕ_i , γ_i , a_w are known.

Deviations to Ideality

Vapor Phase. The fugacity coefficients are taken for NOTHNAGEL et al. (5) correlation.

Liquid Phase. γ and a are derived from the following expression of the excess Gibbs energy.

$$G^E = G^E\ (\text{PITZER}) + G^E\ (\text{BORN}) + G^E\ (\text{EDWARDS})$$

Interactions Between Ions.

$$\frac{G^E\ (\text{PITZER})}{M_w\ n_w\ RT} = f(I) + \sum_{kl} \lambda_{kl}\ m_k\ m_l + \sum_{hkl} \mu_{hkl}\ m_h\ m_k\ m_l \quad (k,l \text{ ions}) \tag{9}$$

Electric Work of Charge.

$$\frac{G^E\ (\text{BORN})}{M_w\ n_w\ RT} = \left(\sum_j L_j\ m_j\right)\left(\frac{V_f + 0{,}5\ V_c}{V_f - V_c}\ \frac{D_w}{D_n} - \frac{V_i + 0{,}5\ V_c}{V_i - V_c}\right) (j \text{ ions}) \tag{10}$$

Interaction Between Neutral Species.

$$\frac{G^E(\text{EDWARDS})}{M_w\ n_w\ RT} = \sum_a \lambda_{aa}\ m_a^2 + \sum_a \mu_{aaa}\ m_a^2 \quad (a \text{ neutral solute}) \tag{11}$$

Activity Coefficients of Ions (molality scale)

$$\ln \gamma_j = \left(\frac{\partial}{\partial n_j}\ \frac{G^E}{RT}\right)_{n_w,\ n_k,\ n_a}$$

$$\ln \gamma_j = \frac{z_j^2}{2}\ \frac{df}{dI} + 2 \sum_k \lambda_{jk}\ m_k + \frac{z_j^2}{2} \sum_{kl} \frac{d\lambda_{kl}}{dI}\ m_k\ m_l$$
$$+ 3 \sum_{kl} \mu_{jkl}\ m_k\ m_l$$
$$+ L_j\ \frac{D_w}{D_n} \sum_a m_a \left[-\frac{\alpha_a}{V_n}\ \frac{V_i + V_c/2}{V_i - V_c} - \frac{1.5\ \bar{v}_a\ V_c}{(V_i - V_c)\ (V_f - V_c)}\right]$$
$$+ \frac{3}{2}\left(\sum_k L_k\ m_k\right)\left[\frac{D_w}{D_n}\ \frac{V_f\ \bar{v}_j^c - V_c\ \bar{v}_j}{(V_f - V_c)^2} - \frac{V_i\ \bar{v}_j^c - V_c\ \bar{v}_j}{(V_i - V_c)^2}\right] \tag{12}$$

Activity Coefficients of Neutral Solutes

$$\ln \gamma_a = \left(\frac{\partial}{\partial n_a}\ \frac{G^E}{RT}\right)_{n_w,\ n_k,\ n_a'}$$

$$\ln \gamma_a = \left(\sum_k L_k\ m_k\right) \frac{D_w}{D_n}\left[-\frac{1.5\ \bar{v}_a\ V_c}{(V_f - V_c)^2} + \left(\frac{V_f + 0.5\ V_c}{V_f - V_c}\right)\left(\frac{D_w}{D_n}\right)\right.$$
$$\left.\left(\frac{\bar{v}_a}{V_n} - \frac{\bar{v}_a + \alpha_a}{V_n}\ \frac{D_w}{D_n}\right)\right] + 2\ \lambda_{aa}\ m_a + 3\ \mu_{aaa}\ m_a^2 \tag{13}$$

Activity of Water

$$\ln a_w = M_w\left[-\sum_j m_j - \sum_a m_a + \frac{1}{M_w}\left(\frac{\partial}{\partial n_w}\ \frac{G^E}{RT}\right)_{n_k,\ n_a}\right] = -M_w\ \phi\left(\sum_j m_j + \sum_a m_a\right)$$

$$\ln a_w = M_w \Big[- \sum_j m_j - \sum_a m_a + f(I) - I \frac{df}{dI} - \sum_{hk} \Big(\lambda_{hk} + I \frac{d\lambda_{hk}}{dI}\Big) m_h m_k$$
$$- 2 \sum_h \sum_k \sum_l \mu_{hkl} m_h m_k m_l$$
$$- \Big(\sum_k L_k m_k\Big) \Big\{ - \frac{\sum_a \alpha_a m_a}{d_w V_n^2} \Big(\frac{D_w}{D_n}\Big)^2 \Big(\frac{V_f + 0.5 V_c}{V_f - V_c}\Big) + \frac{1.5 V_c}{d_w (V_f - V_c)^2}$$
$$\frac{D_w}{D_n} - \frac{1.5 V_c}{d_w (V_i - V_c)^2} \Big\} - \sum_a \lambda_{aa} m_a^2 - 2 \sum_a \mu_{aaa} m_a^3 \Big] \quad (14)$$

List of Symbols

a activity in solution
A apparent molality of acid (mole/kg of water)
A_{DH} DEBYE - HÜCKEL constant

$$A_{DH} = \Big(\frac{2\pi N d_w}{1000}\Big)^{1/2} \Big(\frac{e^2}{D_w kT}\Big)^{3/2} \quad (15)$$

$b = 1.2$
B apparent molality of base (mole/ kg water)
C_1 to C_4 coefficients in equations (20) (22) (29)

C^{ϕ}_{MX} PITZER ternary interaction parameter for salt MX

d_w density of water (kg/dm^3)
D_n dielectric constant of neutral solution (without ions)

$$D_n = D_w \Big(1 + \sum_a \frac{\alpha_a m_a}{V_n}\Big) \quad (16)$$

D_w dielectric constant of water

$$D_w = 305.7 \exp(-\exp(-12.741 + 0.01875\ T) - T/219) \quad (17)$$

e charge of electron ($4,8029 \cdot 10^{-10}$ esu)
$f°$ reference fugacity

$$f(I) = - \frac{A_{DH}}{3} \frac{4I}{b} \ln(1 + bI^{1/2}) \quad (18)$$

$$\frac{df}{dI} = - \frac{2 A_{DH}}{3} \Big[\frac{I^{1/2}}{1 + bI^{1/2}} + \frac{2}{b} \ln(1 + bI^{1/2})\Big] \quad (19)$$

G^E excess GIBBS energy in the definition of electrochemists by reference to the "ideal" solution in molality scale.
H Henry's constant of undissociated acid or base (atm. $kg/mole^{-1}$)

$$\ln H_a = C_1 + C_2/T + C_3 \ln T + C_4\ T \quad (20)$$

I ionic strengh

$$I = \frac{1}{2} \sum_j m_j z_j^2 \quad (21)$$

k BOLTZMANN'S constant $k = 1.38045 \cdot 10^{-16}$ erg K^{-1}

K equilibrium constant in molality scale

$$\ln K = C_1 + C_2/T + C_3 \ln T + C_4 T \tag{22}$$

L_j dimensionless constant (r_j en Å)

$$L_j = \frac{e^2 z_j^2}{2 r_j kT D_w} \cdot 10^8 \tag{23}$$

M_w molecular weight of water 0.018 kg/mole
m molality, mol/kg
n number of mole
N AVOGADRO'S number N = $6.0232 \cdot 10^{23}$ mol^{-1}
P pressure (atm)
Q Combination of PITZER binary ionic interaction parameter $Q = \beta^{(0)}/(\beta^{(0)} + \beta^{(1)})$
$\mathcal{P}$ POYNTING correction
r_j ionic cavity radius (Å)
R gas constant
S sum of PITZER binary ionic interaction parameters $Q = \beta^{(0)} + \beta^{(1)}$
T temperature (K)
v partial molar volume, dm^3/mol
v_j^c volume of ionic cavities for ion j per equivalent (dm^3/mol) r_j en Å

$$v_j^c = \frac{4}{3} \pi N \, r_j^3 \cdot 10^{-27} \tag{24}$$

V_c volume of all ionic cavities (dm^3/kg of water)

$$V_c = \sum_j v_j^c \, m_j \tag{25}$$

V_f volume of real solution (dm^3/kg water)

$$V_f = \frac{1}{d_w} + \sum_j m_j v_j + \sum_a m_a \bar{v}_a \tag{26}$$

V_i volume of ionic solution excluding neutral solutes (dm^3/kg water)

$$V_i = \frac{1}{d_w} + \sum_j m_j \bar{v}_j \tag{27}$$

V_n volume of neutral solution excluding ions (dm^3/kg of water)

$$V_n = \frac{1}{d_w} + \sum_a m_a \bar{v}_a \tag{28}$$

y mole fraction in vapor phase
z_j number of charges of ion j
$\alpha = 2$
α_a dielectric coefficient (dm^3/mol^{-1})
$\beta_{jk}^{(0)}, \beta_{jk}^{(1)}$ parameters in PITZER binary term for interaction between ions of opposite sign j and k obtained from S and Q.
γ activity coefficient (molality scale)
λ_{aa} EDWARDS extended PITZER binary coefficient for interaction of neutral species (β_{aa} in EDWARDS(10) (kg/mol^{-1})

$\lambda_{aa} = C_1 + C_2/T$

λ_{jk} PITZER binary interaction coefficient for ions j.k. of different signs (kg/mol^{-1})

$$\lambda_{jk} = \beta_{jk}^{(0)} + \beta_{jk}^{(1)} \frac{2}{\alpha^2 I} \left[1 - e^{-\alpha\sqrt{I}} (1 + \alpha\sqrt{I}) \right] \quad (30)$$

$$\frac{d\lambda_{jk}}{dI} = \frac{2\beta_{jk}^{(1)}}{\alpha^2 \; I^2} \left[- 1 + e^{-\alpha\sqrt{I}} \left(1 + \alpha\sqrt{I} + \frac{\alpha^2}{2} I\right) \right] \quad (31)$$

all other λ_{jk} are zero.

μ_{aaa} extended PITZER ternary coefficient

$$\mu_{aaa} = - \frac{1}{55.5} (\lambda_{aa} + \frac{1}{166.5}) \quad (32)$$

μ_{hkl} PITZER ternary interaction coefficients ($kg^2 \; mol^{-2}$)

$$\mu_{MMX} = \mu_{MXX} = \frac{C_{MX}^{\phi}}{6} \text{ for 1-1 salts MX} \quad (33)$$

$$\mu_{MMX} = \mu_{MXX} = \frac{C_{MX}^{\phi}\sqrt{2}}{9} \text{ for 1-2 salts MX} \quad (34)$$

$$\mu_{MXY} = \frac{1}{2} (\mu_{MXX} + \mu_{MYY}) \quad (35)$$

all other μ are zero

π_w vapor pressure of water

$$\log_{10} \pi_w = 7.96681 - \frac{1668.21}{T-45.15} \quad (36)$$

ϕ_a fugacity coefficient ϕ osmotic coefficient

Subscripts

a neutral solute

hijkl ionic species

w water

Superscript

∞ infinite dilution

Literature Cited

1. VAN AKEN, A.B. ; DREXHAGE, J.J. ; de SWAAN ARONS, J. Ind. Eng. Chem. Fundam., 1975, 14(3), 154.
2. EDWARDS, T.J. ; NEWMAN, J. ; PRAUSNITZ, J.M. ; A.I.Ch.E.J., 1975, 21(2), 248.
3. PITZER, K.S. ; J. Phys. Chem., 1973, 77(19), 268 ; J. Am. Chem. Soc., 1974, 96(18) , 5701.

4. CRUZ, J. ; RENON, H. ; Ind. Eng. Chem. Fundam., 1979, 18(2), 168.
5. NOTHNAGEL, K.H. ; ABRAMS, D.S. ; PRAUSNITZ, J.M., Ind. Eng. Chem. Proc. Des. Devt., 1973, 12(1), 25.
6. BROWN, I. ; EWALD, A.H. ; Austr. J. Sci. Res. Phys. Ser., 1950, 3, 306.
7. CRUZ, J. ; RENON, H. ; A.I.Ch.E. E. Journal, 1978, 24(5),817.
8. BEUTIER, D. ; RENON, H. ; Ind. Eng. Chem. Proc. Des. Devt., 1978, 17(3), 220.
9. BROMLEY, L.A. ; J. Chem. Therm. , 1973, 4, 669.
10. EDWARDS, T.J. ; MAURER, O. ; NEWMAN, J. ; PRAUSNITZ, J.M. ; A.I.C.h.E. Journal, 1978, 24(6), 966.

RECEIVED January 31, 1980.

9

Sour Water Equilibria

Ammonia Volatility down to PPM Levels; pH vs. Composition; and Effect of Electrolytes on Ammonia Volatility

GRANT M. WILSON, RICHARD S. OWENS, and MARSHALL W. ROE

Wilco Research Company, 488 South 500 West, Provo, Utah 84601

Undesirable sulfur, nitrogen, and oxygen compounds are often encountered in commercial gas production and gas treating facilities relating either to natural sources or to synthetic processes. Water also occurs as condensate in these gas streams or water is brought in contact with gas streams in various processing steps. As a result aqueous waste streams are produced in which undesirable sulfur, nitrogen, and oxygen compounds are present. Their concentrations in the aqueous waste streams depend on their equilibrium concentrations in gas streams from which the aqueous streams are derived. Present and future environmental control restrictions dictate that the concentrations of these undesirable compounds be maintained at very low levels before the streams can be released to the environment. Thus, methods must be developed for controlling these concentrations. One method used in refineries for controlling the concentrations of undesirable compounds in aqueous waste streams is by means of a steam stripper called a sour water stripper where these trace compounds are steam distilled and then condensed as a concentrated product in the condenser of the stripping column. The principal components in these strippers are hydrogen sulfide, carbon dioxide, and ammonia. This concentrated product stream is then further processed for removal of these undesirable compounds. Similar processes undoubtedly will be necessary in existing and new gas production or gas treating facilities. The design of these processes requires data regarding the equilibrium concentrations of undesirable compounds absorbed into various aqueous waste streams, and then equilibrium data are required relating to the removal of these compounds from aqueous waste streams.

This paper reports on measurements made in three areas pertaining to the processing of aqueous waste streams as follows.

A. Vapor-liquid equilibrium measurements on NH_3-H_2O mixtures at 80 and 120°C
B. pH of NH_3-H_2S-CO_2-H_2O mixtures at 25 and 80°C
C. Effect of sodium hydroxide and sodium acetate on NH_3 volatility at 80°C

0-8412-0569-8/80/47-133-187$10.00/0

Measurement Apparatus

A flow-type vapor-liquid equilibrium apparatus shown schematically in Figure 1 was used for the vapor-liquid equilibrium measurements on NH_3-H_2O. The method involves the analysis of an equilibrium nitrogen stream for NH_3 after equilibration with a NH_3-H_2O mixture contained in cylinders 1 and 2 of Figure 1. The partial pressure of NH_3 is then calculated from the vapor mole fraction of NH_3 times the total pressure of the system. Two cylinders were used to saturate the nitrogen stream in order to minimize liquid depletion effects and to insure equilibrium between the gas and liquid phases. Vapor phase analyses were made by absorbing the NH_3 into an aqueous HCl solution which was subsequently titrated and by measuring the amount of nitrogen in the sample by use of a calibrated wet test meter accurate to $\pm$ 1%. Pressures were measured by means of calibrated pressure gauges accurate to $\pm$ 0.1%, and temperatures were measured by means of calibrated thermocouples accurate to $\pm$ 0.05°C. The composition of both the vapor and the liquid samples were determined by potentiometric titration by addition of standardized NaOH solution to samples which initially contained a slight excess of HCl. This method worked very well even at the ppm levels studied in some of the runs. The same titrating solution was used for analyzing both the vapor and the liquid samples of each run thus minimizing errors produced in standardizing the NaOH solution.

Figure 2 gives a schematic of the apparatus used for pH measurements at 25 and 80°C. It consisted of a stoppered Erlenmeyer flask submerged in a temperature bath regulated at either 25 or 80°C. The contents of the flask were stirred by means of a magnetic stirrer coupled to a motor beneath the bath. A pH probe and thermometer were inserted through the stopper at the top of the flask and another hole was stoppered for use in pipeting solution into or out of the flask.

Measurements were made by first calibrating the pH probe using two buffer solutions at pH's of 7 and 10 supplied by Van Labs of Van Waters and Rogers Equipment Company . These solutions have been certified by the National Bureau of Standards to be accurate to $\pm$ 0.01 pH unit at 25°C. This pH calibration was made with the probe inserted in buffer solution at the same temperature as the pH measurements were made.

After calibration the probe was inserted into the flask shown in Figure 2. A concentrated solution of NH_3-H_2S-CO_2-H_2O of measured density was then pipeted into the flask and after temperature and pH equilibration the pH was read. This normally took a period of five minutes for the equilibration process. After reading the pH, the solution was diluted with water by first removing by pipet a portion of the solution in the flask and by subsequent addition of water to replace the solution

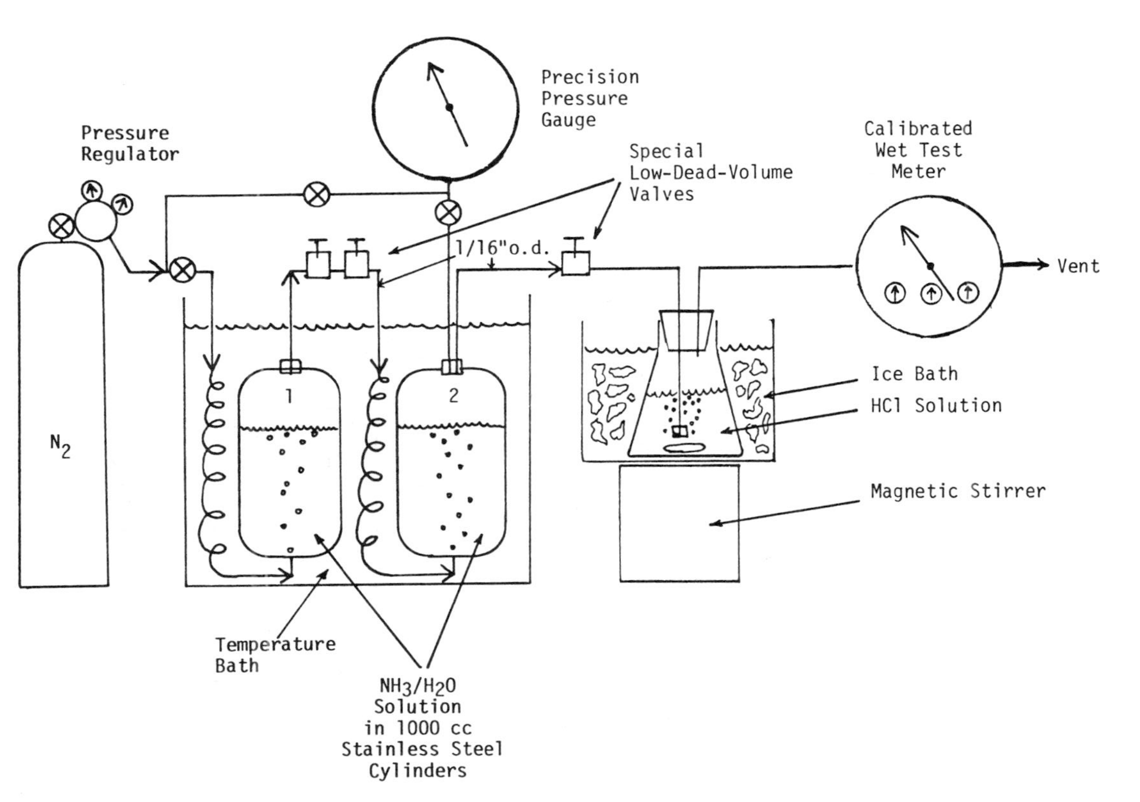

Figure 1. Schematic flow apparatus used for NH_3–H_2O (and electrolyte) vapor–liquid equilibrium measurements

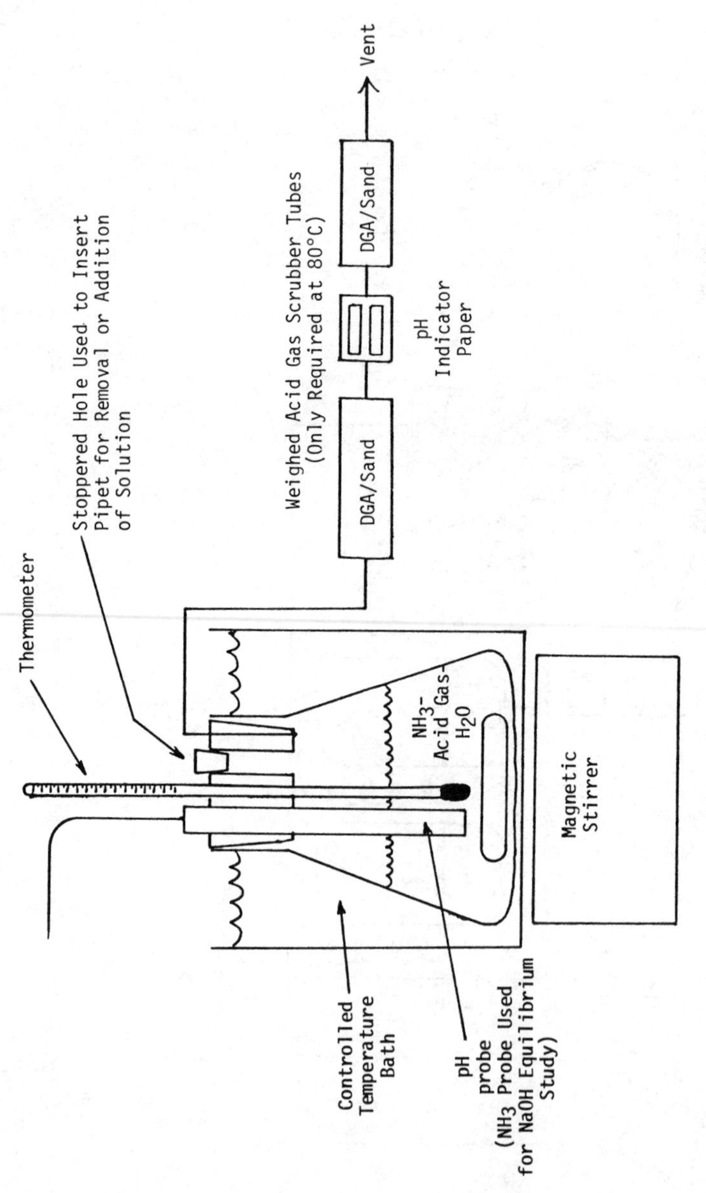

Figure 2. *Schematic of apparatus used for pH (or NH_3 probe) measurements*

removed. Boiled distilled water was used for both the solution preparation and dilution stages of the experiment. After each dilution, the pH was again read.

The concentrated solutions were prepared by bubbling acid gas into a solution of ammonia and water. No gas was allowed to escape so that the increase in the weight of the solution represented the amount of H_2S or CO_2 added. No problems were encountered when adding H_2S by this method, but CO_2 is absorbed very slowly and long periods of agitation of the solution in contact with gaseous CO_2 were necessary to achieve the desired levels. At the higher loadings of acid gas to ammonia, problems were encountered when these solutions were heated to 80°C because the absorbed gases tended to be driven off. This problem was solved by connecting the absorption train composed of diglycolamine (DGA) on sand shown in Figure 2. By this method, any components discharged from the cell were absorbed in these scrubber tubes. The tubes were then weighed and a correction was made in the composition of the solution in the Erlenmeyer flask.

Some problems were encountered with the pH probes used in the pH measurements because the reference electrode is saturated with AgCl. Hydrogen sulfide can react with the AgCl in solution, thus precipitating AgS at the KCl junction between the reference electrode and the solution. In this regard it was found that certain probes work better than others. Our observations are as follows.

1. Some reference electrodes have an exposed filling hole for adding AgCl-saturated KCl. The hole is located on the side of the probe where H_2S can readily enter and contaminate the solution. These electrodes proved entirely unsatisfactory.
2. Some electrodes have the KCl solution in a gel form. These electrodes work fairly well but eventually the gel becomes contaminated with silver sulfide and the probe must be replaced. This type of probe was used for the measurements given in this report. Two probes were actually used for the entire set of measurements.
3. There are also electrodes available which have a ground glass junction between the reference electrode and the solution being measured. These probes would probably be most suitable because they can be dismantled and cleaned. A probe arrived as our measurements were being completed so this type of probe has not yet been tested in our laboratory.

The effects of sodium acetate and sodium hydroxide on NH_3 volatility at 80°C were studied by two methods. Data on the effect of sodium acetate were measured in the same apparatus used to measure the NH_3-H_2O data shown in Figure 1. In this case, sodium acetate was also added to the liquid phase with subsequent

NH_3 partial pressure measurements made in the same manner used for the NH_3-H_2O measurements.

Data on the effect of sodium hydroxide were measured by means of an NH_3 probe supplied by Orion Research Company which operates in a manner analogous to a pH probe except that a membrane is used through which only the NH_3 permeates. Thus the response of the probe is proportional to the activity (or partial pressure) of ammonia.

E.m.f. data on the effect of sodium hydroxide were converted to ammonia partial pressure data using the following equation.

$$f_{NH_3} = f_{NH_3}^{ref.} \cdot 10^{\frac{mv - mv^{ref.}}{\beta T}} \quad (1)$$

where f_{NH_3} = fugacity of NH_3

$f_{NH_3}^{ref.}$ = fugacity of NH_3 in a reference solution of NH_3- H_2O

mv = e.m.f. output of NH_3 electrode in millivolts

$mv^{ref.}$ = output of NH_3 electrode in the reference solution

βT = millivolts per decade, at 80°C T = 70.1 mv/decade

At low pressures the fugacity of a component can be replaced by its partial pressure so that Equation 1 can be approximated as follows.

$$P_{NH_3} = P_{NH_3}^{ref.} \cdot 10^{\frac{mv - mv^{ref.}}{\beta T}} \quad (2)$$

One problem encountered with the NH_3 probe was that its calibration drifted with time, thus requiring frequent recalibration of the probe. For this reason the probe was abandoned when the measurements with sodium acetate were made.

Measurement Results

Vapor-Liquid Equilibrium Data. Vapor-liquid equilibrium measurements on NH_3-H_2O mixtures at 80 and 120°C are summarized in Tables 1 and 2, respectively. At each condition, several vapor and liquid samples were removed for analysis before proceeding to the next condition. These tables summarize the analyses of these samples at each condition and then give an average value for each run condition. The charge analyses are based on analyses of the solution charged to the measurement apparatus at the beginning of each run and liquid analyses are based on samples removed from the apparatus during the run. A comparison of these analyses shows that the liquid analyses are all slightly lower than the charge analysis and that the liquid analyses decrease slightly with each sample. This effect is due to loss of ammonia to the vapor phase in sampling the vapor and thus

TABLE 1. Ammonia-Water Vapor-Liquid Equilibrium Measurements at 80°C by Flow Cell Method

	wt % NH_3 in charge	wt % NH_3 in liquid	Partial Pressure, psia NH_3	Partial Pressure, psia H_2O[a)]	Total Pressure, psia	Volatility Ratio P_{NH_3}/wt % Uncorr.	Volatility Ratio P_{NH_3}/wt % Corr.[b)]
	4.97	4.94	6.88	6.51	13.39	1.39	1.24
		4.90	6.86	6.51	13.37	1.40	1.25
		4.86	6.65	6.51	13.16	1.37	1.22
average		4.90	6.80	6.51	13.31	1.39	1.24
	1.000	.994	1.240	6.80	8.04	1.25	1.23
		.974	1.250	6.80	8.05	1.28	1.26
		.950	1.247	6.80	8.05	1.31	1.29
average		.973	1.246	6.80	8.05	1.28	1.26
	.0993	.0991	.122	6.86	6.98	1.23	1.25
		.0983	.119	6.86	6.98	1.21	1.23
		.0965	.119	6.86	6.98	1.23	1.25
average		.0980	.120	6.86	6.98	1.23	1.25
	100.0 ppm	99.2 ppm	.0111	6.87	6.88	1.12	1.18
		97.7	.0113	6.87	6.88	1.16	1.22
		96.0	.0112	6.87	6.88	1.17	1.23
		94.3	.0110	6.87	6.88	1.17	1.23
average		96.8	.0112	6.87	6.88	1.16	1.22
	10.09 ppm	10.08 ppm	.00108	6.87	6.87	1.07	1.26
		10.04	.00108	6.87	6.87	1.08	1.27
		9.93	.00107	6.87	6.87	1.08	1.27
average		10.02	.00108	6.87	6.87	1.08	1.27

[a)] Actual measurements were made under nitrogen pressure according to conditions summarized in Table 4. The partial pressure of water as given here is based on Raoult's Law which applies at the low NH_3 concentrations given here. By this method the partial pressure of water is given as follows.

$$P_{H_2O} = P^{\circ}_{H_2O}\, x_{H_2O},$$

$P^{\circ}_{H_2O}$ = vapor pressure of pure water

x_{H_2O} = water mole fraction in liquid

The vapor pressure of water was taken from the CRC Handbook, 52nd Ed., page D-147.

[b)] See text for corrections applied.

TABLE 2. Ammonia-Water Vapor-Liquid Equilibrium Measurements at 120°C by Flow Cell Method

	wt % NH_3 in charge	wt % NH_3 in liquid	Partial Pressure, psia[a] NH_3	Partial Pressure, psia[a] H_2O	Total Pressure, psia	Volatility Ratio P_{NH_3}/wt %[b] Uncorr.	Volatility Ratio P_{NH_3}/wt %[b] Corr.
	4.83	4.75	19.0	27.35	46.4	4.00	3.81
	4.68	4.60	18.5	27.39	45.9	4.02	3.83
		4.45	18.2	27.44	45.6	4.09	3.89
average		4.60	18.6	27.39	46.0	4.04	3.85
	.905	.900	3.59	28.52	32.1	3.99	3.97
	.889	.880	3.55	28.53	32.1	4.03	4.01
	.867	.850	3.48	28.54	32.1	4.09	4.07
average		.877	3.54	28.53	32.1	4.04	4.02
	.0955	.0940	.342	28.77	29.1	3.64	3.69
		.0914	.341	28.77	29.1	3.73	3.78
	.0898	.0878	.329	28.77	29.1	3.75	3.80
	.0856	.0840	.315	28.77	29.1	3.75	3.80
average		.0893	.332	28.77	29.1	3.72	3.77
	93.4 ppm	92.2 ppm	.0332	28.79	28.8	3.60	3.77
		90.0	.0332	28.79	28.8	3.69	3.86
		87.8	.0326	28.79	28.8	3.71	3.88
		84.8	.0317	28.79	28.8	3.74	3.91
average		88.7	.0327	28.79	28.8	3.69	3.86
	9.83 ppm	9.72 ppm	.00358	28.80	28.8	3.68	4.24 } appear high
		9.44	.00368	28.80	28.8	3.90	4.49 } appear high
		9.12	.00367	28.80	28.8	4.02	4.63 } appear high
average		9.43	.00364	28.80	28.8	3.86	4.45 } appear high

[a] See footnote "a" at the bottom of Table 2.

[b] See text for corrections applied.

represents slight depletion of the liquid with each vapor sample. The liquid analyses in these tables are averages of analyses made before and after sampling the vapor so the concentrations given represent the average liquid composition during the vapor sampling process.

The ammonia partial pressures given in Tables 1 and 2 are based on the concentration of ammonia found in the vapor stream times the total pressure. The actual pressures applied at each run condition are summarized in Table 3 where the pressures varied from 15 psia at 80°C to 90 psia at 120°C. Because nitrogen was used as a pressurizing fluid, the partial pressure of water and the total pressure excluding nitrogen have been computed in Tables 1 and 2 based on Raoult's law for water as noted at the bottom of Table 1. Raoult's law applies for the partial pressure of water because the activity coefficient of water is virtually unity at the low levels of ammonia used in the liquid phase. Minor effects due to vapor non-ideality have not been applied.

The last two columns of Tables 1 and 2 give equilibrium volatility ratios of ammonia partial pressure divided by the weight percent of ammonia in the liquid phase. The column labeled "Uncorr." gives the volatility ratio computed simply as the partial pressure of ammonia divided by the total weight percent of ammonia in solution. The second column gives a corrected volatility ratio based on the dissociation of ammonia at low concentrations and extrapolation to zero concentration of ammonia. The dissociation correction was applied by dividing the uncorrected volatility ratio by the computed ratio of free ammonia over total ammonia in solution. The ratio of free ammonia over total ammonia was computed from the dissociation constant of ammonia given by Edwards and Prausnitz (1) as follows.

$$NH_4OH \rightarrow NH_4^+ + OH^-$$

Temperature °C	Dissociation Constant
25	1.78×10^{-5}
80	1.66×10^{-5}
120	1.19×10^{-5}

With no other ions present the concentration of free ammonia over total ammonia present is given by the following equation:

$$\frac{(NH_3)_{free}}{(NH_3)_{total}} = 1 - \frac{2k}{4C}\left(\sqrt{1 + 4C/k} - 1\right) \quad (3)$$

where k = dissociation constant of NH_3
C = total concentration of NH_3 moles/Kg water

In addition to this correction the volatility ratio of ammonia

TABLE 3. Measurement Pressures for Ammonia-Water Vapor-Liquid Equilibrium Runs

Temp. °C	NH_3 wt %	NaAc wt %	Measurement Pressure with N_2 psia
80	5	0	30
	1	0	20
	.1	0	15
	100 ppm	0	15
	10 ppm	0	15
	1 wt %	25	20
	1	15	20
	1	5	20
120	5	0	90
	1	0	60
	.1	0	60
	100 ppm	0	60
	10 ppm	0	60

varies due to the solvent effect of free ammonia according to the following equation given in Table 1 of reference 2.

$$\ln(H_{NH_3}) = \ln(H^{\circ}_{NH_3}) + \beta C_{NH_3} \tag{4}$$

where H° = volatility ratio at zero concentration of free ammonia
C_{NH_3} = effect of free ammonia on the volatility
β = temperature dependent constant; $\beta = 131.4/T°R - .1682$
C_{NH_3} = concentration of free ammonia, moles/Kg water

Solving for $\ln(H^{\circ}_{NH_3})$ gives the following equation.

$$\ln(H^{\circ}_{NH_3}) = \ln(H_{NH_3}) - \beta C_{NH_3} \tag{5}$$

or $$H^{\circ}_{NH_3} = H_{NH_3}\exp(-\beta C_{NH_3}) \tag{6}$$

Equations 2 and 5 can be combined to give a total correction as follows.

$$\left(\frac{P_{NH_3}}{wt\ \%\ NH_3}\right)^{\circ}_{Corr.} = \frac{(P_{NH_3}/wt\ \%\ NH_3)_{Uncorr.}\ \exp(-\beta C_{NH_3})}{[(NH_{3_{free}})/(NH_{3_{total}})]} \tag{7}$$

Thus the last column in Tables 1 and 2 corresponds to the volatility ratio of ammonia based on free ammonia and extrapolated to zero concentration of free ammonia. This number should be independent of the concentrations studied at a given temperature, and any variation represents errors in either the measured data or in the applied corrections. The following is a comparison of the averages from each run.

NH_3 Concentration	$(P_{NH_3}/wt\ \%)^{\circ}_{Corr.}$ 80°C	120°C	
5 wt %	1.24	3.85	
1 wt %	1.26	4.02	
.1 wt %	1.25	3.77	
100 ppm	1.22	3.86	
10 ppm	1.27	4.45	(appears high)
Average (Excluding 10 ppm point at 120°C)	1.25 ± 2%	3.88 ± 4%	

This agreement is considered to be quite good when allowance is made for the fact that these are independently measured runs at concentrations varying by a ratio of 5000 to 1! The 10 ppm run at 120°C is probably in error due to a trace contaminant in the vapor samples observed in the potentiometric titration curve. This problem was not observed at 80°C.

The favorable comparison given above shows that the ammonia' volatility ratio can be reliably extrapolated to low concentrations without loss of accuracy. This is important because the bulk of published literature data on the volatility of ammonia are at concentrations of 1% NH_3 or higher.

A comparison of this new volatility ratio with measured literature data is given in Figure 3 where deviation ratios of $P_{NH3(meas)}/P_{NH3(calc)}$ are plotted versus temperature where comparison is made with the SWEQ calculation model of reference 2. This plot shows that these new data are in fair agreement with the calculated values with ratios of about 1.05 at both 80 and 120°C. Fortunately for the authors these data also agree quite well with previously measured data reported by Miles and Wilson (3).

In summary the following can be concluded from the data given in Tables 1 and 2.

1. The volatility of ammonia at ppm levels can be calculated from the ionization constant of ammonia combined with volatility data measured at higher concentrations of ammonia.
2. The volatility of ammonia at 80 and 120°C is about 5% higher than is now predicted by the SWEQ model. This means that the SWEQ model will predict slightly more steam for a given separation, all other effects being equal, than will actually be required.

pH of NH_3-H_2S-CO_2-H_2O Mixtures. pH measurements on NH_3-H_2S-CO_2-H_2O mixtures have been made at 25°C and 80°C as outlined in Part B of the Measurement Program. The results of these measurements are given in Tables 4 to 22. Each table gives the composition of each solution at various dilutions and measured and correlated pH data at each composition. The SWEQ computer program (2) does not accurately predict these pH data, so an empirical correlation of the data was made in order to give comparisons between the data and a smoothing function. This smoothing function is based on the following equilibrium which is assumed to be a function of the concentration of the components in solution.

$$NH_3 + H^+ \longrightarrow NH_4^+ \qquad (8)$$

$$k = \frac{(NH_4^+)}{(NH_3)(H^+)} \qquad (9)$$

In Equation 9, the hydrogen ion concentration is given by the pH measurements; and the ratio of NH_3/NH_4^+ can be approximated from the moles of acid gas per mole of NH_3 assuming all of the acid gas reacts as follows.

$$\begin{array}{ccccccc} NH_3 & + & HA & \longrightarrow & NH_4^+ & + & A^- \\ n_{NH_3} - \alpha & & \alpha - \alpha & & \alpha & & \alpha \end{array} \quad , \quad A^- = HCO_3^- \text{ or } HS^- \qquad (10)$$

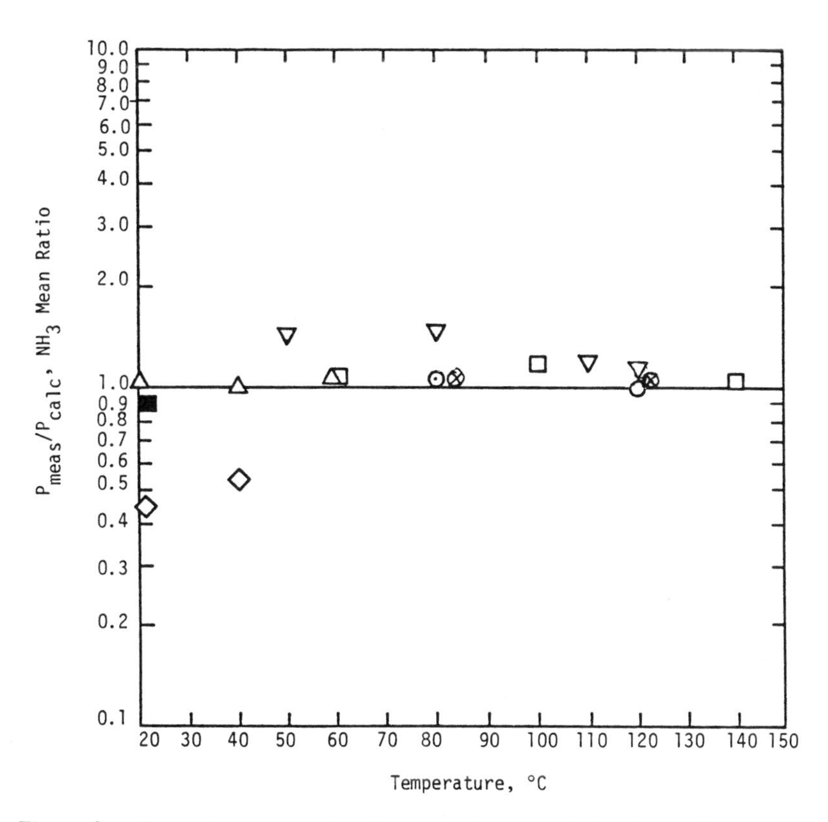

Figure 3. Ammonia mean ratio of measured over calculated partial pressures based on SWEQ correlation (⊗) new data; see *Figure 2 of Ref.* 1 *for the following: (○) Miles & Wilson; (□) Clifford; (△) Van Krevelen NH_3–CO_2; (▽) Cardon & Wilson; (■) Badger & Silver; (◇) Breitenback & Perman)*

TABLE 4. Measured and Correlated pH of Water-Ammonia Mixtures with no Acid Gas Present, 25°C

Ammonia[a)] wt %	Ratio NH_3/NH_4^+ (SWEQ)	pH Meas.	pH Correl.	pH Diff.
.00103	2.82	9.491	9.606	-.115
.00308	5.19	9.807	9.875	-.068
.00925	9.34	10.123	10.137	-.014
.0277	16.5	10.369	10.397	-.028
.111	33.4	10.726	10.733	-.007
.335	58.6	11.014	11.021	-.007
1.01	102	11.283	11.337	-.054
3.05	177	11.597	11.709	-.112
6.17	254	11.883	11.997	-.114
8.30	293	12.066	12.130	-.064
11.21	339	12.272	12.276	-.004
15.20	391	12.588	12.436	.152

[a)] Balance is water

TABLE 5. Measured and Correlated pH of Water-Ammonia Mixtures with 0.253 Mole of H_2S/Mole of NH_3, 25°C

Ammonia[a] wt %	H_2S[a] wt %	Ratio NH_3/NH_4^+ (SWEQ)	pH Meas.	pH Correl.	pH Diff.
.00135	.00068	1.69	9.303	9.385	-.082
.00404	.00205	2.26	9.414	9.517	-.103
.0121	.00612	2.66	9.495	9.604	-.109
.0364	.0184	2.85	9.567	9.643	-.076
.109	.0552	2.91	9.669	9.680	-.011
.328	.166	2.94	9.772	9.732	.040
.985	.499	2.95	9.896	9.818	.078
2.96	1.50	2.95	9.996	9.963	.033
5.97	3.02	2.96	10.083	10.108	-.025
8.90	4.50	2.96	10.155	10.216	-.061
10.50	5.31	2.96	10.196	10.267	-.071
14.11	7.14	2.96	10.230	10.370	-.140

[a] Balance is water

TABLE 6. Measured and Correlated pH of Water-Ammonia Mixtures with 0.491 Mole of H_2S/Mole of NH_3, 25°C

Ammonia[a] wt %	H_2S[a] wt %	Ratio NH_3/NH_4^+ (SWEQ)	pH		
			Meas.	Correl.	Diff.
.00407	.00400	.952	9.040	9.143	-.103
.0122	.0120	1.01	9.080	9.178	-.098
.0366	.0359	1.03	9.174	9.204	-.030
.110	.108	1.04	9.290	9.239	.051
.330	.324	1.04	9.405	9.292	.113
.990	.971	1.04	9.497	9.383	.114
2.97	2.91	1.05	9.599	9.546	.053
5.94	5.83	1.04	9.649	9.697	-.048
9.92	9.73	1.04	9.758	9.852	-.094
13.23	12.98	1.04	9.734	9.958	-.224

[a] Balance is water

TABLE 7. Measured and Correlated pH of Water-Ammonia Mixtures with 0.251 Mole of CO_2/Mole of NH_3, 25°C

Ammonia[a] wt %	CO_2[a] wt %	Ratio NH_3/NH_4^+ (SWEQ)	pH Meas.	pH Correl.	pH Diff.
.00180	.00117	1.66	9.343	9.379	-.036
.00457	.00296	1.98	9.438	9.461	-.023
.0116	.00750	2.18	9.463	9.512	-.049
.0295	.0191	2.27	9.509	9.543	-.034
.0748	.0485	2.29	9.573	9.569	.004
.190	.123	2.29	9.606	9.605	.001
.481	.312	2.25	9.644	9.653	-.009
1.22	.791	2.19	9.735	9.731	.004
1.75	1.13	2.17	9.768	9.775	-.007
5.72	3.71	2.07	10.005	9.988	.017
7.60	4.92	2.06	10.095	10.065	.030
10.1	6.53	2.04	10.210	10.153	.057
13.4	8.65	2.03	10.379	10.256	.123

[a] Balance is water

TABLE 8. Measured and Correlated pH of Water-Ammonia Mixtures with 0.520 Mole of CO_2/Mole of NH_3, 25°C

Ammonia[a)] wt %	CO_2[a)] wt %	Ratio NH_3/NH_4^+ (SWEQ)	pH		
			Meas.	Correl.	Diff.
.000297	.000400	.452	8.579	8.810	-.231
.000754	.00102	.592	8.789	8.929	-.140
.00191	.00257	.695	8.951	9.003	-.052
.00486	.00654	.746	9.028	9.040	-.012
.0123	.0166	.761	9.074	9.060	.014
.0313	.0421	.768	9.096	9.080	.016
.0794	.107	.748	9.105	9.096	.009
.201	.271	.704	9.127	9.112	.015
.661	.890	.597	9.138	9.133	.005
1.31	1.77	.507	9.162	9.146	.016
1.74	2.35	.466	9.163	9.154	.009
2.31	3.11	.428	9.176	9.168	.008
3.06	4.12	.388	9.189	9.184	.005
4.04	5.44	.349	9.210	9.204	.006
5.32	7.16	.314	9.240	9.233	.007
6.97	9.38	.280	9.277	9.268	.009
9.10	12.25	.248	9.329	9.311	.018
11.81	15.89	.220	9.430	9.366	.064

[a)] Balance is water

TABLE 9. Measured and Correlated pH of Water-Ammonia Mixtures with .755 Mole of CO_2/Mole of NH_3, 25°C

Ammonia[a)] wt %	CO_2[a)] wt %	Ratio NH_3/NH_4^+ (SWEQ)	pH		
			Meas.	Correl.	Diff.
.00238	.00463	.278	8.629	8.608	.021
.00579	.0113	.283	8.679	8.623	.056
.0141	.0274	.289	8.709	8.644	.065
.0342	.0664	.285	8.716	8.657	.059
.0828	.161	.271	8.725	8.664	.061
.201	.391	.244	8.718	8.664	.054
.486	.944	.201	8.698	8.650	.048
1.17	2.27	.145	8.641	8.616	.025
2.31	4.49	.0990	8.577	8.574	.003
4.50	8.74	.0632	8.500	8.549	-.049
5.90	11.46	.0513	8.472	8.545	-.073

[a)] Balance is water

TABLE 10. Measured and Correlated pH of Water-Ammonia Mixtures with 0.886 Mole of CO_2/Mole of NH_3, 25°C

Ammonia[a] wt %	CO_2[a] wt %	Ratio NH_3/NH_4^+ (SWEQ)	pH Meas.	pH Correl.	pH Diff.
.000902	.00207	.110	8.265	8.200	.065
.00219	.00501	.119	8.345	8.239	.106
.00532	.0122	.119	8.393	8.247	.146
.0129	.0295	.122	8.400	8.270	.130
.0314	.0719	.118	8.391	8.274	.117
.0761	.174	.113	8.400	8.285	.115
.185	.423	.0996	8.382	8.276	.106
.446	1.02	.0789	8.335	8.245	.090
1.07	2.46	.0521	8.233	8.174	.059
2.11	4.83	.0346	8.131	8.120	.011
2.78	6.37	.0281	8.084	8.093	-.009

[a] Balance is water

TABLE 11. Measured and Correlated pH of Water-Ammonia Mixtures with .124 Mole of CO_2 and .124 Mole of H_2S per Mole of NH_3, 25°C

Ammonia[a)] wt %	CO_2[a)] wt %	H_2S[a)] wt %	Ratio NH_3/NH_4^+ (SWEQ)	pH Meas.	pH Correl.	pH Diff.
0.00298	0.000954	0.000737	2.003	9.248	9.463	-0.215
0.00724	0.00232	0.00179	2.311	9.435	9.531	-0.096
0.0176	0.00563	0.0435	2.494	9.489	9.574	-0.085
0.0427	0.0137	0.0106	2.568	9.476	9.601	-0.125
0.104	0.0332	0.0256	2.623	9.546	9.634	-0.088
0.252	0.0806	0.0623	2.622	9.631	9.670	-0.039
0.612	0.196	0.151	2.613	9.733	9.724	0.009
1.49	0.476	0.367	2.603	9.845	9.811	0.034
3.61	1.16	0.893	2.562	9.975	9.941	0.034
7.24	2.32	1.79	2.556	10.116	10.100	0.016
9.67	3.09	2.39	2.560	10.194	10.185	0.009
12.91	4.13	3.19	2.554	10.298	10.281	0.017

[a)] Balance is water

TABLE 12. Measured and Correlated pH of Water-Ammonia Mixtures with .277 Mole of CO_2 and .277 Mole of H_2S per Mole of NH_3, 25°C

Ammonia[a)] wt %	CO_2[a)] wt %	H_2S[a)] wt %	Ratio NH_3/NH_4^+ (SWEQ)	pH Meas.	pH Correl.	pH Diff.
0.00136	0.000964	0.000753	0.635	8.576	8.961	-0.385
0.00363	0.00257	0.00201	0.704	8.745	9.011	-0.266
0.00967	0.00686	0.00535	0.732	8.877	9.036	-0.159
0.0258	0.0183	0.0143	0.740	8.934	9.055	-0.121
0.0687	0.0488	0.0382	0.730	9.002	9.072	-0.070
0.183	0.130	0.102	0.706	9.085	9.095	-0.010
0.408	0.346	0.271	0.682	9.150	9.127	-0.023
1.296	0.919	0.720	0.587	9.209	9.177	0.032
3.422	2.428	1.900	0.499	9.293	9.273	0.020
6.443	4.571	3.577	0.447	9.376	9.388	-0.012
8.380	6.013	4.653	0.417	9.416	9.443	-0.027
10.87	7.800	6.035	0.399	9.479	9.519	-0.040

[a)] Balance is water

TABLE 13. Measured and Correlated pH of Water-Ammonia Mixtures with 0.375 Mole of H_2S and 0.372 Mole of CO_2 per Mole of NH_3, 25°C

Ammonia[a] wt %	CO_2[a] wt %	H_2S[a] wt %	Ratio NH_3/NH_4^+ (SWEQ)	pH Meas.	pH Correl.	pH Diff.
0.00280	0.00269	0.00210	0.3074	8.542	8.650	-0.108
0.00680	0.00653	0.00509	0.3176	8.598	8.671	-0.073
0.0165	0.0159	0.0124	0.3155	8.643	8.679	-0.036
0.0401	0.0385	0.0300	0.3183	8.686	8.700	-0.014
0.0973	0.0934	0.0729	0.3091	8.732	8.714	0.018
0.0236	0.227	0.177	0.2886	8.768	8.647	0.121
0.0572	0.549	0.428	0.2573	8.774	8.617	0.157
1.38	1.33	1.034	0.2055	8.764	8.746	0.018
3.30	3.17	2.47	0.1514	8.740	8.772	-0.032
5.92	5.69	4.44	0.1145	8.719	8.805	-0.086
7.56	7.27	5.67	0.1003	8.715	8.827	-0.112

[a] Balance is water

TABLE 14. Measured and Correlated pH of Water-Ammonia Mixtures with No Acid-Gas Present, 80°C

Ammonia[a)] wt %	Ratio NH_3/NH_4^+ (SWEQ)	pH Meas.	pH Correl.	pH Diff.[b)]
0.00202	4.4484	8.077	8.138	-.061
0.0121	11.565	8.432	8.636	-.204
0.0273	17.572	8.636	8.881	-.245
0.0614	26.638	8.826	9.143	-.317
0.221	51.020	9.156	9.597	-.441
0.887	102.43	9.526	10.127	-.601
1.79	145.48	9.717	10.396	-.679
3.61	208.46	9.865	10.662	-.797

[a)] Balance is water

[b)] These differences indicate that the NH_3/NH_4^+ ratio from the SWEQ computer program are too high.

TABLE 15. Measured and Correlated pH of Water-Ammonia Mixtures with 0.258 Mole of H_2S/Mole of NH_3, 80°C

Ammonia[a)] wt %	H_2S[a)] wt %	Ratio NH_3/NH_4^+ (SWEQ)	pH Meas.	pH Correl.	pH Diff.
0.00292	0.00151	2.150	7.805	7.840	-.035
0.00709	0.00366	2.504	7.966	7.950	.016
0.0172	0.00889	2.709	8.057	8.045	.012
0.0418	0.0216	2.818	8.208	8.147	.061
0.102	0.0524	2.869	8.323	8.265	.058
0.246	0.127	2.891	8.470	8.400	.070
0.600	0.309	2.898	8.575	8.550	.025
1.46	0.756	2.894	8.675	8.699	-.024
3.58	1.85	2.891	8.777	8.836	-.059

[a)] Balance is water

TABLE 16. Measured and Correlated pH of Water-Ammonia Mixtures with 0.480 Mole of H_2S/Mole of NH_3, 80°C

Ammonia[a)] wt %	H_2S[a)] wt %	Ratio NH_3/NH_4^+ (SWEQ)	pH		
			Meas.	Correl.	Diff.
0.00185	0.00178	.948	7.511	7.472	.039
0.00468	0.00451	1.041	7.618	7.553	.065
0.0119	0.0114	1.085	7.724	7.631	.093
0.0302	0.0291	1.105	7.863	7.722	.141
0.0767	0.0738	1.113	7.952	7.836	.116
0.195	0.187	1.115	8.062	7.975	.087
0.494	0.477	1.111	8.152	8.128	.024
1.26	1.22	1.107	8.235	8.283	-.048
3.21	3.12	1.096	8.324	8.422	-.098

[a)] Balance is water

TABLE 17. Measured and Correlated pH of Water-Ammonia Mixtures with .555 Mole of H_2S/Mole of NH_3, 80°C

Ammonia[a] wt %	H_2S[a] wt %	Ratio NH_3/NH_4^+ (SWEQ)	pH		
			Meas.	Correl.	Diff.
.0935	.104	0.8439	7.809	7.750	.059
.187	.208	0.8453	7.853	7.855	-.002
.375	.416	0.8453	7.874	7.970	-.096
.752	.834	0.8453	7.916	8.088	-.172
1.55	1.70	0.8628	7.985	8.215	-.230
3.07	3.89	0.6293	8.094	8.187	-.093

[a] Balance is water

TABLE 18. Measured and Correlated pH of Water-Ammonia Mixtures with .315 Mole of CO_2/Mole of NH_3, 80°C

Ammonia[a)] wt %	CO_2[a)] wt %	Ratio NH_3/NH_4^+ (SWEQ)	pH Meas.	pH Correl.	pH Diff.
0.00175	0.00142	1.572	7.830	7.687	.143
0.00425	0.00346	1.845	7.927	7.792	.135
0.0103	0.00840	2.011	8.048	7.880	.168
0.0878	0.0714	2.108	8.272	8.118	.154
0.213	0.173	2.075	8.405	8.242	.163
0.515	0.420	1.989	8.496	8.371	.125
1.25	1.02	1.853	8.591	8.491	.100
3.01	2.46	1.665	8.692	8.584	.108
5.98	4.95	1.487	8.830	8.630	.200

[a)] Balance is water

TABLE 19. Measured and Correlated pH of Water-Ammonia Mixture with .429 Mole of CO_2/Mole of NH_3, 80°C

Ammonia[a] wt %	CO_2[a] wt %	Ratio NH_3/NH_4^+ (SWEQ)	pH Meas.	pH Correl.	pH Diff.
0.00244	0.00269	1.164	7.557	7.571	-.013
0.00548	0.00605	1.260	7.601	7.643	-.042
0.0123	0.0136	1.309	7.654	7.712	-.058
0.0278	0.0306	1.327	7.742	7.789	-.047
0.166	0.183	1.297	7.950	8.011	-.061
0.373	0.412	1.243	8.039	8.125	-.086
1.33	1.46	1.079	8.141	8.282	-.141
2.63	2.96	.908	8.210	8.316	-.106
5.13	6.08	.690	8.308	8.293	.015

[a] Balance is water

TABLE 20. Measured and Correlated pH of Water-Ammonia Mixtures with .121 Mole of H_2S and .126 Mole of CO_2 per Mole of NH_3, 80°C

Ammonia[a] wt %	CO_2[a] wt %	H_2S[a] wt %	Ratio NH_3/NH_4^+ (SWEQ)	pH Meas.	pH Correl.	pH Diff.
0.00178	0.000581	0.000430	1.962	7.737	7.783	-.046
0.00401	0.00131	0.000968	2.051	7.794	7.834	-.041
0.00902	0.00294	0.00218	2.675	7.816	7.992	-.176
0.0542	0.0176	0.0131	2.954	8.124	8.196	-.072
0.122	0.0397	0.0295	2.977	8.264	8.306	-.042
0.275	0.0895	0.0664	2.967	8.416	8.429	-.013
0.765	0.252	0.185	2.889	8.609	8.589	.020
2.78	0.925	0.670	2.740	8.787	8.777	.010
5.70	1.90	1.39	2.622	8.901	8.858	.043

[a] Balance is water

TABLE 21. Measured and Correlated pH of Water-Ammonia Mixtures with .188 Mole of H_2S and .225 Mole of CO_2 per Mole of NH_3, 80°C

Ammonia[a)] wt %	CO_2[a)] wt %	H_2S[a)] wt %	Ratio NH_3/NH_4^+ (SWEQ)	pH Meas.	pH Correl.	pH Diff.
0.00439	0.00236	0.00165	1.481	7.579	7.700	-.120
0.00988	0.00530	0.00372	1.573	7.607	7.773	-.166
0.0593	0.0318	0.0223	1.635	7.833	7.962	-.129
0.133	0.0716	0.0502	1.643	8.020	8.074	-.054
0.300	0.161	0.113	1.608	8.158	8.193	-.035
1.08	0.581	0.406	1.522	8.336	8.387	-.051
2.16	1.26	0.813	1.214	8.408	8.406	.002
4.33	2.56	1.65	1.097	8.465	8.463	.002
8.68	5.28	3.50	0.933	8.596	8.480	.116

[a)] Balance is water

TABLE 22. Measured and Correlated pH of Water-Ammonia Mixtures with .363 Mole of H_2S and .328 Mole of CO_2 per Mole of NH_3, 80°C

Ammonia[a] wt %	CO_2[a] wt %	H_2S[a] wt %	Ratio NH_3/NH_4^+ (SWEQ)	pH		
				Meas.	Correl.	Diff.
0.988	0.838	0.717	.4244	7.903	7.851	.052
1.97	1.71	1.45	.3584	7.926	7.889	.037
3.92	3.45	2.93	.2903	7.970	7.897	.073

[a] Balance is water

$$\text{free } NH_3 = n_{NH_3} - \alpha$$

$$NH_4^+ = \alpha$$

$$\frac{\text{free } NH_3}{NH_4^+} = \frac{n_{NH_3} - \alpha}{\alpha} = \frac{1 - \frac{\alpha}{n_{NH_3}}}{\frac{\alpha}{n_{NH_3}}} \tag{11}$$

$$\frac{NH_3}{NH_4^+} = \frac{1 - R}{R}, \quad R = \frac{\alpha}{n_{NH_3}} = \text{moles acid gas per mole } NH_3 \tag{12}$$

This estimated ratio can be improved by using the SWEQ computer program which takes the ionization of NH_3 into account at low concentrations. For example in Table 5, Equation 12 gives a NH_3/NH_4^+ ratio of 2.95. This agrees with the SWEQ program at high concentrations, but at low concentrations the ratio decreases due to the ionization of ammonia. In order to take this ionization effect into account, the SWEQ computer program has been used to calculate the NH_3/NH_4^+ ratios in each table for use in the smoothing function. These calculated ratios are therefore given in each table, and the smoothing function now involves a correlation of the equilibrium constant given by Equation 9 as a function of the concentrations of the components. The resulting equations are the following.

$$\log_{10}(k) = A + \frac{B\sqrt{(1 + R)(\text{ wt \% } NH_3)}}{C + D\sqrt{(1 + R)(\text{wt \% } NH_3)}} \tag{13}$$

where A, B, C, and D = constants

$$R = \frac{NH_4^+}{NH_4^+ + NH_3} \text{ from SWEQ program}$$

$$R = \text{moles acid gas/mole } NH_3$$

$$\text{wt \% } NH_3 = \text{total weight percent of ammonia in soln.}$$

Values of the parameters were found to be as follows.

Temp. °C	Parameter A	B	C	D
25	9.15	.178	1	0
80	7.42	1.36	.9	1

The first parameter, A, is the pK_a of NH_3 which can be compared with literature data; the comparison is as follows.

Temp. °C	pK_a Meas.	Lit.(2,4)	Difference
25	9.15	9.25	-.10
80	7.42	7.84	-.42

From this comparison the agreement is quite good at 25°C. At 80°C the agreement is not as good, but the larger difference is not surprising because both the measurements reported here and measurements in the literature suffer from measurement problems at higher temperatures. However, it would be surprising if the data reported here are off by 0.4 unit because NBS recommended buffer solutions accurate to $\pm$ 0.01 pH unit over the temperature range to 80°C were used. If the data given here are right, it probably means that the dissociation constant of ammonia at 80°C is about 0.65×10^{-5} rather than 1.66×10^{-5} computed from the equation of Edwards and Prausnitz (1).

Differences between measured and correlated pH data are given in each of Tables 4 to 22. No attempt was made to do a least-square fit of the data so in some cases deviations between the correlation and the data can be rather large without any significant error in the measured data. Nevertheless, most of the deviations are less than $\pm$ 0.1 pH unit. One exception is data in Table 14 on the pH of NH_3-H_2O at 80°C with no acid gas present; in this case the calculated pH data depend heavily on the NH_3/NH_4^+ ratio given by the SWEQ model. Differences up to 0.797 pH unit indicate the magnitude of error in the SWEQ model at 80°C. Similar data at 25°C in Table 3 are better predicted.

The data in Tables 4 to 13 at 25°C plot nearly as vertical lines independent of the wt % NH_3 in solution except for pure ammonia. At low concentrations the curves tend to deviate because of the ionization of NH_3 and water. If the amount of ammonia in solution is approximately known, these curves can be used to estimate the moles of acid gas per mole of NH_3 with fairly good accuracy. At 80°C the curves tend to slant more so the amount of ammonia in solution would have to be known more accurately before an estimate of the ratio of acid gas to ammonia could be made. Also we do not recommend the use of a pH probe at 80°C as a control indicator because the response of the probe is more erratic, and precise data are difficult to obtain. The use of an indicator probe at 25°C seems more logical because the output is more stable.

One observation that can be made from the pH data in Tables 4 to 22 is that the pH appears to be affected about equally by either H_2S or CO_2. This result can be seen from an examination of the plots given in Figures 4 and 5. In these figures the pH is plotted versus the moles of acid gas per mole of NH_3 at a fixed concentration of 1.0 wt % NH_3. The curves in Figure 4 correspond to results at 25°C, and the plot in Figure 5 corresponds to results at 80°C. The points plotted in these two figures represent smoothed values obtained from Tables 4 to 22. Data at 25°C in Figure 4 clearly separate into three distinct curves although the curves are close together. This shows that there is some difference between the acid gases, but not a large difference. The points in Figure 5 at 80°C do not resolve into separate curves because there is some scatter in the points.

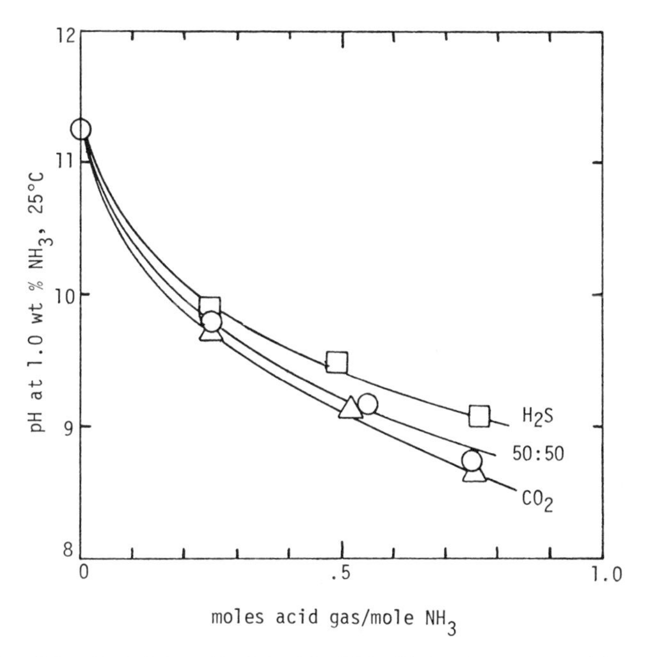

Figure 4. The pH of 1.0 wt % H_2S–CO_2–NH_3–H_2O mixtures vs. acid gas loading at 25°C (smoothed points: (△) 100% CO_2; (□) 100% H_2S; (○) 50:50 H_2S/CO_2)

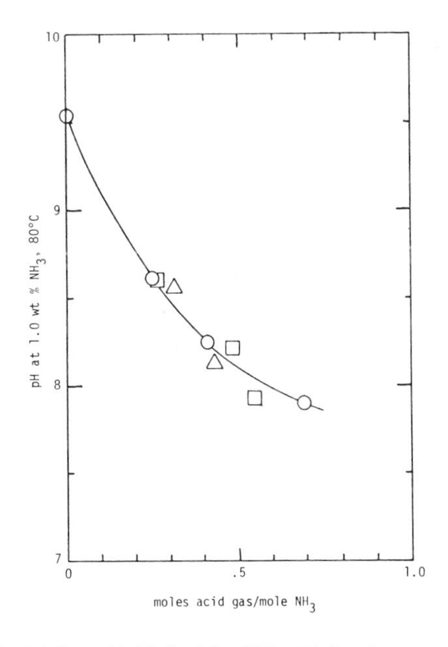

Figure 5. The pH of 1.0 wt % H_2S–CO_2–NH_3–H_2O mixtures vs. acid gas loading at 80°C ((△) 100% CO_2; (□) 100% H_2S; (○) 50:50 H_2S/CO_2)

Within this scatter, the three types of data appear to be correlated by a single curve; thus showing that there is not a great difference in the effect of the acid gases at 80°C.

In summary the following can be concluded regarding the pH data in Tables 4 to 22.

1. The data can be correlated using an equilibrium constant approach with most deviations being less than $\pm$ 0.1 pH unit.
2. The pK_a of ammonia derived from these data agrees with literature data within 0.1 pH unit at 25°C and 0.42 pH unit at 80°C.
3. The molar ratios of H_2S/NH_3 or CO_2/NH_3 have about equal effects on the pH. At 25°C, CO_2 appears to have a slightly greater effect. A similar conclusion at 80°C can be made except that no distinction between the acid gases is visible because of some scatter in the data.
4. The pH probe is more stable at 25°C than at 80°C. For this reason it is recommended that if a pH probe is used as a control sensor in a process that it be used at 25°C so that the response will tend to be more accurate.
5. pH data at 25°C in Tables 4 to 13 can be used to estimate the moles of acid gas per moles of NH_3 in solution with only an approximate knowledge of the total amount of NH_3 in solution.

Sodium Hydroxide and Sodium Acetate Effect on NH_3 Volatility. Ammonia volatility measurements at various concentrations of sodium hydroxide and sodium acetate at 80°C are given in Tables 23 and 24 respectively. Data on the effect of sodium hydroxide were measured using an ammonia probe from Orion Research Company. These measurements were reduced to NH_3 partial pressure measurements using Equation 2 given above. In this case, the reference solution is the composition reached at zero salt concentration at the bottom of Table 23. The ammonia partial pressure above this solution can be inferred from Table 1 where P_{NH_3}/wt % NH_3 at 0.1 wt % NH_3 has a value of 1.23. This information gives the following values for $P^{ref.}_{NH_3}$ and $mv^{ref.}$

$$P^{ref.}_{NH_3} = 0.1005 \times 1.23 = 0.124 \text{ psia}$$

$$mv^{ref.} = -57.3 \text{ mv}$$

These values and the probe output data in Table 23 were substituted into Equation 1 to give the partial pressure data given in Table 23.

The data in Table 24 on the effect of sodium acetate were measured in the same manner as data in Tables 1 and 2, so no assumptions about the ammonia probe behavior were necessary. This was done because readings from the NH_3 probe became erratic after

TABLE 23. Effect of Sodium Hydroxide Electrolyte on Ammonia Volatility at 80°C from Ammonia Probe Data[a)]

wt % in Liquid		NH_3 Probe Output	P_{NH_3}	P_{NH_3}/wt %
NH_3	NaOH	mv	psia	psia
.0800	22.5	-57.1	.123	1.54
.0891	12.7	-57.5	.125	1.40
.0944	6.82	-57.3	.124	1.31
.0974	3.60	-57.2	.124	1.27
.0995	1.27	-57.3	.124	1.25
.1005	0	mv^{ref} = -57.3[b)]	$P_{NH_3}^{ref}$ = .124[c)]	1.23

[a)]Orion Research Company

[b)]This value of -57.3 mv was not directly measured, but is a value obtained by extrapolating mv vs wt % NaOH to zero wt % NaOH.

[c)]This partial pressure for NH_3 was obtained from the NH_3 partial pressure data in Table 2 at 0.1 wt % NH_3. The same ratio of P_{NH_3}/wt % NH_3, uncorr. was used at .1005 wt % NH_3 to give P_{NH_3} as follows.

$$P_{NH_3} = 1.23 \times .1005 = .124 \text{ psia}$$

TABLE 24. Effect of Sodium Acetate Electrolyte on Ammonia Volatility at 80°C from Flow Cell Data

charge, wt % NH_3	charge, wt % NaAc	NH_3 in Liquid wt %	P_{NH_3} [a] psia	P_{NH_3}/wt %
.802	25	.781	1.838	2.35
		.734	1.778	2.42
		.684	1.712	2.50
		.644	1.620	2.51
	average	.711	1.737	2.45
.931	15	.992	1.812	1.83
		.975	1.817	1.86
		.959	1.796	1.87
		.942	1.756	1.86
	average	.967	1.795	1.86
.931	5	.995	1.381	1.39
		.985	1.411	1.43
		.974	1.405	1.44
		.964	1.380	1.43
	average	.980	1.394	1.42
1.00	0	.973	1.246	1.28 average from Table 2

[a] Sampled at 20 psia

the measurements at 80°C on the effect of sodium hydroxide. These data were measured at about 1 wt % NH_3 in the liquid phase instead of 0.1 wt % NH_3 used for the NaOH measurements. For this reason they do not extrapolate to the same volatility ratio at zero concentration of electrolyte. When correction is made for ionization and solvent effects of ammonia, then the two intercepts agree.

From the data in Tables 23 and 24, the following can be concluded.

1. The volatility of ammonia can be significantely affected by high concentrations of dissolved ions in the liquid phase. In sodium acetate the volatility increases by a factor of 1.9 at 25 wt % of salt. In sodium hydroxide the volatility is enhanced to a lesser degree with an increase of 1.25 at 22.5 wt % NaOH. Both electrolytes produce ions with only one electronic charge, but their effects on the volatility of ammonia are significantly different. Thus the effects of various ionic components must be studied individually in order to determine their effect on the volatility of NH_3.
2. At the low ionic concentrations encountered in sour water strippers, the effect of dissolved ions is probably small. Thus at a 1% concentration of sodium acetate the volatility of ammonia only increases about 2.5% due to the salt. This is within the prediction accuracy of the ammonia volatility data and no correction is therefore required. However significant ionic effects could exist in the condenser where high concentrations of the ionic components could exist.

Summary and Conclusions

Ammonia partial pressure data have been determined at concentrations from 10 ppm up to 5 wt % in water at temperatures of 80 and 120°C. The pH of NH_3-H_2S-CO_2-H_2O mixtures have also been measured at 25 and 80°C. Also the effects of sodium hydroxide and sodium acetate on ammonia volatility data have been measured at 80°C. Various conclusions made from the data are as follows.

1. The volatility of NH_3 at ppm levels can be calculated from the ionization constant of ammonia combined with volatility data measured at higher concentrations of ammonia.
2. The volatility of ammonia at 80 and 120°C is about 5% higher than is now predicted by the SWEQ model.
3. The volatility of ammonia can be significantly affected by high concentrations of dissolved ions in the liquid phase. Unfortunately the effect varies depending on the types of ions in solution. Sodium acetate produces a

much larger enhancement in the volatility than is produced by sodium hydroxide.

4. At the low ionic concentrations encountered in sour water strippers, the effect of dissolved ions is probably small. A 1% concentration of sodium acetate only enhances the volatility of ammonia by about 2.5%. High ionic concentrations in the condenser may produce bigger effects.
5. Measured pH data on NH_3-H_2S-CO_2-H_2O mixtures can be correlated using an equilibrium constant approach with most deviations being less than $\pm$ 0.1 pH unit.
6. The pK_a of ammonia derived from the pH data agrees with literature data within 0.1 pH unit at 25°C. At 80°C the agreement is not as good with a difference of 0.42 pH unit.
7. H_2S and CO_2 have about equal effects on the pH of H_2S-CO_2-NH_3-H_2O mixtures with CO_2 showing a slightly greater effect at 25°C.
8. pH measurements at 25°C tend to be more reliable than data at higher temperatures because the output from the probe is more stable at 25°C. For this reason it is recommended that if a pH probe is used as a control sensor in a process that it be used at 25°C.
9. Measured pH data on H_2S-CO_2-NH_3-H_2O mixtures at 25°C can be used to estimate the moles of acid gas per mole of ammonia in solution with only an approximate knowledge of the total amount of ammonia in solution.
10. Certain types of reference electrodes are found to be more suitable than others when determining the pH of solutions containing H_2S. Recommendations regarding various electrodes are given in the text of the paper.

Acknowledgment

This work was supported by the Gas Processors Association and the American Petroleum Institute.

The conscientious work of Kent C. Wilson in doing many of the potentiometric titrations and Katie Toepke for typing this report is appreciated.

Literature Cited

1. Edwards, T. J. and Prausnitz, J. M. A.I.Ch.E. Journal, 1975, 21, (2), 248-259.
2. Wilson, G. M. "A New Correlation of NH_3, CO_2, and H_2S Volatility Data from Aqueous Sour Water Systems", American Petroleum Institute, February 1978.
3. Miles, D. H. and Wilson, G. M. "Vapor-Liquid Equilibrium Data for Design of Sour Water Strippers", Annual Report to the American Petroleum Institute for 1974, October 1975 (Data in this report are also summarized in reference 2).
4. CRC Handbook, 51st Edition, page D-122 (1970-1971).

RECEIVED February 27, 1980.

10

Predicting Vapor-Liquid-Solid Equilibria in Multicomponent Aqueous Solutions of Electrolytes

JOSEPH F. ZEMAITIS, JR.

OLI Systems, Inc., 15 James Street, Morristown, NJ 07960

It is evident from the title of this symposium that as a result of recent requirements to reduce pollutant levels in process wastewater streams, improved techniques for predicting the vapor-liquid-solid equilibria of multicomponent aqueous solutions of strong and/or weak electrolytes are needed. In addition to the thermodynamic models necessary for such predictions, tools have to be developed so that the engineer or scientist can use these thermodynamic models correctly and with relative ease.

Within the past few years the advances made in hydrocarbon thermodynamics combined wtih increased sophistication in computer software and hardware have made it quite simple for engineers to predict phase equilibria or simulate complex fractionation towers to a high degree of accuracy through software systems such as SSI's PROCESS, Monsanto's FLOWTRAN, and Chemshare's DISTILL among others. This has not beem the case for electrolyte systems.

For processes involving electrolytes either directly or indirectly, the techniques used have relied heavily on correlations of limited data which are often embedded into design techniques where inexact extrapolations are the rule rather than the exception. Hampering the improvement of design tools to a level comparable to that for hydrocarbons have been several restrictions such as:

o The lack of good simple data for strong and/or weak electrolytes.

o Until recently, there was no need to reduce pollutant levels to the ppm range.

o A lack of understanding of electrolyte thermodynamics with regards to the need to satisfy the ionic equilibria and electroneutrality expressions for the particular set of species of interest. This one point has negated many design approaches proposed in the past.

0-8412-0569-8/80/47-133-227$05.00/0

- The lack of a suitable thermodynamic framework for electrolytes over a wide range of concentrations and temperatures.

- Mistaken approaches to data gathering, leading to the collection of considerable data in doped solutions to hold ionic strength constant.

Recognizing these problems while forseeing the need for a new tool useful for aqueous systems by users in a wide variety of applications we began in 1974 to forge the proper tools. We were fortunate to start at that time and not five years sooner since it was during the early 1970's when a resurgence in work in the area of electrolyte thermodynamics started being published by several of our symposium's distinguished participants.

In this paper I will discuss the tools we have developed and are continuing to develop and enhance. In particular, I will describe the basic thermodynamic framework which we have synthesized and how this framework is implemented into a computer software system that:

- Predicts the vapor-liquid-solid equilibrium of multi-component aqueous systems.

- Allows the user to expand the system's data bank through a series of programs to reduce raw data and regress to fit the thermodynamic framework of the system.

- Allows the user to further expand the data available through a series of estimating and/or extrapolating programs when little or no data are available.

- Solves complex systems and predicts results as a function of temperature or concentrations in order to develop phase diagrams or study the effects of various levels of certain constituents.

- Incorporates the thermodynamic framework and associated data estimation techniques into multistage multicomponent fractionation tower software for the design and/or simulation of sour water strippers, nitric acid towers, amine scrubbers or any other tower where ionic equilibria and/or reactions occur.

To illustrate the power of the general purpose tools we have developed, I will describe the application of our software to two systems. First, since most of the participants in this session are using it as an example, and because it is important, I will give some of our results for the NH_3-CO_2-H_2O system. Secondly, to illustrate the prediction of combined vapor-liquid-solid equilibria and the effect of high ionic strengths, I will present some of our results for an interesting system, that of $FeCl_2$-HCl-H_2O.

In referring to the computer software system which I developed, the acronym ECES will be used. ECES signifies the Equilibrium Composition of Electrolyte Systems.

Thermodynamic Framework

In developing the thermodynamic framework for ECES, we attempted to synthesize computer software that would correctly predict the vapor-liquid-solid equilibria over a wide range of conditions for multicomponent systems. To do this we needed a good basis which would make evident to the user the chemical and ionic equilibria present in aqueous systems. We chose as our cornerstone the law of mass action which simply stated says: "The product of the activities of the reaction products, each raised to the power indicated by its numerical coefficient, divided by the product of the activities of the reactants, each raised to a corresponding power, is a constant at a given temperature."

If the user writes a simple mass balance for the solubility equilibrium or ionic equilibrium such as - the reaction of b moles of B with c moles of C has come to equilibrium with the product of d moles of D and e moles of E, then the mass balance is given by

$$bB + cC = dD + eE \tag{1}$$

and the law of mass action states at equilibrium

$$a_D^d \, a_E^e \, / \, a_B^b \, a_C^c = K \tag{2}$$

where K is the thermodynamic equilibrium constant for the temperature of interest. The thermodynamic equilibrium constant can be readily calculated at any temperature if we can evaluate the standard free energy of formation for the reaction expressed by equation (1) at the temperature desired.

Such computations can be done rather easily if we have available information on the standard free energy of formation, the heat of formation, and standard heat capacities for each of the constituents of our equilibrium equation. A considerable amount of this type of data is available in compilations published by the National Bureau of Standards (1) and others. Such information has been stored within a database in the ECES system. Then, using user input in the form of equation (1), ECES writes the expression for computing the thermodynamic equilibrium constant as a function of temperature to a file where it will eventually become part of a program to solve the many equilibria that might describe a complex system.

In order to describe fully the system, methods for evaluating the activities of the various species in equation (2) must be used. For an electrolyte solution the activity of an individual ionic species is given by

$$a_i = \gamma_i m_i \tag{3}$$

where the activity of the species is based on the hypothetical ideal solution of unit molality. The molality of the ion i is defined by m_i equal to the number of gram-moles of ion in 1000 grams of water.

For a pure salt dissolved in water it is not feasible to determine the activity coefficient, γ_i, for the cation and anion separately so that the mean activity coefficient concept has been defined for a salt's cation-anion pair as

$$\gamma_{\pm} = (\gamma_{+}^{\nu+} + \gamma_{-}^{\nu-})^{\frac{1}{\nu+ + \nu-}} \tag{4}$$

The mean activity coefficient is the standard form of expressing electrolyte data either in compilations of evaluated experimental data such as Hamer and Wu (2) or in predictions based on extensions to the Debye-Huckel model of electrolyte behavior. Recently several advances in the prediction and correlation of mean activity coefficients have been presented in a series of papers starting in 1972 by Pitzer (3), Meissner (4), and Bromley (5) among others.

As the basis for predicting ionic activity coefficients we chose to adopt an empirical modification of Bromley's (5) extension of the Debye-Huckel model. The mean activity coefficient of a pure salt in water is given by

$$\log \gamma_{\pm} = \frac{-A|z^{+}z^{-}|I^{1/2}}{1 + I^{1/2}} + \frac{(0.06+0.6B)|z^{+}z^{-}|}{\left(1 + \frac{1.5I}{|z^{+}z^{-}|}\right)^{2}} + BI + CI^{2} + DI^{3} \tag{5}$$

where A is the Debye-Huckel parameter, a function of temperature, and B, C, D are empirical coefficients which are functions of temperature and of the salt in question. We chose the Bromley form due to the fact that Bromley's version of equation (5) which neglected the polynomial terms established a basis for predicting the coefficient B for salts where no data was available that would give reasonable results to ionic strengths, I, of about 6 molal.

The typical system for which the equilibrium composition is desired however does not contain a single salt in solution but more usually the equivalent of several salts in solution. In addition, the activities required in equilibrium expressions arising from the law of mass action are single ion activities or in general, single ion activity coefficients. And, we are interested in the ionic activity coefficeint of each species in a multicomponent system.

As our basis we chose the Bronsted-Guggenheim (6) equation for the mean activity coefficient $\gamma_{R,X}$ of the electrolyte in a solution of several electrolytes having cation R and anion X. The mean activity coefficient $\gamma_{R,X}$ is given by

$$\log\gamma_{R,X} = \frac{-A\,|z^+z^-|\,I^{1/2}}{1 + I^{1/2}} + \frac{\nu^+}{\nu^+ + \nu^-} \Sigma_{x^1}\,\beta_{R,x^1}\, m_{x^1} + \frac{\nu^-}{\nu^+ + \nu^-} \Sigma_{R^1}\,\beta_{R^1,x^1}\, m_{R^1} \qquad (6)$$

If we take the logarithm of equation (4) and apply the rules of logs we obtain

$$(\nu^+ + \nu^-)\log\gamma_{R,x} = \nu^+ \log R + \nu^- \log\gamma_x \qquad (7)$$

Which if we combine with equation (6) we can develop a definition of the activity coefficient of a single ion in a multi-ion solution as

$$\log_R = \frac{-A|z^+|\,I^{1/2}}{1+I^{1/2}} + \Sigma_{x^1}\,\beta_{R,x^1}\, m_{x^1} \qquad (8)$$

A common failing of earlier work on multicomponent solutions was to consider the $\beta_{R,x}$ to be a constant for each salt. As Bromley (5) pointed out, the equations developed from the Bronsted-Guggenheim approach to multicomponent equations are only exact if $\beta_{R,x}$ is constant, the basic equations are reasonable exact if $\beta_{R,x}$ is a slowly varying function of concentration and Harned's Rule (7) for mixed electrolytes through an empirical correlation does show such of an effect.

For ECES as our basis of developing a concentration dependence for the interaction coefficient we equated equation (5) to the alternate expression

$$\log\gamma_{Rx} = \frac{-A|z^+z^-|\,I^{1/2}}{1 + I^{1/2}} + \beta_{Rx} I \qquad (9)$$

and developed the following expression for the interaction coefficient of a cation-anion pair:

$$\beta_{Rx} = \frac{(0.06 + 0.6B)|z^+z^-|}{\left(1 + \frac{1.5\,I}{|z^+z^-|}\right)^2} + B_{Rx^1} + C_{Rx} I + D_{Rx} I^2 \qquad (10)$$

which is used in Bromley's form of the Bronsted-Guggenheim equation

$$\log \gamma_{Rx} = \frac{-A|z^+z^-| I^{1/2}}{1 + I^{1/2}} + \frac{\sqrt{R}}{\sqrt{R} + \sqrt{x}} \sum_{x^1} \beta_{R,x^1} \overline{z_{R,x^1}}^2 m_{x^1}$$

$$+ \frac{\sqrt{x}}{\sqrt{R} + \sqrt{x}} \sum_{R^1} \beta_{R^1,x} \overline{z}_{R^1,x}^2 m_{R^1} \qquad (11)$$

where $z_{R,x} = \frac{|z_R + z_x|}{2}$. This is our basic expression for the activity coefficient of an ion in a multicomponent solution, the result for the activity coefficient of a cation is given by

$$\log\gamma_R = \frac{-A|z_R^2| I^{1/2}}{1 + I^{1/2}} + \sum_{x^1} \beta_{R,x^1} \overline{z_{R,x^1}}^2 m_{x^1} \qquad (12)$$

where the β_{R,x^1} are given by equation (10) and are functions of concentration and temperature.

The activity of the solvent (water) in a solution of pure electrolyte dissolved in water can be computed by application of the Gibbs-Duhem equation:

$$-55.51 \; d(\ln a_w) = (\nu_+ + \nu_-)m \; d(\ln(\gamma m)) \qquad (13)$$

which can be integrated upon substitution of equation (5) to give:

$$-55.51 \ln (a_w) = \frac{2 I}{|z^-z^+|} + \frac{4.606}{|z^+z^-|} \; -A|z^+z^-|[(1+I^{1/2})$$

$$-2 \ln (1 + I^{1/2}) - 1/(1 + I^{1/2})] + \frac{(0.06 + .6B) I|z^+z^-|}{0.75}$$

$$\frac{1 + 3 I/(|z^+z^-|)}{(1 + 1.5I/(|z^+z^-|))^2} - \frac{\ln (1 + 1.5I/|z^+z^-|)}{1.5I/|z^+z^-|}$$

$$+ B I^2/2 + 2C I^3/3 + 3D I^4/4 \qquad (14)$$

For multicomponent salt solutions, the integration of equation (13) can be quite horrendous and Meissner and Kusik (8) proposed a simplication which is exact for solutions containing ions of the same charge magnitude and is a first order approximation for solutions of ions of various charge magnitude. This formulation was adopted into the ECES framework particularly after evaluations of various expressions by Sangster and Lenzi (9) showed the suitability of this approach. Meissner and Kusik's (8) expression for the computation of the activity of water as adopted for use in our thermodynamic framework becomes

$$\ln(a_w)_{mix} = \frac{1}{x_I y_I} \sum_x Rx^1 \ln(a_w)_{Rx^1} + \sum_{R^1} R^1x \ln(a_w) R^1x$$

where $$x_I = \sum_{R^1} z^2_{R^1} m_{R^1} \text{ and } y_I = \sum_{x^1} z^2_{x^1} m_{x^1} \qquad (15)$$

For non-electrolytes in solutions of electrolytes the prediction of activity coefficients for these species is not nearly as advanced. Most predictions are variations of the well-known Setschenow equation.

$$\log \gamma_i = \log \frac{Si^\circ}{Si} = K_s C_s \qquad (16)$$

where Si° is the solubility of the non electrolyte in pure water, Si the solubility in the salt solution, K_s is a parameter which is dependent on the particular salt and C_s is the salt solution concentration. Since little information is available for the prediction of K_s, except for recent attempts using scaled particle theory or empirical data as summarized in the review article by Long and McDevit (10), or for multicomponent effects, attempts to develop correlations or predictions of the activity coefficient of a non-electrolyte in a multicomponent system follow approaches similar to those outlined by Edwards, Newman and Prausnitz (11).

Within the ECES framework the activity coefficient of a non-electrolyte is given by the following expression:

$$\ln(\gamma i) = \beta_i I + 2 \sum_x \beta_x m_x$$

where β_i is an empirical molecule interaction parameter and β_x is an empirical ion interaction parameter. These parameters are found by either data evaluation or proprietary correlations. At present no temperature or concentration dependencies are incorporated into these interaction parameters as they are into similar ion-ion interaction parameters.

I have outlined the basic thermodynamic framework we adopted and in many cases where we have extended the work of several of the researchers present at this symposium. The particular structure was chosen for several reasons including the following:

- In most cases, state of the art techniques are used.

- The particular structure is well-suited for imbedding into a sophisticated tool which will be easy to use.

- Using the activity coefficient concept makes the resulting model for a particular system easy for an engineer or scientist to understand.

- Particular formulations were chosen since data regression where required will be relatively straight forward.

- The formulations chosen also enabled the development of parameter estimating techniques for the usual case of insufficient pure salt data.

Certain weaknesses are obvious, these include:

- For ion interactions, application of the Bronsted assumption, that is only interactions between oppositely charged ions.

- For ion activity coefficients, currently no effects of non electrolytes on the ion activity coefficients.

- For ion activity coefficients, no high order interactions.

- For water activity predictions in multicomponent solutions, the same objections as applied to ions as well as a first order approximation to unsymmetrical electrolytes.

- For non electrolytes, a simple temperature independent first order effect with respect to ion-molecule interactions only.

Despite these apparent weaknesses, within the context of a general purpose system for predicting the vapor-liquid-solid equilibria of multicomponent aqueous solutions, ECES as a tool succeeds remarkably well as will be seen in a few illustrations after the following description of the software structure and use.

Overall Structure of the ECES System

In developing a tool for describing electrolyte equilibria we defined as our objectives those given in the introduction to this paper and summarized as:

- The ability to expand the databank through experimental data analysis and regression

- The ability to estimate data and build a suitable databank when little or no experimental data is available

- The ability to solve complex systems for specific parameter studies (phase diagrams)

- The ability to predict vapor-liquid-solid equilibria in multistage multicomponent systems (i e., strippers, etc.)

All of these objectives are to be met in a computer software system that is user oriented.

To do this, we developed a computer software system composed of several major program blocks. These programs and their functions are:

Program Builder - This block of programs takes the user description of the multicomponent system including all chemical and ionic equilibria of interest and either:

a) writes a complete model description suitable for solving a single stage flash where the user species inlet concentrations of species and temperature and pressure.

b) writes a complete set of subroutines for linking to a fractionator program designed to handle electrolyte systems.

Which ever path is followed, the thermodynamic framework which I earlier described is used as the basis along with data obtained from a large data-base created by using the Data Regression Program Block or Data Estimation Block, to finally describe the system.

Program Generator (ASAP) - This block of programs utilizes the complete model description created by the program generator to automatically write and compile a FORTRAN program to solve various forms of single stage flash problems or parameter study problems.

Case Run Block - This program group is used to solve the resulting program for the Program Generator. Imbedded in it are heuristics to take user input data and solve the complex models by a sophisticated Newton-Raphson technique.

FRACHEM Block - This block of programs is linked to the subroutine created by the Program Generator to solve multistage electrolyte towers which are not limited to being in complete chemical equilibria. The FRACHEM Block and applications are described separately in a recent paper (12).

Data Entry Block - This group of programs is used to evaluate and regress experimental data into the form suitable for use by the Program Builder Block.

Data Estimation Block - This group of programs allows the user to create species entries for electrolyte species where little or no data is available. Unfortunately, for many experimental systems of interest there is often little or no suitable data on the pure salt and its behavior in aqueous solution. Within this block we have incorporated several proprietary correlations that draw on and extend parameter estimating procedures as proposed by Bromley (5), Edwards, Newman and Prausnitz (11), Meissner and Kusik (8), Criss and Cobble (13). These programs range from estimate procedures for Bromley B interaction parameters, through temperature dependence estimations and extrapolation techniques incorporating the work of Meissner and Kusik (8).

In the next section of this paper, I have looked at the application of ECES system to two problems and briefly describe the user input and ECES results.

Using the ECES System to Predict Vapor Liquid Equilibria (NH_3-CO_2-H_2O)

An important system in industrial pollution control is the NH_3-CO_2-H_2O system. This equilibria represents a good test for the ECES system because data is available on the vapor pressures of NH_3 and CO_2 over various compositions for a range of temperatures. Furthermore, there is a lack of data on some of the pure salts that are inherent in this system and data estimates are necessary. To complete this test, this system involves both weak and strong electrolytes.

To use the ECES Program Builder, the user creates computer input as shown in Figure 1. The input consists of three parts: The names of the components in the inlet stream to the flash unit, the names of the species present in all the outlet streams - vapor, aqueous, or solids, and the chemical and/or ionic equilibria of interest. From this input the Program Builder Block using the imbedded thermodynamic framework described earlier writes the model description as given in Figure 2.

Upon examining this model which is written in essentially textbook notation, the following equations may be found:

Model Eq. 1 - Expression for B_{RX} referred to in activity coefficient equations as SE(1,x) where x is a code number for a particular cation-anion pair.

2 - Expression for $ln(aw)_{RX}$, referred to in eq. 8 as SE(z,x)

3 - Expression for X_I, referred to in eq. 8 as SE(3)

4 - Expression for Y_I

5 - Expression for Debye-Huckel term

6 - Summation of $\beta_i m_i$ for use in activity coefficient of non-electrolytes

7,9,10,11,15,19, and 21 - Equilibrium expressions developed from user input of mass law balance expressions

8 - Expression to calculate activity of water

12,13,16,17,20, and 22 - Expressions for predicting the activity coefficients of individual ions.

14 and 18 - Expressions for predicting the activity coefficients of the molecules in solution

24 -Definition of ionic strength

25 -Definition of Electroneutrality

26 and 27 - Elemental balances written automatically until phase rule satisfied

28 -Balance on gas phase - low pressure version ideal gas behavior assumed

These expressions were developed automatically from the user input and the embedded thermodynamic framework. In addition a separate routine was written to compute the equilibrium constants as a function of temperature and the B,C, and D coefficients for each salt pair. If in the associated ECES database, certain species were not already present, a message such as

NH4NH2CO2PPT NOT IN DATABASE
HNH2CO2PPT NOT IN DATABASE

would be printed as indeed was the case during the first attempt to create this model. This message indicates that there is no prediction of the activity coefficient of pure NH4NH2CO2 in water, for example. The user has two options - either to continue in which case the $_{RX}$ for these salt pairs will be computed by the expression

$$\frac{0.06}{(1+1.5I)^2}$$

or to either obtain data or use the Estimation Block to obtain parameter estimates to be stored in the data bank. The latter was adopted here.

After the model is built, the program can be generated and compiled. At execution time, the user has considerable flexibility and we chose to predict the bubble point pressure for a fixed temperature and specified total system composition in order to compare some of our results with the data of Otsuku (14). Figure 3 presents the results for a system composed of 10.14 wt% CO_2 and NH_3 at a temperature of 80° where the %CO_2 in the CO_2 and NH_3 was varied.

Of considerable interest would be the ECES results at the 35% CO_2 point in which we can see the predicted activity coefficients of the species and the concentrations of the species on solution.

Calculated Concentrations (molality)	
NH_4^+	0.9269
H^+	3.514×10^{-9}
OH^-	1.560×10^{-4}
HCO_3^-	.2072
CO_3^{2}	.01923
NH_3aq	2.672
CO_2aq	4.672×10^{-4}

Activity Coefficients	
NH_3aq	1.094
CO_2aq	1.116
NH_4^+	.4986
OH^-	.6521
H^+	.6764
HCO_3^-	.5005
$NH_2CO_2^-$	.4874
CO_3^{-2}	.1460

Activity	
H_2O	0.9738

```
INPUT
H2OIN
NH3IN
CO2IN
SPECIES
H2OVAP
NH3VAP
CO2VAP
NH3AQ
CO2AQ
NH4ION
OHION
HION
HCO3ION
CO3ION
NH2CO2ION
EQUILIBRIUM
H2OVAP = H2O
NH3VAP = NH3AQ
CO2VAP = CO2AQ
NH3AQ + H2O = NH4ION + OHION
CO2AQ + H2O = HION + HVO3ION
HCO3ION = HION + CO3ION
NH2CO2ION + H2O = NH3AQ + HCO3ION
H2O = HION + OHION
```

Figure 1. Input to ECES to describe NH_3–CO_2–H_2O problem

```
1  (.06+0.6*B(I1))*Z(I1)/(1+1.5/Z(I1)*I)**2+B(I1)+C(I1)*I+D(I1*I*
   *2
2  -1/55.51*(2*I/Z(I1)+4.606/Z(I1)*(-CON1*Z(I1)*((1+I**0.5)-2*ALOG
   (1+I**0.5)-./(1+I**0.5))+(0.06+0.6*B(I1))*I*Z(I1)*((1+3*I/Z(I
   1))/(( 1+1.5*I/Z(I1))**2)-(ALOG(1.5*I/Z(I1)+1))/(1.5*I/Z(I1)))
   /0.75+ B(I1)*(I**2)/2+2*C(I1)*(I1)*(I**3)/3+3*D(I1)*(I**4)/4))
3  1*NH4ION+1*HION
4  1*OHION+1*HCO3ION+4*CO3ION+1*NH2CO2ION
5  -CON1*I**0.5/(1+I**0.5)
6  0.01132*NH4ION-0.0388*HION+0.09425*OHION+0.02616*HCO3ION+0.1252
   8*CO 3ION+0.04492*NH2CO2ION
7  KH2OVAP-AH2O/PH2O=0
8  ALOG(AH2O)-1/(SE(3)*SE(4))*(NH4ION*OHION*SE(2,1)+NH4ION*HCO3ION
   *SE(2,2)+NH4ION*4*CO3ION*SE(2,3)+NH4ION*NH2CO2ION*SE(2,4)+HION
   *HCO 3ION*SE(2,5))-1/(SE(3)*SE(4))*(HION*4*CO3ION*SE(2,6)+HION
   *NH2C O2ION *SE(2 ,7))=0
9  KNH3VAP-ANH3AQ*NH3AQ/PNH3=0
10 KCO2VAP-ACO2AQ*CO2AQ/PCO2=0
11 KNH3AQ-1/(AH2O(*ANH4ION*NH4ION*AOHION*OHION/(ANH3AQ*NH3AQ)=0
12 ALOG10(ANH4ION)-(1)**2*SE(5)-(SE(1,1)*OHION+SE(1,2)*HCO3ION N+S
   E(A+SE( 1,3) *CO3ION*3**2/4+SE(1,4)*NH2CO2ION)=0
13 ALOG10(AOHION)-(1)**2*SE(5)-(SE(1,1)*NH4ION)=0
14 ALOG(ANH3AQ)+.0084*I-2*SE(6)=0
15 KCO2AQ-1/(AH2O)*AHION*HION*AHCO3ION*HCO3ION/(ACO2AQ*CO2AQ)=0
16 ALOG10(AHION)-(1)**2*SE(5)-(SE(1,5)*HCO3ION+SE(1,6)*CO3ION*3**2
   /4+S E(1,7)*NH2CO2ION)=0
17 ALOG10(AHCO3ION)-(1)**2*SE(5)-(SE(1,2)*NH4ION+SE(1,5)*HION)+0
18 ALOG(ACO2AQ)-0.01300*I-2*SE(6)=0
19 KHCO3ION-ACO3ION*AHION/AHCO3ION*CO3ION*HION/HCO3ION=0
20 ALOG10(ACO3ION)-(2)**2*SE(5)-(SE(1,3)*NH4ION*3**2/4+SE(1,6)*
   HION*3**2/4)=0
21 KNH2CO2ION-1/(AH2O)*ANH3AQ*NH3AQ*AHCO3ION*HCO3ION/(ANH2CO2ION*N
   H2CO2ION)=0
22 ALOG10(ANH2CO2ION)-(1)**2*SE(5)-(SE(1,4)*NH4ION+SE(1,7)*HION)=0
23 KH2O-AOHION*AHION*OHION*HION/AH2O=0
24 I-0.5*(SE(3)+SE(4))=0 N*HION/AH2O=0
25 NH4ION+HION-OHION-HCO3ION-2*CO3ION-NH2CO2ION=0
26 3*H2OIN+4*NH3IN+3*CO2IN-H2)/55.5*(4*NH3AQ+3*CO2AQ+5*NH4ION+2*OH
   ION+H ON+5*HCO3ION+4*CO3ION+6*NH2CO2ION)-V/PT*(3*PH2O+4*PNH3+3
   *PCO2 )-3* H2O=0
27 2*H2OIN-2*H2O+3*NH3IN-H2O/55.5*(3*NH3AQ+4*NH4ION+OHION+HION+HCO
   3ION +2*NH2CO2ION)-V/PT*(2*PH2O+3*PNH3)=0
28 PT-PH2O-PNH3-PCO2=0
29 (HCO3ION+CO2AQ+CO3ION+NH2ION)*H2O/55.5V/PT*PCO2-CO2IN=0
```

Figure 2. Model summary for single stage flash of NH_3–CO_2–H_2O system (low pressure)

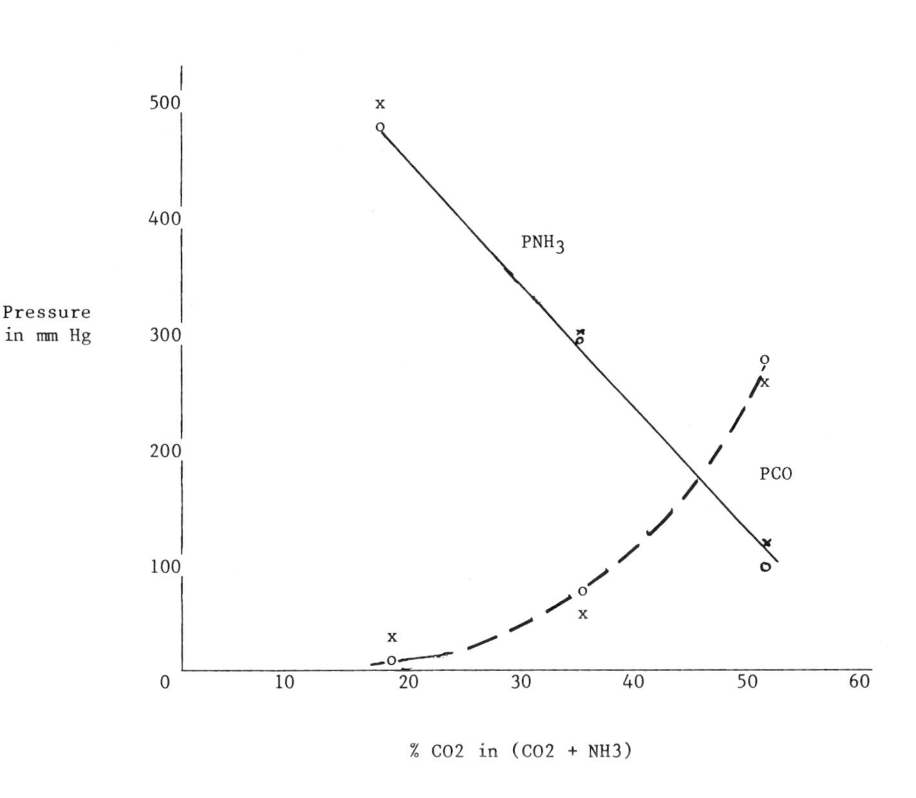

Figure 3. Comparison of ECES prediction (×) vs. data of Otsuku (○) (NH_3–CO_2–H_2O equilibrium at 80°C, 10.14 wt % (CO_2 + NH_3))

This illustrates that the CO_2 in solution is predominately in the form of carbamate ion and as CO_2 concentration is increased we finally reach a point where sufficient aqueous CO_2 is present for the CO_2 vapor pressure to increase markedly with fraction of total CO_2 in solution. Similar input can be used to create subroutines for a fractionator program and this system is described for such a case in a recent paper (12).

Using the ECES System for Vapor-Liquid-Solid Equilibria ($FeCl_2$-HCl-H_2O)

As an example of a different form of problem more common to the basic chemical industry, let us consider the case where concentrated solutions of ferrous chloride are the by-products of an industrial process. In this particular example at different points of the process cycle, two questions arose: the first being a need to know the vapor pressure of hydrochloric acid over mixtures of $FeCl_2$-HCl-H_2O and the second, the solubility of $FeCl_2$ as a function of HCl concentration.

To use the ECES system, activity coefficient data for $FeCl_2$ had to be developed. A recent paper by Susarev et al (15) presented experimental results of the vapor pressure of water over ferrous chloride solutions for temperatures from 25 to 100°C and concentrations of 1 to 4.84 molal. This data was entered into the ECES system in the Data Preparation Block with a routine VAPOR designed to regress such data and develop the interaction coefficients B, C, D of our model. These results replaced an earlier entry which was based on more limited data. All other data for studying the equilibria in the $FeCl_2$-HCl-H_2O system was already contained within the ECES system.

To develop the ECES model the following input was entered by the user:

```
INPUT
H2OIN
FEIICL2
HCLIN
SPECIES
H2OVAP
HCLVAP
FEIICL2.4H2O
HION
CLION
FEIIION
EQILIBRIUM
H2OVAP = H2O
HCLVAP = HION + CLION
FEIICL2.4H2O = FEIIION + 2CLION + 4H2O
```

Since the basic model generated by the Program Builder Block looks similar in many ways to the previously illustrated model, we have not shown it here. One difference in this case is that we have an expression for solid-liquid equilibria which is written from the last mass balance expression entered by the user and is then written by the Program Builder as:

```
KFEIICL2.4H2O - AFEIIION*FEIIION*ACLION**2*CLION**2AH2O**4 = 0
```

For convenience within ECES the ferrous species is designated FEII while the ferric species is designated FEIII, which at times results in having up to four i's in a row. In testing the ECES model, solubility data of $FeCl_2$.2H2O as a function of HCL concentration were available in a paper by Schemmel (16). Typical results of the ECES prediction compared to the experimentally determined solubility at 40°C are:

$FeCl_2$ - HCL-H_2O @ 40°C solid phase $FeCl_2.4H_2O$

HCl molality	Measured FeCl2 molaltity	ECES predicted FeCl2 molality
0.0	5.415	5.420
2.328	4.197	4.088
4.522	2.966	2.921

As can be seen less than 2% error in this multicomponent system occurred when using ECES. This system is quite different than the NH_3-CO_2-H_2O system since we are dealing only with strong electrolytes. For example, the second datum point predicted by ECES give the following results for the concentrations, activity coefficients and water activity in the aqueous phase.

Concentrations

Ionic Strength	14.59 molal
FEIIION	4.088
HION	2.328
CLION	10.504

Activity Coefficients

H^+	7.739
Cl^-	3.693
FE^{++}	1.319

Activity

a_{H2O}	0.549

Within the Parameter Study Block, the algorithm is designed to test whether solubility has been exceeded or not. As a result the same model generated by ECES can be used to predict HCl vapor pressures over unsaturated solutions of FeCl2-HCl-H2O without modifying the basic program created by ECES. In a paper by Chen (17) some limited experimental data was presented on the vapor pressure of HCl over ferrous chloride system.

HCl Conc Moles/liter	FeCl2 Conc Moles/liter	Temp °C	PHCl (EXP) mm Hg	PHCl (ECES) mm Hg
5.43	1.50	60	14.6	15.8
5.43	1.50	70	25.9	29.2
5.43	1.50	80	32.1	51.6
5.43	1.50	90	79.6	80.2

A large discrepancy appears at the 80°C datum point. However, this appears to be an error in the published data since some recent experimental work (unpublished) to resolve this difference measured a vapor pressure of approximately 47 mm Hg which is close to the ECES prediction. These results for the FeCl2-HCl-H2O system were just a subset of a more complex model involving several additional species. Because of the proprietary nature of the work, we cannot disclose the results at this time, but needless to say, the performance of the ECES thermodynamic framework was excellent in predicting multicomponent phase equilibria over a wide range of operating temperatures with ionic strengths up to approximately 30 molal for the extensions of the FeCl2-HCl-H2O system.

Conclusions and Significance

With the current advances made in our understanding of electrolyte systems, as illustrated by the papers and discussions of this symposium, a complete design tool can be synthesized with a consistent thermodynamic framework which gives to the engineer or scientist capabilities similar to those that have been available for many years in the hydrocarbon processing industry. There are tremendous data gaps however when working with electrolyte solutions and as a result many estimation techniques need to be developed to temporarily fill in the holes. Additionally, our thermodynamic framework does not have the total strength it needs yet. This symposium hopefully will show directions to be pursued to strengthen that framework.

Besides the difference in the expressions for activity coefficients and other thermodynamic properties from those published and used by the hydrocarbon processing industries, it is more important to realize the need to describe the ionic and chemical equilibria that are to be satisfied in aqueous systems. Most of our participants in this symposium have presented results as good as if not better than those predicted by ECES, however in order to obtain such predictions, detailed models have to be written and coded.

Our fundamental contribution is the marrying of several different technologies to develop a general purpose software system capable of predicting vapor-liquid-solid equilibrium in both single and multistaged systems over a wide range of conditions. We have combined concepts of automatic program generation, advanced numerical techniques, a generalized thermodynamic framework, data analysis and estimation routines into a software system that can be used to answer many of the industrial problems arising concerning electrolytes. Currently the ECES system is being used in such diverse areas as food processing, petroleum production, pollution control, and of course in the design and simulation of aqueous based chemical processes. Furthermore the system is evolving towards a more comprehensive and easier to use tool as a result of the feedback from its users as more and more applications are tried.

Literature Cited

1. Wagman et al, "Selected Values of Chemical Thermodynamic Properties", National Bureau of Standards, 1968; Technical Note 270-3,4,5,...

2. Hamer, W.J.;Wu, Y.C.,"Osmotic Coefficients and Mean Activity Coefficients of Uni-univalent Electrolytes in Water at 25°C", J. Phys. Chem., 1972, 1, 1047.

3. Pitzer, K.S., "Thermodynamics of Electrolytes I Theoretical Basis and General Equations", J. Phys. Chem., 1973, 77, 268
4. Meissner, H.P.; Kusik, C.L., "Activity Coefficients of Strong Electrolytes in Multicomponent Aqueous Solutions", AIChE J., 1972, 18, 294

5. Bromley, L.A., "Thermodynamic Properties of Strong Electrolytes in Aqueous Solutions",AIChE J,1973, 19, 313

6. Guggenheim, E.A.; Turgeon, J.C., "Specific Interaction of Ions",Trans Faraday Society, 1955, 51, 747

7. Harned, H.S.; Owen, B.B., "The Physical Chemistry of Electrolyte Solutions", 3ed, Reinhold Publishing Company, New York, 1958

8. Meissner, H.P.; Kusik, C.L., "Aqueous Solutions of Two or More Strong Electrolytes-Vapor Pressures and Solubilities", I&EC Process Des. Develop., 1973, 12, 205

9. Sangster, J.; Lenzi,F., "On the Choice of Methods for the Predictions of the Water-activity and Activity Coefficients for Multicomponent Aqueous Solutions", Can. J. Chem. Eng., 1974, 52, 392

10. Long, F.A.; McDevit, W.F., "Activity Coefficients of Nonelectrolyte Solutes in Aqueous Salt Solutions", Chem. Reviews 1952, 51, 119

11. Edwards, T.J.,; Newman, J.; Prausnitz, J.M., "Thermodynamics of Aqueous Solutions Containing Volatile Weak Electrolytes", AIChE 1975, 21, 248

12. Zemaitis, J.F., "Counter Current Stage Separation with Chemical and Ionic Equilibrium and/or Reaction", to be published in Computer Applications to Checmical Engineering Process Design and Simulation, ACS Symposium Series, 1979-1980

13. Criss, C.M.; Cobble, J.W., "Thermodynamic Properties of High Temperature Aqueous Systems. IV Entropies of the Ions up to 200°C and the Correspondence Principles", JACS, 1964, 86, 5385

14. Otsuku, E., "Measured Vapor Liquid Equilibria of NH3-CO2-H2O", Kogyo Kagaku Zasohi, 1960, 63, 1214

15. Susarev, M.P. et al, "Vapor Pressure of Water Over Ferrous Chloride Solutions", Zhurnal Pukladnoi Khimii, 1976, 49, 1045

16. Schemmel, F.A., "Ternary System Ferrous Chloride-Hydrogen Chloride-Water", JACS, 1952, 74, 4689

17. Chen, E.C.; McGuire, G.; Lu, H.Y., "Vapor-Liquid Equilibria of the Hydroclhoric Acid Ferrous Chloride-Water System", J. Chem. Eng. Data, 1974, 15, 233

RECEIVED January 31, 1980.

11

Effects of Aqueous Chemical Equilibria on Wet Scrubbing of Sulfur Dioxide With Magnesia-Enhanced Limestone

CLYDE H. ROWLAND, ABDUL H. ABDULSATTAR, and DEWEY A. BURBANK

Bechtel National, Inc., P.O. Box 3965, San Francisco, CA 94119

In wet scrubbing of SO_2 from boiler flue gas by limestone slurry, the concentration of dissolved species in the scrubbing liquor that can react with incoming SO_2 gas is very low, about one to two m-mole/l. This is far below the SO_2 make-per-pass in the scrubber, typically about 10 m-mole of SO_2 absorbed per liter of liquor for one pass through the scrubber. Therefore, the SO_2 absorption rate is largely dependent upon the slow rate of limestone dissolution into the liquor passing through the scrubber.

Addition of magnesia to the scrubbing liquor increases the concentration of two dissolved sulfite species, $SO_3^=$ and $MgSO_3^o$, with $CaSO_3^o$ remaining constant. This increase in dissolved sulfite concentration makes the SO_2 absorption rate more dependent on the very fast liquid phase reactions of the basic sulfite species with the strong dibasic acid $SO_2(aq)$:

$$\text{base} + \text{dibasic acid} \longrightarrow \text{monobasic acid}$$

$$SO_3^= + SO_2(aq) + H_2O \longrightarrow 2\ HSO_3^-$$

$$MgSO_3^o + SO_2(aq) + H_2O \longrightarrow 2\ HSO_3^- + Mg^{++}$$

$$CaSO_3^o + SO_2(aq) + H_2O \longrightarrow 2\ HSO_3^- + Ca^{++}$$

Conversion of $SO_2(aq)$ to HSO_3^- by the above reactions in the scrubber facilitates the absorption of more $SO_2(g)$ into the liquor as $SO_2(aq)$, thus increasing SO_2 removal.

A chemical model for magnesia wet scrubbing of SO_2 was previously verified with experimental data for aqueous solutions of magnesium, SO_2, and sulfate over the temperature range 15-60°C for dissolved magnesium concentrations as high as 1200 m-mole/l (1). The present chemical model for limestone wet scrubbing with magnesia enhancement is applicable for liquors containing dissolved magnesium, calcium, SO_2, sulfate, and chloride, where the molality of chloride is less than twice that of magnesium. The limestone/magnesia scrubbing model

0-8412-0569-8/80/47-133-247$05.25/0

consists of the thermodynamic equilibria among the important chemical species, the mass balances for total magnesium and chloride species, the electroneutrality relationship, and equations for calculating significant activity coefficients.

Formulation and Verification of Chemical Model

Specified and Calculated Variables. The variables that are specified as input to the limestone/magnesia chemical model are the scrubber temperature, the liquor pH, the total dissolved magnesium concentration, the total dissolved chloride concentration, and the fractional degree of saturation of the liquor with calcium sulfite ($CaSO_3 \bullet 1/2\ H_2O$) and gypsum ($CaSO_4 \bullet 2H_2O$) solids. An equilibrium computer program containing the equations in the model calculates the equilibrium SO_2 partial pressure; the total dissolved concentrations of calcium, sulfite, SO_2 species (sulfite plus bisulfite), and sulfate; and the concentrations of the individual species.

The 12 individual species that are calculated by use of the chemical model are: $SO_2(g)$, $SO_2(aq)$, HSO_3^-, $SO_3^=$, $SO_4^=$, Mg^{++}, Ca^{++}, $CaSO_3^o$, $MgSO_3^o$, $MgSO_4^o$, $CaSO_4^o$, and Cl^-. The liquor pH is used to determine the ratios of the activities of $SO_2(aq)$, HSO_3^- and $SO_3^=$, but the hydronium ion (H^+) and hydroxide ion (OH^-) concentrations are negligibly small, and can be excluded from the ionic balance. As in previous work (2,3), chloride is assumed not to form complexes, so that all chloride is present as chloride ion, Cl^-. Other species, such as bicarbonate ion, bisulfate ion, and the calcium and magnesium complexes of bicarbonate and hydroxide, have negligibly small concentrations in limestone/magnesia scrubbing liquors.

Equilibria, Mass Balances, and Activity Coefficients. Nine aqueous thermodynamic equilibria apply, one each for the formation of $SO_2(g)$, $SO_2(aq)$, HSO_3^-, $MgSO_3^o$, $MgSO_4^o$, $CaSO_3^o$, $CaSO_4^o$, $CaSO_3 \bullet 1/2\ H_2O(s)$, and $CaSO_4 \bullet 2H_2O(s)$. These equilibria are shown below. The constants for these equilibria at 25°C and 1 atmosphere total pressure are given in Table I. The temperature dependence of these constants was presented previously (3). Limestone wet scrubbers operate at essentially ambient pressure, so that no pressure correction of the constants is required. The value of 3.5×10^{-7} for the solubility product of $CaSO_3 \bullet 1/2\ H_2O(s)$ was derived from limestone/magnesia scrubber inlet liquor analyses obtained at the Environmental Protection Agency (EPA) Shawnee test facility (9), and may include kinetic, as well as thermodynamic, interaction between slurry solids and liquor.

Table I

Equilibrium Constants at 25 °C and 1 Atmosphere

Species Formed	Equilibrium Constant	Reference
$SO_2(g)$	1.23*	(4)
$SO_2(aq)$	0.0130	(4)
HSO_3^-	6.24×10^{-8}	(5)
$MgSO_3^0$	1.2×10^{-3}	(3)
$MgSO_4^0$	5.6×10^{-3}	(6)
$CaSO_3^0$	4.0×10^{-4}	(3)
$CaSO_4^0$	4.9×10^{-3}	(7)
$CaSO_3 \cdot 1/2\ H_2O(s)$	3.5×10^{-7}	**
$CaSO_4 \cdot 2H_2O(s)$	2.4×10^{-5}	(8)

* Henry's constant for SO_2 is the inverse of this value, or 0.813 atm/ (g-mole/l).

** Derived from limestone/magnesia scrubber inlet liquor analyses obtained during 1976 at the EPA Shawnee Alkali Scrubbing Test Facility.

$$SO_2(aq) \rightleftharpoons SO_2(g)$$

$$HSO_3^- + H^+ \rightleftharpoons SO_2(aq) + H_2O$$

$$SO_3^= + H^+ \rightleftharpoons HSO_3^-$$

$$Mg^{++} + SO_3^= \rightleftharpoons MgSO_3^0$$

$$Mg^{++} + SO_4^= \rightleftharpoons MgSO_4^0$$

$$Ca^{++} + SO_3^= \rightleftharpoons CaSO_3^0$$

$$Ca^{++} + SO_4^= \rightleftharpoons CaSO_4^0$$

$$Ca^{++} + SO_3^= + 1/2\ H_2O \rightleftharpoons CaSO_3 \cdot 1/2\ H_2O(s)$$

$$Ca^{++} + SO_4^= + 2\ H_2O \rightleftharpoons CaSO_4 \cdot 2H_2O(s)$$

Excluding activity coefficients, three relationships are required in addition to the nine thermodynamic equilibria in order to calculate concentrations of the 12 unknown species. These relationships are the mass balances for magnesium and chloride, and the electroneutrality equation.

The activity coefficients that are significant in determining the concentrations of the unknown species are the activity coefficients of Mg^{++}, Ca^{++}, $SO_3^=$, $SO_4^=$, and HSO_3^-. The activity coefficient of Ca^{++} is assumed to be equal to that of Mg^{++}, and that of $SO_3^=$ equal to that of $SO_4^=$. The method for calculating the activity coefficients of Mg^{++}, $SO_4^=$, and HSO_3^- was presented previously (1). The activity coefficients of the neutral dissolved species and the fugacity coefficient of $SO_2(g)$ are assumed equal to unity, as in (1).

Verification of Model for Solutions Containing Calcium and Chloride. The ability of the chemical model to predict equilibrium SO_2 partial pressure and the relationship of liquor pH to liquor composition was previously verified for solutions containing magnesium, SO_2, and sulfate (1). In order to test the model for liquors also containing calcium and chloride, measured and predicted values of dissolved calcium concentration have been compared for 39 experimental data points, as shown in Figure 1. Of these data, 26 are for Shawnee (9) limestone/magnesia scrubber inlet liquors in contact with calcium sulfite and gypsum slurry solids at 50°C, and the remaining 13 are laboratory (10) data for aqueous magnesium sulfate and chloride solutions saturated with gypsum at 25°C in the absence of sulfite. The data cover a magnesium concentration range of 6-650 m-mole/l, a chloride range of 0-470 m-mole/l, and a chloride-to-magnesium range of 0-1.4 mole/mole. The chemical model explains 72 percent of the variation in the calcium data

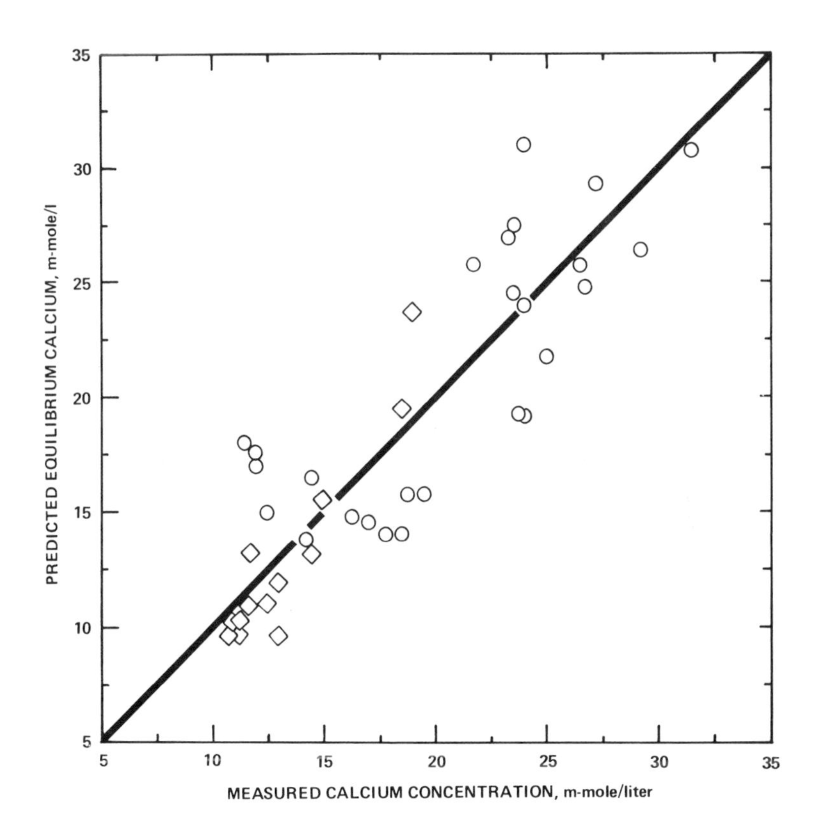

Figure 1. Predicted vs. measured calcium concentration ((○) Shawnee (9) prototype scrubber data, 50°C; (◇) laboratory (10) $MgSO_4$/gypsum data, 25°C)

(correlation coefficient of 0.85) with a standard error of estimate of 3.1 m-mole/l of calcium.

Correlations for Equilibrium SO_2 Partial Pressure and Liquor Composition

In order to facilitate calculations and increase the usefulness of the chemical model, the computerized model has been used to obtain simplified correlations suitable for hand calculations. These correlations predict equilibrium SO_2 partial pressure and liquor composition as functions of liquor pH and the total dissolved concentrations of magnesium and chloride. The correlations apply for:

- Typical scrubbing temperature of 50 °C
- Liquor saturated with both calcium sulfite and gypsum
- Liquor pH range of 4.5-6.0
- Dissolved magnesium concentration range of 80-1200 m-mole/l
- Dissolved chloride range of 0-400 m-mole/l
- Dissolved chloride molarity less than 1.4 times that of magnesium

The correlations, Equations 1-4, are used with the mass balances, Equations 5-6, to calculate the SO_2 partial pressure and the total dissolved concentrations of sulfite, bisulfite, calcium, SO_2 (sulfite plus bisulfite), and sulfate.

Correlations:

$$\log pSO_2 = 10.44 - 2\,pH + 0.33 \log Mg - 0.39\, Cl/Mg + 0.76\, C \quad (1)$$

$$Sulfite = 1.22 + 0.0349\, C\, (Mg - 0.44\, Cl) \quad (2)$$

$$\log (bisulfite) = 4.96 - pH + 0.47 \log Mg - 0.37\, Cl/Mg + 0.75\, C \quad (3)$$

$$Ca = 4.3 + \frac{9.1}{C\,(1 - Cl/2\,Mg)} \quad (4)$$

where $C = 1 - (6-pH)^2\,(0.242 - 0.062 \log Mg)$

Mass Balances:

$$SO_2 = sulfite + bisulfite \quad (5)$$

$$SO_3 = Mg + Ca - sulfite - (1/2)\,(bisulfite + Cl) \quad (6)$$

where Mg, Ca, SO_2, SO_3, and Cl are the total dissolved concentrations (m-mole/l) of magnesium, calcium, sulfur dioxide (including sulfite and bisulfite), sulfate, and chloride species, respectively; and pSO_2 is the equilibrium partial pressure of SO_2 in ppm (10^{-6} atm).

Dissolved sulfite concentrations predicted from Equation 2 agree with the computerized chemical model to within a relative standard error of 0.5 percent. Equations 1, 3, and 4 have relative standard errors of less than three percent for prediction of SO_2 partial pressure, bisulfite (HSO_3^-) concentration, and total calcium concentration, respectively.

Figure 2 is a plot of the correction parameter C as a function of magnesium concentration and pH. Note that for the typical range of limestone scrubber inlet liquor pH, 5.2 to 6.0, C has a narrow range of 0.92-1.00. This narrow range for C justifies the use of a constant value, C = 0.97, in Equations 1-4 for pH values between 5.2 and 6.0.

Equilibrium SO_2 Partial Pressure. Figure 3 is a plot of Equation 1 for zero chloride concentration, showing the very strong effect of pH and the relatively weak effect of magnesium concentration on the equilibrium SO_2 partial pressure, pSO_2. To a first approximation, pSO_2 is proportional to the square of the hydronium ion (H^+) concentration, so that pSO_2 increases by a factor of nearly 100 (actually about 80) as the pH decreases by one unit.

An increase in magnesium concentration from 100 to 1000 m-mole/l increases pSO_2 by about a factor of two. Most limestone scrubbing systems with magnesium enhancement would operate midway between these magnesium concentrations, about 200-400 m-mole/l.

The main purpose of magnesia addition to a limestone wet scrubbing system is to facilitate high SO_2 removal. For a wet scrubber that cleans flue gas from a utility coal-fired boiler, the scrubber inlet gas SO_2 concentration is typically about 700 ppm by volume per one weight percent sulfur in the fired coal. For sub-bituminous coal having only 0.7 weight percent sulfur content, the inlet SO_2 concentration is about 500 ppm, and, for example, the outlet SO_2 has to be less than 50 ppm to achieve 90 percent removal. In order to avoid serious inhibition of mass transfer because of SO_2 back-pressure, the equilibrium SO_2 partial pressure should be about four or more times lower than the actual SO_2 partial pressure in the gas. Thus, Figure 3 indicates that for this low-sulfur coal system, the scrubber inlet pH should be at least 5.5, and the outlet pH at least 5.0. These conditions require a moderate-to-high limestone stoichiometry, about 1.3 mole of limestone added to the scrubber system per mole of SO_2 absorbed, for a well-designed scrubber system (9).

For high-sulfur bituminous coal containing 4.0 weight percent sulfur, the scrubber inlet SO_2 concentration is about 2800 ppm, and the required outlet SO_2 for 90 percent removal is 280 ppm. SO_2 back-pressure should not be a limiting factor for a scrubber inlet pH of at least 5.0 with an outlet pH of at least 4.6. Thus, limestone stoichiometries as low as 1.1 are feasible with high-sulfur coal and a well-designed scrubber system.

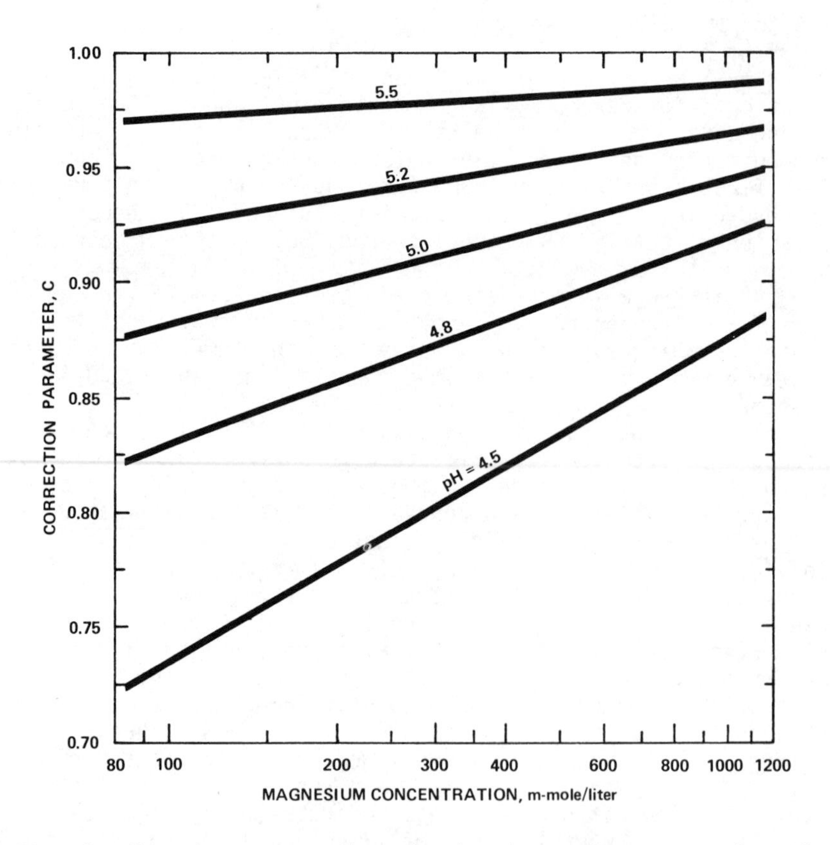

Figure 2. Correction parameter C *as a function of magnesium concentration and pH*

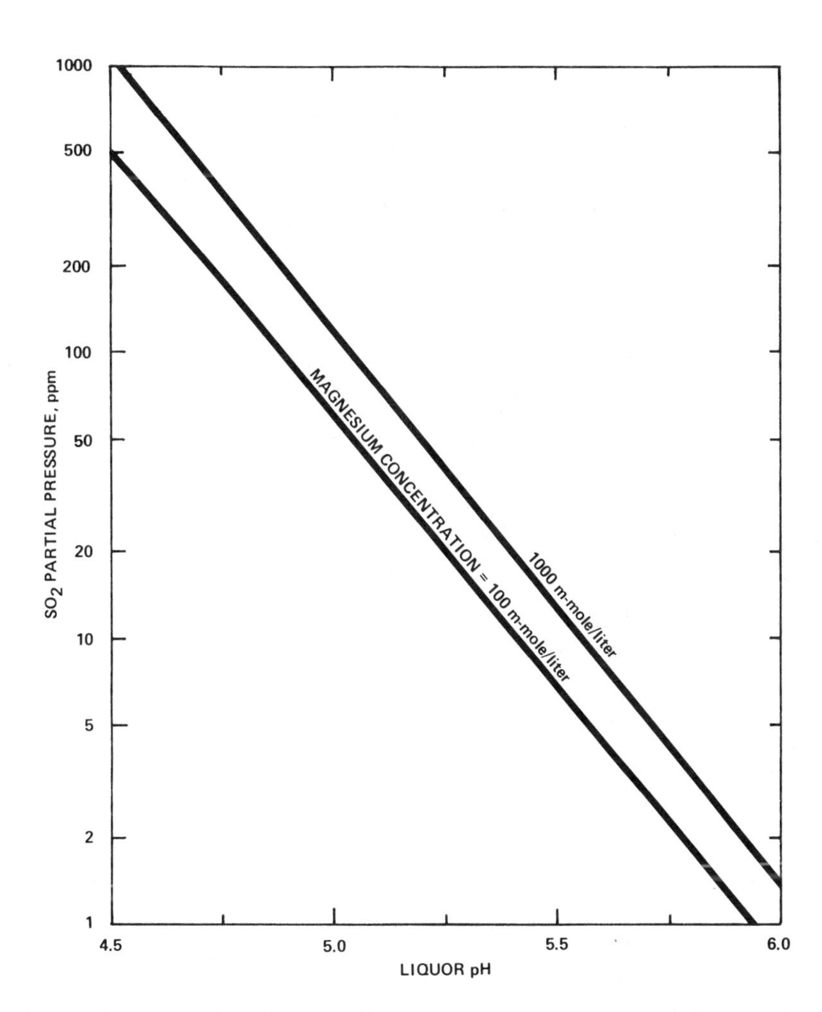

Figure 3. Equilibrium SO_2 partial pressure as a function of pH and magnesium concentration for liquors saturated with calcium sulfite and gypsum at 50°C with no chloride

Dissolved Sulfite Concentration. Figure 4 is a plot of Equation 2 for a pH of 5.5 presenting the effects of dissolved magnesium and chloride concentrations on the dissolved sulfite ($SO_3^=$, $MgSO_3^o$, $CaSO_3^o$) concentration. Dissolved sulfite is important because it is the base that converts the strong dibasic acid SO_2 (aq) to the weaker monobasic acid HSO_3^-, thus facilitating SO_2 absorption.

In the absence of chloride, the dissolved sulfite increases linearly with magnesium concentration from 1.2 m-mole/l of sulfite without magnesium additive to 15 m-mole/l at 400 m-mole/l dissolved magnesium, and to more than 40 m-mole/l of sulfite at 1200 m-mole/l magnesium concentration. For a limestone wet scrubber, each liter of liquor absorbs about 10 m-mole of SO_2 in one pass through the scrubber. For a typical magnesium concentration of 200-400 m-mole/l, the corresponding sulfite concentration is 8-15 m-mole/l, so that the reactant stoichiometry is about one mole of dissolved sulfite in the liquor entering the scrubber per mole of SO_2 absorbed.

The negative linear effect of chloride on sulfite concentration causes magnesium chloride, $MgCl_2$, to be largely ineffective in scrubbing SO_2. A negative coefficient of 0.50, rather than 0.44, for the chloride-to-magnesium ratio in Equation 2 would correspond to no effect of $MgCl_2$ concentration on dissolved sulfite and hence no effect on scrubber liquor reactivity with SO_2. This approximation is sufficiently accurate that effective magnesium, defined as magnesium concentration minus half the chloride concentration, has been successfully used as a dependent variable for correlation of SO_2 removal by limestone wet scrubbing (9,11).

Dissolved Concentrations of Calcium and SO_2 Species. The equilibrium dissolved concentrations of total calcium and SO_2 (sulfite plus bisulfite) species are important because comparison of these equilibrium concentrations with actual measured values determines the degree of gypsum saturation, and hence the potential for gypsum scale formation in the scrubber. As a first approximation, the fraction gypsum saturation of a scrubber liquor, having specified pH and specified concentrations of magnesium and chloride, is proportional to the measured calcium concentration, and inversely proportional to the measured SO_2 concentration.

Figure 5 is a plot of Equation 4 for a pH of 5.5 showing the increase in equilibrium calcium concentration with increasing chloride-to-magnesium ratio. For a liquor saturated with gypsum and calcium sulfite, the dissolved calcium concentration increases from 12 m-mole/l without chloride to 17 m-mole/l for a molar chloride-to-magnesium ratio of 0.5, and to 35 m-mole/l of calcium for a chloride-to-magnesium ratio of 1.4. For a given magnesium concentration, an increase in chloride decreases the other major anions, the SO_2 and sulfate species, thus increasing

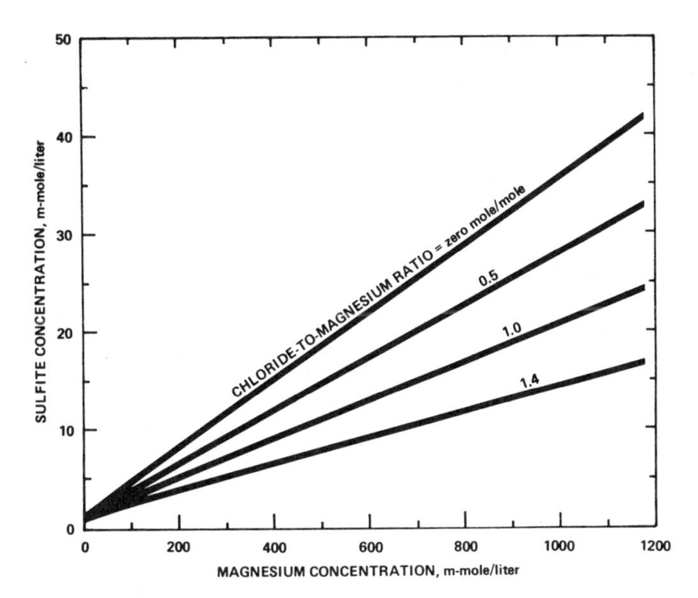

Figure 4. Dissolved sulfite concentration as a function of magnesium concentration and chloride-to-magnesium ratio for liquors saturated with calcium sulfite and gypsum at pH 5.5 and 50°C

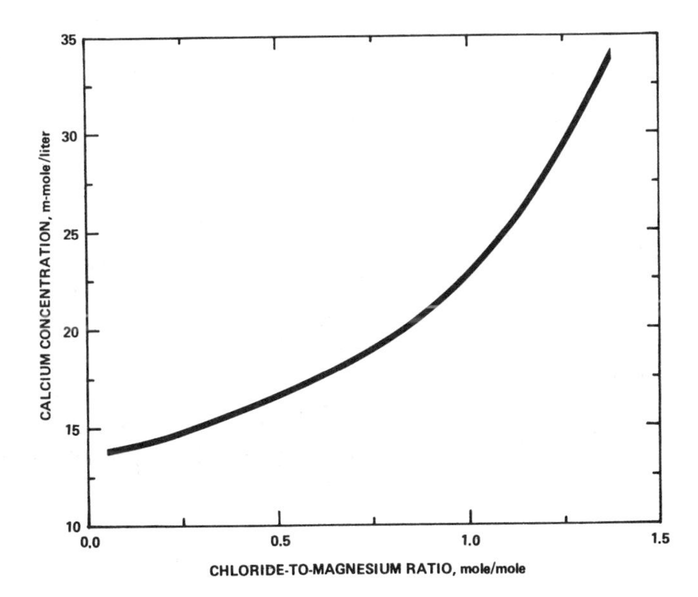

Figure 5. Dissolved calcium concentration as a function of chloride-to-magnesium ratio for liquors saturated with calcium sulfite and gypsum at pH 5.5 and 50°C

calcium concentration according to the solubility products for calcium sulfite and gypsum.

Figure 6 presents the strong effects of both pH and magnesium concentration on the equilibrium concentration of dissolved SO_2 species, where chloride is absent. The figure is derived from Equations 2, 3, and 5.

The dissolved SO_2 concentration decreases by a factor of about two for a pH increase of 0.5. This results from a strong decrease in bisulfite concentration with increasing pH; the sulfite species increase slightly as pH rises.

Dissolved SO_2 increases by about a factor of two as magnesium increases from 100 to 300 m-mole/l, and by a factor of four as magnesium increases from 100 to 1000 m-mole/l.

Equations for Calculating the Degree of Gypsum Saturation

For a liquor of known pH and magnesium and chloride concentrations, the degree of gypsum saturation can be determined by measurement of either the total dissolved calcium or the total dissolved SO_2 (sulfite plus bisulfite). The chemical model has been used to obtain correlations for gypsum saturation, presented below. The correlations, Equations 7 and 9, are valid for a typical scrubbing temperature of 50 °C, and for the same ranges of pH, magnesium, and chloride as for Equations 1-4.

Gypsum Saturation From Measurements of Dissolved Calcium. Equation 7 below can be used with Equation 4 to calculate gypsum saturation from measurements of total dissolved calcium.

$$\text{Fraction gypsum saturation} = 1 + (1.12/C)\,[(Ca)_M/Ca - 1] \qquad (7)$$

where $(Ca)_M$ is the measured total concentration of dissolved calcium species (m-mole/l), and Ca is the total dissolved calcium concentration for a liquor saturated with gypsum, obtained from Equation 4. Calculation of fraction gypsum saturation by use of Equations 4 and 7 gives results that agree with the chemical model to within a standard error of estimate of 0.05 fraction saturation for saturations between 0.5 and 2.0.

For a typical limestone scrubber inlet liquor pH range of 5.2-6.0, and for liquors having a chloride-to-magnesium ratio of 0.2 mole/mole or less, the following simplified equation can be used to determine gypsum saturation from calcium measurements:

$$\text{Fraction gypsum saturation} = \frac{(Ca)_M}{12.4} - 0.13 \qquad (8)$$

Equation 8 has a standard error of estimate of 0.07 fraction gypsum saturation over the saturation range of 0.5-2.0. The equation is useful for monitoring the actual gypsum saturation

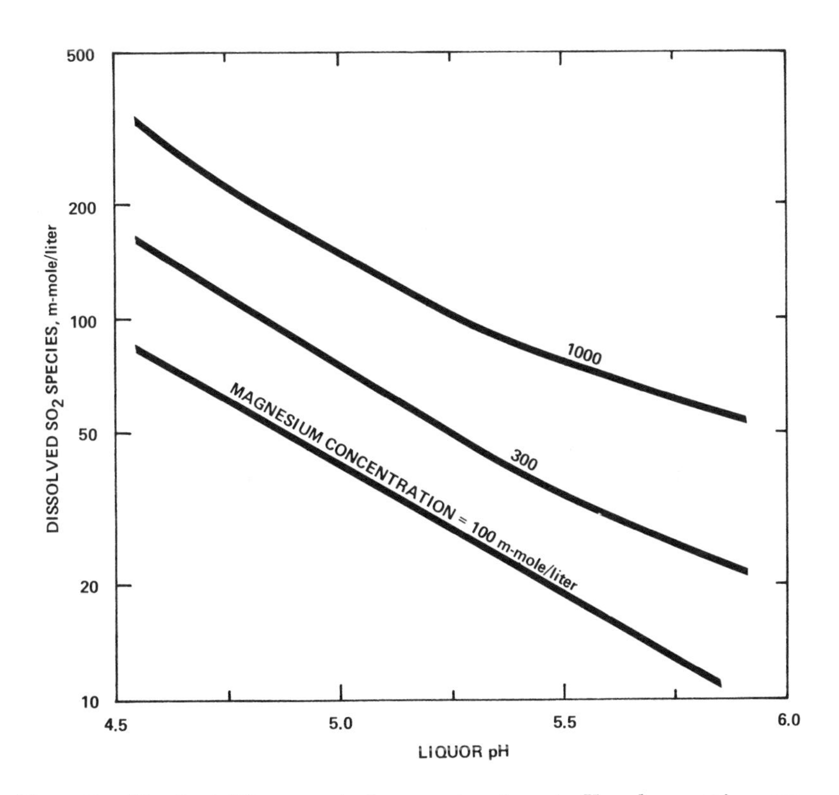

Figure 6. Dissolved SO_2 concentration as a function of pH and magnesium concentration for liquors saturated with calcium sulfite and gypsum at 50°C with no chloride

for an operating unit where computational resources are limited.

Gypsum Saturation from Measurements of Dissolved SO_2. Use of measurements of dissolved calcium to determine gypsum saturation is relatively easy from a computational standpoint; use of measurements of dissolved SO_2 is more difficult. However, wet chemical analyses for calcium are frequently subject to interference by high concentrations of magnesium. For installations where a quick and reliable analysis for calcium is not available, the use of dissolved SO_2 is preferred, and the following correlation applies:

$$\text{Fraction gypsum saturation} = R\,(1 + D \log R) \tag{9}$$

where $D = 5.92 - 3.98\,C - 0.62 \log (Mg - 1/2\,Cl)$

and R = ratio of calculated (Equations 2, 3, and 5) total dissolved SO_2 for a saturated liquor to measured total dissolved SO_2 in the actual liquor.

Use of Equations 2, 3, 5, and 9 to determine gypsum saturation gives results that agree with the chemical model to within a standard error of estimate of 0.04 fraction gypsum saturation for saturations of 0.5-2.0.

Enhancement Effect of Magnesia-Induced Dissolved Sulfite on SO_2 Removal

The chemical model has been used to correlate the enhancement effect of magnesia-induced dissolved sulfite on SO_2 removal by limestone wet scrubbing. The correlated data were obtained at the 10-MW equivalent EPA Alkali Scrubbing Test Facility during 1976 (9), using a Turbulent Contact Absorber, or TCA (12). The TCA contained three beds of 1.5-inch diameter nitrile foam spheres, held in place by four retaining grids having about 70 percent open area. Slurry was distributed over the beds, countercurrent to the gas flow, by low-pressure (about 3 psig) nozzles located above the top bed. The action of gas and liquid partially fluidized the sphere beds, thus improving gas-liquid contacting and enhancing mass transfer.

A total of 81 tests were conducted, 44 without magnesia addition, and 37 with magnesia. The tests were designed to evaluate the effects of magnesia addition, slurry flow rate to the scrubber, scrubber inlet liquor pH, and total height of spheres on SO_2 removal. Gas velocity has no significant effect on SO_2 removal for the range tested.

The operating conditions were:

Dissolved sulfite concentration = 1.2 (no magnesium), 7-9, or 12-16 m-mole/l
Magnesium concentration = 0-600 m-mole/l
Chloride concentration = 50-500 m-mole/l
Chloride-to-magnesium ratio with magnesia additive = 0.1-1.2 mole/mole
Slurry flow rate = 20, 30, or 40 gpm/ft^2

Gas velocity through scrubber = 8-12 ft/sec
Scrubber inlet liquor pH = 5.0-6.1
Total height of spheres = zero or 15 inches
Scrubber inlet gas SO_2 concentration = 2200-3600 ppm

The following correlation for prediction of SO_2 removal fits the experimental data:

$$\text{Fraction } SO_2 \text{ Removal} = 1 - \exp\left[-\ 7.4 \times 10^{-4}\ L^{0.80} \exp\left(0.035\ H + 0.76\ pH + 0.053\ S\right)\right] \quad (10)$$

where S is the dissolved sulfite concentration ($SO_3^=$, $MgSO_3^0$, $CaSO_3^0$) in m-mole/l calculated by Equation 2, L is the slurry flow rate to the scrubber in gpm/ft^2, H is the total height of spheres in inches and the pH is that of the scrubber inlet liquor.

Equation 10 explains 95 percent of the variation in the data for SO_2 removal with a standard error of estimate of 3.2 percent SO_2 removal. Values of SO_2 removal predicted by Equation 10 are plotted against the corresponding measured values in Figure 7.

The effect of the concentration of dissolved sulfite, the reactive base, on SO_2 removal in Equation 10 can be represented as:

$$N \propto \exp\left(0.053\ S\right) \quad (11)$$

where

$$N = \text{number of mass transfer units} = \ln\left(\frac{1}{1\text{-fraction } SO_2 \text{ removal}}\right)$$

The coefficient of 0.053 on dissolved sulfite concentration has a standard error of 0.0025; i.e., the t-test value for the coefficient is 0.053/0.0025 = 21.

Figure 8 shows the effects of dissolved sulfite concentration and slurry flow rate on SO_2 removal as predicted by Equation 10 for the TCA with 15 inches of spheres and a scrubber

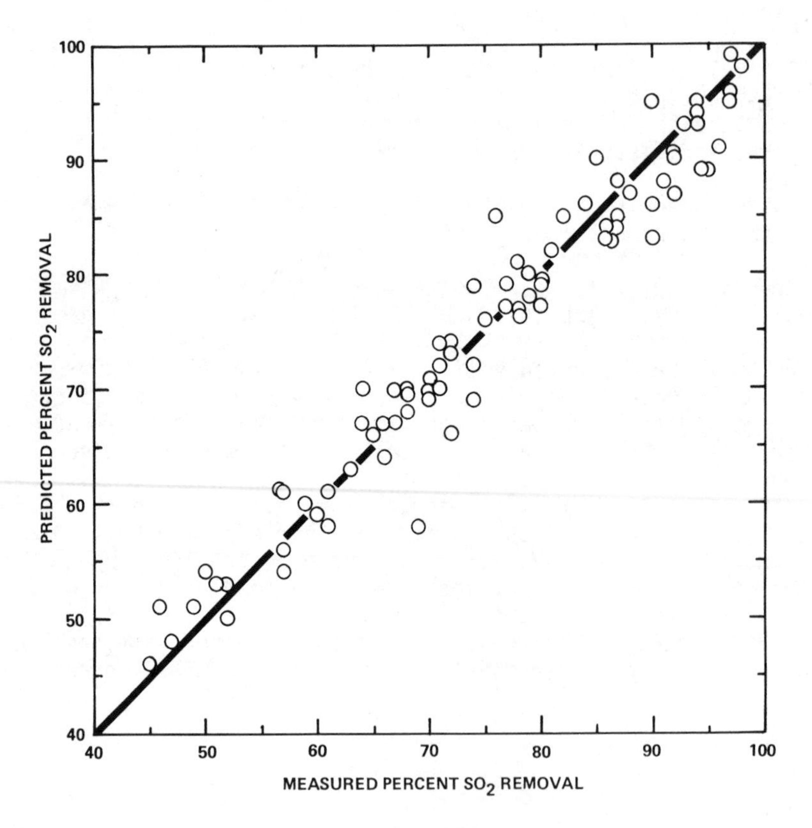

Figure 7. Predicted (Equation 10) vs. measured SO_2 removal for the Shawnee (9) 10-MW turbulent contact absorber

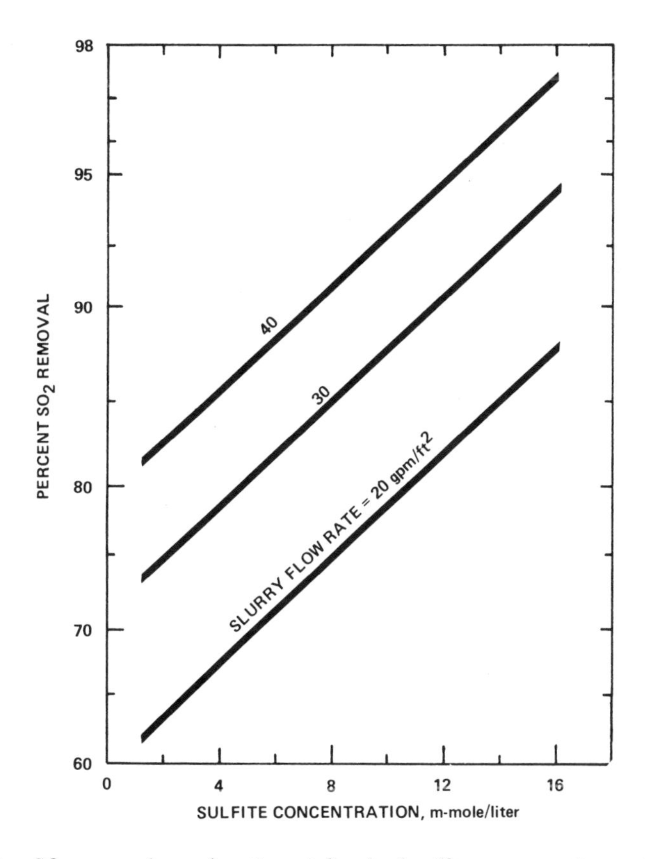

Figure 8. SO_2 removal as a function of dissolved sulfite concentration and slurry flow rate for the Shawnee (9) TCA with 15 in. of spheres at an inlet liquor pH of 5.5

inlet liquor pH of 5.5. For a slurry flow rate of 30 gpm/ft^2, an increase in dissolved sulfite concentration from 1.2 m-mole/l to 10 m-mole/l increases SO_2 removal from 73 to 87 percent, thus reducing SO_2 emission by a factor of two. A further sulfite increase to 16 m-mole/l increases SO_2 removal to 94 percent, decreasing SO_2 emission by an additional factor of two.

In the absence of chloride, these sulfite concentrations of 1.2, 10, and 16 m-mole/l are achieved for dissolved magnesium concentrations of zero, 250, and 450 m-mole/l, respectively (see Figure 4).

Figure 8 indicates that the addition of 10-12 m-mole/l of dissolved sulfite to the liquor has an enhancement effect equivalent to a doubling of the slurry flow rate. According to Equation 10, such an increase in dissolved sulfite is also equivalent to the addition of 15 inches of spheres, or to a scrubber inlet liquor pH increase of 0.7 units, nearly the entire allowable pH range.

An advantage of expressing the magnesia enhancement effect on SO_2 removal in terms of dissolved sulfite concentration is that the sulfite species are the basic species that react with incoming SO_2 gas. This allows a comparison of the scrubbing abilities of sulfite/sulfate liquors containing different dissolved cations, such as calcium, magnesium, and sodium, in terms of the sulfite concentration obtainable with each cation. A limestone wet scrubbing system with calcium-based liquor has only about 1.2 m-mole/l of dissolved sulfite. This calcium system requires a liquid-to-gas ratio (L/G) of about 50 gallons per thousand actual cubic feet at 50 °C, equivalent to 30 gpm/ft^2 of slurry flow rate at a gas velocity of 10 ft/sec, to achieve 70-75 percent SO_2 removal at 2000-3000 ppm inlet SO_2 (see Figure 8). A magnesia-enhanced limestone system having about 15 m-mole/l of dissolved sulfite achieves 90-95 percent SO_2 removal at the same L/G of 50 gal/mcf. A concentrated dual alkali system using sodium sulfite/sulfate scrubbing liquor containing about 150 m-mole/l of dissolved sulfite obtains 90-95 percent SO_2 removal at a much lower L/G of 10 gal/mcf (13,14). The actual selection of scrubber system chemistry and mechanical design is, however, very site specific and dependent upon complex economic considerations.

Summary and Conclusions

- A chemical model has been developed that predicts equilibria for wet scrubbing of SO_2 by limestone slurry, with magnesia additive to enhance SO_2 removal. The model is implemented by an equilibrium computer program.

- Simplified correlations, derived from the computerized model and applicable for a typical scrubber temperature of 50 °C, can be used to facilitate calculations. For specified pH, magnesium concentration, and chloride concentration, the correlations determine equilibrium SO_2 partial pressure and equilibrium concentrations of calcium, sulfite, bisulfite, and sulfate. For actual scrubbing liquors, the correlations also determine the degree of gypsum saturation. The correlations agree with the chemical model to within a relative standard error of estimate of less than three percent.

- The enhancement effect of magnesia on SO_2 removal is caused by increased concentrations of basic sulfite species, $SO_3^=$ and $MgSO_3^o$, that react with the strong acid $SO_2(aq)$ to form bisulfite ion, HSO_3^-.

- The concentration of basic sulfite species increases linearly with magnesium and decreases linearly with chloride. Magnesium chloride, $MgCl_2$, is almost completely ineffective in increasing the sulfite concentration.

- SO_2 removal is strongly correlated with the concentration of dissolved basic sulfite species in the scrubbing liquor. The enhancement effect of sulfite on the number of mass transfer units in a scrubber is expressed by a simple exponential relationship.

- Dissolved sulfite concentration is a convenient basis for comparing limestone/magnesia scrubbing to other types of flue gas wet scrubbing systems.

- SO_2 partial pressure has a strong negative relationship to pH, increasing by about a factor of 80 as the pH decreases by one unit. This relationship of SO_2 partial pressure to pH should be considered in the design of installations where low-sulfur coal is burned.

- The degree of gypsum saturation of a limestone/magnesia scrubbing liquor can be calculated from either dissolved calcium or dissolved SO_2 (sulfite plus bisulfite) analyses. Use of dissolved calcium permits the use of simpler correlations, but the accuracy of calcium measurements in the presence of magnesium is dependent upon the analytical method used.

Abstract

A chemical model previously developed for magnesia wet scrubbing of SO_2 has been extended to magnesia-enhanced limestone scrubbing of SO_2. These magnesia-enhanced limestone systems use scrubbing slurry containing magnesium-based liquor and calcium solids. For such systems at 50°C having dissolved magnesium concentrations of up to 1.2 M, the chemical model has been used to develop correlations for prediction of equilibrium SO_2 partial pressure, equilibrium concentration of dissolved sulfite species that react with SO_2, and the degree of gypsum saturation of actual scrubbing liquors.

Data for magnesia-enhanced limestone scrubbing of SO_2 by a Turbulent Contact Absorber were obtained from the 10-MW equivalent EPA Alkali Scrubbing Test Facility. These data and the chemical model have been used to predict the enhancement effect of magnesia-induced dissolved sulfite concentration on SO_2 removal.

For liquors saturated with both calcium sulfite and gypsum, the concentration of dissolved sulfite species increases linearly with dissolved magnesium concentration, decreases linearly with dissolved chloride concentration, and is only weakly affected by pH.

Scrubber inlet liquor pH should be maintained at 5.5 or higher for low-sulfur coal to avoid excessively high SO_2 back-pressure.

Acknowledgment

This paper utilizes data that were acquired as part of the Environmental Protection Agency's Shawnee Wet Scrubbing Test Program, and is presented with the support and encouragement of John E. Williams, Shawnee Project Officer, EPA Industrial Environmental Research Laboratory, Research Triangle Park, NC.

Literature Cited

1. Rowland, C.H.; Abdulsattar; A.H. Env. Sci. Tech., 1978, 12 (10), 1158.

2. Garrels, R.M.; Christ, C.L. "Solutions, Minerals, and Equilibria"; Harper and Row: New York, 1965.

3. Lowell, P.S.; Ottmers, D.M.; Strange, T.I.; Schwitzgebel, K.; De Berry, D.W. "A Theoretical Description of the Limestone Injection-Wet Scrubbing Process"; Final Report for EPA Contract No. CPA-22-69-138 with Radian Corporation, 1970.

4. Johnstone, H.F.; Leppla, D.W. J. Am. Chem. Soc., 1934, 56, 2233.

5. Tartar, H.V.; Garretson, H.H. J. Am. Chem. Soc., 1941, 63, 808.

6. Nair, V.S.K.; Nancollas, G.H. J. Chem. Soc. (London), 1958, 3706.

7. Bell, R.P.; George, J.H.B. Trans. Faraday Soc., 1953, 49, 619.

8. Latimer, W.M. "Oxidation Potentials"; Prentice Hall: New York, 1952.

9. Bechtel Corporation, "EPA Alkali Scrubbing Test Facility: Advanced Program, Third Progress Report"; EPA Report 600/7-77-105, September 1977.

10. Borgwardt, R.H. "Limestone Scrubbing at EPA Pilot Plant"; Progress Report No. 13, November 1973.

11. Rowland, C.H.; Bell, N.E.; Leivo, C.C.; Head, H.N. "Predicting SO_2 Removal by Limestone/Lime Wet Scrubbing: Correlations of Shawnee Data"; presented at 71st Annual APCA Meeting, Houston, June 25-30, 1978.

12. Universal Oil Products, Air Correction Division. "UOP Wet Scrubbers"; Bulletin No. 608, 1971.

13. La Mantia, C.R. "Final Report: Dual Alkali Test and Evaluation Program, Volume III. Prototype Test Program - Plant Scholz"; EPA Report 600/7-77-050c, May 1977.

14. Kaplan, N. "Summary of Utility Dual Alkali Systems"; presented at EPA Flue Gas Desulfurization Symposium, Las Vegas, March 5-8, 1979.

RECEIVED February 14, 1980.

12

Sulfur Dioxide Vapor Pressure and pH of Sodium Citrate Buffer Solutions with Dissolved Sulfur Dioxide

GARY T. ROCHELLE

Department of Chemical Engineering, University of Texas at Austin, Austin, TX 78712

Aqueous scrubbing followed by steam stripping is a potentially attractive method of desulfurizing stack gas with the production of concentrated SO_2 (1, 2). SO_2 is absorbed from stack gas containing 500 to 5000 ppm SO_2 by an aqueous solution at 30° to 60°C. The solution is regenerated by stripping with steam at 80° to 150°C. Liquid water is easily condensed from the stripper overhead vapor, leaving concentrated SO_2. This process has not received commercial acceptance because it generally requires an excessive amount of steam for stripping.

Sodium citrate was recognized as a potential aqueous absorbent for absorption/stripping as early as 1934 (3, 4). It has recently reappeared in work by the U. S. Bureau of Mines (5), in process development sponsored by Peabody, Inc., and in a process offered by Flakt, Inc. (6). This paper reports on work which is part of a development program on absorption/stripping sponsored by the Electric Power Research Institute.

Buffers in the pH range of 3.5 to 5.5 provide for reversible SO_2 absorption as bisulfite (HSO_3^-) by the acid/base reaction:

$$SO_2(g) + H_2O \rightleftharpoons H^+ + HSO_3^-$$

In a perfectly-buffered solution the SO_2 vapor pressure will be directly proportional to the total concentration of SO_2 and bisulfite, giving a linear equilibrium relationship. In simple alkali sulfite solution without added buffer, the equilibrium relationship is highly nonlinear, because H^+ accumulates as SO_2 is absorbed. Under these conditions is it not possible to carry out reversible SO_2 absorption/stripping in a simple system, resulting in greater steam requirements than expected with a linear equilibrium relationship. Weak acid buffers such as sodium citrate have been proposed to "straighten" the equilibrium relationship and thereby reduce ultimate steam requirements (1, 2, 7). Citrate buffer is attractive because it is effective over a wide range, from pH 2.5 to pH 5.5 in concentrated solutions.

Johnstone, et al, (2) found that the ratio of SO_2 vapor

0-8412-0569-8/80/47-133-269$05.75/0

pressure to H_2O vapor pressure (P_{SO_2}/P_{H_2O}) over weak acid buffer solutions was generally independent of temperature. The reaction of HSO_3^- with H^+ to give gaseous SO_2 has an enthalpy change of about 10.8 kcal/g-mole (8). The extraction of H^+ from a weak acid usually requires negligible enthalpy change. Therefore the net enthalpy change for desorption of gaseous SO_2 from a weak acid is about 10.8 kcal/g-mole, almost equal to the enthalpy of water vaporization:

$$HSO_3^- + H^+ \rightarrow SO_2(g) + H_2O(1) \qquad \Delta H = 10.8 \text{ kcal/g-mole}$$

$$HA \rightarrow H^+ + A^- \qquad \Delta H \approx 0$$

$$HSO_3^- + HA \rightarrow SO_2(g) + H_2O(1) + A^- \qquad \Delta H = 10.8 \text{ kcal/g-mole}$$

It is expected that sodium citrate solution will behave as a typical weak acid buffer. Both solution pH and P_{SO_2}/P_{H_2O} should be independent of temperature. Under these conditions, the steam requirements will generally be independent of the stripper operating pressure and temperature (1, 2).

This work was carried out to confirm minimal temperature dependence of P_{SO_2}/P_{H_2O} over sodium citrate solutions and to determine the dependence of P_{SO_2}/P_{H_2O} on solution composition. Measurements of pH as a function of temperature and solution composition have been performed in order to separate the effects of the specific buffer on P_{SO_2}/P_{H_2O}. Design calculations are presented to estimate the steam requirements on typical applications.

Experimental Methods

Solution Preparation. Solutions were prepared from reagent grade citric acid monohydrate, sodium citrate dihydrate, $NaHSO_3$, Na_2SO_4, NaCl, and standardized NaOH solution. Hydroquinone (0.1 wt %) was added to inhibit oxidation of solutions with $NaHSO_3$.

The $NaHSO_3$ was analyzed by iodine titration and was typically 97-98% of the expected SO_2 content. Several of the solutions used for vapor/liquid equilibrium experiments were analyzed for total SO_2 and found to contain 5 to 10% less than the nominal concentration. Nominal concentrations were used in presenting and analyzing the data, unless noted otherwise. Therefore, correlated values of P_{SO_2} may be 5 to 10% low for a given solution composition.

pH Measurements. A Ag/AgCl combination electrode was used for all measurements at 25°C. A thalamid combination electrode was used at 55° and 95°C. It was conditioned in pH 4.00 buffer at 55° or 95°C for at least 24 hours before use. The electrodes were standardized at the measurement temperature by phthalate buffer at pH 4.00 and phosphate buffer at pH 7.00. Response of the electrodes to citrate buffers required 15 to 30 minutes for a stable reading. In later experiments secondary citrate buffers were used to standardize the electrodes and thereby reduce equilibrium time.

Dynamic Saturation. Dynamic saturation was the primary method used for vapor-liquid equilibrium (VLE) determinations. Typical apparatus is shown in figure 1. N_2 gas was sparged through a solution of known, relatively constant composition, then analyzed for SO_2. Typically 1 to 5 liters of N_2 or N_2/SO_2 was sparged through a coarse, fritted-glass dispersion tube submerged 7 to 15 cm in 125 to 300 ml of solution in a single-stage glass or stainless steel saturator. Gas leaving the saturator was assumed to be in equilibrium with the solution. Comparable results were obtained in experiments using three saturators in series and in experiments using SO_2 absorption rather than desorption. The saturator was maintained ± 1°C. Most data at 25° to 55°C were taken at atmospheric pressure. Data at 55° to 150°C were taken at 4.4 to 12 atm.

SO_2 content of the gas was determined by sparging into a 125 ml gas washing bottle containing a known amount of I_2 in acetate buffer at pH 4-5.5 with a starch indicator. N_2 flow was measured by a wet test meter.

Water content of the gas leaving the saturator was estimated using a modified Raoult's law. Constants for vapor pressure lowering were obtained from Weast (8). Constants for $NaH_2Citrate$, $Na_2HCitrate$, $Na_3Citrate$, and $NaHSO_3$ were assumed to be equal to those for NaH_2PO_4, Na_2HPO_4, Na_3PO_4, and NaCl, respectively. The activity of water was assumed to be independent of temperature and is given by:

$$(1 - a_{H_2O})760 = 30.0\,[Na_3Cit] + 23.5\,[Na_2HCit] + 21.0\,[NaH_2Cit] + 25.2\,([NaHSO_3] + [NaCl]) + 25.0\,([Na_2SO_4] + [Na_2S_2O_3]) \quad (1)$$

The lowest water activity encountered in these experiments was about 0.88 (2 M Citrate). The vapor pressure of water over the solution is given by:

$$p_{H_2O} = a_{H_2O} P^{\circ}_{H_2O}$$

where $P^{\circ}_{H_2O}$ is the vapor pressure of pure water at the saturator temperature.

Assuming that the vapor leaving the saturator is an ideal gas the SO_2 vapor pressure is given by:

$$P_{SO_2} = \frac{n_{SO_2}}{n_{SO_2} + n_{N_2}}(P_T - P_{H_2O})$$

where n_{SO_2} and n_{N_2} are the moles of SO_2 and N_2 and P_T is the total pressure. Under most conditions P_{H_2O} was less than 10% of P_T. Therefore, P_{SO_2} is relatively insensitive to errors in P_{H_2O}.

SO_2 Electrode. A gas-sensing SO_2 electrode marketed by Ionics, Inc. was used to provide additional VLE data at 25°C as a function of composition. Aqueous SO_2 equilibrates across a polymeric membrane with a filling solution containing about 0.1 M $NaHSO_3$. Ionic species do not diffuse across the membrane. A small combination glass electrode measures the pH of the filling solution. The SO_2 activity (P_{SO_2}) is proportional to the activity of H^+ (10^{-pH}), because the bisulfite activity is constant:

$$HSO_3^- + H^+ \rightleftharpoons SO_2(g) + H_2O$$

$$P_{SO_2} = K\ a_{H^+} a_{HSO_3^-}$$

The SO_2 electrode has a linear response to P_{SO_2} as low as 10^{-5} atm.

Water also diffuses across the polymer membrane to a limited extent. Therefore the electrode response is unstable and unreliable if there is a significant difference between the osmotic pressure of the filling solution and the unknown solution. To partially alleviate this problem,data were taken with filling solutions containing 0, 1.0, and 2.0 M additional KCl.

The response of the SO_2 electrode is actually read as voltage, (mV). Two constants are needed to convert this to P_{SO_2}:

$$\log P_{SO_2} = c + \frac{V}{d}$$

We have used the theoretical value of d at 25°C which is 59.16 mV. The constant c must be determined by measurement of a known solution each day the electrode is used. Alternatively, the electrode can be used to provide relative P_{SO_2} for a series of measurements.

pH Behavior

A semiempirical correlation of pH measurements with 156 solutions gave the following relationship:

$$pH = a_1 + a_2 f + a_3 [\text{Anion}]_T^{0.5} \tag{1}$$

$$a_1 = 2.77 \pm 0.05$$

$$a_2 = 3.60 \pm 0.05$$

$$a_3 = -0.53 \pm 0.03$$

The total concentration of anions is given by:

$$[\text{Anion}]_T = [\text{Citrate}] + [Na_2SO_4] + [NaCl] + [NaHSO_3]$$

Neglecting H^+ and assuming that SO_2 is present as bisulfite, the fraction neutralization is defined as:

$$f = \frac{[Na^+] - [SO_2] - [Cl^-] - 2[SO_4^{=}]}{3[\text{Citrate}]}$$

This correlation includes solutions over the following range of conditions:

$$f = 0.20 - 0.833$$

$$[\text{Citrate}] = 0.05 - 2.0\ M$$

$$[Na_2SO_4] = 0 - 1.5\ M$$

$$[NaHSO_3] = 0 - 1.0\ M$$

$$[NaCl] = 0 - 1.9\ M$$

The standard deviation of pH prediction is 0.12 pH units.

In these solutions pH is more strongly correlated with total anion concentration than with ionic strength. Thus 1 $\underline{M}$ Na_2SO_4 and 1 $\underline{M}$ NaCl have about the same effect on the pH of a solution at a given fraction neutralization. Figure 2 shows pH at 50% neutralization as a function of anion concentration in the solutions which are primarily citrate, Na_2SO_4, or NaCl, as well as in mixed solutions.

The effect of solution anionic concentration is probably related to effects on activity coefficients and ion pair formation of more highly charged buffer species. In more concentrated solutions, the activity of the highly charged species is reduced by both ionic strength and ion pair formation. The effect on less charged, acidic species is less. Therefore, as solutions become more concentrated, the activity of basic species is reduced relative to that of acidic species, and at a given fraction

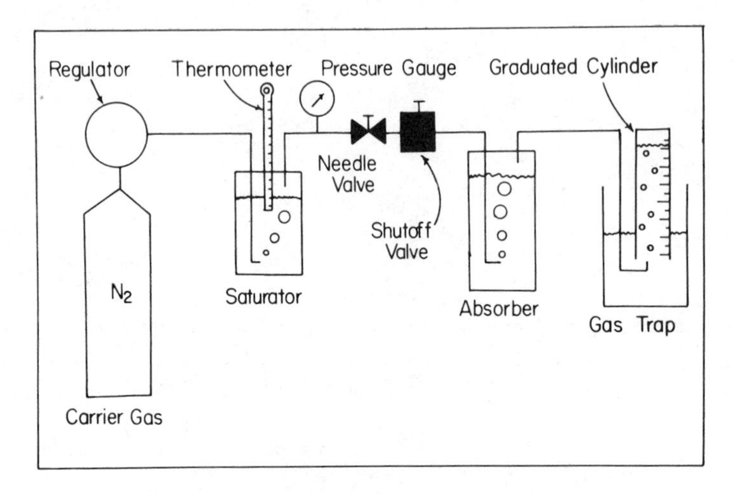

Figure 1. Apparatus for dynamic saturation

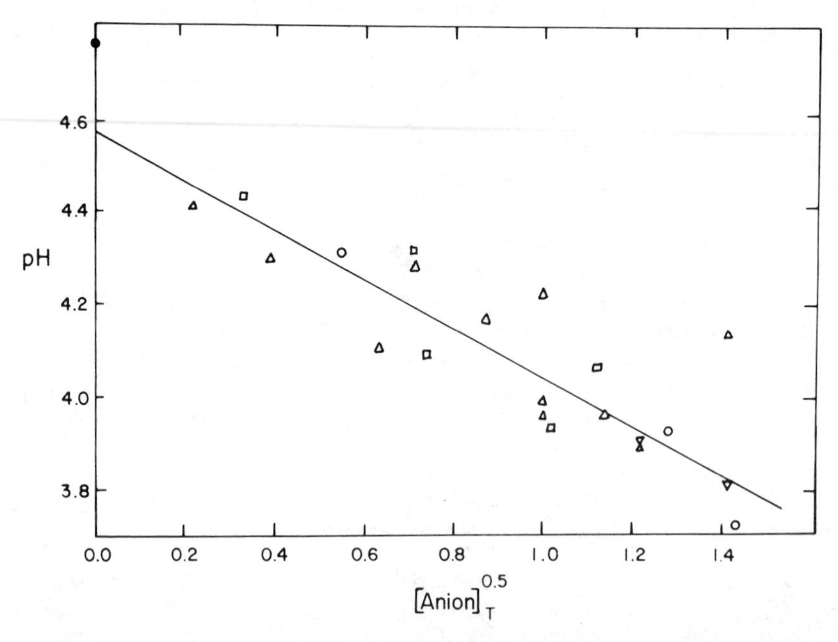

Figure 2. Dependence of pH on total anion concentration ((●) infinite dilution; (○) NaCl; (□) Na_2SO_4; (△) sodium citrate; (▽) mixed solution)

neutralization the pH decreases. In concentrated solutions polyvalent anions such as $SO_4^=$ and citrate$^{\equiv}$ probably form ion pairs such as $NaSO_4^-$ and $Na_2citrate^+$. Therefore the effective ionic strength and Na^+ concentration available for ion pair formation varies with total anion concentration rather than ionic strength.

The effect of fraction neutralization on pH is illustrated in figure 3 for a solution of constant anionic concentration. This corresponds to titrating citric acid with NaOH. The titration curve is very nearly linear from pH 2.2 to about pH 5.5 with a slope of 3.60. The effects of the three functional buffer groups of citric acid are smeared so that no S-shape or inflection points are apparent.

As shown in figure 4, titration with HCl or absorption of SO_2 as bisulfite results in a different dependence of pH on fraction neutralization because the total anion concentration is increased. The slope of pH versus f is typically greater than 4.0 and is modeled by equation (1).

As shown in Table I, the ΔH values of the buffer reactions corresponding roughly to K_{a1}, K_{a2}, and K_{a3} (16.7, 50.0, and 83.3% neutralization, respectively) all have absolute values less than 2.0 kcal/g-mole. The reactions corresponding to K_{a2} and K_{a3}, which are most relevant for SO_2 absorption/stripping, had ΔH values of -0.7 and +0.1 kcal/g-mole, respectively. The careful data of Bates and Pinching (9) in dilute citrate solutions give ΔH values between 25° and 50°C of -0.01, -0.002, and +0.03 kcal/g-mole, respectively for K_{a1}, K_{a2}, and K_{a3}. Because overall temperature effects in SO_2 absorption/stripping are on the order of 10 kcal/g-mole, we can neglect the enthalpy change of the buffer reaction.

SO_2 Vapor Pressure

SO_2 vapor pressure was determined in eight series of experiments (Tables II and III) with a total of about 80 solutions over the following range of conditions:

f = 0.40 - 0.80

[Citrate] = 0.2 - 2.0 M

$[NaHSO_3]$ = 0.025 - 1.0 M

$[Na_2SO_4]$ = 0 - 1 M

$[Na_2S_2O_3]$ = 0 - 0.6 M

T = 25 - 168°C

A semiempirical correlation of the data gives:

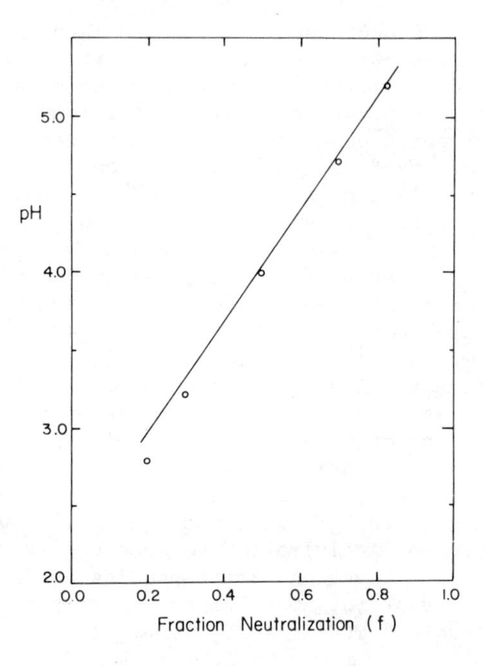

Figure 3. Dependence of pH on fraction neutralization—titration with NaOH

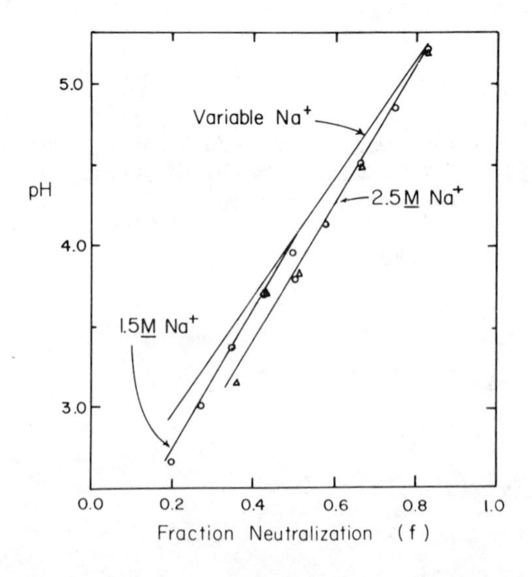

Figure 4. Dependence of pH on fraction neutralization—titration with HCl (△) or SO_2 (○)

Table I: Heat of Reaction of H^+ with Concentrated Sodium Citrate Buffers

Fraction neutralization[a]	Solution composition (M)	ΔH[d] (kcal/g-mole)
0.167	0.30 NaCl	- 1.6
	1.50	+ 0.2
	0.60 Citrate	- 2.8
	3.00	- 3.8[c]
	Mean	- 2.0 ± 1.7
0.500	0.30 NaCl	- 0.9
	1.50	- 0.4
	0.50 Na_2SO_4	+ 0.2
	0.15 Citrate	- 0.7
	0.50	- 1.5[c]
	0.75	- 0.1
	1.00	- 1.0[c]
	2.00	- 0.9[c]
	Mean	- 0.7 ± 0.5
0.833	0.30 NaCl	+ 0.7
	1.50	+ 0.3
	0.50 Na_2SO_4	+ 0.4
	0.0667 Citrate	+ 0.7
	0.333	- 0.1
	0.50	- 0.2[c]
	1.00	- 0.4[c]
	2.00	- 0.9[c]
	Mean	+ 0.1 ± 0.6

[a]fraction neutralization $= \dfrac{[Na^+] - [Cl^-] - 2[SO_4^=]}{3[\text{Citrate}]}$

[b]Chloride and sulfate solutions contain 0.00667 M Citrate. All solutions contain NaOH as indicated by fraction neutralization

[c]ΔH calculated from pH values at 25° and 95°C

[d]ΔH calculated from pH values at 25° and 55°C unless noted

Table II: SO_2 Vapor Pressure Obtained by Dynamic Saturation at 25° to 158°C, Series 1

Solution Composition (M) Citrate	SO_2	f	T (°C)	P_{SO_2}/P_{H_2O} x 10^3	Calc P_{SO_2} / Meas P_{SO_2}
0.5	0.20	0.533	27	14.8	
			55	12.1	
			55	14.7	
			75	14.2	
			95	14.0	
			95	12.4	
			115	14.3	
			135	13.2	
			mean	13.7 ± 1.0	1.26
1.0	0.25	0.583	24	14.6	
			55	18.8	
			75	16.3	
			95	16.5	
			115	17.1	
			135	15.7	
			mean	16.5 ± 1.4	1.06
2.0	0.80	0.700	55	31.7	
			75	34.8	
			95	36.6	
			115	33.6	
			135	35.9	
			mean	34.5 ± 1.9	0.93
0.5[a]	0.20	0.533	25	22.2	
			168	19.0	
			mean	20.6 ± 2.3	0.85
1.0	0.164[b]	0.612	24	8.7	
			51	10.6	
			92	10.8	
			117	11.4	
			148	10.6	
			mean	10.4 ± 1.0	0.86

[a] 0.05 M $Na_2S_2O_3$ was also added to the solution

[b] As analyzed by iodine titration, nominal SO_2 concentration was 0.2 M

Table III: SO_2 Vapor Pressure at 25°C Obtained by Dynamic Saturation or the SO_2 Electrode

Solution Composition (M) Citrate	SO_2	f	pH	a_{H_2O}	P_{SO_2}/P_{H_2O} x 10^3	Calc P_{SO_2} / Meas P_{SO_2}
Series 2, Dynamic Saturation						
0.5	0.20 a	0.533	4.60	0.970	14.5	1.27
2.0	1.00	0.667	4.38	0.910	77.6	0.69
1.0	0.50	0.667	4.32	0.968	25.0	0.79
2.0	0.60	0.567	4.01	0.923	70.7	0.90
1.0	0.30	0.567	4.04	0.961	26.6	0.90
0.5	0.15	0.567	4.22	0.980	8.9	1.10
2.0	0.20	0.467	3.85	0.938	38.7	1.07
1.0	0.10	0.467	3.90	0.968	17.7	0.91
0.5	0.05	0.467	3.94	0.984	7.0	0.96
2.0	0.20	0.800	5.12	0.922	3.2	1.07
1.0	0.10	0.800	5.11	0.961	1.1	1.18
2.0	0.30	0.617	4.39	0.930	22.9	0.90
1.0	0.15	0.617	4.37	0.965	7.8	1.02
0.5	0.075	0.617	4.45	0.982	4.3	0.76
					mean	0.97 ± 0.16
Series 3, Dynamic Saturation						
2.0	0.2	0.52	4.24		20.7	1.35
1.0	0.1	0.52	4.22		7.3	1.48
0.5	0.05	0.52	4.33		3.01	1.49
1.0	0.9	0.52	4.17		83.8	1.41
1.0	0.1 f	0.52	4.22		9.88	1.39
1.0	0.1 b	0.52	4.15		9.06	1.38
2.0	0.2	0.74	4.99		3.15	1.68
1.0	0.1	0.74	4.96		1.31	1.56
0.5	0.05	0.74	4.99		0.76	1.10
1.0	0.90	0.74	4.83		14.9	1.51
1.0	0.10 f	0.74	4.90		2.42	1.07
1.0	0.10 b	0.74	4.89		1.53	1.56
					mean	1.42 ± 0.18

... continued

Table III: Continued

Solution Composition (M) Citrate	SO_2	f	pH	a_{H_2O}	P_{SO_2}/P_{H_2O} x 10^3	Calc P_{SO_2} / Meas P_{SO_2}
Series 4, SO_2 Electrode with 1 M KCl						
0.50	0.40	0.400	3.81		97.1	1.03
0.50	0.20	0.533	4.33		16.1	1.07
0.50	0.10	0.600	4.60		4.7	1.06
0.50	0.05	0.633	4.73		1.93	0.99
0.50	0.025	0.650	4.83		0.95	0.87
1.00	0.40	0.533	4.09		37.6	1.13
1.00	0.20	0.600	4.34		11.1	1.09
1.00	0.10	0.633	4.47		4.5	1.03
1.00	0.05	0.650	4.53		2.0	0.98
1.00	0.025	0.658	4.57		1.0	0.91
					mean	1.02 ± 0.08
Series 5, SO_2 Electrode with 2 M KCl						
2.0	1.0	0.667	4.38		55.2	0.97
1.0	0.5	0.667	4.32		22.4	0.88
2.0	0.6	0.567	4.01		60.5	1.05
1.0	0.3	0.567	4.04		24.4	0.98
2.0	0.20	0.467	3.85		35.4	1.18
1.0	0.10	0.467	3.90		15.8	1.02
1.0	0.10	0.800	5.11		1.3	1.00
2.0	0.30	0.617	4.39		21.0	0.98
1.0	0.15	0.617	4.37		8.4	0.94
					mean	1.00 ± 0.08
Series 6, SO_2 Electrode with 2 M KCl						
1.0	0.02	0.493	4.00		2.72	0.95
1.0	0.065	0.478	3.94		10.2	0.93
1,0	0.11	0.463	3.85		16.3	1.14
1.0	0.155	0.448	3.80		27.9	1.04
1.0	0.20	0.438	3.71		43.9	0.97
1.0	0.05	0.650	4.47		2.14	0.93
1.0	0.1625	0.613	4.37		10.0	0.88
1.0	0.275	0.575	4.16		22.0	0.93
1.0	0.3875	0.538	4.08		39.3	1.00
1.0	0.500	0.500	3.90		65.9	1.05
1.0	0.05	0.650	4.35		2.47	1.03

... continued

Table III: Continued

Solution Composition ($\underline{M}$) Citrate	SO_2	f	pH	a_{H_2O}	P_{SO_2}/P_{H_2O} x 10^3	Calc P_{SO_2} / Meas P_{SO_2}
Series 6, continued						
1.0	0.05	0.650	4.40		2.34	0.97
1.0	0.275	0.575	4.12		22.0	0.93
1.0	0.08	0.807	4.99		1.03	0.96
1.0	0.26	0.747	4.75		5.38	0.98
1.0	0.44	0.687	4.51		14.5	1.02
1.0	0.62	0.627	4.27		33.0	1.03
1.0	0.80	0.567	4.01		65.0	1.11
					mean	0.99 ± 0.07
Series 7, Dynamic Saturation, 17 experiments						
1.0	0.45 c	0.483			58.0 ± 4.1	1.22 ± 0.09
Series 8, Dynamic Saturation, SO_2 analyzer						
1.6	0.17 i	0.506	3.92		36.3	0.67
1.0	0.17	0.510	3.84		30.1	0.67
0.4	0.17	0.525	3.81		25.4	0.60
0.2	0.17	0.550	3.82		22.7	0.50
1.0	0.17 f	0.510	3.82		39.3	0.65
1.0	0.17 d	0.510	3.81		33.9	0.68
1.0	0.17	0.510	3.72		40.3	0.63
0.2	0.17	0.550	3.83		36.2	0.49
					mean	0.61 ± 0.08

a +0.2 $\underline{M}$ $Na_2S_2O_3$

b +0.6 $\underline{M}$ $Na_2S_2O_3$

c +0.05 $\underline{M}$ Na_2SO_4

d +0.5 $\underline{M}$ Na_2SO_4

e +0.6 $\underline{M}$ Na_2SO_4

f +1.0 $\underline{M}$ Na_2SO_4

g +1.5 $\underline{M}$ Na_2SO_4

h +1.0 $\underline{M}$ NaCl

i Determined by iodine titration. Nominal concentration was 0.2 $\underline{M}$ SO_2.

$$\log K_c = a_4 + a_5 f + a_6 [\text{Anion}]_T^{0.5} \tag{2}$$

$$a_4 = 0.46 \pm 0.075$$

$$a_5 = -3.27 \pm 0.12$$

$$a_6 = 0.26 \pm 0.04$$

where

$$K_c = \frac{P_{SO_2}}{P_{H_2O}[SO_2]_T}$$

The correlation includes temperature dependence as the calculated value of water vapor pressure over the solution (P_{H_2O}).

For the entire set of data, the standard deviation for prediction of log K_c is 0.011 or about 27% of K_c. However, there is evidence of systematic error in series 3 and series 8, and from run to run in series 1. This error does not appear to be associated with variables such as solution composition or temperature. It may have resulted from errors in standardization of iodine solution or in calibration of the SO_2 analyzer (for series 8). This systematic error affects the usefulness of these series in determining the constant a_4, but they still provide valuable data for use in determining dependence of the equilibrium on temperature, fraction neutralization, and total anion concentration.

The first series of data includes five solutions with measurements by dynamic saturation at 25° to 168°C. As shown in Table II the ratio P_{SO_2}/P_{H_2O} is essentially independent of temperature. Yet from 25° to 150°C, P_{SO_2} increases a factor of 150. For any given solution the standard deviation of P_{SO_2}/P_{H_2O} over the temperature range was generally less than 10%. However, the error of estimate of K_c by equation (2) is as high as 26%, suggesting systematic errors from one experiment to the next.

Series 2 by dynamic saturation and series 3, 4, and 5 by the SO_2 electrode were intended primarily to show the effect of fraction neutralization. Data collected by the SO_2 electrode have been calibrated by the correlation (equation (2)) so that they contribute to a_5 and a_6, the slope terms, but not to a_4, the intercept. Figure 5 shows the ratio of calculated and measured values of P_{SO_2} as a function of the fraction neutralization. There is no consistent trend with fraction neutralization, so the correlating equation is adequate. However, series 3 is high and series 8 is low, suggesting systematic errors. Each of series 2 through 8 are internally consistent, with standard deviations around a mean error less than 17%.

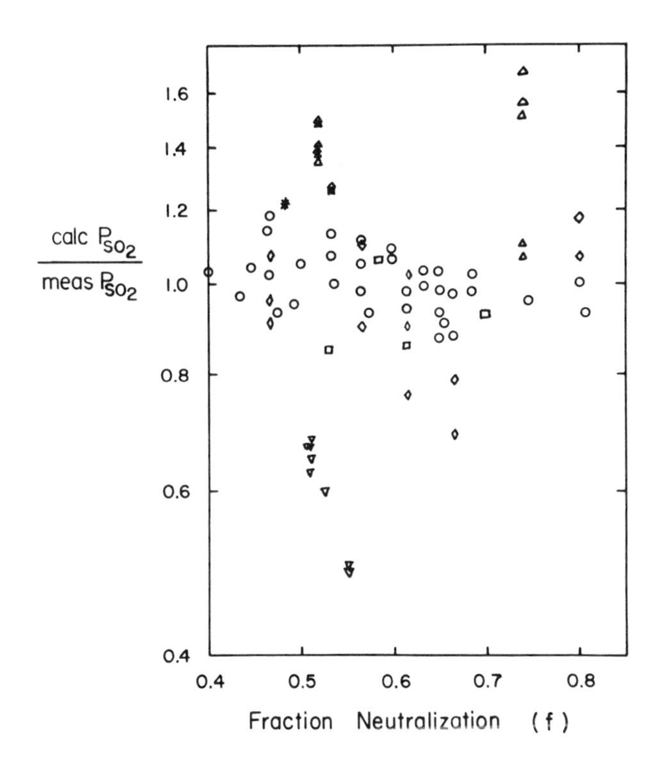

Figure 5. Relative error of P_{SO_2} prediction vs. fraction neutralization (series: 1 (□); 2 (◇); 3 (△); 4,5,6 (○); 7 (*); 8 (▽))

Series 7 included 17 experiments with a single solution composition but with a variable number of saturators and absorbers in the dynamic saturation apparatus. The results had a standard deviation of about 8%.

Series 8 in combination with earlier series was intended to provide data on the effects of total anion concentration. The results are internally consistant with the correlation, having a standard deviation of about 15% around the mean error. However the measured values of P_{SO_2} were about 40% lower than the general correlation. An SO_2 analyzer, rather than iodine titration, was used to determine SO_2 gas concentration from the saturator. The analyzer was calibrated with dry SO_2/N_2 span gas. In later experiments it was shown that humid gas gives a lower analyzer response. With constant fraction neutralization increased anionic concentration increases P_{SO_2} because pH decreases faster than effective bisulfite activity.

Equations (1) and (2) can be combined to give an explicit prediction of P_{SO_2} as a function of pH:

$$\log K_c = (a_1 + a_4) + (a_2 + a_5)f + (a_3 + a_6)\ [\text{Anion}]_T^{0.5} - \text{pH}$$

$$a_1 + a_4 = 3.23$$

$$a_2 + a_5 = 0.33$$

$$a_3 + a_6 = -0.27$$

The dependence on fraction neutralization and total anion concentration should reflect the extent to which the bisulfite activity is not proportional to total dissolved SO_2. As expected, the dependence on f is quite small, since dissolved SO_2 is present primarily as bisulfite at pH 3.5 to 5.0. The effect of anion concentration is in the direction expected since bisulfite activity would be reduced by ion pairing in more concentrated solutions.

Design Calculations for Absorption/Stripping

The primary characteristic of an acid/base buffer affecting steam requirements is the temperature dependence of P_{SO_2}/P_{H_2O}. Since P_{SO_2}/P_{H_2O} of the citrate system is independent of temperature, the ideal minimum steam requirement (moles H_2O/moles SO_2) of a simple stripper is equal to the ratio, P_{SO_2}/P_{H_2O}, of the humidified inlet stack gas (1). This steam requirement is independent of the stripper temperature, but assumes that the stripper feed is preheated to its boiling point, that there are an infinite number of stages in the absorber and stripper, and that the equilibrium curve is linear.

The second characteristic of equilibrium data affecting steam requirements is the nonlinearity of P_{SO_2}/P_{H_2O} versus $[SO_2]$. That nonlinearity is quantified primarily by the dependence of K_c on f. Figure 6 shows predicted and measured values of P_{SO_2}/P_{H_2O} versus $[SO_2]$ for 1 M citrate solutions with 1.5,2.0,and 2.5 M Na^+.

Performance of a simple absorption/stripping system can be determined by use of a McCabe-Thiele diagram. Equilibrium for both stripper and absorber can be represented as a single line when plotting P_{SO_2}/P_{H_2O} versus $[SO_2]$. Material balance gives operating lines for the absorber and stripper. The slope of the absorber operating line is the ratio of liters of circulating solution to moles of H_2O in the saturated stack gas. The slope of the stripper operating line is the ratio of liters of circulating solution to moles of steam. The steam requirement in moles per mole of SO_2 absorbed is equal to the inverse of the ratio, P_{SO_2}/P_{H_2O}, at the top of the stripper.

Figure 7 illustrates how minimum steam requirements can be estimated for 1 M citrate with 2.0 molar Na^+ in a simple absorption/stripping system to remove 90% of the SO_2 from stack gas at 55°C containing 3000 ppm SO_2. The operating lines for the absorber and stripper are straight, assuming that the H_2O vapor rate is constant throughout the absorber and throughout the stripper. With an infinite number of stages the absorber is pinched at the top and bottom. Using live steam, the stripper pinches in the middle because of the nonlinearity of the equilibrium curve. The gas leaving the stripper would have P_{SO_2}/P_{H_2O} equal to 0.0173, giving a minimum steam requirement of 57.8 moles H_2O/mole SO_2 (16.3 kg/kg). If the equilibrium were linear the minimum steam requirement would be 50 moles H_2O/mole SO_2. Thus, nonlinearity increases the steam requirement by a factor of 1.16.

Figure 8 illustrates the performance of a system with three equilibrium stages in the absorber and six in the stripper. The actual steam requirement is 147 moles/mole SO_2 (41.3 kg/kg). The use of a finite number of stages increases the steam requirement a factor of 2.5 from the case of infinite stages with a nonlinear equilibrium.

Table IV gives minimum steam requirement (infinite stages) at several different solution capacities. The factor attributable to equilibrium nonlinearity increases as more SO_2 is absorbed, because the buffer capacity is consumed to a greater extent. Any capacity for SO_2 absorption can be achieved by varying Na concentration (pH) in the solution. At low pH ([Na] = 1.5 M) the solution capacity for SO_2 absorption is small, but the nonlinearity factor is also small (1.05). Solution capacity can be increased by operating at higher pH ([Na] = 2.5 M), but nonlinearity is more severe (1.32).

As shown by case 3 in Table IV, the minimum steam requirement in an optimized system is not sensitive to the magnitude of P_{SO_2} over the solution, but only to its dependence on temperature

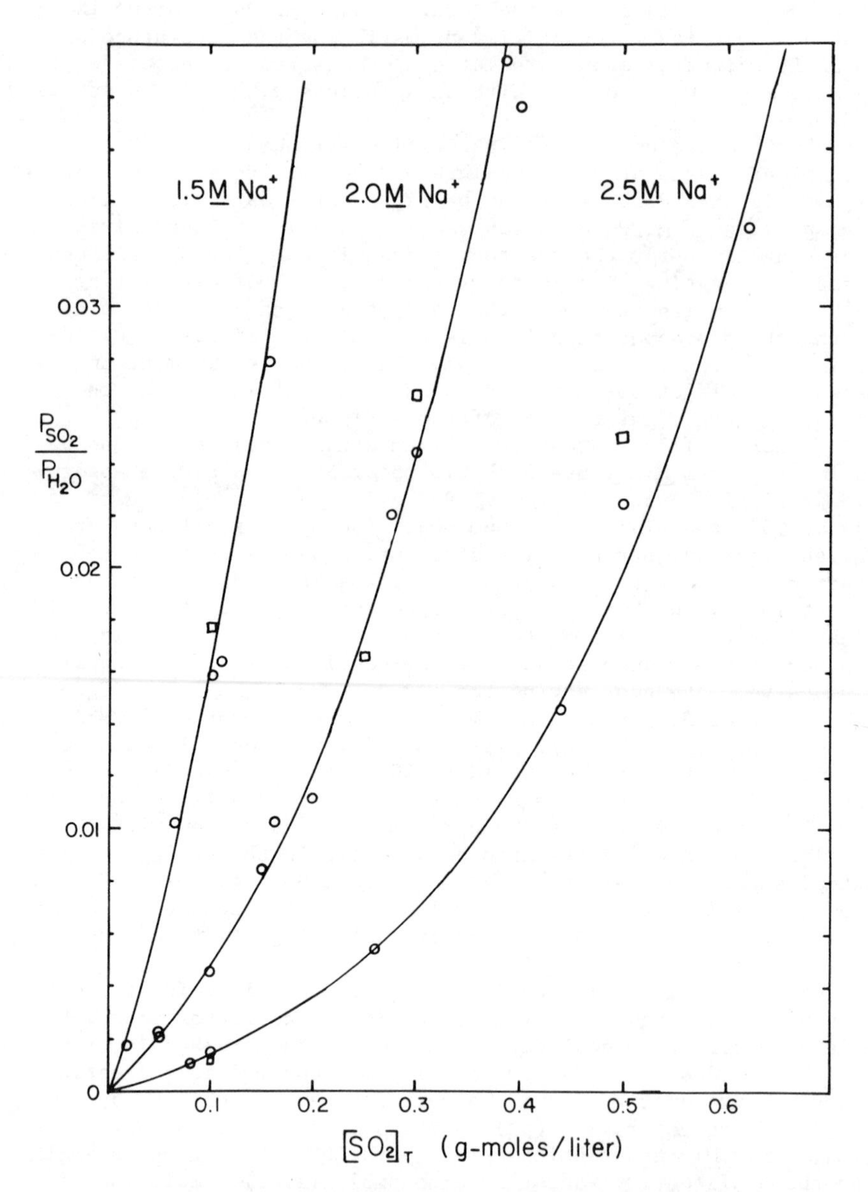

*Figure 6. Dependence of P_{SO_2} on solution composition—1.0*M *citrate, 25°C ((○) SO_2 electrode; (□) dynamic saturation)*

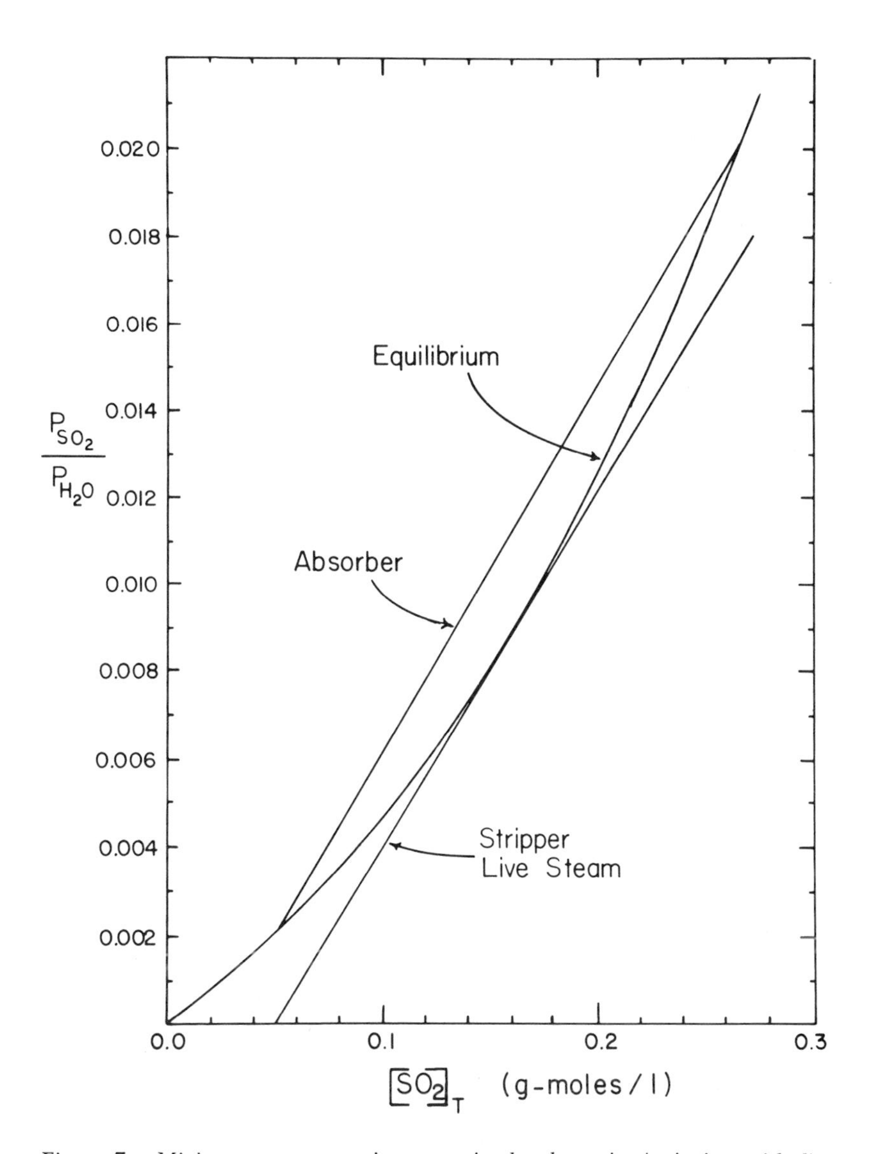

Figure 7. Minimum steam requirement, simple absorption/stripping with live steam, 3000 ppm SO_2 in at 55°C, 90% removal, 1.0M citrate, 2.0M Na

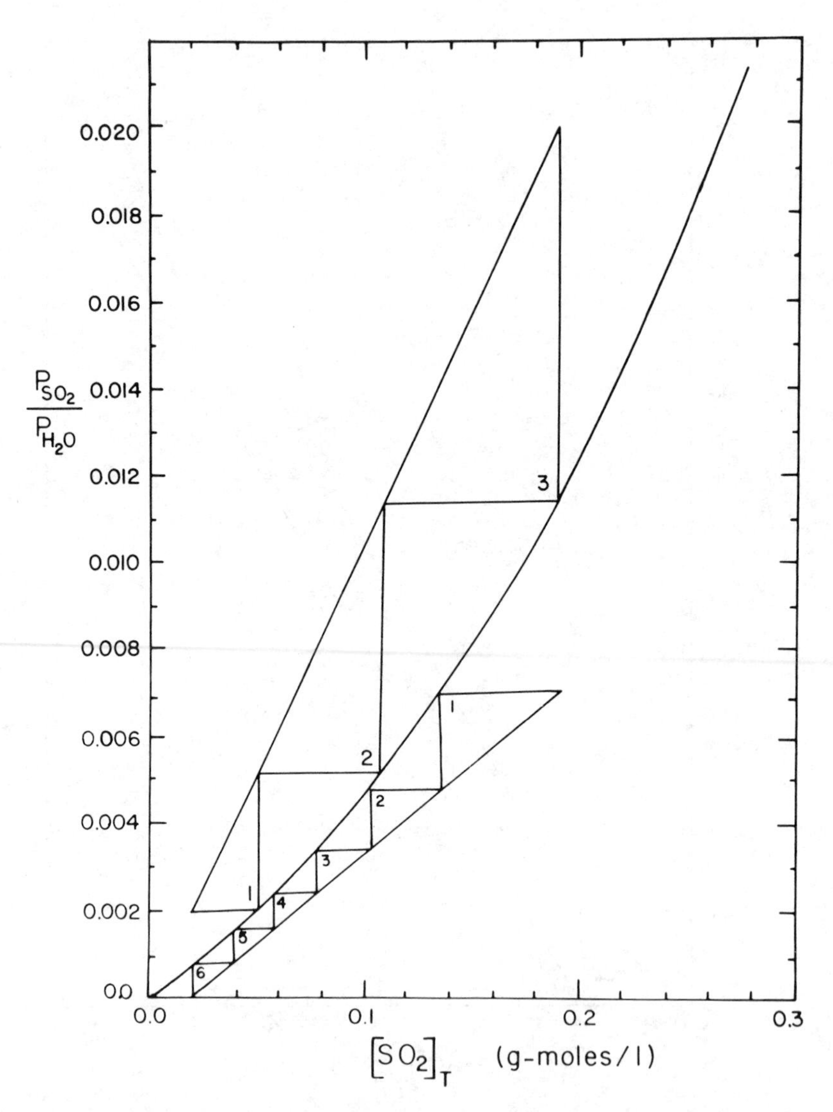

*Figure 8. Actual steam requirements, simple absorption/stripping with live steam, 3000 ppm SO_2 in at 55°C, 90% removal, 1.0*M *citrate, 2.0*M *Na*

and SO_2 solution concentration. If the constant a_4 in the equation for K_c is 0.30 rather than 0.46, the minimum steam requirement increases by only 3%.

Table IV: Minimum steam requirement, inlet P_{SO_2}/P_{H_2O} = 0.02, 90% SO_2 removal, infinite stages, 1.0 M Citrate

[Na] (M)	Steam Requirement Moles H_2O/Mole SO_2	Nonlinearity Factor	Capacity Moles SO_2/liter
1.5	52.4	1.05	0.103
2.0	57.8	1.16	0.173
2.0[a]	59.5	1.19	0.262
2.5	65.8	1.32	0.368

[a]Value of a_4 taken to be 0.30 rather than 0.46.

Conclusions

1. Temperature dependence of pH and P_{SO_2}/P_{H_2O} over sodium citrate buffer solutions is insignificant.
2. Composition dependence of pH and P_{SO_2}/P_{H_2O} can be represented as a function of fraction neutralization and total anion concentration.
3. Actual steam requirement with typical stack gas should be about 41 kg/kg SO_2.
4. Optimized steam requirement is relatively insensitive to solution pH. Solution capacity for SO_2 absorption can reasonably vary from 0.1 to 0.4 g-moles SO_2/liter.
5. The SO_2 gas sensing electrode is an effective tool for vapor/liquid equilibrium at room temperature.

Nomenclature

a	Activity
a_1, a_2, a_4, a_5	Constants in correlation of pH and P_{SO_2}
a_3, a_6	Constants with anionic concentration in correlation of pH and P_{SO_2}, $M^{0.5}$
c	Intercept for calibration of SO_2 electrode.
[Citrate]	Total concentration of citric acid and its anions, M
d	Slope for calibration of SO_2 electrode, mV
f	Fraction neutralization
K	Equilibrium constant for SO_2 absorption as bisulfite, atm M^{-2}
K_{a1}, K_{a2}, K_{a3}	Equilibrium constants for dissociation of citric acid, M
K_c	Dependent variable in correlation of SO_2 vapor pressure, M^{-1}

M	Molarity, gmol/liter
n	Number of g-moles
P	Pressure or partial pressure, atm
P_{SO_2}	Vapor or partial pressure of SO_2, atm
$P^{o}_{H_2O}$	Vapor pressure of pure water, atm
P_{H_2O}	Vapor or partial pressure of H_2O, atm
$[SO_2]$	Total concentration of S^{+4} species (mostly bisulfite), M
[]	Concentration, M

Abstract

SO_2 vapor pressure (P_{SO_2}) was measured by dynamic saturation and by a gas-sensing SO_2 electrode over solutions containing 0.5 to 2.0 M sodium citrate at pH 3.5 to 5 with up to 1 M $NaHSO_3$, Na_2SO_4, and NaCl. P_{SO_2} was measured at 25° to 168°C; pH at 25° to 95°C. Both pH and the vapor pressure ratio P_{SO_2}/P_{H_2O} were independent of temperature. The composition and temperature dependence of the data are correlated by the semiempirical expressions:

$$pH = 2.77 + 3.60\ f - 0.53\ [Anion]_T^{0.5}$$

$$\log \frac{P_{SO_2}}{P_{H_2O}\,[SO_2]_T} = 0.46 - 3.27\ f + 0.26\ [Anion]_T^{0.5}$$

where f is the fraction neutralization of the citrate buffer. The steam requirement for simple absorption/stripping with 90% removal of SO_2 from stack gas containing 3000 ppm SO_2 at 55°C was estimated to be about 40 kg/kg SO_2.

Acknowledgements

This work was primarily supported by contracts TPS 77-747 and RP 1402-2 with the Electric Power Research Institute. Curtis Cavanaugh, Richard Ulrich, Pui Lin, and Michael Ragsdale have contributed experimental data as research assistants.

any information, apparatus, method, or process disclosed in this report may not infringe privately owned rights; or,

b. Assumes any liabilities with respect to the use of, or for damages resulting from the use of, any information, apparatus, method or process disclosed in this report.

Literature Cited

1. Rochelle, G. T., "Proceedings: Symposium on Flue Gas Desulfurization - Hollywood, FL", EPA-600/7-78-058b, p. 902 (1977).
2. Johnstone, H. F., H. F. Read, and H. C. Blankmeyer, Ind. Eng. Chem., 30, 101 (1938).
3. Boswell, M. C., U. S. Patent 1,972,074, Sept. 4, 1934.
4. Applebey, M. P., J. Soc. Chem. Ind. Trans., 56, 139 (1937).
5. Nissen, W. I., D. A. Elkins, and W. A. McKinney, "Proceedings: Symposium on Flue Gas Desulfurization - New Orleans", EPA-600/2-76-136b, p. 843 (1976).
6. Farrington, J. F. and S. Bengtsson, "The Flakt-Boliden Process for SO_2 Recovery", presented at AIME annual Meeting, February 19, 1979.
7. Johnstone, H. F., Ind. Eng. Chem., 39, 3896 (1935).
8. Weast, R. C., "Handbook of Chemistry and Physics", 54th ed., CRC Press, pp. D-70, E-1 (1973).
9. Bates, R. G. and G. D. Pinching, J. Am. Chem. Soc., 71, 1274 (1949).

RECEIVED January 31, 1980.

THERMODYNAMICS OF SYNTHESIS GAS AND RELATED SYSTEMS

13

Principles of Coal Conversion

JOSEPH F. McMAHON

Foster Wheeler Energy Corporation, Livingston, NJ 07039

Direct use of coal as a primary fuel is often the most efficient and economic method of utilizing this important energy resource. In many cases, however, certain undesirable properties of coal make direct utilization difficult. Coal is a solid and requires more effort to handle, measure and control than gases or liquids. Coal is usually contaminated with ash and other undesirable components and has widely variable chemical and physical properties. As a result, there is often a need to convert coal into more convenient and cleaner forms of energy and products. Before considering the basic principles of coal conversion, some important characteristics of fossil fuels will be reviewed.

Fossil Fuel Characteristics

Carbon and hydrogen are the two elements in fossil fuels of primary importance with respect to energy and chemical products. The hydrogen to carbon ratios (H/C) of light hydrocarbon gases, petroleum liquids and coal are compared in Figure 1. Methane, the principal component of natural gas, has the highest H/C ratio of the hydrocarbon series. Petroleum fractions have H/C ratios ranging from about 2.3 for naphtha through 1.5 for distillate oils. Coal has the lowest H/C ratio, ranging from about 0.8 for bituminous coal to about 0.4 or lower for anthracite coal and cokes.

The chemical energy content of fossil fuels generally parallels the H/C ratio as shown in Table 1. Methane has the highest heating value of the hydrocarbon series, corresponding to its high H/C ratio while coal has the lowest heating value. The normal physical state of fossil fuels also parallels the H/C ratio. Methane is a gas at ambient temperature and pressure conditions. Petroleum fractions are mobile liquids except for the heaviest fraction which can be solid at ambient temperature. Coal is a solid material.

0-8412-0569-8/80/47-133-295$05.00/0

Table 1
Heating Values Of Selected Fossil Fuels

Fuel	kJ/kg
Methane	53,800
Naphtha	46,500-48,800
Crude Oil	44,200-46,500
Resid	39,500-41,900
Coal*	27,900-32,600

*maf

Methane (natural gas) and petroleum are clearly more convenient to handle and use than coal. This situation led to widespread industrial use of these fuels for electric power production, industrial and residential fuel requirements and chemicals production. Recently, however, it has become apparent that more use must be made of coal to conserve supplies of natural gas and petroleum despite the difficulties in handling and use. There are large reserves of coal in many parts of the world, many near centers of population and industrial activity.

Basic Methods of Coal Conversion

Conversion of coal to more desirable forms of energy such as:

- pipeline gas
- synthesis gas
- fuel gas
- liquid hydrocarbons
- chemicals

generally requires considerable technical effort. There are many variations in technical processes that have been developed for converting coal to secondary fuels. These processes, however, all have the common objective of either removing carbon from or adding hydrogen to coal to improve its original H/C ratio. The choice of method and the extent of improvement in H/C ratio depends on the type of product required and considerations of cost. Removal of carbon can be accomplished by

- pyrolysis
- gasification

while processes for hydrogen addition include

- liquefaction
- hydrogasification

Carbon Removal Processes

Pyrolysis. Conversion of coal by pyrolysis involves heating coal to a temperature of 500 to 700°C. Gases and liquids are evolved from the coal at these temperatures, leaving char which has a lower H/C ratio than the original coal:

$$\text{coal} \xrightarrow[\text{heat}]{} \text{gas} + \text{liquid} + \text{char}$$

Pyrolysis processes differ primarily in the means used to supply the required heat to the coal. Invariably, the heat is generated by burning a portion of the coal with an oxygen-containing gas. In some cases, a circulating stream of char is used to carry the heat to the fresh coal. Alternately, an inert solid is used as heat carrier although, in this case, means must be provided to separate heat carrier from product char.

Gasification. The extent of carbon removal from coal by pyrolysis is relatively limited. As a result, the yields of secondary fuels having increased H/C ratio are not large. Essentially complete control of the amount of carbon removed can be achieved, however, by complete gasification of coal with an oxygen-containing gas and steam:

$$\text{coal} + O_2 + H_2O \longrightarrow H_2 + CO + CH_4 + CO_2$$

Major components of the gaseous product are hydrogen, carbon monoxide and variable amounts of methane and light hydrocarbons, as well as carbon dioxide. Carbon is removed from the system by separating carbon dioxide from the gas by well known gas scrubbing processes. Some additional hydrogen over that contained in the original coal is introduced by reaction of steam with the coal. The result is a hydrogen-carbon monoxide gas mixture having an effective H/C ratio of two or higher. This gas mixture, often referred to as snythesis gas, can be further processed to pipeline gas, liquid hydrocarbons, or chemicals such as methanol and ammonia.

The common feature of coal gasification processes is that coal is contacted in a gasifier with an oxygen-containing gas and steam at a temperature of at least 700°C. The main types of gasifiers are:

- moving bed (Figure 2)
- fluidized bed (Figure 3)
- entrained flow (Figure 4)

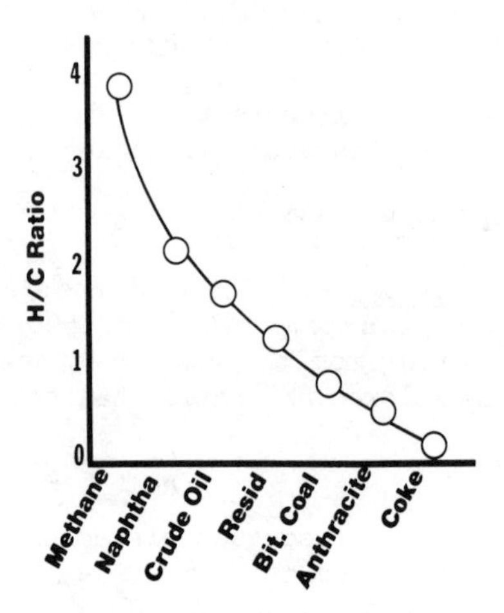

Figure 1. H/C ratios for various fossil fuels

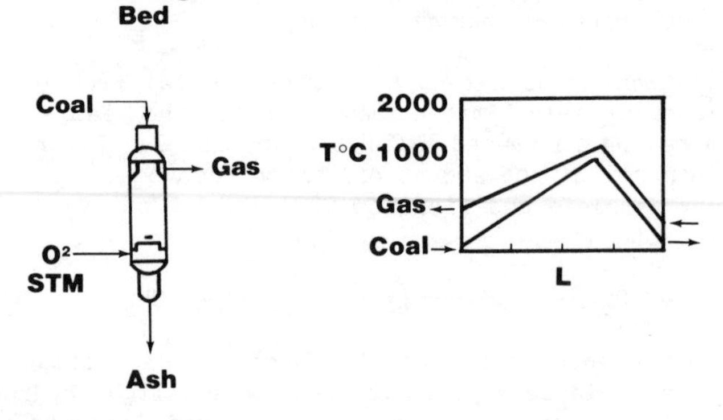

Figure 2. Schematic of moving-bed coal gasifier with typical reaction temperature profiles

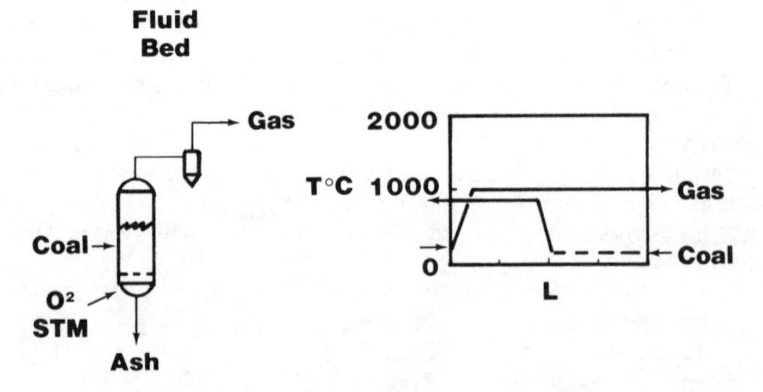

Figure 3. Schematic of fluidized-bed coal gasifier with typical reactor temperature profiles

In the moving bed gasifier, lump size coal flows downward through the gasifier, countercurrent to ascending hot gases. Coal flows successively through drying, devolatilization, gasification and combustion zones, each zone operating at increasingly higher temperatures up to and beyond the melting point of the ash. Ash is withdrawn from the bottom of the gasifier. Oxygen or air, together with steam, are introduced into the gasification zone and flow upward through the gasifier.

In the fluidized bed gasifier, crushed coal is introduced into a fluidized bed of char together with oxygen or air and steam. Coal undergoes drying, devolatilization, gasification and combustion at essentially constant temperature of about 1000°C because of the rapid mixing characteristics of fluidized beds.

In the entrained flow gasifier, pulverized coal flows cocurrently with oxygen and steam through a reaction zone at temperatures up to 1800°C. The coal is entrained as a dilute suspension in the flowing gases. Coal drying, devolatilization, gasification and combustion occur very rapidly in the high temperature reaction zone. Entrained flow gasifiers may be single stage or two stage, the latter involving introduction of coal into a first stage where it is rapidly devolatilized by hot gases from a second stage. Char is recovered from the first stage and gasified with steam and oxygen or air in the second stage.

The overall process of coal gasification is endothermic and heat must be supplied to the system. Up to 40% of the heating value of the coal may be used for this purpose, depending on the specific gasification process and coal used. Countercurrent processes usually produce the highest ratio of chemical to sensible heat in the product gas because of relatively low gas outlet temperature. Cocurrent single stage entrained flow processes produce the lowest ratio of chemical to sensible heat in the product gas; fluidized bed processes are intermediate in this respect. Heat is usually supplied by partial combustion of char in the gasifier with oxygen or air.

The properties of coal ash, particularly softening temperature, have an important effect on gasification processes. Ash can be discharged in "dry" form, in which case the maximum temperature in the gasifier cannot exceed the ash softening point. In moving bed processes, the maximum temperature occurs in the combustion zone and steam is added for temperature control. Higher temperatures can be allowed if the ash is removed as sintered agglomerates or as a liquid. In these cases, increased gas production and a greater extent of gasification can be achieved compared to dry ash operation.

Hydrogen Addition Processes

Liquefaction. (Figure 5) Coal liquefaction involves addition

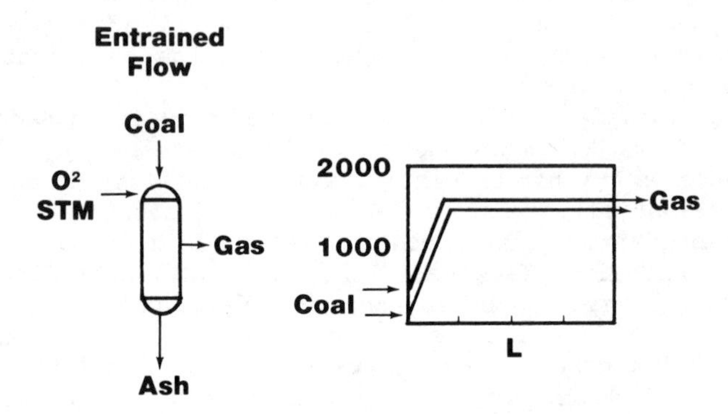

Figure 4. Schematic of entrained flow coal gasifier with typical reactor temperature profiles

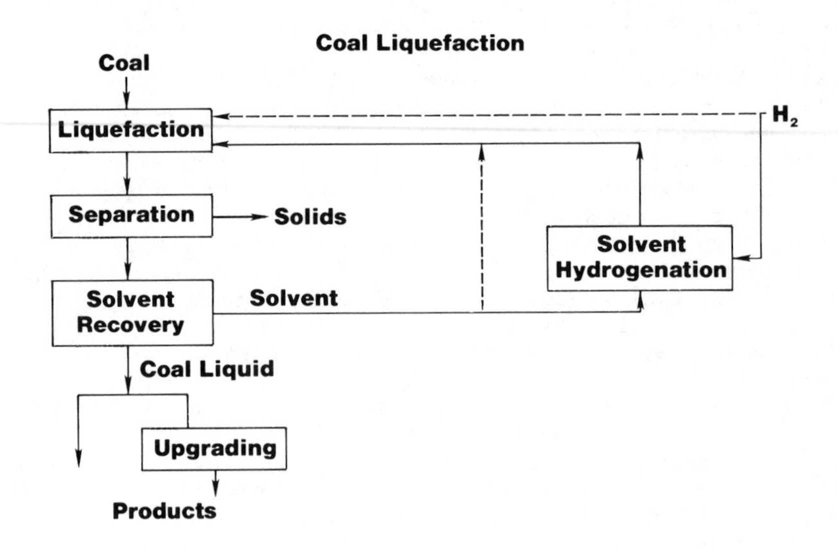

Figure 5. Generalized flowchart for coal liquefaction processes

of hydrogen to coal to produce secondary fuels of increased H/C ratio. It has been known for many years that heating a mixture of coal and hydrogen donor solvent to temperatures of 400 to 500°C results in the production of liquids of increased H/C ratio compared to the original coal. Hydrogen donor solvents are compounds such as tetralin or mixtures of similar compounds containing labile hydrogen atoms. At elevated temperatures, these hydrogen atoms of the solvent react with coal, breaking down the three dimensional coal structure into fragments which dissolve in the solvent. The molecular weight of these fragments decreases as the amount of hydrogen reacted increases. Contaminants such as sulfur, nitrogen and oxygen are eliminated as hydrogen sulfide, ammonia and water.

The principle of donor hydrogen reaction with coal has been applied in various ways in processes for coal liquefaction. In one application, hydrogen donor solvent is generated from the coal itself. The solvent, usually a distillate fraction of the coal liquid product, is hydrogenated and recycled to the coal liquefaction reaction.

In a variation of this technique, molecular hydrogen is added to the liquefaction reactor together with the solvent. Hydrogen uptake is usually limited to about two weight percent of the coal and the liquefied coal product has a melting point of about 150 to 200°C. In a further variation, a portion of the high boiling liquid product is recycled to the reactor. The ash contained in this stream exerts a catalytic effect on the hydrogen donor reaction. Hydrogen uptake is increased and a full boiling range liquid product is produced containing naphtha, gas oil and heavy fuel oil fractions.

Catalytic coal liquefaction processes do not specifically use hydrogen donor solvents although coal is introduced into the liquefaction reactor as a slurry in a recycle liquid stream. Catalyst is used as a powder or as granules such as pellets or extrudates. If powdered catalyst is used, it is mixed with the coal/liquid stream entering the reactor. Pelleted catalyst can be used in fixed bed reactors if precautions are taken to avoid plugging with solids or in fluidized bed reactors. In the latter case, the reacting system is actually a three phase fluidized bed, that is, catalyst particles and coal solids, as well as liquid, are fluidized by gas.

Hydrogenation of coal is a highly exothermic reaction corresponding to a heat evolution of about 15 kilojoules per cubic metre of hydrogen reacted. Means must be provided to remove this heat from the reaction zone so that the reaction temperature can be maintained in the optimum range. This is usually accomplished by injecting coal liquid as quench into various sections of the reactor.

Hydrogasification. Hydrogasification of coal involves reaction of hydrogen with coal carried out at elevated temperatures under high partial pressure of hydrogen. The objective is to add sufficient hydrogen to coal to produce methane as the major product. It has been found that many types of coal can be hydrogasified if the coal is heated rapidly to reaction temperatures. Even under favorable conditions, however, conversion to methane is not complete and aromatics such as benzene are made as by-products.

Hydrogasification of coal is also a very exothermic reaction. One means of absorbing the heat of reaction is to use a fluidized bed reactor and inject hydrogen and coal reactants at sufficiently low temperature so that the sensible heat required to heat coal and hydrogen to reaction temperature is equivalent to the heat of reaction.

Efficiency Considerations

Heat recovery efficiency is a consideration of major importance in the conversion of coal to secondary fuels. This parameter is defined as the percent of the heating value of the coal used which is recovered as heating value in the desired secondary fuel. Heat recovery efficiency which can be attained in a coal conversion process depends firstly on the theoretical chemical and thermodynamic requirements of the process, and secondly on the practical realization of the process. The first factor determines the theoretical maximum heat recovery efficiency that can be obtained under ideal circumstances. The second factor determines the extent to which the practical process approaches the theoretical ideal.

Several aspects of the concept of theoretical heat recovery efficiency can be understood by considering an idealized conversion of coal to a secondary fuel having a high H/C ratio, such as methane. In the following discussion, it is assumed that the conversion reactions proceed to completion at a temperature of 15°C and a pressure of 1 atmosphere although, of course, this cannot be realized in practice. Coal is assumed to have the idealized chemical formula of $C_{10}H_8$.

Several important chemical reactions for the conversion of coal to methane are shown in Table 2. Steam conversion involves the reaction of coal with steam to produce hydrogen and carbon monoxide. Hydrogen conversion is a reaction in which coal and hydrogen react to form methane. Oxygen conversion produces hydrogen and carbon monoxide by partial oxidation of coal. Methanation involves a reaction in which methane and water are produced from carbon monoxide and hydrogen. The water gas shift reaction between carbon monoxide and steam produces carbon dioxide and hydrogen.

Several of these reactions can occur simultaneously or can be used consecutively to produce methane from coal. Methanation and water gas shift reactions occur simultaneously in varying degrees with steam and oxygen conversion reactions. Hydrogen required for hydrogen conversion can be produced by steam or oxygen gasification.

Table 2
Coal Conversion Chemical Reaction

Steam Conversion

$$C_{10}H_8 + 10\ H_2O \longrightarrow 10\ CO + 14\ H_2$$

Hydrogen Conversion

$$C_{10}H_8 + 16\ H_2 \longrightarrow 10\ CO + 4\ H_4$$

Oxygen Conversion

$$C_{10}H_8 + 5\ O_2 \longrightarrow 10\ CO + 4\ H_2$$

Methanation

$$CO + 3\ H_2 \longrightarrow CH_4 + H_2O$$

Water Gas Shift

$$CO + H_2O \longrightarrow CO_2 + H_2$$

The potential of these reactions for methane production can be compared in terms of theoretical yields and heat recovery efficiencies. Theoretical methane yield is defined by the chemical equations. Theoretical heat recovery efficiency is defined as the percent of the higher heating value of the coal which is recovered in the form of methane product. These idealized parameters provide a measure of the ultimate capability of conversion systems and are useful for evaluating actual conversion processes.

Table 3 shows the theoretical methane yields and heat recovery efficiencies for

- steam conversion-methanation
- hydrogen conversion
- oxygen conversion-methanation

Table 3
Theoretical Methane Yield and Heat Recovery Efficiency

Feedstock	H:C Wt. Ratio	CH_4 Yield m^3/kg	Heat Recovery Efficiency %
	Steam Conversion - Methanation		
$C_{10}H_8$ (coal)	0.067	1.11	100
	Hydrogen Conversion		
$C_{10}H_8$	0.067	1.84	92
	Hydrogen Conversion - H_2 Supply by Oxygen Conversion		
$C_{10}H_8$	0.067	0.86	81
	Oxygen Conversion - Methanation		
$C_{10}H_8$	0.067	0.65	61

Steam conversion/methanation has a theoretical heat recovery efficiency of 100%. Hydrogen conversion has a theoretical efficiency of about 90%; if the production of hydrogen by steam conversion is taken into account, however, the theoretical efficiency drops to 81%. Oxygen conversion/methanation has a theoretical efficiency of only 61% which is the lowest of the conversion systems.

Practical conversion processes can only approach the theoretical efficiencies shown in Table 3. The coal conversion reactions do not proceed to completion at ambient temperatures within practical time limitations. As a result, a portion of the coal feedstock must be burned to supply heat so that the reactions can be carried out at elevated temperatures and pressure where the rates of conversion are rapid. In practical systems, this additional heat can only be partially recovered. Consequently, practical conversion processes have actual heat recovery efficiencies of about 60-70% for production of high H/C ratio products. Production of secondary fuels having somewhat lower H/C ratio, i.e. about 2.0, permits attainment of heat recovery efficiencies of 70 to 80%.

Although conversion processes result in the loss of a significant part of the chemical heat energy of the original coal, secondary fuels are produced which are clean and are more convenient to handle. The favorable characteristics of these clean secondary fuels justify in many cases the cost of plant and energy required for conversion.

RECEIVED February 14, 1980.

14

Thermodynamic Data for Synthesis Gas and Related Systems

Prediction, Development, Evaluation, and Correlation

GRANT M. WILSON

Wilco Research Company, 488 South 500 West, Provo, UT 84601

The problems associated with new synthesis gas processes are far greater than problems associated with gas processing plants or refineries because of water, salt, sludge, ammonia, and cresols present in the process streams. These problems need to be resolved rapidly if we are to meet our energy needs of the next decade. For this reason we cannot wait for data on one plant to be used in designing another one as has been done in refinery development. Therefore we must use available basic data, pilot plant data, and our capabilities involving mathematical tools and computers to develop reliable processes without having to learn from the failures of prior processes. When basic mathematical models predict pilot plant results, then one has considerable faith in using the model to predict full-scale plant operation. The advantages of the mathematical models are that the effects of varying stream rates, temperatures, and pressures can be predicted with considerable reliability by simply establishing that existing pilot plant data agree with the models. This means that when differences between pilot plant data and the models exist that they have to be reconciled. In general, the cost of developing the necessary basic data and models is considerably less than the cost of operation of pilot plants or the cost associated with inefficiency of operation of full-scale plants. Unfortunately the development of basic data and mathematical models takes time, so full scale plants have to be built in situations where the necessary basic data are not adequate. When this occurs, enough inefficient over-design has to be built into the process to compensate for uncertainties in the basic data. In these situations, the process of development of new basic data should continue because the results will be useful when operating problems occur in the new plant or when designing additional plants. Thus our immediate and long range needs are for basic data which can be used in mathematical model developments. If this information is developed rapidly, the overall cost to our economy will be considerably less than if the data are developed slowly. Subsequent sections of this paper discuss the magnitude of work that needs to be done,

0-8412-0569-8/80/47-133-305$05.00/0

how this is to be accomplished, and identifies some of the areas requiring development.

The Magnitude of Work to be Done

Existing processes for producing oil and gas products have required the development of phase behavior and other thermodynamic data on light hydrocarbons, heavy hydrocarbons, and the acid gases CO_2 and H_2S. For this reason a lot of basic data are available on these systems; but there is still a lot we don't know such as how to characterize the behavior of hydrocarbon fractions containing numerous paraffin, naphthene, and aromatic components. Additional basic data on these systems would help to improve the efficiency of these existing processes.

The picture becomes considerably more complicated when the additional components of synthesis gas processes are added. The various types of components involved are portrayed in Figure 1 where each type of component is indicated by a circle. The components of existing oil and gas processes, shown on the right-hand portion of the figure, primarily involve the light hydrocarbons C_1, C_2, C_3, C_4, and C_5; the oil fractions: C_6+; and the acid gas components: composed primarily of H_2S and CO_2.

The development of basic modeling data involves the development of interaction data between components in each type and interaction data between types. In the case of oil and gas components these involve interactions between three different types of compounds or three interactions between types of groups. The work required to develop data on these systems has been very large and has involved a time span of many years. But the work has been necessary, and much could have been saved in plant costs and operating efficiency if the data had been developed faster. Three additional circles have been added at the left of Figure 1 to represent the additional components involved in the production of synthesis gas. These involve the light gases: H_2 and CO, with N_2, O_2, and Ar as minor components; water and ammonia with amines as minor components; and cresols and other organic components. These three additional types of components produce a total of 15 combinations of interactions between the various types of components or 12 additional interactions. Thus the additional work to be done could be as much as four to five times the amount already done on oil and gas components.

Some examples of new problems encountered in synthesis gas production are the following.

1. Ammonia. Ammonia interferes with existing acid gas removal processes because it can pass on through the scrubbers and then solidify on cyrogenic surfaces or it can go with the acid gases and poison the sulfur conversion catalysts. If ammonia is absorbed into an aqueous stream, then this aqueous stream must be

further processed to remove the ammonia in order to avoid its release as an aqueous waste-stream pollutant.

2. Cresol-Type Components. Large amounts of cresol-type components are produced in synthesis gas processes. These components concentrate in aqueous streams and represent a serious pollution threat because of their toxicity. For this reason they must be recovered from any aqueous waste stream before leaving a plant.
3. Water. Water is a major component in synthesis gas processes as it is the main source of hydrogen. Thus the phase behavior of water with H_2, CO, light hydrocarbons, heavy hydrocarbons, etc. needs to be known in order to design processes which will handle streams containing significant amounts of water. Since water is an extremely polar component this presents new problems in predicting the thermodynamic properties of mixtures. Existing equations of state are not adequate so new methods to handle these mixtures must be developed.

The solution of these problems will involve additional processing which would not be encountered in existing oil and gas processes; thus data for all three classes of components mentioned above are essential to the successful development of synthesis gas and other coal conversion processes. Additional problems which are equally as serious may appear as these processes are further developed, thus considerable thought must be given to the orderly development of data relating to these new processes.

Methods of Data Development

The orderly development of basic data involves a research cycle shown schematically in Figure 2. This cycle involves four stages as follows.

1. Plan new measurements; predict or guess order of magnitude of experimental result.
2. Make measurements
3. Evaluate data
4. Correlate data

This cycle will be most successful when each of the stages are carefully carried out. Failure in any one stage can negate the results of subsequent stages, thus causing errors and inefficiency. The final product, useful for process design, are the correlations produced as a reult of the research cycle. The rapid development of these correlations thus requires that attention be given to each of the stages of the research cycle. In some situations it has appeared that much work has been spent on only one or two stages without much attention to the other stages.

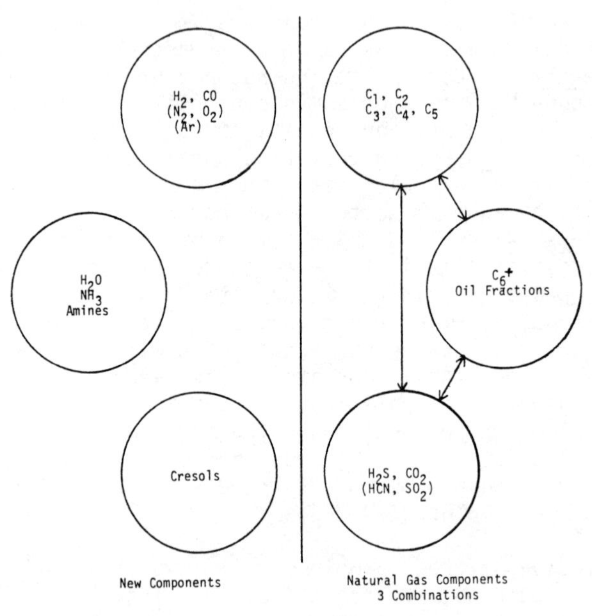

Figure 1. Synthesis gas components

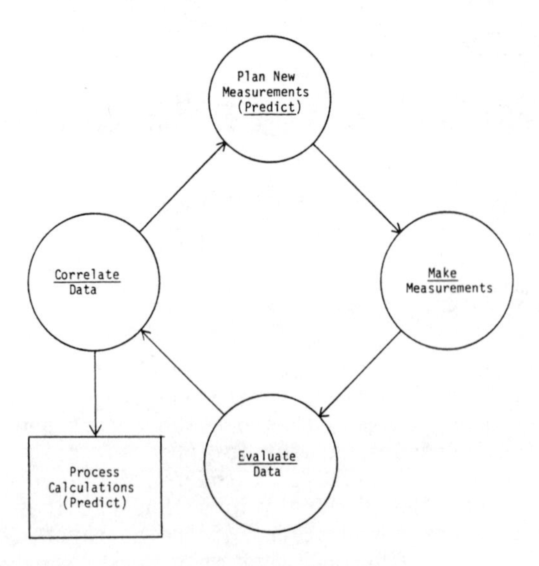

Figure 2. ***Re****-search cycle*

For example, much effort can be spent correlating existing data when in reality more basic data measurements are needed. This lack of emphasis on certain stages produces inefficiency in the research process which impedes the development of improved correlations; thus attention must be given to each stage.

The first stage of the research cycle is the planning and predicting stage. Planned measurements must be coordinated with industry needs so that correlations can be produced in areas of need. After the measurements are planned, and the better one can predict results; the more successful the measurements will be. This occurs because predicted data help to identify possible areas of measurement problems. For example, a different analytical procedure might be used to determine the concentration of a trace component than for a major component. Predicted data help to identify when this will occur. Generally speaking, the more that is known--the more successful will be the experiment. Information useful in predicting data can be found in existing physical property compilations, data books, equations of state, group contribution methods, correlations, etc. In many cases, interactions between the components have to be "guessed" in order to make a prediction; but this is better than no estimate at all.

The second stage of the research cycle is the measurement stage. This is probably the most difficult stage because a successful experiment involves the simultaneous operation of many different types of measurement and control equipment. The failure of just one part negates the results of the entire experiment. The probability of this happening is quite high because a malfunction in the apparatus can re-occur or new malfunctions can appear. In some cases, these malfunctions can go undetected until an examination of the results shows that they are entirely inconsistent with predicted data.

The third stage of the research cycle is the evaluation stage. It is important that this stage be done by the same group doing the measurements so that any differences between measured and predicted data can be reconciled. Too often data are measured and reported without being evaluated. When this occurs, the effort of an entire research cycle can be lost. But when immediate evaluation is made, it is easy to go back and modify the experiment in order to confirm or remedy the data; thus significantly improving the efficiency of the research cycle.

The fourth and most useful stage of the research cycle is the correlation stage because this is where results appear that can be used in process design calculations. Because of their immediate utility, there is a tendency on the part of individual companies to over-emphasize the correlation stage without enough emphasis on the other stages of the research cycle. Thus data tend to be over correlated with the same basic problems in all of the correlations due to lack of input data. Nevertheless, correlations are a necessary and important stage of the research cycle.

When possible, new data should be integrated into existing correlations; thus making it possible to immediately use the results in process calculations. New correlation methods present problems even though they may be more accurate because the new correlation must be added to the design calculation procedure before it can be used. This usually causes a time lag which often prevents the new correlation from being used for long time periods. Thus new methods should only be used when old methods simply are not suitable. When this occurs, it is helpful to try to anticipate the needs of industry so that useful correlation methods capable of a wide range of problems can be adopted.

Mixtures with Polar Components

Rather than attempt to discuss all areas where data are needed, only the problem of predicting the thermodynamic properties of mixtures containing polar components will be discussed here. Existing methods based on equations of state appear to be adequate for engineering purposes when only non-polar and slightly polar components are present. Various versions of the Redlich-Kwong equation, BWR equation, and variations of these equations such as the Peng-Robinson equation of state are now in use. These equations of state have proved to be very useful for predicting phase behavior and thermodynamic data on hydrocarbon systems. Unfortunately these equations of state appear incapable of correlating the behavior of water-hydrocarbon systems without using separate interaction coefficients for the aqueous and hydrocarbon phases. The use of separate interaction coefficients represents an immediate solution to the problem for current process calculations when only water is present, but it does not represent a long-range solution to handle situations where significant concentrations of water-soluble components are also present including regions near the upper consolute temperature of water-hydrocarbon systems. When this occurs, the same interaction coefficients must apply over the entire composition range.

The basic features of equations of state are not complicated when they are expressed as PV/RT versus density. Figure 3 is a sample plot for methanol. These curves are characteristic of all fluids, and equations of state only differ in their ability to accurately predict these curves. The actual curves are relatively simple and they change only slightly from one material to another; for this reason, simple equations of state such as the Redlich-Kwong equation have been about as successful as the BWR equation. The simplicity of the actual curves is often hidden because the data are not usually plotted as PV/RT versus density. More often the data are plotted as PV/RT versus pressure shown in Figure 4, or pressure versus volume shown in Figure 5. Both of these plots obscure the real simplicity shown in Figure 3.

The real problem with equations of state is in the accurate prediction of mixture behavior. This is somewhat analogous to

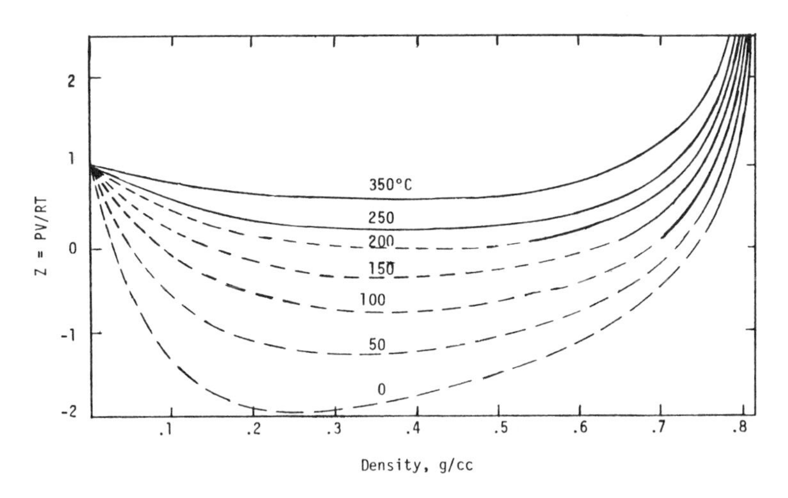

Figure 3. Compressibility factor of methanol

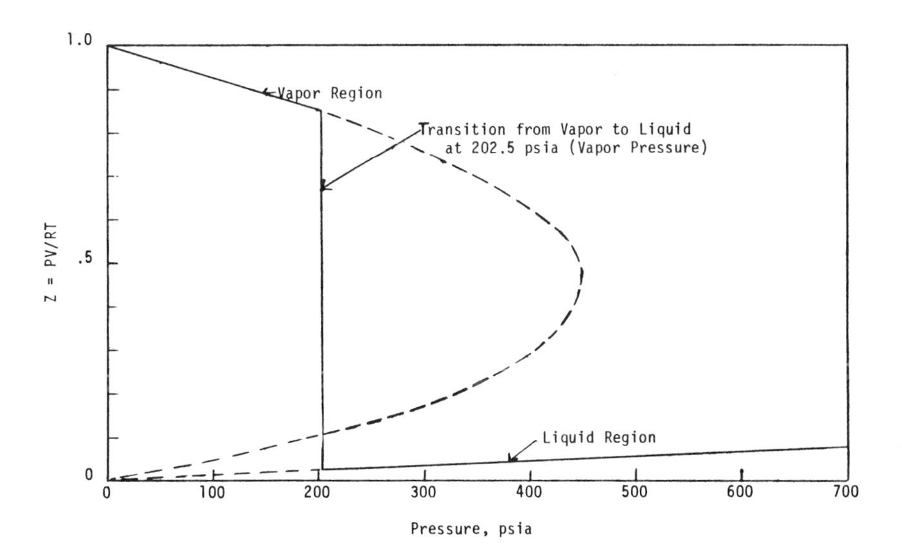

Figure 4. Compressibility factor of methanol at 302°F

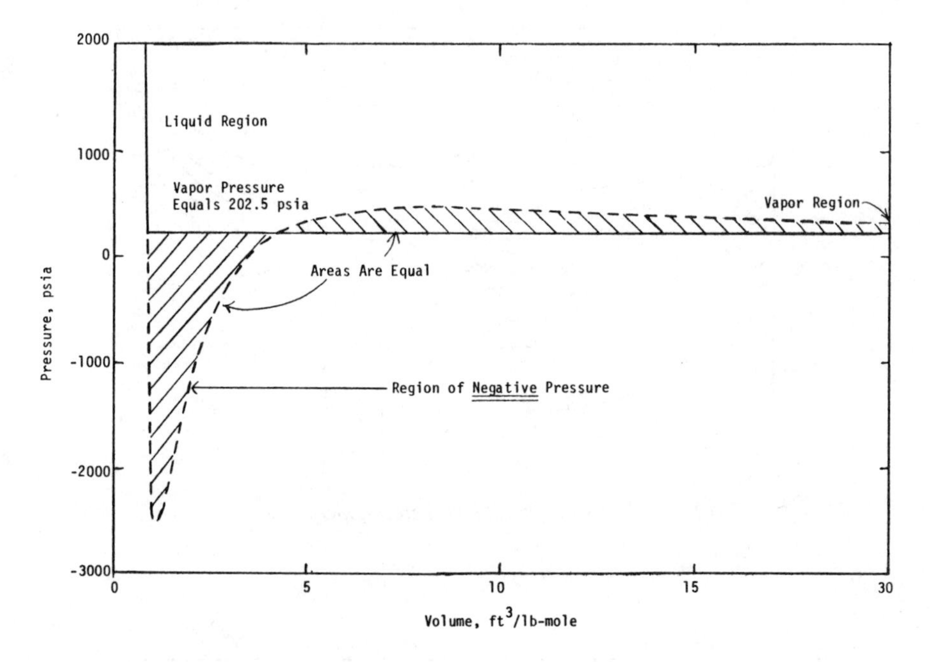

Figure 5. P–V isotherm of methanol at 302°F

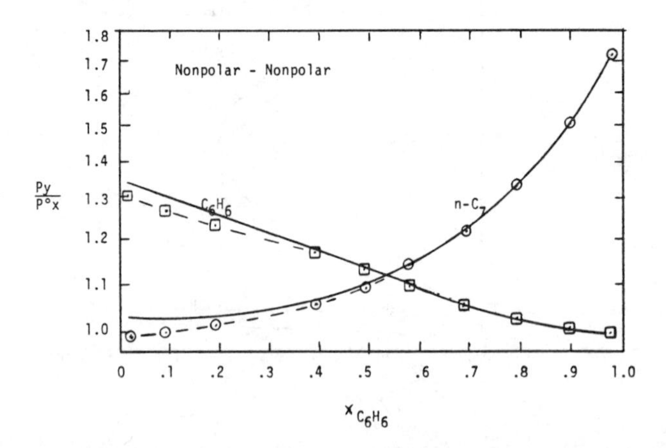

*Figure 6. Redlich–Kwong correlation of nonpolar–nonpolar benzene-*n *heptane at 60°C (data of I. Brown and A. H. Ewald,* Austral. J. Sci. Res., (A), 4, *198 (1951))*

the problem of predicting activity coefficients of non-ideal mixtures. If this problem were solved, the rest would be relatively easy.

To illustrate the problem with mixtures, the Redlich-Kwong equation has been used to correlate literature data on three types of mixtures as follows.

Type of Mixtures	Components
Nonpolar-Nonpolar	benzene/n-heptane
Polar-Polar	ethanol/water
Nonpolar-Polar	n-hexane/methanol

In each case pure component parameters were chosen to fit the pure component vapor pressure and liquid density, and a binary interaction parameter was chosen to fit the total pressure curve at the 50:50 mole percent composition in the liquid phase. If the equation of state is adequate for representing these mixtures, then this should be sufficient information for predicting data at other compositions. The results are given in Figures 6 to 8. Figure 6 shows that data on nonpolar-nonpolar benzene/n-heptane are fairly well represented by the Redlich-Kwong equation. This isn't surprising because the Redlich-Kwong equation is widely used for mixtures of this type. Figure 7 shows that data on polar-polar ethanol-water are predicted with some degree of accuracy. This result is a bit surprising and suggests that the equation may be suitable for correlating mixture data when the components only differ slightly in polarity; regardless of whether the components are polar or nonpolar. Figure 8 shows that data on nonpolar-polar n-hexane/methanol are not predicted very well at all. Thus the most serious problems occur with mixtures of nonpolar and polar components. One generality can be drawn from examining Figures 7 and 8; namely that the activity coefficient of the polar component at low concentrations is higher than predicted and the activity coefficient of the nonpolar component is lower than predicted. One can show by examining other mixtures that this trend consistently occurs; thus this gives a clue that adequate equations can be developed which will compensate for the systematic prediction error of the Redlich-Kwong equation. This systematic nature of the main error tends to indicate that only minor changes in the Redlich-Kwong equation might be necessary to yield accurate predictions for these systems. In addition, it may be possible to develop other relatively simple equations of state which will predict this behavior. This discussion shows that there are tests which can be applied in order to determine the adequacy or inadequacy of equations of state when components of differing polarity are present in a mixture. However no attempt is made here to suggest what changes may be necessary in order to adequately modify the equations of state or to develop new equations of state.

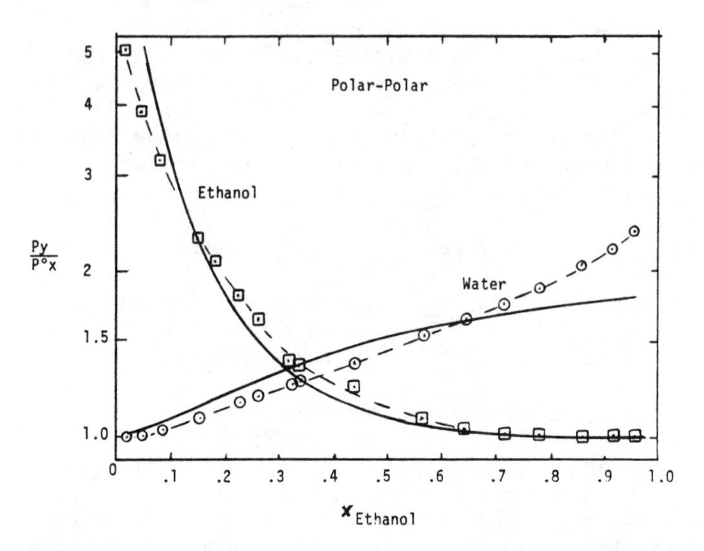

Figure 7. Redlich–Kwong correlation of polar–polar ethanol–water mixture at 150°C (data of F. Barr-David and B. F. Dodge, J. Chem. Eng. Data, 4, *107 (1959))*

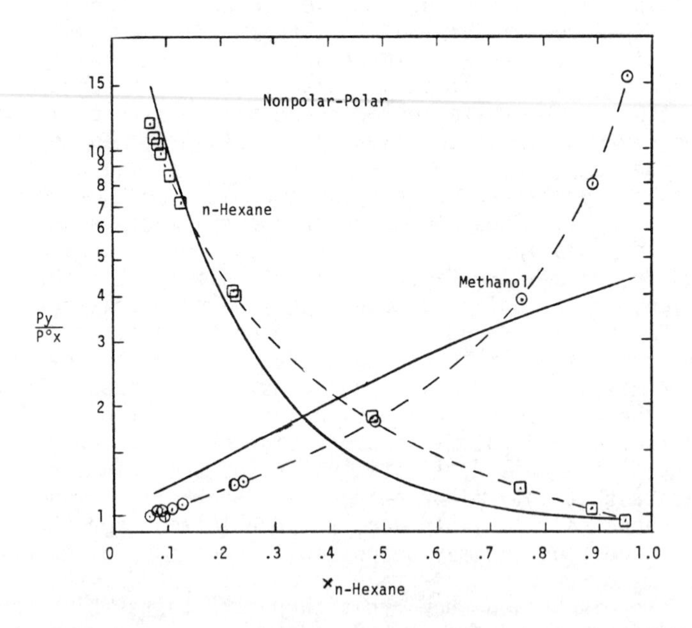

Figure 8. Redlich–Kwong correlation of nonpolar–polar n-*hexane–methanol mixture at 45°C (data of J. B. Ferguson,* Z. Phys. Chem., 36, *1123 (1932))*

One area not discussed so far is the problem of thermodynamic properties of mixtures containing both electrolyte and nonelectrolyte components. It is the belief of the author that this problem will not be completely resolved until equations of state are developed to handle these systems. We will only make progress in this direction when we recognize this as a problem and work toward solving the problem. Anything short of this also falls short in solving the ultimate problems of these mixtures.

Summary

This discussion can probably be best summarized by saying that the magnitude of work necessary for the development of thermodynamic data on synthesis gas processes is very large because six different types of compounds are involved rather than three types involved in oil and gas processes. These additional types introduce twelve new interactions between the various types of compounds compared with three for oil and gas systems. Thus the amount of work could be as much as five times as much as work already done on oil and gas systems.

In order to develop the necessary data rapidly it will be necessary for all research groups to cooperate.

Abstract

The problems associated with new synthesis gas processes are far greater than problems associated with gas processing plants or refineries because of water, salt, sludge, ammonia, and cresols present in the process streams. This paper attempts to identify the magnitude of the problems and methods for solving these problems. The problem of predicting the thermodynamic properties of nonpolar-polar mixtures by means of equations of state is also identified as an area needing study.

RECEIVED February 26, 1980.

15

The Gas Processors Association Research on Substitute Gas Thermodynamics

M. A. ALBRIGHT—Phillips Petroleum Company, Trailer #2, Bartlesville, OK 74004

D. A. FORBES—Continental Oil Company, P.O. Box 1267, Ponca City, OK 74601

L. D. WIENER—Mobil Research and Development Corporation, Box 900, Dallas, TX 75221

Since 1974, The Gas Processors Association (GPA) has been sponsoring research on substitute gas thermodynamics and other water containing systems. GPA is an unincorporated trade association with 156 corporate members. All of the members either process natural gas or are studying and/or producing substitute gas or both. GPA's thermodynamics research budget averages about $250,000 per year. About one third of that is presently directed towards systems involving water. Research areas involving water include substitute gas, sour water stripping, sour gas treating, and gas-solid hydrate equilibria.

Processes Involved

A. In the substitute gas area, there are three processing steps involving water. They are the reaction step, the quench step, and the water treating step. The system involved and the data needed for process design in each step are:

1. Reaction Step

   Temperature: 640-1925 K (700-3000 F)
   Pressure: 100-17,250 kPa (1-170 Atm)
   Needed Properties: Enthalpies, P-V-T relationships, free energies of formation
   Compounds of Concern: H_2O, H_2, CO, CO_2, CH_4, N_2

2. Quench Step

   Temperature: 310-640 K (100-700 F)
   Pressure: 100-17,250 kPa (1-170 Atm)
   Needed Properties: Enthalpies, P-V-T relationships, L-L-V phase equilibria (water-hydrocarbon-gas), and dew points
   Compounds of Concern: H_2O, H_2, CO, CO_2, N_2, CH_4, H_2S, trace (NH_3, HCN, COS, CS_2, phenols, aromatics, fatty acids, tars, oxides of nitrogen and sulfur)

0-8412-0569-8/80/47-133-317$05.00/0

3. Water Treating Step

Temperature: 280-590 K (60-600 F)
Pressure: 100-7,100 kPa (1-70 Atm)
Needed Properties: Solubilities, L-V phase equilibria, enthalpies, heats of adsorption
Compounds of Concern: H_2O, NH_3, CO_2, CO, aromatics, phenol, fatty acids, H_2S, COS, CS_2, HCN, oxides of nitrogen and sulfur

B. Sour gas treating involves the removal of the acid gas components CO_2 and H_2S from natural gas. Most ways of doing this involve water solutions. Treating is normally at near ambient temperatures and at pressures to 7100 kPa (70 Atm). The treating of high acid gas content natural gas is becoming more important as the value of natural gas increases.

C. Sour water treating is found in natural gas processing, refining, and the substitute fuel processes. Cmmpounds present normally include a few to all those listed in A. 3. above. In addition, the water pH is normally controlled with caustic soda or other salt. In the past, processing has been near ambient temperatures and pressures. Future operations may, however, be at elevated temperatures and pressures.

D. Gas-solid hydrate may occur in high pressure gas pipelines and processing facilities. They occur at temperatures below 285 K (55 F) and phase boundary pressures near 4100 kPa (40 Atm).

GPA Projects Involving Water

GPA research is proposed, financed, and conducted on an individual project basis. The GPA research strategy is normally to measure only sufficient data to allow parameter determination in models. GPA seldom attempts definitive system studies. GPA research is directed by an Enthalpy Steering Committee and a Phase Equilibrium Committee. Dr. L. D. Wiener chairs the Enthalpy Committee. Dr. K. H. Kilgren chairs the Equilibrium Committee. Both committees are divisions of Technical Section F chaired by M. A. Albright.

Completed and current GPA research areas include:

A. Enthalpy Projects

Project 742

"Calorimetric Determination of Enthalpies of Binary and Multicomponent Mixtures of Components Encountered in SNG Processes"

1. Systems: H_2-CH_4; H_2-CO (1)
   Temperature: 283-477 K (50-400 F)
   Pressure: 100-17,300 kPa (1-170 Atm)
   Investigator: J. M. Berryman (P-V-T, Inc.)

2. System: NH_3, H_2O, H_2S, CO_2
   Temperature: 283-477 K (50-400 F)
   Pressure: 100-17,300 kPa (1-170 Atm)
   Investigator: John Cunningham (BYU)

Project 772

"Experimental Determination of Enthalpy Departures of Well Defined Simulated Natural Gas/Water Mixtures." In addition, densities and dew points are being measured.

1. Systems: Natural gas - 1.5, 5, and 10% water
   Sour Natural gas - 1.5, 5, and 10% water.
   Gas phase measurements only
   Temperatures: 283-477 K (50-400 F)
   Pressures: 100-34,500 kPa (1-340 Atm)
   Investigator: Toby Eubank (Texas A&M)

Project 773

"Calorimetric Enthalpy Data for Binary Syngas Mixtures"

1. Systems: CH_4, CO, H_2, CO_2, N_2 binaries with H_2O
   Temperature: 310-1090 K (100-1500 F)
   Pressure: 100-20,700 kPa (1-205 Atm)
   Investigator: John Cunningham (BYU)

Project 791

"Extension of the Peturbed Hard Chain Correlation (Statistical Mechanical Theory of Fluids)" (2, 5). Extend the PHC program under development to include additional compounds including water. This work is an attempt to combine good correlations for phase equilibrium, enthalpy, entropy, and density into a single model.

B. Phase Equilibria

Project 758

"Phase Equilibria and Modeling for Gaseous Components With Water"

1. Bibliography and Data Compilation (4)
   Investigator: John Erbar (Oklahoma State)

2. System: H_2-H_2O (3)
   Temperature: 365-590 K (200-600 F)
   Pressure: 1378-11,035 kPa (13.6-108.9 Atm)
   Investigator: Will Devaney (P-V-T, Inc.)

3. System: NH_3-CO_2-H_2S-H_2O with and without salts (6)
   Temperature: 298-393 K (77-248 F)
   Pressure: Saturation
   Investigator: Grant Wilson (Wilco Co.)

4. Systems: CO-H_2O; N_2-H_2O; H_2S-H_2O
   Includes phase densities
   Work in Progress
   Investigator: Dick Hall (Texas A&M)

5. System: CH_4-CO_2-H_2S-H_2O
   Work in Progress
   Investigator: Don Robinson (Alberta)

6. System: H_2-CO-H_2O
   Work in Progress
   Investigator: Grant Wilson (Wilco Co.)

Project 775

"Allowable Water Content of Low Temperature Gas Streams and NGL"

1. Systems: Natural gas and its constituents with water hydrate
   Temperature: Below 273 K (32 F)
   Pressures: Phase Boundaries (psuedo)
   Investigator: Riki Kobayashi (Rice)

When all of the above multi-year projects are completed, the estimated cost is $752,000.

Future Plans

Planned experimental work involving water systems include phase equilibria on sour water streams at higher temperatures and pressures. This work will be applicable to gas-water separations in natural gas, substitute gas, and refinery processes.

Also planned are enthalpy measurements on selected multicomponent systems which will include water in some cases.

Other Work

In addition to the projects involving water mentioned above, GPA has several projects involving phase equilibria and enthalpies of the other substitute gas constituents with the natural gas components.

The API Subcommittee for Technical Data is sponsoring phase equilibria work by Grant Wilson (Wilco Co.) on water non-hydrocarbon/hydrocarbon systems. The first system will be n-octane, ethylbenzene, and ethylcyclohexane as binaries with water and as ternaries with hydrogen sulfide as the third component.

ABSTRACT

Since 1974, The Gas Processors Association (GPA) has been sponsoring research on substitute gas thermodynamics. Enthalpy and vapor-liquid-equilibrium data required for the process design of all types of substitute gas are being measured. Known properties must be extended to much higher pressures and temperatures than previously found in commercial processes. Technically, the problem has been the addition of new compounds to the natural gas compounds whose pure and mixture properties are well known. These new compounds include water, hydrogen, carbon monoxide, polycyclic aromatics, undefined heavy mixtures, and trace compounds. Among the trace compounds are ammonia, hydrogen cyanide, the gaseous oxides of sulfur and nitrogen, hydrogen sulfide, and many others. The GPA program is carefully integrated with other programs (American Petroleum Institute, Electric Power Research Institute, etc.) to avoid duplication of effort.

LITERATURE CITED

1. Berryman, J. M.; Dulaney, W. E.; Eakin, B.; Bailey, N. L.; "Enthalpy Measurements on Synthetic Gas Systems: Hydrogen-Methane, Hydrogen-Carbon Monoxide"; Research Report 37; Gas Processors Association, Tulsa, OK

2. Donahue, M. D.; Praushitz, J. M.; "Statistical Thermodynamics of Solutions in Natural Gas and Petroleum Refining"; Research Report 26, Gas Processors Association, Tulsa, OK

3. DeVaney, W.; Berryman, J. M.; Kao, P-L; Eakin, B.; "High Temperature Measurements for Substitute Gas Components"; Research Report 30, Gas Processors Association, Tulsa, OK

4. Erbar, J.; Pendergraft, P.; Marston, M.; Gonzales, M.; Rice, V.; "Literature Survey for Synthetic Gas Components - Thermodynamic Properties"; Research Report 36, Gas Processors Association, Tulsa, OK

5. Hohman, E. C.; "A Preliminary Version of the PHC Equation of State Computerized for Engineering Calculations"; Research Report 38, Gas Processors Association, Tulsa, OK

6. Wilson, G. M.; Owens, R. S.; Roe, M. W.; "Sour Water Equilibria: Ammonia Volatility Down to ppm Levels, Effect of Electrolytes on Ammonia Volatility - pH vs. Composition"; Research Report 34, Gas Processors Association, Tulsa, OK

RECEIVED January 31, 1980.

16

Overview

Gas Supply Research Program at the Gas Research Institute

A. FLOWERS and J. C. SHARER

American Gas Association, 1515 Wilson Blvd., Arlington, VA 22209

The paper deals briefly with the Gas Research Institute and its research in alternative sources of gas. As a not-for-profit organization, the Gas Research Institute plans, finances, and manages applied and basic research and technological development programs associated with gaseous fuels. These programs are in the general areas of production, transportation, storage, utilization and conservation of natural and manufactured gases and related products. Research results, whether experimental or analytical, are evaluated and publicly disseminated.

Since the proved reserves of conventional natural gas have declined in recent years, the need for new supply options was of primary importance in the 1979 research program. Forty-four projects are being undertaken this year to further the development of four new sources of gas supply. They are:

- o Unconventional Natural Gas
- o Substitute Natural Gas from Fossil Fuels
- o Substitute Natural Gas from Biomass
- o Nonfossil Hydrogen

GRI'S HISTORY

Until the organization of the Gas Research Institute, almost all cooperative research in the gas industry was carried out under the auspices of the American Gas Association (A.G.A.), the trade association of a wide cross section of the regulated gas distribution and transmission companies.

With the 1973-74 oil embargo, the gas industry realized that a major national effort would be needed to assure adequate, secure, and environmentally acceptable supplies of all forms of energy.

The concept of GRI was based on the recommendation of an ad hoc committee of the Boards of A.G.A. and the Interstate Natural Gas Association of America (INGAA), the trade association of the interstate pipeline companies. GRI was incorporated in Illinois as a not-for-profit scientific research corporation

0-8412-0569-8/80/47-133-323$05.00/0

on July 8, 1976, following the approval of these recommendations by the A.G.A. and INGAA Boards. At the beginning of 1977, GRI began recruiting charter members. Charter membership dues and contributions by A.G.A. and INGAA provided initial funds, and committees composed of gas industry executives and consultants from outside the gas industry assisted in organizing GRI.

GRI's progress was given great impetus when the former Federal Power Commission (FPC) proposed a rule change that would allow advance approval of R&D programs developed under a set of carefully drawn guidelines by organizations which derive financial support from companies under FPC jurisdiction. The proposed regulations were adopted by the FPC in June 1977, and GRI has since been operating under them, as promulgated by FPC's successor, the Federal Energy Regulatory Commission (FERC).

To provide an objective basis for its program, GRI established an effective planning methodology which integrates cost-benefit and state-of-the-art studies of relevant technologies with the expert judgment of four advisory bodies appointed by and reporting to the Board of Directors. Members of these advisory bodies participate directly in the planning process, and are especially sensitive to the broad national interest. The culmination of the planning process is the five-year plan and annual program that is submitted yearly to the FERC for approval. Most simply stated, GRI's program is designed to identify and pursue those scientific and technological opportunities that best meet the needs of the gas consumers served by the nation's regulated pipeline and distribution companies.

THE OVERALL GRI R&D PROGRAM

In 1978, GRI administered about 60 contracts comprising most of the former research programs of the A.G.A. These were the utility research program, funded at $9.7 million, and the coal gasification program, funded in the past at an annual rate of $10 million by the gas industry and $20 million by the DOE. A.G.A. continued to raise funds from its members during all of 1978 for utility research and for the first half of 1978 for coal gasification research. GRI raised or obtained commitments from the former coal gasification subscribers to cover the approximately $5 million gas industry share for the second half of 1978.

With approval by the FERC of the 1978 R&D program, GRI began to negotiate and let contracts for a supplemental program, largely in the areas of unconventional natural gas supply and gas conservation. Since the FERC funding mechanism did not become effective until June 1 and cash flow from interstate sales and transportation services did not start until late summer, GRI received only approximately $6 million for the FERC approved 1978 R&D program and was not able to place contracts for all of its intended program during the 1978 calendar year.

In 1979, GRI will administer over approximately 150 contracts, all of which are being funded by the FERC funding mechanism. The total GRI R&D budget is about $36 million and can be broken down by percentage among the 5 operating divisions as shown in Table I.

TABLE I - PERCENTAGES OF GRI R&D BUDGET BY DIVISION

	1979	1980
Gas Supply	56%	45%
Efficient Utilization	27%	32%
Planning (Economic & Systems Analysis)	6%	6%
Environment and Safety	6%	9%
Basic Research	5%	8%
	100%	100%

GRI sumbitted its 1980 Program to FERC on June 4, 1979. This plan is in the approval process at the time this paper was written. The total R&D budget request was about $50 million and can be broken down amongst the operating divisions as shown in Table I.

THE GAS SUPPLY PROGRAM

The proven reserves of so called "conventional" natural gas have declined in recent years and the need for additional supplies and new supply options have been identified as a high priority requirement to benefit the gas consumers. Therefore, the GRI Gas Supply Program has been established to identify, evaluate, and develop new gas supplies that will guarantee abundant quantities of gaseous fuels for gas consumers in the future. To fulfill this objective the GRI Gas Supply Program has been divided into the following subprograms:

- Unconventional Natural Gas
- SNG from Coal
- SNG from Oil Shale
- SNG from Biomass
- Hydrogen

These subprograms have been developed to provide for near, mid- and long-term gas options. It is essential that a well-conceived, properly managed, gas supply program having a high priority and a high funding level, be maintained such that multiple gas options can be adequately investigated. Prioritization to a "single gas supply option" at this time will not yield a cost effective research program and will not be in the best interest of the gas consumer.

Near term options include primarily the unconventional natural gas resources (Western tight gas sands, Eastern Devonian gas shales, geopressured aquifers, and methane from coal deposits). Mid-term options include coal gasification (peat included), in-situ coal gasification and biomass. Long-term options include hydrogen. In evaluating and prioritizing these programs, options that do not make technical or economic sense are eliminated. For example, SNG from Oil Shale will not receive funding in 1980 because no real advantages have been identified in oil shale gasification over coal. In a similar manner, budget allocations have been determined by not only impacts on gas supplies but also on the funding requirements for developing these options in a cost effective manner.

To maximize the output from our program and to fulfull our stated objectives in the most timely and cost effective manner, it is essential to be informed with respect to the DOE and industry programs, coordinate with these programs where possible, and co-fund and co-manage projects that will satisfy the needs of the GRI program. GRI has attempted to coordinate all subprograms within the Gas Supply Division through discussion with appropriate DOE personnel. GRI believes that our program planning with DOE has produced a well conceived coordinated effort from both the GRI and DOE standpoint.

1.1 UNCONVENTIONAL NATURAL GAS

Research to date has shown that a significant resource base exists in what is commonly called Unconventional Natural Gas Resources. These resources differ in geological formation and geographical location and are typically categorized as follows:

- o Western Tight Gas Sands
- o Eastern Devonian Gas Shales
- o Methane from Coal Deposits
- o Geopressured Aquifers
- o Gas Hydrates.

Potential resource bases and the economic benefits of using the resource are being determined for unconventional natural gas sources. Numerous assessments have been performed by DOE, the gas industry, and other groups. The ranges of resource estimates from these assessments are summarized below.

UNCONVENTIONAL GAS RESOURCE ESTIMATES

RESOURCE	ESTIMATED TOTAL RESOURCE IN-PLACE TCF*	RECOVERABLE RESOURCE TCF*
Western Gas Sands	50 - 600	23 - 313
Eastern Gas Shales	75 - 700	10 - 504
Methane from Coal Deposits	72 - 860	15 - 487
Geopressured Methane	3,000 - 50,000	150 - 2,000
Gas Hydrates	450 - 30 X 10^6**	?

* Trillion Cubic Feet

**Represents possible total world supply

There is considerable disagreement over both the amount of the resources in place and the economics of their recovery. However, there is general agreement that:

- o There are considerable quantities of gas to be recovered even at conservative estimates.
- o The differences in resource estimates and the uncertainties in the economics of recovery show the need for R&D.

We have initiated projects in 1979 that are coordinated with DOE. In 1979, GRI will complete the development of detailed subprogram plans including specification of those tasks that will be co-funded with DOE. Core samples, stimulation, and production data will be collected from wells in western tight sand formations and geopressured zones. Laboratory data on the effects of gas fracturing on specific Devonian shale formations will be collected. These data collection activities will better characterize the geologic structure of gas-containing formations and will define the extent to which new technology development is necessary. Specific technology developments initiated in 1979 include testing of diagnostic techniques to determine stimulation effects, determination of flow and corrosion properties of geopressured brines, and evaluation of turbodrill hardware for use in coal seams.

1.2 SNG FROM COAL

The production of synthetic natural gas (SNG) from coal is considered one of the major alternatives for augmenting the tightening supplies of natural gas in the United States. Coal (bituminous, sub-bituminous, lignite and peat) represents 70% to 80% of the remaining fossil fuel reserves in the United States and is, therefore, the most logical material for conversion to SNG. Of course, it has long been possible to manufacture SNG from coal by a proven commercial process (Lurgi), but significant improvements are possible. This subprogram aims to develop modern coal gasification processes with dramatic improvements in either technical simplicity, higher efficiency, or lower cost.

The SNG from coal subprogram is broken down into three project areas:

- o Gasification Processes
- o Associated Technologies
- o *In-Situ* Gasification

A large pilot-plant research program was started in 1971 to develop several coal gasification concepts which utilize modern engineering techniques. This effort was funded by DOE (ERDA, OCR) and GRI (A.G.A.) on a 2/3 to 1/3 basis, respectively. It included several coal gasification processes that were ready for pilot plant scale testing at that time. This joint program has completed its eighth year of operations, and all the gasification processes in the original Joint Program starting in 1971, except two, have been terminated or successfully completed. The two remaining processes are BI-GAS and HYGAS. BI-GAS, because of technical problems and lack of good economic potential, was dropped at the end of 1978 from the Joint Program at the request of GRI. HYGAS Steam/Oxygen has been the most successful of the processes in the Joint Program. The HYGAS Pilot Plant will be operated through 1980 to support a demonstration plant design effort funded by DOE.

The continuing search for technical and economic improvements in the coal gasification areas has provided incentives to develop newer process concepts starting in 1979. The Joint Program for 1979 includes processes in the Process Development Unit (PDU) stage of development which promise technical and/or economic improvements. Gasification processes included in the 1979 program are:

a. HYGAS Steam/Oxygen
b. PEATGAS Process
c. Exxon Catalytic Process
d. Westinghouse Fluid Bed Process
e. Rockwell Hydrogasification Process
f. Bell Aerospace High Mass Flux Process

The HYGAS project was dropped from the Joint Program on June 30, 1979. The remaining processes received several independent critical evaluations during 1979. These evaluations pointed out that the Rockwell process cannot claim distinct advantages over earlier technology. Therefore, it will not be included in our 1980 program.

In critical review of the SNG from Coal subprogram, it became evident that research efforts directed to the operations upstream and downstream of the gasifier must be substantially increased to achieve optimization of the overall coal gasification plant and to achieve maximum cost reductions.

The present state-of-the-art technology for the conversion of raw gas exiting the gasifier into pipeline quality SNG is not very satisfactory. Enormous volumes of gas are cooled, reheated, cooled again, reheated again, and cooled for a third time. Steam is added to the gas, and condensate is subtracted twice (earlier and later). This series of operations is complex and costly. This area offers lots of opportunity for spectacular process improvement with the potential of high economic payoff.

Materials of construction are becoming more critical with development of the newer processes and unit operations. Therefore, an expanded materials research program is essential to the successful development of a coal gasification industry.

Engineering evaluation has become extremely important to the SNG from Coal subprogram because of the need to critically assess the technical and/or economic impacts of the research projects being funded. In order to select processes having a greater potential for technical and/or economic advantages over other processes at the earliest possible time and to maximize prioritization within the SNG from Coal subprogram, increased efforts in engineering evaluations was deemed essential.

In-situ coal gasification appears to have the potential of producing SNG at a lower cost than above-the-ground processing. For the first time, a process is being developed that has shown some potential of being developed into a workable process. Therefore, this process deserves further development to determine if large-scale underground coal gasification is technically feasible and to develop sufficient data for a good economic evaluation.

The Lawrence Livermore Laboratory in-situ coal gasification project has the objective of producing synthesis gas that can be upgraded to SNG. An initial two-day oxygen burn during 1978 indicated that oxygen gasification appeared feasible. The work plan for 1979 will develop additional data with a longer duration oxygen burn in a shallow coal bed using directional drilling to assure a positive connection between the injection and production wells. If the 1979 effort is successful, the next step in this development is to verify the 1979 results in a deep coal bed.

1.3 SNG FROM OIL SHALE

Oil Shale is second to coal as the most abundant potential source of raw material for supplemental gas supplies. The DOE development program on oil shale is focused on the production of liquids by thermal retorting, either by above-ground or in-situ processing. GRI has been funding the development of a hydrogasification process that can handle both Eastern and Western shales to produce a range of gaseous or high-grade liquid fuels depending on the operating conditions selected. The PDU program is now nearing completion and the next step would be the pilot plant testing where a substantial amount of funding is required.

GRI considers SNG from Oil Shale to have lower priority then coal since coal is far advanced in the development cycle. In reviewing developments to date, no real advantages have been determined in oil shale gasification over coal. Therefore, GRI decided that SNG from Oil Shale will be held at the PDU level until the coal program is fully developed or whenever the resumption of development of the shale program is warranted.

1.4 SNG FROM BIOMASS

A very promising long-range solution to the problem of fossil-fuel depletion is to convert a major source of continuously renewable carbon to SNG. The greatest potential sources of this carbon are water- and land-based biomass produced from ambient carbon dioxide and solar energy by photosynthesis. Biomass is defined as all growing organic matter (such as plants, trees, algae, and organic wastes) and, it is perpetually renewable. The production of SNG from low-cash-value, high-fuel-value biomass would offer a major, controllable, nonpolluting, storable resource of fossil-fuel substitutes.

The SNG from Biomass subprogram is broken down into three project areas:

- o Land-Based Biomass
- o Water-Based Biomass
- o Wastes.

Land-based biomass has a potential of providing from 7 to 11 quads/year of SNG if all available marginal land suitable for the growing of crops could be utilized. However, the doubtful availability of land and water for growing land-based biomass specifically for SNG production gives this supply option a lower priority than water-based biomass. The initial stage of the project concentrates on identifying natural species that will produce a maximum quantity of gas.

The Marine Biomass Program has the overall objective of developing an integrated system for the production of methane gas from marine biomass on a commercial scale that will provide a major contribution to the nation's gas supply. Giant brown kelp grows naturally along the coast of California. It is commercially harvested by two chemical companies for use as a food additive and animal feed supplement. In the Marine Biomass Program, sponsored by the Gas Research Institute, Department of Energy, and New York State ERDA, the kelp will be grown and cultivated in the open ocean on an artificial structure with fertilizer being supplied by mechanically upwelled, nutrient-rich, deep ocean water. The kelp is then mechanically harvested and converted to methane by the anaerobic digestion process.

A test farm has been deployed about 5 miles off the coast of California and has been in operation for about 9 months. Laboratory experiments are also underway involving kelp pretreatment, post-treatment and conversion to methane utilizing both sewage based and marine based inocula.

The upper limit of energy potentially capable of being produced from a marine or ocean based biomass has not yet been determined, but it is estimated that the potential contribution to the long-term U.S. energy supply could be at least equal to today's natural gas consumption or 20 quads per year.

Our enthusiasm for the program is based on the following assessments:

1. A virtually unlimited potential exists for growing a huge biomass resource in the ocean.
2. No scientific breakthroughs are required to commercialize this concept.
3. Preliminary studies indicate gas costs could be competitive with other SNG sources.
4. The biomass is a renewable resource with no apparent negative environmental impacts.

Another source of non-fossil carbon that can be used to produce SNG is organic wastes. The growing environmental and pollution problems caused by the generation of organic wastes in the United States provide an opportunity to combine waste recovery with the production of SNG. Considering the amount of organic solids economically available for conversion, a potential of about 1 to 1.5 quads of SNG per year could be produced. Industry and government have been funding R&D efforts in this area for several years. GRI has given this supply option a low priority.

1.5 HYDROGEN

Hydrogen is of interest as a means to deliver gaseous fuel from non-fossil primary energy resources such as nuclear reactors, or high temperature solar collectors. It is believed that hydrogen may phase into the energy market at such a time when fossil-based fuels either become too expensive or environmentally unsatisfactory. Hydrogen and biomass are the only two potentially visible options at the present time for the gas industry if that does take place.

Hydrogen is used today as a unique industrial chemical in petroleum processing and in the synthesis of ammonia and methanol, and other organic chemicals. The world wide production of hydrogen has increased by three orders of magnitude in the last four decades. At present, the amount of U.S. energy consumed to produce industrial hydrogen is about 1.4 quads/year which is more than 1% of the total national energy use and is expected to increase to about 5 quads/year by the year 2000. Most of this hydrogen is produced by steam reforming natural gas or light oils.

Hydrogen can be considered an insurance policy for the gas industry. The time frame for which hydrogen becomes economically viable, is, at this stage, unknown. Long range research programs, at low funding levels can identify the viability of hydrogen producing schemes. At this stage of research, GRI is only interested in the production of hydrogen. The uncertainty of the economics of producing hydrogen from water is the key problem to the implementation of utilizing hydrogen as a gaseous fuel. Other areas such as transportation, storage, distribution and utilization are not being investigated by GRI. Research in these areas should be delayed until large-scale, economical, production techniques appear feasible.

The Hydrogen subprogram has been divided in two project areas:

- Thermochemical Hydrogen
- Electrolytic Hydrogen

Thermochemical Hydrogen is concerned with the production of hydrogen as a gaseous fuel through a closed loop thermochemical process for water-splitting. There are two ongoing projects in this area.

The second project area involves the optimization of water electrolysis technology.

RECEIVED January 31, 1980.

17

Application of the PFGC-MES Equation of State to Synthetic and Natural Gas Systems

M. MOSHFEGHIAN, A. SHARIAT, and J. H. ERBAR

School of Chemical Engineering, Oklahoma State University, Stillwater, OK 74078

The need for methods of accurately describing the thermodynamic behavior of natural and synthetic gas systems has been well established. Of the numerous equations of state available, three--the Soave-Redlich-Kwong (SRK) (19), the Peng-Robinson (PR) (18) and the Starling version of the Benedict-Webb-Rubin (BWRS) (13, 20)--have satisfied this need for many hydrocarbon systems. These equations can be readily extended to describe the behavior of synthetic gas systems. At least two of the equations (SRK and PR) have been further extended to describe the thermodynamic properties of water-light hydrocarbon systems.

All of these equations suffer from at least one common deficiency--they require that the critical properties of all components in the system be defined. This requirement extends to any undefined component (C_{6}+, crude oil, heavy tar fractions, etc.) which may be present in the system. Prediction of the critical properties of these compounds is at best an art. Changing the critical temperature of an undefined fraction present in quantities less than one mol percent by 10°C can change the predicted dew point of a natural gas system by 35 bar. Would anyone pretend that our current (or future methods) of estimating the critical properties of an undefined component are accurate to ±10°C?

Since most synthetic and natural gas systems will contain some amount (however small) of heavy undefined components, we have been searching for improved methods of predicting critical properties and an equation of state which does not use critical constants (or quasi critical constants) to determine the parameters for the equation. Development of improved critical property prediction methods appears to be a waste of time. Wilson and Cunningham (6) have presented an equation--the Parameters From Group Contributions (PFGC) equation of state which satisfies our needs. As the name implies, the parameters in this equation of state are estimated by group contribution rather

0-8412-0569-8/80/47-133-333$06.75/0

than relying on the critical properties. This ability offers a real advantage when dealing with undefined compounds. The groups (fragments) in these compounds can be determined by advanced analytical methods such as $C^{13}NMR$ spectroscopy. Reduced reliance on the less refined but standard analytical procedures, i.e., specific gravity, average boiling point, UOPK, etc., will surely result in improved predictions of the thermodynamic properties of synthetic and natural gas systems.

Our recent research efforts have been directed to developing an improved and extended set of parameters for the various groups used in the PFGC equation of state (16). We have dubbed our version of this equation the PFGC-MES. A thorough evaluation of the ability of this equation to predict the thermodynamic properties of hydrocarbon systems has been a part of this development process. A companion project has been the extension of the equation to describe the behavior of hydrocarbon-water-alcohol-nonhydrocarbon systems. This paper gives some of the results of our work on hydrocarbon-water-methanol systems. We also report some of our work on extending the SRK equation of state to hydrocarbon water systems.

Theory

The basic equations used to predict the thermodynamic properties of systems for the SRK and PFGC-MES are given in Tables I and II, respectively. As can be seen, the PFGC-MES equation of state relies only on group contributions--critical properties, etc., are not required. Conversely, the SRK, as all Redlich-Kwong based equations of states, relies on using the critical properties to estimate the parameters required for solution.

Wilson and Cunningham developed the PFGC equation of state on the basis of the following definition of the Helmholtz free energy

$$\frac{A}{RT}^{PFGC} = \frac{A}{RT}^{FH} + \frac{A}{RT}^{W}$$

Using the modified hole theory and the appropriate mathematical manipulations, the set of equations shown in Table II resulted.

Application of either equation of state to the prediction of the thermodynamic properties of hydrocarbon systems is straightforward once the appropriate parameters are available. The ability of the SRK to describe the phase behavior of light hydrocarbons is well known. Moshfeghian, et al. (16) have reported that the PFGC-MES equation of state gives similar results for these systems except in the critical region.

While not immediately apparent, the PFGC-MES equation of state for the compressibility factor, z, cubic in behavior. Because of the complexity of the formulation an iterative solution procedure is used to determine the appropriate root for the

TABLE I
SOAVE REDLICH KWONG EQUATIONS

$$K_i = \frac{\phi_i^L}{\phi_i^V}$$

$$\ln \phi_i = \frac{b_i}{b}(Z-1) - \ln(Z-B) - \frac{A}{B}\left[\frac{2(ac\alpha)_i}{(ac\alpha)} - \frac{b_i}{b}\right]\ln\left(1+\frac{B}{Z}\right)$$

$$-\frac{\Delta H}{RT} = \left[\frac{A}{B} - \frac{\beta}{Rb}\right]\ln\left(1+\frac{B}{Z}\right) + 1 - Z$$

$$\frac{\Delta S}{R} = \Sigma x_i \ln \phi_i - \frac{\Delta H}{RT} + \Sigma x_i \ln x_i + \ln\frac{P}{P^0}$$

$$Z^3 - Z^2 + (A - B - B^2)Z - AB = 0.0$$

$$A = \frac{(ac\alpha)P}{R^2T^2}; \quad B = b\frac{P}{T}; \quad b = \Sigma x_i b_i$$

$$b_i = 0.08667\frac{RT_{ci}}{P_{ci}}; \quad a_{ci} = 0.42747\frac{R^2T_{ci}^2}{P_{ci}}; \quad \alpha_i^{0.5} = 1 + m_i(1 - T_{ri}^{0.5})$$

$$(ac\alpha) = \sum_i\sum_j x_i x_j a_{ci}^{0.5} a_{cj}^{0.5} \alpha_i^{0.5} \alpha_j^{0.5}(1-k_{ij})$$

$$(ac\alpha)_i = \sum_{j=1}^{n} x_j a_{ci}^{0.5} a_{cj}^{0.5} \alpha_i^{0.5} \alpha_j^{0.5}(1-k_{ij})$$

$$\beta = \sum_{i=1}^{n}\sum_{j=1}^{n} x_i x_j \left[\frac{a_{ci}^{0.5} a_{cj}^{0.5} \alpha_j^{0.5} m_i}{2T_{ci}T_{ri}} + \frac{a_{cj}^{0.5} a_{ci}^{0.5} \alpha_i^{0.5} m_j}{2T_{ci}T_{rj}}\right](1-k_{ij})$$

TABLE II
PFGC EQUATIONS

$$\frac{Pv}{RT} = Z = 1 - \frac{sv}{b}\ln\left(1-\frac{b}{v}\right) - s + b\left(\frac{c}{b_H}\right)\sum_k^g \psi_k\left(\frac{b - b\sum_n^g \psi_n \tau_{nk}}{v - b + b\sum_n^g \psi_n \tau_{nk}}\right)$$

$$\frac{\mu_i}{RT} = s_i\left(\frac{v}{b}-1\right)\ln\left(1-\frac{b}{v}\right) + 1 - \frac{sb_i}{b}\left[\frac{v}{b}\ln\left(1-\frac{b}{v}\right)+1\right]$$

$$+ \ln\left(\frac{RT}{v}\right) - \left(\frac{c}{b_H}\right)\left\{\sum_k^g\left[m_{ik}\, b_k \ln\left(\frac{v - b + b\sum_n^g \psi_n \tau_{kn}}{v\,\tau_{kk}}\right)\right]\right.$$

$$\left. + b\sum_k^g\left[\psi_k\left(\frac{-b_i + \sum_n^g m_{in} b_n \tau_{kn}}{v - b + b\sum_n^g \psi_n \tau_{kn}}\right)\right]\right\}$$

$$\frac{\Delta H}{RT} = (Z-1) + \left(\frac{c}{b_H}\right)\frac{b^2}{T}\left\{\sum_k^g \psi_k\left(\frac{\sum_n^g \psi_n\left(\frac{d\tau_{kn}}{dT}\right)}{v - b + b\sum_n^g \psi_n \tau_{kn}}\right)\right\}$$

$$b = \sum_i^c x_i b_i \qquad b_i = \sum_k^g m_{ik} b_k \qquad s = \sum_i^c x_i s_i \qquad s_i = \sum_k^g m_{ik} s_k$$

$$\tau_{kn} = e^{-E_{kn}/T} \qquad E_{kn} = K_{kn}\left[E_k + E_n\right]/2.0$$

$$E_k = E_k^{(0)} + E_k^{(1)}\left(\frac{283.2}{T} - 1\right) + E_k^{(2)}\left[\left(\frac{283.2}{T}\right)^2 - 1\right]$$

equation of state. Once the proper root (liquid or vapor) has been determined, solution of the remaining equations for chemical potential and isothermal effect of pressure on enthalpy is straight forward.

The parameters b, s, and τ appearing in the equations are group dependent parameters. Standard linear mixing rules are used to calculate the b and s terms. The τ term is also a function of temperature; a group interaction parameter is used to account for non-ideal group interactions. Parameters for the various groups are not reported here since these parameters are not yet finalized. Checking and adjustment of the parameters is being continued to assure that the best possible results are achieved over the broadest possible temperature-pressure-composition ranges. Values of the group parameters will be published when the final "polishing" of the parameters is completed.

The mixing rules of both equations of state had to be modified to predict the phase behavior of water hydrocarbon systems. This modification consisted mainly of defining different binary interaction parameters for the various phases present--i.e., one interaction parameter per binary pair for the vapor phase and hydrocarbon rich liquid; another binary interaction parameter was defined for the water rich liquid phase. In addition, water phase binary interaction parameter usually had to be temperature dependent to achieve good agreement between predicted and experimental hydrocarbon solubilities in the water rich phase. The temperature dependence of the water phase binary interaction parameter k_{12}^2, appears to be nearly linear with temperature for most systems. A typical set of SRK based k_{ij}^2's of the methane-water system are shown in Figure 1. This kind of temperature dependence was observed for all binary pairs when using the SRK equation of state. The PFGC-MES equation of state required temperature dependent k_{ij}^2's for only a few groups (molecules). The PFGC-MES k_{ij}^2's were also linear in temperature.

For ternary and higher order mixtures, we have usually assumed that the interaction parameters for the non-water binary pairs in the water rich phase are identical to the vapor (hydrocarbon rich liquid phase) interaction parameters. Some work has been done on changing all water phase interaction parameters; we concluded that predicted results were not improved enough to warrant the expenditure of time required to develop the additional parameters. A third interaction parameter for the hydrocarbon rich liquid could also be determined. Again, our work indicated that little improvement resulted from using this third parameter. Additional work is being done on both points.

The experimental data for water hydrocarbon systems are relatively limited. Consequently, a generalized correlation was developed to estimate the equation for the temperature dependent k_{ij}^2 term for those compounds for which no data are available. This generalized correlation was developed only for the SRK equation of state. The variation in the slopes of the k_{ij}^2

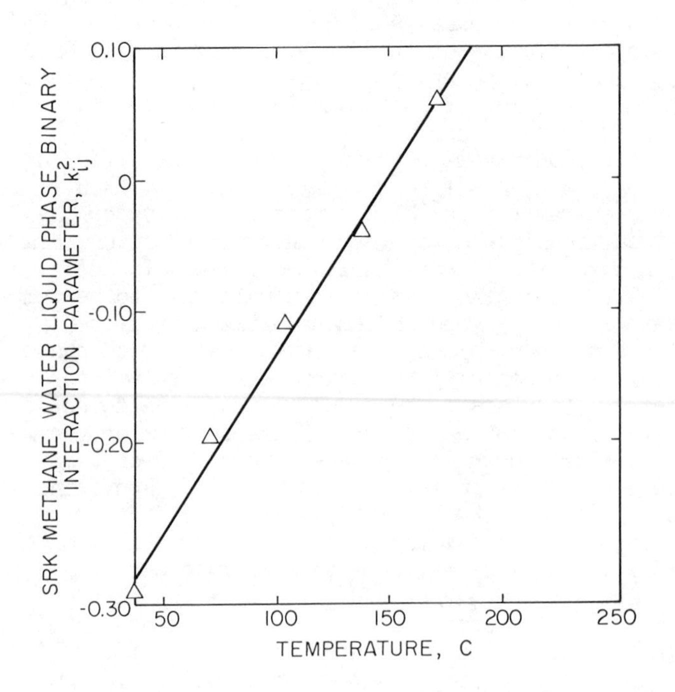

Figure 1. Effect of temperature on methane k^2_{ij}

equation for the various homologous hydrocarbon series and selected specific properties are considered. The quality of this correlation is unknown--no data are available to evaluate it.

Note that a similar correlation did not have to be developed for the PFGC-MES equation of state. Adequate data were available to develop (or estimate) group parameters for all the molecules considered in this work. The predicted water solubilities of the higher molecular weight components appear to be reasonable, but no data are available to check the quality of the predictions.

A similar strategy was used to develop the PFGC-MES equation of state parameters for describing the behavior of methanol hydrocarbon acid gas water systems. Multiple phase binary interaction parameters were used as required. Again, these second phase binary interaction parameters were usually not temperature dependent.

Model Validation

The parameters for both equations of state were developed using the available binary water data. The quality of agreement between predicted and experimental results should be good to excellent for these systems. Results for the methane-water system are shown in Figure 2; these results are typical of the results for the hydrocarbon systems. For pressures up to 350 bars both the SRK and PFGC-MES give excellent predictions of the concentrations of the various phases including the water rich and hydrocarbon rich liquid phase. In fact, the predictions from both equations of state are coincident for all temperatures except 38 and 71°C. At pressures above 350 bars, the SRK continues to perform well, while the quality of PFGC-MES predictions are not very good. Similar results were noted for all systems except the CO_2-water and H_2S-water systems. At temperatures below about 100°C and pressures greater than 70 bars significant errors in the predicted values were observed for these systems. At other conditions, excluding this region, agreement commensurate with the hydrocarbon results was found.

After the available binary data had been used to determine the various parameters, the available multicomponent (ternary and higher order systems) were used to validate both equations of state. Some of the results of the validation are given in Tables III through V and Figure 3. The hydrocarbon compositions are summarized in Table IV. Again, the available data are skimpy and some are of doubtful quality. As a general rule, both equations of state give excellent agreement with the experimental values up to 350 bars. Above this pressure limit, the SRK is superior to the PFGC-MES equation.

Validation of the PFGC-MES on methanol-water-hydrocarbon-acid gas systems was practically impossible. Many investigators have reported data for the C_5 - C_8 paraffin aromatic naphthenic

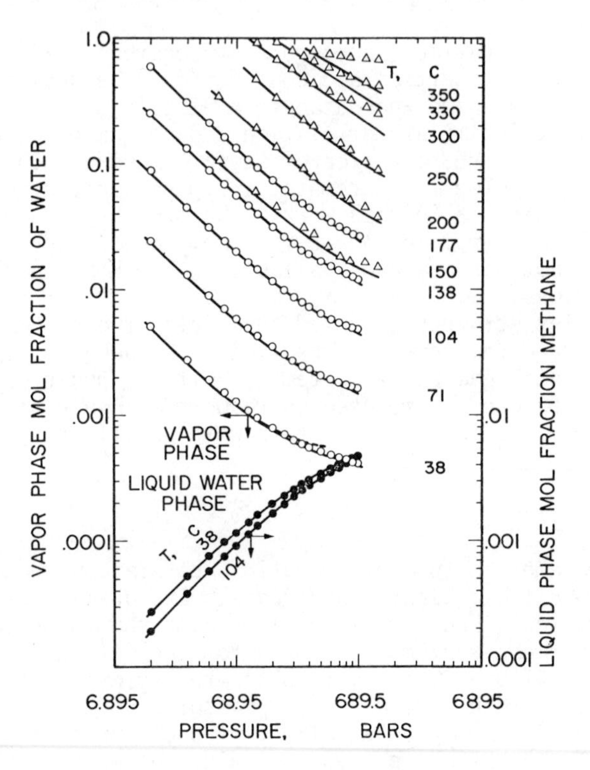

Figure 2. Comparison of predicted and experimental phase solubilities for methane–water systems (phase: liquid—(●) (17); vapor—(○) (17); (△) (21,22); (——) SRK; (– – –) PFGC–MES)

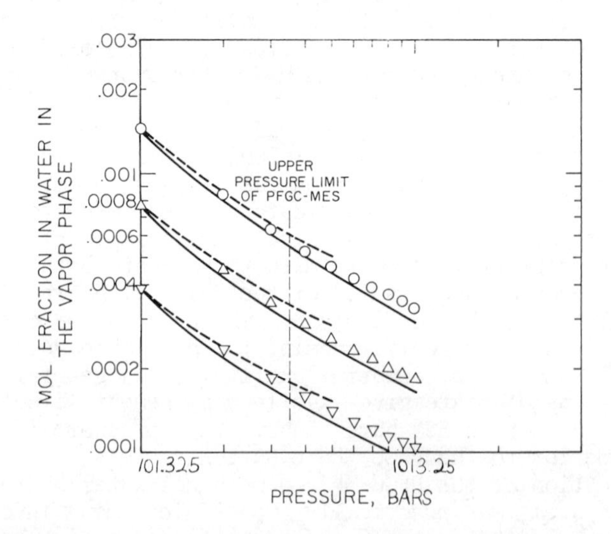

Figure 3. Comparison of predicted and experimental N_2–H_2 mixture vapor phase water concentrations (system composition: 75% H_2/25% N_2 (molar); from Ref. 3—(○) 50°C; (△) 37.5°C; (▽) 25°C; predicted—(——) SRK; (– – –) PFGC–MES)

TABLE III

SUMMARY OF COMPARISON OF
TERNARY HYDROCARBON (NON-HYDROCARBON)
WATER SYSTEMS
(From References 2 & 5)

Component 1	Component 2	mol% Range Component 2	Temperature Range, °C	Pressure Range, Bars	Absolute Avg Error in Predicted Water Content SRK	PFGC*
CH_4	C_2H_6	8.27/50.0	38/71	69.90/138.9	4.22	4.33
CH_4	CO_2	11.32/20.22	38/71	69.90/138.9	5.52	7.70
CH_4	H_2S	8.30/29.0	54/71	24.75/138.9	4.55	5.25
CH_4	C_3H_8	45.90	105	72.20/678.0	4.40	2.10
CH_4	nC_4H_{10}	25.60/39.35	105	75.35/685.0	3.68	4.04
CH_4	nC_5H_{12}	14.7	105	70.80/690.5	1.88	3.16

* PFGC-MES maximum evaluation parssure = 350.00 bars

TABLE IV

EXPERIMENTAL MULTICOMPONENT MIXTURE COMPOSITIONS
(From Reference 2)

	System			
	1	2	3	4
N_2	-	.0100	.148	-
C_1	.8851	.9436	.7379	.8690
CO_2	-	.0060	.003	.0210
C_2	.0602	.0264	.066	.0740
C_3	.0318	.0096	.039	.0220
iC_4	.0046	-	.0035	.0051
nC_4	-	.0044	.0048	.0043
iC_5	.0098	-	-	-
nC_5	.0085	-	-	.0025
nC_6	-	-	-	.0018
nC_7	-	-	-	.0013

TABLE V

COMPARISON OF PREDICTED AND EXPERIMENTAL WATER CONTENTS OF MULTICOMPONENT SYSTEMS (From Reference 2)

System No.	Temperature Range, °C	Pressure Range Bars	Absolute Avg Error in Predicted Water Content of Vapor Phase	
			SRK	PFGC-MES*
1	37.8/65,6	68.95/137.90	4.25	4.03
2	37.8	68.95/137.90	7.32	4.18
3	104.6	71.71/680.17	2.01	5.70
4	104.6	70.88/689.13	1.18	2.22

* pressure limited to 350 bars

range hydrocarbon-methanol systems; the data available for lighter hydrocarbons is sparse. The binary data for MeOH-other component systems is generally well represented; some of these results have been reported earlier (16). The results for two binary systems are reported here-- CH_3OH-CO_2 and CH_3OH-H_2S (Figures 4 and 5). The predicted bubble pressures agree with the experimental values within about 5%. Unfortunately, only one set of experimental vapor phase compositions was available (Katayama). The absolute average error in the predicted vapor phase composition for carbon dioxide was less than 1%. Predicted and experimental bubble point pressures for a four component system--nitrogen, carbon dioxide, hydrogen sulfide and methanol--are reported in Table VI. The average error in the predicted bubble point pressures is about 5%. Though not shown, the absolute average error in the nitrogen vapor phase concentration is about 8%. Similar evaluations have been made for hydrocarbon systems when possible. Most of these evaluations are based on proprietary data--results very similar to the values shown here have been obtained.

Predictions of the vapor volume and enthalpy departure for a water containing natural gases were compared with data being generated by Hall and coworkers at Texas A&M under GPA sponsorship (10). Both equations of state performed well; the average error in the predicted volume was less than one percent (abs) and the absolute average error in the predicted enthalpy departure was about four KJ/kg.

Based on our evaluations, we believe that the SRK equation of state gives reasonable predictions of water hydrocarbon behavior at any pressure up to about 700 bars and over the temperature range 0°C to 300°C. The upper pressure limit of the PFGC-MES appears to be about 350 bars while the valid temperature range is about the same as the SRK.

Sample Problems

A sample calculation has been made to illustrate the practical application of the PFGC-MES equation of state.

The problem is a typical pipeline transporting a water saturated natural gas. The pipeline conditions are such that hydrates could form in the pipeline. Methanol will be used to depress the hydrate formation temperature to an acceptable level (a 15°C hydrate depression was used).

Three different approaches to this problem can be developed. These are:

1. Conventional pipeline calculations in which "dry" hydrocarbon flashes are performed to determine the hydrocarbon liquid formation; the liquid water condensed is estimated using one of the available natural gas water content charts (1, 15), and the Hammerschmidt equation (11) and a graphical correlation are used to

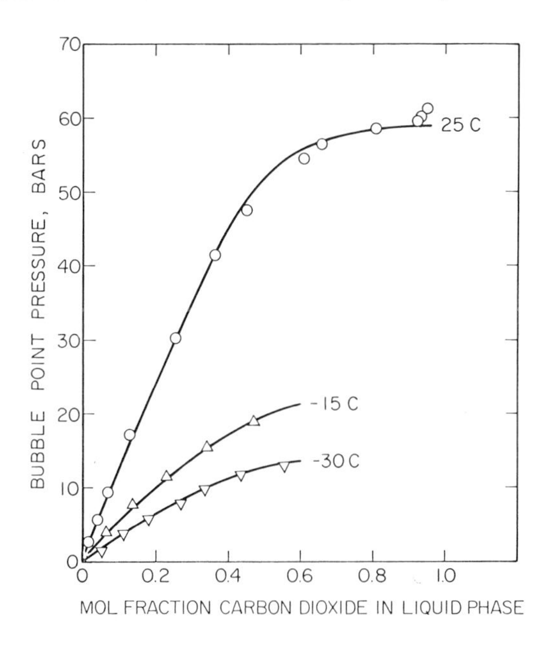

Figure 4. Comparison of predicted and experimental bubble point pressures for methanol–carbon dioxide system ((○) (14); (△, ▽) (24); (——) predicted)

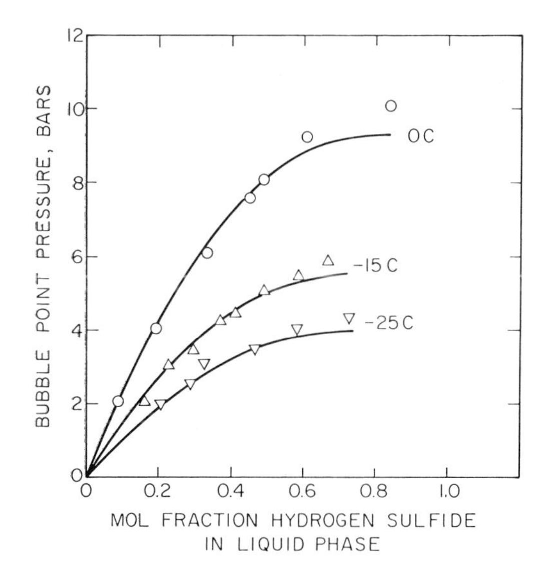

Figure 5. Comparison of predicted and experimental bubble point pressures for methanol–hydrogen sulfide system ((○, △, ▽) (24); (——) predicted)

TABLE VI

COMPARISON OF PREDICTED AND EXPERIMENTAL BUBBLE POINT PRESSURE FOR THE NITROGEN-CARBON DIOXIDE, HYDROGEN SULFIDE, METHANOL SYSTEM AT -15°C
(From Reference 9)

Experimental Bubble Point Pressure, bars	Experimental Values Liquid Phase Mol Fraction				Predicted Bubble Point Pressure bars
	N_2	CO_2	H_2S	MeOH	
9.49	.003	.069	.001	.927	10.28
9.88	.003	.067	.001	.929	10.17
20.16	.007	.083	.001	.909	20.31
28.27	.010	.070	.001	.919	26.71
35.06	.013	.065	.002	.920	33.95

estimate the methanol requirements (4).

2. Perform "wet" hydrocarbon flashes to estimate the hydrocarbon and water liquid formation and estimate the methanol requirements using the above approach.
3. Perform "wet"-methanol-hydrocarbon flashes to estimate the liquid water plus methanol and hydrocarbon phases; the methanol concentration is adjusted to satisfy the Hammerschmidt equation prediction for the desired hydrate formation temperature depression.

Calculations for all three cases have been performed for the system described in Tables VII and VIII and Figure 6. In this case the raw feed gas was flashed at 66°C and 138 bars with sufficient water to assure that the gas leaving the separator was water saturated. Each of the calculational philosophies described above was used to predict the phase behavior of the systems at each pressure temperature point in the pipeline. The results of these calculations are summarized in Tables IX through XI and Figures 7 through 10.

Several factors become immediately apparent after analysis of these results:

1. The type of flash calculation--"dry", "wet" or "wet" plus methanol--has no practical effect on the predicted hydrocarbon liquid formation.
2. The conventional technology of pipeline calculations does not normally predict the carbon dioxide or hydrocarbon content of the water rich phase. These can be estimated by other graphical correlations.
3. Both "wet" flash calculations predict a higher concentration of water in the vapor phase than the graphical correlations. The presence of methanol reduces the predicted water content of the vapor phase.
4. The presence of methanol in the liquid water phase substantially increases the predicted carbon dioxide solubility in that phase.
5. Methanol has little effect on the predicted solubility of hydrocarbons in the liquid water phase.
6. All three approaches predict about the same methanol requirements to give the specific hydrate formation temperature depression.

Most of these effects were expected; three deserve further comment.

Most of the available water content charts are applicable only to sweet lean natural gases. Moore, et al. (15) have developed a set of charts which are based on the Heidemann (8, 12) version of the SRK water prediction. The system used in this study contains nearly 6.0% carbon dioxide. The acid gases cause increased water solubility in the vapor phase. Our calculations simply verify these observations.

The "wet"-methanol-hydrocarbon flash predicts the complete phase distribution of methanol at each point in the system.

TABLE VII

FEED COMPOSITION

Component	Mol Percent
N_2	2.76
CH_4	76.95
CO_2	5.30
C_2H_6	4.06
C_3H_8	1.94
iC_4H_{10}	0.43
nC_4H_{10}	0.76
iC_5H_{12}	0.39
nC_5H_{12}	0.33
Lt. Aromatics (C_6H_6)	0.42
C_{7+} ($nC_{10}H_{22}$)	3.75
H_2O	2.91

TABLE VIII

PROCESS/PIPELINE CONDITIONS

Primary Separator at 66°C and 138.00 Bars

Pipeline Pressure Temperature Profile

	P, Bars	T, C
Inlet from separator	138.00	66
	124.00	38
	110.00	27
	96.50	21
	82.70	16
Discharge from pipeline	68.95	16

TABLE IX

RESULTS BASED ON
DRY HYDROCARBON FLASH

P/T	Hydrocarbon Liq	Water kg/Mm^3(N)*		Methanol	
Bars/C	m^3/Mm^3(N)*	Liquid	Vapor	Water Rich Liquid wt%	Vapor Phase kg/Mm^3(N)
138.00/66	-	-	2 036	-	1 354
124.00/38	18.86	1 476	560	-	-
110.00/27	25.66	1 901	331	-	-
96.50/21	29.14	1 780	256	-	-
82.70/16	32.28	1 846	190	-	-
68.95/16	31.38	1 831	205	24.75	740

* 10^6 normal cubic meters

TABLE X

RESULTS BASED ON
WET HYDROCARBON FLASH

P/T	Hydrocarbon Liq	Water kg/Mm (N)*		Methanol	
Bars/C	m^3/Mm^3 (N)*	Liquid	Vapor	Water Rich Liquid wt%	Vapor Phase kg/Mm^3 (N)*
138.00/66	-	-	2 142	-	1 354
124.00/38	18.86	1 469	673	-	-
110.00/27	25.60	1 732	410	-	-
96.50/21	29.08	1 817	325	-	-
82.70/16	32.11	1 886	256	-	-
68.95/16	31.33	1 861	281	24.75	740

* 10^6 normal cubic meters

TABLE XI

RESULTS BASED ON
WET-METHANOL-HYDROCARBON FLASH

P/T	Hydrocarbon Liq	Water kg/Mm (N)*		Methanol	
Bars/C	m^3/Mm^3 (N)*	Liquid	Vapor	Water Rich Liquid wt%	Vapor Phase kg/Mm^3 (N)*
138.00/66	-	-	2 143	-	1 339
124.00/38	18.88	1 536	607	16.22	1 013
110.00/27	25.53	1 787	356	20.40	819
96.50/21	29.04	1 868	275	22.68	711
82.70/16	32.14	1 930	213	24.78	601
68.95/16	31.28	1 910	233	24.77	606

* 10^6 normal cubic meters

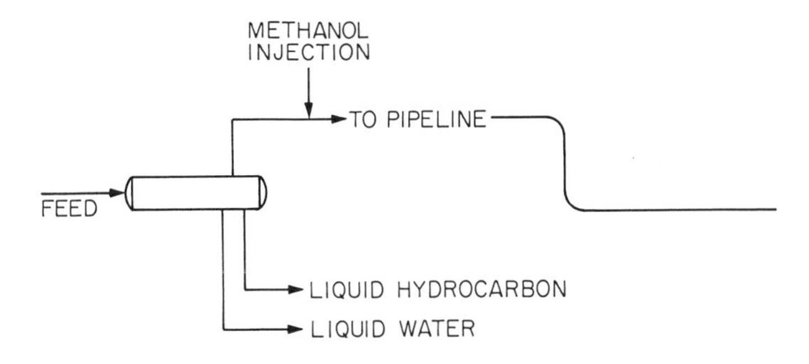

Figure 6. Diagram of process

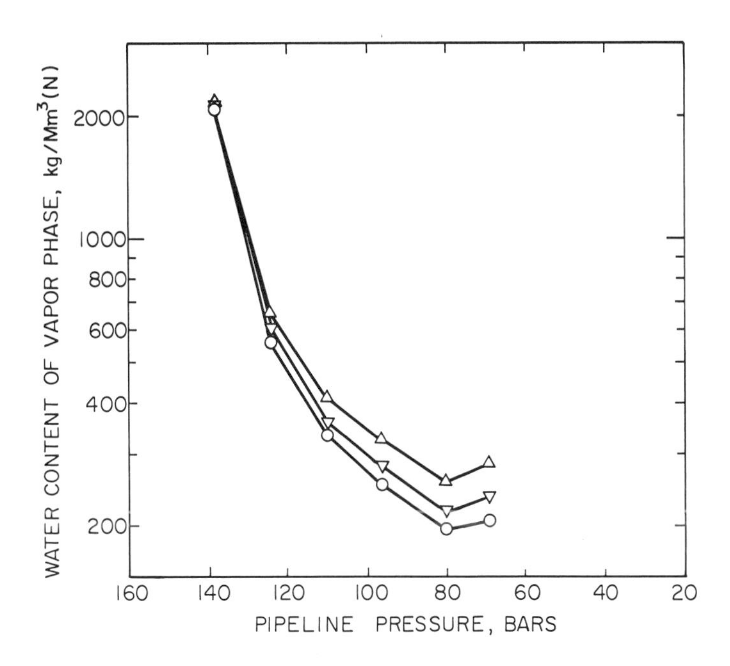

Figure 7. Vapor phase water concentration profile ((○) conventional technology; (△) wet hydrocarbon flash; (▽) wet methanol hydrocarbon flash)

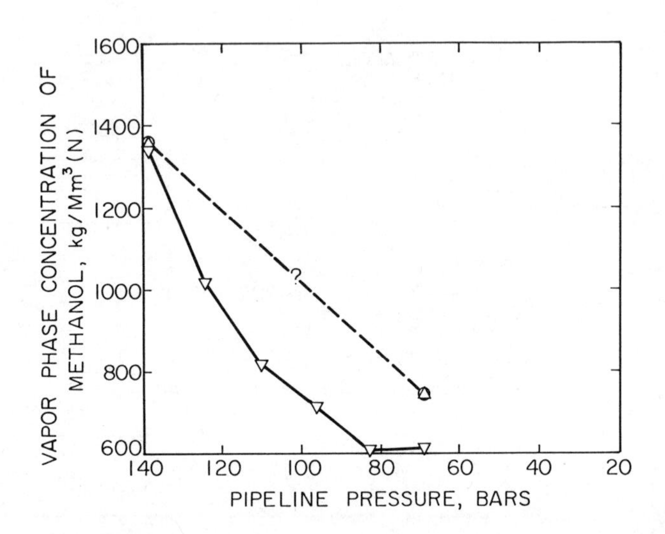

Figure 8. Vapor phase methanol concentration profile ((○) dry hydrocarbon flash; (△) wet hydrocarbon flash; (▽) wet methanol hydrocarbon flash)

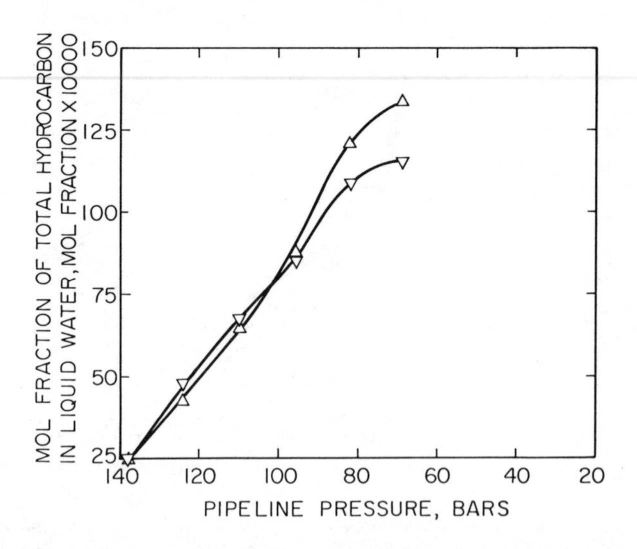

Figure 9. Effect of methanol on predicted hydrocarbon solubility in water phase ((△) wet hydrocarbon flash; (▽) wet methanol hydrocarbon flash)

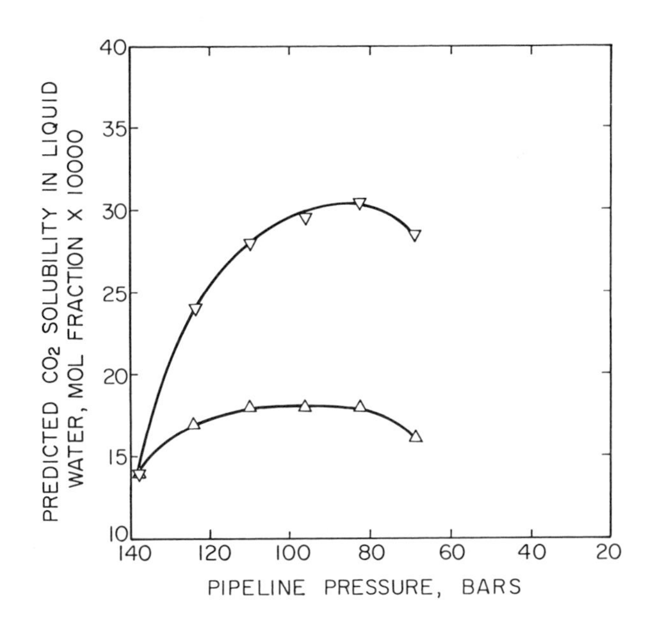

Figure 10. Effect of methanol on predicted CO_2 solubility in water phase ((△) wet hydrocarbon flash; (▽) wet methanol hydrocarbon flash)

Anomolous pressure-temperature profiles in a pipeline can cause unusual behavior of the water condensation profile in the line. If this happens, the methanol requirements predicted on the basis of terminal pipeline conditions could be low.

Probably the most important feature of the "wet" methanol hydrocarbon flash results in the enhanced carbon dioxide solubility in the liquid water phase. This would probably mean potentially more severe corrosion problems in the pipeline and associated equipment. The design and operation of the corrosion prevention system and program can be substantially improved.

Current and Future Work

We believe that the SRK equation of state has been pushed to its limits. Some improvements in its ability to describe the behavior of hydrocarbon water-other components systems can probably be made. Some of our earlier work indicated that the vapor liquid behavior of selected organic water systems could be reasonably well described (7, 23). Unfortunately, the results of this work could not be extended beyond the range of data used in the fitting process.

Our major efforts are being devoted to extension of the capabilities of the PFGC-MES equation of state. We are or soon will be developing parameters for:

1. The glycols used in natural/synthetic gas dehydration processes.
2. The physical absorbents for acid gas removal from natural and synthetic gases.
3. The aromatic molecules which contain nitrogen, sulfur and oxygen that commonly occur in coal liquefaction processes.
4. The sulfur containing components which occur in coal gasification/liquefaction systems such as, carbonyl sulfide, methyl and ethyl mercaptans, etc.
5. The halogenated refrigerants.
6. Other organic chemicals.

In addition, we are evaluating the ability of the equation of state to predict the vapor liquid equilibria for systems appropriate to each group of compounds. We recognize that these objectives represent a very ambitious project; we hesitate to speculate about the completion date or quality of the final results.

Summary

The ability of the SRK equation of state to reliably predict the vapor phase water content of natural and synthetic gas systems has been demonstrated. In addition, the ability of the PFGC-MES equation to describe the phase behavior of hydrocarbon, acid gas, methanol, water systems has been described. Both equations of state appear to give predictions of the phase behavior of the subject systems that are completely adequate for routine engineering design/evaluation work provided the limits of the correlation are not exceeded.

We believe that the PFGC equation of state approach will be the most fruitful new route to predicting phase behavior of the diverse systems encountered in the natural gas/petroleum/coal liquefaction gasification process industry. We commend it to your attention.

ACKNOWLEDGEMENT

The authors greatly appreciate the financial support of the Oklahoma State University College of Engineering and the Mobil Research Foundation which made this work possible. We also appreciate the generous grant of computer time from CEC Inc.

Nomenclature List

A - parameter in SRK equation of state
a_c - parameter in SRK equation of state
$a_c\alpha$ - mixture parameter in SRK equation of state
α - parameter in SRK equation of state
B - parameter in SRK equation of state
β - term in SRK equation of state based enthalpy departure equation
b - parameter in SRK or PFGC-MES equation of state
c/b_H - arbitrary constant in PFGC equation of state
ΔH - isothermal effect of pressure on enthalpy
ΔS - isothermal effect of pressure on entropy
E - parameter in PFGC-MES equation of state
k - binary interaction parameter for SRK or PFGC-MES equation of state
K - vapor liquid K value = y/x
m - component characteristic parameter for SRK equation of state or number of groups in molecule i
P - pressure, absolute units
P° - reference pressure for entropy calculations, absolute units
Pc - critical pressure, absolute units
ψ - volume fraction of group
R - gas law constant
T - temperature, absolute units
τ - parameter in PFGC-MES equation of state
Tc - critical temperature, absolute units
v - molar volume
x - liquid phase (or phase) mol fraction
y - vapor mol fraction
z - compressibility factor

subscripts

i - any component
k - any group
n - any group

superscripts

2 - second phase binary interation parameter
(0) - coefficient of T^0
(1) - coefficient of T^1
(2) - coefficient of T^2

Summation counters

k - number of groups in systems or molecule
c - number of components
n - number of components in SRK equation of state

REFERENCES

1. ______________, Engineering Data Book, 9th edition, GPSA, 1972.
2. ______________, IGT Research Bulletin No. 8, Reproduced by Socony Mobil Oil Co., Inc., FRL. with Institute of Gas Technology permission, 1955.
3. Bartlett, E. P., JACS, January 11, 1927.
4. Campbell, J. M., "Gas Conditioning and Processing," Vol. 2, Campbell Petroleum Series, Norman, OK (1978).
5. Campbell, J. M., Private Communication (1979).
6. Cunningham, J. and G. M. Wilson, Proceedings of the 54th GPA National Meeting, Denver, Colorado (1974).
7. Erbar, J. H., Invited paper presented at Polish Academy of Sciences Symposium, Jablonna, Poland, November (1975).
8. Evelein, K. A., R. G. Moore and R. A. Heidemann, Ind. Eng. Chem. Process Des. Dev., Vol. 15, p. 423, 1976.
9. Ferrell, J. K., R.W. Rousseau and D. B. Bass, "The Solubility of Acid Gases in Methanol," EPA Report - 600/7-79-097 (April, 1979).
10. Hall, K. R., Private Communication (1979).
11. Hamerschmidt, E. G., Western Gas 9, 9-11, 29, December (1933).
12. Heidemann, R. A., AIChE J, Vol. 20, p. 847, 1974.
13. Hopke, S. W. and G. J. Lin, Paper presented at 76th National Meeting Tulsa, OK (1974).
14. Katayama, T., O. Kazunari, G. Mackawa, M. Goto and T. Nagano, Journal of Chem. Eng. of Japan, Vol. 8, p. 89 (1975).
15. Moore, R. G., R. A. Heidemann and E. Wichert, Energy Processing/Canada, p. 40, January, 1979.

16. Moshfeghian, M., A. Shariat and J. H. Erbar, "Application of the PFGC Equation of State to Gas Processing Systems," presented at the National AIChE meeting, Houston, Texas, April 5, 1979.
17. Olds, R. H., B. H. Sage and W. N. Lacey, Ind. Eng. Chem., Vol. 34 p. 1223, 1942.
18. Peng, D. Y. and D. B. Robinson, Canadian Journal of Chemical Engineering, Vol. 54, p. 595, December, 1976.
19. Soave, G., Chem. Eng. Sci., Vol. 27, p. 1197, 1972.
20. Starling, K. E. and M. S. Han, Hydrocarbon Processing, Vol. 51, No. 5, p. 129 (1972).
21. Sultanov, R. G., Skripka, V. G. and A. Y. Namiot, Gazovaia Promyshlennost, Vol. 16, No. 4, p. 6, 1971.
22. Sultanov, R. G., Skripka, V. G. and A. Y. Namiot, Gazovaia Promyshlennost, Vol. 17, No. 5, p. 6, 1972.
23. Vellendorf, G., Master's Thesis, Oklahoma State University (1976).
24. Yorizane, M., Sada Moto, H. Masuoka, and Y. Eto, Kogyo Kagoku. Zashins, Vol. 72, p. 2174 (1969).

RECEIVED January 31, 1980.

18

Considerations Affecting Observation of Interaction Second Virial Coefficients by Chromatography

K. N. ROGERS, S. H. TEDDER[1], J. C. HOLSTE, P. T. EUBANK, and K. R. HALL

Chemical Engineering Department, Texas A&M University, College Station, TX 77843

Its precise basis in statistical mechanics makes the virial equation of state a powerful tool for prediction and correlation of thermodynamic properties involving fluids and fluid mixtures. Within the study of mixtures, the interaction second virial coefficient occupies an important position because of its relationship to the interaction potential between unlike molecules. On a more practical basis, this coefficient is useful in developing predictive correlations for mixture properties.

Several techniques are available for measuring values of interaction second virial coefficients. The primary methods are: reduction of mixture virial coefficients determined from $P\rho T$ data; reduction of vapor-liquid equilibrium data; the differential pressure technique of Knobler et al.(1959); the Burnett-isochoric method of Hall and Eubank (1973); and reduction of gas chromatography data as originally proposed by Desty et al.(1962). The latter procedure is by far the most rapid, although it is probably the least accurate.

Besides speed, the chromatography experiment has the advantage that relatively "nasty" systems present minimal problems. Coal chemicals in light carrier gases or H_2S systems are good examples. Such considerations led us to develop a chromatography apparatus with the specific goal of obtaining interaction second virial coefficients. A vast literature was available to guide us in this effort. The more noteable references include the book by Littlewood (1970) and review articles by Conder (1968), Locke (1976) and Kobayashi et al.(1967). Papers which specifically address determination of the interaction virial coefficient, among other properties, include those by Everett (1965), Laub and Pecsok (1974), Young and coworkers (1966, 1967, 1968, 1968, 1969). Conder and Purnell (1968, 1968, 1969, 1969) provide an excellent overview of the total experiment.

We have chosen the water - carbon dioxide system for this study for several reasons. Relatively little literature exists for this important system. This mixture is important in combus-

[1] Current address: AMOCO Production Company, Tulsa, OK

0-8412-0569-8/80/47-133-361$05.00/0

tion studies, ordinance design, and tertiary oil recovery. Finally, it is a difficult system to study and we reasoned that it would demonstrate equipment capability and inadequacy.

Analysis of Experiment

The analysis providing interaction second virial coefficients from chromatography rests upon three principal assumptions: 1) vapor-liquid equilibrium exists in the column; 2) the solute (component 1) is soluble in both the carrier gas (component 2) and the stationary liquid phase (component 3); 3) the carrier gas and stationary liquid are insoluble. Under assumption #1, we can write

$$\hat{f}_i^v = \hat{f}_i^\ell \tag{1}$$

where $\hat{f}_i$ is the fugacity of component i in a solution and the superscripts indicate vapor and liquid. Taking the vapor side of the equation, we write

$$\hat{f}_1^v = y_1 P \hat{\phi}_1 \tag{2}$$

where y_i is the vapor-phase composition, P is the pressure, and $\hat{\phi}_i^v$ is the vapor-phase fugacity coefficient of component i. This equation also serves to define $\hat{\phi}_i^v$. Standard thermodynamic derivations, such as those of Smith and Van Ness (1975), provide the relationship

$$\ell n \hat{\phi}_i^v = \int_{nV}^{\infty} \left\{ \left[\frac{\partial (nZ)}{\partial n_i} \right]_{T,nV,n_j} - 1 \right\} \frac{d(nV)}{(nV)} - \ell n Z \tag{3}$$

where nV is the total volume, n_i is number of moles of component i and Z is the compressibility factor. Note that assumption #3 forces the vapor (and the liquid) to be binary mixtures. From the virial equation, we obtain

$$Z = 1 + (y_1^2 B_{11} + y_2^2 B_{22} + 2y_1 y_2 B_{12})/V + (y_1^3 C_{111} + 3y_1^2 y_2 C_{112} + 3y_1 y_2^2 C_{122} + y_2^2 C_{222})/V^2 + \dots \tag{4}$$

where B_{ij} and C_{ijk} are the second and third virial coefficients. Using Equation 4 multiplied by n, we obtain by differentiation

$$\left\{ \left[\frac{\partial (nZ)}{\partial n_1} \right]_{T,\, nV, n_2} - 1 \right\} = (2y_1 B_{11} + y_2 B_{12})/V + 3(y_1^2 C_{111} + 2y_1 y_2 C_{112} + y_2^2 C_{122})/V^2 + \dots \tag{5}$$

Substituting Equation 5 into 3 and integrating provides

$$\ln\hat{\phi}_1^v = 2(y_1B_{11} + y_2B_{12})/V + 1.5(y_1C_{111} + 2y_1y_2C_{112} + y_2^2C_{122})/V^2 + \ldots - \ln Z \qquad (6)$$

which, in turn, we substitute into Equation 2.

Next, we examine the liquid side of Equation 1 and write as a definition

$$\hat{f}_1^\ell = x_1 f_1^\ell \gamma_1^\ell = \hat{f}_{1r}^\ell(\hat{f}_1^\ell/\hat{f}_{1r}^\ell) = \hat{f}_{1r}^\ell \exp\left[\int_{P_r}^{P} \frac{\bar{V}_1^\ell}{RT} dP\right] \qquad (7)$$

where γ_i^ℓ is the activity coefficient of component i and subscript r denotes a reference state. In this case, we take the reference pressure, P_r, to be the vapor pressure of the stationary liquid at the column temperature (in most cases, including this one, $P_r \simeq 0$). Now, we further define

$$\hat{\phi}_1^f = \hat{f}_{1r}^\ell / x_1 P_1^{vap} \qquad (8)$$

where x_1 is the liquid phase composition and P_1^{vap} is the vapor pressure of the solute. Therefore, substituting Equation 8 into 7 we obtain

$$\hat{f}_1^\ell = x_1 P_1^{vap}\, \hat{\phi}_1^f \exp\left[\int_{P_r}^{P} \frac{\bar{V}_1^\ell}{RT} dP\right] \qquad (9)$$

and, upon substitution of Equations 2 and 9 into 1, we have

$$K_1 \equiv y_1/x_1 = \frac{P_1^{vap}\, \hat{\phi}_1^f \exp\left[\int_{P_r}^{P} \frac{\bar{V}_1^\ell}{RT} dP\right]}{P\hat{\phi}_1^v}$$

For the experiment we propose to use, we may further assume that the solute is present at infinite dilution in both the liquid and vapor phase. As a result, Equation 9 becomes

$$K_1^\infty = \frac{P_1^{vap}(\hat{\phi}_1^f)^\infty \exp\left[\int_{P_r}^{P} \frac{(\bar{V}_1^\ell)^\infty}{RT} dP\right]}{P(\hat{\phi}_1^v)^\infty} \qquad (10)$$

Using Equation 6, we obtain at infinite dilution

$$(\hat{\phi}_1^v)^\infty \equiv \lim_{y_1 \to 0} \hat{\phi}_1^v = \exp\,(2B_{12}/V_2 + 1.5\ C_{122}/V_2^2 + \ldots - \ell n Z_2) \tag{11}$$

which, in combination with Equation 10, becomes

$$\ell n\left(\frac{K_1^\infty P}{Z_2 P_1^{vap}}\right) = \ell n(\hat{\phi}_1^f)^\infty + \int_{P_r}^{P} \frac{(\bar{V}_1^\ell)^\infty}{RT}\,dP - 2B_{12}\left(\frac{P}{Z_2RT}\right) - 1.5C_{122}\left(\frac{P}{Z_2RT}\right)^2 - \ldots \tag{12}$$

Now, recalling that $P_r = P_3^{vap} \simeq 0$, we can approximate the partial molar volume by

$$(\bar{V}_1^\ell)^\infty \simeq (\bar{V}_1^\ell)_r^\infty + \left[\frac{\partial(\bar{V}_1^\ell)_r^\infty}{\partial P}\right]_T P + \ldots \tag{13}$$

or for Equation 12

$$\int_0^P \frac{(\bar{V}_1^\ell)^\infty}{RT}\,dP = \frac{(\bar{V}_1^\ell)^\infty}{RT} P + 1/2 \left[\frac{\partial(\bar{V}_1^\ell)_r^\infty}{\partial P}\right]_T P^2 + \ldots \tag{14}$$

Finally, we obtain

$$\ell n\left(\frac{K_1^\infty P}{Z_2 P_1^{vap}}\right) = \ell n(\hat{\phi}_1^f)^\infty + \left[(\bar{V}_1^\ell)^\infty - \frac{2B_{12}}{Z_2}\right]\left(\frac{P}{RT}\right) + 1/2 \left\{\left[\frac{\partial(\bar{V}_1^\ell)_r^\infty}{\partial P}\right] - \frac{3C_{122}}{Z_2^2}\right\} \cdot \left(\frac{P}{RT}\right)^2 + \ldots \tag{15}$$

A plot of $\ln(K_1^\infty P/Z_2 P_1^{vap})$ vs (P/RT) will produce $\ln(\hat{\phi}_1^f)^\infty$ as the intercept, $(\bar{V}_1^\ell)_r^\infty - 2B_{12}/Z_2$ as the slope and so forth. In practice, experimental precision will mask any curvature and the effect of Z_2 on the slope, and our working equation is

$$B_{12} = \frac{(\bar{V}_1^\ell)_r^\infty - \text{slope}}{2} \tag{16}$$

Also, at infinite dilution, the Henry's law constant for solute i in solvent j is

$$H_{12} = (\hat{f}_1^v/y_1)^\infty = P(\hat{\phi}_1^v)^\infty = \frac{P}{Z_2} \exp\left[\frac{2B_{12}}{Z_2}\left(\frac{P}{RT}\right) + \ldots\right] \tag{17}$$

$$H_{13} = (\hat{f}_1^\ell/x_1)^\infty = P_1^{vap}\,(\hat{\phi}_1^f)^\infty \exp\left[(\bar{V}_1^\ell)_r^\infty\left(\frac{P}{RT}\right) + \ldots\right] \tag{18}$$

To utilize the analysis we must resolve three difficulties: 1) defining P (we assume constant pressure but there is a pressure drop across the column), 2) establishing K_1^∞, and 3) determining $(\bar{V}_1^\ell)_r^\infty$. We can use an average pressure for the experiment, but we must choose between distance and time averages. Current wisdom favors a time average based upon

$$P = P_{out}\, J_n^m \tag{19}$$

where P_{out} is the outlet pressure for the column and

$$J_n^m = \frac{n}{m}\left[\frac{(P_{in}/P_{out})^m - 1}{(P_{in}/P_{out})^n - 1}\right] \tag{20}$$

where P_{in} is the column inlet pressure. For our experiment we would obtain the following corrections for P_{out}

distance average = 1.121

J_2^3 = 1.125

J_3^4 = 1.129

or 1.125 $\pm$ 0.004 which is within our experimental error so we use J_2^3.

To obtain K_1^∞, inspect Figure 1 which plots the flowing phase concentration against the stationary phase concentration of component 1. In the infinite dilution limit, we have

$$C_{1f}^\infty = \beta C_{1s}^\infty \tag{21}$$

and using the overall phase concentrations

$$(C_{1f}^\infty/C_f)\ C_f = \beta(C_{1s}^\infty/C_s)C_s \tag{22}$$

which is by definition

$$y_1^\infty\ C_f = \beta x_1^\infty\ C_s \tag{23}$$

and the distribution coefficient is

$$K_1^\infty = (y_1/x_1)^\infty = \beta C_s/C_f \tag{24}$$

Analysis of chromatography experiments reveals that

$$(\beta C_s/C_f)(n_f/n_s) = t_{inert}/\ (t_{solute} - t_{inert}) \tag{25}$$

where t_{inert} is the time required to detect an air peak in the column and $t_{solute} - t_{inert}$ is the retention time. K_1^∞ then becomes

$$K_1^\infty = \frac{Z_2^a RT^a (m/MW)_3}{P^a \dot{V}^a (t_{solute} - t_{inert})} \tag{26}$$

where m is the mass and MW the molecular weight of the stationary phase, $\dot{V}^a$ is the observed volumetric flowrate (from a bubble column, for example), and superscript a denotes ambient conditions. We should note that usually MW_3 is unknown so all the definitions change slightly to reflect a mass rather than a mole basis.

Finally, we come to the "Achilles heel" of the experiment, $(\bar{V}_1^\ell)_r^\infty$. The common assumption for this variable is to use the volume of saturated liquid solute (assume an ideal solution). This assumption can be ridiculous when $MW_3 >> MW_1$ as is the usual case. Figure 2 illustrates this assertion. A more realistic approach is to use statistical mechanics to provide an approximator as suggested by Flory et al.(1964), Flory (1965) and Abe and Flory (1965).

We suggest a third alternative. This procedure requires fiduciary data for B_{1u}, the interaction second virial coefficient for solute 1 and an unspecified carrier gas (a good choice would be helium). The B_{1u} should come from a precise experiment such as the Burnett-mixing or pressure change method mentioned in the introduction. Running the 1-u system through the chromatograph would allow use of Equation 16 in rearranged form

$$(\bar{V}_1^{\ell})_r^{\infty} = 2B_{1u} + \text{slope} \tag{27}$$

The value of $(\bar{V}_1^{\ell})_r^{\infty}$ is independent of carrier gas so it can be used in Equation 16 with component 2 or any other carrier gas. This should provide the most accurate value of $(\bar{V}_1^{\ell})_r^{\infty}$ for use in this experiment.

Experimental Procedure. Figure 3 presents a schematic diagram of the apparatus. The chromatograph is a Varian Aerograph Series 1400 with a thermal conductivity detector and an associated Varian CDS 111 Chromatography Data System (integrator). We modified the chromatograph in two ways: 1) we installed a fine-adjust for temperature, and 2) we replaced the 2% flow controller supplied with the instrument with an Analabs HGS-187 Flow Controller which has 0.3% precision. The temperature adjustment was necessary to permit satisfactory reproducibility, $\pm$ 0.02°C. An error analysis of the experiment indicated that accuracy was especially sensitive to flow control so we obtained the better controller to bring this measurement in line with our other observables.

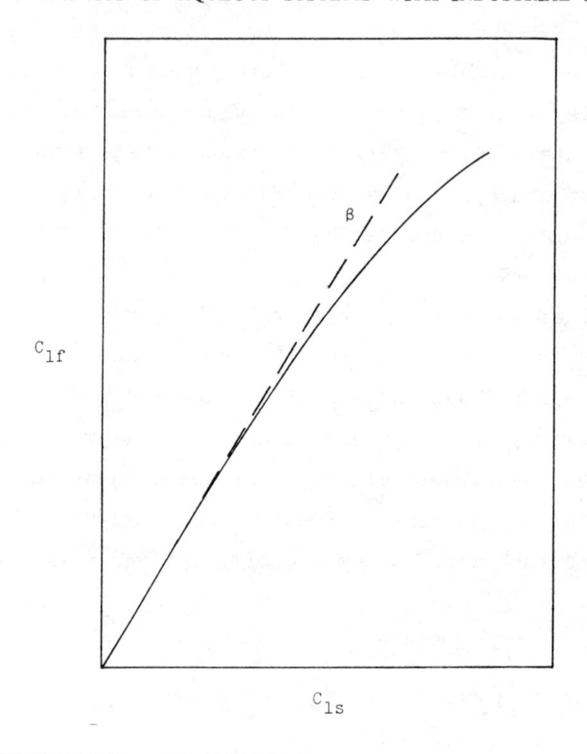

Figure 1. Representative adsorption isotherm

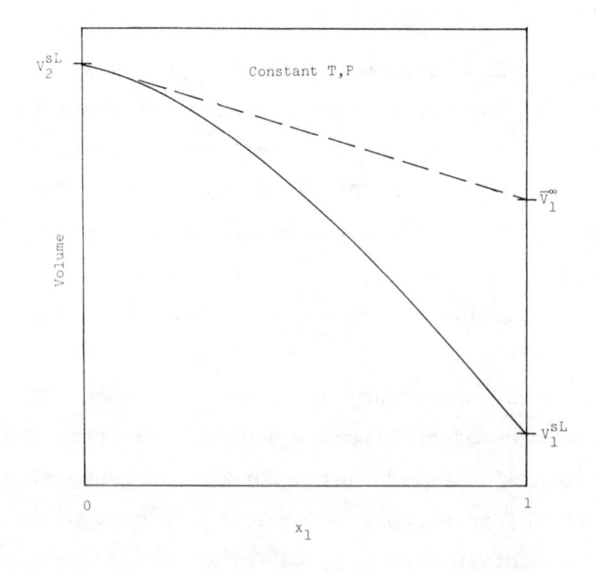

Figure 2. Relationship between saturated liquid volume and partial molar volume

We measured flow rate with a Varian P/N 29-000086-00 Soap Bubble Meter. The pressure gauges were Heise bourdon-tube type with 0.1 psia divisions. The column was 16 feet of 1/4 inch copper tubing packed with Carbowax 20M loaded to 20% on Fluoropak 90.

The experimental procedure is approximately standard chromatography. First, we set the desired column temperature and adjust the detector and injector temperatures to values about 30°C higher. Next we adjust the inlet and outlet pressure and measure the flow rate. Then, we set the detector current and attenuation to give maximum sensitivity with an acceptable baseline (typically the current is 50 mA with an attenuation of 1). For injections, we draw 1/2 the desired air into the syringe followed by the sample followed by the remaining 1/2 of the desired air. We establish the desired amount of air by trial and error to obtain a proper peak size. We also vary the sample size (from 1 to 2 μℓ) between runs at a given P and T. This allows extrapolation of peak areas to zero sample size as shown in Figure 4. The flow rate within the column must be constant for an isotherm. We record the following parameters: ambient pressure and temperature, inlet pressure, outlet pressure, column temperature, retention time for both air and water, water peak area, ambient flow rate, regulator pressure, sample sizes, detector current and temperature, injector temperature, and attenuation.

We collected data for at least five outlet pressures along each isotherm to improve the statistical analysis of the data. The outlet pressures ranged from 0 to 32 psig, but a wider range would have reduced scatter considerably. We varied the inlet pressure to achieve an approximately constant flowrate (20 cc/min) in the column. This flowrate resulted from a compromise between speed and accuracy. The higher the flowrate, the faster the experiment, but too fast a flowrate and the system did not approach equilibrium. We selected 20 cc/min by testing flowrates varying from 10 cc/min to 50 cc/min; rates from 10 cc/min to 20 cc/min gave essentially identical results while rates above 25 cc/min showed significant deviations.

Finally, we must be certain we are observing vapor-liquid equilibrium in the column not vapor-solid equilibrium. Braun and Guillet (1976) reported that a discontinuity in a plot of $\ell n\ (V_g^o)$ *vs* 1/T (where V_g^o is the retention volume at 273.15 K) indicated a phase transition. We calculated V_g^o from the following relationship

$$V_g^o = \frac{\dot{V}^a(t_{solute} - t_{inert})}{m_3}\left[\frac{P^a Z^o T^o}{J_2^3\ P_{out}\ Z^a T^a}\right] \tag{28}$$

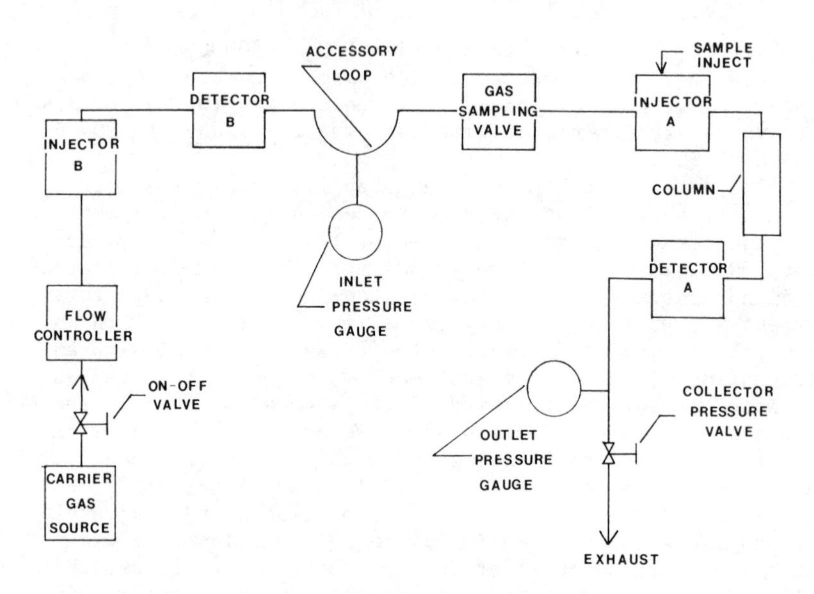

Figure 3. Schematic of modified GLC apparatus

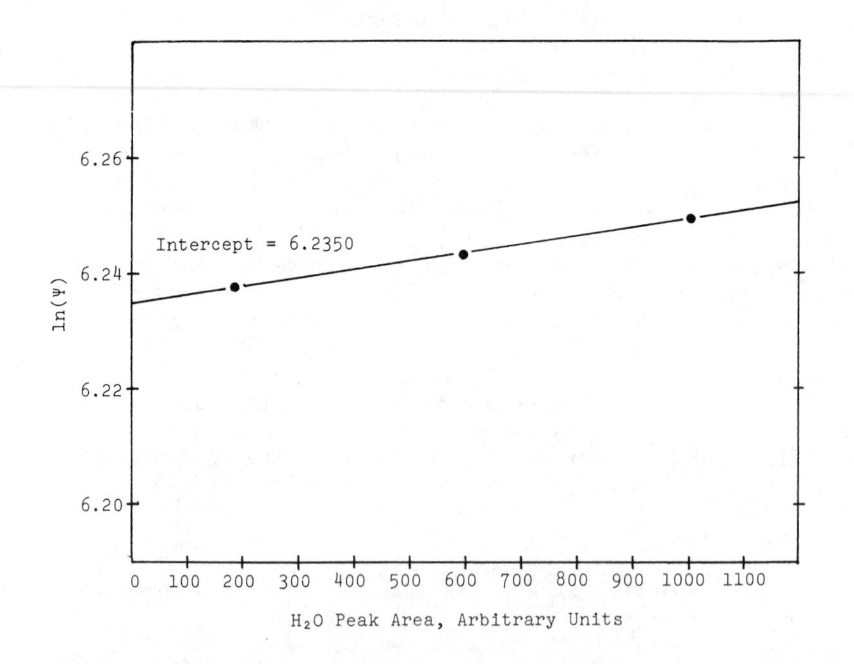

Figure 4. Extrapolation to zero peak area (operating conditions: $T \simeq 150°C$*;* $P_{out} \simeq 30$ *psig)*

where superscript o indicates 273.15K values. Figure 5 indicates that a phase transition occured in our experiment between 75°C and 125°C.

Results. Our experimental observations appear in Table I and Figure 6. We have defined Ψ as $K_1^{\infty}P/Z_2P_1^{vap}$. The slopes from Figure 6 inserted into Equation 16 provide values for B_{12}. Unfortunately, we were not able to find the fiduciary data necessary to evaluate $(\bar{V}_1^{\ell})_r^{\infty}$ and must leave the calculation unfinished. The ideal solution assumption is totally inadequate in this particular experiment.

We were able to determine a surface fit for the data using the expression

$$\ell n\ \Psi = (A + BT^2) + C\ \exp(DT^2)\ [P/RT] \qquad (29)$$

Figures 7 and 8 illustrate the behavior of the intercepts and slopes from Figure 6 corresponding to the functional forms of Equation 29. The error bars on Figures 7 and 8 represent one standard deviation as determined from isothermal fits. The intercepts have deviations on the order of 0.5% which is consistent with an apparatus analysis. The slopes, however, have much larger uncertainties ranging up to 15%. Increasing the pressure range would greatly reduce this large and important error.

Table II presents our final results. We only report the four isotherms which were definitely vapor-liquid and we present both isothermal analysis and surface analysis values.

Conclusions. Our work indicates that it may be possible to obtain good values of B_{12} by chromatography, but only with painstaking attention to every detail. It is clear from our experience why values measured by this technique are generally regarded as not top quality.

It is essential to verify experimentally each assumption. In this experiment, we used five different flowrates to assure ourselves that the column did approach equilibrium: 20 cc/min was satisfactory, 25 cc/min was already questionable. The solute was soluble in both phases as evidenced by peak separation on the chromatogram. The stationary liquid was not soluble in the carrier gas as evidenced by no long time peak on the chromatogram. The carrier gas was slightly soluble in the stationary liquid. We checked this by injecting CO_2 into the carrier and we did obtain a peak-however, analysis showed that $K_{CO_2} \simeq 70{,}000$ therefore the CO_2 concentration in the stationary liquid was insignificant. We assure infinite dilution values by performing multiple injections and extrapolating to zero peak area. This procedure is also essential in the experiment.

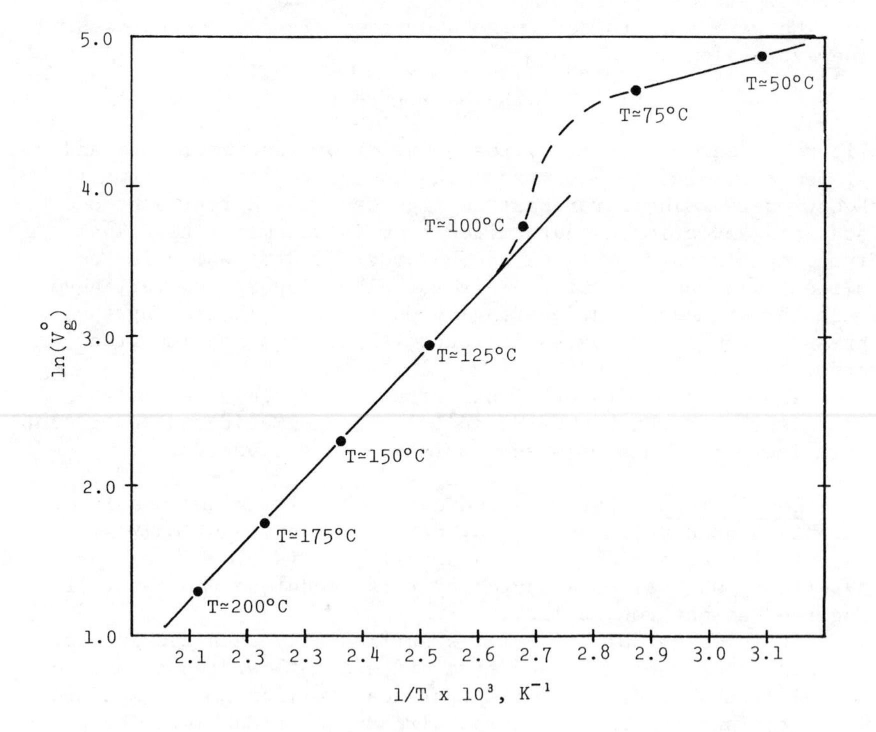

Figure 5. Phase transition indicator

Table I. Calculated Values of ln(Ψ) and P/RT

Avg. T, K	Avg. F_c, cc/min	Avg. P_{out}, psig	Infinite Dilution ln(Ψ)	Weighting Factor	Avg. P/RT x 10 , g-mole/cc
398.16	10.61	0.00	6.2587	172.45	3.2640
	14.99	0.00	6.2535	951.25	3.3678
	20.09	0.00	6.2475	1303.32	3.4842
	20.11	10.32	6.2898	3106.94	5.6470
	20.26	14.00	6.3114	164.60	6.4115
	19.97	19.80	6.3217	1012.60	7.6162
	20.67	29.30	6.3452	534.89	9.6155
423.40	17.04	0.00	6.1578	458.94	3.2212
	19.86	7.20	6.1958	88.57	4.7240
	19.33	15.21	6.2030	64.58	6.2567
	20.53	21.33	6.2344	527.72	7.5164
	21.12	28.22	6.2350	1483.59	8.8571
448.24	17.21	0.00	6.0759	253.84	3.1032
	20.10	7.10	6.1096	115.01	4.4916
	19.36	15.10	6.1249	46.87	5.9501
	20.94	21.75	6.1169	62.79	7.2335
	20.82	29.50	6.1380	52.20	8.6619
473.21	20.53	0.00	5.9826	103.46	3.0627
	21.41	7.55	6.0266	11.76	4.0112
	25.92	15.90	6.0388	70.37	5.9837
	20.51	23.56	6.0184	61.58	7.2002
	20.90	31.24	6.0295	31.23	8.5602

Table II. Values of the Slope and Intercept Predicted for each Isotherm

Average Temperature, K	Slope, cm^3/g-mole		Intercept	
	Isotherm Fit	Surface Fit	Isotherm Fit	Surface Fit
398.16	1607	1611	6.198	6.197
423.43	1342	1300	6.120	6.124
448.24	1115	1040	6.046	6.048
473.21	928	820	5.962	5.967

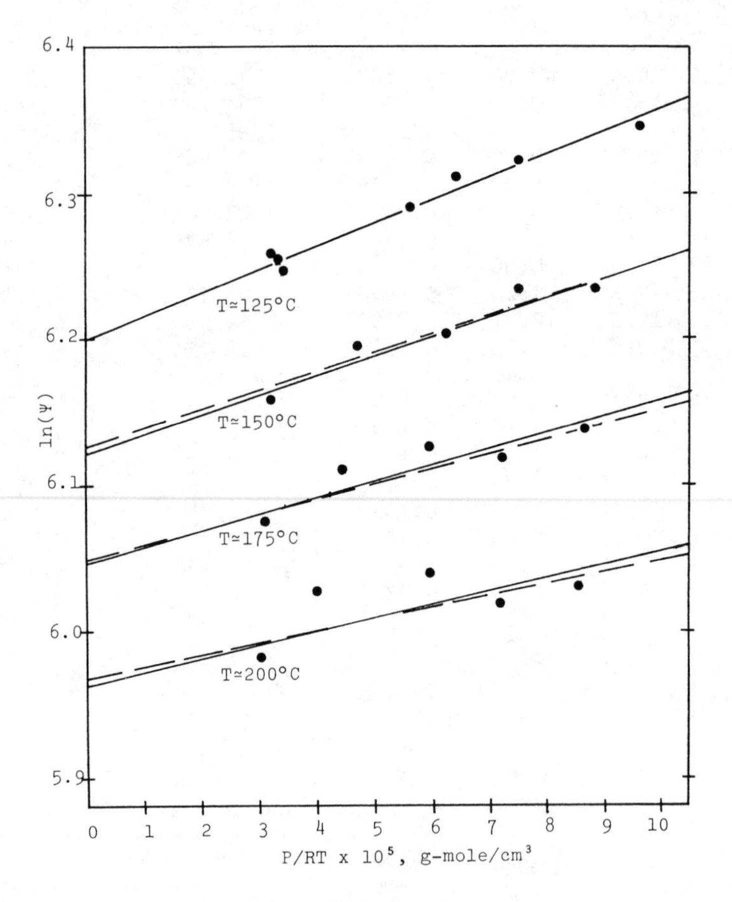

Figure 6. Experimental data for determining slopes ((——) isotherm fit; (- - -) surface fit; surface fit indistinguishable at T ≃ 125°C)

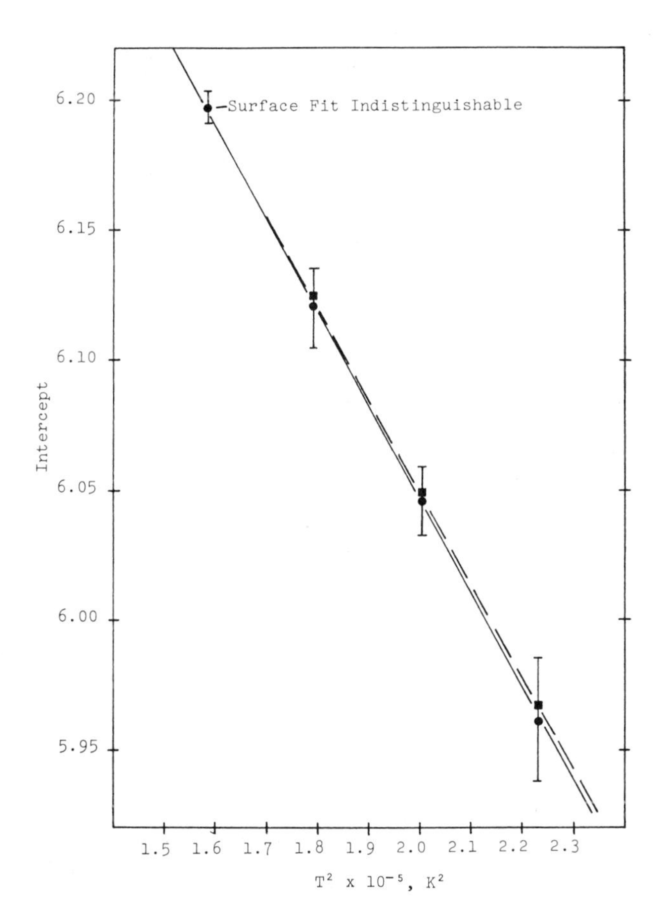

Figure 7. Predicted intercept as a function of temperature ((— ● —) isotherm fit; (– – – ■ – – –) surface fit)

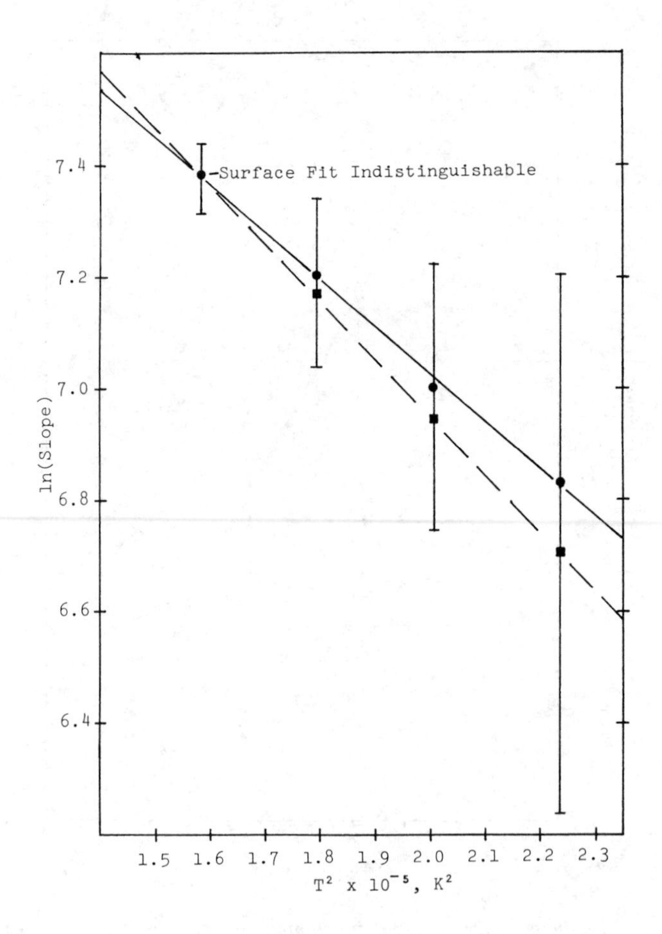

Figure 8. Predicted slope as a function of temperature ((—●—) isotherm fit; (– –■– –) surface fit)

A preliminary experimental error analysis indicated that the flowrate control and, to a lesser degree, the temperature control would be critical. It is necessary to change the off-the-shelf flow controllers for commercial chromatographs and desirable to change the temperature controller.

The value used for $(\bar{V}_1^{\ell})_r^{\infty}$ is crucial for final result accuracy. The ideal solution assumption will range from bad to ridiculous. For polar systems, such as the present one, the <u>major</u> <u>effect</u> comes from $(\bar{V}_1^{\ell})_r^{\infty}$ in the slope and, unless a very precise value is available, the experiment is worthless! We have proposed a new method for obtaining $(\bar{V}_1^{\ell})_r^{\infty}$ using fiducial data which should produce acceptable values of this quantity. Using this procedure could reduce the B_{12} error in simple systems by one order-of-magnitude and in complex systems by many orders-of-magnitude. Certainly $(\bar{V}_1^{\ell})_r^{\infty}$ can be determined to about 1 % using the new technique.

It is essential to use at least five injections spaced over as wide a range as possible to obtain each value of Ψ. It is also essential to observe the values of Ψ over as wide a pressure range as possible. These precautions will greatly reduce scatter in the slope required by Equation 16.

Finally, we have presented a new analysis which is not found in the literature. This analysis is more sensitive to the various effects discussed above and therefore should produce better values in the long run.

<u>Acknowledgements</u>. Financial support for this project came in part from the National Science Foundation (Grants ENG 77-01070 and 76-00692), the Petroleum Research Fund (Grant 10057-AC7-C), The Gas Research Institute (Grant 5014-363-0118), The Texas Engineering Experiment Station and the Gas Processors Association (Projects 772 and 758). We gratefully acknowledge their assistance.

<u>Literature Cited</u>

Abe, A.; Flory, P.J.,; "The Thermodynamic Properties of Mixtures of Small, Nonpolar Molecules," <u>JACS</u>, <u>87</u>, 1838(1965).

Braun, J.M.; Guillet, J.E.; "Study of Polymers by Inverse Gas Chromatography," Adv. Poly. Sci., <u>21</u>, 107(1976).

Conder, Jr.; Purnell, J.H.; "Gas Chromatography at Finite Concentration," Pt. 1: <u>Trans Far. Soc.</u>, <u>64</u>, 1505(1968)/ Pt. 2: <u>Trans. Far. Soc.</u>, <u>64</u>, 3100(1968)/ Pt. 3: <u>Trans. Far. Soc.</u>, <u>65</u>, 824(1969); Pt. 4: <u>Trans. Far. Soc.</u>, <u>65</u>, 839(1969).

Conder, J.R.; "Physical Measurement by Gas Chromatography," Adv. Anal. Chem. Instrum., 6, 209(1968).

Desty, D.H./ Goldup, A.; Luckhurst, G.R.; Swanton, W.T.; "The Effect of Carrier Gas and Column Pressure on Solute Retention," Gas Chromatography 1962, Butterworths (London), 67(1962).

Everett, D.H.; "Effect of Gas Imperfections on G.L.C. Measurements," Trans. Far. Soc., 61, 1637(1965).

Flory, P.J./ Statistical Thermodynamics of Liquid Mixtures," JACS, 87, 1833(1965).

Flory, P.J.; Orwoll, R.A.; Vrij, A.; "Statistical Thermodynamics of Chain Molecule Liquids," JACS, 86, 3515(1964).

Hall, K.R.; Eubank, P.T.; "Experimental Technique for Direct Measurement of Interaction Second Virial Coefficients," J. Chem. Phys., 59, 709(1973).

Knobler, C.M.; Beenakker, J.J.M.; Knapp, H.E.P.; "The Second Virial Coefficient of Gaseous Mixtures at 90K," Physica, 25, 909(1959).

Kobayashi, R./ Chappelear, P.S.; Deans, H.A.; "Physico-Chemical Measurements by Gas Chromatography," I&EC, 59, 63(1967).

Laub, R.J.; Pecsok, R.L.; "Determination of Second-Interaction Virial Coefficients by Gas-Liquid Chromatography," J. Chrom., 98, 511(1974).

Littlewood, A.B.; Gas Chromatography, Academia Press (London), 2nd ed. (1970).

Locke, D.C.; "Physicochemical Measurements Using Chromatography," Adv. Chrom., 14, 87(1976).

Tsonopoulous, C.; "Second Virial Cross Coefficients: Correlation and Prediction of k_{ij}," Adv. in Chem., in press(1979).

Young, C.L.; Cruickshank, A.K.B.; Windsor, M.L.; The Use of GLC to Determine Activity Coefficients and Second Virial Coefficients of Mixtures," Proc. Roy. Soc., A295, 271(1966).

Young, C.L.; Windsor, M.L.; Thermodynamics of Mixtures from Gas-Liquid Chromatography," J. Chrom., 27, 355(1967).

Young, C.L.; Cruickshank, A.J.B.; Gainey, B.W.; "Activity Coefficients of Hydrocarbons C_4 to C_8 in n-Octadecane at 35 °C," Trans. Far. Soc., 64, 337(1968).

Young, C.L.; Gainey, B.W.; "Activity Coefficients of Benzene in Solutions of n-Alkanes and Second Virial Coefficients of Benzene + Nitrogen Mixtures," Trans. Far. Soc., 64, 349(1968).

Young, C.L.; Cruickshank, A.J.B.; Gainey, B.W.; Hicks, C.P.; Letcher, T.M.; Moody, R.W.; Gas-Liquid Chromatographic Determination of Cross-Term Second Virial Coefficients using Glycerol,: Trans. Far. Soc., 65, 1014(1969).

RECEIVED January 31, 1980.

19

Critical Points In Reacting Mixtures

ROBERT A. HEIDEMANN

University of Calgary, Calgary, Alberta, Canada T2N 1N4

Processes in which chemical reaction and phase equilibria are simultaneously of significance present a considerable challenge to the thermodynamicist. The challenge is both to develop models which are suitable to describe the mixtures and to find computational procedures which permit analysis of equilibrium behavior.

The presence of water in synthesis gas mixtures along with light components, such as carbon monoxide or hydrogen, has the effect that phase separations may persist even under extreme conditions of temperature and pressure. The need exists to demonstrate that these phase separations, perhaps with simultaneous reaction equilibrium, can be described by models capable of some accuracy.

The present paper deals with one aspect of this problem; the calculation of phase separation critical points in reacting mixtures. The model employed is the Soave-Redlich-Kwong equation of state (1), which is typical of several equations of state (2, 3) which have relatively recently come into wide use as phase equilibrium models for light gas mixtures, sometimes including water and the acid gases as components (4, 5, 6). If the critical point contained in the equation of state (perhaps even for the mixture at reaction equilibrium) can be found directly, the result will aid in other equilibrium computations.

Other workers have reported computations of reaction equilibria in mixtures described by equations of state (7, 8). Only occasionally have non-ideal mixtures with phase separations been tackled (9, 10), and no previous computations of critical points in reacting mixtures appear in the literature.

Stoichiometry

The algebra required to express the stoichiometry of chemically reacting systems is well established. The species present in the reacting mixture are A_i, i = 1, N. The reactions are denoted by

0-8412-0569-8/80/47-133-379$05.00/0

$$\sum_{i=1}^{N} \nu_{ij} A_i = 0; \quad j = 1, R. \tag{1}$$

The ν_{ij} are stoichiometric coefficients in the R independent reactions. The progress of each of the independent reactions is indicated by a reaction extent ε_j, j=1, R. The mole number of species A_i in a reacting mixture is given by

$$n_i = n_{i_o} + n_{T_o} \sum_{j=1}^{R} \nu_{ij} \varepsilon_j \tag{2}$$

where n_{i_o} is the mole number in a reference mixture (with ε_j=0, j=1, R) and n_{T_o} is the total number of moles in the reference mixture,

$$n_{T_o} = \sum_{i=1}^{R} n_{i_o} \tag{3}$$

The reaction extents, ε_j, are dimensionless numbers in this formulation. They are restricted to values such that all the n_i are non-negative.

When all the species remain in a single homogeneous phase, the mole fractions in the phase are given by

$$y_i = (y_{i_o} + \sum_{j=1}^{R} \nu_{ij} \varepsilon_j)/(1 + \sum_{j=1}^{R} \varepsilon_j \Delta\nu_j) \tag{4}$$

where

$$\Delta\nu_j = \sum_{i=1}^{N} \nu_{ij} \tag{5}$$

Critical Points

In classical thermodynamics, there are many ways to express the criteria of a critical phase. (Reid and Beegle (11) have discussed the relationships between many of the various equations which can be used.) There have been three recent studies in which the critical points of multicomponent mixtures described by pressure-explicit equations of state have been calculated. (Peng and Robinson (12), Baker and Luks (13), Heidemann and Khalil (14)) In each study, a different statement of the critical criteria and a different computational algorithm has been employed. In each case, however, the critical point is required to satisy two algebraic equations.

The formulation of Heidemann and Khalil (14) is used here, with some modifications proposed by Michelsen (15, 16) which improve computational speed. Reference (14) contains adequate detail to permit the computations to be reproduced. The elements of the method are given here.

Heidemann and Khalil (14) find the critical point of a mixture of given composition by solving for the critical temperature and volume; the critical pressure is found from the equation of state.

A critical point must satisfy two conditions. The first defines the limit of stability. In the procedure of Heidemann and Khalil, this condition takes the form that the determinant of matrix Q is zero, where the elements of Q are

$$q_{ij} = n_T(T/100)(\partial \ell n f_i/\partial n_j)_{T,V} \tag{6}$$

For the Soave-Redlich Kwong equation, the fugacity derivatives are

$$n_T(\partial \ell n f_i/\partial n_j)_{T,V} = \frac{\delta_{ij}}{y_i} + \frac{2b_{ij}}{v - b} + \frac{\beta_i \beta_j}{(v - b)^2} + \frac{a\beta_i\beta_j}{RTb} \cdot \frac{1}{(v + b)^2} + \frac{B_1}{RTb^2} \cdot \frac{1}{(v + b)} + \frac{B_2}{RTb^3} \ell n \left(\frac{v + b}{v}\right) \tag{7}$$

Coefficients a and b are both evaluated as quadratics in the mole fractions;

$$b = \tfrac{1}{2} \sum_j \sum_i y_i y_j (b_{ii} + b_{jj})(1 - c_{ij}) \tag{8}$$

$$a = \sum_j \sum_i y_i y_j \sqrt{a_{ii} a_{jj}}(1 - k_{ij}). \tag{9}$$

This introduces two "interaction parameters" per binary pair. The pure component coefficients, a_{ii} and b_{ii}, are evaluated from critical data and the acentricity, as proposed by Soave in his original paper (1). The pure component a_{ii} varies with reduced temperature so as to match vapor pressure. (Soave's recently revised expression for a_{ii} (17) has not been used.)

Other symbols in equation (7) are defined as follows:

$$\alpha_i = 2\sum_j y_j a_{ij} \tag{10}$$

$$\beta i = 2\sum_j y_j b_{ij} - b \tag{11}$$

$$\delta_{ij} = \begin{cases} 0, & i \neq j \\ 1, & i = j \end{cases} \tag{12}$$

$$\gamma_{ij} = 2b_{ij} - \beta_i - \beta_j \tag{13}$$

$$B_1 = 2a\beta_i\beta_j - b(\alpha_i\beta_j + \alpha_j\beta_i + a\gamma_{ij}) \tag{14}$$

and

$$B_2 = -B_1 - 2a_{ij}b^2 \tag{15}$$

The first of the two critical criteria is that

$$Q \equiv \det(Q) = 0 \tag{16}$$

When this equation is satisfied, there is a vector $\Delta\overline{n}$

$$\Delta\overline{n} \equiv (\Delta n_1, \Delta n_2, \ldots)^T \tag{17}$$

which satisfies

$$Q\cdot\Delta\overline{n} = \overline{0} \tag{18}$$

Heidemann and Khalil (14) use as the second critical criterion that the cubic form C is zero where,

$$C = \left[(v-b)/2b\right]^2 \sum\sum\sum n_T{}^2(\partial^2 \ell n f_i/\partial n_k \partial n_j)\Delta n_i \Delta n_j \Delta n_k \tag{19}$$

(In (14) the second derivatives of $\ell n fi$ are given in detail for the SRK equation.) The $\Delta\overline{n}$ in this equation is the normalized vector which satisfies (18).

Michelsen (15, 16) has pointed out that the evaluation of the cubic form in (19) can be performed very economically. Typical terms in (19) can be reduced as follows:

$$\sum\sum\sum \Delta n_i \Delta n_j \Delta n_k = (\sum \Delta n_i)^3 \tag{20}$$

$$\sum\sum\sum \beta_i \Delta n_i \Delta n_j \Delta n_k = (\sum \beta_i \Delta n_i)(\sum \Delta n_i)^2 \tag{21}$$

$$\sum\sum\sum a_{ij} \Delta n_i \Delta n_j \Delta n_k = (\sum\sum a_{ij} \Delta n_i \Delta n_j)(\sum \Delta n_i) \tag{22}$$

Because of these and similar identities the triple sum in (19) can be replaced by terms involving, at most, double sums.

The two equations, (16) and

$$C = 0, \tag{23}$$

together define the critical point.

Evaluating Critical Points

It was found most convenient, as the first stage in finding a critical point for a given mixture, to solve for the temperature which makes Q equal to zero at a given volume. The temperature - volume pair gives one point on the stability limit.

Corresponding to this point, there is a vector $\Delta\bar{n}$ which satisfies equation (18). The second stage in the critical point calculation is to solve for and normalize this vector. The vector $\Delta\bar{n}$ is then employed to evaluate the cubic form.

The cubic form evaluated in this way is a function of volume alone (since the temperature and $\Delta\bar{n}$ are fixed by the volume). To find the critical point requires to find a volume at which the cubic form is zero.

In Heidemann and Khalil (14) additional detail is given about numerical procedures which were found effective in solving these equations. The overall solution strategy, as described above, requires nested one-dimensional searches; the critical volume is found by solving (23), but at each volume (16) must be solved for the temperature. The multiplier (T/100) in each term of the matrix Q and the multiplier $[(v-b)/2b]^2$ in the cubic form were introduced to improve the behavior of the numerical methods.

Water-Gas Shift Mixtures

The specific reaction examined in this paper is the water-gas shift reaction

$$H_2O + CO \rightarrow H_2 + CO_2 \tag{24}$$

As this reaction proceeds, beginning with some initial mixture, each of the four mole fractions can be represented as a function of a single reaction extent, ε.

At every ε the mixture critical point (if indeed, the mixture has a critical point) can be found using the procedure described above.

Binary Critical Points

Before describing variations in the critical points in the four-component water-gas shift mixture it is instructive to examine the critical points in the various binary mixtures. There are six binary pairs to consider.

The calculated critical points of the binary pairs, particularly the critical pressures, are quite sensitive to the values used for the interaction parameters in the mixing rules for a and b in the equation of state. One problem in undertaking this study is that no data are available on the critical lines of any of the binary pairs except for CO_2 - H_2O. Even for CO_2 - H_2O, two sets of critical data available (18, 19) are in poor quantitative agreement, though they present the same qualitative picture of the critical phenomena.

Because of the uncertainty as to the data and because of the sensitivity to the parameters it should be understood that calculated critical points reported in this paper need not represent actual behavior, even of the binary pairs.

Most of the interaction parameters employed were taken from other studies (20, 21), and are reportedly obtained by minimizing errors in the match of phase equilibrium data. However, in (21), the SRK equation employed was slightly different from that used here. The parameters for CO_2 - H_2O were chosen because they had been shown to give a critical line which is qualitatively correct. The H_2O - CO interaction parameter is the value given in (20) for H_2S - CO. For H_2O - H_2, k_{ij} was taken to be -0.25 in the absence of any literature studies.

Table I. INTERACTION PARAMETERS

System	k_{ij}	c_{ij}	Reference
H_2O - CO_2	0.0	-0.03	5
H_2O - CO	0.0603	0	-
H_2O - H_2	-0.25	0	-
CO_2 - CO	-0.064	0	21
CO_2 - H_2	-0.3570	0	20
CO - H_2	0.1000	0	20

The calculated critical lines for the binary pairs are shown in Figure 1. All these lines are discontinuous, indicating high density phase separations. For each binary pair the principal part of the critical line begins at the critical point for the component with the higher critical temperature. There is a second branch of each of the critical lines, beginning at the critical point of the component with the lower critical temperature, which terminates on intersecting a liquid-liquid-vapor three-phase line.

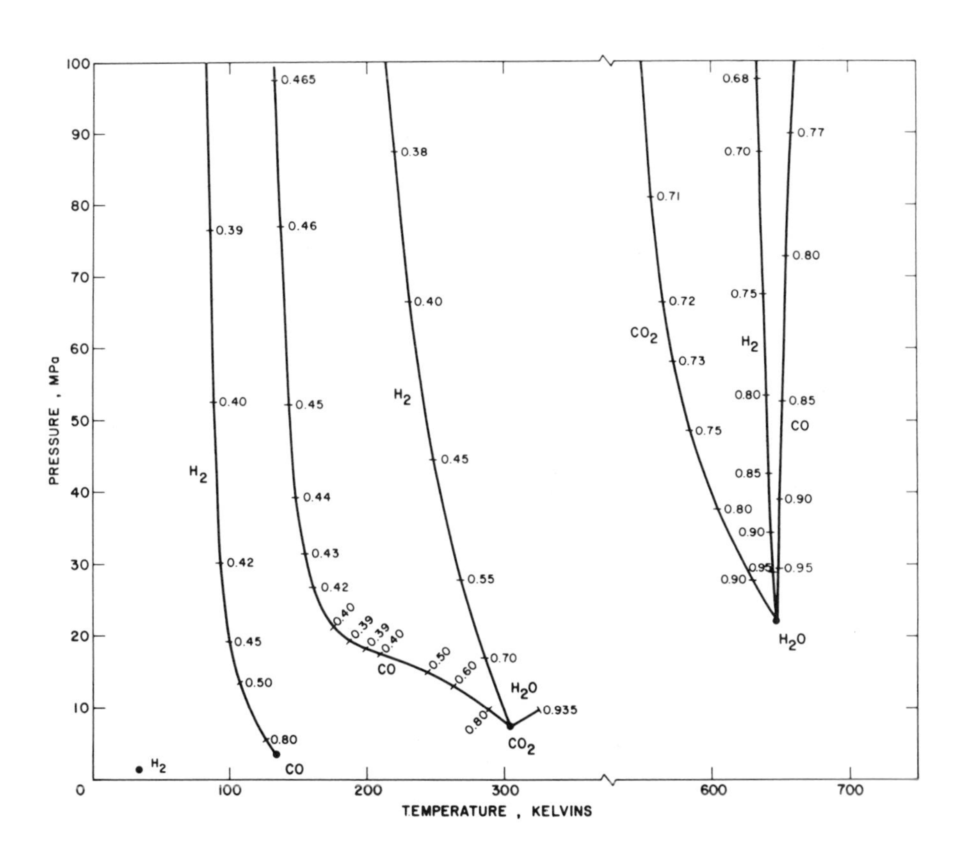

Figure 1. Critical lines in water–gas shift binary pairs

In Figure 1 the low temperature branches of the critical lines have been omitted, except for the CO_2 - H_2O binary. The reason for the omission is that these lines are extremely short and are not of much interest.

The mole fraction of the component whose critical point is the origin of the critical line has been indicated as a parameter along each of the lines. It should be noted from these numbers that there are compositions in each of the binary pairs for which there is no critical point. It should also be noted that some CO_2 - CO mixtures have two critical points.

It is most significant that certain binary mixtures have no critical point. With this in mind it is not surprising that certain four-component mixtures have no critical point.

The general shape of the binary critical lines dictates the shape of critical lines in the reacting mixtures.

Water-Gas Shift Critical Lines

Mixtures rich in the three lighter components have no critical point unless the water content is less than a few percent. The more interesting behavior is in mixtures rich in water.

The mixture mole fractions reached during the reaction depends on the initial composition. The sorts of mixtures studied here are generated by beginning with water and carbon monoxide, the water in excess.

Critical points have been calculated for such mixtures with various initial H_2O/CO ratios. The results are shown in Figure 2. In this figure, the line of zero conversion is the critical line for H_2O - CO binary. For mixtures with initial H_2O/CO ratios greater than about 4, it is possible to find critical points for all conversions. When the initial ratio is less, this is no longer possible. Lines of 25%, 50% and 75% conversion are also shown in Figure 2.

The critical points in these mixtures are all at pressures higher than the critical pressure of water; many are at temperatures higher than the water critical temperature. The mixture critical points indicate that high density phase separations persist to extreme conditions of temperature and pressure.

Reaction Equilibrium

The necessary condition for equilibrium in the chemical reactions of equation (1) is that

$$\ln K_j = \sum_i \nu_{ij} \ln f_i; \quad j = 1, R. \tag{25}$$

The equilibrium constants, K_j, may be evaluated as functions of temperature using readily available thermochemical data.

A procedure was developed that permitted location of points

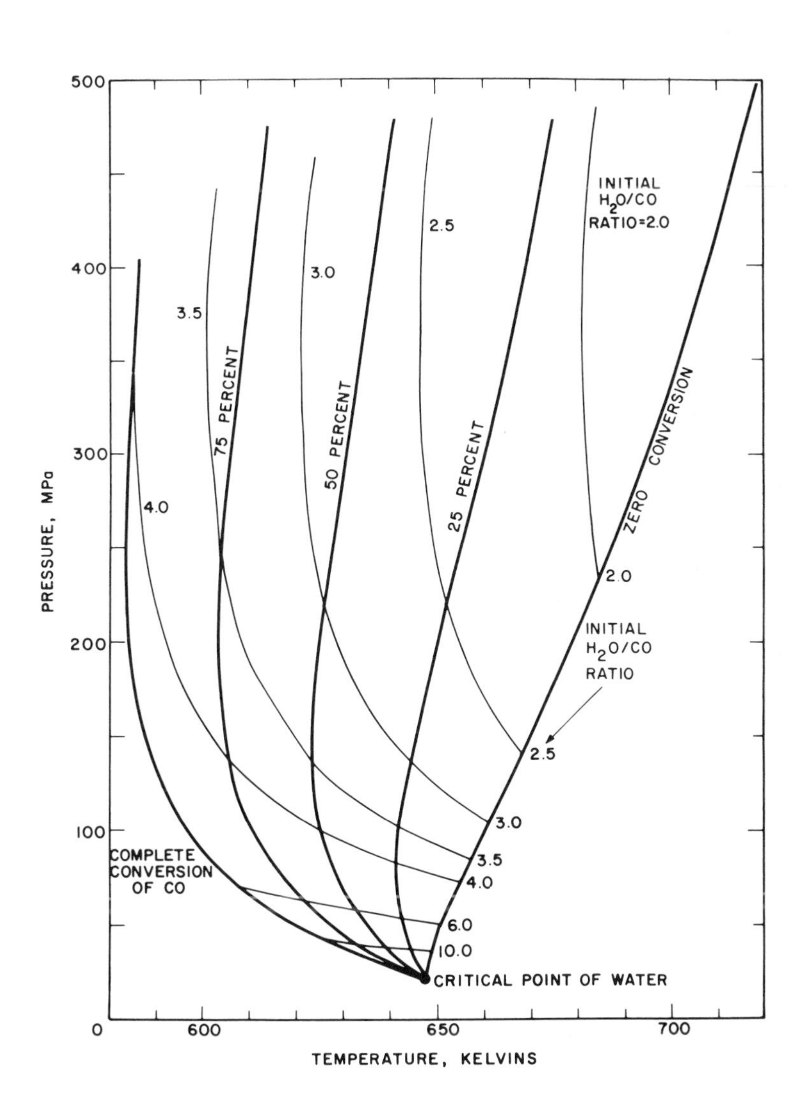

Figure 2. Critical lines in reacting mixtures of CO with excess H_2O

which are critical points and which obey equation (25) for the water gas shift reaction. The necessary thermochemical data were taken from Reid, Prausnitz and Sherwood (22). The fugacities of the four reacting species were evaluated using the SRK equation of state.

For every initial H_2O/CO ratio, the mixture mole fractions, hence the critical temperature and volume, are determined by the reaction extent ε. The equilibrium constant is calculated at the critical temperature. The fugacities are calculated also at the critical condition for the given ε. The function F, defined as

$$F(\varepsilon) = -\ell n K + \ell n (f_{CO_2} f_{H_2})/(f_{CO} f_{H_2O}) \qquad (26)$$

hence varies with ε alone.

Although it has been shown that thermodynamic models which imply phase separations can create difficulties with uniqueness in solving the reaction equilibrium equations (10, 23, 24), there proved to be only one solution to equation (26) under the conditions studied. It is conceivable that more than one critical point could be found for some reacting mixtures at certain reaction extents (two critical points are indeed indicated in Figure 1 for some CO_2 - CO mixtures), in which case $F(\varepsilon)$ will not be a single-valued function. This possibility was not explored.

In the mixtures with initial H_2O/CO ratios greater than 4, unique critical points were found for all ε covering the range from zero to 100% conversion of CO. In this range $F(\varepsilon)$ varies from $-\infty$ to $+\infty$, hence must cross zero at some ε. There was only one such zero crossing, however, indicating that non-uniqueness, if it occurs, will arise only because of the branching of $F(\varepsilon)$.

Calculated Critical and Equilibrium Points

The Newton-Raphson procedure was used to find ε satisfying $F(\varepsilon) = 0$. Iterations began at high conversion and the derivative $dF/d\varepsilon$ was found by numerical differentiation. Convergence was obtained in 5 iterations, with 10 critical point evaluations, in about 10 seconds. The computer used was the University of Calgary Honeywell HIS-Multics system.

Results of these calculations are given in Table 2 and in Figure 3. Also shown are results of calculations made assuming the mixture to be a perfect gas at the calculated critical temperature. (The pressure has no effect on the calculated ideal gas conversion in this reaction.)

Also calculated and presented in Table 2 is the quantity K_ϕ, defined by

$$K = K_\phi (p_{CO_2} p_{H_2} / p_{CO} p_{H_2O}) \qquad (27)$$

which provides an indication of the effect due to non-ideal gas

Table 2

CRITICAL POINTS AT REACTION EQUILIBRIUM

initial mol H_2O / mol CO	T_c,K	P_c,MP_a	fraction CO converted	$\ell n K$	K_ϕ	ideal gas CO conversion
10	626.1	42.48	0.9770	2.964	4.21	0.9943
8	620.0	50.50	0.9711	3.037	4.48	0.9932
6	608.7	70.20	0.9606	3.176	5.15	0.9918
5	599.0	98.54	0.9504	3.301	6.03	0.9910
4.5	592.9	133.11	0.9416	3.382	6.89	0.9905
4.25	590.2	165.9	0.9349	3.418	7.53	0.9901
4.0	589.2	223.8	0.9252	3.432	8.31	0.9895
3.9	589.8	259.8	0.9200	3.423	8.63	0.9890
3.8	591.5	308.3	0.9137	3.401	8.94	0.9884
3.7	594.6	375.1	0.9060	3.359	9.19	0.9875

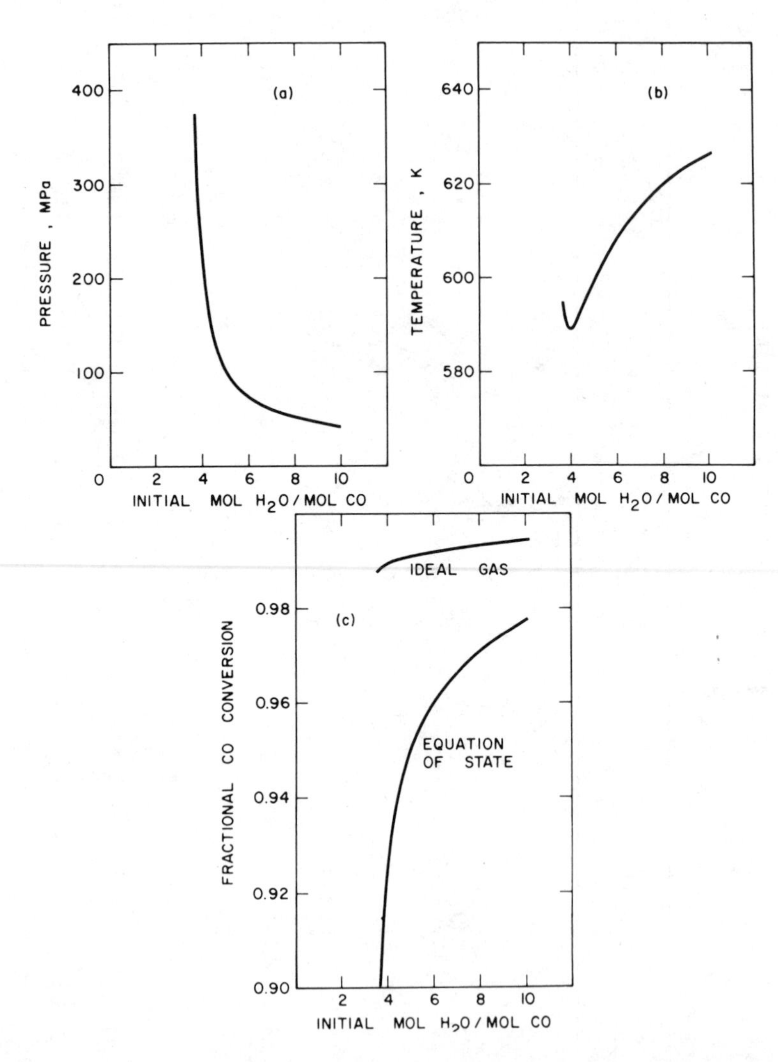

Figure 3. Critical and reaction equilibrium points: (a) critical pressure; (b) critical temperature; (c) conversion of CO

behavior. In the case of the water-gas shift reaction at the critical point, K_ϕ is greater than 1.0 indicating that equilibrium conversions will be less than would be calculated assuming perfect gases. It is the non-ideality of water which most affects K_ϕ but none of the other species could be regarded as ideal gases at the temperatures and pressures of the critical points.

The equilibrium conversions of CO achieved at the critical point are shown in Figure 3.c. The conversions attainable are well below the values corresponding to ideal gas behavior at the critical temperature.

Conclusions

The study described above for the water-gas shift reaction employed computational methods that could be used for other synthesis gas operations. The critical point calculation procedure of Heidemann and Khalil (14) proved to be adaptable to the mixtures involved. In the case of one reaction, it was possible to find conditions under which a critical mixture was at chemical reaction equilibrium by using a one dimensional Newton-Raphson procedures along the critical line defined by varying reaction extents. In the case of more than one independent chemical reaction, a Newton-Raphson procedure in the several reaction extents would be a candidate as an approach to satisfying the several equilibrium constant equations, (25).

Since synthesis gas mixtures are capable of showing more than one critical point, there is a possibility of finding more than one solution to the simultaneous equations defining reaction equilibria and the critical point. This possibility was not encountered in the present study.

Acknowledgment

This work was supported by a grant from the National Science and Engineering Research Council of Canada.

Literature Cited

1. Soave, G. Chem. Eng. Sci., 1972, 27, 1197.
2. Wilson, G.M., 1969, paper presented at the 65th National AIChE Mtg., Cleveland.
3. Peng, D.-Y.; Robinson, D.B. Ind. Eng. Chem. Fundamentals, 1976, 15, 59.
4. Heidemann, R.A. AIChE J, 1974, 20, 847.
5. Evelein, K.A.; Moore, R.G.; Heidemann, R.A. Ind. Eng. Chem. Process Des. Dev., 1976, 15, 423.
6. Peng, D.-Y.; Robinson, D.B. Canadian J. of Chem. Engng, 1976, 54, 595.
7. van Zeggeren, F.; Storey, H.S. "The Computation of Chemical Equilibria", Cambridge University Press, Cambridge, 1970.

8. Danes, F.E.; Geana, D. Revista de Chimie, 1979, 30, Nr. 3, 244.
9. Dluzniewski, J.H.; Adler, S.B.; Ozkardesh, H.; Barner, H.E., 1973, paper presented at the 75th National AIChE Mtg., Detroit.
10. Heidemann, R.A. Chem. Eng. Sci., 1978, 33, 1517.
11. Reid, R.C.; Beegle, B.L. AIChE J., 1977, 23, 726.
12. Peng, D.-Y.; Robinson, D.B. AIChE J., 1977, 23, 137.
13. Baker, L.E.; Luks, K.D., 1978, paper SPE 7478 presented at the 53rd Annual Conference, Soc. Pet. Eng. of the AIME, Houston.
14. Heidemann, R.A.; Khalil, A.M., 1979, paper presented at the 86th National AIChE Mtg., Houston.
15. Michelsen, M.L., personal communication, 1979.
16. Michelsen, M.L. Fluid Phase Equilibria (in press).
17. Soave, G. Chem. Eng. Sci, 1979, 34, 225.
18. Takenouchi, S.; Kennedy, G.C. Amer. J. Sci., 1964, 262, 1055.
19. Todheide, K.; Franck, E.U. Z. Phys. Chem. (Frankfort am Main), 1963, 37, 387.
20. Oellrich, L.; Plocker, U.; Prausnitz, J.M.; Knapp, H. Chem. Ing. Tech., 1977, 12, 955.
21. Graboski, M.S.: Daubert, T.E. Ind. Eng. Chem. Process Des. Dev., 1978, 17, 448.
22. Reid, R.C.; Prausnitz, J.M.; Sherwood, T.K. "Properties of Gases and Liquids", 3rd Edition, McGraw-Hill, New York, 1977.
23. Caram, H.S.; Scriven, L.E. Chem. Eng. Sci., 1976, 31, 163.
24. Othmer, H.G. Chem. Eng. Sci, 1976, 31, 993.

RECEIVED January 31, 1980.

20

Two- and Three-Phase Equilibrium Calculations for Coal Gasification and Related Processes

D.–Y. PENG and D. B. ROBINSON

Department of Chemical Engineering, University of Alberta, Edmonton, Alberta, Canada T6G 2G6

The gasificiation of coal, shale-oil, or other lower grade hydrocarbon base stocks inevitably leads to the production of process streams which contain a very wide range of paraffinic, naphthenic, aromatic and olefinic hydrocarbons in the presence of associated non-hydrocarbons such as hydrogen, nitrogen, carbon dioxide, hydrogen sulfide and ammonia. These streams are often contacted with water at process conditions which normally lead to either gas - water - rich liquid equilibrium or gas - water - rich liquid - hydrocarbon rich liquid equilibrium. The processing conditions and stream compositions which may lead to the formation of these different phases and the distribution of the components between phases are of great importance to the design engineer. For this reason the establishment of reliable procedures for predicting the behavior of these mixtures in the one-, two-, and three-phase regions is a matter of considerable importance.

In an earlier paper (1), the authors presented an efficient procedure for predicting the phase behavior of systems exhibiting a water - rich liquid phase, a hydrocarbon - rich liquid phase, and a vapor phase. The Peng-Robinson equation of state (2) was used to represent the behavior of all three phases. It has the following form:

$$P = \frac{RT}{v-b} - \frac{a(T)}{v(v+b) + b(v-b)} \qquad (1)$$

where $a(T) = a_c \alpha$

$$a_c = 0.45724 \frac{R^2 T_c^2}{P_c}$$

$$\alpha^{\frac{1}{2}} = 1 + \kappa(1 - T_R^{\frac{1}{2}}) \qquad (2)$$

0-8412-0569-8/80/47-133-393$05.50/0

$$\kappa = 0.37464 + 1.54226\omega - 0.26992\omega^2 \tag{3}$$

$$b = 0.07780 \frac{RT_c}{P_c}$$

For mixtures

$$a = \sum_i \sum_j x_i x_j (1-\delta_{ij}) a_i^{\frac{1}{2}} a_j^{\frac{1}{2}} \tag{4}$$

$$b = \sum x_i b_i \tag{5}$$

Although the calculated phase compositions for the hydrocarbon - rich liquid phase and the vapor phase showed excellent agreement with the experimental data, the calculated hydrocarbon contents of the aqueous liquid phase was consistently several orders of magnitude lower than the reported experimental values. It was speculated that additional temperature - dependent interaction parameters would be required to bring the predicted values and the experimental results into quantitative agreement; nevertheless, no attempt was made at that time to try to accomplish this.

In this study, it has been possible to devise a procedure which can be used to generate reliable phase compositions for both the hydrocarbon - rich phase and the aqueous phase over a wide range of temperature and pressure. Moreover, the calculation procedure has been successfully applied to non-hydrocarbon - water systems with quantitative results.

Calculation Procedure. With the exception of two significant modifications, the calculation procedure used in this study was basically the same as that used previously.

The first modification concerns the use of Eqn. (2) for water. When developing the original correlation for $\alpha^{\frac{1}{2}}$ and κ as expressed by Eqn. (2) and (3), water was not included as one of the components, and consequently the predicted vapor pressures for water were not as good as expected. Thus in order to correlate the vapor pressure of water more accurately over the entire temperature range, Eqn. (2) was modified for this compound at temperatures where $T_R^{\frac{1}{2}} < 0.85$ as follows:

$$\alpha^{\frac{1}{2}} = 1.0085677 + 0.82154 (1-T_R^{\frac{1}{2}}) \tag{6}$$

At temperatures where $T_R^{\frac{1}{2}} \geq 0.85$, Eqn. (2) still applies.

The second modification concerns the correlation of the composition of the aqueous liquid phase. In order to accomplish this, a temperature - dependent interaction parameter was used for the aqueous liquid phase and the previous temperature - independent parameter was used for the non-aqueous liquid phase and the vapor phase. Thus for the aqueous - liquid phase Eqn. (4) becomes

$$a = \sum_i \sum_j x_{B_i} x_{B_j} (1-\tau_{ij}) a_i^{1/2} a_j^{1/2} \quad (7)$$

where τ_{ij} is a temperature - dependent interaction parameter.

The introduction of this parameter for each aqueous binary pair means that the interaction between the water molecule and the gas molecule in the aqueous liquid phase is much different from that in the nonaqueous phases. For all the aqueous binaries which have been examined in this study, the temperature - dependent interaction parameters take on negative values at ambient temperature and monotonically increase as temperature increases. This indicates that the attraction energy between the water molecular and the other molecules decreases as the temperature increases.

Non-Hydrocarbon - Water Binaries

Of the many possible non-hydrocarbon - water binary systems which are related to substitute gas processes, the data on only the water binaries containing H_2S, CO_2, N_2, and NH_3 were used in this study. The treatment of hydrogen, a quantum gas, is different from that of the other gases. A separate paper will deal with the correlation of the data on hydrogen mixtures.

Hydrogen Sulfide - Water System. The data of Selleck et al. (3) were used to evaluate the interaction parameters for the hydrogen sulfide - water system. The data include the composition of both phases at temperatures from 100°F to 340°F and pressures from 100 to 5000 psia in the coexisting vapor and aqueous liquid - hydrogen sulfide - rich liquid - vapor equilibrium locus.

A single, constant interaction parameter has been determined for the hydrogen sulfide - rich phases. This determination was based on the three-phase pressure - temperature locus. While investigating the three-phase region, it was noted that the three-phase locus and the composition of the hydrogen sulfide - rich phase were rather insensitive to the temperature - dependent aqueous phase interaction parameter. Furthermore, the composition of the aqueous phase was relatively independent of the constant interaction parameter. For these reasons, the solubility of hydrogen sulfide in the aqueous liquid was correlated at the same time as the parameter was being determined for the hydrogen sulfide - rich phases.

The calculated and experimental gaseous and liquid phase compositions are shown in Figures 1 and 2 respectively.

Carbon Dioxide - Water System. The data of Wiebe and Gaddy (4, 5, 6) were used exclusively in this study to determine the interaction parameters for the carbon dioxide - water binary system. These data cover the temperature and pressure range from 12°C to 100°C and from 25 atm to 700 atm respectively. As with the H_2S - H_2O system, a constant interaction parameter has been obtained for the gaseous phase and the carbon dioxide - rich

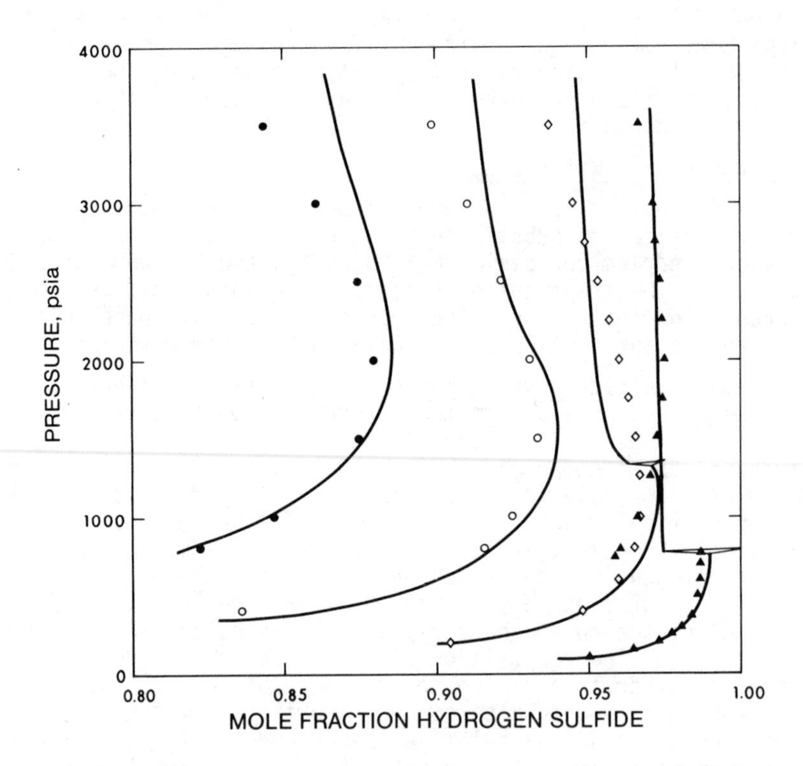

Figure 1. Experimental and predicted vapor phase compositions for the hydrogen sulfide–water system ((——) P–R prediction; data from Ref. 3: *(●) 340°F; (○) 280°F; (◇) 220°F; (▲) 160°F)*

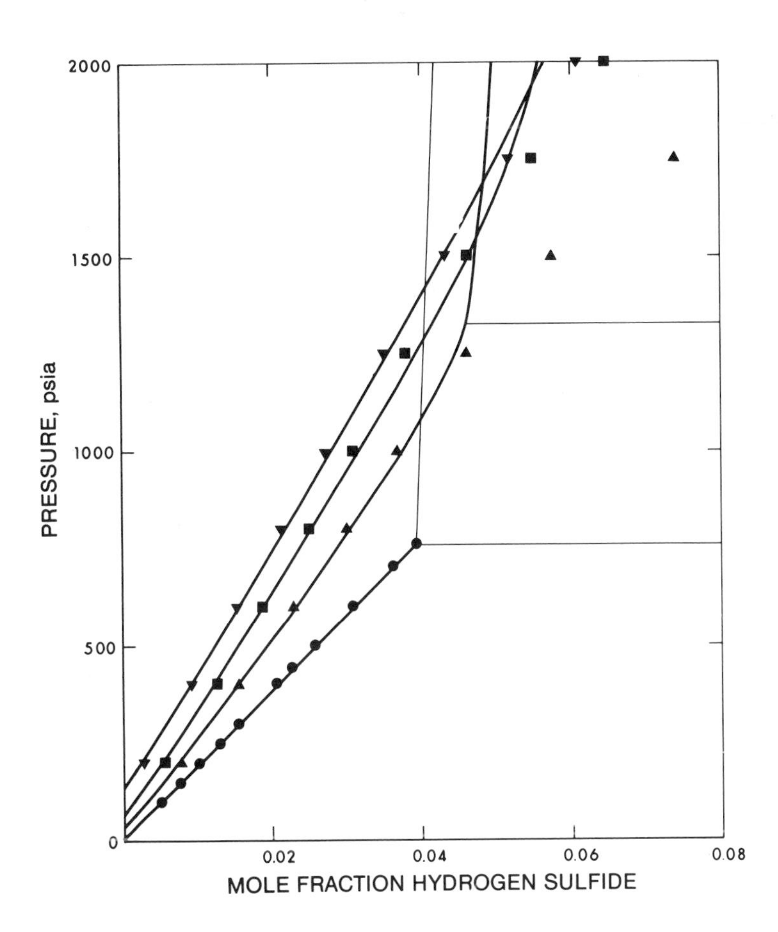

Figure 2. Experimental and predicted aqueous liquid phase compositions for the hydrogen sulfide–water system ((——) P–R prediction; data from Ref. 3: (●) 160°F; (▲) 220°F; (■) 280°F; (▼) 340°F)

liquid phase. At each temperature, the solubility of carbon dioxide in water can be correlated accurately through the whole pressure range using one interaction parameter for the aqueous phase. The equilibrium aqueous liquid and vapor compositions for this binary at two temperatures are shown in Fig. 3.

Malinin (7), Todheide and Franck (8) and Takenouchi and Kennedy (9) reported equilibrium data for this system at temperatures up to 350°C and pressures to 3500 bars. However, the vapor phase data of these authors do not always agree with each other. The aqueous phase data have been used to extend the temperature - dependent interaction parameter to 300°C.

Nitrogen - Water System. The interaction parameters for the nitrogen - water system have been evaluated using the data of Wiebe and Gaddy (10), Paratella and Sagramora (11), Rigby and Prausnitz (12)and O'Sullivan and Smith (13). As with the two previous systems, only one constant interaction parameter was necessary to correlate the vapor phase composition while the interaction parameter for the aqueous liquid phase increased monotonically with temperature. A comparison of the calculated and experimental vapor phase and aqueous liquid phase compositions is given in Table I.

Ammonia - Water System. Interaction parameter for the ammonia - water system was obtained using the data of Clifford and Hunter (14) and of Macriss et al. (15). A single - valued parameter was capable of representing the composition of the liquid phase reasonably well at all temperatures, however, the calculated amount of water in the vapor phase in the very high ammonia concentration region was somewhat lower than the data of Clifford and Hunter and Macriss et al. Edwards et al. (16) have applied a new thermodynamic consistency test to the data of Macriss et al and have concluded that the data appear to be inconsistent and that the reported water content of the vapor phase is too high.

The experimental data and the calculated results are given in Fig. 4.

Hydrocarbon - Water Binaries

The interaction parameters for binary systems containing water with methane, ethane, propane, n-butane, n-pentane, n-hexane, n-octane, and benzene have been determined using data from the literature. The phase behavior of the paraffin - water systems can be represented very well using the modified procedure. However, the aromatic - water system can not be correlated satisfactorily. Possibly a differetn type of mixing rule will be required for the aromatic - water systems, although this has not as yet been explored.

Methane - Water System. Interaction parameters were generated for the vapor phase and the aqueous liquid phase for the methane -

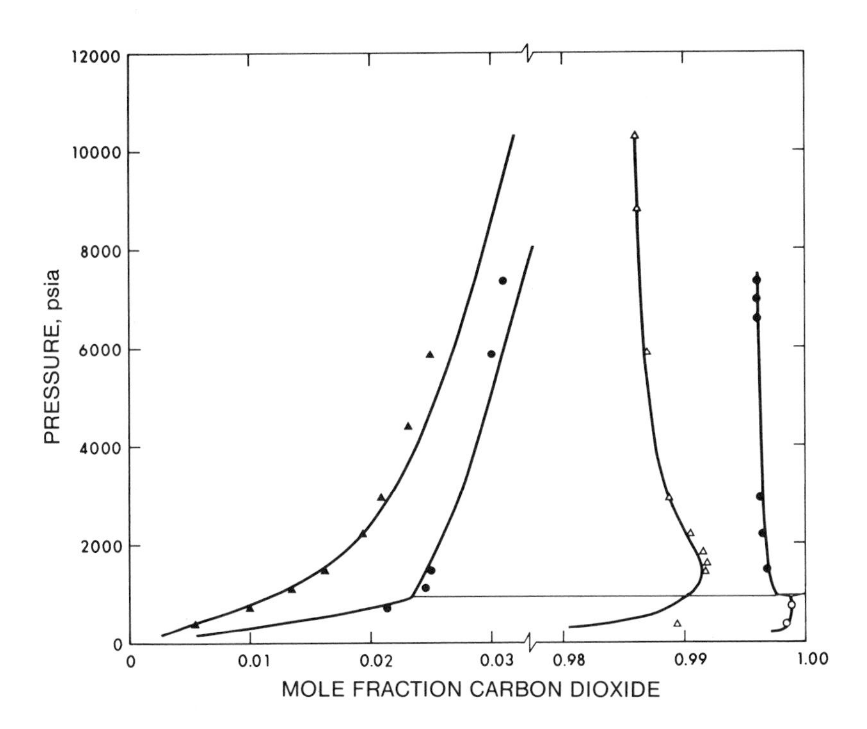

Figure 3. Experimental and predicted vapor and aqueous liquid phase compositions for the carbon dioxide–water system ((——) P–R prediction; data from Ref. 6: liquid—(▲) 167°F; (●) 77°F; vapor—(△) 167°F; (○) 77°F)

TABLE I. Experimental and Calculated Aqueous Liquid and Vapor Phase Compositions for the Nitrogen - Water System.

Pressure, atm.	$x^*_{N_2} \times 10^3$		$y^*_{H_2O} \times 10^3$	
	Expt.	Calc.	Expt.	Calc.
		T = 25°C		
22.20			1.529	1.502
25	0.280	0.278		
30.50			1.149	1.123
38.19			0.941	0.919
50	0.542	0.537		
100	1.015	1.004		
200	1.812	1.795		
300	2.455	2.458		
500	3.558	3.555		
800	4.909	4.869		
1000	5.720	5.604		
		T = 50°C		
20.81			6.260	6.190
25	0.219	0.220		
36.93			3.680	3.640
50	0.436	0.428		
59.04			2.420	2.410
75.99			1.956	1.952
100	0.812	0.810		
200	1.470	1.470		
300	2.034	2.032		
500	2.982	2.968		
800	4.181	4.084		
1000	4.900	4.701		

Table I - continued.

Pressure, atm.	$x^*_{N_2} \times 10^3$		$y^*_{H_2O} \times 10^3$	
	Expt.	Calc.	Expt.	Calc.
		T = 75°C		
25	0.204	0.203		
41.66			10.09	10.12
50	0.397	0.398		
60.35			7.21	7.25
88.55			5.23	5.23
100	0.760	0.760		
200	1.390	1.395		
300	1.936	1.942		
500	2.872	2.859		
800	4.052	3.948		
1000	4.747	4.544		
		T = 100°C		
25	0.214	0.206		
50	0.415	0.410		
56.42			19.94	19.94
78.44			15.03	14.89
100	0.792	0.792		
100.o9			12.19	12.09
200	1.462	1.470		
300	2.042	2.060		
500	3.044	3.052		
800	4.294	4.223		
1000	5.003	4.857		

* Mole Fraction

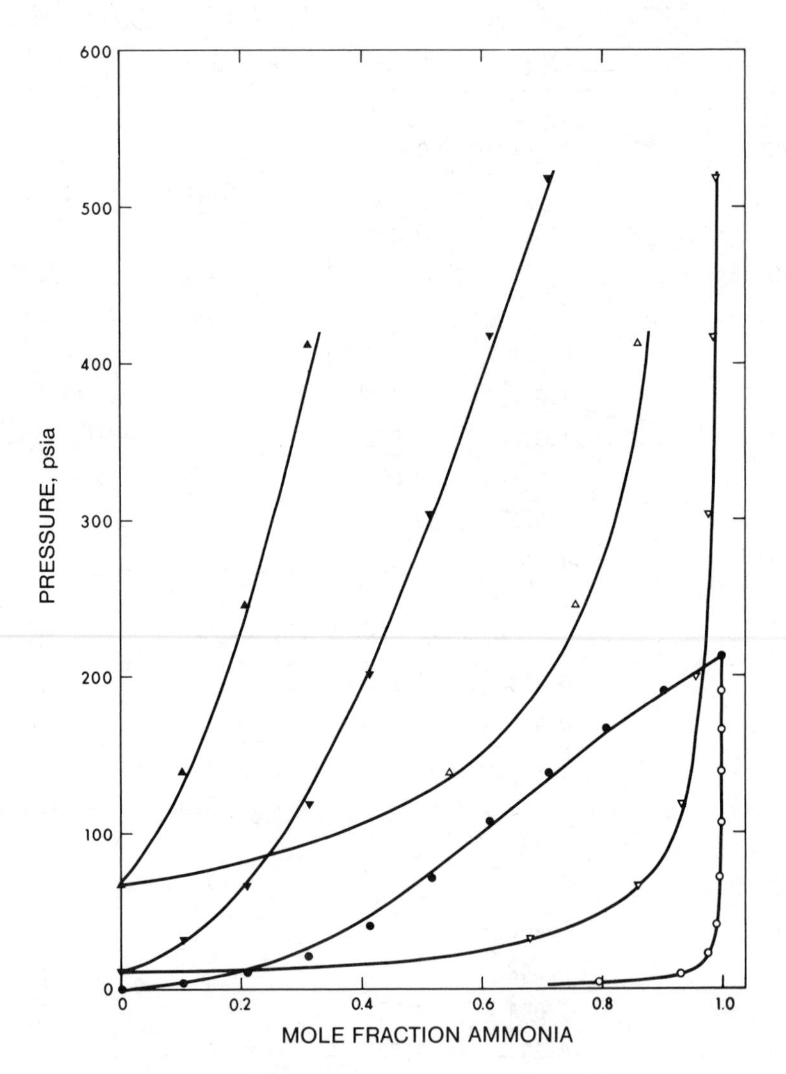

Figure 4. Experimental and predicted vapor and liquid phase compositions for the ammonia–water system ((——) P–R prediction; data from Ref. 15: *liquid—(▲) 300°F; (▼) 200°F; (●) 100°F; vapor—(△) 300°F; (▽) 200°F; (○) 100°F)*

water binary system using experimental data reported by Sultanov et al. (17, 18), Olds et al. (19), and Culberson and McKetta (20).

The vapor-phase mole fractions of water of Olds et al. (19) can be represented very well using the Peng-Robinson equation of state in conjunction with a constant interaction parameter over the temperature range from 100°F to 460°F. The same interaction parameter can be used to reproduce the data of Sultanov et al. (18) up to 300°C with good results. However, at higher temperatures the calculated water content in the vapor phase deviated somewhat from their data.

The temperature - dependent interaction parameters were determined from 77°F to 680°F using the data of Culberson and McKetta (20) and of Sultanov et al. (18). This parameter increases with temperature and appears to converge to the value of the constant parameter used for the vapor phase as the critical temperature of water is approached.

The experimental and calculated results for this binary system at 250°C are presented in Figure 5.

Ethane - Water System. The data used for the determination of the interaction parameters for the ethane - water binary are those of Culberson and McKetta (21), Culberson et al. (22) and Reamer et al. (23).

A constant interaction parameter was capable of representing the mole fraction of water in the vapor phase within experimental uncertainty over the temperature range from 100°F to 460°F. As with the methane - water system, the temperature - dependent interaction parameter is also a monotonically increasing function of temperature. However, at each specified temperature, the interaction parameter for this system is numerically greater than that for the methane - water system. Although it is possible for this binary to form a three-phase equilibrium locus, no experimental data on this effect have been reported.

Figure 6 illustrates the calculated and experimental equilibrium phase compositions at 220°F for this binary system.

Propane - Water System. The interaction parameters for the propane - water system were obtained over a temperature range from 42°F to 310°F using exclusively the data of Kobayashi and Katz (24). This is because among the available literature on the phase behavior of this binary system, their data appear to give the most extensive information. A constant interaction parameter was obtained for the propane-rich phases at all temperatures. The magnitude of the temperature - dependent interaction parameter for this binary was less than that for the ethane - water binary at the same temperature. Azarnoosh and McKetta (25) also presented experimental data for the solubility of propane in water over about the same temperature range as that of Kobayashi and Katz but at pressures up to 500 psia only. However, a different set of temperature - dependent parameters

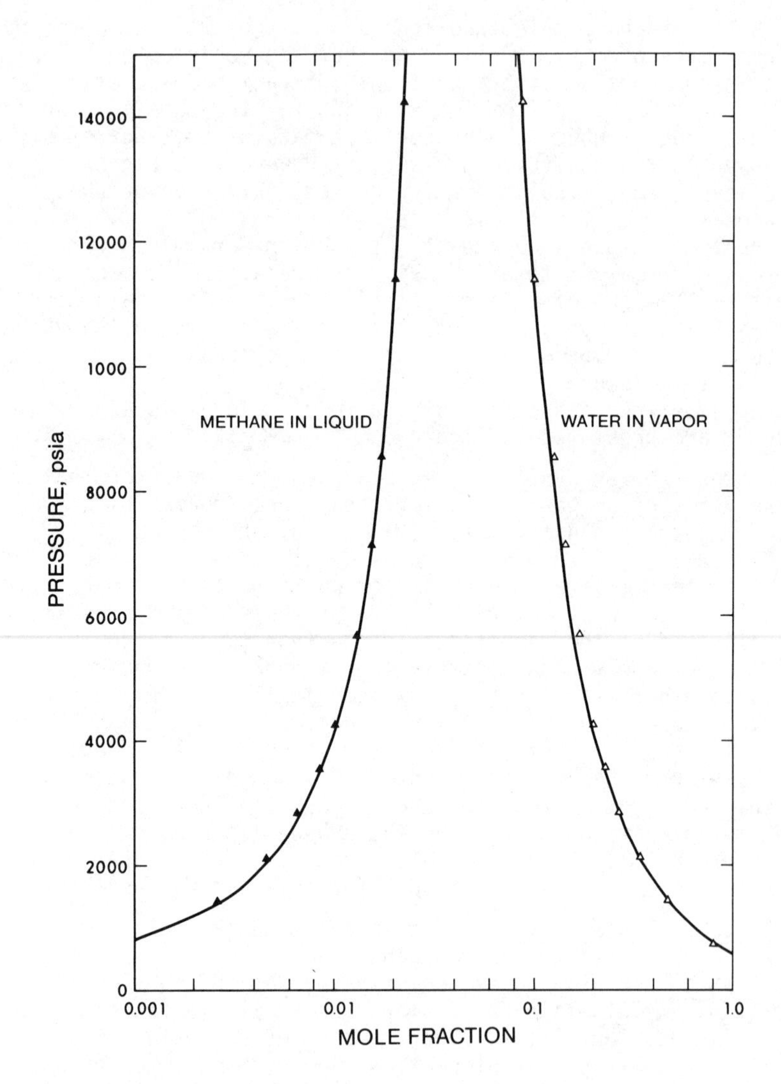

Figure 5. Experimental and predicted vapor and liquid phase compositions for methane–water system at 250°C ((——) P–R prediction; (△) (17); (▲) (18))

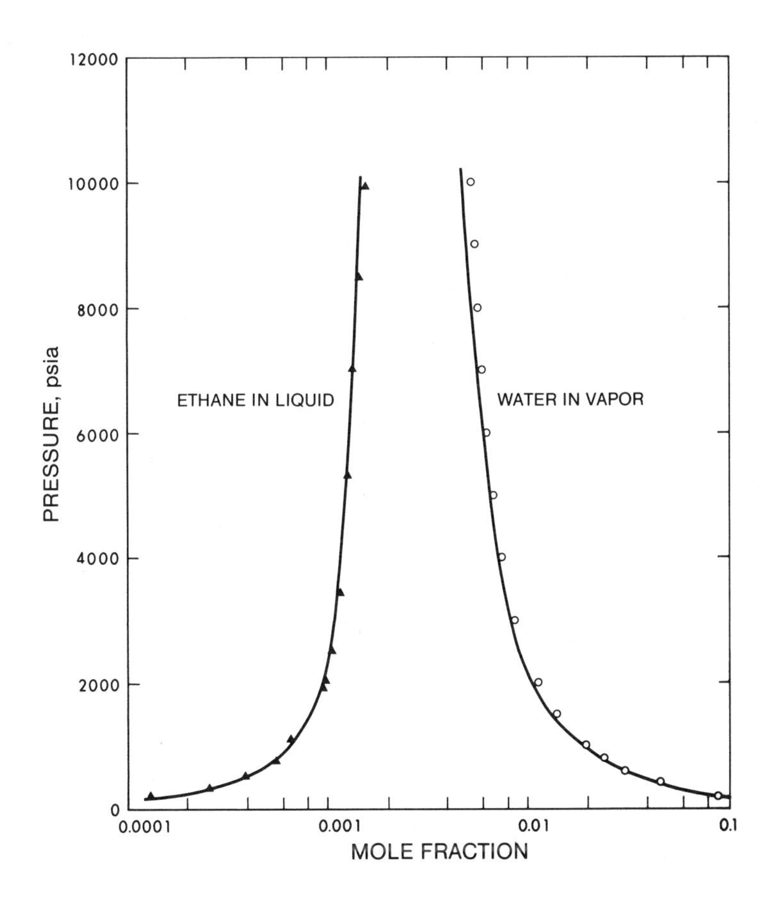

Figure 6. Experimental and predicted vapor and liquid phase compositions for the ethane–water system at 220°F ((——) P–R prediction; (▲) (21); (○) (23))

would be required to accurately correlate their results.

The water content of the propane - rich phases in the aqueous liquid - propane liquid - vapor region are illustrated in Figure 7.

n-Butane - Water System. Reamer et al. (26) have determined the concentration of water in the n-butane - water system in the vapor and the n-butane liquid phases in the three-phase region. Reamer et al. (27) have published experimental data for the solubility of n-butane in water and of water in n-butane in the two-phase region covering a temperature range from 100°F to 460°F and at pressures up to 10,000 psia. LeBreton and McKetta (28) have presented the results of an experimental study on the solubility of n-butane in water at four temperatures but at pressures up to only 1000 psia. While the reported three-phase pressures from these two sources agree very well, the data on the solubility of n-butane in water show marked differences. The solubility values presented by LeBreton and McKetta are consistently lower than those reported by Reamer et al. In view of the fact that the data of Reamer et al. covered a much broader range of both temperature and pressure, their data were used for determining the interaction parameters for this system.

As with the first three paraffin - water systems, only a constant parameter was required to correlate the hydrocarbon - rich phases although a temperature - dependent parameter was necessary to fit the aqueous - liquid phase data.

The equilibrium composition of the n-butane - water binary in the three-phase region are illustrated in Figure 8.

n-Pentane - Water System. Scheffer (29) has presented the three-phase locus for a mixture of i-pentane and n-pentane over a temperature range from 150°C to 187.1°C. However, no compositional measurements were reported. Namiot and Beider (30) reported the solubility of n-pentane in water at three temperatures along the three-phase locus. Interaction parameters for the n-pentane - water system were determined using these data.

n-Hexane - Water System. The n-hexane - water system is the lightest paraffin - water binary where the vapor pressure locus of the hydrocarbon intersects that for pure water. The experimental phase behavior data available in the literature for this system cover a wide range of temperature and pressure. Unfortunately these data do not corroborate each other and noticeable discrepancies exist. The data of Scheffer (31), Rebert and Hayworth (32), and Sultanov and Skripka (33) were employed in determining the interaction parameter for the hydrocarbon - rich phases. A unique value for this interaction parameter could not be obtained because of the discrepancies among the data. However, a tentative value, based on the extrapolation of the values for other paraffin - water interaction parameters, has been assigned to the constant interaction parameter. The interaction parameters for the aqueous liquid phase were determined using the data of

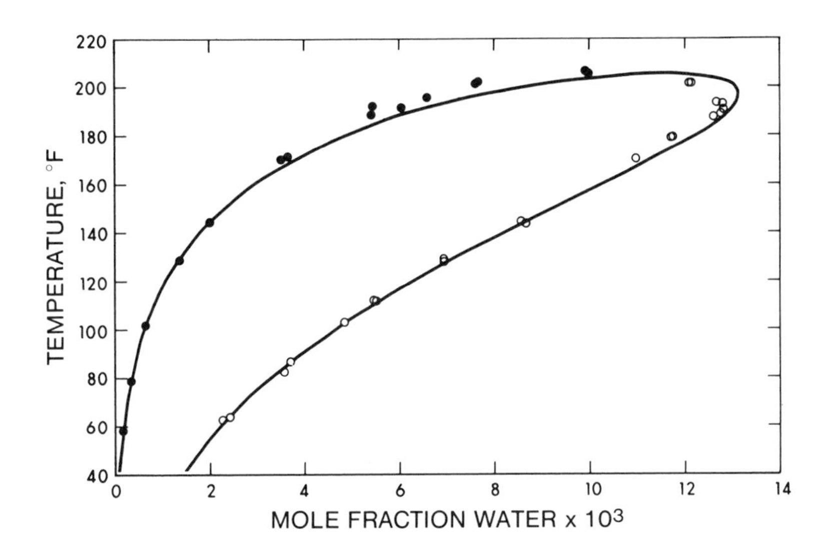

Figure 7. Experimental and predicted water content of propane liquid and vapor phases for the propane–water system along the three-phase locus ((——) P–R prediction; data from Ref. 24: (○) vapor; (●) liquid)

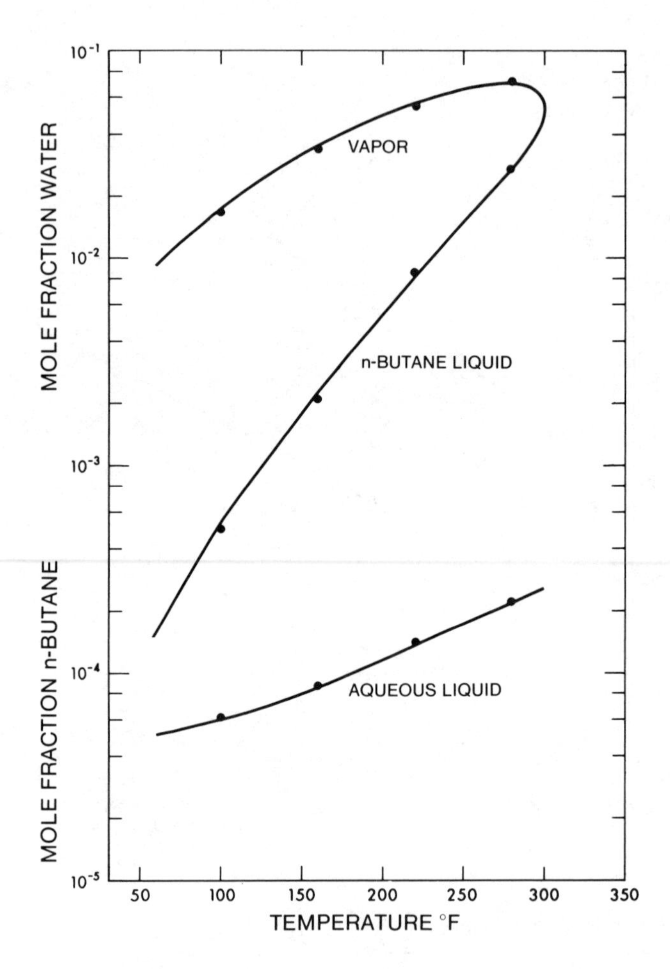

Figure 8. Experimental and predicted water content of n-*butane and vapor phases for the* n-*butane–water system along the three-phase locus ((——) P–R prediction; (●) (27))*

Kudchadker and McKetta (34). Their solubility data in the vapor - liquid region are believed to be in error probably due to their incorrect procedure of smoothing the raw data. However, their data in the liquid - liquid region seem to be acceptable. The data of Rebert and Hayworth (32) were used to extend the temperature - dependent interaction parameters to temperatures above the critical point of n-hexane.

Other Hydrocarbon - Water Systems. Interaction parameters were generated for the benzene - water system. The data used were those of Scheffer (31), Rebert and Kay (35), and Connolly (36). As with the alkane - water systems, the interaction parameters for the aqueous liquid phase were found to be temperature - dependent. However, the compositions for the benzene - rich phases could not be accurately represented using any single value for the constant interaction parameter. The calculated water mole fractions in the hydrocarbon - rich phases were always greater than the experimental values as reported by Rebert and Kay (35). The final value for the constant interaction parameter was chosen to fit the three phase locus of this system. Nevertheless, the calculated three-phase critical point was about 9°C lower than the experimental value.

Interaction parameter was also generated for the hydrocarbon - rich phases of the n-octane - water system. The data of Kalafati and Piir (37) were used. There were no data available for the water - rich liquid phase for this binary.

Experimental solubility data are available for some higher alkane - water systems (see, for example, Skripka et al., (38)). However, these data either cover only a very limited temperature range or contain results for one phase only. No attempt has been made to determine the interaction parameters for water - hydrocarbon systems where the hydrocarbon is larger than n-octane.

The temperature - dependent interaction parameters determined for several alkane - water systems are plotted in Figure 9. The values for the hydrogen sulfide - carbon dioxide -, and nitrogen - water binaries are given in Figure 10. It can be seen that a systematic trend exists for these parameters. The interaction parameter increases with the size of the molecule and furthermore it appears to converge rapidly as the carbon number increases. At a given temperature and pressure, the solubility of alkanes in water generally decreases as the molecular weight of the hydrocarbon increases. The amount of n-octane and heavier hydrocarbons dissolved in water streams resulting from synthetic gas processes is believed to be insignificant. The calculation of the solubility of these compounds in water under these conditions by using extrapolated values from interaction parameters of lighter paraffin - water binaries probably will not cause large errors.

Three-Phase Loci. Figure 11 shows the three-phase loci for the alkane - water systems. No experimental three-phase data were available in the literature for the ethane - water binary.

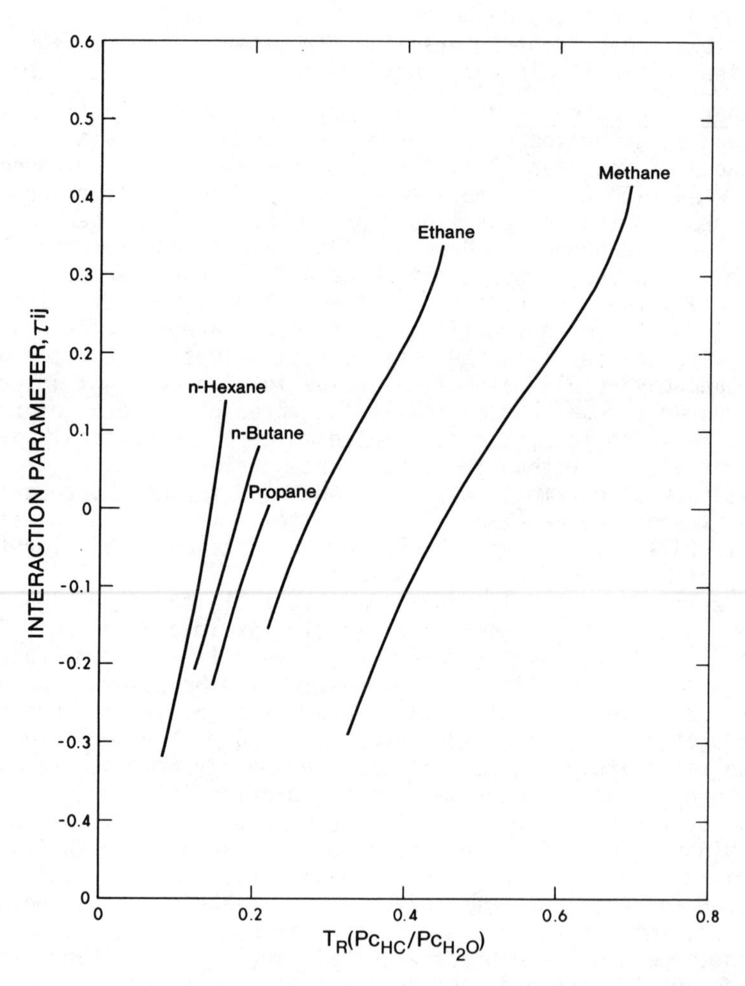

Figure 9. Temperature-dependent interaction parameters for selected paraffin–water binary systems

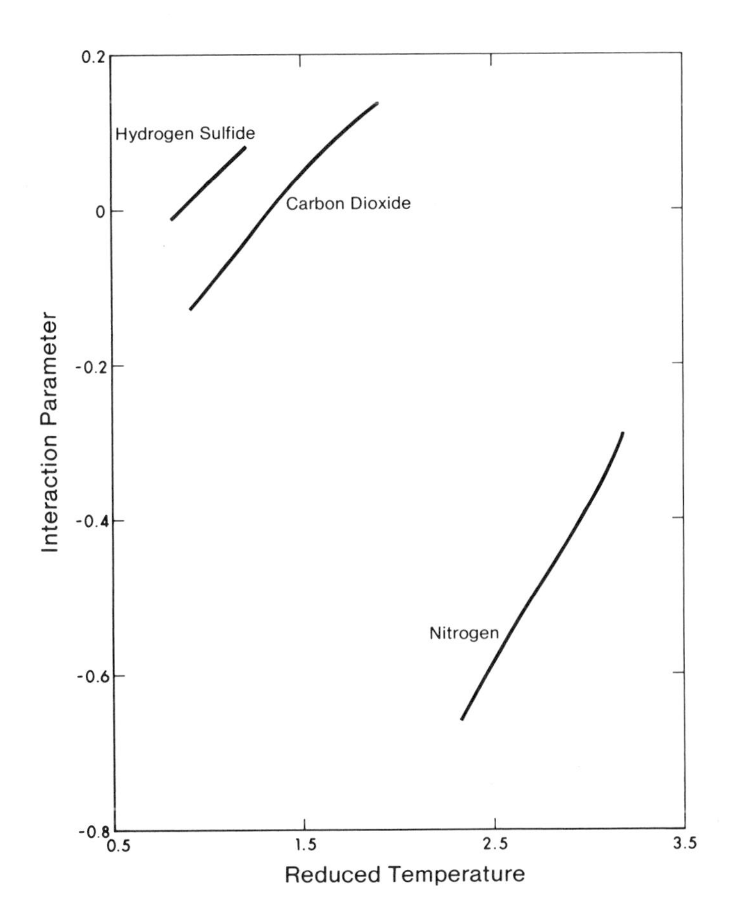

Figure 10. Temperature-dependent interaction parameters for nitrogen, carbon dioxide, and hydrogen sulfide with water

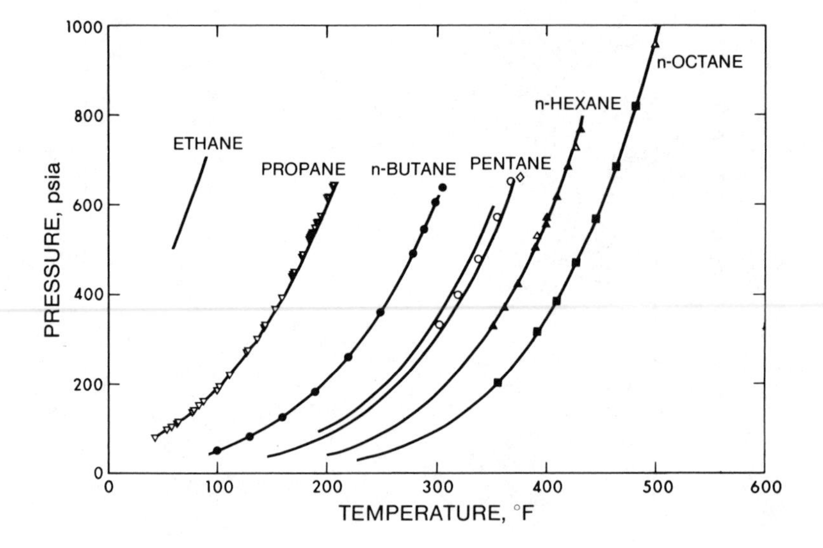

Figure 11. Experimental and predicted three-phase loci for selected paraffin–water binary systems ((——) P–R prediction; (▽) (24); (●) (27); (◇) (39); (○) (29); (▲) (31); (△) (33); (■) (37))

Nevertheless, a calculated locus is included for completeness and to indicate the possible region of three-phase equilibrium.

As was mentioned earlier, the three-phase data reported by Scheffer (29) for pentane - water were for a "binary" composed of water and a mixture of i-pentane and n-pentane. As shown in the figure, these data are bounded by the calculated loci of the i-pentane - water and n-pentane - water systems.

Conclusion. The mixing rule for use with the Peng-Robinson equation of state has been modified to include temperature - dependent interaction parameters. Both the constant and the temperature - dependent interaction parameters covering a wide range of temperatures have been determined for hydrocarbon - water systems including methane - water, ethane - water, propane - water, n-butane - water, and n-hexane - water and non-hydrocarbon - water systems including hydrogen sulfide - water, carbon dioxide - water, nitrogen - water, and ammonia - water. The inclusion of these temperature - dependent parameters has greatly improved the accuracy of predictions of three-phase and two-phase equilibrium for systems involving water.

Acknowledgement. The financial support received from the National Science and Engineering Research Council of Canada is sincerely appreciated.

Abstract

Two-constant equation of state phase behavior calculations for aqueous mixtures often require the use of temperature dependent binary interaction parameters. The methods used for evaluating these parameters for some of the typical aqueous binary pairs found in coal gasification and related process streams are described. Experimental and predicted phase compositions based on these methods are illustrated for aqueous pairs containing CO_2, H_2S, NH_3, and other gases.

Literature Cited

1. Peng, D.-Y., Robinson, D.B., Can. J. Chem. Eng., 1976, 54, 595.
2. Peng, D.-Y., Robinson, D.B., Ind. Eng. Chem. Fundam., 1976, 15, 59.
3. Selleck, F.T., Carmichael, L.T., Sage, B.H., Ind. Eng. Chem., 1952, 44, 2219.
4. Wiebe, R., Gaddy, V.L., J. Am. Chem. Soc., 1939, 61, 315.
5. Wiebe, R., Gaddy, V.L., J. Am. Chem. Soc., 1940, 62, 815.
6. Wiebe, R., Gaddy, V.L., J. Am. Chem. Soc., 1941, 63, 475.
7. Malinin, S.D., Geokhimiya, 1959, 3, 235.
8. Todheide, K., Franck, E.U., Z. Physik. Chem. (Frankfurt), 1963, 37, 387.
9. Takenouchi, S., Kennedy, G.C., Am. J. Sci., 1964, 262, 1055.
10. Wiebe, R., Gaddy, V.L., Heins, C., J. Am. Chem. Soc., 1933, 55, 947.

11. Paratella, A., Sagramora, G., Ricerca Sci., 1959, 29, 2605.
12. Rigby, M., Prausnitz, J.M., J. Phys. Chem. 1968, 72, 330.
13. O'Sullivan, T.D., Smith, N.O., J. Phys. Chem. 1970, 74, 1460.
14. Clifford, I.L., Hunter, E., J. Phys. Chem. 1933, 37, 101.
15. Macriss, R.A., Eakin, B.E., Ellington, R.T., Huebler, J., Research Bulletin, No. 34, American Gas Association, 1964.
16. Edwards, T.J., Newman, J., Prausnitz, J.M., Ind. Eng. Chem. Fundam., 1978, 17, 264.
17. Sultanov, R.G., Skripka, V.G., Namiot, A. Yu., Gazovaya Promyshlennost, 1971, 16 (4), 6.
18. Sultanov, R.G., Skripka, V.G., Namiot, A. Yu., Gazovaya Promyshlennost, 1972, 17 (5), 6.
19. Olds, R.H., Sage, B.H., Lacey, W.N., Ind. Eng. Chem., 1942, 34, 1226.
20. Culberson, O.L., McKetta, J.J., Petrol. Trans. AIME, 1951, 192, 223.
21. Culberson, O.L., McKetta, J.J., Petrol. Trans. AIME, 1950, 189, 319.
22. Culberson, O.L., Horn, A.B., McKetta, J.J., Petrol. Trans. AIME, 1950, 189, 1.
23. Reamer, H.H., Olds, R.H., Sage, B.H., Lacey, W.N., Ind. Eng. Chem. 1943, 35, 790.
24. Kobayashi, R., Katz, D.L., Ind. Eng. Chem., 1953, 45, 440.
25. Azarnoosh, A., McKetta, J.J., Petroleum Refiner. 1958, 37, (11), 275.
26. Reamer, H.H., Olds, R.H., Sage, B.H., Lacey, W.N., Ind. Eng. Chem., 1944, 36, 381.
27. Reamer, H.H., Sage, B.H., Lacey, W.N., Ind. Eng. Chem., 1952, 44, 609.
28. LeBreton, J.G., McKetta, J.J., Hydrocarbon Processing & Petrol. Refiner. 1964, 43, (6), 136.
29. Scheffer, F.E.C., Proc. Kon. Akad. Wet. Amsterdam, 1914, 17, 834.
30. Namiot, A. Yu., Beider, S. Ya., Khim. Tekhnol. Topliv. Masel 1960, 5, (7), 52.
31. Scheffer, F.E.C., Proc. Kon. Akad. Wet. Amsterdam, 1913, 16, 404.
32. Rebert, C.J., Hayworth, K.E., A.I.Ch.E.J., 1967, 13, 118.
33. Sultanov, R.G., Skripka, V.G., Zh. Fiz. Khim., 1972, 46, (8), 2170.
34. Kudchadker, A.P., McKetta, J.J., Hydrocarbon Processing & Petrol. Refiner. 1961, 40, (9), 231.
35. Rebert, C.J., Kay, W.B., A.I.Ch.E.J., 1959, 5, 285.
36. Connolly, J.F., J. Chem. Eng. Data, 1966, 11, 13.
37. Kalafati, D.D., Piir, A.E., Zh. Fiz. Khim., 1972, 46, (2), 325.
38. Skripka, V.G., Hubkina, G.F., Boksha, O.A., Zh. Fiz. Khim., 1974, 48, (3), 781.
39. Roof, J.G., J. Chem. Eng. Data, 1970, 15, 301.

RECEIVED January 31, 1980.

21

Phase Equilibrium Calculations by Equation of State for Aqueous Systems with Low Mutual Solubility

M. BAUMGAERTNER, R. A. S. MOORWOOD, and H. WENZEL

Lehrstuhl fuer Technische Chemie, Universitaet Erlangen-Nuernberg, Egerlandstrasse 3, D-8520 Erlangen, Fed. Rep. of Germany

At present there are two fundamentally different approaches available for calculating phase equilibria, one utilising activity coefficients and the other an equation of state. In the case of vapour-liquid equilibrium (VLE), the first method is an extension of Raoult's Law. For binary systems it requires typically three Antoine parameters for each component and two parameters for the activity coefficients to describe the pure-component vapour pressure and the phase equilibrium. Further parameters are needed to represent the temperature dependence of the activity coefficients, therebly allowing the heat of mixing to be calculated. The equation-of-state method, on the other hand, uses typically three parameters p_c, T_c and ω for each pure component and one binary interaction parameter k_{12}, which can often be taken as constant over a relatively wide temperature range. It represents the pure-component vapour pressure curve over a wider temperature range, includes the critical data p_c and T_c, and besides predicting the phase equilibrium also describes volume, enthalpy and entropy, thus enabling the heat of mixing, Joule-Thompson effect, adiabatic compressibility in the two-phase region etc. to be calculated.

In view of the obvious advantages of using an equation of state, it is perhaps surprising that the activity-coefficient approach is still in widespread use and the object of current research. This is because equations of state connot be generally applied to systems in which hydrogen bonding occurs. Such systems represent a large proportion of those of industrial interest and also include most systems exhibiting liquid-liquid equilibrium (LLE).

There are, however, some published examples of equations of state being applied to associating substances. Heidemann (1) used the Redlich-Kwong equation as modified by Wilson (2) to calculate aqueous hydrocarbon systems. Similar calculations were done by Peng and Robinson (3) using their own equation of state. In both

0-8412-0569-8/80/47-133-415$05.00/0

cases the representation of the organic phase was good but the aqueous phase was incorrect by some orders of magnitude. Evelein, Moore and Heidemann (4) in 1976 used Soave's modification of the RK equation to calculate the systems H_2O-H_2S and H_2O-CO_2 over a wide temperature and pressure range; they were able to represent VLE,LLE and gas-gas equilibrium but had to use two adjustable binary parameters.

In 1977 De Santis et al. (5) as well as Heidemann et al. (6) calculated the gas-phase fugacities in the systems H_2O-air and H_2O-N_2-CO_2 by equation of state; in these calculations the liquid phase was not included. One of the authors (7) showed in 1978 that aqueous systems with some inert gases and alkanes as well as H_2S and CO_2 could be represented by an equation of state if the molecular weight of water was artificially increased. An extension of this method applied to alcohols was found to be only partially successful. Gmehling et al. (8) treated polar fluids such as alcohols, ketones and water as monomer-dimer mixtures using Donohue's equation of state (9); various systems including water-methanol and water-ethanol were succussfully represented.

The presence of hydrogen bonding in water is shown by its anomalous thermophysical properties and has been confirmed by more recent methods of investigating its structure such as NMR, neutron scattering, dielectric relaxation, ultrasonic absorption, IR and Raman spectroscopy etc.

Many models for the structure of water have been proposed but little concensus exists as to their validity. For example, Vinogradov (14) reports that in a comparison of twenty published models it was found that the predicted concentrations of broken hydrogen bonds at 0 ^{o}C varied between 2.5 % and 70 %. The models mainly fall into two categories: association models which assume that water is a mixture of specified polymer types (e.g. Eucken's monomer-diver-tetramer-octamer model (11)) and continuum models which assume that water is a continuous but mobile network of molecules linked by hydrogen bonds, in which it is meaningless to consider individual water polymers (e.g. the models of Bernal (12) and Pople (13)). The various experimental methods of investigating water give contradictory results: Vinogradov concluded that I.R. spectroscopy tends to support the continuum model while Raman spectroscopy indicates the presence of individual polymers; NMR measurements suggest that water polymers of high molecular weight do not occur. Thus the structure of water remains uncertain. For association models, however, most authors assume low concentrations of monomers in liquid water at 0 ^{o}C. There are no firm indications as to the nature and concentrations of the higher polymers present in the liquid phase; virial-coefficient measurements suggest, however, that the vapour phase is predominantly monomeric.

Scope

The object of this work was to extend the field of application of the equation-of-state method. The method was applied to aqueous systems in conjunction with a model that treats water as a mixture of a limited number of polymers, an approach similar to that previously adopted for the carboxylic acids (2). Association is calculated by the law of mass action; corrections for non-ideal behaviour are made by means the equation of state. A major problem of the method is the large number of parameters needed to describe the properties and concentrations of the polymers together with their interaction with molecules of other substances. The Mecke-Kemptner model (15) (also known as the Kretschmer-Wiebe model (16) and experimental values for hydrogen-bond energies were used for guidance in fixing these parameters.

Method of Calculation

The equation of state. A recently published equation of state (17) was used:

$$p = \frac{RT}{V - b} - \frac{a}{V^2 + (1 + 3\omega)\ bV - 3\omega b^2} \qquad (1)$$

where ω is Pitzer's acentric factor and a and b at the critical point are functions of the critical properties P_c and T_c. The temperature dependence of the parameter a is similar to that given by Soave's expression (18), and b is constant. Thus a pure component is characterised by the three quantities P_c, T_c and ω.

The following mixing rules were used:

$$\bar{a} = \sum_i \sum_j y_i y_j a_{ij} \text{ with } a_{ij} = (1 - k_{ij}) \sqrt{a_{ii} a_{jj}} \quad i\ j$$

$$\bar{b} = \sum_i y_i b_i \qquad (2)$$

$$\bar{\omega} = \sum_i y_i \omega_i$$

where y_i are the vapour-phase mole fractions. Analogous expressions for the liquid phase were used with x_i substituted for y_i.

k_{ij} is an interaction parameter to be fitted to the equilibrium compositions of a binary system.

Eqn. (1) has the advantage over other commonly used equations of the van der Waal's type of giving an improved representation of the vapour pressure and molar liquid volume.

Also, like all cubic equations of state, Eqn. (1) requires relatively little computing time for calculating thermophysical properties.

Physical equilibrium. When two phases are in equilibrium, the chemical potentials μ_i for each component i must be equal in each phase. The K-factors $K_i = y_i/x_i$ can be obtained from this condition, using the relationship:

$$\frac{\partial \mu_i}{\partial p} = V_i \tag{3}$$

where V_i is the partial molar volume of component i, which can be calculated by the equation of state. Several papers (e.g.19,20) show for similar equations of state how the expression for K_i can be derived.

Chemical equilibrium. The polymer concentrations in the ideal gas phase of water are related to the association equilibrium constants according to the expression:

$$k^o_{pi} = \frac{y_i}{y_1^{n_i}\, p^{n_i-1}} \tag{4}$$

where p is the pressure and y_i the mole fraction of component i, with y_1 referring to the monomer. Component i is defined as being formed by the association of n_i monomers. The temperature dependence of the equilibrium constant k^o_{pi} is given by

$$\ln k^o_{pi} = \frac{\Delta H_{ass}}{RT} - \frac{\Delta S_{ass}}{R} \tag{5}$$

where the association enthalpy ΔH_{ass} and entropy ΔS_{ass} are regarded as independent of temperature. Cross-dimers may also occur between the water molecules and those of the non-aqueous component for which the expression corresponding to Eqn. (4) is

$$k^o_{p_{cross}} = \frac{y_{cross}}{y_1\, y_{inert}\, p} \tag{6}$$

The compositions in the gas phase follow from the contraint

$$\sum_i y_i + y_{inert} + y_{cross} = 1 \tag{7}$$

where only one inert component and one cross-dimer are considered. If the equilibrium constants and y_{inert} are specified, all the concentrations in Eqn. (7) may be expressed in terms of y_1 using Eqns. (4) and (6). Eqn. (7) is then solved for y_1 using Newton's method. Once y_1 is found, all other compositions follow from Eqn. (4) and Eqn. (6). Concentrations under non-ideal conditions such as in the liquid phase or in a compressed gas phase are calculated using expressions of the type $\int V_i dp$ which, as is described elsewhere (20), can be evaluated by means of the equation of state.

Combined physical and chemical equilibrium. Vapour-liquid equilibria were determined in this work by performing dew-point calculations. The procedure is:

(1) Choose an arbitrary value of temperature T and inert concentration
(2) Estimate a dew-point pressure P
(3) Calculate the association equilibrium in the vapour phase at T and P, as previously outlined.
(4) From the conditions of physical equilibrium, calculate the K-factors $K_i = y_i/x_i$
(5) Calculate the liquid concentrations $x_i = \frac{y_i}{K_i}$ and form the sum $S = \sum_i x_i$. Modify P using a Newton routine and repeat steps (3) to (5) until S = 1.

Because of the physical equilibrium, the association in the liquid phase is determinded by that in the vapour phase. Therefore no additional association constants are required for the liquid phase. In the case of liquid-liquid equilibrium calculations, an analogous procedure was adopted using convergence test (5), with y_i referring to the second liquid phase.

Determination of the Parameters of Pure Water

The models tested in more detail in this work were the 1-2-3, 1-2-3-4, 1-2-4-8 and 1-2-4-8-12 models, where the numbers refer to the types of polymers proposed. The 1-2-4-8-12 model gave the best results, the parameters of which are shown in Table 1. The interaction parameters k_{ij} of Eqn.(2) between the different water polymers were all set to zero.

The parameters were optimised to represent the critical data p_c and T_c, the vapour pressure from room temperature up to the critical point and the saturated vapour and liquid molar volumes. Account was also taken of the representation of the binary systems H_2O-CH_4 to C_4H_{10}. Initially the method of Powell (22) was used but later the 'Simplex Method' of Nelder and Mead (23) was found to be more reliable for this problem.

A comparison of calculated and experimental values of vapour pressure and saturated liquid and vapour volumes at various temperatures is shown in Table 2. The water composition in the liquid phase and the corresponding degree of association, $DA = x_1 + 2x_2 + 4x_3 + 8x_4 + 12x_5$, are given at various temperatures in Table 3. Except in the critical region, the vapour composition is virtually completely monomer. For example, at room temperature, the monomer and dimer concentrations are $y_1 = 0.9988$ and $y_2 = 0.0012$ respectively. At 200 ^{o}C the values are $y_1 = 0.9814$

Table 1

Parameters for the Assumed Species in Water

Species of water	T_c / K	P_c / bar	ω	ΔH_{ass} / kcal/mol	ΔS_{ass} / e.u.
Monomer	618.5	271.4	0.1617		
Dimer	743.1	146.3	0.2715	5.86	25.9
Tetramer	939.8	85.4	0.4188	19.33	71.9
Octamer	1216.9	51.5	0.6104	48.83	176.5
Dodecamer	1492.3	43.7	0.7536	74.86	281.4

Table 2

Comparison of Calculated and Experimental Data of Vapour Pressure p_s and Saturated Liquid and Gas Volumes, V_L and V_G

Experimental Data by Canjar (37)

T / K	$\Delta P/P$ x 100	P / bar	$V_{L_{exp}}$ / ml	$V_{L_{calc}}$ / ml	$V_{G_{exp}}$ / l	$V_{G_{calc}}$ / l
273.2	4.3	0.0061	18.017	17.72	3718.	3720.
294.3	1.2	0.0250	18.06	17.88	976.1	976.0
322.0	-0.56	0.1167	18.22	18.05	228.6	228.7
344.3	-0.92	0.3269	18.43	18.24	86.93	87.06
366.5	-0.69	0.7947	18.70	18.45	37.83	37.97
373.2	-0.54	1.013	18.80	18.52	30.14	30.28
399.8	0.25	2.443	19.22	18.85	13.23	13.34
455.4	2.1	10.55	20.37	19.78	3.326	3.402
499.8	3.0	26.31	21.66	20.88	1.369	1.427
555.4	3.1	66.36	24.18	23.07	0.5229	0.5698
599.8	2.6	123.2	27.78	26.06	0.2475	0.2916
633.2	1.7	186.7	34.30	30.06	0.1254	0.1744

Table 3

Calculated Compositions and Degree of Association of Liquid Water at some Temperatures

degree of ass	T / K	monomer	dimer	tetramer	octamer	dodecamer
		mole fraction				
5.3	273.2	0.362	0.031	0.127	0.349	0.132
4.1	373.2	0.395	0.057	0.245	0.254	0.050
2.7	477.6	0.508	0.088	0.303	0.094	$0.65 \cdot 10^{-2}$
1.9	577.6	0.663	0.106	0.216	0.014	$0.20 \cdot 10^{-3}$
1.6	622.0	0.736	0.105	0.155	$0.45 \cdot 10^{-2}$	$0.24 \cdot 10^{-4}$
1.2	647=T_c	0.866	0.082	0.052	$0.25 \cdot 10^{-3}$	$0.11 \cdot 10^{-6}$

and y_1 = 0.9814 and y_2 = 0.0179. All other concentrations are less than 10^{-3}

Mixtures of Water and Inert Components

The critical data and ω values used for inert components were those given by Ambrose (24). The interaction parameters between the water and the inert component were found by performing a dew-point calculation as described above but with the interaction parameter k_{ij} rather than P_s taken as the iteration variable. As the water model included five components (indexed 1 to 5), there were five values of $k_{i,6} = k_{6,i}$, i = 1,5, where index 6 refers the inert. For simplicity the following assuptions were made:

$$k_{2,6} = \frac{1}{2} k_{1,6} \;;\; k_{3,6} = k_{4,6} = \frac{1}{8} k_{1,6} \;;\; k_{5,6} = \frac{1}{12} k_{1,6} \quad (8)$$

so that $k_{1,6}$ was the only independent parameter to be fitted. The systems calculated and the corresponding values of $k_{1,6}$ are shown in Table 4, which also gives references to the tables and figures in which the experimental and computed results are compared. The concentrations given are analytical concentrations, i.e. the concentrations that substances would appear to have, if they are considered to exist only as monomers (Figures 1-6).

Certain figures for VLE are not given; in most cases the results were similar to those given at other temperatures. In the case of the systems H_2O-CH_4 and H_2O-C_2H_6 at T = 344.4 K and T = 377.6 K agreement was less good: for example, in the H_2O - CH_4 system at T = 344.3 K and P = 100 bar the calculated analytical concentration is y_{H_2O} = 0.0025 compared with an experimental value of 0.0044. The corresponding values for the H_2O-C_2H_6 system are 0.0012 and 0.0043 respectively. In the H_2O-C_3H_8 and H_2O-C_4H_{10} systems, however, agreement is excellent in all cases including the liquid-liquid equilibria at lower temperatures. With butane, deviations occur at 477 K, but the rather unusual behaviour of the liquid solubility line is well represented.

Various liquid-liquid equilibria were calculated and the results are given in Table 5. The calculations were performed with $k_{1,6}$ set at 0.5 (method 1) and alternatively with $k_{1,6}$ fitted to correctly predict the water solubility in the organic phase at various temperatures (method 2). The table shows that in some cases divergences between experimental and calculated results are considerable. However, as McAuliffe (34) has demonstrated, divergences between experimental results of different workers are also large. Diagrams for the aqueous systems with gases H_2, CO and O_2 are not given as data for the solubility of water in these gases were not available to us. Also, the

Table 4

Interaction Parameters $k_{1,6}$ at Various Temperatures

Systems	Type	$\frac{T}{K}$	$k_{1,6}$	Fig.	/ Table
H_2O-CH_4	VLE	344.3	1.13	-	-
		377.6	1.01	-	-
		444.3	0.57	1	-
H_2O-C_2H_6	VLE	344.3	0.93	-	-
		377.6	0.89	-	-
		444.3	0.67	-	-
H_2O-C_3H_8	VLE, LLE	344.3	0.70	2	-
	VLE	377.6	0.67	-	-
	VLE	427.6	0.55	3	-
H_2O-C_4H_{10}	LLE	293.2	0.50	-	5
	LLE	310.9	0.50	-	5
	VLE, LLE	344.3	0.50	4	5
	VLE, LLE	377.6	0.50	5	5
	VLE, LLE	410.9	0.48	-	5
	VLE	477.6	0.37	6	-
H_2O-C_5H_{12}	LLE	293.2	0.50	-	5
H_2O-C_6H_{14}	LLE	293.2	0.44	-	5
		338.7	0.44	-	5
		394.3	0.23	-	5
H_2O-C_6H_{12}	LLE	293.2	0.54	-	5
H_2O-C_6H_6	LLE	283.2	0.39	-	5
		293.2	0.41	-	5
		303.2	0.42	-	5
		313.2	0.43	-	5
		323.2	0.43	-	5
		333.2	0.43	-	5
H_2O-C_7H_{16}	LLE	293.2	0.42	-	5
H_2O-C_8H_{18}	LLE	293.2	0.40	-	5
H_2O-H_2	VLE	313.2	-0.014	-	-
		373.2	-0.026	-	-
		423.2	-0.027	-	-
H_2O-CO	VLE	323.2	0.099	-	-
		373.2	-0.002	-	-
		423.2	-0.011	-	-
H_2O-O_2	VLE	373.2	1.57	-	-
		398.2	1.40	-	-
		436.0	1.08	-	-
H_2O-N_2	VLE	323.2	0.52	7	-
		373.2	0.47	-	-
		513.2	-0.030	-	-
H_2O-air	VLE	323.2	-	8	-

Table 5

Experimental and Calculated Analytical Compositions for some Liquid-Liquid Equilibria. X, y Mole Fraction in Aqueous and Organic Phase Respectively

Systems	$\frac{T}{K}$	$\frac{P}{bar}$	y_{exp}	X_{exp}	method 1: with $k_{1,6}$ = 0,5 y_{calc}	X_{calc}	method 2: with $k_{1,6}$ fitted to organic phase X_{calc}	$k_{1,6}$
$H_2O.C_4H_{10}$	293.2	1.013	$2.00.10^{-4}$ [35]	$1.91.10^{-5}$ [34]	$3.05.10^{-4}$	$3.13.10^{-5}$	$2.16.10^{-5}$	0.57
	310.9	4.137	$5.00.10^{-4}$ [26]	$6.20.10^{-5}$ [26]	$7.35.10^{-4}$	$4.73.10^{-5}$	$3.28.10^{-5}$	0.57
	344.3	13.79	$2.10.10^{-3}$ "	$8.80.10^{-5}$ "	$3.03.10^{-3}$	$8.79.10^{-5}$	$5.57.10^{-5}$	0.60
	377.6	20.68	$8.50.10^{-3}$ "	$1.40.10^{-4}$ "	$1.01.10^{-2}$	$1.40.10^{-4}$	$1.05.10^{-4}$	0.56
	410.9	34.47	$2.68.10^{-2}$ "	$2.20.10^{-4}$ "	$3.15.10^{-2}$	$1.99.10^{-4}$	$1.47.10^{-4}$	0.57
$H_2O-C_5H_{12}$	293.2	1.013	$3.44.10^{-4}$ [35]	$9.63.10^{-6}$ [34]	$3.33.10^{-4}$	$3.01.10^{-6}$	$3.12.10^{-6}$	0.50
$H_2O-C_6H_{14}$	293.2	1.013	$5.30.10^{-4}$ "	$1.99.10^{-6}$ "	$3.66.10^{-4}$	$2.87.10^{-7}$	$4.50.10^{-7}$	0.44
	338.7	1.013	$3.80.10^{-3}$ [38]	-	$2.81.10^{-4}$	$1.75.10^{-6}$	$2.70.10^{-6}$	0.44
	394.3	1.013	$4.50.10^{-2}$ "	-	$1.81.10^{-2}$	$7.84.10^{-4}$	$4.17.10^{-4}$	0.23
$H_2O-C_6H_{12}$	293.2	1.013	$4.20.10^{-4}$ [35]	$1.18.10^{-5}$ [34]	$5.30.10^{-4}$	$1.77.10^{-5}$	$1.39.10^{-5}$	0.54
$H_2O-C_6H_6$	283.2	1.013	$1.95.10^{-3}$ [39]	$3.77.10^{-4}$ [39]	$8.93.10^{-4}$	$1.64.10^{-4}$	$3.42.10^{-4}$	0.39
	293.2	1.013	$2.48.10^{-3}$ "	$4.04.10^{-4}$ "	$1.30.10^{-3}$	$2.02.10^{-4}$	$3.78.10^{-4}$	0.41
	303.2	1.013	$3.22.10^{-3}$ "	$4.39.10^{-4}$ "	$1.85.10^{-3}$	$2.45.10^{-4}$	$4.23.10^{-4}$	0.42
	313.2	1.013	$4.11.10^{-3}$ "	$4.76.10^{-4}$ "	$2.58.10^{-3}$	$2.91.10^{-4}$	$4.65.10^{-4}$	0.43
	323.2	1.013	$5.48.10^{-3}$ "	$5.20.10^{-4}$ "	$3.55.10^{-3}$	$3.40.10^{-4}$	$5.33.10^{-4}$	0.43
	333.2	1.013	$7.06.10^{-3}$ "	$5.78.10^{-4}$ "	$4.81.10^{-3}$	$3.93.10^{-4}$	$5.91.10^{-4}$	0.43
$H_2O-C_7H_{16}$	293.2	1.013	$7.01.10^{-4}$ [35]	$5.27.10^{-7}$ [34]	$4.14.10^{-4}$	$2.63.10^{-7}$	$5.51.10^{-8}$	0.42
$H_2O-C_8H_{18}$	293.2	1.013	$9.00.10^{-4}$ "	$1.04.10^{-7}$ "	$4.79.10^{-4}$	$2.52.10^{-8}$	$6.86.10^{-9}$	0.40
$H_2O-(C_2H_5)_2O$	273.2	1.013	0.0429 [39]	0.0311 [39]	0.0459	0.0190		
	283.2	1.013	0.0514 "	0.0235 "	0.0525	0.0171		
	293.2	1.013	0.0570 "	0.0176 "	0.0592	0.0158		
	303.2	1.013	0.0690 "	0.0135 "	0.0658	0.0152		

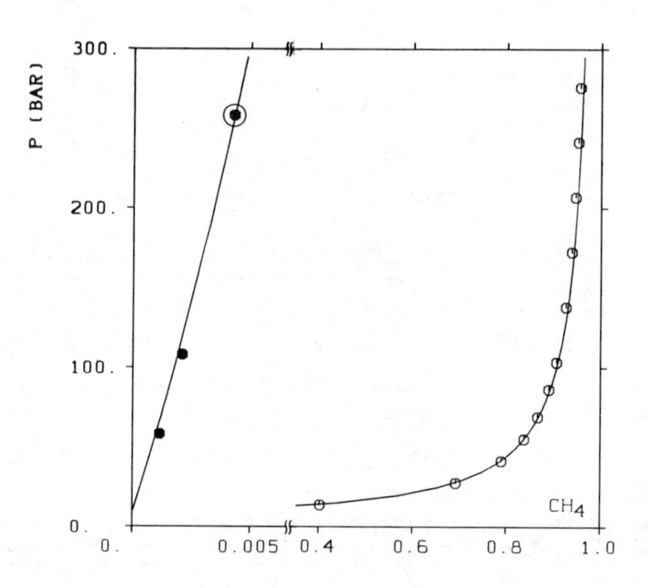

Figure 1. VLE in the system H_2O–methane at 444.3°K, $k_{1,6} = 0.57$; experimental data: (●) (25); (○) (26); (○) fitting point; (——) calculated

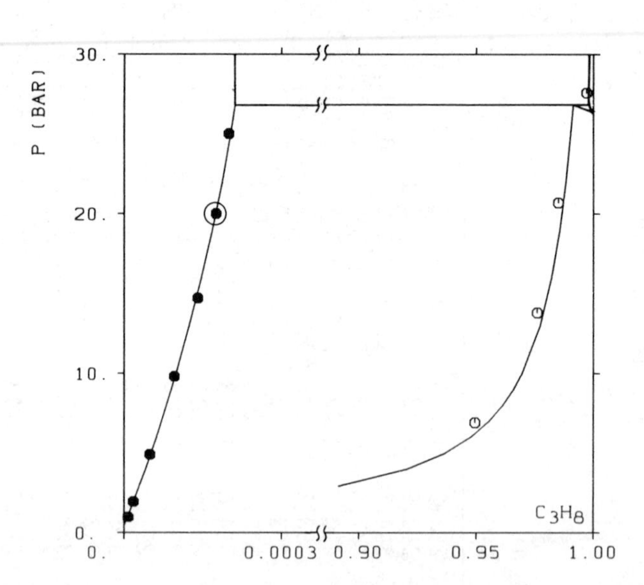

Figure 2. VLE and LLE in the system H_2O–propane at 344.3°K, $k_{1,6} = 0.70$; experimental data: (●) (25); (○) (27); (○) fitting point; (——) calculated

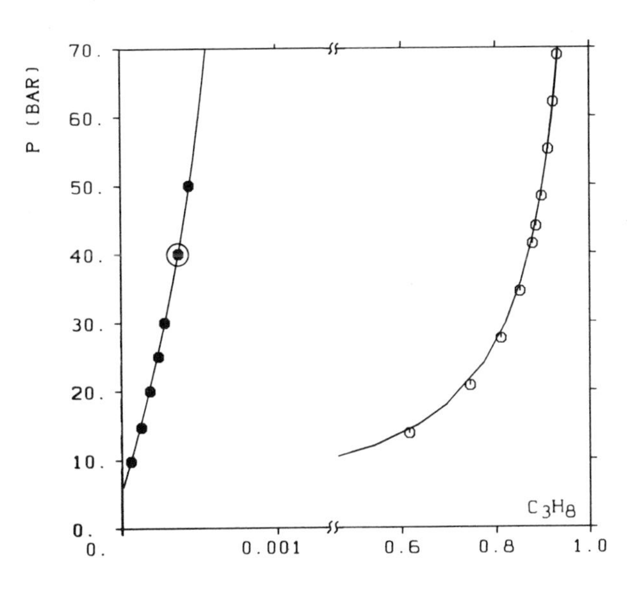

Figure 3. VLE in the system H_2O–propane at 427.6°K, $k_{1,6}$ = 0.55; experimental data: (●) (25); (○) (27); (◎) fitting point; (——) calculated

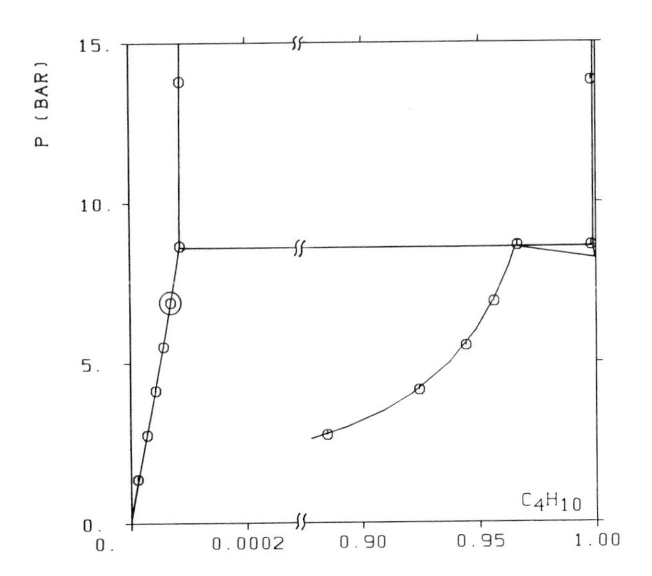

Figure 4. VLE and LLE in the system H_2O–n-butane at 344.3°K, $k_{1,6}$ = 0.50; experimental data from Ref. 26; (◎) fitting point; (——) calculated

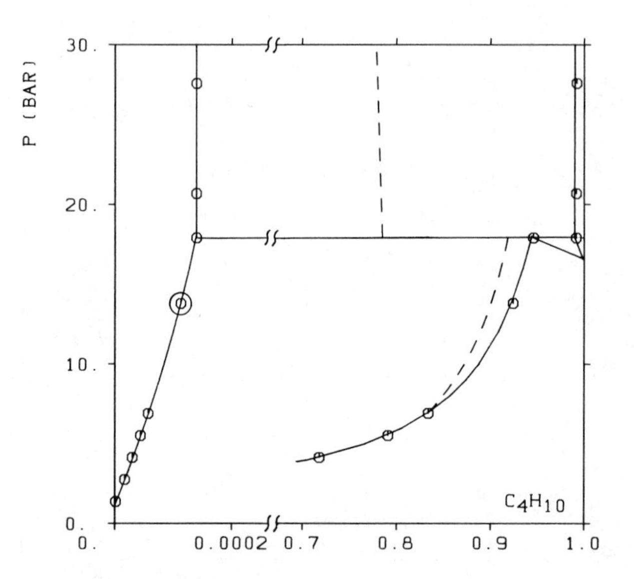

Figure 5. VLE and LLE in the system H_2O–n-butane at 377.6°K, $k_{1,6} = 0.50$; experimental data from Ref. 26; (○) fitting point; (——) calculated; (– – –) calculated with water taken as monomer

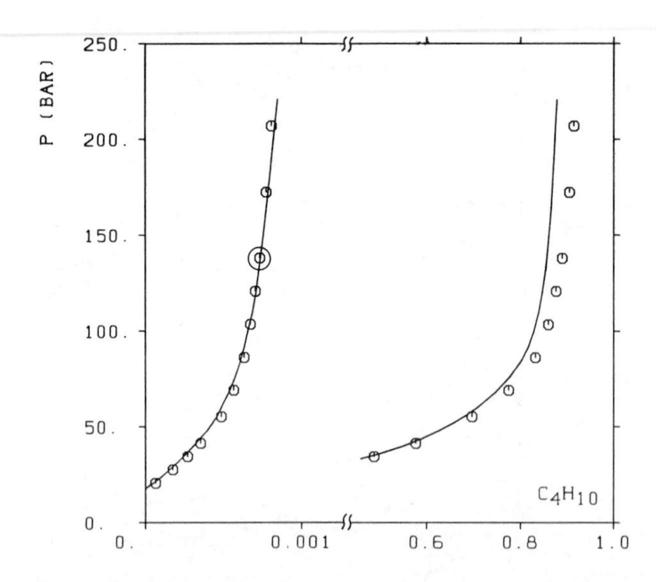

Figure 6. VLE in the system H_2O–n-butane at 477.6°K, $k_{1,6} = 0.37$; experimental data from Ref. 26; (○) fitting point; (——) calculated

solubility of these gases in water follows Henry's law closely, behaviour which the equation of state has no difficulty in representing. On the other hand, data do exist for the solubilities of water in compressed nitrogen. Calculated values agree well with experimental values as shown in Fig. 7 at 323 K. Agreement is also good at T = 373 K but less so at 513 K, a result previously found when a different method of calculation was used (7). Fig. 8 shows the solubility of H_2O in compressed air at T = 323.2 K. The agreement is good considering that k_{ij}-values used were those for the systems H_2O - N_2 and H_2O - O_2 without further fitting.

Mixtures with active Compounds

The substances H_2S, CO_2 and $(C_2H_5)_2O$ (diethyl ether) were treated as slightly active as there are indications that with these compounds cross-association occurs in liquid water. In the case of diethyl ether this conclusion is supported by the considerably higher mutual solubility for the H_2O-$(C_2H_5)_2O$ system compared with water-alkane systems as shown in Table 5. To keep the number of parameters low, the following procedure was adopted in accounting for cross-association:

(1) Only cross-dimers (indexed 7) between monomers and those of the active component were considered
(2) P_c and ω of the cross-dimer were estimated using approximate rules found by analogy with the hydrocarbons.
(3) T_c and $k^o_{p,cross}$ were fitted by trial and error to the binary systems considered
(4) The interaction parameters $k_{i,7}$, i = 1,5 between the cross-dimer and all the water polymers were taken as zero and those between water and the active component, $k_{1,6}$, i = 1,5, were estimated by analogy with the interaction parameters found for previously investigated systems.

The parameters used for the active and cross components are shown in Table 6. ΔH_{ass} and ΔS_{ass} were found by determining $k^o_{p,cross}$ at two temperatures. The resulting mole concentrations of cross-dimers have maximum values of around 0.1 at the lower temperatures considered. The H_2O - H_2S system at 444.3 K is shown in Fig. 9; a similar result was obtained at T = 411 K. The curvature of the equilibrium lines at higher pressures is not correctly represented but the results are greatly improved compared with models that desregard association. A liquid-liquid equilibrium is found at T = 311 K in agreement with experiment, the three-phase pressure being P = 25.3 bar compared with 26.9 bar experimentally.

Figs.10 and 11 show results for the H_2O - CO_2 system. Comparable results were obtained at T = 373 K, 423 K and 473 K. At 298 K a liquid-liquid equilibrium was found at p = 64.0 bar, for which no experimental data are available. Heidemann's (4) calcu-

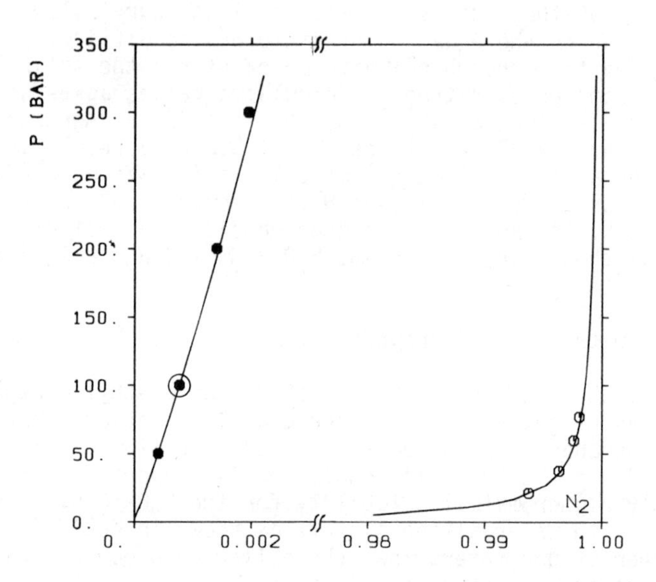

Figure 7. VLE in the system H_2O–nitrogen at 323°K, $k_{1,6} = 0.52$; experimental data: (●) (25); (○) (28); (○) fitting point; (——) calculated

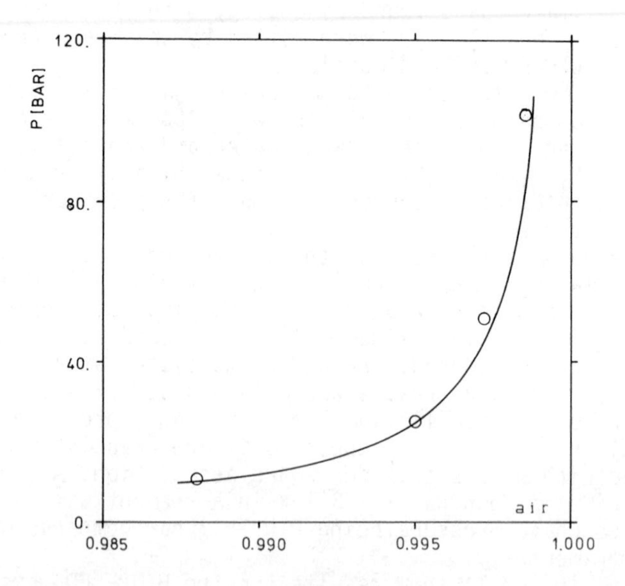

Figure 8. Solubility of water in air at 323.2°K; experimental data from Ref. 36; (——) calculated

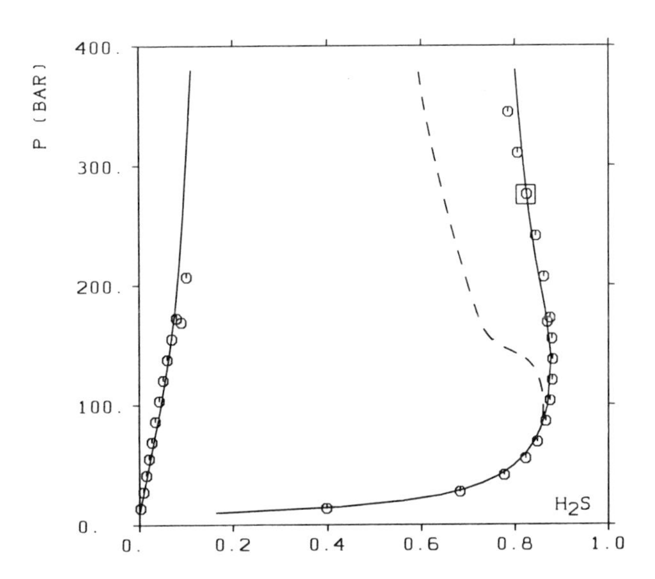

Figure 9. VLE in the system H_2O–H_2S at 444.3°K, $k_{1,6}$ = 0.50; experimental data from Ref. 26; (——) calculated with cross dimers; (– – –) calculated with water taken as a monomer and without cross dimers; (□) fitting point

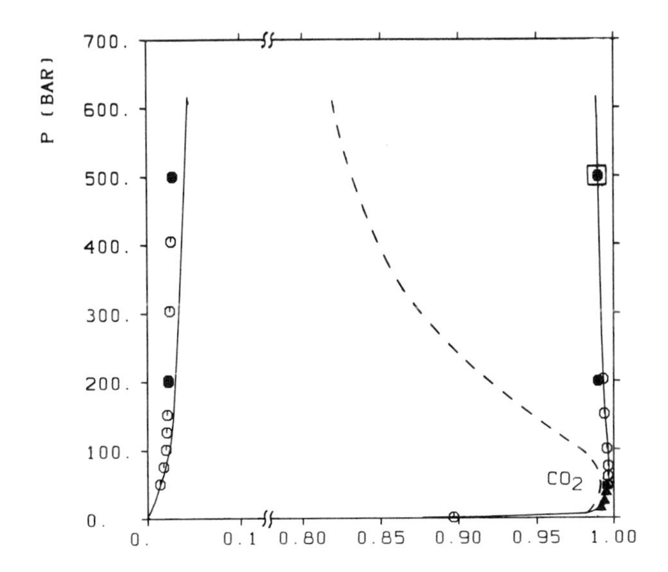

Figure 10. VLE in the system H_2O–CO_2 at 323.2°K, $k_{1,6}$ = 0.50; experimental data: (●) (30); (○) (32); (▲) (33); (——) calculated with cross dimer; (– – –) calculated without cross dimer and water taken as a monomer; (□) fitting point

Table 6

Parameters of Assumed Cross-Dimers for Various Systems

Systems	T_c^+ / K	P_c^+ / bar	ω^+	ΔH_{ass} / kcal/mol	ΔS_{ass} / e.u.
H_2O - CO_2	486.0	74.77	0.25	1.17	14.6
H_2O - H_2S	540.	82.26	0.18	3.90	19.6
H_2O - $(C_2H_5)O$	708.7	44.75	0.338	6.14	25.9

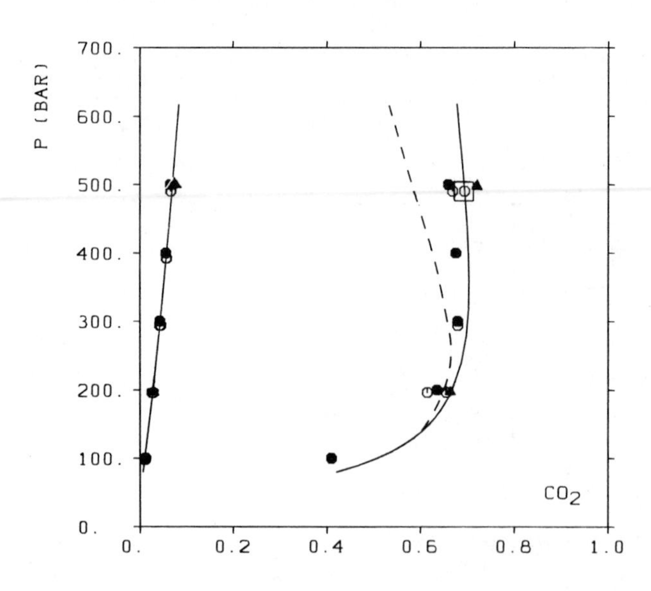

Figure 11. VLE in the system H_2O–CO_2 at 523.2°K, $k_{1,6}$ = 0.50; experimental data: (●) (29); (○) (31); (▲) (30); (□) fitting point; (——) calculated with cross dimers; (– – –) calculated without cross dimers and water taken as a monomer

lations also gave a similar equilibrium at this temperature. At 573 K larger deviations were found, the calculated critical pressure of the system being too high.

Table 5 includes the result of calculations on the H_2O-$(C_2H_5)_2O$ system. The solubility of water in the organic phase as well as its temperature dependence is well represented. This system exhibits the unusual property of having opposite signs for the temperature depence of the solubilities in the two phases.

Conclusions

It has been shown that the selected water model in conjunction with the equation of state provides a uniform method to calculate VLE and LLE of aqueous systems over a wide temperature and pressure range. The remaining discrepancies could possibly be eliminated with modified rules for the interaction parameters. The system H_2O - CO_2, H_2O - H_2S and H_2O - $(C_2H_5)_2O$ can be described by assuming cross-association. The particular temperature dependence of the solubility for diethyl ether was reproduced by the calculation without making it the object of a fitting process. This suggests that the method might be able to describe systems with both an upper and lower critical solution point.

Further work is needed to be able to predict the properties of the cross-dimers and thus reduce the number of adjustable parameters in cases where cross-association must be considered. More complex associated species may have to be postulated to improve the results.

The proposed model for water would appear to be a reasonable approximation in the light of some of the present experimental evidence. Despite the large number of parameters involved, it was found that the degree of association in liquid water could only be selected within a limited range. If it was less than 4 at 20 $^{\circ}C$, serious deviations appeared in the phase-equilibrium calculations, especially for LLE; if it was greater than 6, it was found to be difficult to reproduce the vapour pressure curve of water. Some results (H_2O - C_410_{10} at 477 K and H_2O - CH_4 at 344 K and 377 K) suggest that the degree of association should be somewhat smaller at lower temperatures and greater at higher temperatures than in the suggested model. So far this could not be achieved without worsening the representation of the vapour pressure curve for water. There also appears to be limited freedom in deciding at what point the proposed sequence of polymers should terminate. Apart from the 1-2-4-8 model, which came closest to the proposed model in the accuracy of its predictions, all the other investigated models gave inferior results for all water-alkane systems tested. Therefore, it appears that, the method is able to provide a new way of obtaining information on the extent of association in water.

Symbols

p	pressure
T	absolute temperature
V	molar volume
R	molar ideal gas constant
p_c	critical pressure
T_c	critical temperature
a, b	parameters in the equation of state that characterise each component
ω	Pitzer factor
k_{ij}	binary interaction parameters
x_i	mole fraction (concentration of component i in the liquid phase (VLE) or the aqueous phase (LLE)
y_i	mole fraction (concentration) of component in the vapour phase (VLE) or the non-aqueous phase (LLE)
y_{inert}	mole fraction (concentration) of the non-aqueous component
y_{cross}	mole fraction (concentration) of the cross-dimer
k_i	K-factor of component i
V_i	partial molar volume of component i
k^o_{pi}	association factor of component i under ideal-gas conditions
$k^o_{p\ cross}$	association factor of the cross-dimer under ideal-gas conditions
n_i	number of water monomers that must associate to form a polymer of component i
ΔH_{ass}	standard enthalpy of formation of a component
ΔS_{ass}	standard entropy of formation of a component

Acknowledgement

This work was supported by the "Bundesministerium fuer Forschung and Technologie" and the "Deutsche Forschungsgemeinschaft" of the Federal Republic of Germany.

Abstract

Using a recent equation of state of the van der Waals type developed to describe non-polar components, a model is presented which considers water as a mixture of monomers and a limited number of polymers formed by association. The parameters of the model are determined so as to describe the pure-component properties (vapour pressure, saturated volumes of both phases) of water and the phase equilibria (vapour-liquid and/or liquid-liquid) for binary systems with water including selected hydrocarbons and inorganic gases. The results obtained are satisfactory for a considerable variety of different types of system over a wide range of pressure and temperature.

Literature Cited

1. Heidemann, R. A., AIChE Journal, 1974, 20, 847
2. Wilson, G. M., Adv. Cryog. Eng., 1964, 9, 168
3. Peng, D. Y. and Robinson, D. B., Can. J. Chem. Eng. 1976, 54, 595
4. Evelein, K. A., Moore, R. G and Heidemann, R. A., Ind. Eng. Chem. Proc. Des. Dev., 1976, 15, 423
5. De Santis, R. and Marrelli, L., Can. J. Chem. Eng., 1977, 55, 712
6. Heidemann, R. A. and Prausnitz, J. M., Ind. Eng. Chem., 1977, 16, 375
7. Wenzel, H. and Rupp, W., Chem. Eng. Sci., 1978, 33, 683
8. Gmehling, J., Lind, D. D. and Prausnitz, J. M., Chem. Eng. Sci., 1979, 34, 951
9. Donohue, M. D. and Prausnitz, J. M., AIChE Journal, 1978, 24, 849
10. Roentgen, W. K., Ann. Phys. (Wied), 1892, 45, 91
11. Eucken, A., Z. Elektrochem., 1948, 52, 264
12. Bernal, I. D. and Fowler, R. H., J. Chem. Phys., 1933, 1, 516
13. Pople, J. A., Proc. Roy. Soc., 1951, A 205, 1 3
14. Vinogradov, S. N. and Linnell, R. H., Hydrogen Bonding, New York, 1971
15. Mecke, R. and Kemptner, H., Z. phys. Chem., 1940, B 46, 229
16. Kretschmer, C. B. and Wiebe, R., J. Chem. Phys., 1954, 22, 1697
17. Schmidt, G. and Wenzel, H., submitted to Chem. Eng. Sci., Aug. 1979
18. Soave, G., Chem. Eng. Sci., 1972, 27, 1197
19. Wenzel, H. and Peter, S., Chem.-Ing.-Techn., 1971, 43, 856
20. Peter, S. and Wenzel, H., Chem.-Ing.-Techn., 1973, 45, 573
21. Baumgaertner, M., Rupp, W. and Wenzel, H., I. Chem. E. Symposium Ser.,1979, 56, 1.2/31
22. Powell, M. J. D., Comp. J., 1965, 7
23. Nelder, J. A. and Mead, R., Comp. J., 1965, 7, 308

24. Ambrose, D. and Townsend, R., NPL Report, 1978
25. Landolt-Boernstein, Bd. IV/4cl: Gleichgewicht der Absorption von Gasen in Flüssigkeiten, Berlin 1976
26. Sage, B. H. and Lacey, W. N., Some Properties of the Lighter Hydrocarbons, Hydrogen Sulfide and Carbon Dioxide, A.P.I., New York, 1955
27. Kobayashi, R. and Katz, D. L., Ind. Eng. Chem., 1953, 45, 440
28. Rigby, M. and Prausnitz, J. M., J. Phys. Chem., 1968, 72, 330
29. Takenouchi, S. and Kennedy, G, C., Am. J. Sci., 1964, 262, 1055
30. Toedheide, K. and Franck, E. V., Z. Phys. Chem. N. F., 1963, 37, 387
31. Malinin, S. D., Geochemistry, 1959, 3, 292
32. Wiebe, R. and Gaddy, V. L., J. Am. Chem. Soc., 1940, 62, 815; 1941, 63, 475
33. Coan, C. R. and King, A. D. Jr., J. Am. Chem. Soc., 1971, 93, 1857
34. Mc Auliffe, C., J. Phys. Chem., 1966, 70, 1267
35. Black, C., Joris, G. G. and Taylor, H. S., J. Chem. Phys., 1948, 16, 537
36. Hyland, R. W. and Wexler, A., J. Res. Nat. Bur. Std. A. Phys. Chem., 1973, 77A, 115, 133
37. Canjar, L. N. and Manning, F. S., Thermodynamic Properties and reduced Correlations for Gases, Gulf Publ. Comp., Houston, 1967
38. API Technical Data Book, American Petroleum Inst., Div. of Refining, Washington D.C., 1970
39. Stephen, H. and Stephen, S., Solubilities of Inorganic and Organic Compounds, Oxford, 1963

RECEIVED January 31, 1980.

Excess Enthalpies of Some Binary Steam Mixtures

C. J. WORMALD and C. N. COLLING
School of Chemistry, The University, Bristol BS8 1TS, UK

Despite the importance of mixtures containing steam as a component there is a shortage of thermodynamic data for such systems. At low densities the solubility of water in compressed gases has been used (1,2) to obtain cross term second virial coefficients B_{12}. At high densities the phase boundaries of several water + hydrocarbon systems have been determined (3,4). Data which would be of greatest value, pVT measurements, do not exist. Adsorption on the walls of a pVT apparatus causes such large errors that it has been a difficult task to determine the equation of state of pure steam, particularly at low densities. Flow calorimetric measurements, which are free from adsorption errors, offer an alternative route to thermodynamic information. Flow calorimetric measurements of the isothermal enthalpy-pressure coefficient ϕ_p (5) extrapolated to zero pressure yield the quantity $\phi_o = B - TdB/dT$ where B is the second virial coefficient. From values of ϕ_o it is possible to obtain values of B without recourse to pVT measurements.

As with pure steam the properties of binary steam mixtures can be obtained from flow calorimetric measurements of the enthalpy of the mixture. With steam + n-alkane binaries, for which the enthalpies of both components are known, it is more sensible to measure the excess enthalpy directly rather than measure the large total enthalpy of the mixture to determine a small excess quantity. Extrapolation of the excess enthalpy H_p^E at pressure p to zero pressure yields $H_o^E = x_1x_2p(2\phi_{12} - \phi_{11} - \phi_{22})$ and from this quantity B_{12} for a steam + n-alkane interaction can be obtained.

0-8412-0569-8/80/47-133-435$05.00/0

Outline of the Flow Calorimetric Apparatus

A mixing calorimeter suitable for measurements at low densities is described in the literature (6), and this calorimeter has been used to make measurements on steam mixtures at temperatures around 373 K. Although high pressure calorimeters have been described (7,8,9) none of these designs is suitable for work at high temperature. Details of the calorimeter used in this work will be published shortly (10). The flow system is shown in outline in figure 1. The system is pressurized with nitrogen which enters at 1. Pumps 2 and 3 supply liquid components to flash boilers 4 where the liquids are vaporized. The vapours mix in calorimeter 5 which is contained in a pressurized vessel immersed in a fluidized bed thermostat. The fall in temperature produced on mixing is sensed by four platinum resistance thermometers and a heater in the centre of the calorimeter is used to obtain isothermal conditions. The mixture is next passed into a total enthalpy boil-off calorimeter 6 where the enthalpy change of the fluid vaporizes some n-pentane. The n-pentane vapour is condensed at 7 and the rate of boil-off is measured using calibrated bulbs 8. The liquid condensate is collected in vessel 9. The apparatus has been used to make measurements at pressures up to 16 MPa and at temperatures up to 698 K. With any flow mixing calorimeter it is important to test for the presence of heat leaks. This can be done by measuring the enthalpy of mixing at constant composition, temperature and pressure over a wide range of flow rate. Tests on our calorimeter were done on steam + nitrogen at $x = 0.5$. The results of measurements at 698 K and 12.3 MPa are shown in figure 2, and demonstrate that even under these extreme conditions heat leaks are negligible. The results also show the good reproducibility (one percent) of which the apparatus is capable.

Measurements have so far been made on mixtures of steam + hydrogen, nitrogen, argon, methane, carbon-dioxide, n-hexane, n-heptane, benzene and cyclohexane. The measurements cover the range 373 to 698 K at pressures from 0.1 MPa to saturation or 12.5 MPa. The only exception to this is steam + carbon dioxide for which the measurements extend up to 5.5 MPa. The accuracy of the measurements is around ±2 percent.

Results for the Mixture Steam + n-Heptane

Some results for the mixture steam + n-heptane at $x = 0.5$ are shown in figure 3. The results for steam + n-hexane, + cyclohexane, and + benzene are similar. The measurements at 548 and 598 K are above the critical temperature of n-heptane (540 K) and below that of steam (647 K). The measurements at 648 and 698 K are above the critical temperature of both components. All the results which are below the critical temperature of one of the components show a maximum and terminate at the saturation pressure

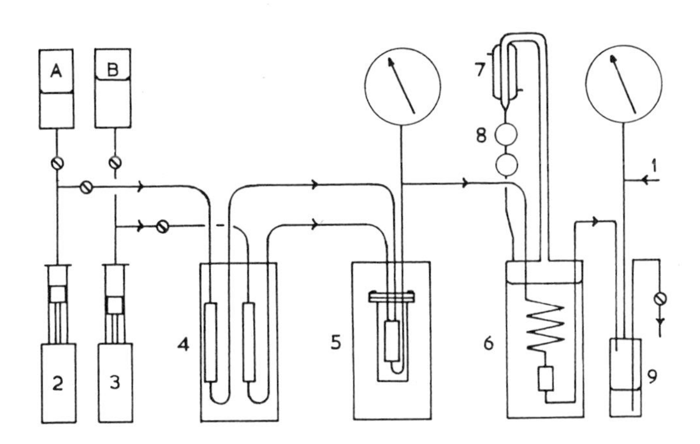

Figure 1. Diagram of the flow calorimetric apparatus

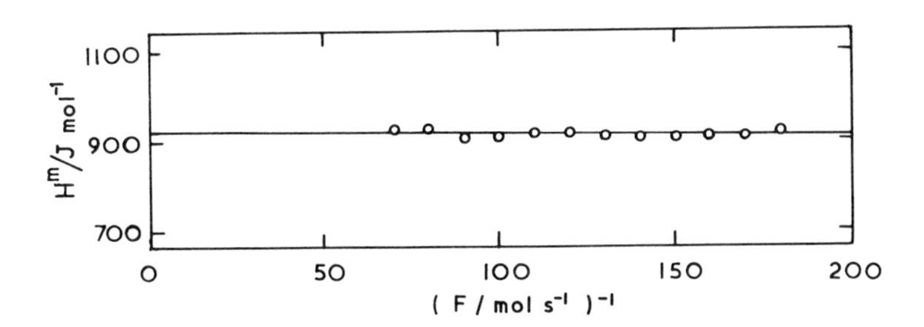

Figure 2. Test of the flow mixing calorimeter on steam + nitrogen at x = *0.5 (measurements were made at 698 K and 12.3 MPa)*

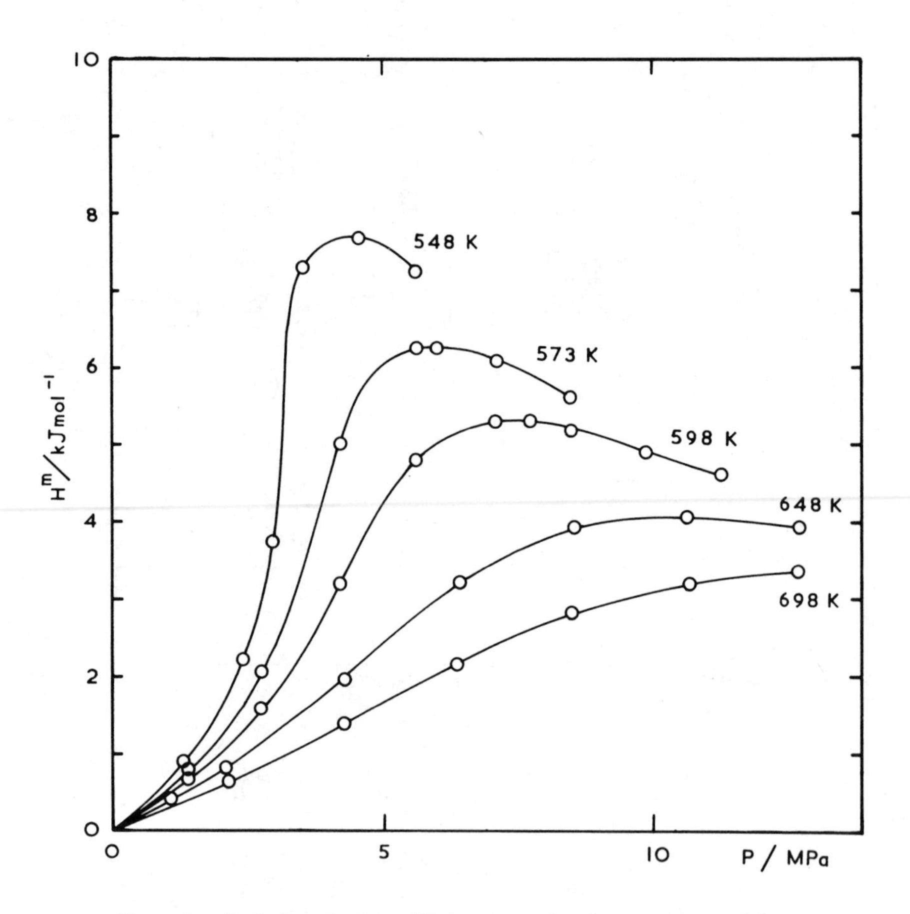

Figure 3. Enthalpy of mixing H^m *for steam* + n-*heptane at* $x = 0.5$

of the subcritical component. Analysis of the results at low pressures can be done using the virial equation of state. Following Lambert (11) we separate the second virial coefficient B into a physical B^o and a chemical contribution

$$B = B^o - RTK \tag{1}$$

where K is the equilibrium constant for dimer formation. The excess enthalpy H^E of a steam (1) and n-heptane (2) mixture can be written (12)

$$H^E = p[\phi_m - x_1(\phi_1^o + K\Delta H) - x_2\phi_2]$$

$$-(p^2/RT)[B_m\phi_m - x_1(B_1^o - RTK)(\phi_1^o + K\Delta H) - x_2B_2\phi_2]$$

where

$$\phi_m = x_1^2(\phi_1^o + K\Delta H) + 2x_1x_2\phi_{12} + x_2^2\phi_2$$

$$B_m = x_1^2(B_1^o - RTK) + 2x_1x_2B_{12} + x_2^2B_2 \tag{2}$$

and ΔH is the enthalpy of dimer formation. B_1^o can be taken as the second virial coefficient of methyl fluoride which has the same dipole moment as steam (1.85D). B_{12} and ϕ_{12} can be calculated from B_1^o and B_2 using a Kihara-Stockmayer potential. With $\Delta H = -16.426$ kJ mol^{-1} and $K = 0.385$ MPa^{-1} (at 298 K) equation (2) fits H^E for steam + n-alkane mixtures at low densities to within experimental error. Figure 4 shows the fit to the results at standard atmospheric pressure. To fit the results at higher pressures requires further virial coefficients, and the method runs into difficulties.

Analysis of the Results at High Pressures

The enthalpy of mixing H^m is given by the equation

$$H^m = H_M^* - x_1H_1^* - x_2H_2^* \tag{3}$$

where H_1^* and H_2^* are the residual enthalpies of components 1 and 2 and H_M^* is the residual enthalpy of the mixture. For fluids which are slightly polar we might expect the Peng-Robinson (P-R) equation of state (13) to give a reasonable estimate of H^m. While we would not expect it to work well for mixtures containing steam it is instructive to see what it gives. The residual enthalpy H^* is given by

$$H^*(V,T) = \frac{RTV}{V-b} - \frac{aV}{V(V+b) + b(V-b)} - RT - a(1+\kappa)^2 \ln\left[\frac{V + (1-2^{\frac{1}{2}})b}{V + (1+2^{\frac{1}{2}})b}\right] \tag{4}$$

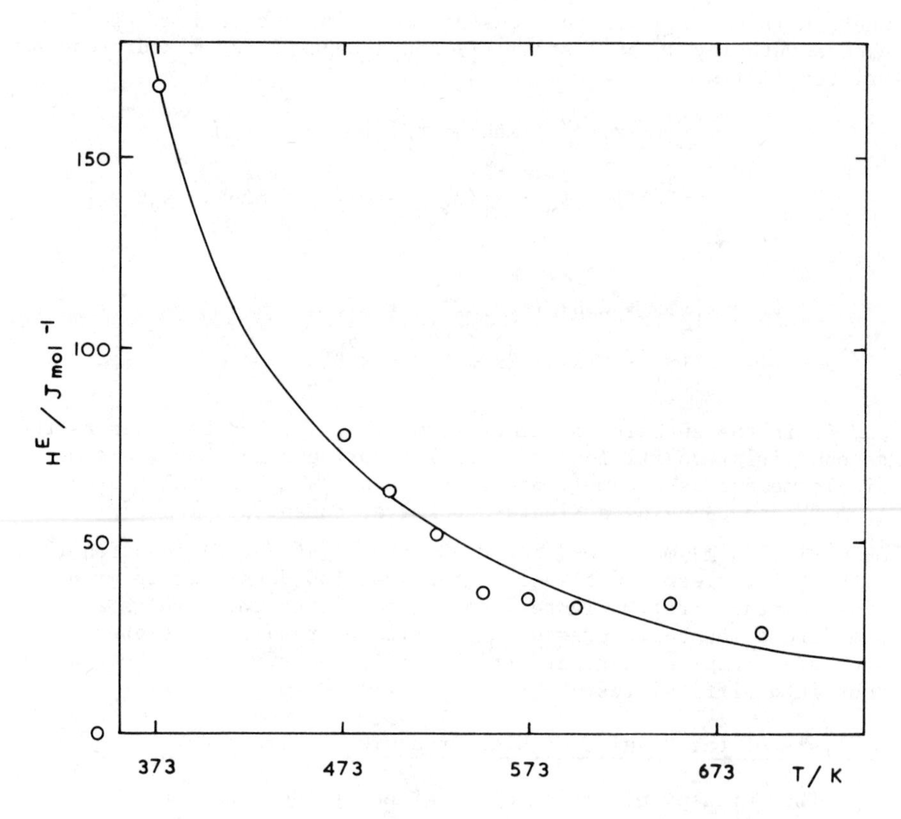

Figure 4. Excess enthalpy H^E *of steam* + n*-heptane at* x = *0.5 and standard atmospheric pressure ((——) calculated using Equation 2)*

To calculate H_M^* we use the mixing rules

$$a_M = \sum_i \sum_j x_i x_j a_{ij} \quad ; \quad b_M = \sum_i x_i b_i$$

$$a_{ij} = k_{ij}(a_i a_j)^{\frac{1}{2}} \quad ; \quad \kappa_M = \sum_i x_i \kappa_i \tag{5}$$

Putting $k_{ij} = 1$, and using criticality conditions to calculate a and b for steam + n-heptane, the enthalpies of mixing are found to be only about half as big as the experimental results and show only the right qualitative behaviour. It is interesting to see if a value of k_{ij} which brings the calculated enthalpies into agreement with experiment can be found. The best k_{ij} turns out to be -0.3, and curves calculated using this value are shown in figure 5. To obtain a fit to the results at 548 K requires $k_{ij} = -0.5$, and to fit the results at 698 K requires $k_{ij} = 0.0$. Now the P-R equation is not a good fit to the residual enthalpies of either steam or n-heptane. At temperatures up to 623 K values of the enthalpy of n-heptane are available. If P-R parameters are chosen to fit the residual enthalpies of steam and n-heptane, it is found that the best value of k_{ij} is -0.2. A temperature dependent value of k_{ij} is still required, although the change of k_{ij} with temperature is less than was required when pure component enthalpies were obtained from criticality conditions. The large negative values of k_{ij} clearly indicate that a different approach to the calculation of mixture properties is needed.

The Separated Associated Fluid Interaction Model for Polar + Nonpolar Mixtures

Woolley (14) has developed the equation of state for an associated fluid in terms of the formation of dimer, trimer, tetramer, etc. characterised by equilibrium constants K_2, K_3, K_4, etc. The molecules in the model have no size, and are simply points between which interactions occur. Lambert (11) developed a similar approach for low density gases, and regarded the observed second virial coefficient as the sum of a "physical term" which he obtained from the Berthelot equation of state and which implies that the molecules have finite size and attractive forces, and a "chemical" term characterised by a dimerisation constant K_2. The inclusion of the physical term ensures that B can be positive at high temperatures without K_2 having to change sign. A disadvantage of the Woolley treatment is that extension to mixtures of associated fluids is impossible, as cross term virial coefficients cannot be calculated. The inclusion of a "physical

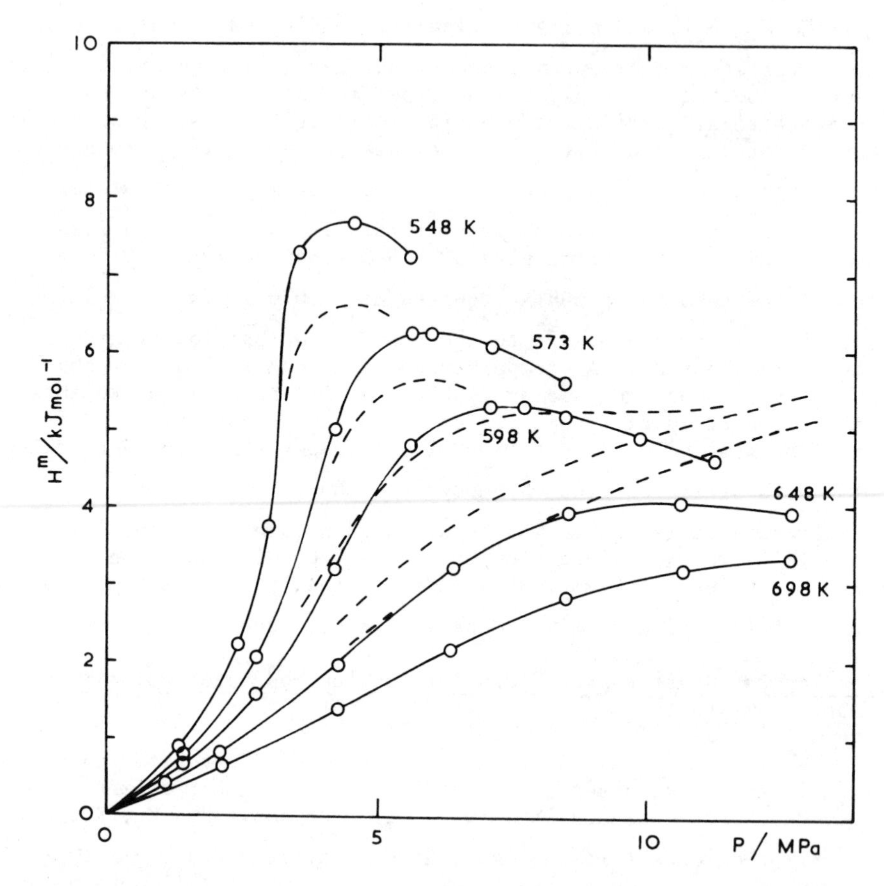

Figure 5. Comparison of the Peng–Robinson equation using $k_{ij} = -0.3$ *with the results for steam +* n*-heptane ((– – –) calculated using Equations 3 and 4)*

term" in the virials overcomes this problem. With steam in mind Vukalovitch (15) developed a model similar to that of Woolley, in which he regarded each cluster as obeying the van der Waals equation of state. As there is no easy way of assigning van der Waals parameters a and b to the clusters, the Vukalovitch equations are almost unusable.

The model presented here develops these ideas and introduces features which make the calculation of mixture properties simple. For a polar fluid with approximately central dispersion forces together with a strong angle dependent electrostatic force we may separate the intermolecular potential into two parts so that the virial coefficients, B, C, D, etc. of the fluid can be written as the sum of two terms. The first terms B^o, C^o, D^o, etc, arise from dispersion forces and may include a contribution arising from the permanent dipole of the molecule. The second terms contain equilibrium constants K_2, K_3, K_4, etc. which describe the formation of dimer, trimer, etc. by hydrogen bonding. The first three virial coefficients for the polar fluid can be written

$$B = B^o - K_2(RT)$$

$$C = C^o - (2K_3 - 4K_2^2)(RT)^2$$

$$D = D^o - (3K_4 - 18K_2K_3 + 20K_2^3)(RT)^3 \qquad (6)$$

We can in principle calculate B^o, C^o, D^o, etc. from a suitable pair potential, and so obtain K_2, K_3 and K_4. For gas mixtures at low densities where B and C terms are sufficient this approach can be used successfully. However, the calculation of D^o and higher coefficients becomes prohibitively difficult, and for mixtures the calculation of cross coefficients is an additional problem.

An alternative approach is possible. Just at the coefficients B, C, D, etc. define the thermodynamic properties of the real fluid so coefficients B^o, C^o, D^o, etc. define thermodynamic properties for a hypothetical fluid which we will call the primary fluid. The primary fluid can be regarded as having the properties which the real fluid might have in the absence of association. It is assumed that when secondary interactions such as hydrogen bonding are imposed on the primary fluid the real fluid will be simulated. This assumption is an acceptable approximation at low densities, but is unlikely to hold at high densities where the addition of hydrogen bonds may produce new structural features. At moderate densities we can make the not unreasonable approximation that any property which is a function f of the virial coefficients can be separated into two contributions

$$\underset{\text{real fluid}}{f(B, C, D \ldots)} = \underset{\text{primary fluid}}{f(B^o, C^o, D^o \ldots)} + \underset{\text{secondary equilibria}}{f(K_2, K_3, K_4 \ldots)} \qquad (7)$$

Our present concern is to find a model for steam which can be used to calculate mixture properties. An estimate of B^o, the second virial coefficient which steam might have in the absence of association, could be obtained by selecting a molecule which has about the same dispersion force as steam, such as argon, taking the Lennard-Jones 12-6 parameters which fit B for argon together with the dipole moment of steam 1.85 D, and using the Stockmayer potential to calculate values of B^o. If we do this it is found that the calculated values of B^o lie close to the second virial coefficient of methyl fluoride which also has a dipole moment of 1.85 D, and which would also be a reasonable model for unassociated steam. The following four steps are now taken.

1. The primary fluid is replaced by the real fluid methyl fluoride. This removes the need to evaluate $f(B^o,C^o,D^o,etc.)$, as the thermodynamic function X required is simply that for methyl fluoride.
2. The contribution of the secondary equilibria $f(K_2,K_3,K_4 \ldots)$ is obtained by subtracting the thermodynamic property of the primary fluid from that of the real fluid.

$$X(\text{association}) = X(\text{real fluid}) - X(\text{primary fluid}) \qquad (8)$$

3. An equation of state which will represent the thermodynamic property X_2 of the non polar component of the mixture is chosen. This same equation of state is used to fit the property of the primary fluid, component 1.
4. Using combining rules appropriate to the equation of state the thermodynamic property X_M of the non polar component + primary fluid mixture is calculated. The change in property X on forming the mixture X^m is given by an equation similar to 3

$$X^m = X_M - x_1X(\text{primary fluid}) - x_1X(\text{association}) - x_2X_2 . \qquad (9)$$

Equation 9 was used to calculate H^m for steam + n-heptane as follows. The Peng-Robinson equation with parameters obtained from criticality conditions was used to calculate the residual enthalpy H_1^* of methyl fluoride. Peng-Robinson parameters for n-heptane were obtained by fitting to the residual enthalpy of the fluid at temperatures below the critical, and by using criticality conditions at higher temperatures. The mixing rules given in equation 4 with k_{ij} = 1 were used to calculate H_M. As hydrogen bonding occurs only for H_2O + H_2O interactions and not for H_2O + n-heptane interactions H^*(ass) makes no contribution to H_M^*. At the temperatures at which mixing experiments had been done, the residual enthalpy of steam calculated from steam tables was fitted to polynomial equations in powers of the pressure, and H^* (association) was obtained by subtracting H_1^*.

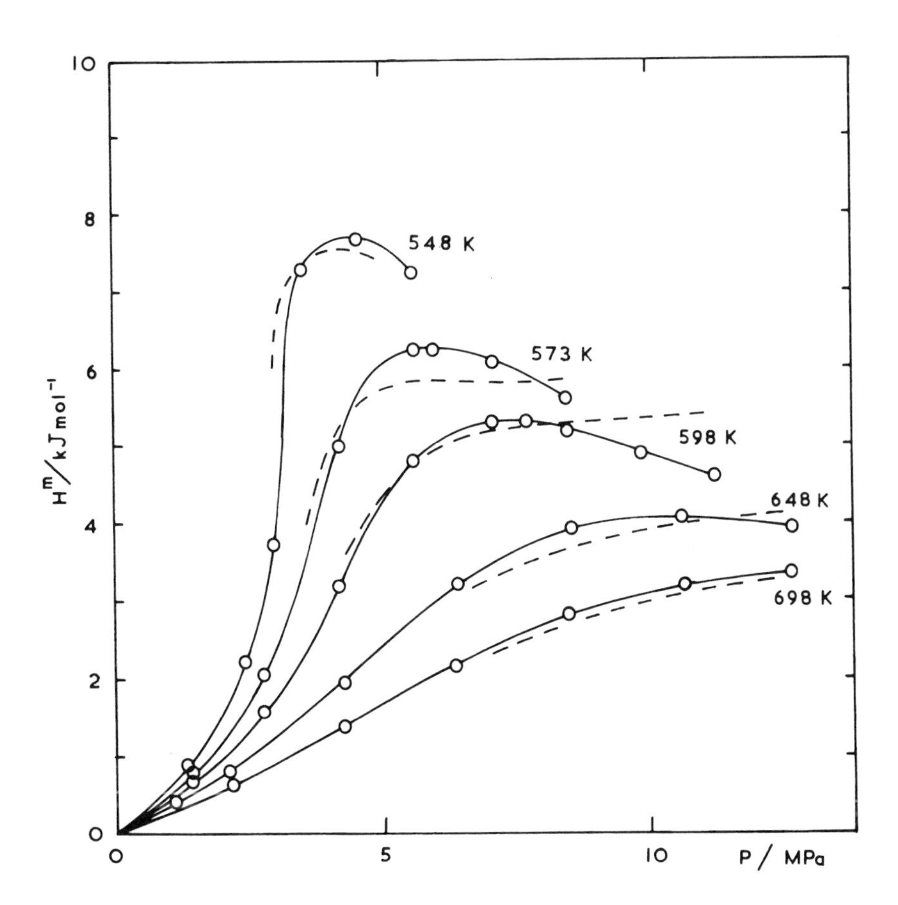

Figure 6. Comparison of the Separated Associated Fluid Interaction Model using no adjustable parameters with the results for steam + n-heptane ((– – –) calculated from the model)

H^m for steam + n-heptane calculated by the above method is shown by the dashed lines in figure 6. Considering the simplicity of the model and the fact that no adjustable parameters have been used, agreement with experiment is remarkable. For mixtures of steam + n-hexane, benzene and cyclohexane agreement with experiment is much the same. At low densities the model reproduces the curvature of the lines through the results better than the virial equation of state. The method fails to fully reproduce the downward turn of the experimental curves at pressures near saturation, but does marginally better in this region than the P-R equation with $k_{ij} = -0.3$. At supercritical temperatures the model seems to work well. For steam + methane, + nitrogen, + argon, the model gives values of H^m which are too large, and does no better than the virial equation of state.

The choice of methyl fluoride as the primary fluid for steam is evidently reasonable but is not necessarily the best. There is no reason why T^c, P^c and ω for the primary fluid should not be treated as adjustable parameters so that a hypothetical primary fluid which gives best agreement with results on all the above steam + hydrocarbon mixtures can be defined. There are clearly many ways in which the model can be modified.

As it is formulated above, the model can be used without further modification for the estimation of the thermodynamic properties of mixtures of steam with hydrocarbons in the C_4 to C_8 range. Below the critical temperature of steam the model can be used at pressures up to 10 MPa with reasonable confidence. At temperatures above the critical point of steam the model can be used at higher pressures.

Acknowledgment

We are grateful to the British Gas Corporation for financial support which made this work possible.

Abstract

Flow calorimetric measurements of the excess enthalpy of a steam + n-heptane mixture over the temperature range 373 to 698 K and at pressures up to 12.3 MPa are reported. The low pressure measurements are analysed in terms of the virial equation of state using an association model. An extension of this approach, the Separated Associated Fluid Interaction Model, fits the measurements at high pressures reasonably well.

Literature Cited

1. Rigby, M; Prausnitz, J.M. J.Phys.Chem. (1968), 72, 330.
2. Coan, C.R.; King, A.D. J.Amer.Chem.Soc. (1971), 93, 1857.

3. Bröllos, K.; Peter, K.; Schneider, G.M. Ber.Bunsengesellschaft Phys.Chem. (1970), 74, 682.
4. Robert, C.J.; Kay, W.B. A.I.Ch.E.J. (1959), 5, 285.
5. Pocock, G.; Wormald, C.J. Faraday Trans.I (1975), 71, 705.
6. Wormald, C.J. J.Chem.Thermodynamics (1977), 9, 901.
7. Wormald, C.J.; Lewis, K.L.; Mosedale, S.E. J.Chem.Thermodynamics (1977), 9, 27.
8. Lee, J.I.; Mather, A.E. J.Chem.Thermodynamics (1970), 2,881.
9. Hejmadi, A.V.; Katz, D.L.; Powers, J.E. J.Chem.Thermodynamics (1971), 3, 483.
10. Wormald, C.J.; Colling, N. J.Chem.Thermodynamics (1980),
11. Lambert, J.D.; Roberts, G.A.H.; Rowlinson, J.S.; Wilkinson, V.J. Proc.Roy.Soc.A. (1949), 196, 113.
12. Richards, P; Wormald, C.J. J.Chem.Thermodynamics (1980).
13. Peng, D.-Y.; Robinson, D.B. Ind.Eng.Chem.Fundamentals (1976), 15, 59.
14. Woolley, H.W. J.Chem.Phys. (1953), 21, 236.
15. Vukalovitch, M.P. "Thermodynamic Properties of Water and Steam", 6th Edn. (1958).

RECEIVED January 31, 1980.

PROPERTIES OF AQUEOUS SOLUTIONS— THEORY, EXPERIMENT, AND PREDICTION

23

Thermodynamics of Aqueous Electrolytes at Various Temperatures, Pressures, and Compositions

KENNETH S. PITZER

Department of Chemistry and Lawrence Berkeley Laboratory,
University of California, Berkeley, CA 94720

Early in this century there was great interest in the apparently anomalous properties of aqueous electrolytes. The anomaly concerned the limiting behavior at low concentration. While several investigators contributed substantially to the resolution of this problem, it was the classic work of Debye and Hückel (1) which provided a simple yet adequate explanation of the effect on thermodynamic properties of the long-range electrostatic forces between ions in solution. The experimental work of that era tended to emphasize dilute solutions at room temperature. While Debye and Hückel recognized the short-range repulsive forces between ions by assuming a hard-core model, the statistical mechanical methods then available did not allow a full treatment of the effects of this hard core. Only the effect on the electrostatic energy was included--not the direct effect of the hard core on thermodynamic properties.

As is often the case, after the intense activity of the 1920's, the investigation of aqueous electrolytes proceeded at a more relaxed pace. But careful and systematic experimental research continued in this area and was summarized by Harned and Owen (2) and by Robinson and Stokes (3) in their excellent monographs. The latter volume contains in the appendix a comprehensive set of tables of the osmotic and activity coefficients of the common inorganic solutes at 25°C and at concentrations up to 6 M in most cases.

Subsequently major theoretical advances were made, principally by Mayer, in creating an adequate statistical mechanical theory in which both long-range electrostatic forces and short-range forces of whatever origin were properly considered. Friedman has contributed greatly to further theoretical advances and will discuss recent work in this symposium. Also important are the Monte Carlo calculations of Card and Valleau (4). The writer has published (5) an elementary review of these theoretical advances; more advanced reviews are available by Friedman (6) and by Andersen (7). The result is that the properties of univalent aqueous electrolytes based on the hard core and other simple

0-8412-0569-8/80/47-133-451$05.00/0

models (8) are now known from reliable theory up to a concentration of 1 or 2 M. While the actual interionic potentials of mean force doubtless differ somewhat from the models treated, the resulting thermodynamic properties will be similar. Thus we have good theoretical guidance in selecting forms of equations for semi-empirical use. For more highly charged ions or a solvent with substantially lower dielectric constant the situation is somewhat more complex, but the theory is reasonably satisfactory and is being further improved.

From the thermodynamic viewpoint, the basic statistical theory is still too complex to provide useful working equations, but it does suggest forms of equations with some purely theoretical terms, and other terms including parameters to be evaluated empirically. In general, the theoretical terms arise from the electrostatic interactions which are simple and well-known while the empirical terms relate to short-range interionic forces whose characteristics are qualitatively but not quantitatively known from independent sources. But, as we shall see, this division is not complete - there are interactions between the two categories.

In recent years there has been a resurgence of experimental research on aqueous inorganic electrolytes emphasizing the broader domain of high temperatures or high pressures or both. Also many organic solutes have been investigated at room temperature. Thus most of the pure aqueous electrolytes likely to be of engineering interest have been investigated at room temperature, a substantial number have been studied over the 0 - 50°C range, and a smaller but increasing number at high pressures and at temperatures to 300°C or occasionally higher.

Most practical systems, however, are mixtures rather than pure electrolytes. The experimental measurement of a wide variety of mixtures over closely spaced grids of composition would be very burdensome. It is here, for mixed electrolytes, that theory, confirmed by a limited number of experiments, is particularly valuable. The electrostatic interactions are all simply defined; also the short-range forces between a pair of ions of different sign are the same in a mixed electrolyte as in the pure electrolyte comprising that pair of ions. Thus a proper definition of terms will allow the evaluation of all of these effects in mixtures from information on the various pure electrolytes. It is for the effect of short-range forces between ions of the same sign that new terms arise for mixtures. But ions of the same sign repel one another and are unlikely to be so close together that their short-range forces have a large effect. Indeed Bronsted (9) postulated that these differences among short-range interactions among ions of the same sign could be ignored and Guggenheim (10) developed detailed equations on that basis. Kim and the writer (11) found that such differences were not completely negligible for mixtures of singly charged ions or for 2-1 charged mixtures but that they were very small. Also these difference terms can be evaluated from existing measurements on simple mixtures of the most important ions. Thereafter, for complex mixtures of practical importance, all of the important terms are known and only very small terms must be neglected from lack of information.

From the preceding paragraphs it is clear that the capacity to calculate the properties of a variety of mixed electrolytes depends on an adequate theoretical structure within which the available experimental data can be organized. Thus the primary emphasis for the remainder of this paper will be the description of this structure of semi-empirical equations. The array of substances for which experimental data are available will be described in general terms but there is not sufficient space to list results in detail. Also a severe test of predictions for mixed electrolytes will be reported.

But before turning to the detailed consideration of electrolytes of moderate concentration, it is interesting to note the properties of a few systems which exist as liquids from pure fused salts to dilute aqueous solutions.

Miscible Electrolytes

There are two systems for which the vapor pressure and thereby the activity of water has been measured over the full range of composition from fused salt to dilute solution in water. In each case the salt is a simple mixture of approximately equal molal proportions. The system $(Ag,T\ell)NO_3$ was measured at 98°C by Trudelle, Abraham, and Sangster (12) while $(Li,K)NO_3$ was measured in the most concentrated range at 119°C by Tripp and Braunstein (13) and over the remainder of the range at 100°C by Braunstein and Braunstein (14). These results are shown on Figure 1 which also includes similar data for several systems of large but limited solubility. The composition variable is the mole fraction on an ionized basis, i.e., $x_1 = n_1/(n_1 + \nu n_2)$ where n_1 and n_2 are moles of water and salt, respectively, and ν is the number of ions in the salt. On this basis Raoult's law applies in the very dilute range, with the Debye-Hückel correction applicable as the concentration increases.

The similarity of the curves on Figure 1 to those for nonelectrolyte solutions is striking. The dashed line representing $a_1 = x_1$ can be called "ideal-solution behavior" for these systems, as it is for nonelectrolytes, but it is realized that a statistical model yielding that result would be more complex for the ionic case. Also the Debye-Hückel effect is a departure from this ideal behavior. Nevertheless, it seems worthwhile to explore the use for these systems of the simple equations for nonelectrolytes. One of the simplest and most successful had its origin in the work of van Laar (15) and has been widely used since. Prausnitz (16) discusses this and related equations as well as the contributions of Margules, Hildebrand, Scatchard, Guggenheim, and others to this topic. For the activity of either component, referenced to the pure liquid, one has

$$\ln a_1 = \ln x_1 + w_1 z_2^2 \tag{1a}$$

$$\ln a_2 = \ln x_2 + w_2 z_1^2 \tag{1b}$$

$$z_1 = n_1/[n_1 + \nu n_2(b_2/b_1)] \tag{1c}$$

$$z_2 = \nu n_2/[n_1(b_1/b_2) + \nu n_2] \tag{1d}$$

$$w_2 = w_1(b_2/b_1). \tag{1e}$$

Note first that if (b_1/b_2) is unity, z_1 and z_2 reduce to the mole fractions x_1 and x_2. Then one has the even simpler equations

$$\ln a_1 = \ln x_1 + w x_2^2 \tag{2a}$$

$$\ln a_2 = \ln x_2 + w x_1^2. \tag{2b}$$

In either equations (1) or (2) the non-ideality parameter w (sometimes written w/RT) arises from the difference between the intermolecular attraction of unlike species as compared to the mean of the intermolecular attraction for pairs of like species. The second parameter in equation (1), (b_1/b_2), is sometimes ascribed to the ratio of the volumes of the molecules or to the ratio of molal volumes in the liquid, although in some systems, especially metallic solutions, equation (1) is still quite satisfactory but (b_1/b_2) departs greatly from the ratio of molal or atomic volumes. For fused salt-water mixtures it seems best to regard (b_1/b_2) as a freely adjustable parameter and subsequently to compare the values with ratios of molal volumes.

Equation (1) was fitted to the two systems remaining liquid over the full range of composition with the results $w_1 = 1.02$, $(b_1/b_2) = 0.50$ for $(Ag,T\ell)NO_3$-H_2O and $w_1 = -0.89$, $(b_1/b_2) = 1.2$ for $(Li,K)NO_3$-H_2O. Water is component 1 and the salt component 2. For the latter system the simpler equation (2) serves almost as well with $w = -0.80$ (this implies $b_1/b_2 = 1.0$). The calculated curves based on equation (1) are compared with the experimental data in Figure 2.

These results shown in Figures 1 and 2 demonstrate the similarity of the effects of short-range forces on the properties of nonelectrolytes and concentrated electrolytes. One finds both positive and negative deviations from ideality and these effects may be ascribed to the difference between the intermolecular potential energy of attraction of unlike species to the mean of the corresponding potentials for pairs of like molecules. Previous discussion of these systems has focused on the hydration of the positive ion as the dominant effect, but we see in Figure 1 that

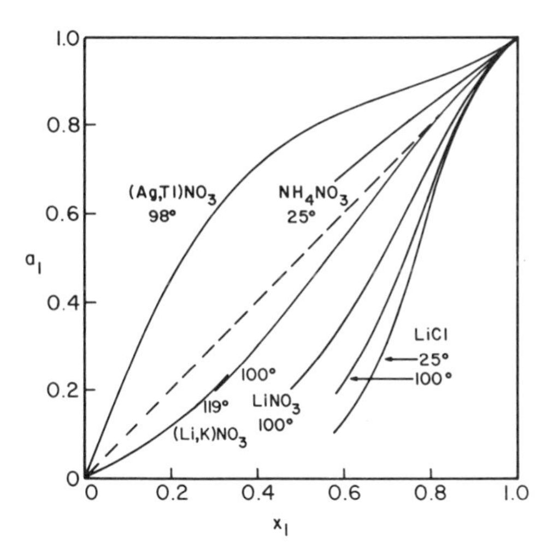

Figure 1. Activity of water for water–salt solutions over a very wide range of composition

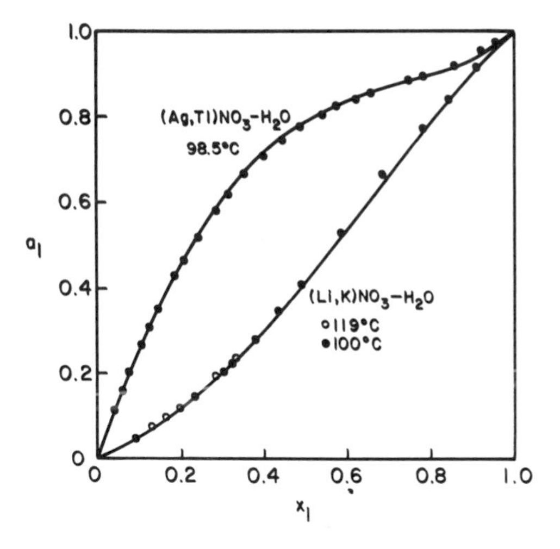

Figure 2. Comparison of the calculated and experimental activity of water for water–salt solutions over the full range of composition

the shift from nitrate to chloride is comparably important. Just as for nonelectrolytes, one must consider all intermolecular forces in electrolytes.

With experimental data for the activity of water, one can, of course, integrate the Gibbs-Duhem equation to obtain the activity of the salt, over the same range in composition, without the use of any model or semi-empirical equation. But equation (1) appears to fit so well that its use is very convenient. As presented, the constant of integration is evaluated for the pure-liquid reference state for each component. Thus equation (1b) gives the activity of the salt in relation to the pure fused salt. Since this form is obtained by integrating the Gibbs-Duhem equation over composition to the fused salt, $x_1 = 0$, $x_2 = 1$, the Debye-Hückel range is avoided and no error from that source is introduced.

If the fused salt does not exist at the temperature of interest, one normally uses the infinitely dilute solute standard state. While these equations can easily be converted to that basis, the results are not immediately useful for two reasons: (1) Debye-Hückel effects are significant in the dilute range and are not considered, and (2) the usual composition scale for the solute standard state is molality rather than mole fraction. Both of these problems have been overcome, and the more complex relationships are being presented elsewhere (17). However, for most purposes, the virial coefficient equations for electrolytes are more convenient and have been widely used. Hence our primary presentation will be in those terms.

Virial Coefficient Equations for Electrolytes

A very effective method of representing the properties of non-ideal gases is by use of a series in increasing powers of density or concentrations. The coefficients, called virial coefficients, are unambiguously related to a particular number of molecules. Thus the first term relates to individual molecules and is the ideal gas law. The second virial coefficient arises from binary intermolecular forces and may be either positive or negative as repulsive or attractive forces predominate. The third virial coefficient arises from triple interactions, etc. The MacMillan-Mayer (18) solution theory established that a formally similar treatment applied to solutes in a solvent provided the intermolecular potentials are replaced by potentials of mean force in that solvent. For electrolytes one must recognize the long-range character of coulombic forces which prevents their inclusion in the virial series. But as Mayer (19) and others have shown, one may combine a Debye-Hückel term for electrostatic effects with a virial series for the effects of short range forces. In this case, however, the virial coefficients depend on the ionic strength as well as the temperature and other properties. These theoretical principles were used by the writer (20) to establish the form of an equation for electrolyte properties in which the virial coefficients are evaluated empirically. Virial type equations were used earlier for electrolytes (10, 21, 22) but without recognition of the ionic strength dependence of the second virial coefficient. The basic equation is postulated for the excess Gibbs energy from which other functions can be obtained from appropriate derivatives.

$$G^{ex}/n_w RT = f(I) + \sum_i \sum_j \lambda_{ij}(I)\, m_i m_j + \sum_i \sum_j \sum_k \mu_{ijk}\, m_i m_j m_k. \qquad (3)$$

Here G^{ex}/n_w is the excess Gibbs energy per kilogram of solvent and m_i, m_j, etc., are the molalities of the various ions or neutral solutes present. The long-range electrostatic forces lead to the Debye-Hückel term f(I) where I is the ionic strength. Short-range interparticle-potential effects are taken into account by the virial coefficients λ_{ij} for binary interactions, μ_{ijk} for ternary, etc. As noted above, electrostatic effects lead to an ionic strength dependence on λ for ionic interactions. For μ this is neglected; also μ is omitted if all ions are of the same sign. While fourth virial coefficients could be added, they do not appear to be needed for most applications. Indeed the third virial coefficients are so small that they can often be omitted at moderate concentration (I up to about 2).

The derivative equations for osmotic and activity coefficients, which are presented below, were applied to the experimental data for wide variety of pure aqueous electrolytes at 25°C by Pitzer and Mayorga (23) and to mixtures by Pitzer and Kim (11). Later work (24-28) considered special groups of solutes and cases where an association equilibrium was present (H_3PO_4 and H_2SO_4). While there was no attempt in these papers to include all solutes for which experimental data exist, nearly 300 pure electrolytes and 70 mixed systems were considered and the resulting parameters reported. This represents the most extensive survey of aqueous electrolyte thermodynamics, although it was not as thorough in some respects as the earlier evaluation of Robinson and Stokes (3). In some cases where data from several sources are of comparable accuracy, a new critical evaluation was made, but in other cases the tables of Robinson and Stokes were accepted.

In addition to the activity and osmotic coefficients at room temperature, the first temperature derivatives and the related enthalpy of dilution data were considered for over 100 electrolytes (26, 29). The data for electrolytes at higher temperatures become progressively more sparse. Quite a few solutes have been measured up to about 50°C (and down to 0°C). Also, over this range, the equations using just first temperature derivatives have some validity for rough estimates in other cases. But the effects of the second derivative (or the heat capacity) on activity coefficients at higher temperatures is very substantial.

Sodium chloride has been studied much more thoroughly at high temperature than any other electrolyte. The osmotic coefficient measurements of Liu and Lindsay (30) and various types of measurements of Federov and associates (31) are particularly noteworthy. A preliminary effort to represent all of the data on sodium chloride by virial coefficient equations has been published (32) and a revision is being completed (33).

Over the wide range of temperature to 300°C and concentration to 10 M, sodium chloride shows only very moderate and slow changes in its properties; the principal change is the increase in the Debye-Hückel parameter (34) which lowers both activity and osmotic coefficients. The osmotic coefficient for NaCl at high temperature is shown on Figure 3; the curves were calculated by the virial coefficient equation (33). The behavior of a typical 2-1 electrolyte, $MgCl_2$, is shown on Figure 4, which gives both the experimental values of Holmes, Baes, and Mesmer (35) and their curves from a virial coefficient equation. Only a few additional salts have been studied extensively at high temperatures, although solubility information yields less a complete picture for several others (36). However, the general patterns of behavior are simple enough that one can make estimates under some circumstances based on detailed data at room temperature for the solute of interest and high temperature data for other solutes of the same valence type.

The importance of the virial-coefficient equations is especially great for mixed electrolytes. Of the needed virial coefficients for a complex mixture such as sea water, most are determined by the pure electrolyte measurements and all the others of any significance are determined from data on simple mixtures such as NaCl-KCl, NaCl-$MgCl_2$, NaCl-Na_2SO_4, etc., which have been measured. The effect of the terms obtained from mixtures is very small in any case and these terms can be ignored for all but the most abundant species.

A very severe test of these virial-coefficient equations for the sea-water-related Na-K-Mg-Ca-Cl-SO_4-H_2O system has been made by Harvie and Weare (37) who calculated the solubility relationships for most of the solids which can arise from this complex system. There are 13 invariant points with four solids present in the system Na-K-Mg-Cl-SO_4-H_2O and the predicted solution compositions in all 13 cases agree with the experimental values of Braitsch (38) substantially within the estimated error of measurement. In particular, Harvie and Weare found that fourth virial coefficients were not required even in the most concentrated solutions. They did make a few small adjustments in third virial coefficients which had not previously been measured accurately, but otherwise they used the previously published parameters.

There are also many less severe tests (11) of predictions for mixed electrolytes which illustrate the accuracy to be expected in various cases. Thus it is well-established that the virial coefficient equations for electrolytes yield reliable predictions of

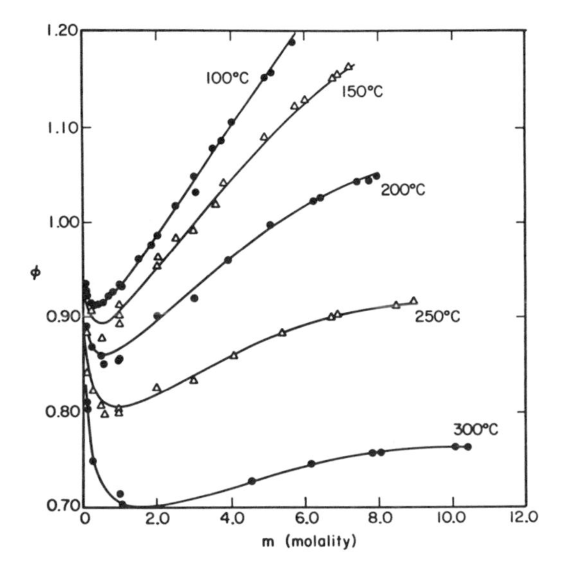

Figure 3. Osmotic coefficient for sodium chloride solutions at various temperatures

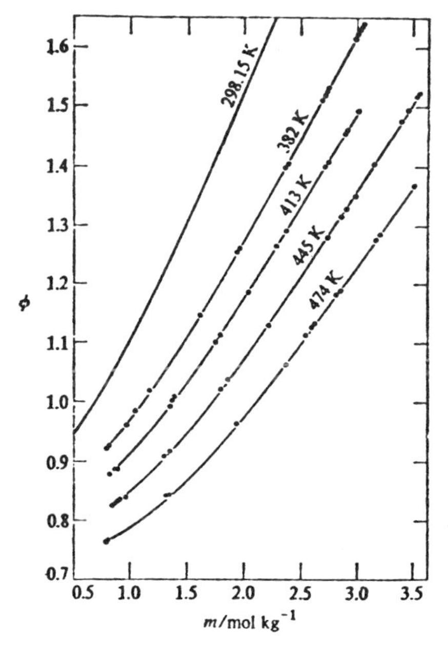

Journal of Chemical Thermodynamics

Figure 4. Osmotic coefficient for magnesium chloride solution at various temperatures (35)

mixed-electrolyte properties provided the coefficients measurable for the pure electrolyte components are known. The predictions are more accurate if the additional coefficients measurable from simple mixtures are also known but their effect is usually very small.

The working equations for osmotic and activity coefficients, derived from equation (3) are given as equations (4) and (5), respectively. The various secondary relationships are defined in several additional equations stated and briefly described thereafter. Additional details and derivations of equations for the entropy, the heat capacity, and other related functions can be found in various published papers (11, 20, 23-29, 32-34).

$$(\phi-1) = \frac{2}{(\sum_i m_i)}\left\{-\frac{A_\phi I^{3/2}}{I+bI^{1/2}} + \sum_c\sum_a m_c m_a (B^\phi_{ca} + ZC_{ca}) + \frac{1}{2}\sum_c\sum_{c'} m_c m_{c'}(\Theta^\phi_{cc'} + \sum_a m_a \psi_{cc'a}) + \frac{1}{2}\sum_a\sum_{a'} m_a m_{a'}(\Theta^\phi_{aa'} + \sum_c m_c \psi_{aa'c})\right\}. \quad (4)$$

Although one wishes activity coefficients for neutral combinations of ions, it is convenient to use equations for single-ion activity coefficients which can then be combined appropriately.

$$\ln \gamma_M = z_M^2 F + \sum_a m_a(2B_{Ma} + ZC_{Ma}) + \sum_c m_c(2\Theta_{Mc} + \sum_a m_a \psi_{Mca}) + \frac{1}{2}\sum_a\sum_{a'} m_a m_{a'} \psi_{aa'M} + |z_M| \sum_c\sum_a m_c m_a C_{ca} \quad (5a)$$

$$\ln \gamma_X = z_X^2 F + \sum_c m_c(2B_{cX} + ZC_{cX}) + \sum_a m_a(2\Theta_{Xa} + \sum_c m_c \psi_{Xac}) + \frac{1}{2}\sum_c\sum_{c'} m_c m_{c'} \psi_{cc'X} + |z_X| \sum_c\sum_a m_c m_a C_{ca}. \quad (5b)$$

Here m_c is the molality of cation c with charge z_c and correspondingly for anion a. Sums over c or a cover all cations or anions, respectively. B's and Θ's are measurable combinations of λ's whereas C's and ψ's are combinations of the μ's in Equation (3). Note that the Θ's and ψ's are zero and these terms disappear for pure electrolytes.

The electrostatic function f must contain the Debye-Hückel limiting law with the parameter

$$A_\phi = (1/3)(2\pi N_o d_w/1000)^{1/2}(e^2/DkT)^{3/2}$$

but it proves empirically advantageous to take an extended form. Among alternatives, the form found best was

$$f(I) = -4A_\phi I b^{-1} \ln(1-bI^{1/2}) \qquad (6)$$

with b = 1.2 chosen for all electrolytes in water. At 25°C the value of A_ϕ is 0.391. The appropriate derivative gives the term in Equation (4) for ϕ. For the activity coefficient it is convenient to define

$$F = -A_\phi[I^{1/2}/(1+bI^{1/2}) + (2/b)\ln(1+bI^{1/2})] + \sum_b \sum_a m_c m_a B'_{ca}$$

$$+ \frac{1}{2}\sum_c \sum_{c'} m_c m_{c'} \Theta'_{cc'} + \frac{1}{2}\sum_a \sum_{a'} m_a m_{a'} \Theta'_{aa'} \qquad (7)$$

which includes both the Debye-Hückel term with A_ϕ and certain derivatives of the second virial terms.

The second virial coefficients, B_{MX}, are functions of ionic strength. Again an empirical choice was made among theoretically plausible forms for B_{MX} and the following was chosen with $\beta^{(0)}$ and $\beta^{(1)}$ parameters fitted to the data for each solute.

$$B^\phi_{MX} = \beta^{(0)}_{MX} + \beta^{(1)}_{MX}\, e^{-\alpha I^{1/2}} \qquad (8a)$$

$$B_{MX} = \beta^{(0)}_{MX} + \beta^{(1)}_{MX}\, g(\alpha I^{1/2}) \qquad (8b)$$

$$B'_{MX} = \beta^{(1)}_{MX}\, g'(\alpha I^{1/2})/I, \qquad (8c)$$

where the functions g and g' are given by

$$g(x) = 2[1 - (1+x)e^{-x}]/x^2 \qquad (9a)$$

$$g'(x) = -2[1 - (1+x+\tfrac{1}{2}x^2)e^{-x}]/x^2 \qquad (9b)$$

with $x = \alpha I^{1/2}$. In Equations (8) the form given is for 1-1 and 1-2 electrolytes for which the value $\alpha = 2$ was chosen emperically. For higher valence types, such as 2-2 electrolytes, where there is a tendency toward ion-pairing, an additional term is added and Equations (8) become

$$B_{MX} = \beta_{MX}^{(0)} + \beta_{MX}^{(1)} e^{-\alpha_1 I^{1/2}} + \beta_{MX}^{(2)} e^{-\alpha_2 I^{1/2}} \tag{10a}$$

$$B_{MX} = \beta_{MX}^{(0)} + \beta_{MX}^{(1)} g(\alpha_1 I^{1/2}) + \beta_{MX}^{(2)} g(\alpha_2 I^{1/2}) \tag{10b}$$

$$B'_{MX} = \beta_{MX}^{(1)} g'(\alpha_1 I^{1/2})/I + \beta_{MX}^{(2)} g'(\alpha_2 I^{1/2})/I. \tag{10c}$$

In this case the values of $\alpha_1 = 1.4$ and $\alpha_2 = 12.0$ are assigned.

The parameters, C_{MX}, are related to the tabulated parameters of Pitzer and Mayorga (23), C^{ϕ}_{MX}, as follows,

$$C_{MX} = C^{\phi}_{MX}/2|z_M z_X|^{1/2} \tag{11}$$

also the quantity Z is defined to be

$$Z = \sum_i m_i|z_i| = 2\sum_c m_c z_c. \tag{12}$$

The mixed electrolyte terms in Θ and ψ account for differences among interactions between ions of like sign. The defining equations for the second virial coefficients, Θ_{ij}, are given by Equations (13),

$$\Theta^{\phi}_{ij} = \theta_{ij} + {}^E\theta_{ij}(I) + I\,{}^E\theta'_{ij}(I) \tag{13a}$$

$$\Theta_{ij} = \theta_{ij} + {}^E\theta_{ij}(I) \tag{13b}$$

$$\Theta'_{ij} = {}^E\theta'_{ij}(I) \tag{13c}$$

θ_{ij}, a single parameter for each pair of anions or each pair of cations, is the only adjustable parameter in Equations (13).

The terms ${}^E\theta_{ij}(I)$ and ${}^E\theta'_{ij}(I)$ account for the electrostatic effects of unsymmetrical mixing. Equations for calculating these terms were derived by Pitzer (39); this effect was discovered by Friedman (6). The important features of ${}^E\theta_{ij}(I)$ and ${}^E\theta'_{ij}(I)$ are that they depend only on the charges of the ions i and j and the total ionic strength. They do not constitute additional parameterization. ${}^E\theta_{ij}(I)$ and ${}^E\theta'_{ij}(I)$ are zero when the ions i and j are of the same charge. Although these terms are important for 1-3 mixtures, such as HCl-$AlCl_3$, they did not appear to be really needed for simple 1-2 mixtures. However, Harvie and Weare (37) have found these special electrostatic terms to be important for the $CaSO_4$-$NaCl$ system and some more complex mixtures involving singly and doubly charged ions.

Table I

Parameters for virial coefficient equations at 25°C

M	X	$\beta^{(0)}_{MX}$	$\beta^{(1)}_{MX}$	$\beta^{(2)}_{MX}$	C^{ϕ}_{MX}
Na	Cl	.07650	.2264	--	.00127
Na	SO_4	.01958	1.1130	--	.00497
K	Cl	.04835	.2122	--	-.00084
K	SO_4	.04995	.7793	--	0
Mg	Cl	.35235	1.6815	--	.00519
Mg	SO_4	.22100	3.3430	-37.25	.025
Ca	Cl	.31590	1.6140	--	-.00034
Ca	SO_4	.20000	2.650	-57.70	0

Table II

Parameters for mixed electrolytes with the virial coefficient equations (at 25°C)

i	j	k	θ_{ij}	ψ_{ijk}
Na	K	Cl	-.012	-.0018
		SO_4		-.010
Na	Mg	Cl	.07	-.012
		SO_4		-.015
Na	Ca	Cl	.07	-.014
		SO_4		-.023
K	Mg	Cl	.0	-.022
		SO_4		-.048
K	Ca	Cl	.032	-.025
		SO_4		0
Mg	Ca	Cl	.007	-.012
		SO_4		.05
Cl	SO_4	Na	.02	.0014
		K		0
		Mg		-.004
		Ca		0

It would burden this paper excessively to list the parameters for all known electrolytes, even at room temperature. As examples giving the pattern of magnitudes as well as values for widely appearing salts, the values for the Na-K-Mg-Ca-Cl-SO_4-H_2O system are listed in Tables I and II. Most of these are taken from Pitzer and Mayorga (23) or Pitzer and Kim (11) but a few were revised in later work including that of Harvie and Weare (37). It is apparent that the pure-electrolyte parameters in Table I are much larger than those for mixing of ions of the same sign in Table II. Also the second virial coefficients are much larger than the third virial coefficients in Table I.

Conclusions

It is shown that the properties of fully ionized aqueous electrolyte systems can be represented by relatively simple equations over wide ranges of composition. There are only a few systems for which data are available over the full range to fused salt. A simple equation commonly used for nonelectrolytes fits the measured vapor pressure of water reasonably well and further refinements are clearly possible. Over the somewhat more limited composition range up to saturation of typical salts such as NaCl, the equations representing thermodynamic properties with a Debye-Hückel term plus second and third virial coefficients are very successful and these coefficients are known for nearly 300 electrolytes at room temperature. These same equations effectively predict the properties of mixed electrolytes. A stringent test is offered by the calculation of the solubility relationships of the system Na-K-Mg-Ca-Cl-SO_4-H_2O and the calculated results of Harvie and Weare show excellent agreement with experiment.

Acknowledgements

This research was supported by the Office of Basic Energy Sciences, U. S. Department of Energy, under Contract No. W-7405-Eng-48. I thank Dr. John H. Weare for sending me his results in advance of publication.

References

1. Debye, P. and Hückel, E., Physik. Z. (1923) 24, 185, 334; (1924) 25, 97.

2. Harned, H. S. and Owen, B. B., "The Physical Chemistry of Electrolyte Solutions," 3rd Ed. Reinhold Publishing Co., New York (1958).

3. Robinson, R. A. and Stokes, R. H., "Electrolyte Solutions," Butterworths, London, 1955, revised 1959.

4. Card, D. N. and Valleau, J. P., J. Chem. Phys. (1970) 52, 6232.

5. Pitzer, K. S., Accounts of Chemical Research, (1977) 10, 371.

6. Friedman, H. L. "Ionic Solution Theory," Interscience (1962).

7. Anderson, H. C., "Improvements upon the Debye-Hückel Theory of Ionic Solutions," in "Modern Aspects of Electrochemistry," No. 11, B. E. Conway and J. O'M. Bockris, Eds., Plenum Press, New York, 1975.

8. Ramanathan, P. S. and Friedman, H. L., J. Chem. Phys. (1971) 54, 1086.

9. Bronsted, J. N., Kgl. Dan. Vidensk. Selsk., Mat. Fys. Medd., (1929) 4, (4); J. Am. Chem. Soc., (1922) 44, 877; (1923) 45, 2898.

10. Guggenheim, E. A., Phil. Mag. [7] (1935) 19, 588; also Guggenheim, E. A. and Turgeon, J. C., Trans. Faraday Soc., (1955) 51, 747.

11. Pitzer, K. S. and Kim, J. J., J. Am. Chem. Soc. (1974) 96, 5701.

12. Trudelle, M-C., Abraham, M. and Sangster, J., Can. J. Chem. (1977) 55, 1713.

13. Tripp, T. B. and Braunstein, J., J. Am. Chem. Soc. (1954) 73, 1984.

14. Braunstein, H. and Braunstein, J., J. Chem. Thermodynamics (1971) 3, 419.

15. van Laar, J. J., Z. physik. Chem. (1906) 72; (1910) 723.

16. Prausnitz, J. M., "Molecular Thermodynamics of Fluid-Phase Equilibria," Prentice-Hall, Inc., Englewood Cliffs, NJ (1969).

17. Pitzer, K. S., in preparation for publication.

18. McMillan, W. G. and Mayer, J. E., J. Chem. Phys., (1945) 13, 276.

19. Mayer, J. E., J. Chem. Phys. (1950) 18, 1426; see also Reference 6.

20. Pitzer, K. S., J. Phys. Chem. (1973) 77, 268.

21. Scatchard, G., J. Am. Chem. Soc. (1968) 90, 3124.

22. Scatchard, G., Rush, R. M. and Johnson, J. S., J. Phys. Chem. (1970) 74, 3786.

23. Pitzer, K. S. and Mayorga, G., J. Phys. Chem. (1973) 77, 2300; J. Solution Chem. (1974) 3, 539.

24. Pitzer, K. S. and Silvester, L. F., J. Solution Chem., (1976) 5, 269.

25. Pitzer, K. S., Roy, R. N., and Silvester, L. F., J. Am. Chem. Soc. (1977) 99, 4930.

26. Pitzer, K. S., Peterson, J. R., and Silvester, L. F., J. Solution Chem. (1978) 7, 45.

27. Pitzer, K. S. and Silvester, L. F., J. Phys. Chem. (1978) 82, 1239.

28. Downes, C. J. and Pitzer, K. S., J. Solution Chem. (1976) 5, 389.

29. Silvester, L. F. and Pitzer, K. S., J. Solution Chem. (1978) 7, 327.

30. Liu, C. and Lindsay, W. T., J. Solution Chem. (1972) 1, 45; J. Phys. Chem. (1970) 74, 341.

31. Puchkow, L. V., Styazhkin, P. S., and Federov, M. K., J. Appl. Chem. (USSR) (1977) 50, 1004; Atonov, N. A., Gilyarov, V. N., Zarembo, V. I., and Federov, M. K., ibid, (1976) 49, 120; and other papers there cited.

32. Silvester, L. F. and Pitzer, K. S., J. Phys. Chem. (1977) 81, 1822.

33. Pitzer, K. S., Bradley, D. J., Rogers, P. S. Z., and Peiper, J. C., paper at Honolulu meeting of the A.C.S. (1979), to be published.

34. Bradley, D. J. and Pitzer, K. S., J. Phys. Chem., (1979) 83, 1599.

35. Holmes, H. F., Baes, C. F., Jr., and Mesmer, R. E., J. Chem. Thermo. (1978) 10, 983.

36. For example, Marshall, W. L., J. Phys. Chem. (1967) 71, 3584.

38. Braitsch, O., "Salt deposits: their origin and composition," Springer Verlag (1971).

39. Pitzer, K. S., J. Solution Chem. (1975) 4, 249.

RECEIVED January 31, 1980.

24

Current Status of Experimental Knowledge of Thermodynamic Properties of Aqueous Solutions

RANDOLPH C. WILHOIT

Thermodynamics Research Center, Texas A&M University, College Station, TX 77843

The current state of knowledge of aqueous solutions is the result of all that has been learned in the past. These studies have formed a major part of modern science since its beginning. Theories of aqueous electrolytes have played a major role in the history of chemistry.

We can recognize four main periods in the history of the study of aqueous solutions. Each period starts with one or more basic discoveries or advances in theoretical understanding. The first period, from about 1800 to 1890, was triggered by the discovery of the electrolysis of water followed by the investigation of other electrolysis reactions and electrochemical cells. Developments during this period are associated with names such as Davy, Faraday, Gay-Lussac, Hittorf, Ostwald, and Kohlrausch. The distinction between electrolytes and nonelectrolytes was made, the laws of electrolysis were quantitatively formulated, the electrical conductivity of electrolyte solutions was studied, and the concept of independent ions in solutions was proposed.

The second period, from 1890 to around 1920, was characterized by the idea of ionic dissociation and the equilibrium between neutral and ionic species. This model was used by Arrhenius to account for the concentration dependence of electrical conductivity and certain other properties of aqueous electrolytes. It was reinforced by the research of Van't Hoff on the colligative properties of solutions. However, the inability of ionic dissociation to explain quantitatively the properties of electrolyte solutions was soon recognized.

The theory proposed by Debye and Huckel dominated the study of aqueous electrolytes from around 1920 to near the end of the 1950's. The Debye-Huckel theory was based on a model of electrolyte solutions in which the ions were treated as point charges (later as charged spheres), and the solvent was considered to be a homogeneous dielectric. Deviations from ideal behaviors were assumed to be due only to the long range electrostatic forces between ions. Refinements to include ion-ion pairing and ion

0-8412-0569-8/80/47-133-467$06.75/0

hydration were eventually added. Most of the experimental work by physical chemists during this period was done to discover the limiting concentration dependence of various thermodynamic and transport properties at infinite dilutions and to compare such results to the predictions of the Debye-Huckel theory. It was found that the Debye-Huckel theory does predict the correct limiting laws for all properties which have been sufficiently studied. However, for some properties; i.e., partial molal enthalpy, the concentration must be very low. The failure of the Debye-Huckel theory and its various modifications to account for properties of solutions more concentrated than a few tenths molal led to a stalemate in further understanding of such solutions during the 1940's and the 1950's.

The prevalence of water in many industrial processes has led to the accumulation of a large body of experimental data on aqueous solutions of both electrolytes and nonelectrolytes. It is commonly recognized that aqueous solutions are usually highly non-ideal, but until recent years, no theoretical explanation was available for aqueous solutions of nonelectrolytes.

The past fifteen years or so have seen a decided resurgence in theories of aqueous solutions. Although these theories are not the result of any single event, the main components of this final and fourth period can be recognized.

The newer ideas incorporate specific interactions between water molecules and between water and solute molecules. They also make extensive use of the accumulated knowledge on the intermolecular structure and order of liquid water and of the geometrical constraints which they imply. The new developments have been partly inspired by the investigation of biochemists and biophysicists on the effect of water and aqueous solutes on the tertiary structure of biological macromolecules. Terms such as "hydrophobic bonding" and "structure making" and "structure breaking" solutes originate from these studies. The spectacular rise in the study of molten salts during the past twenty years has also had an effect. Finally, the prevalence of computers in recent years has made the use of more realistic, and thus more complicated, models possible. Computer simulations of all kinds of fluids have given important clues for the behavior of solutes in water. It is now apparent that the new theories not only are beginning to provide an understanding of moderately concentrated aqueous solutions of electrolytes but also of nonelectrolytes and electrolyte-nonelectrolyte solutions as well.

Figure 1 lists the major industrial processes that use thermodynamic data and the kind of data that are relevant. The design and use of distillation columns are the largest consumers of thermodynamic data. The feed stream to most industrial columns contains at least several components. The predication of the operating characteristics of such columns requires complicated calculations, and much computer software has been written

I. Separation Processes
   A. Distillation
      Vapor-liquid equilibrium compositions, K-values, activity coeff., etc., azeotrope temperature and composition, enthalpy and heat capacity, heats of vaporization
   B. Solvent extraction
      Activity coefficients, distribution coefficients, equilibrium constants for reactions
   C. Ion-exchange
      Activity ceofficients, surface and absorption effects, equilibrium constants of reactions
   D. Crystallization from solution
      Phase diagrams, solubility
   E. Osmosis
      Activity of solvent, volumetric data, properties at high pressure
II. Heat and Mass Transport
   Enthalpy and heat capacity, volumetric properties, surface tension, transport properties
III. Water Treatment
   Equilibrium constants for many reactions among electrolytes and non-electrolytes, and solubility data under a wide range of conditions
   A. Boiler feed
   B. Waste water
   C. Sea water
IV. Electrochemical Processes
   Thermochemical data, oxidation-reduction potentials, activity coefficients, surface properties
   A. Electrolysis
   B. Electroplating
   C. Storage cells and fuel cells
   D. Corrosion
V. Chemical Manufacturing
   Thermochemical data
VI. Metallurgy
   A. Hydrometallurgy
      Solubility, vapor pressure, distribution coefficient, reaction equilibria
   B. Electrometallurgy
      Electrode potentials, equilibrium constants

Figure 1. Industrial processes that use thermodynamic data

for this purpose. To separate components which have similar boiling points or which form azeotropes, it is becoming increasingly common to feed additional components to the column. If the additional substance is volatile and migrates primarily to the vapor stream, the process is called azeotropic distillation. If the additional substance stays in the residue, the process is called extractive distillation of aqueous solutions. The separations of various alcohols, ethylene glycol, acetic acid, acetone, and nitric acid from water is an example in which extractive distillation has been used or proposed. References (108) and (126) describe processes.

Solvent extraction is a major industrial technique. The usual objective is to selectively remove one or more solutes from a complex mixture. Selectivity usually depends on strong specific solvent-solute interactions or on the formation of complexes between ions and ligands. Thus solvent extraction systems are likely to include a number of chemical reactions and to exhibit large deviations from ideality. The design of liquid extraction processes may require many kinds of data. References (31, 32, 55, 61, 81, and 118) are concerned specifically with solvent extraction.

The treatment of boiler feed water is a specialized topic which depends strongly on empirical and proprietary methods. The problem of removing objectionable material from waste water streams has become acute in recent years. Petroleum refineries, coal processing plants, and many chemical manufacturing operations produce large volumes of waste water. Because of government regulations and the increased need for energy efficiency, practices used in the past may no longer be acceptable. Several aspects of this problem have already been discussed in this symposium. The Thermodynamic properties of sea water have been extensively studied during the past decade. This has occurred because of their applications to oceanography and because sea water is used in some industrial operations. References (21, 37, 42, 48, 49, and 100) describe sea water. References (37, 45, 48, 64, 108, 117, 118, 126, and 133) emphasize various industrial applications for the thermodynamics of aqueous solutions. References (19, 45, 48, 64, 58, 99, 119, 120, 121, 12, 141, and 142) report data at temperatures above the normal boiling point.

Theoretical aspects of aqueous solutions are being discussed by other speakers at this session. Kruss (112) has written a useful introduction to both theoretical and experimental studies of solutions. Additional reviews of modern theory may be found in references (76, 80, 82, 87, 90-101, 111, 114, 116, 119, 120, and 121). By an extension of the Debye-Huckel equation, Pitzer and his co-workers (129, 130, 131, 134, 135, 141, 143, and 144) have developed a popular mathematical model of the thermodynamic properties of aqueous electrolyte solutions. However, it is my intention to provide a guide for the rapid location of numerical values of thermodynamic properties of aqueous solutions of importance to industry.

For data on aqueous solutions, the engineer must still rely primarily, either directly or indirectly, on experimental measurements. Figure 2 summarizes the principal techniques for measuring thermodynamic properties. It is important to note that none of these are new. As most of these techniques have been used regularly for nearly eighty years and some for nearly one hundred years, each one identified in Figure 2 has a long history of development and refinement. All have participated in the general improvement in accuracy and sophistication in laboratory instruments over the years. Shifts in the relative amounts of usage of different techniques have occurred during this period. Many of these procedures are described in reference (136).

The measurement of the composition of phases in mutual equilibrium has many direct applications. However, it is common to reduce such data to some form of Gibbs energy. These forms include excess Gibbs energy of mixing, the chemical potential relative to some standard state, activities, activity coefficients, or Gibbs energy of transfer from one solvent to another. These calculations depend on the fact that the chemical potential of any component is the same in all phases in mutual equilibrium. Phase equilibrium measurements at different temperatures allow the calculation of the corresponding enthalpy, entropy, and heat capacity.

Most data on vapor-liquid equilibria in which at least two components are volatile, have been obtained by the use of some kind of recirculating ebulliometer. The liquid mixture is boiled in a closed system until a steady state is reached. Then the temperature is noted and samples of the liquid and vapor phases are withdrawn for analysis. The constraints of mass balance and of the Gibbs-Duhem equation impose functional dependence among the chemical potentials of the components of the system. Thus much effort has been devoted to devising methods of testing data obtained by ebulliometers for thermodynamic consistency. It is possible to calculate the chemical potentials of the components in a system consisting of liquid and vapor phases at equilibrium from observations on only the total vapor pressure as a function of composition. This technique avoids the thermodynamic consistency problem, as well as many other difficulties, with the use of ebulliometers. Vapor pressures of mixtures are usually measured by a static technique in a sealed system. The calculation of chemical potentials from total vapor pressure data is much more difficult than it is from ebulliometric data, but the ready availability of computer algorithms for this calculation has eliminated this disadvantage for binary solutions.

The isopiestic method has been used frequently to measure the vapor pressure of aqueous solutions of nonvolatile solutes. In this technique, two or more solutions are placed in separate cups and stored in a sealed container. All the cups are open to a common vapor space. Air is removed from the container and

I. Phase Equilibria – Gibbs Energy and Related Properties as Functions of Composition
   A. Vapor-liquid
      1. Ebulliometry
      2. Total vapor pressure (direct and isopiestic)
   B. Freezing point
   C. Solubility
   D. Osmotic pressure
II. Calorimetry – Enthalpy and Heat Capacity
   A. Thermochemistry (heats of reaction)
   B. Solution, mixing, dilution
   C. Heat capacity
III. Volumetric Measurements
   A. Density and volume
   B. Sound velocity (adiabatic compressibility)
IV. Electrochemical Measurements – Gibbs Energy and Related Quantities for Chemical Reactions
   A. Reversible cell potentials
   B. Polarography
   C. Electrical conductivity
V. Equilibrium Constants of Chemical Reactions – Gibbs Energy, Enthalpy, Entropy for Reactions
   A. Potentiometric
   B. Spectrophotometric
   C. Thermometric

Figure 2. Techniques for measuring thermodynamic properties

it is shaken gently for a period of time in a carefully thermostatted environment. Water distills back and forth between the different solutions until they all attain the same vapor pressure. Generally several days are required to reach equilibrium. The solutions are then removed and their concentrations are measured. One of the solutions is a standard whose vapor pressure as a function of temperature has been established by other methods. Thus the vapor pressures of all the other solutions are also determined at the equilibrium concentration.

Roughly half of the data on the activities of electrolytes in aqueous solutions and most of the data for nonelectrolytes, have been obtained by isopiestic technique. It has two main disadvantages. A great deal of skill and time is needed to obtain reliable data in this way. It is impractical to measure vapor pressures of solutions much below one molal by the isopiestic technique because of the length of time required to reach equilibrium. This is generally sufficient to permit the calculation of activity coefficients of nonelectrolytes, but the calculation for electrolytes requires data at lower concentrations, which must be obtained by other means.

From a thermodynamic standpoint, freezing point measurements and isopiestic measurements are similar since both yield directly the activity of the solvent. When done carefully, freezing point data can generate activity coefficient values at concentrations down to 0.001 molal. During the first half of this century, much activity coefficient data was obtained from freezing point measurements. However, the popularity of this technique has decreased and is seldom used for aqueous solutions at the present time.

The design and operation of solution calorimeters is an extensive topic. Reference (125) reviews modern calorimetry and identifies earlier discussions. The thermometric titration type of calorimeter has been perfected during the past fifteen or twenty years. It is especially useful for measuring heats of reaction that take place in several steps. The availability of advances in thermometry has had a major effect on calorimetry. New types of thermometers include sensitive and reliable thermistors and quartz crystal thermometers.

Reversible cell potentials have been the source of much thermodynamic data on aqueous electrolytes. In recent years, this technique has been extended to nonaqueous solutions and to molten salt systems. Its use for aqueous solutions, relative to other techniques, has decreased. Various ion specific electrodes have been developed in recent years. These are used primarily in analytical chemistry and have not produced much thermodynamic data.

Figures 3 and 4 illustrate some trends in the publication of experimental data from 1931 to 1976. They graph the number of articles appearing during certain selected years for the forty-four year period. Articles published before 1967 were identified

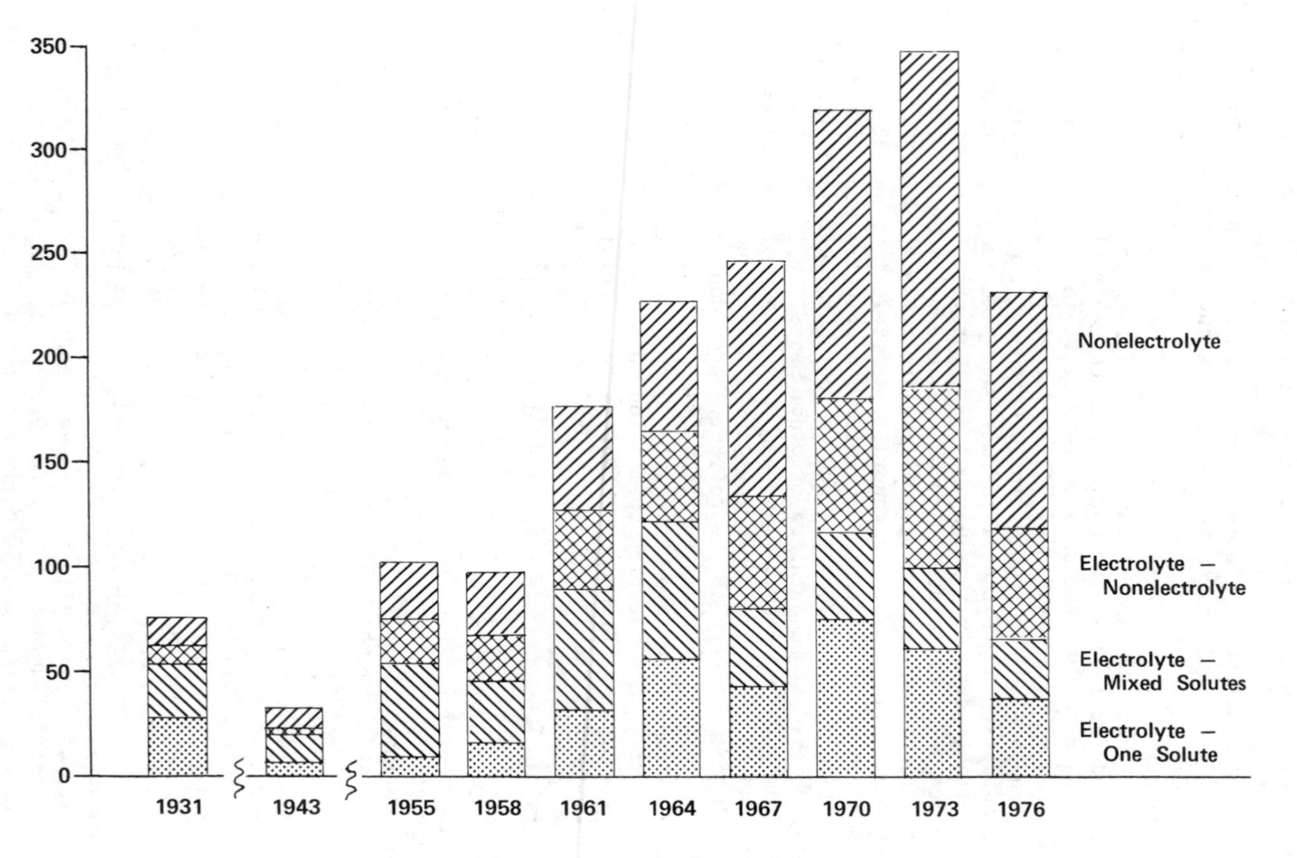

Figure 3. Number of articles published on aqueous systems

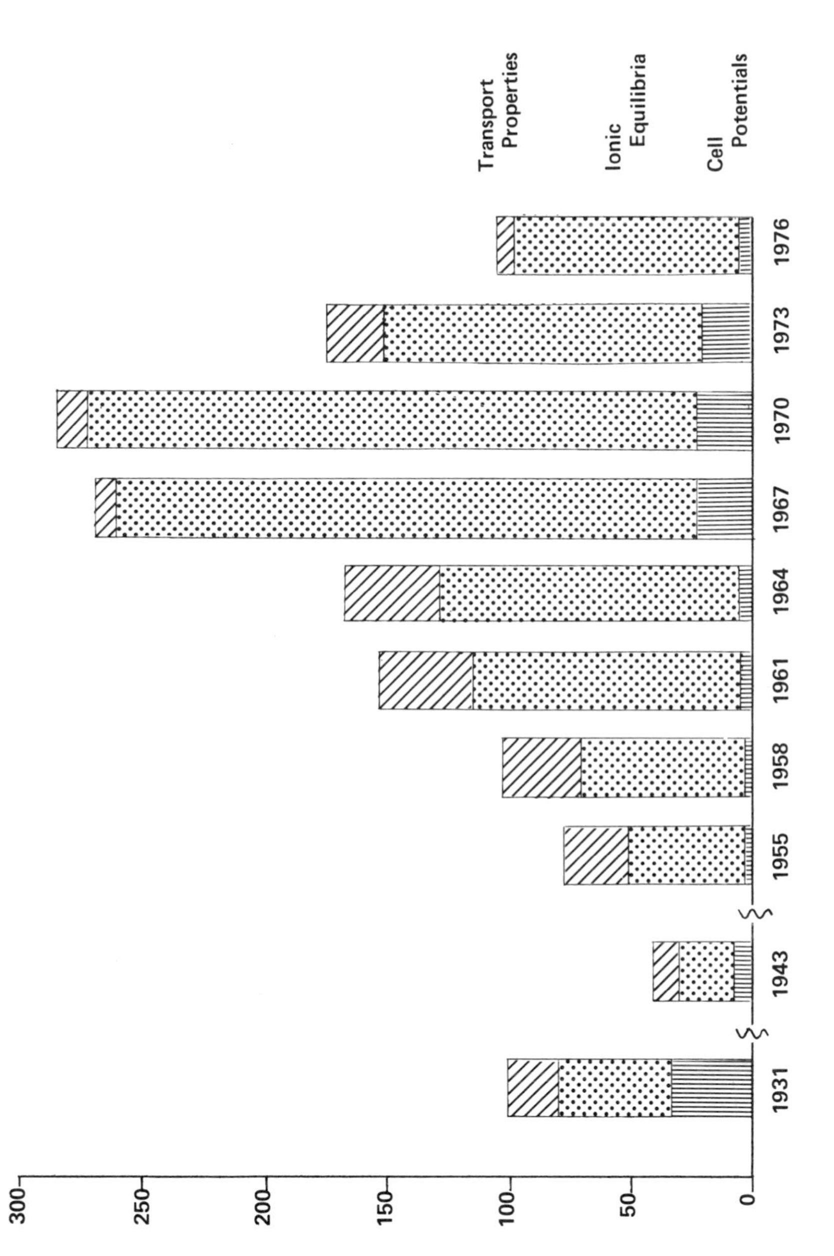

Figure 4. Number of articles published on aqueous systems

by a scan of appropriate sections of Chemical Abstracts. From 1967 to 1976, they were identified by searching the bibliographic section of the Bulletin of Chemical Thermodynamics (1). At this time, only two volumes of the Bulletin have been issued since 1976. Since at least three volumes are needed to collect all references for a particular year, the count for 1976 is probably about 20% low. Only those papers which reported quantitative measurements of equilibrium thermodynamic properties of well-defined systems were counted. Aqueous systems were defined as those in which water is a major component. Studies on colloids or polymers were not included.

The overall increase in rate of publication reflects the general trend in scientific literature. However the slump during the 1940-1960 period is obvious, and a peak during the early part of the 1970's is suggested. Figure 3 graphs the measurements derived from phase equilibrium studies and from measurements of properties of a single phase treated as a system of independent components. They are classified according to the type of components other than water. The figure shows the rapid increase in studies of nonelectrolytes and in electrolyte-nonelectrolyte systems during the past fifteen years. Figure 4 shows the number of articles which report data on electrode and cell potentials, on thermodynamic properties of ionic reactions, and on transport properties. It points out the great popularity of ionic equilibrium studies during the period of 1967-1970.

The largest increase in experimental measurements on aqueous solutions has been in those designed to furnish information on molecular interactions and order. These techniques, along with the kinds of information which can be derived from them, are outlined in Figure 5. Although the principles behind all these techniques have been known for many years, advances in instrumentation and in data collection have encouraged their widespread application to solutions of all kinds. The use of mass spectrometry to study interactions between isolated solvent and solute molecules has been perfected largely within the past ten years. This topic is reviewed in reference (113).

The list of 145 references at the end of this chapter has been collected to help those looking for numerical values of thermodynamic properties of aqueous solutions for industrial applications. The emphasis has been primarily on items published since 1964, although a few older ones of special utility have been included.

The first group consists of eleven bibliographies on various aspects of the thermodynamics of aqueous solutions. Extensive bibliographies may also be found in many other references in the list. The Bulletin of Chemical Thermodynamics (formerly the Bulletin of Thermodynamics and Thermochemistry), reference (1), deserves special mention. This document is issued annually under the sponsorship of the International Union of Pure and Applied Chemistry (IUPAC). It began publication under the

I. Spectroscopic Techniques
   A. Infrared and Raman – vibrational energy levels
      Vibration-translation energy transfer, solute-solvent interaction, H-bonds, ion pairs
   B. Nuclear Magnetic Resonance – nuclear-electronic interactions
      Molecular conformations, solute-solvent interactions, chemical reactions
II. Ultrasonic Absorption – Relaxation Phenomena Involving T and P Changes
   Molecular conformations, solute-solvent interactions, chemical reactions
III. Dielectric Relaxation – Relaxation Phenomena Involving Electric Moment Changes
IV. Scattering Phenomena
   A. X-Ray – electron density
      Structure and order at molecular level, radial distribution function
   B. Light – density and concentration fluctuations
      Orientation relaxations, long range order
   C. Neutron – nuclear position
      Molecular rotation, diffusion, chemical equilibria
V. Ion-Molecule Reactions in the Gas Phase – Mass Spectra
   Thermochemistry of ion-solvent reactions, proton affinity
VI. Computer Simulations – Testing of Models
   A. Monte-Carlo
   B. Molecular dynamics

Figure 5. Techniques for studying intermolecular forces and structure

editorship of H. A. Skinner in 1955. Professor Edgar Westrum served as editor from 1965 to 1976, and Dr. Robert D. Freeman has served since that time. It was not comprehensive or widely circulated until 1963. It is an international cooperative project. For the past fifteen years, the Office of Standard Reference Data of the National Bureau of Standards has supplied a major financial subsidy for the work.

Each issue of the Bulletin consists of two major parts. One part presents brief descriptions of work on thermodynamics and thermochemistry in progress at the time, which are submitted by research investigators from around the world. The other part consists of a substance-property index and a bibliography of published literature. This latter part is separated into four main sections: Organic substances, organic mixtures, inorganic compounds and mixtures, and biological systems. Since 1970, the substance-property index has been assembled at the National Bureau of Standards by a computerized system from data supplied by tapes prepared by cooperating institutions. The Bulletin also contains news items and other information about developments in chemical thermodynamics. Details about the contents and instructions for ordering copies can be obtained from Professor Freeman (see reference (1)).

The second group of citations identifies compilations of numerical data. Additional specialized tables can also be found in some of the references listed in the third and fourth groups. References (13) and (14) are the last two volumes of a four volume compilation of properties of mixtures prepared by J. Timmermans. They contain a large compilation of various properties of aqueous solutions collected from all the previous literature. They are neither complete nor selective, however.

References (20, 22, 23, 24, 29, and 74) comprise the series of Technical Notes 270 from the Chemical Thermodynamics Data Center at the National Bureau of Standards. These give selected values of enthalpies and Gibbs energies of formation and of entropies and heat capacities of pure compounds and of aqueous species in their standard states at 25 ^{o}C. They include all inorganic compounds of one and two carbon atoms per molecule. They also list enthalpies of formation at a series of concentrations for many solutes in water.

There are four other major projects which are conducting continuing and systematic compilations of evaluated thermodynamic data on aqueous solutions. One of these is the Electrolyte Data Center of the National Bureau of Standards. Publications of this and related groups at NBS are cited in references (10, 11, 17, 25, 52, 59, 63, 71, 72, 73, 122, and 139). This work will be described more fully by Dr. Staples in the fifth lecture of this session.

The Deutsche Gesellschaft fur Chemisches publishes a series of data compilations under the DECHEMA series. It is intended primarily for engineering applications. Reference (50) cites tables for aqueous nonelectrolyte solutions from this series.

Another continuing project on aqueous nonelectrolytes is the International Data Series B. The editor is Dr. J. A. Larkin of the National Physical Laboratory in England. It is published in the form of supplements of loose-leaf sheets and issued at irregular intervals of a few months. Each sheet is prepared and submitted by an author or authors. The data are presented in a standard format and must follow certain rules with regard to kind of properties, style, units, and kind of auxiliary information to be included. Standard table formats have been designed for each kind of property included. Each table is reviewed by an editor specially selected from an international panel.

The data may be the result of original measurements not previously published, data which have been published elsewhere but have been modified to conform to the requirements of IDS B, or compilations gathered from several sources. Thus the International Data Series combines the function of a scientific research journal with those of a data compilation.

The first issue of IDS Series B appeared in 1978 and contained 66 data sheets. Two issues containing about the same number of data sheets are planned for 1979.

IDS Series B is a companion project to the International Data Series A. Series A reports data on nonaqueous organic mixtures. It is published by the Thermodynamics Research Center at Texas A&M University in College Station, Texas. The Executive Officer is Professor Bruno J. Zwolinski. The general objectives and manner of processing data are quite similar for the two series of IDS. Series A started publication in 1973 and about 750 data sheets have been issued. Other sections, or series, of IDS are planned, including a series on electrolytes, one on metal alloys, and one on mixtures of special importance to industry. The editor-in-chief of IDS Series A is Henry V. Kehiaian of Marseille, France.

The final major continuing project is the Solubility Data Project, which is sponsored by the International Union of Pure and Applied Chemistry (IUPAC). The project has been underway for several years, and the results are appearing as a series of volumes of the "Solubility Data Series". The collection, compilation, and evaluation of data are being carried out by a large staff of experts in each area on a world-wide basis. The first eight volumes have been announced for 1979. References (67, 68, 69, and 70) are scheduled for publication in 1979 and 1980. Dr. A. S. Kertes of Israel is editor-in-chief of the series.

Special mention should be made of recently published volumes of the Landolt-Bornstein Tables, references (35) and (51). These contain a large amount of data on aqueous solutions presented in a compact form. Reference (58) cites a new handbook on the thermodynamic properties of inorganic compounds. It gives tables of enthalpy, Gibbs energy, entropy, and heat

	Electrolytes	Electrolyte Non-electrolyte	Non-electrolyte
Phase Equilibria			
Gas Solubility		37,64,123,124, 127	44,54,88,102, 111, 138
Vapor-Liquid	13,14,18,19, 21,57,64,99	108,126	14,18,27,34,35, 50,56,57
Condensed Phases	13,14,28,36, 41,43,67,70	37	14,15,37,43,44, 56,68,99,104, 111,137
Gibbs Energy & Derived Properties	13,14,25,26, 40,42,52,53, 57,58,59,65, 66,71,72,73, 87,119,122, 128,130,131, 134,140	12,64,95,108, 129	14,15,56,57, 128,140
Enthalpy & Heat Capacity	13,14,17,20, 22,23,24,29, 33,42,48,51, 74,94,120, 141,144		14,17,20,22,23, 24,29,51,56,74
Volumetric Properties	13,14,21,41, 49,51,57,75, 91		14,51,56,57
Transport Properties	13,14,21,38, 48,57,76	108	14,38,57
Thermochemical and Chemical Equilibria	16,20,22,23, 24,29,33,34, 43,45,47,55, 58,61,62,63, 64,69,74,80, 85,115,117		20,22,23,24,29, 43,74
Electrode & Cell Potentials	30,43,60		
Distribution Coefficients	31,32,55,61		31,32,55,61

Note: Numbers refer to references cited in Bibliography.

Figure 6. Index to sources of compiled or correlated properties

capacity over a range of temperatures for both pure compounds and ions in solution.

The third section of the reference list identifies treatises and reviews which deal primarily with the theoretical aspects of aqueous solutions. Many of them contain some numerical data as well. The section entitled Miscellaneous Reports cites references which give useful methods of correlating and predicting data and also discussions of special and related topics.

Figure 6 is an index to numerical compilations of data according to the class of property and type of solute. Although an impressive number of compilations have been published in the past fifteen years, only a small fraction of the scientific literature on aqueous solutions has been covered. In spite of the increasing volume of data on mixtures of electrolyte and nonelectrolyte solutes which is being produced, there are few comprehensive compilations on such systems. Accordingly, the engineer who needs thermodynamic data on aqueous solutions very often must still spend a great deal of time searching the primary literature.

BIBLIOGRAPHY ON THERMODYNAMIC PROPERTIES OF AQUEOUS SOLUTIONS BIBLIOGRAPHIES

1. Freeman, R. D., Ed. "Bulletin of Thermodynamics and Thermochemistry"; Thermochemistry Inc., University of Oklahoma, Stillwater, Oklahoma, 1955-1978.
2. Horvath, A. L. "Reference Literature to Solubility Data Between Halogenated Hydrocarbons and Water"; J. Chem. Doc., 1972, 12, 163.
3. Wichterle, I.; Linek, J.; Hala, E., Eds. "Vapor-Liquid Equilibrium Data Bibliography"; Elsevier Scientific Publishing Co., Amsterdam, 1973.
   Covers literature on electrolytes and non electrolytes to 1972
4. Hawkins, D. L. "Bibliography on the Physical and Chemical Properties of Water, 1969-1974"; J. Sol. Chem., 1975, 4, 621.
5. Horvath, A. L. "Reference Literature to the Critical Properties of Aqueous Electrolyte Solutions"; J. Chem. Inf. & Computer Sci., 1975, 15, 245.
   71 solutes indexed by compound name
6. Wichterle, I.; Linek, J.; Hala, E., Eds. "Vapor-Liquid Equilibrium Data Bibliography Supplement 1"; Elsevier Scientific Publishing Co., Amsterdam, 1976.
   Covers literature from 1973 to 1975
7. Armstrong, G. T.; Goldberg, R. N. "An Annotated Bibliography of Compiled Thermodynamic Data Sources for Biochemical and Aqueous Systems (1930-1975)"; Nat. Bur. Stand. Special Publ. No. 454, September 1976.
   Equilibrium, enthalpy, heat capacity and entropy data
8. Hicks, C. P., Ed., Chapter 9 "A Bibliography of Thermodynamic Quantities for Binary Fluid Mixtures in Specialist Periodical Report. Chemical Thermodynamics"; Vol. 2, McGlashan, M. L.; Senior Reporter, The Chemical Society, Burlington House, London, 1978.
9. Wisniak, J.; Tamir, A. "Mixing and Excess Thermodynamic Properties. A Literature Source Books"; Elsevier Scientific Publishing Co., New York, 1978.
   A review of correlations is given in the introduction. The bibliography includes inorganic and organic solutes and electrolytes and non-electrolytes
10. Smith-Magowan, D.; Goldberg, R. N. "A Bibliography of Sources of Experimental Data Leading to Thermal Properties of Binary Aqueous Electrolyte Solutions"; Nat. Bur. Stand. Special Publ. No. 537, March 1979.
11. Goldberg, R. N.; Staples, B. R.; Nuttall, R. L.; Arbuckle, R. "A Bibliography of Sources of Experimental Data Leading to Activity or Osmotic Coefficients for Polyvalent Electrolytes in Aqueous Solution"; Nat. Bur. Stand. Special Publ. No. 485, July 1979.

COMPILATIONS OF DATA

12. Long, F. A.; McDevit, W. F. "Activity Coefficients of Nonelectrolyte Solutes in Aqueous Salt Solutions"; Chem. Rev., 1952, 51, 119.

13. Timmermans, J. "Systems with Metallic Compounds" in "The Physico-Chemical Constants of Binary Systems in Concentrated Solutions"; Vol. 3, Interscience Publishers, Inc., New York, 1960.

A compilation of various thermodynamic and transport properties, including aqueous electrolyte and non-electrolyte solutions

14. Timmermans, J. "Systems with Inorganic and Organic or Inorganic Compounds (Excepting Metallic Derivatives)" in "The Physico-Chemical Constants Binary Systems in Concentrated Solutions"; Vol. 4, Interscience Publishers, Inc., New York, 1960.

15. Francis, A. W. "Critical Solution Temperatures"; Advances in Chemistry Series No. 31, American Chemical Soc., Washington, D.C., 1961.

16. Kortum, G.; Vogel, W.; Andrussow, K. "Dissociation Constants of Organic Acids in Aqueous Solution"; Pure Appl. Chem., 1961, 1, 190.

17. Parker, V. B. "Thermal Properties of Aqueous Uni-Univalent Electrolytes"; NSRDS-NBS 2, U.S. Department of Commerce, National Bureau of Standards, Washington, D.C., April 1965.

18. Hala, E.; Wichterle, I.; Polak, J.; Boublik, T. "Vapor-Liquid Equilibrium Data at Normal Pressures"; Pergamon Press, New York, 1968.

Tables and correlating equations for 27 binary and ternary aqueous systems

19. Lindsay, W. T., Jr.; Liu, C. T. "Vapor Pressure Lowering of Aqueous Solutions at Elevated Temperatures"; Office of Saline Water, U.S. Government Printing Office, Cat. No. I 1.88:347, 1968.

20. Wagman, D. D.; Evans, W. H.; Parker, V. B.; Halow, I.; Bailey, S. M.; Schumm, R. H. "Selected Values of Chemical Thermodynamic Properties. Tables for the First Thirty-Four Elements in the Standard Order of Arrangement"; Nat. Bur. Stand. Tech. Note No. 270-3, January 1968.

21. Fabuss, B. M.; Korosi, A. "Properties of Sea Water and Solutions Contaning Sodium Chloride, Potassium Chloride, Sodium Sulfate and Magnesium Sulfate"; Office of Saline Water, Report No. 384, U. S. Department of the Interior, 1968.

Tables and correlating equations for density, vapor pressure, thermal conductivity and viscosity of binary and ternary solutions

22. Wagman, D. D.; Evans, W. H.; Parker, V. B.; Halow, I.; Bailey, S. M.; Schumm, R. H. "Selected Values of Chemical Thermodynamic Properties. Tables for Elements 35 through 53 in the Standard Order of Arrangement"; Nat. Bur. Stand. Tech. Note No. 270-4, May 1969.
23. Wagman, D. D.; Evans, W. H.; Parker, V. B.; Halow, I.; Bailey, S. M.; Schumm, R. H.; Churney, K. "Selected Values of Chemical Thermodynamic Properties. Tables for Elements 54 through 61 in the Standard Order of Arrangement"; Nat. Bur. Stand. Tech. Note No. 270-5, March 1971.
24. Parker, V. P.; Wagman, D. D.; Evans, W. H. "Selected Values of Chemical Thermodynamic Properties. Tables for the Alkaline Earth Elements (Elements 92 through 97 in the Standard Order of Arrangement)"; Nat. Bur. Stand. Tech. Note No. 270-6, November 1971.
25. Hamer, W. J.; Wu, Y. C. "Osmotic and Mean Activity Coefficients of Uni-Univalent Electrolytes in Water at 25°C"; J. Phys. Chem. Ref. Data, 1972, 1, 1047.
Numerical data and coeffficients in smoothing equations for 80 compounds
26. Liu, C. T.; Lindsay, W. T., Jr. "Thermodynamics of Sodium Chloride Solutions at High Temperatures"; J. Sol. Chem., 1972, 1, 45.
27. Horsley, L. H. "Azeotropic Data-III"; Advances in Chemistry Series 116, American Chemical Society, Washington, D.C., 1973.
28. Zdanovskii, A. B.; Solov'eva, E. F.; Lyakhovskaya, E. I. "Three Component Systems. Books 1 and 2" in "Handbook of Experimental Data on Solubility of Multicomponent Water-Salt Systems"; Vol. 1, Khimiya, Leningrad Otd., Leningrad, USSR, 1973.
29. Schumm, R. H.; Wagman, D. D.; Bailey, S.; Evans, W. H.; Parker, V. B. "Selected Values of Thermodynamic Properties. Table for the Lanthanide (Rare Earth) Elements (Elements 62 through 76 in the Standard Order of Arrangement)"; Nat. Bur. Stand. Tech. Note No. 270-7, April 1973.
30. Meites, L.; Zuman, P. "Organic, Organometallic Biochemical Substances" in "Electrochemical Data"; Part 1, John Wiley and Sons, New York, 1974.
Oxidation-reduction potentials, polarographic data, etc.
31. Marcus, Y.; Kertes, A. S.; Yamir, E., Compilers, Introduction and Part 1 "Organophosphorus Extractants" in "Equilibrium Constants of Liquid-Liquid Distribution Reactions"; IUPAC Analytical Chemistry Division, Butterworths, London, 1974.
32. Kertes, A. S.; Marcus, Y.; Yamir, E., Compilers, Part 2 "Alkylammonium Salt Extractants" in "Equilibrium Constants of Liquid-Liquid Distribution Reactions"; IUPAC Analytical Chemistry Division, Butterworths, London, 1974.

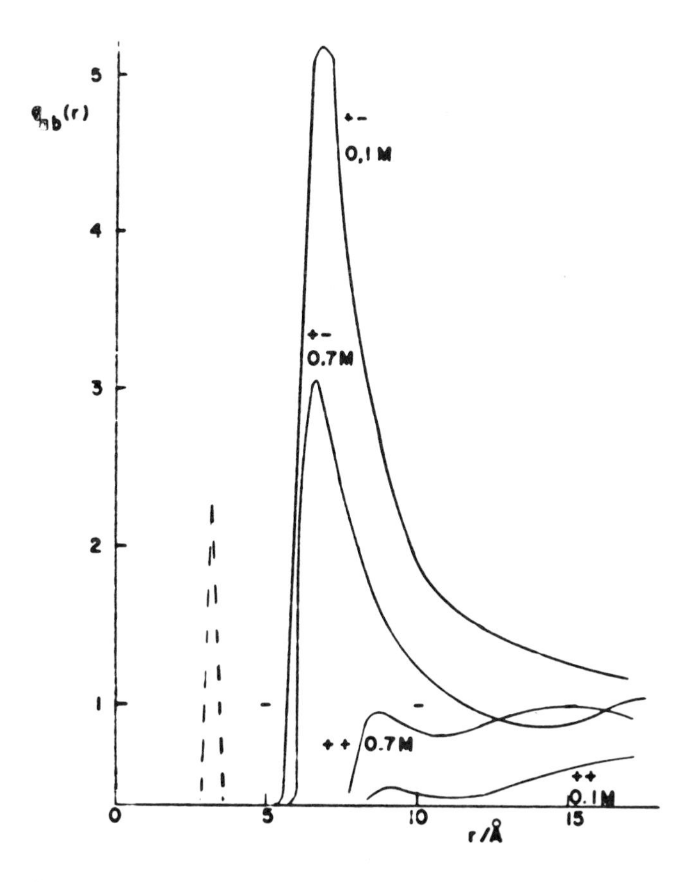

Figure 1. Correlation functions in $NiCl_2(aq)$ at 0.1M and 0.7M.

The solid curves below 7 Å are calculated accurately for a model (Section 4) that fits the osmotic coefficient data. The curves above 7 Å are merely schematic, showing in exaggerated form the oscillations that appear in g_{ab} at large r when the concentration is large, even for the models in Section 4. The dashed curve indicates the location and intensity of the peak in $g_{+-}(r)$ identified in aqueous $NiCl_2$ in neutron diffraction and EXAFS studies, as reviewed in Section 5.

2. HAMILTONIAN MODELS AT VARIOUS LEVELS

The word 'model' now is so often used to mean <u>any</u> set of approximations that it is convenient to use the term Hamiltonian model to mean a physical model. The model's Hamiltonian specifies the forces acting upon each particle in each possible configuration of the system, i.e. each set of locations of all of the particles. This may be done at several levels. (5, 8)

At the deepest level of interest here, the Schroedinger level, the particles are the nuclei and electrons making up the solution of interest. The Hamiltonian is well known. The laws of quantum mechanics must be used to calculate the wave functions which carry the information about the structure of the system.

Among the functions one can, at least in principle, calculate at the Schroedinger level is the Born-Oppenheimer (BO) potential surface, the potential of the forces among the nuclei assuming that at each nuclear configuration the time-independent Schroedinger equation is satisfied. We may think of this as the 'electron-averaged' potential. Such an N-body potential U_N often may be adequately represented as a sum of pair potentials

$$U_N(\underset{\sim}{r}_1, \underset{\sim}{r}_2, \ldots, \underset{\sim}{r}_N) = \sum_{\text{pairs}} u_{ij}(r_{ij}) \tag{3}$$

where the particle indices in the pair potentials may pertain to monatomic particles or to molecules; in the latter case the coordinates must in general include orientational and other internal coordinates as well as the center-to-center distance r_{ij} .

The program of calculating the BO-level potentials from Schroedinger level cannot often be carried through with the accuracy required for the intermolecular forces in solution theory. (9) Fortunately a great deal can be learned through the study of BO-level models in which the N-body potential is pairwise additive (as in Eq. (3)) and in which the pair potentials have very simple forms. (2,3,6) Thus for the hard sphere fluid we have, with σ=sphere diameter,

$$\begin{aligned} u(r) &= \infty \quad \text{if} \quad r<\sigma \\ &= 0 \quad \text{if} \quad r>\sigma \end{aligned} \tag{4}$$

while for the 6-12 fluid we have

$$u(r) = e_0[(\sigma/r)^{12} - 2(\sigma/r)^6] \tag{5}$$

The 6-12 potential is only qualitatively like the realistic potentials that can be derived by calculations at Schroedinger level for, say, Ar-Ar interactions. But it requires careful and detailed study to see how real simple fluids (i.e. one component fluids with monatomic particles) deviate from the behavior calculated from the 6-12 model. Moreover the principal structural features of simple fluids are already quite realistically given by the hard sphere fluid.

The lesson, that drastically simple Hamiltonian models are adequate to generate quite realistic fluid properties and hence to understand the structure of fluids, can be reinforced by many other examples. For the present Symposium the most important may be the Stillinger-Rahman series of studies of a BO-level

Hamiltonian model for water. (10,11)

In a McMillan-Mayer level model (MM-level) for a solution, the particles are the solute particles (i.e. the ions with positive, negative, or zero charge). The ion-ion potentials can, in principle, be generated by calculations in which one averages over the solvent coordinates in a BO-level model which sees the solvent particles. (4,5,12) Pairwise additivity (we use overbars for solvent-averaged potentials)

$$\bar{U}_N(r_1, r_2,\ldots,r_N) = \sum_{\text{pairs of ions}} \bar{u}_{ij}(r_{ij}) \tag{6}$$

is not so accurate or realistic as at BO-level, but is perhaps realistic enough for solutions in which the ionic concentrations do not much exceed 1M .

The simplest model for ionic solutions at the MM-level is the primitive model, (13,14)

$$\bar{u}_{ij} = u_{ij}^{HS} + e_i e_j/\varepsilon r \tag{7}$$

where u_{ij}^{HS} is the potential given in Eq. (4), with $\sigma \rightarrow \sigma_{ij}$, and where ε is the dielectric constant of the solvent medium. This is implicitly the model studied by Debye and Hückel, and in most later studies of ionic solution theory as well. Their well known result for the potential of average force,

$$w_{ij}(r) = u_{ij}^{HS}(r) + e_i e_j e^{-\kappa r}/\varepsilon r , \tag{8}$$

where κ^{-1} is the Debye shielding length, is not very accurate, even for this model, compared to some of the later results. Some of the advances are incorporated in the system of equations for the thermodynamic excess properties developed by K. S. Pitzer. (15)

In Section 4 we discuss some more refined MM-level models.

3. MM-LEVEL $\bar{u}_{+-}(r)$ FROM BO LEVEL

Until very recently there was no information about how an MM pair potential should look, based upon calculations from the deeper BO level. In the simplest BO level model for an ionic solution the solvent molecules are represented as hard spheres with centered point dipoles and the ions as hard spheres with centered charges. Now there are two sets of calculations, (16,17) by very different approximation methods, for this model where all of the spheres are 3Å in diameter, where the dipole moments are near 1 Debye, and where the ions are singly charged. The temperature is 25° and the solvent concentration is about 50M, corresponding to a liquid state. The dielectric constant of the model solvent is believed to be near 9.6.

As shown in Fig. 2 the primitive model with ε=9.6 is

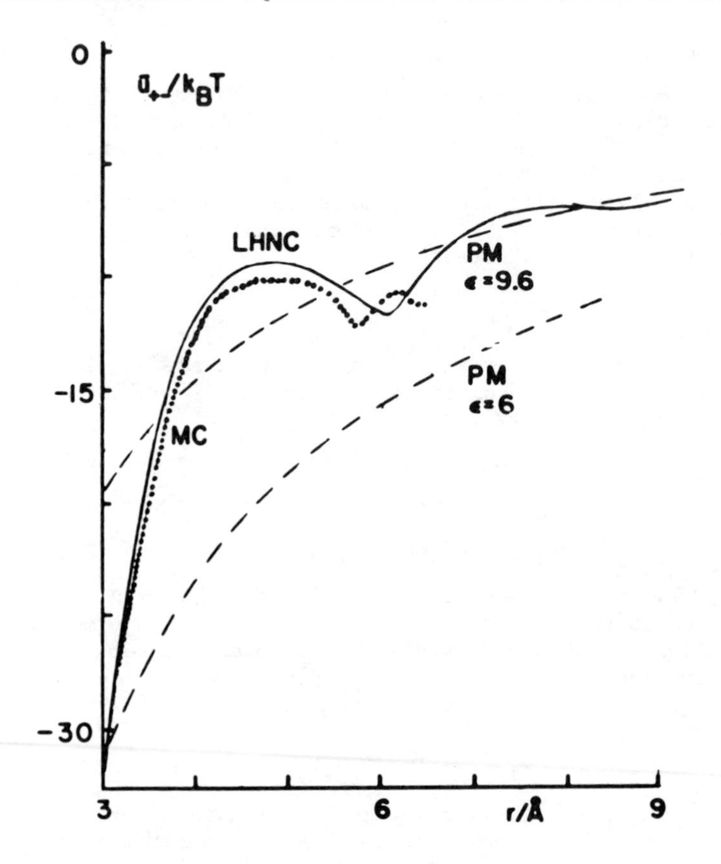

Figure 2. Solvent-averaged potential for charged hard-sphere ions in a dipolar hard-sphere solvent. MC approximation by Patey and Valleau (16) and LHNC approximation by Levesque, Weis, and Patey (17). Also shown are the primitive model functions for solvent dielectric constants 9.6 and 6.

only very close to the calculated $\bar{u}_{+-}(r)$ for $r>7Å$. At $r=3Å$, corresponding to a configuration in which the ions are in contact, the deviation of $\bar{u}_{+-}/k_BT$ from the primitive model is about 11 units. About 1 unit is expected on the basis of so-called liquid structure effects. Thus if the charges and dipole moments were made vanishingly small one would find $u_{+-}/k_BT \simeq -1$. The remaining 10 units is associated with the poor shielding of the ionic charges by the solvent dipoles when the ions are in contact. Indeed the contact value observed for $\bar{u}_{+-}/k_BT$ corresponds to a primitive model with $\varepsilon \simeq 6$ and the same ion sizes as the actual model, as shown in the figure.

The small- r behavior in Fig. 2 is quite surprising from the following point of view. For real 1-1 electrolytes in solvents with $\varepsilon \simeq 10$, e.g. tetrabutylammonium picrate in ethylene dichloride, the mass action constant for forming +- ion pairs, as estimated from the conductivity data, is rather close to what may be calculated from the primitive model, with a realistic ion size parameter, using the theory of Bjerrum (18). However the mass action constant for forming ion pairs estimated from the non-primitive model curve in Fig. 2 is larger than the primitive model result by about three orders of magnitude. This comparison suggests that the particular BO-level model treated in Fig. 2 is not very realistic. One does not yet know what additional features are required to improve the model.

Another method of calculation developed by Adelman and applied by him to a model for a 1-1 electrolyte in water gives much smaller deviations from the primitive model. (19)

We turn to $w_{ij}(r)$ in the case that both i and j are hydrophobic solute particles in water. The so-called hydrophobic bond (20) corresponds to a well in $w_{ij}(r)$ and a peak in the corresponding $g_{ij}(r)$. From a BO-level model for a solution of two Ne atoms in water, with the Bjerrum-like ST2 model for water (10), Geiger, Rahman, and Stillinger (11) find that their calculations suggest that $\bar{u}_{Ne,Ne}(r)$ has a minimum at an r value that corresponds to a solvent-separated hydrophobic bond. In a more detailed and specialized calculation along similar lines but for a model for Xe in ST2 water, Pangali, Rao, and Berne (21) find that $\bar{u}_{Xe,Xe}(r)$ has two wells, one near 4Å that corresponds to the usual idea of contact hydrophobic bonds and one near 7Å that corresponds very closely to the type of solvent-separated hydrophobic bond that one can get if the water near the hydrophobic species has the characteristic hydrogen bonded structure of water in the clathrate hydrates. Moreover the $\bar{u}_{Xe,Xe}(r)$ potential function is in excellent agreement with that calculated by Pratt and Chandler (22) by means of a very different approximation method and also a somewhat different model. Thus Pratt and Chandler are able to finesse the specification of the water-water pair potential in the BO-level model by introducing, at an appropriate point in the theory, the experimental (x-ray) distribution function (essentially $g_{OO}(r)$) for pure water.

The earlier rather complicated evidence for clathrate structures enforced by hydrophobic pairs (from EPR lineshape phenomena for paramagnetic hydrophobic solutes, (23)) and for two states of the hydrophobic bond (from thermodynamic excess functions (24,25)) is provided with a detailed background by these important theoretical developments.

4. REFINED MM-LEVEL MODELS

Extensive studies have been made of MM-level models with the potential (26)

$$\bar{u}_{ij}(r) = COR_{ij} + CAV_{ij} + e_i e_j/\varepsilon r + GUR_{ij} \qquad (9)$$

The COR term represents the short range repulsion between the ions, either by a power law $(\sigma_{ij}/r)^9$ or by an exponential function $e^{-r/\sigma_{ij}}$. Its parameters are calculated from the known sizes, e.g. crystal radii, of particles i and j. The CAV term, which varies like $1/r^4$, accounts for a particular dielectric effect, something like the one in the salting- out theory of Debye and McAulay. (27) It turns out to be numerically unimportant even for Setchenow coefficients. The Coulomb term is well known. The GUR term is intended to account for an effect pointed out some time ago by Gurney (28) and by H. S. Frank. (29) Thus if there is a region (the cosphere) around each ion within which the solvent is modified by the ion, then the force between two ions will have a contribution from the free energy changes when the cospheres overlap. This idea is quantified in the form (26)

$$GUR_{ij}(r) = A_{ij}V_{mu}(r)/V_w \tag{10}$$

where V_w is the molecular volume of the solvent and $V_{mu}(r)$ is the mutual volume of the cospheres of i and j when the distance between their centers is r. The amplitude of the effect, A_{ij}, the only parameter in Eq. (9) that is adjusted to fit the solution data, has the significance of the free energy change per molecule of water displaced from the overlapping cospheres. Most often the cospheres are assumed to include one molecular layer of solvent.

The osmotic coefficients calculated from Eq. (9) can be brought into good agreement with solution data up to about 1M for aqueous solutions of alkali (26) and alkaline earth halides, (30) tetraalkyl ammonium halides, (31) mixed electrolytes, where the Harned coefficients are measured, (32) and electrolyte-non electrolyte mixtures, where Setchenow coefficients are measured. (33) Equally good results have been found for excess enthalpies and volumes when these have been attempted.

Contributing to A_{ij} are, in addition to the solvent structural effects explicitly considered, contributions from dielectric saturation, from the liquid structure effects one has even in simple fluids, from solvent-mediated dispersion interactions of the ions, from charge-polarizability interactions of the ions, and so on. It is difficult to tell a-priori which effects are dominant or how big they are. However the collection of A_{ij} coefficients has characteristics that are consistent with the first named effect being dominant.

We find, from the models that fit the experimental data, that A_{ij}/k_BT is most often in the range from -0.3 to 0. It varies quite regularly with variation of species in a manner that is consistent with the conclusion that the important effects of ion size are explicitly accounted for in Eq. (9). (33,34) The most striking regularity is that if i and j are both species that are all or largely hydrophobic then A_{ij} tends to be nearly independent of species. This behavior makes it relatively easy to predict the thermodynamic excess functions of aqueous solutions of hydrophobic solutes with a certain accuracy, which doubtless could be improved by further work. (35)

One of the further refinements which seems desirable is to modify Eq. (9) so that it has wiggles (damped oscillatory behavior). Wiggles are expected in any realistic MM-level pair potential as a consequence of the molecular structure of the solvent; (2,3,10,11,21,22) they would be found even for two hard sphere solute particles in a hard-sphere liquid or for two $H_2{}^{18}O$ solute molecules in ordinary liquid H_2O, and are found in simulation studies of solutions based on BO-level models. In ionic solutions in a polar solvent another source of wiggles, evidenced in Fig. 2, may be associated with an oscillatory non-local dielectric function $\varepsilon(r)$. (36) These various studies may be used to guide the introduction of wiggles into Eq. (9) in a realistic way.

5. NON-THERMODYNAMIC DATA

Electrical conductivities are much easier to measure than thermodynamic coefficients and, of course, are affected by the solvent-averaged ion-ion forces. Using charged hard sphere models refined by incorporation of a Gurney-like term, Ebeling and coworkers (37,38) and Justice and Justice (39,40) have adjusted Gurney parameters for aqueous alkali halides and some other systems to fit the conductivity data. Quite remarkably the Gurney parameters adjusted to fit the conductivity data are very well correlated with the Gurney parameters adjusted to fit the thermodynamic excess functions. (Fig. 3) On this basis one could use conductivity data to find Gurney parameters in cases in which the thermodynamic data were incomplete and then use the results to predict thermodynamic excess functions which otherwise would be poorly known. A useful and significant correlation between the Gurney parameters and the solvation free energies also has been reported. (38)

Next we describe some work that is directed at the question, does the agreement with experiment of model potentials like the one in Eq. (9) imply that the models are right? It seems unlikely that the answer is affirmative because of the great variety of equations which, over the years, have been reported to give precision fits to the thermodynamic excess function data; apparently there is relatively little information in these data. More directly, we find that we can change some important aspects of the models within the scope of Eq. (9) and still fit the data for thermodynamic excess functions. Typical examples are given below.

For the rate constant k_{ab} of an activation controlled reaction of a solute particle of species a with one of species

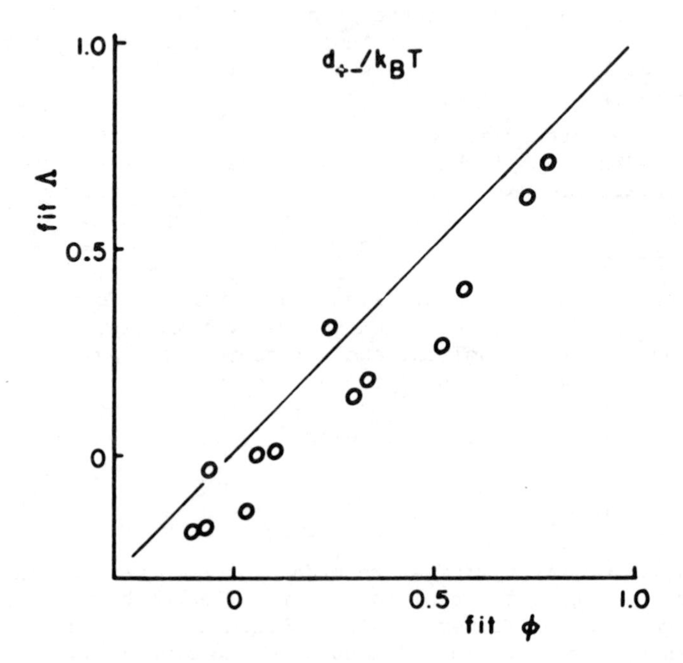

Figure 3. Gurney parameters for + − pairs determined by fitting data for osmotic coefficient φ and conductivity Λ (38). The line in the figure represents ideal correlation. The data are for alkali halides, except fluorides, in water. The parameter d_{+-} is defined for a simpler MM-level model than Equation 9; in Ref. 38 it is reported that $d_{+-}/k_BT = 0.75 + 3.6\ A_{+-}/k_BT$. These correlations have been found by Justice and Justice as well (39, 40).

b one has

$$k_{ab} = \int g_{ab}(r)\hat{k}_{ab}(r)\ d^3r \qquad (11)$$

where $\hat{k}_{ab}(r)$ is the rate when the distance between the reactants is r . Comparing calculated and experimental k_{ab} is a much better way to test models than comparison of the thermodynamic excess functions because k_{ab} depends only on g_{ab} whereas the thermodynamic excess functions depend on all of the solute-solute g_{ij} , weighted by the concentration product c_ic_j . (34,41) While the theory for ordinary chemical kinetics is not sufficiently advanced so that one knows enough about $\hat{k}_{ab}(r)$ for this purpose, the situation is much better for spin relaxation induced by the dipolar spin-spin interaction, as in the process

$$^7Li^+(m) + Ni^{2+} = {}^7Li^+(m') + Ni^{2+} \quad ,$$

where $m \rightarrow m'$ represents a change in nuclear spin state of the ^7Li, in this case due to collision with the paramagnetic Ni^{2+}. The rate constant is determined by measuring T_1 of ^7Li as a function of Ni^{2+} concentration in aqueous solutions of LiCl and $NiCl_2$, with added $MgCl_2$ to keep the ionic strength fixed. The rate constant is calculated for models based on Eq. (9) for each one of the following cases.

A. Ni^{2+} and Li^+ both are rigidly hydrated so that the ionic radii used in Eq. (9) are respectively, 0.7+2.76 Å and 0.6+2.76Å .

B. Ni^{2+} is rigidly hydrated but Li^+ is not; its radius in Eq. (9) is only 0.6 Å.

C. Neither Ni^{2+} nor Li^+ is rigidly hydrated.

All three models may be fit to the thermodynamic data for aqueous solutions of $NiCl_2$, LiCl, and their mixtures by adjusting their Gurney parameters. Thus, it is difficult to tell from the thermodynamic data whether one model is more realistic than the others. However only model B is in satisfactory agreement with the spin relaxation data. (41)

At this time diffraction data for ion-ion distributions in aqueous solutions of moderate concentration are beginning to become available. In aqueous $NiCl_2$ solutions very refined neutron diffraction studies indicate that the Ni^{2+}-Cl^- pair correlation function has a peak near 3.1Å under conditions in which the Cl^- does not penetrate the $Ni(H_2O)_6^{2+}$ unit. (42) It is reported that EXAFS studies give the same result. (43) While the information is most welcome it is puzzling because a geometrical calculation indicates that the closest center to center distance for the Ni^{2+} and a Cl^- that does not penetrate the hydration shell is closer to 3.9Å. (7)

6. CORRESPONDING STATES FOR IONIC SYSTEMS

It has been proposed to define a reduced temperature T_r for a solution of a single electrolyte as the ratio of k_BT to the work required to separate a contact +- ion pair, and the reduced density ρ_r as the fraction of the space occupied by the ions. (44) The principal feature on the T_r, ρ_r corresponding states diagram is a coexistence curve for two phases, with an upper critical point as for the liquid-vapor equilibrium of a simple fluid, but with a markedly lower reduced temperature at the critical point than for a simple fluid (with the corresponding definition of the reduced temperature, i.e. the ratio of k_BT to the work required to separate a van der Waals pair.) In the case of a plasma, an ionic fluid without a solvent, the coexistence curve is for the liquid-vapor equilibrium, while for solutions it corresponds to two solution phases of different concentrations in equilibrium. Some non-aqueous solutions are known which do unmix to form two liquid phases of slightly different concentrations. While no examples in aqueous solution are known, the corresponding states principle for ionic fluids suggests that solutions of even 1-1 electrolytes in water might unmix above 300°C in some concentration range. It is difficult to be more precise because the coexistence curve apparently is sensitive to details in the pair potentials. (36)

7. MIXING COEFFICIENTS

In a symposium concerning industrial applications one expects solutions of single electrolytes to take second place in importance behind solutions of mixed electrolytes. The latter have scarcely been mentioned in this review. To redress the balance we recall here an interesting question regarding the mixing coefficient g_1 that appears in the expansion of the excess free energy of mixing two electrolytes with a common ion at constant molal ionic strength I.

$$\Delta_m G^{ex}(y,I) = y(1-y)I^2 WRT[g_0(I) + (1-2y)g_1(I) + \ldots] \quad (12)$$

Here y is the fraction of the ionic strength due to one of the electrolytes and W is the mass (kg) of solvent in the mixture. It is well known (4) that higher order limiting laws, depending upon only the charge type and ε, determine the behavior of g_0 at very low I. Recently it was found experimentally by Cassel and Wood (45) that in mixtures of different charge types at low I the coefficient g_1 (actually $\partial(g_1/T)/\partial(1/T)$) seemed to be diverging as $I \to 0$ and, moreover, that the species-dependence of the divergence indicated that the divergence was a limiting law phenomenon. Calculations by two approximation methods (46) showed that g_1 would indeed appear to diverge in the range studied by Cassel and Wood, and that the effect was indeed due to the long range Coulomb interactions, but the calculated and experimental coefficients disagreed in both sign and magnitude! On the other hand, more recent measurements in a different system by Khoo and coworkers (47) seem consistent with the theory, although the comparison has yet to be carefully developed.

8. ACKNOWLEDGEMENT

We are grateful for the support of the present work by the National Science Foundation.

LITERATURE CITED

1. Hill, T. L. "Statistical Mechanics," McGraw-Hill, New York, 1956.
2. Hansen, J. P. and McDonald, I. R., "Theory of Simple Liquids," Academic Press, N.Y., 1976.
3. Watts, R. O. and McGee, I. J. "Liquid State Physics," J. Wiley and Sons, N.Y., 1976.

4. Friedman, H. L., "Ionic Solution Theory," Interscience Publishers, N.Y., 1962.

5. Friedman, H. L. and Dale, W. D. T., in "Modern Theoretical Chemistry, Vol. V, Statistical Mechanics," Plenum Press, N.Y., 1977, Editor, Berne, Bruce J., pp. 85-135.

6. Andersen, H. C., Ann. Revs. Phys. Chem., 1975, 26, 145.

7. Friedman, H. L., Jour. Electrochem. Soc., 1977, 124, 421C.

8. Friedman, H. L., in "Protons and ions involved in fast dynamic phenomena," Elsevier, N.Y., 1978. Laszlo, P., editor.

9. Lie, G. C., Clementi, E., and Yoshimine, N., J. Chem. Phys., 1976, 64, 2314.

10. Rahman, A. and Stillinger, F. H., J. Chem. Phys., 1971, 55, 3336.

11. Geiger, A., Rahman, A., and Stillinger, F. H., J. Chem. Phys., 1979, 70, 263.

12. McMillan, W. G. and Mayer, J. E., J. Chem. Phys., 1945, 13, 276.

13. Friedman, H. L., J. Chem. Phys., 1960, 32, 1134.

14. Stell, G. R. and Høye, J., Faraday Discuss. Chem. Soc., 1977, 64, 16.

15. Pitzer, K. S., Accts. Chem. Res., 1977, 10, 371.

16. Patey, G. N. and Valleau, J. P., J. Chem. Phys., 1975, 63, 2334.

17. Levesque, D., Weis, J. J., and Patey, G. N., Physics Letters, 1978, 66A, 115.

18. Kraus, C. A., J. Phys. Chem., 1956, 60, 129.

19. Adelman, S. and Chen, J.-H., J. Chem. Phys., 1979, 70, 4291.

20. Kauzmann, W., Adv. Protein Chem., 1959, 14, 1.

21. Pangali, C., Rao, M., and Berne, B. J., J. Chem. Phys., 1979, 71, 2975.

22. Pratt, L. and Chandler, D., J. Chem. Phys., 1977, 67, 3683.

23. Jolicoeur, C. and Friedman, H. L., Ber. Bunsenges., 1971, 75, 248.

24. F. Franks in "Water, A Comprehensive Treatise," Plenum Press, N.Y.. 1974, Vol. 4, Ch. 1. Franks, F., editor.

25. Clarke, A. H., Franks, F., Pedley, M. D., and Reid, D. S., Faraday Trans. Chem. Soc. I, 1977, 73, 290.

26. Ramanathan, P. S. and Friedman, H. L., J. Chem. Phys., 1971, 54, 1086.

27. Debye, P. and McAulay, J., Physik. Z., 1925, 26, 22.

28. Gurney, R. W., "Ionic Processes in Solution," Dover, N.Y., 1953.

29. Frank, H. S., in "Chemical Physics of Ionic Solutions, edited by Conway, B. E. and Barradas, R. G., Wiley, N.Y., 1966, Z. Physik Chem. (Leipzig), 1965, 228, 364.

30. Friedman, H. L., Smitherman, A., and De Santis, R., J. Solution Chem., 1973, 2, 59.

31. Ramanathan, P. S., Krishnan, C. V., and Friedman, H. L., J. Solution Chem., 1972, 1, 237.

32. Friedman, H. L. and Ramanathan, P. S., J. Phys. Chem., 1970, 74, 3756.

33. Krishnan, C. V. and Friedman, H. L., J. Solution Chem., 1974, 3, 727.

34. Friedman, H. L., Krishnan, C. V., and Hwang, L. P., in "Structure of Water and Aqueous Solutions," Luck, W., editor, Verlag Chemies Gmbh, 1974, p. 169.

35. Rossky, P. J. and Friedman, H. L., J. Phys. Chem., 1979, submitted.

36. Dogonadze, R. R. and Kornyshev, A. A., J. Chem. Soc. Faraday Trans., 1974, 70, 2.

37. Bich, E., Ebeling, W., and Krienke, H., Z. Phys. Chem., Leipzig, 1976, 257, 549.

38. Wiechert, H., Krienke, H., Feistel, R., and Ebeling, W., Z. Phys. Chem. Leipzig, 1978, 259, 1057.

39. Justice, J.-C. and Justice, M.-C., Faraday Discuss. Chem. Soc., 1977, 64, 265.

40. Justice, M.-C. and Justice, J.-C., Pure and Applied Chem., 1979, 51, 1681.

41. Hirata, F., Holz, M., Hertz, H. G., and Friedman, H. L., to be submitted.

42. Soper, A. K., Neilson, G. W., Enderby, J. E. and Howe, R. A., J. Phys. C, 1977, 10, 1793.

43. Sandstrom, D. R., J. Chem. Phys., 1979, 71, 2381.

44. Friedman, H. L. and Larsen, B., J. Chem. Phys., 1978, 69, 92.

45. Wood, R. H. and Cassel, R. B., J. Phys. Chem., 1974, 78, 1924.

46. Friedman, H. L. and Krishnan, C. V., J. Phys. Chem., 1974, 78, 1927.

47. Khoo, K. H., Lim, T. K., and Chan, C. Y., Faraday Trans. Chem. Soc. I, 1978, 74, 2037.

RECEIVED January 31, 1980.

Activity Coefficients, Ionic Media, and Equilibrium Constants

R. M. PYTKOWICZ

Oregon State University, Corvallis, OR 97330

In this work the concepts of ionic medium, effective ionic strength and free versus total activity coefficients are examined. Then they are applied to the study of permissible and incorrect translations of equilibrium constants from one medium to another.

Ionic Media

Ionic media are solutions of background electrolytes which are concentrated enough so that the activity coefficients of the electrolytes of interest do not change during processes which are occurring. Typical ionic media are a 1 m $HClO_4$ or $NaClO_4$ solution and seawater.

Let us consider the dissolution-precipitation process in seawater in the following example. The normal concentrations of calcium and of carbonate in the near-surface oceanic waters are about $[Ca^{2+}] = 0.01$ and $[CO_3^{2-}] = 2 \times 10^{-4}$ M. The $CaCO_3$ in solution is metastable and roughly 200% saturated (1). Should precipitation occur due to an abundance of nuclei, $[CO_3^{2-}]$ will drop to 10^{-4} M but $[Ca^{2+}]$ will change by no more than 2%. Therefore, the ionic strength of the ionic medium seawater will remain essentially constant at 0.7 M. The major ion composition will also remain constant. We shall see later what the implications are for equilibrium constants.

It is important to realize that one must take into consideration ion association of the ionic media electrolytes because it affects the effective ionic strength (2). This in turn changes the activity coefficients of the ions under study. The effective ionic strength, I_e, of a 2-1 electrolyte CA_2 which associates is given by

$$I_e = 0.5\{4[C^{2+}]_F + [A^-]_F + [CA^+]\} \quad (1)$$

where the subscript F refers to free species. We shall see that I_e plays a key role in equilibria.

0-8412-0569-8/80/47-133-561$05.00/0

Electrolytes which often are considered to be completely dissociated but in effect are not are NaCl and $NaClO_4$ (3).

Ion-Pairing

In the preceeding section mention was made of ion association (ion-pairing) which, for the purposes of this paper, will refer to coulombic entities with or without cosphere overlap. Experimental support for ion-pairing has come from sound attenuation (4), Raman spectroscopy (5) and potentiometry (2, 3). Credibility has resulted from the model of Fuoss (6) applied by Kester and Pytkowicz (7).

Our method at present is not based upon theoretical models or departures from ideal behavior. It consists in the use of potentiometric determinations and literature values of activity coefficients, starting with HCl-$HClO_4$ electrolyte mixtures and with the assumption that $HClO_4$ is completely dissociated since the association constant pK = 7 is extremely small in this case. A comparison of experimental results with those calculated from the Fuoss (6) theory is presented in Table I. The theory is only valid approximately so that the order of magnitude agreement is fairly good, except in the cases of $MgCO_3{}^0$ and $CaCO_3{}^0$. Stoichiometric association constants K* are then obtained from the activity coefficients, expressions for K*, and from equations for the conservation of mass. The latter express the total concentration of a given ion as the sum of the concentrations of the free ion and of the ion-pairs. Values of K* and of the activity coefficients of free ions in ionic media depend only upon the effective ionic strength as is shown later.

Table I.
A comparison of stoichiometric association constants calculated from the Fuoss (6) model with Debye radii and from the measurements of Johnson and Pytkowicz (3).

Ion Pair	Calculated	Measured
$NaSO_4{}^-$	2.1	2.02
$MgSO_4{}^0$	4.6	10.2
$CaSO_4{}^0$	6.5	10.8
$NaHCO_3{}^0$	0.46	0.28
$MgHCO_3{}^+$	2.8	1.62
$CaHCO_3{}^+$	2.4	1.96
$NaCO_3{}^-$	2.0	4.25
$MgCO_3{}^0$	3.9	112.3 (?)
$CaCO_3{}^0$	4.9	279.6 (?)
$NaCl^0$	0.43	0.321
$MgCl^+$	2.5	1.91
$CaCl^+$	2.3	2.24

From the free activity coefficients and values of K* obtained in binary solutions it is possible then to calculate total (stoichiometric) activity coefficients in more complex solutions.

Our results lead to good predictions of activity coefficients in multicomponent systems from data measured in simple solutions. Also, they yield values similar to those of Pitzer and Kim (8) as is shown in Table II.

Table II.
A comparison of trace mean activity coefficients calculated by the methods of Johnson and Pytkowicz (3) and Pitzer and Kim (8).

System A/B	A (1)	A (2)	A (3)
$KCl/NaCl$	0.620	0.636	0.626
$KCl/CaCl_2$	0.639	0.628	0.662
$KCl/MgCl_2$	0.649	0.636	0.674
$NaCl/CaCl_2$	0.671	0.643	0.682
$NaCl/MgCl_2$	0.682	0.650	0.694
$CaCl_2/MgCl_2$	0.461	0.460	0.461

(1) Experimental
(2) Johnson and Pytkowicz
(3) Pitzer and Kim

For an artificial seawater consisting of Na^+, K^+, Mg^{2+}, Ca^{2+} and Cl^- (HCO_3^- and CO_3^{2-} missing) we obtain as examples that Na^+ is 85.1% and Ca^{2+} is 45.7% free.

Activity Coefficients

The ion-pair concept leads to two types of activity coefficients; $\gamma_{\pm T}$ for free plus dissociated species and $\gamma_{\pm F}$. These γ's are related by

$$a = \gamma_{\pm F}^2[F] = \gamma_{\pm T}^2[T] \quad (2)$$

where [F] and [T] are concentrations and a is the activity (2).

$\gamma_{\pm F}$ is a construct which includes effects such as those due to hydration, ion-cavity interactions, hard core, etc., but not to ion-pairing. Pytkowicz and Kester (2) and Johnson and Pytkowicz (3) have shown that $\gamma_{\pm F}$ depends only upon the I_e of the medium while $\gamma_{\pm T}$ is also a function of the composition.

It is important in equilibrium calculations to select the proper equation for activity coefficients when the authors do not actually measure them or if measured values in the medium of interest are not directly available. Much of what I say is known but some new concepts are entered into what follows.

Some measurements in single electrolyte solutions are available at concentrations of the order of 10^{-3} m although most of the data has been obtained in the range 0.1 - 6.0 m for cases when the solubility of the salt was not exceeded.

In single electrolyte solutions the Debye-Hückel limiting equation works up to 10^{-3} m. Its well known extended form used at m > 10^{-3} is

$$\log \gamma_{\pm T} = - \frac{|Z_C Z_A| A_D I^{0.5}}{1 + B_D a_D I^{0.5}} \quad (3)$$

however, it is entirely empirical because Frank and Thompson (9) showed that the ionic atmosphere is then coarse-grained. The sphere of radius $1/\kappa$ may contain only one or a fraction of an ion in contrast to the requirements of the theory. If equation (3) is used then a_D has to be calculated by curve-fitting to data for the electrolyte of interest. This fit to experimental data means that $\gamma_{\pm T}$ rather than $\gamma_{\pm F}$ is being determined.

In terms of empirical equations, the following one provides good fits from 10^{-3} to 1.0 m solutions according to our work. The Davies equation (10)

$$\log \gamma_{\pm} = - \frac{Z_C Z_A \; A_C I^{0.5}}{1 + B_C I^{0.5}} + C_C I + D_C I^{1.5} + E_C I^2 \quad (4)$$

in contrast to the above equations, is a predictive tool that works well up to I $\cong$ 0.1. Note should be made, however, to the effect that, if ion-pairing or complexing occurs, then the Davies equation only works if $\gamma_{\pm}$ is corrected for association. Thus, this equation yields $(\gamma_{\pm})_T = (\gamma_{\pm})_F$ when there is no association and $(\gamma_{\pm})_F$ when ion-pairing occurs.

Several approaches are available in the case of mixed electrolyte solutions. The Guntelberg equation can be used at very high dilutions to avoid the ambiguity in the meaning of a_D, the distance of closest approach, when several electrolytes are present. This equation is empirical and has fewer terms than the Debye-Huckel extended equation. I found it to yield poor agreement with experimental results even at m = 0.01 for NaCl at 25°C ($\gamma_{\pm}$ calc = 0.8985 and $\gamma_{\pm}$ exp = 0.9024). For the Davies equation for m = 0.20 one obtains $\gamma_{\pm}$ calc = 0.752 and $\gamma_{\pm}$ exp = 0.735 also for NaCl at 25°C.

An extended form of the Debye-Hückel equation is the hydration one of Robinson and Stokes (11). It contains two adjustable parameters, a_D and h, where h is related to the hydration number. It can be fitted to $\gamma_{\pm}$ for several electrolytes for concentrations in excess of 1 m. Their equation has the valuable feature of describing not only the salting-in but also the salting-out part of the $\gamma_{\pm}$ versus m curve. It should be noted, however, that the arbitrary allocation of non-hydration terms to the Debye-Hückel equation for the estimate of a_D at high concentrations weakens the theoretical validity of the hydration equation. Furthermore, it is an oversimplification to only consider the nonspecific coulombic and the hydration terms.

Note that, unless otherwise indicated, I am using I_e and $(\gamma_\pm)_T = (\gamma_\pm)_{exp}$ in this review. Furthermore, the simple equations presented so far do not include terms for the cosphere overlap, the hard core term, and ion-cavity interactions.

Next, predictive equations for activity coefficients in mixed electrolyte solutions, based upon results in simpler ones, will be mentioned. The work of Brønsted (12) and of Guggenheim (13) led to the specific interaction equation

$$\log \gamma_{\pm B} = - \frac{Z_C Z_A \; A_D I^{0.5}}{1 + I^{0.5}} + [2xB_{MX} + (B_{NX} + B_{MY})(1 - x)]m \quad (5)$$

$x/(1 - x)$ is the molal ratio of B to C where B is MX and C is NY. A similar condition applies to $\log \gamma_{\pm C}$. One can derive Harned's rule from this expression although it can also be done from the ion pairing model (2). The equations work well for I up to 0.1 m and permit the calculation of $\gamma_{\pm B}$ and $\gamma_{\pm C}$ in a mixture of the two electrolytes from data obtained in single electrolyte solutions. The method suffers, however, because the B coefficients are not allowed to vary with the ionic strength.

The statistical thermodynamic approach of Pitzer (14), involving specific interaction terms on the basis of the kinetic core effect, has provided coefficients which are a function of the ionic strength. The coefficients, as the stoichiometric association constants in our ion-pairing model, are obtained empirically in simple solutions and are then used to predict the activity coefficients in complex solutions. The Pitzer approach uses, however, a first term akin to the Debye-Hückel one to represent nonspecific effects at all concentrations. This weakens somewhat its theoretical foundation.

As most other methods, including our ion-pairing model which was described earlier, the Pitzer approach is empirical in practice. The interaction coefficients in this case are determined by curve fitting in single electrolyte solutions.

In the equations developed by Reilly and Wood (15) from the cluster integral model (16), $\gamma_\pm$ is calculated in complex solutions from excess properties of single salt solutions. Note that the cluster integral approach is based upon terms which represent the contributions of pair-wise ion interactions in various types of clusters to the potential interaction energy. Then, the partition function and the excess properties of the solution can be evaluated. The procedure is akin to the virial expansion in terms of clusters.

The more recent statistical thermodynamic work approaches the problem from the standpoint of the radial distribution function (17). In using this method, Ramanthan and Fridman (18), in contrast to the method of Pitzer, made the Gurney cosphere overlap, the ion-cavity repulsion, and repulsive core potential explicit. The curve fitting was done by means of the Gurney term.

The most fertile approaches so far, from the standpoint of predicting $\gamma_\pm$ in complex solutions from data in simple ones, have been that of Pitzer (14), and the ion-pair approach of Pytkowicz and Kester (2) and of Johnson and Pytkowicz (3). The lattice model of Pytkowicz and Johnson (19) is, at this time, an explanatory rather than a predictive tool.

The only three methods which do not require curve-fitting at present are the use of the Debye-Hückel limiting law and of the equations of Guntelberg and of Davies. Unfortunately these equations are of value only in very dilute and simple solutions.

Equilibrium Constants

Let us examine the following types of constants now in use and their relationships to the thermodynamic constants $K^{(t)}$:

a) Apparent dissociation constants which are of practical use in ionic media (20)

$$K_{HA}' = \frac{ka_H[A^-]_T}{[HA]} = K_{HA}^{(t)} \frac{(\gamma_{HA})_T}{(\gamma_A)_T} \tag{6}$$

b) Stoichiometric solubility products, such as

$$K_{sp}' = [Ca^{2+}]_T[CO_3{}^{2-}]_T = K_{sp}^{(t)}/(\gamma_{\pm CaCO_3})_T^2 \tag{7}$$

c) Free solubility product

$$K_{sp}'' = [Ca^{2+}]_F[CO_3{}^{2-}]_F = K_{sp}^{(t)}/(\gamma_{\pm CaCO_3})_F^2 \tag{8}$$

d) Stoichiometric association constant for ion-pairs

$$K^* = \frac{[CaCO_3{}^0]}{[Ca^{2+}]_F[CO_3{}^{2-}]_F} = K^{*(t)} \frac{(\gamma_{\pm CaCO_3})_F^2}{\gamma_{CaCO_3}} \tag{9}$$

e) Affinity constants for complexes

$$\beta = \frac{[PbCl_2]}{[Pb]_F[Cl^-]_F^2} = \beta^{(t)} \frac{(\gamma_{\pm PbCl_2})_F^3}{\gamma_{PbCl_2{}^0}} \tag{10}$$

$(\gamma_\pm)_F$ depends only upon the effective ionic strength of the medium. Thus, K_{sp}'', K^*, and β can be determined in one ionic medium such as $HClO_4$ and applied in another one such as seawater, provided that I_e is the same in the two solutions.

K_{HA}' and K_{SP}', on the other hand, must be determined and applied in the same medium because γ_T depends upon the composition of the solution. Note that ion-pairing models are required for both classes of constants to ascertain that I_e is indeed the same.

Literature Cited

1. Pytkowicz, R.M., J. Geol., 1965, 73, 196-199.
2. Pytkowicz, R.M.; Kester, D.R., Am. J. Sci., 1969, 267, 217-229.
3. Johnson, K.; Pytkowicz, R.M., Am. J. Sci., 1978, 278, 1428-1447.
4. Fisher, F.H., Science, 1967, 157, 823.
5. Daly, F.P.; Brown, C.W.; Kester, D.R., J. Phys. Chem., 1972, 76, 3664-3668.
6. Fuoss, R.M., J. Am. Chem. Soc., 1958, 80, 5059-5061.
7. Kester, D.R.; Pytkowicz, R.M., Mar. Chem., 1975, 3, 365-374.
8. Pitzer, K.S.; Kim, J.J., Am. Chem. Soc. Jour., 1974, 96, 5701.
9. Frank, H.S.; Thompson, P.T., J. Chem. Phys., 1959, 31, 1086-1095.
10. Davies, C.W., "Ion Association", Butterworths, London, 1962.
11. Robinson, R.A.; Stokes, R.H., "Electrolyte Solutions", Butterworths, London, 1968.
12. Brønsted, J.N., J. Am. Chem. Soc., 1922, 44, 877-898.
13. Guggenheim, E.A., Philos. Mag., 1935, A, 588-643.
14. Pitzer, K.J., Phys. Chem., 1973, 77, 268-277.
15. Reilly, P.J.; Wood, R.H., J. Phys. Chem., 1969, 73, 4292-4297.
16. Friedman, H.L., "Ionic Solution Theory", Interscience, New York, 1962.
17. Vaslow, F., "Water and Aqueous Solutions", R.A. Horne, ed., Interscience, New York, 1972, pp. 465-518.
18. Ramanathan, P.S.; Friedman, H.L., J. Chem. Phys., 1971, 54, 1086-1099.
19. Pytkowicz, R.M.; Johnson, K., "Activity Coefficients in Electrolyte Solutions", R.M. Pytkowicz, ed., CRC Press, Coral Gables, Florida, In press.
20. Pytkowicz, R.M.; Ingle, S.E.; Mehrback, C., Limnol. Oceanogr., 1974, 19, 665-669.

RECEIVED February 11, 1980.

Flow Calorimetry of Aqueous Solutions at Temperatures up to 325°C

R. H. WOOD and D. SMITH–MAGOWAN
University of Delaware, Newark, DE 19711

1) Need

The fact that we are gathered together at a conference on "Thermodynamics on Aqueous Systems with Industrial Application" indicates the importance of thermodynamic data on aqueous solutions. In particular, there is a great need for data on high temperature aqueous systems. Because of the experimental difficulties, there are relatively few measurements on these systems and yet they are of very great industrial importance.

As the temperature of water approaches the critical temperature, 374°C, it becomes a low dielectric constant solvent with the dielectric constant approaching one tenth its value at 25°C. This is about the same dielectric constant as 1,1 dichloroethane at 25°C. Because of this large change in solvent properties there are corresponding very large changes in the properties of electrolytes dissolved in water. The few data available indicate[1-4] that the heat capacities of dilute aqueous solution become very large and negative as the temperature approaches the critical temperature and the data presented below will show that the relative apparent molal enthalpy, L_{ϕ}, becomes very large and positive at higher temperatures. These very large changes in thermodynamic properties make it very difficult to predict high temperature properties of solutions from low temperature measurements and increase the need for accurate measurements.

2) Information Obtained

Calorimetric measurements, when combined with the normally available room temperature thermodynamic properties, give values for free energy, enthalpy, heat capacity and even volume at high temperatures.

We have been actively developing two types of calorimeters which will operate at elevated temperatures and pressures. One type is a heat of mixing calorimeter to measure enthalpies of dilution in order to obtain differences in partial molal enthalpy

0-8412-0569-8/80/47-133-569$05.00/0

or heat contents:

$$\Delta H_D = -(L_{\phi \underline{m}_f} - L_{\phi \underline{m}_i})$$

This property simply considered is the first temperature derivative of the free energy or activity and can be used to obtain osmotic coefficients and activity coefficients by the relationships:

$$\phi(\underline{m},\mathcal{T}) - \phi(\underline{m},T_2) = \{\underline{m}^{\frac{1}{2}}/2\nu R\}\int_{\mathcal{T}}^{T_2} (\frac{\partial L_\phi}{\partial \underline{m}^{\frac{1}{2}}}) d(\frac{1}{T})$$

and

$$\ell n\gamma(\underline{m},T) = [\phi(\underline{m},T)-1] + \int_0^m (\phi(\underline{m}',T)-1)d\ell n\underline{m}'$$

where $\mathcal{T}$ is a reference temperature. We have also developed a heat capacity calorimeter for these extreme conditions.

The heat capacity is the second temperature derivative of the free energy and can be used to calculate the temperature dependence of equilibria by the relationship:

$$\frac{\Delta G(T,P)}{T} = \frac{\Delta G(\mathcal{T},P)}{\mathcal{T}} + \Delta H(\mathcal{T},P)[\frac{1}{T} - \frac{1}{\mathcal{T}}] + \int_{\mathcal{T}}^{T} \int_{\mathcal{T}}^{T'} (\Delta C_p \, dT'') \, d(\frac{1}{T'})$$

Similarly the temperature dependencies of the relative apparent molal heat content can be determined from the heat capacity by:

$$L_\phi(m,T_2) = L_\phi(m,\mathcal{T}) + \int_{\mathcal{T}}^{T_2} (C_{p\phi}(m,T) - C_p^{\circ}(0,T))dT$$

These calorimeters can be used to determine these thermal properties throughout a wide range of pressures. The pressure dependence can be used to calculate volumetric properties by means of the relationships:

$$(\frac{\partial H}{\partial P})_T = -T(\frac{\partial V}{\partial T})_P + V$$

$$(\frac{\partial C_p}{\partial P})_T = -T(\frac{\partial^2 V}{\partial T^2})_P$$

3) Advantages

In a flow calorimeter the thermodynamic properties are measured in a flowing stream contained in a small diameter tubing. Flow calorimetric techniques have been used for many years at room temperature because of their speed and convenience[5-9]. For operations at high temperatures or with a high vapor pressure solvent, the advantages of using flow calorimetric techniques are overwhelming. In the first place there is no vapor space so there is no necessity for corrections for the change in the vapor space composition. Because of the small diameter tubing used, relatively thin walled tubing is strong enough to contain the high pressures and as a result the calorimeter can be constructed for very rapid thermal response. A third advantage is that experiments can be run consecutively without cooling and reloading the calorimeter. All that is necessary is to start pumping into the calorimeter from room temperature the fluids for the next measurement. In the following section we will show that these advantages of flow calorimetric techniques can be realized in practice for measurements on high temperature aqueous solutions by discussing the operation of several instruments that have been constructed in our laboratory.

4) Measurements of $\Delta_m H$

The first attempt in our laboratory to apply flow techniques to high temperature operation was the construction by Dr. E.E. Messikomer of a flow, heat-of-mixing calorimeter[10]. Unfortunately, because the thermopiles used in this instrument did not work above 100°C the instrument was limited to this temperature. However, the results were encouraging because they showed that very rapid and accurate thermodynamic data could be obtained and that the operation of the calorimeter was as easy at 100°C as it was at room temperature.

Because of the encouraging results obtained with the first calorimeter, Dr. James Mayrath build a new version which was successfully operated up to 200°C[11]. A schematic of this calorimeter is given in figure 1. Basically it is a heat-flow calorimeter in which the two liquids to be mixed are pumped at room temperature into a counter current heat exchanger, after which they are equilibrated with an aluminum calorimetric block. Next the two liquids are mixed, and the heat generated by the mixing process is extracted from the flowing stream by a series of thermopiles which can measure the heat extracted. Using several heat extractors guaranteed that most of the heat generated is measured by the thermopiles and the sum of the voltage on the thermopiles is then a measure of the rate of heat production.

A diagram of a typical run is given in figure 2 which shows the power generated by mixing of magnesium chloride with water at 200°C. In this calorimeter a heat of dilution takes 30 minutes and from an initial base line it takes about 15 minutes for the calorimeter to reach a steady state. The sensitivity of this calorimeter was equivalent to being able to detect a 2×10^{-4}K

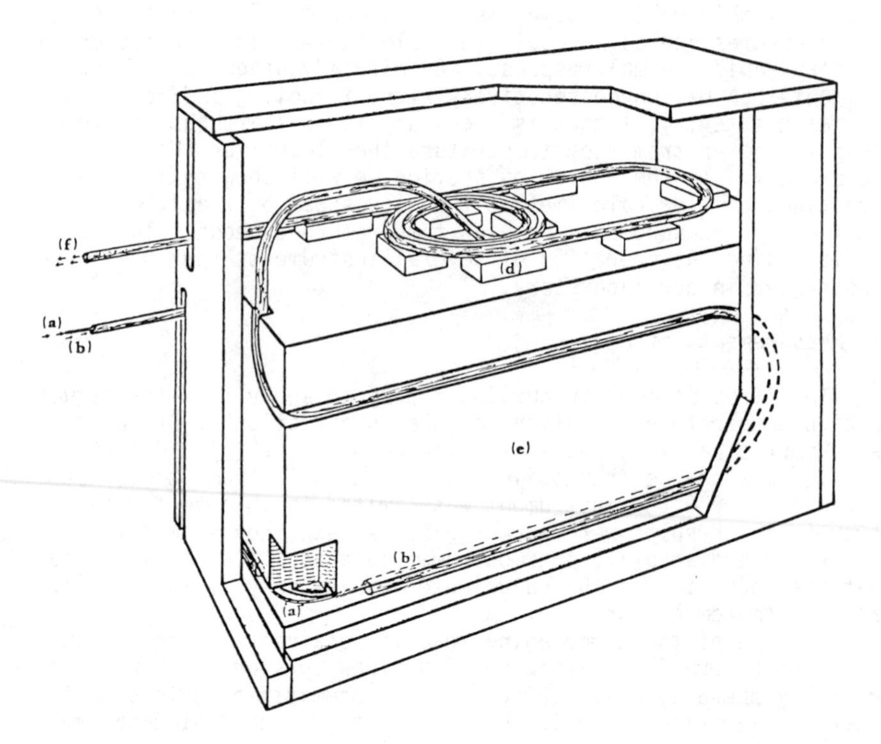

Figure 1. Flow heat of mixing calorimeter: (a and b) solutions to be mixed; (c) calorimetric block; (d) thermopiles for detecting heat flow; (e) exit for mixture

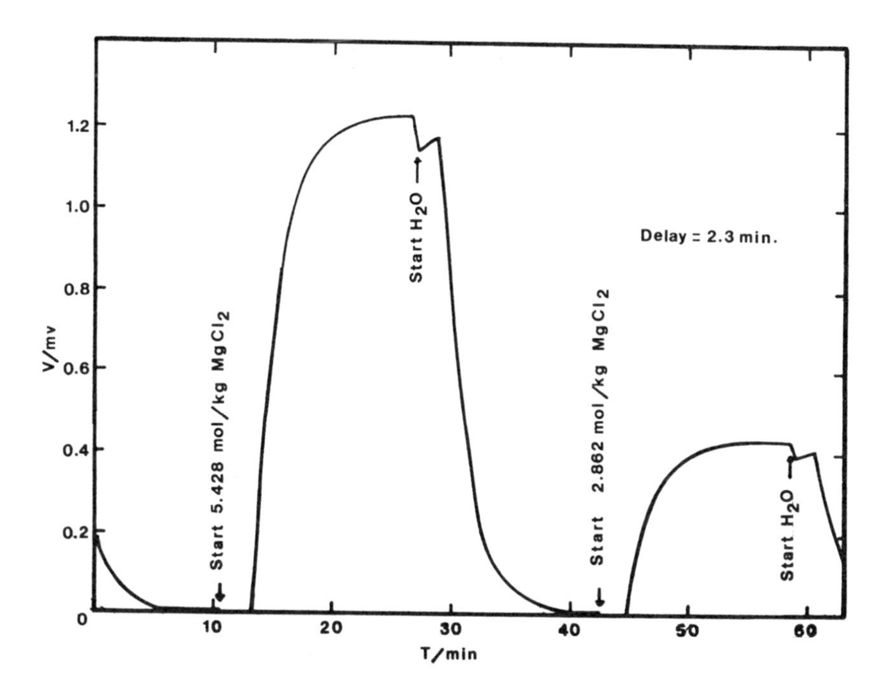

Figure 2. Results of an enthalpy of dilution run on aqueous $MgCl_2$ at 473 K (thermopile voltage vs. time—maximum 1.23 ± 0.002; Q = 1.129 watts; $\triangle$T = 12.8 K)

temperature rise on mixing the calorimetric fluids. Some typical results from this calorimeter are shown in figures 3 and 4. It should be noted that the heat of dilution of magnesium chloride at 200°C is 40 kJ mol^{-1} and that this is close to the heat of reaction of H^+ with OH^- at room temperature. Thus, the heat effects in water at 200°C are extremely large and changing rapidly with temperature. This is a further reminder of the difficulty of predicting the properties of high temperature aqueous solutions from their properties at room temperatures.

5) Heat Capacity Measurements

The success of the mixing calorimeters further encouraged us to construct a flow, heat-capacity calorimeter using the basic design principals of Patrick Picker et al[8]. A schematic diagram of our heat capacity calorimeter is given in figure 5. The operation of the instrument is as follows: Water is pumped with a high pressure liquid chromatography pump through a 1.5 mm outside diameter hastalloy tube through a heat exchanger and onto a copper calorimetric block where it is equilibrated at the reference temperature. The fluid in the tube is then heated about 2°K and the resulting temperature rise detected by a thermistor. The stream then leaves the calorimetric block through the counter current heat exchanger and goes to a sample injection valve which allows a sample loop (containing 10 ml of the solution to be measured) to be interjected into the flow stream. The stream then goes through a second counter current heat exchanger and into an identical clorimetric unit with heater and thermistor to detect the temperature rise. After leaving the block through the counter current heat exchanger the solution exits the system through a back pressure regulator. In operation the pump and heaters are turned on with water flowing through both calorimetric units and a wheatstone bridge containing the two thermistors in opposite arms is balanced. After a steady state is reached the sample loop valve is opened and the sample solution flows through the second calorimetric unit. When the sample hits the heater and thermistor on the second unit there is a change in heat capacity and a consequent change in temperature. The heater on this unit is adjusted to rebalance the thermistor bridge and thus to keep the temperature rise exactly the same. The ratio of the power applied to the heater with water flowing and with the sample flowing is then the ratio of the volumetric heat capacities of the two solutions. The ratio of the mass flows through the calorimetric units in the two cases is just the ratio of the densities of water and of the solution to be measured at the temperature of the sample loop. This can be easily shown using the assumptions that 1) at constant composition, the mass flow is independent of temperature 2) at constant temperature, the volumetric flow is independent of the composition and 3) the heat and volume of mixing at the interface between the two solutions pro-

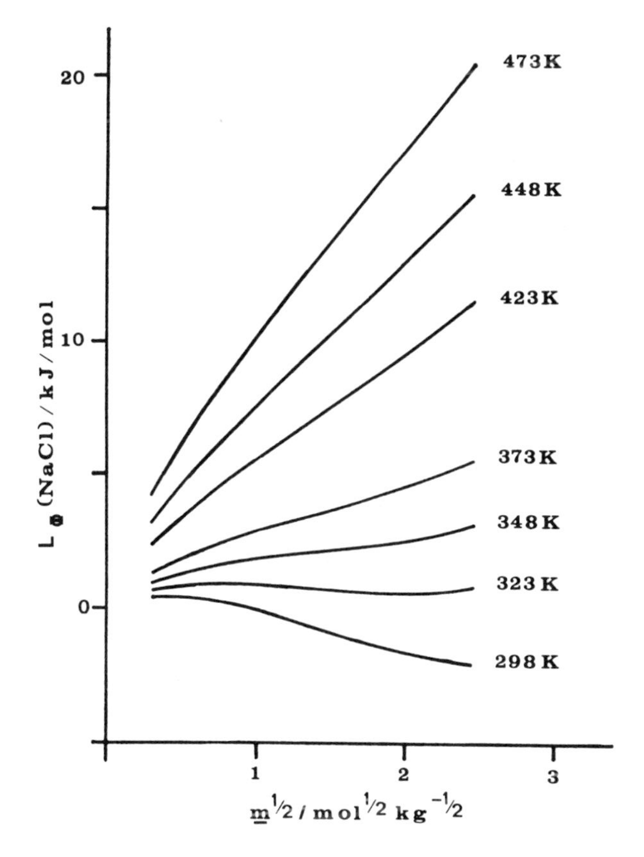

Figure 3. Apparent molal enthalpy of aqueous NaCl as a function of molality and temperature

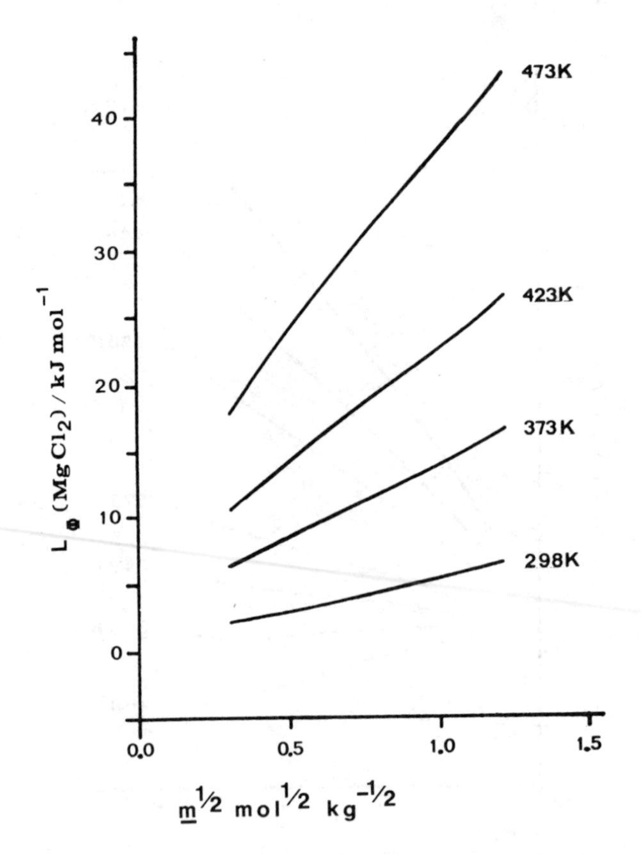

Figure 4. Apparent molal enthalpy of aqueous $MgCl_2$ as a function of molality and temperature

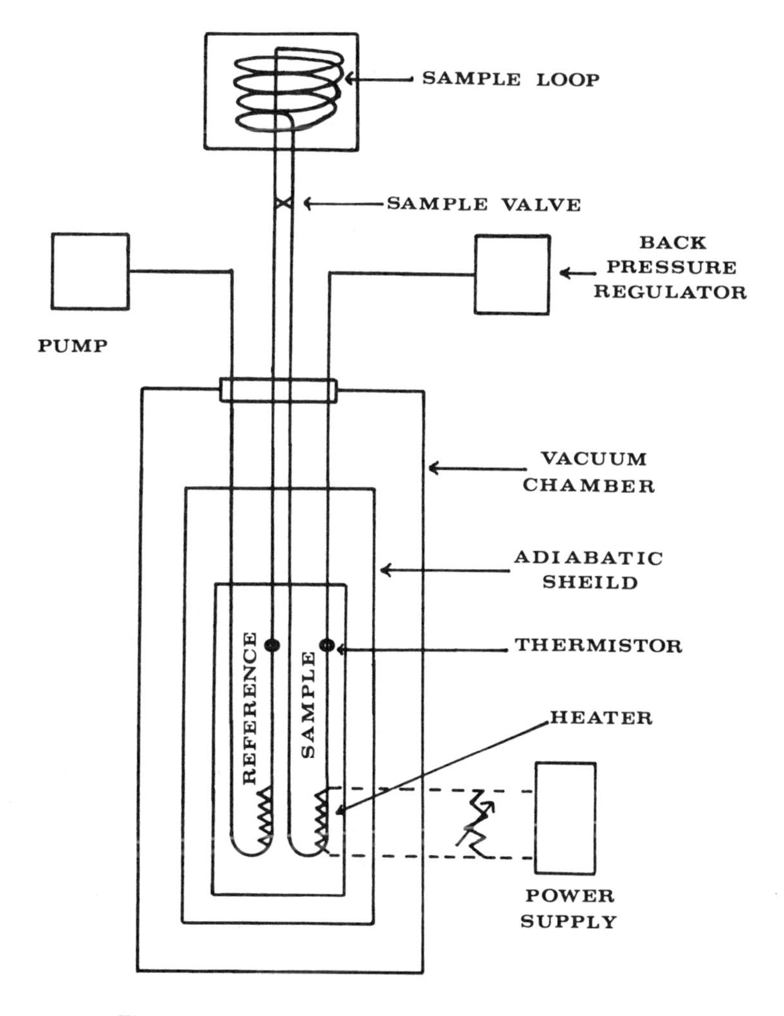

Figure 5. Schematic of flow, heat-capacity calorimeter

duces a negligible change in volume. Experience with the Picker instrument has shown that this latter assumption is quite accurate. The resulting equation for the heat capacities of the two solutions is

$$\frac{C_{p,2}}{C_{p,1}} = \frac{\dfrac{P_2}{\rho_1}}{\dfrac{P_1}{\rho_2}}$$

Where P_1 and P_2 are the powers in the heater with solutions 1 and 2 flowing and ρ_1 and ρ_2 are the densities of the solutions at the temperature of the sample loop.

The resulting instrument has a response time for changes of electrical power input of 50 seconds for 99% response, a sensitivity to a change in power of about 0.005% and a proven capability of operating with this sensitivity up to temperatures as high as 325°C.

The results of some measurements on sodium chloride solutions are given in figure 6. The difficulty of predicting properties of solutions at high temperatures is emphasized by these results. At low concentrations there is a very large change in heat capacity as temperature increases which is not present at higher concentrations. Indeed, the heat capacity differences at 25°C seems almost negligible compared with the changes found with changing temperature at low molality and with changing molality at high temperature.

It is difficult to compare these results with the previous results of other authors since we chose to make measurements along an isobar which was accessible at all temperatures. This allows convenient data reduction. Other authors have generally measured along the saturated water vapor pressure curve with consequent complications in data reduction. While at present, only this one isobar has been systematically investigated, a few measurements of the pressure dependence at 320K and 572K have been made. These results show that at all temperatures the pressure dependence of the heat capacity is appreciable and needs to be more fully evaluated, since the present body of volumetric data is not sufficiently precise to make these corrections accurately. In fact, our very limited results at different pressures suggest that the precision of this procedure may be sufficient to supplant volumetric determinations at high temperature and pressures.

6) Conclusion

These preliminary results show that the promise of flow calorimetric techniques for investigating the thermodynamic properties of high temperature aqueous solutions has been realized. Although there are many experimental difficulties in adapting

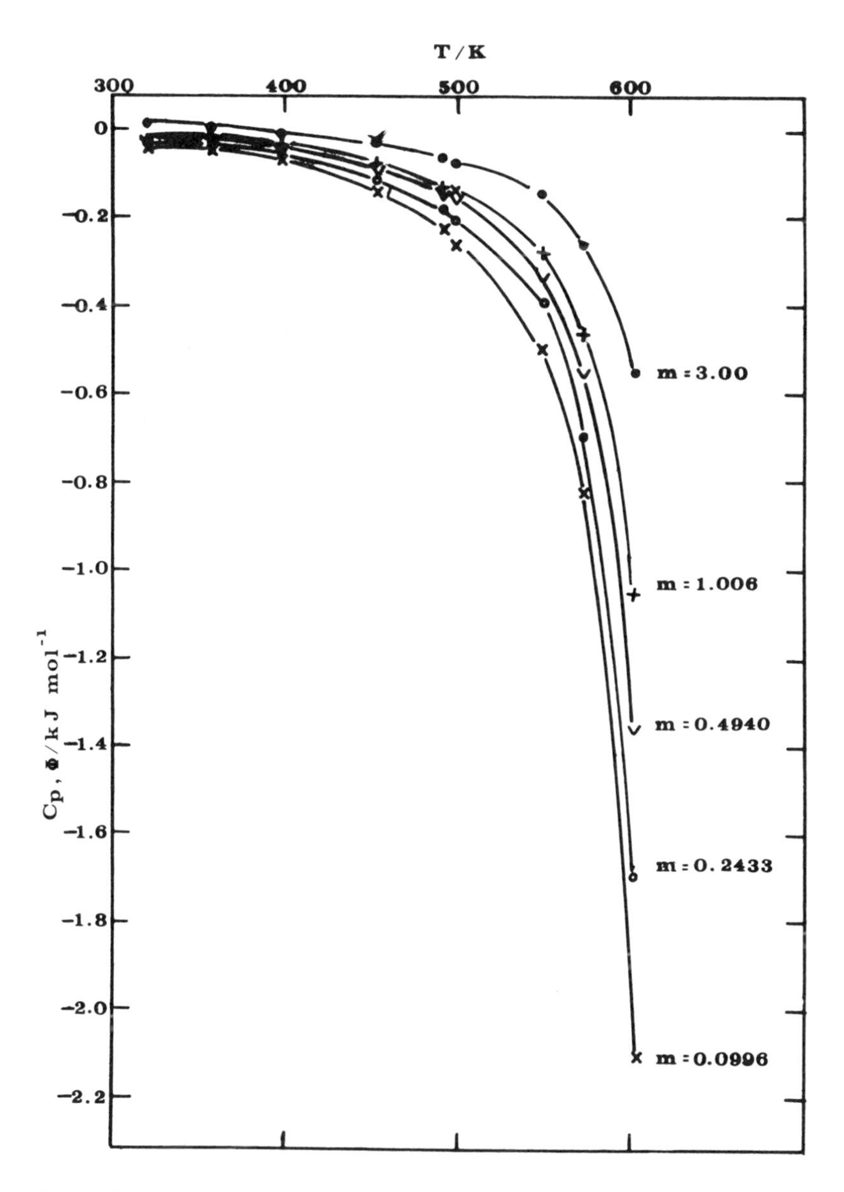

Figure 6. Apparent molal heat capacity of aqueous NaCl at 177 bars as a function of molality and temperature

these techniques to high temperature operation, these problems have now been solved and we have a rapid and sensitive measuring instrument. As a result we can now get down to the business of exploring aqueous solution chemistry at temperatures up to 325°C.

Literature Cited

1. a) Gardner, W.L.; Mitchell, R.E.; Cobble, J.W. J. Phys. Chem., 1969, 73, 2025.
   b) Cobble, J.W. and Murray, Jr., R.C. Discussions Faraday Soc., 1977, 64.
2. C-T. Liu and W.T. Lindsay. J. Soln. Chem., 1972, 1, 45.
3. Likke, S. and Bromley, L.A. J. Chem. Eng. Data, 1973, 18, 189.
4. Puchkov, L.V.; Styazhkin, P.S.; Fedorova, M.K. J. Applied Chem. (Russ.), 1976, 49, 1268.
5. Monk, P. and Wadsö, I. Acta Chem. Scand. 1968, 22, 1842.
6. Picker, P.; Jolicoeur, C.; Desnoyers, J.E. J. Chem. Thermodynamics, 1969, 1, 469.
7. Gill, S.J. and Chen, Y.-J. Rev. Sci. Instruments, 1972, 43, 774.
8. Picker, P.; Fortier, J.-L.; Philip, P.R; Desnoyers, J.E. J. Chem. Thermodyn., 1971, 3, 631.
9. Elliot, K.; Wormald, C.J. J. Chem. Thermodynamics, 1976, 8, 881.
10. Messikomer, E.E.; Wood, R.H. J. Chem. Thermodynamics, 1975, 7, 119.
11. Mayrath, J.E., "A Flow Microenthalpimetric Survey of Electrolyte Solution Thermodynamic Properties from 373K to 473K" Ph.D. Dissertation, University of Delaware, June, 1979.

RECEIVED January 30, 1980.

Table III

Water + Gas
Thermodynamics of Solution, 298.15 K

Gas	$\Delta G°/kJ\ mol^{-1}$	$\Delta H°/kJ\ mol^{-1}$	$\Delta S°/J\ K^{-1}mol^{-1}$	$\Delta C_p°/J\ K^{-1}mol^{-1}$
He	29.42	-0.688	-101.0	116
Ne	29.05	-3.868	-110.4	157
Ar	26.25	-12.24	-129.1	167
Kr	24.80	-15.63	-136	201
Xe	23.42	-19.39	-143.6	228
Rn	21.56	-21.33	-144	291
H_2	27.69	-4.07	-107	138
N_2	28.12	-10.30	-129	206
O_2	26.48	-12.06	-129	200

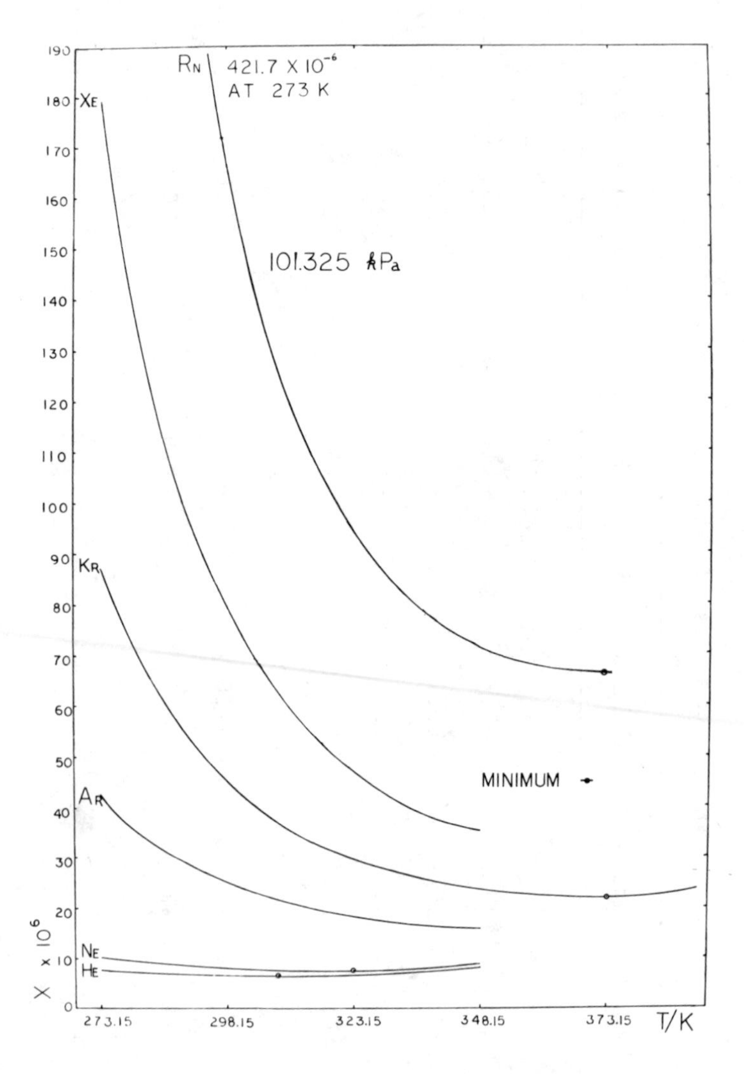

Figure 1. Solubility of the noble gases in water. Mole fraction solubility at 1 atm noble gas partial pressure vs. temperature. The curves are based on the equations of Battino (1–5).

Table IV. Polarizability and Molar Volume at 273.15 K and 1 atm (101.325 kPa).

Gas	$\overline{V}/dm^3\ mol^{-1}$	$\alpha_o x\ 10^{24}/cm^3\ molec^{-1}$
Helium	22.4263	0.2036
Neon	22.4256	0.3926
Hydrogen	22.4225	0.8023
Ideal	22.41383	
Nitrogen	22.4035	1.734
Argon	22.3933	1.6264
Oxygen	22.386	1.561
Krypton	22.3517	2.4559
Radon	22.290	-
Xenon	22.2659	3.9989

Polarizabilities from E. A. Moelwyn-Hughes, Physical Chemistry, 1961

one atmosphere gas partial pressure was estimated from the data and the value was taken to equal the inverse of the Henry constant. Some high pressure solubility data at temperatures below 350 K were included.

The estimation of the solubility at one atmosphere gas pressure was made by one of two procedures. If the solubility was measured at only one pressure at a given temperature, Henry's law was used, and the inverse of Henry's constant was calculated as $X_2(1\ atm) = 1/H_{2,1} = X_2/P_2$. The procedure works well at moderate gas partial pressures, but at higher gas partial pressures of 25 atm or more the procedure often appears to give low solubility values. However, it is the only practical procedure when the solubility was measured at only one pressure. When solubility values were measured at several pressures at a given temperature, the data were fitted by a linear regression to an empirical function $X_2/P_2 = a + bP_2$ to obtain the unit pressure solubility value. In some cases a quadratic rather than a linear function of pressure was used.

After all of the data were tabulated on a given gas, the one atmosphere partial pressure values estimated from the moderate to high pressure measurements up to 600 K were combined with the selected data of Battino in the 273 to 350 K temperature range in a linear regression to obtain the constants of equation (2). The equation parameters are listed in Table V.

The evaluation of three constants was usually

adequate to represent the data. For several of the gases four constants were evaluated. Because of the scatter of the data in the 350 to 600 K temperature interval the four constant equation is not a statistical improvement. However, if the four constant equation gave a noticeably better fit to the data in the 273-350 K temparature interval, the four constant equation was used.

Battino's [1-5] evaluations contain the recommended equations to represent the gas solubility from 273.15 to 350 K. The equation parameters in Table V are for a tentative equation for the calculation of values in the 350 to 600 K temperature range.

Table V. Parameters for the Equation $\ell n\ x_{gas} = \ell n\ (1/H_{2,1}) = A_1 + A_2/(T/100) + A_3 \ln(T/100) + A_4(T/100)$ for the Temperature Interval 350 to 600 K

System	A_1	A_2	A_3	A_4
Helium + Water	-41.7476	43.0404	14.1325	
Neon + Water	-21.2647	16.3471	-0.1417	1.4195
Argon + Water	-80.8022	105.6778	40.6348	-3.2257
Krypton + Water	-61.1802	82.3244	21.5757	
Xenon + Water	-66.2269	92.5765	23.5567	
Hydrogen + Water	15.2051	-33.3273	-28.9664	5.5265
Nitrogen + Water	-55.0165	69.2199	18.7292	
Oxygen + Water	-54.0411	68.8961	18.5541	

Figures 2 to 9 show the solubility values used in the linear regressions and three curves. The curves were calculated from Battino's recommended equation for data between 273-350 K (solid line ―――――), the tentative three constant equation from all of the data (dash-dot line-•-•-•), and the tentative four constant equation from all of the data (dashed line -------).

The extrapolation of Battino's recommended equations above about 350 K is not recommended. However, for several of the systems, especially helium + water and krypton + water, the extrapolated equation represents the higher temperature data surprisingly well. The extrapolation of Battino's equation for the neon +

water system is least reliable. It may indicate the low temperature data on the neon + water system is not as reliable as for the other gases.

Some specific comments on each of the systems follow.

Helium + water. Battino selected solubility data from nine papers for the 273-348 K region (1). We have added values calculated from the data of Potter and Clynne (6) and from Wiebe and Gaddy (7). The solubility value which was calculated from the data of Wiebe and Gaddy at 590 K, was not used in the linear regression. The data and curve are shown in Figure 2. Only one curve is shown. Battino's recommended equation for the solubility data below 348 K and the equation for the entire data set differ by only a fraction of a percent. The curve for the four constant equation is not shown.

Neon + water. The only solubility data above 350 K are the data of Potter and Clynne (6). These were combined with Battino's selected data (1) for the linear regression. Figure 3 shows the extrapolation of Battino's equation, which is much too high, and the curves for both the three and four constant fits to the entire data set. Values of the parameters for the four constant equation are given in Table V. The four constant equation gives a better fit to the data at the low temperature than does the three constant equation. Of the five noble gas + water systems, the neon + water system is the only one for which the Potter and Clynne values are lower than Battino's selected values near 350 K temperature where the data sets overlap.

Argon + water. This is the only system for which a detailed comparison can be made between the solubility data of Potter and Clynne (6) and the selected data of Battino (3). Potter and Clynne made numerous measurements between the temperatures of 298 and 365 K as well as measurements at temperature up to 561 K on the argon + water system. At temperatures of 298, 303, 308, and 313 K Potter and Clynne's one atm argon solubility values show an average deviation of 0.20 percent from Battino's recommended values. However, at temperatures of 323, 338, and 353 K Potter and Clynne's values are higher than the recommended values by 5, 14, and 22 percent, respectively. The discrepancy between the solubility values determined by Potter and Clynne's high pressure method and the recommended values determined by atmospheric methods is disturbing. It does not appear that the discrepancy can be resolved with information presently

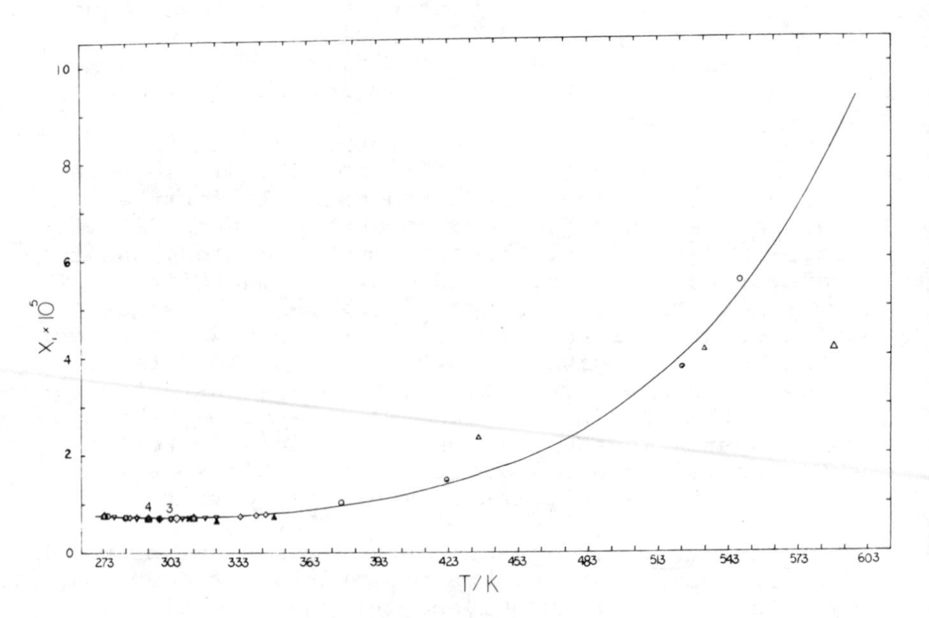

Figure 2. Helium + water—mole fraction solubility at 1 atm helium partial pressure vs. temperature. At temperatures above 353 K: (○) (6); (▲,△) (7). The 590 K value of Wiebe and Gaddy was not included in the linear regression.

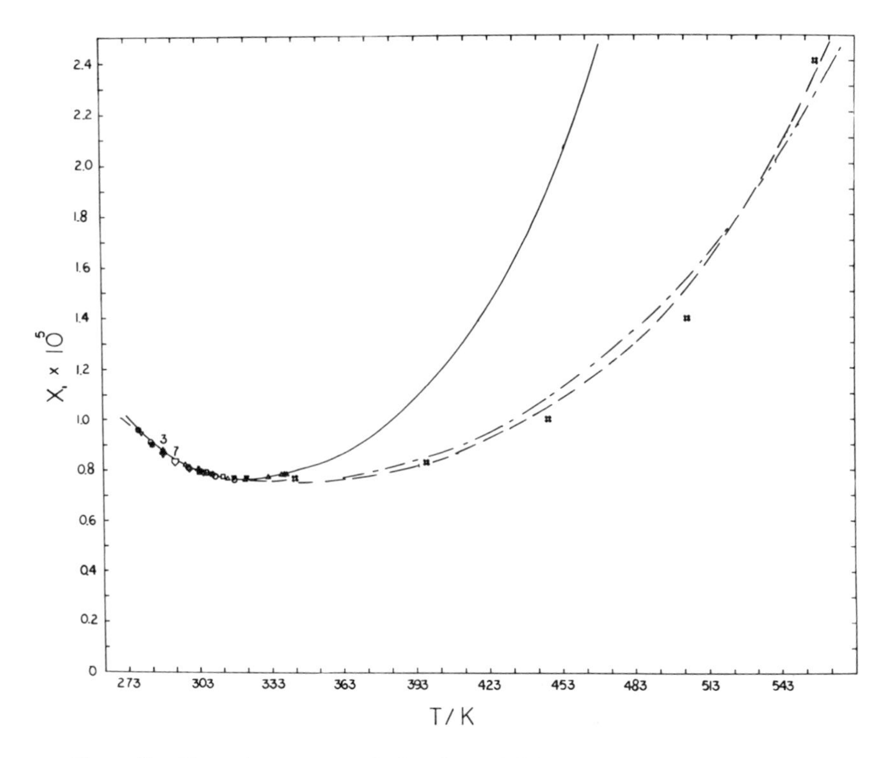

Figure 3. Neon + water—mole fraction solubility at 1 atm neon partial pressure vs. temperature: (+) (6)

available, but that more experimental work in the temperature region where the atmospheric pressure and high pressure methods overlap is needed. One point of concern is that most of the atmospheric pressure techniques are Ostwald methods in which the solubility is measured at a total pressure of gas + solvent vapor of one atmosphere. The experimentally measured Ostwald coefficient is converted to a one atmosphere mole fraction solubility with the assumptions that the gas is ideal, the Ostwald coefficient is independent of pressure, and that Henry's law is obeyed. In the presence of nearly one-half atmosphere of water vapor (at 353 K) these assumptions may not be valid as customarily accepted.

Sisskind and Kasarnowsky (8) measured one high pressure solubility value for the system at 273 K. The one atmosphere argon value estimated from their work and the values from Potter and Clynne's work are shown in Figure 4 along with the curves of Battino's equation, and the three and four constant linear regression of all the data except the Sisskind and Kasarnowsky value. Parameters for the four constant equation are given in Table V for use in the tentative equation for solubility values in the 350-600 K temperature range.

Krypton + water. Battino (2) selected 30 solubility values from 3 papers for the basis of his 273-353 K recommended equation. Both Potter and Clynne (6), and Anderson, Keeler, and Klach (9) report the solubility of kyrpton in water at higher temperatures, but quite different pressure conditions. Potter and Clynne worked at krypton pressures between 4 and 6 atmospheres. Anderson *et al*. worked with radioactive krypton at pressures of 0.4 to 53.2 x 10^{-4} psia in the presence of oxygen gas and water vapor at pressures between 110 and 2015 psia. Although the experimental conditions were vastly different, the krypton mole fraction solubilities at one atmosphere pressure estimated from the two papers agree reasonably well (Figure 5). No effort was made to apply corrections for non-ideal gas behavior. The two sets of data were combined with Battino's selected low temperature atmospheric pressure data in a linear regression. Figure 5 shows only the solubility values from the high temperature experiments (6,9) and the lines for Battino's low temperature equation, and the three and four constant equations from all of the data. The three constants are given in Table V for the tentative equation to calculate solubilities in the 350 to 600 K temperature interval.

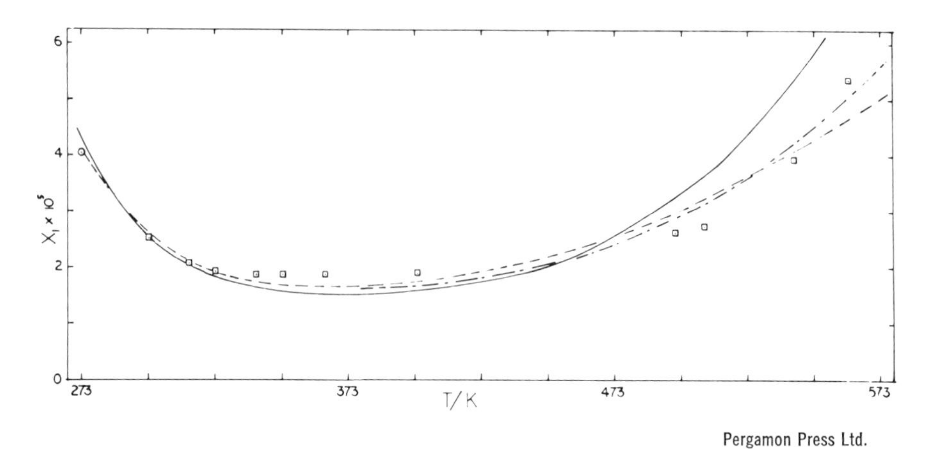

Figure 4. Argon + water—mole fraction solubility at 1 atm argon partial pressure vs. temperature: (○) (8); (□) (6); figure from Ref. 3

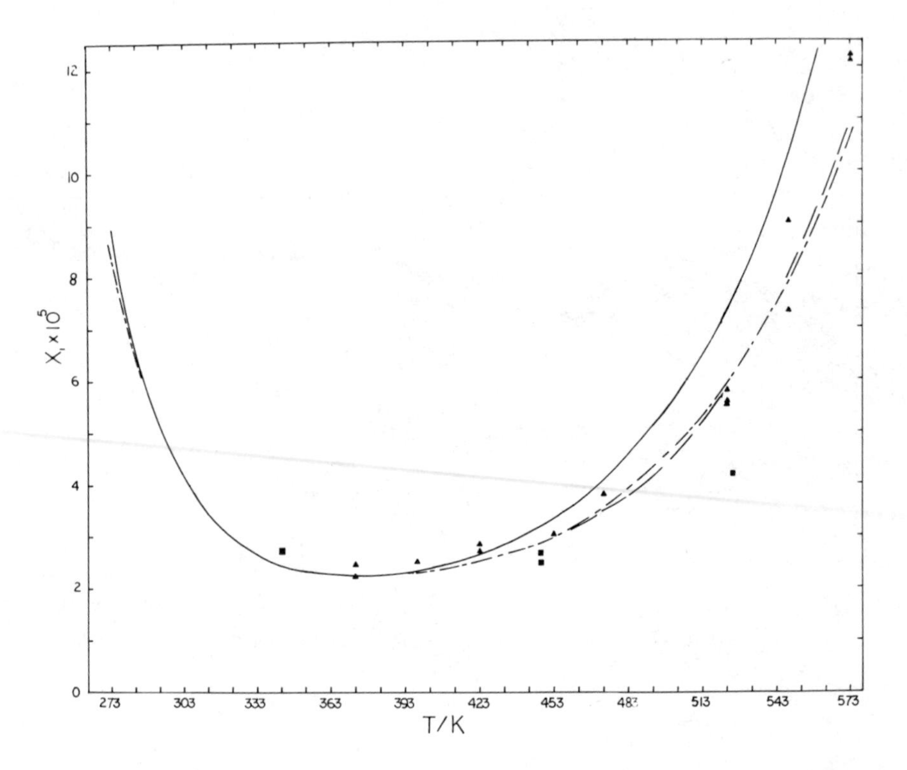

Figure 5. Krypton + water—mole fraction solubility at 1 atm krypton partial pressure vs. temperature: (■) (6); (▲) (9)

Xenon + water. The solubility data of Potter and Clynne (6) and of Stephan, Hatfield, Peoples and Pray (10) were used to estimate the mole fraction solubility at one atmosphere xenon pressure at the higher temperatures. The two sets of data were combined with the 20 solubilities selected from five papers by Battino (2) in a linear regression. Figure 6 shows the data, and Battino's equation and the three constant equation. The three constants for the tentative equation for use between the temperatures of 350 and 600 K are in Table V. The Stephan et al. solubility value at 574 K was not included in the regression.

Hydrogen + water. Battino (4) selected 69 solubility values from nine papers that reported measurements between temperatures of 273 and 348 K. The mole fraction solubilities at one atmosphere partial pressure of hydrogen at the higher temperatures were estimated from the data of Wiebe and Gaddy (11), Pray, Schweichert, and Minnich (12), and Stephan, Hatfield, Peoples and Pray (10). The data from Pray, Schweichert and Minnich were combined with Battino's selected data in a linear regression to obtain the tentative four constant equation for the hydrogen solubility in water between 350 and 600 K (Figure 7 and Table V).

Nitrogen + water. Battino (4) selected 74 solubility values from nine papers that reported measurements between the temperatures of 273 and 348 K. Solubility data measured at higher temperatures from seven other papers (10, 12-17) were considered. As can be seen in Figure 8, many of the values give low atmospheric pressure solubilities when compared with Battino's selected low temperature data. The values estimated from the work cited in references (12-14,17) (large filled circles on Figure 8) were selected to be combined with Battino's recommended low temperature values in a linear regression. The parameters for the three constant equation are given in Table V.

Oxygen + water. Battino's recommended four constant equation from an earlier work (5) was used to represent the low temperature (273-348 K) mole fraction oxygen solubility values. The data determined at the Battelle Memorial Institute laboratories in the early 1950's (10,12) were used to estimate the one atmosphere oxygen pressure solubilities at higher temperatures. The data sets were combined in a linear regression to obtain the parameters of the three constant tentative equation for the solubility between 350 and 600 K (Table V, Figure 9).

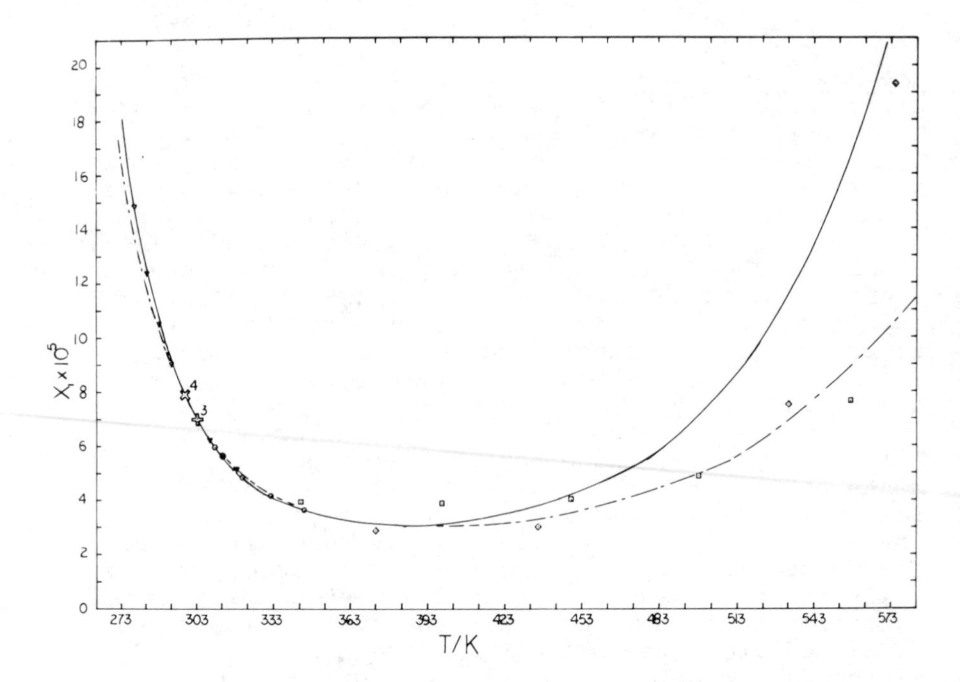

Figure 6. Xenon + water—mole fraction solubility at 1 atm xenon partial pressure vs. temperature: (□) (6); (◇) (10); the value of Stephan et al. at 574 K was not included in the linear regression

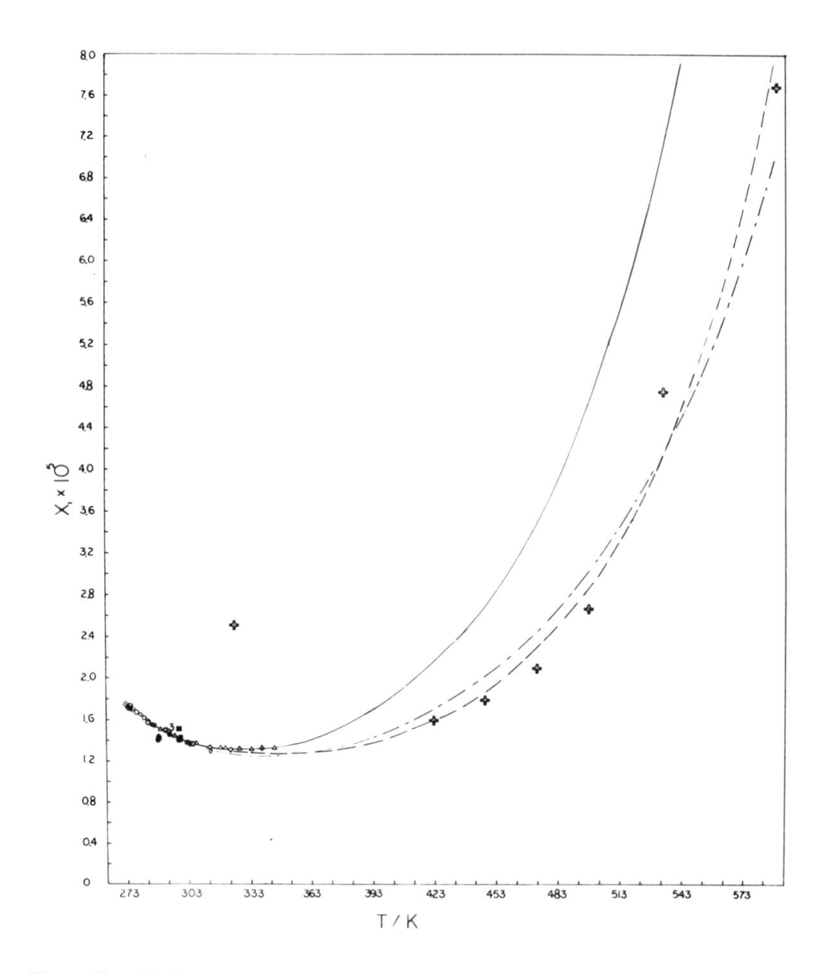

Figure 7. Hydrogen + water—mole fraction solubility at 1 atm hydrogen partial pressure vs. temperature: (+) (12); value at 323 K not used in the linear regression

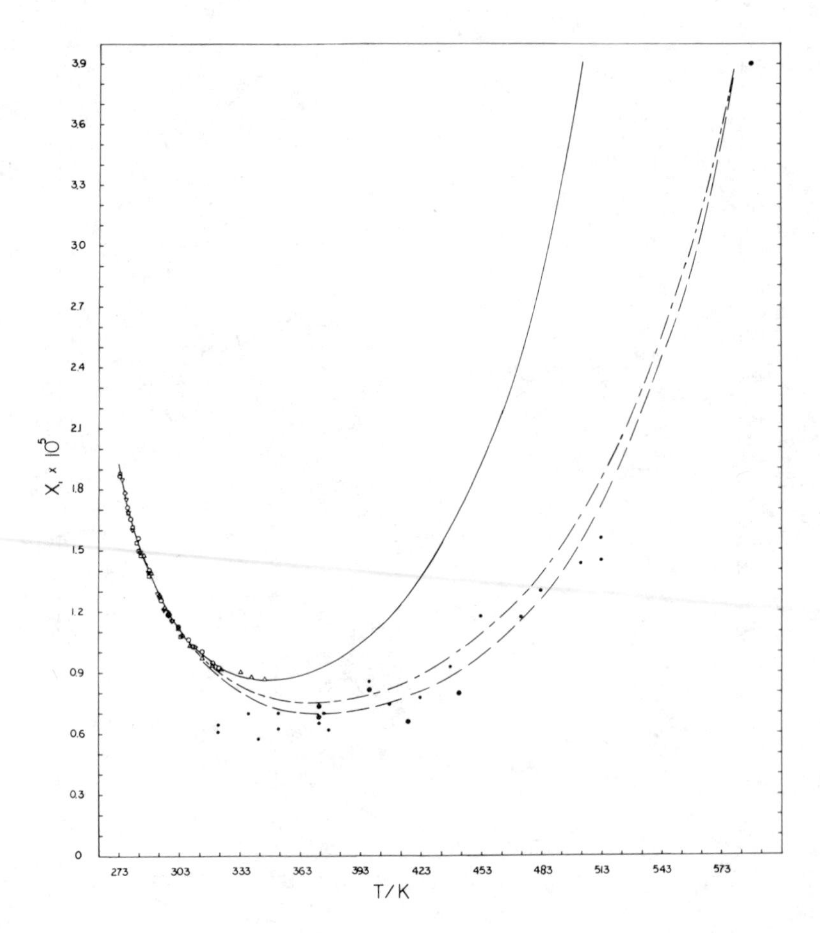

Figure 8. Nitrogen + water—mole fraction solubility at 1 atm nitrogen partial pressure vs. temperature: (·) (10, 15, 16); (●) (12, 13, 14, 17) (values used in the linear regression)

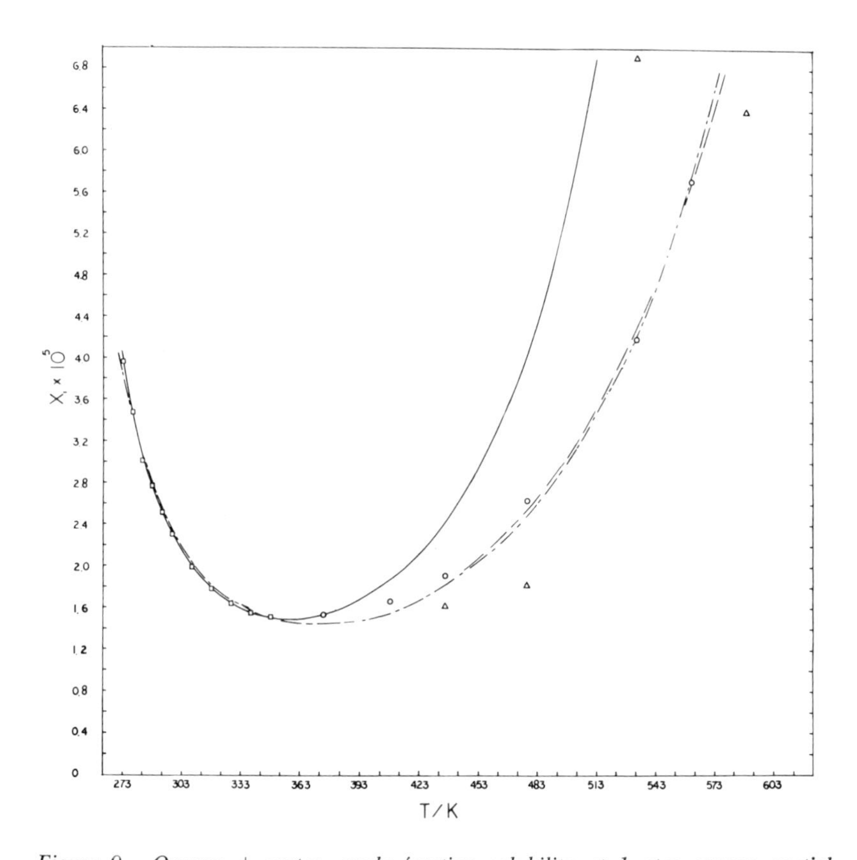

Figure 9. Oxygen + water—mole fraction solubility at 1 atm oxygen partial pressure vs. temperature: (○) (10); (△) (12)

There are other important papers on the behavior of the gas + water systems. Lentz and Franck (18) made a study of the argon + water system to establish the existence of gas-gas immiscibility. The data were presented in graphical form. Franck, Lentz and Welsch (19) studied the xenon + water phase boundary conditions at temperatures up to 635 K. There are other papers on the solubility of hydrogen, nitrogen, and oxygen in water. Particularly noteworthy are the results of W. L. Sibbitt and co-workers (20) determined in the early 1950's. Their published results are in graphs. We are trying to obtain copies of the theses which may contain numerical data.

The effect of pressure on gas solubility

The tentative equation summarized in Table V allows the calculation of the solubility at one atmosphere gas partial pressure which is numerically equal to the inverse of Henry's constant (equation 1). Although Henry's law may be adequate up to moderate pressures, it requires some corrections for the solubilities at higher pressures. Table VI summarizes some approaches that have been used to correlate solubility pressure isotherms. These have been discussed in many places including references [21 and 22].

Table VI. Effect of Pressure on Gas Solubility (Constant Temperature)

Herny's Law $$H_{2,1} \equiv \lim_{x_2 \to 0} \frac{f_2}{x_2}$$

Empirical $$x_2 = ap + bp^2$$

$$x_2/p = a + bp$$

Krichevsky-Kasarnovsky equation

$$\ln(f_2/x_2) = \ln H_{2,1} + \bar{v}_2^\infty (p - p_1^s)/RT$$

Krichevsky-Ilinskaya equation

$$\ln(f_2/x_2) = \ln H_{2,1} + (x_1^2 - 1)A/RT + \bar{v}_2^\infty (p - p_1^s)/RT$$

The Krichevsky-Kasarnowsky equation takes into account both the temperature and pressure dependence of Henry's constant. The volume term in the equation is the partial molal volume of the gas in the liquid

Table VII

Partial Molal Volumes of Gases in Water/298 K

Gas	$V_m/cm^3\ mol^{-1}$	Comments
He	15 - 17.5	High pressure solubility data
	29.7	Volume/hydrostatic pressure
	78.4 ± 1.9	Density, gas saturated water, 20 - 100 atm
Ar	31.7-32.3	Three different methods
	55.2 ± 0.7	Density, gas saturated water, 9.7 - 95.8 atm
	31.71 ± 0.43	Recommended. Tiepel & Gubbins, 1972
Kr	38 ± 5	4.70 atm Density, gas saturated
	32 ± 2	12.00 atm water, 4.7 - 20.1 atm,
	31.2 ± 1.4	15.20 atm Popov & Drakin, 1974
H_2	19 - 21.5	High pressure gas solubility
	24 - 26	Dilatometry
	25.20 ± 0.56	Recommended. Tiepel & Gubbins, 1972
N_2	32.5-33.5	Several methods
	37 - 40	Micropyknometer, dilatometry.
O_2	∿25	Early dilatometry
	30 - 32	By several methods
	30.38 ± 0.97	Recommended. Tiepel & Gubbins, 1972

at infinite dilution. Many tests of the equation show that the volume term required to make the equation fit the experimental data is usually appreciably smaller than the directly measured partial molal volume. Although the equation works well for many systems it often fails under conditions of high pressure and high solubility.

When the Krichevsky-Kasarnowsky equation fails it may be because of either changing activity coefficient of the solute gas with composition, changing partial molal volume of the gas with pressure, or both. The Krichevsky-Ilinskaya equation takes into account the variation in the activity coefficient of the solute gas with mole fraction by means of a two-suffix Margules equation.

Goldman has studied the effect of nonclassical behavior on the solubility of helium, hydrogen, and neon in simple solvents and benzene (23). He has discussed the Henry's constant of water as a solute (24). So far his results do not apply to water as a solvent.

The partial molal volumes of gases in water are needed to apply the Krichevsky-Kasarnowsky and the Krichevsky-Ilinskaya equations. A survey of the available experimentally measured partial molal volumes is given in Table VII. The results of Tiepel and Gubbins (25) seem especially reliable. The recent results of Popov and Drakin (26) usually appear to be much too high, possibly because of Popov and Drakin depended on literature solubility values for the concentration to be used in their calculation of the partial molal volume from the density data.

A more detailed analysis of the effect of pressure on the solubility of gases is planned for the future.

Acknowledgment

We wish to express our appreciation to colleagues of the IUPAC Solubility Data Project, Rubin Battino and Colin L. Young for access to their data collections.

Literature Cited

1. Battino, R. *Solubility Data Series*, Kertes, A. S., Editor, *Helium and Neon, Volume 1*, Clever, H. L., Editor, Pergamon Press, Ltd., Oxford and New York, 1979, pp. 1-4, 124 - 126.
2. Battino, R. *Solubility Data Series*, Kertes, A. S., Editor, *Krypton, Xenon, and Radon, Volume 2*, Clever, H. L., Editor, Pergamon Press, Ltd., Oxford and New York, 1979, pp. 1-3, 134-136, 227-229.
3. Battino, R. *Solubility Data Series*, Kertes, A. S., Editor, *Argon, Volume 4*, Clever, H. L., Editor, Pergamon Press, Ltd., Oxford and New York, 1980, pp. 1-7, 20-21.
4. Battino, R. *Solubility Data Series, Volumes on Hydrogen, Oxygen, and Nitrogen*, in preparation.
5. Wilhelm, E.; Battino, R.; Wilcock, W. J. *Chem. Rev.* 1977, *77*, 219-262.
6. Potter, R. W. II; Clynne, M. A. *J. Solution Chem.* 1978, *7*, 837-844.
7. Wiebe, R.; Gaddy, V. L. *J. Am. Chem. Soc.* 1935, *57*, 847-851.
8. Sisskind, B.; Kasarnowsky, I. *Z. anorg. Chem.* 1931, *200*, 279-286.
9. Anderson, C. J.; Keeler, R. A.; Klach, S. J. *J. Chem. Eng. Data*, 1962, *7*, 290-294.
10. Stephan, E. L.; Hatfield, N. S.; Peoples, R. S.; Pray, H. A. H. Battelle Memorial Laboratory Report, BMI-1067, 1956.
11. Wiebe, R.; Gaddy, V. L. *J. Am. Chem. Soc.* 1934, *56*, 76-79.
12. Pray, H. A.; Schweichert, L. E.; Minnich, B. H. *Ind. Eng. Chem.*, 1952, *44*, 1146-1151.
13. Goodman, J. B.; Krase, N. W. *Ind. Eng. Chem.* 1931, *23*, 401-404.
14. Wiebe, R.; Gaddy, V. L.; Heins, C. *Ind. Eng. Chem.* 1932, *24*, 823-825, 927.
15. Saddington, A. W.; Krase, N. W. *J. Am. Chem. Soc.* 1934, *56*, 353-361.
16. Smith, N. O.; Kelemens, S.; Nagy, B. *Geochim. Cosmochim. Acta* 1962, *26*, 921-926.
17. O'Sullivan, T. D.; Smith, N. O.; Nagy, B. *Ibid.* 1966, *30*, 617-619, *J. Phys. Chem.* 1970, *74*, 1460-1466.
18. Lentz, H.; Franck, E. U. *Ber. Bunsenges. Phys. Chem.* 1969, *73*, 28-35.
19. Franck, E. U.; Lentz, H.; Welsch, H. *Z. Phys. Chem. N. F.*, 1974, *93*, 95-108.

20. Zoss, L. M.; Suciu, S. N.; Sibbitt, W. L. *Trans. Am. Soc. Mech. Engrs.* 1954, *76*, 69-71.
21. Battino, R.; Clever, H. L. *Chem. Rev.* 1966, *66*, 395-463.
22. Prausnitz, J. M. *Molecular Thermodynamics of Fluid Phase Equilibria* Prentice-Hall, Inc., Englewood Cliffs, NJ, 1969, Chapter 8.
23. Goldman, S. *J. Phys. Chem.* 1977, *81*, 608-614.
24. Goldman, S.; Krishnan, T. R. *J. Solution Chem.* 1976, *5*, 693-707.
25. Tiepel, E. W.; Gubbins, K. E. *J. Phys. Chem.* 1972, *76*, 3044-3049.
26. Popov, G. A.; Drakin, S. I. *Zh. Fiz. Khim.* 1974, *48*, 631-634.

RECEIVED January 31, 1980.

Correlation of Thermodynamic Properties of Aqueous Polyvalent Electrolytes

BERT R. STAPLES

Electrolyte Data Center, National Bureau of Standards,
National Measurement Laboratory, Washington, DC 20234

Studies of problems in the area of water quality control, the application of geothermal energy, the desalination of water, sewage treatment, industrial applications, and bioengineering all must treat aqueous solutions containing ionic species. A significant need for reliable quantitative data on the properties of aqueous solutions has become apparent in recent years, particularly with the development of large scale models that attempt to simulate complex aqueous ecosystems (Morel and Morgan, 1972 and Zemaitas, 1975). Reliable quantitative data must often be selected from discordant results of various experimenters. A critical evaluation is the best approach to providing "standard" sets of data.

Critical evaluations of activity and osmotic coefficient data were undertaken early in the 1930-1940 period by Harned and Owen (1958) and by Robinson and Stokes, (1965). Wu and Hamer (1968) evaluated activity and osmotic coefficient data for a series of electrolytes but their work on polyvalent electrolytes was not completed. Their work on the 1:1 electrolytes was published in 1972. The evaluation of polyvalent electrolyte data has been continuing in the Electrolyte Data Center at the National Bureau of Standards, and this paper will summarize the methods used in evaluating data for over 100 aqueous polyvalent electrolytes. Models and the associated correlating equations will be discussed as well as the methodology for the correlating schemes.

Models and Correlating Equations

In 1923, the Debye-Hückel limiting law was derived and it has served as an excellent model for simple salts at very low concentrations. The limiting form of this theory can be derived in several ways which should also give correct results at moderate concentrations. The mathematics involved in proceeding beyond the limiting law stage are so formidable that the theories have been of very limited usefulness in the experimental range of concentrations.

Friedman (1962) has used the cluster theory of Mayer (1950) to derive equations which give the thermodynamic properties of electrolyte solutions as the sum of convergent series. The first term in these series is identical to and thus confirms the Debye-Hückel limiting law. The second term is an IℓnI term whose coefficient is, like the coefficient in the Debye-Hückel limiting law equation, a function of the charge type of the salt and the properties of the solvent. From this theory, as well as from others referred to above, a higher order limiting law can be written as

$$\ln \gamma_{\pm} = - |z_+ z_-| A_m I^{1/2} - \frac{(\sum_i \nu_i z_i^3)^2}{\nu \sum_i (\nu_i z_i^2)} A_m^2 I \ln I. \tag{1}$$

where $\gamma_{\pm}$ is the mean ionic activity coefficient, z the ionic charge, A_m, the Debye-Hückel constant on the molality scale, I is the ionic strength and ν_i is defined in equation (2). For symmetrical electrolytes the coefficient of the IℓnI term is zero.

Higher terms involve direct potentials corresponding to the forces between sets of ions and become mathematically very difficult.

Pitzer et al (1972, 1973, 1974, 1975, 1976) have proposed a set of equations based on the general behavior of classes of electrolytes. Pitzer (1973) writes equations for the excess Gibbs energy, ΔG^{ex}, the osmotic coefficient ϕ, and the activity coefficient $\gamma_{\pm}$ for single unassociated electrolytes as

$$(\Delta G^{ex}/n_w RT) = f^{Gx} + m^2 (2\nu_M \nu_X) B_{MX}^{Gx} + m^3 [2(\nu_M \nu_X)^{3/2}] C_{MX}^{Gx} \tag{2}$$

$$\phi - 1 = |z_M z_X| f^{\phi} + m \left(\frac{2\nu_M \nu_X}{\nu}\right) B_{MX}^{\phi} + m^2 \frac{2(\nu_M \nu_X)^{3/2}}{\nu} C_{MX}^{\phi} \tag{3}$$

$$\ln\gamma = |z_M z_X| f^{\gamma} + m \left(\frac{2\nu_M \nu_X}{\nu}\right) B_{MX}^{\gamma} + m^2 \frac{2(\nu_M \nu_X)^{3/2}}{\nu} C_{MX}^{\gamma} \tag{4}$$

where ν_M and ν_X are the numbers of M and X ions in the formula $M_{\nu_M} X_{\nu_X}$ and z_M and z_X are their respective charges in electronic units; also $\nu = \nu_M + \nu_X$, while n_w is the number of kg of solvent and m is the conventional molality. The functions f, B_{MX}, and C_{MX} are virial terms and the reader is referred to Pitzer's publications for details concerning the model equations and definitions of these other quantities.

In addition, several other forms of correlating equations give comparable fits to the experimental data. One equation uses the higher order limiting law, followed by an empirical polynomial in the square-root of molality. Similarly, another equation uses the Debye-Hückel limiting law with B set equal to zero, followed by an empirical polynomial in the square-root of molality. Both of these have been discussed in detail elsewhere (Staples and Nuttall, 1977).

Criteria for selecting correlating equations should include an adequate description of the experimental data over a wide range of concentrations. Not only should they reproduce the data well, but they should take into account the very dilute region because they are used to evaluate the integral in the Gibbs-Duhem relation. Thus we endeavor to include the Debye-Hückel limiting law as the first term. What is appropriate at slightly higher concentrations is difficult to determine. Correlations are carried out at NBS using a variety of equations but the tabulated and recommended values are generally based on the empirical form of an equation used previously by Hamer and Wu (1972) and Lietzke and Stoughton (1962):

$$\ln \gamma_{\pm} = \frac{-|z_+ z_-| A_m I^{1/2}}{1 + BI^{1/2}} + Cm + Dm^2 + Em^3 + \ . \ . \ . \tag{5}$$

The ionic strength I is given by $I = 1/2\sum m_i z_i^2$, and B is used to denote $B_m \cdot a$, with a being the distance of closest approach in the Debye-Hückel expression. The constants, B, C, D, E, etc. are empirical.

The osmotic coefficient and excess Gibbs energy can be expressed in terms of the same parameters by

$$\phi = 1 + \frac{|z_+ z_-| A_m}{B^3 I} [-(1 + BI^{1/2}) + 2\ln(1 + BI^{1/2}) + 1/(1 + BI^{1/2})] + (1/2)Cm + (2/3)Dm^2 + (3/4)Em^3 + \ldots, \tag{6}$$

and

$$\Delta G^{ex} = \nu mRT \frac{|z_+ z_-| A_m}{B^3 I} [(2-BI^{1/2})BI^{1/2} - 2\ln(1+BI^{1/2})] + (1/2)Cm + (1/3)Dm^2 + (1/4)Em^3 + \ . \ . \ . \tag{7}$$

Values for the parameters are determined by a least squares fit of experimental data using eq (5) for experiments such as galvanic cells measurements that measure solute activity and thus γ/γ_{ref} values, and eq (6) for experiments such as vapor pressure measurements that measure solvent activity and thus ϕ values. All the original data are used in a single fitting program to determine the best values for the parameters. A detailed description of the evaluation procedure has been illustrated for the system calcium chloride-water (Staples and Nuttall, 1977), and calculations deriving activity data from a variety of experimental technique measurements have also been described.

Experimental Techniques

Most determinations of activity and osmotic coefficients of an electrolyte solution are based on these experimental techniques:

(1) isopiestic or vapor pressure equilibrium
(2) vapor pressure lowering
(3) freezing-point depression
(4) boiling-point elevation
(5) vapor pressure osmometry measurements
(6) electromotive force (emfs) of galvanic cells without liquid junction
(7) emfs of galvanic cells with transference
(8) diffusion measurements
(9) solvent extraction measurements
(10) ultracentrifuge measurements.

The first five measure the activity of the solvent and the last five measure the activity of the solute. The boiling point method is generally not included in evaluations for two reasons: there are little data from these measurements or the thermal data are not adequate to apply an accurate correction to obtain an activity at 298 K.

Vapor pressure osmometry in principle, should produce valid results at moderate molalities (a few hundredths to a few $mol{\cdot}kg^{-1}$) but the real accuracy and precision of this method have not yet been properly demonstrated and/or documented (Goldberg, Nuttall, and Staples, 1979).

Solvent extraction measurements and ultracentrifuge measurements have been reported only in single instances but agreement with other methods has been observed for $UO_2(NO_3)_2$ (Goldberg, 1979) and $BaCl_2$ (Goldberg and Nuttall, 1978), respectively.

The primary sources of data used in our critical evaluations are to be found in a recent compilation by Goldberg, and others, (1977). Additional data sources concerning the thermal properties of electrolytes have also been published (Smith-Magowan and Goldberg, 1979). Results of correlating data from a variety of experimental sources generally indicate agreement among a wide range variety of these experiments.

Correlations of Activity and Osmotic Coefficient Data

Activity and osmotic coefficient data derived from ten experimental methods have been critically evaluated and correlating equations have been formulated for more than 100 aqueous polyvalent electrolyte systems at 298 K. Evaluations for the major reference solutions KCl and NaCl (Hamer and Wu, 1972), and $CaCl_2$ (Staples and Nuttall, 1977) have been published; that for H_2SO_4 (Staples, 1980) is available on request.

About a dozen recent publications have appeared in the literature or are in press. These describe other evaluations for the alkaline earth halides (Goldberg and Nuttall, 1978); the bi-univalent compounds of iron, nickel and cobalt (Goldberg, Nuttall, and Staples, 1979); lead, copper, manganese and uranium (Goldberg, 1979). Several other publications are in preparation (Goldberg, 1980a, 1980b).

Generally, agreement has been found between our correlations and those of Pitzer, and others (1972, 1973, 1974, 1975, 1976) and Rard, and others (1976, 1977). Many of our correlations agree fairly well with Robinson and Stokes, (1965) and Harned and Owen, (1958) but in most cases a much larger data base and more recent measurements have been incorporated into the evaluations. It has been observed that agreement with Pitzer's equations is found below moderate concentrations (several molal), but often deviate at higher concentrations where the Pitzer equations do not contain enough parameters to account for the behavior of the activity (or osmotic) coefficient.

The agreement with Pitzer's equations for sulfuric acid is reasonably good up to nearly 5 $mol \cdot kg^{-1}$. This agreement is depicted in Fig. 1, where the square symbols show values from Pitzer's equations, the crosses are experimental results, and the solid line is my evaluation (Staples, 1980).

Figure 2 shows the osmotic coefficient as a function of the molality for the alkaline earth chlorides (Goldberg and Nuttall, 1978). Figure 3 illustrates the osmotic coefficients for the iron, cobalt, and nickel chlorides (Goldberg, Nuttall, and Staples, 1979). Periodic trends are observed.

Other Thermodynamic Properties

Other thermodynamic properties of aqueous solutions are being evaluated. A recent publication reports values calculated for the association constants of aqueous ionic species at 298 K for alkaline earth salts (Staples, 1978).

In addition, the critical evaluation of enthalpies of dilution and solution, as well as evaluations of heat capacities have been initiated. These evaluations will allow calculations and correlations of activity and osmotic coefficients as a function of temperature and composition.

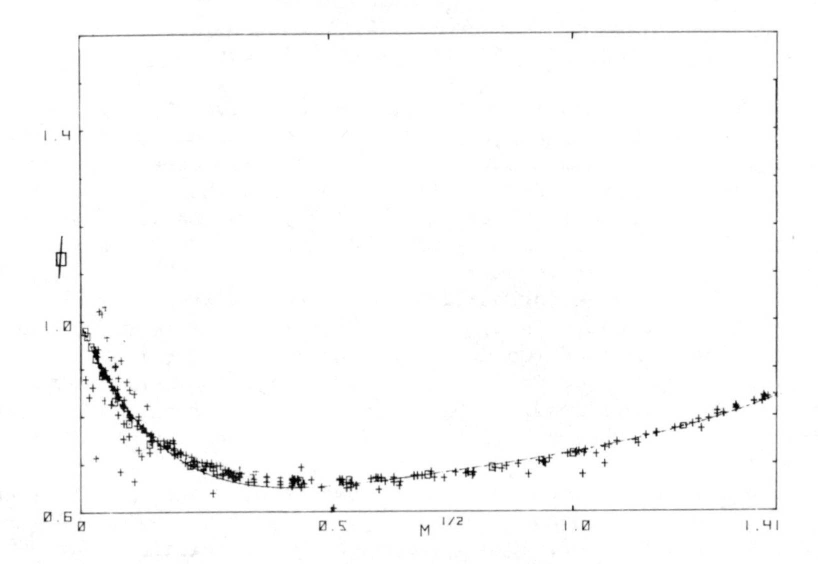

Figure 1. Osmotic coefficient of aqueous sulfuric acid (up to 2m) at 298 K as a function of the square-root of molality ((□) Pitzer evaluation; (+) experimental data)

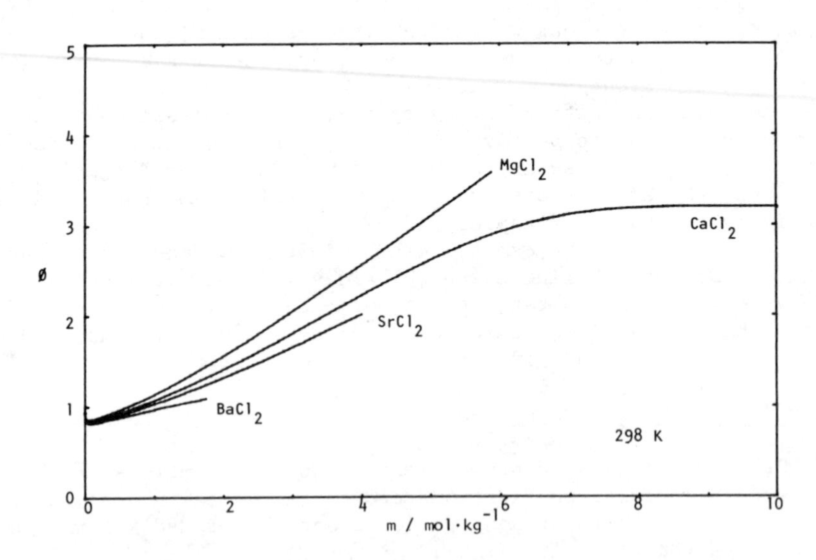

Figure 2. Osmotic coefficient of aqueous alkaline earth chlorides as a function of the molality

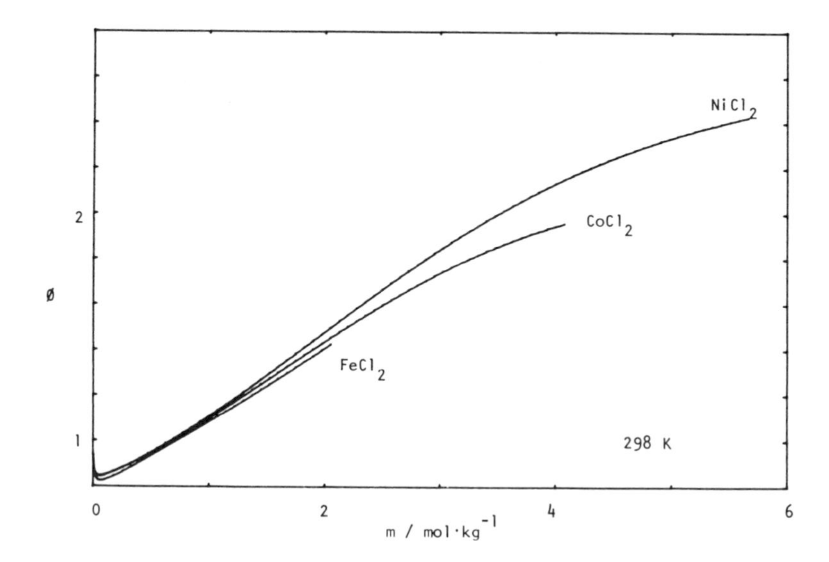

Figure 3. Osmotic coefficient of aqueous chlorides of iron, cobalt, and nickel as a function of the molality

Acknowledgements

The Electrolyte Data Center is supported by the Office of Standard Reference Data, and in part by the Department of Energy. My personal thanks also go to the authors of many articles, Ralph L. Nuttall and Robert N. Goldberg.

Abstract

The techniques used in the critical evaluation and correlation of thermodynamic properties of aqueous polyvalent electrolytes are described. The Electrolyte Data Center is engaged in the correlation of activity and osmotic coefficients, enthalpies of dilution and solution, heat capacities, and ionic equilibrium constants for aqueous salt solutions.

The substances include all inorganic salts for which data are available. The range of concentrations extend from very dilute to saturation. The Debye-Hückel theoretical slopes are used as the basis of correlations in the very dilute region and an empirical equation in powers of $m^{1/2}$ is used to extend the correlating equation to high concentrations.

Current results for sample correlations will be compared and observed periodic trends, and trends according to charge-type, will be discussed.

Literature Cited

Debye, P., and L. Hückel, (1923). *Physik. Z.*, 24, 185.

Friedman, H. L., (1962). *Ionic Solution Theory*, Interscience Publishers.

Goldberg, R. N., (1979). "Evaluated Activity and Osmotic Coefficients for Aqueous Solutions: Bi-univalent Compounds of Lead, Copper, Manganese, and Uranium" manuscript in press, *J. Phys. Chem. Ref. Data*.

Goldberg, R. N., (1980a). "Evaluated Activity and Osmotic Coefficients for Aqueous Electrolyte Solutions: Thirty-Seven Uni-bivalent Systems," manuscript in preparation.

Goldberg, R. N., (1980b). "Evaluated Activity and Osmotic Coefficients for Aqueous Solutions: Bi-univalent Compounds of Zinc and Cadmium, and Ethylene bis(trimethylammonium) Chloride and Iodide," manuscript in review.

Goldberg, R. N. and R. L. Nuttall, (1978). *J. Phys. Chem. Ref. Data*, 7, 263.

Goldberg, R. N., R. L. Nuttall and B. R. Staples, (1979). "Evaluated Activity and Osmotic Coefficients for Aqueous Solutions: $FeCl_2$ and the Bi-univalent Compounds of Nickel and Cobalt," manuscript in press, *J. Phys. Chem. Ref. Data.*

Goldberg, R. N., B. R. Staples, R. L. Nuttall and R. Arbuckle, (1977). "A Bibliography of Sources of Experimental Data Leading to Activity or Osmotic Coefficients for Polyvalent Electrolytes in Aqueous Solution", Nat. Bur. Stand. (U.S.) Spec. Publication 485, U.S. Gov't. Printing Office, Washington, D.C.

Hamer, W. J. and Y. C. Wu, (1972). *J. Phys. Chem. Ref. Data,* 1, 1047.

Harned, H. S. and B. B. Owen, (1958). The Physical Chemistry of Electrolytic Solutions, 3rd ed., Reinhold Pub. Corp., N.Y.

Lietzke, N. H. and R. W. Stoughton, (1962). *J. Phys. Chem.,* 66, 508.

Mayer, H. E., (1950). *J. Chem. Phys.,* 18, 1426.

Morel, F. and J. J. Morgan, (1972). *Environ. Sci. Technol.,* 6, 58.

Pitzer, K. S., (1972). *J. Chem. Soc., Faraday Trans. II,* 68, 101.

Pitzer, K. S., (1973). *J. Phys. Chem.,* 77, 268.

Pitzer, K. S., (1975). *J. Solution Chem.,* 4, 249.

Pitzer, K. S. and J. J. Kim, (1974). *J. Am. Chem. Soc.,* 96, 5701.

Pitzer, K. S. and G. Mayorga, (1973). *J. Phys. Chem.,* 77, 2300.

Pitzer, K. S. and G. Mayorga, (1974). *J. Solution Chem.,* 3, 539.

Pitzer, K. S. and L. F. Silvester, (1976). *J. Solution Chem.,* 5, 269.

Rard, J. A., A. Habenschuss and F. H. Spedding, (1976). *J. Chem. Eng. Data,* 21, 374.

Rard, J. A., A. Habenschuss and F. H. Spedding, (1977). *J. Chem. Eng. Data,* 22, 180.

Robinson, R. A. and R. H. Stokes, (1965). "Electrolyte Solutions," 2nd ed., revised, Butterworth and Co., London, (5th impression).

Smith-Magowan, D. and R. N. Goldberg, (1979). "A Bibliography of Experimental Data Leading to Thermal Properties of Binary Aqueous Electrolyte Solutions," Nat. Bur. Stand. (U.S.) Spec. Pub. 537, U.S. Gov't. Printing Office, Washington, D.C.

Staples, B. R., (1978). *Environ. Sci. and Technol.*, 12, 339.

Staples, B. R., (1980). "Activity and Osmotic Coefficients of Aqueous Sulfuric Acid," in press, *J. Phys. Chem. Ref. Data*.

Staples, B. R., and R. L. Nuttall, (1977). "The Activity and Osmotic Coefficients of Aqueous Calcium Chloride at 298.15 K", *J. Phys. Chem. Ref. Data*, 6, 385.

Wu, Y. C. and W. J. Hamer, (1969). "Osmotic Coefficients and Mean Activity Coefficients of a Series of Uni-Bivalent and Bi-Univalent Electrolytes in Aqueous Solutions at 25°C," National Bureau of Standards Report (NBSIR) 10052, Part XIV, 83 p.

Zemaitis, J., (1975). "Equilibrium Compositions", *Industrial Research*, November issue.

RECEIVED January 31, 1980.

28

Current Trends in the Fundamental Theory of Ionic Solutions

HAROLD L. FRIEDMAN

Department of Chemistry, S.U.N.Y., Stony Brook, NY 11794

The basic questions in theory for this symposium concern the role of the forces among the constituent ions and solvent molecules in determining the thermodynamic properties of the solutions. Also there are qualitatively new thermodynamic features in some of the less well known regimes of composition and temperature.

The forces among the ions and solvent molecules are not well known so one commonly starts with approximations for these basic functions, i.e. with Hamiltonian models. Currently there is intense activity in applying new powerful methods of statistical mechanics to ionic solution models and it is already possible to compare some features of the results as calculated by different techniques.

The study of McMillan-Mayer level models, in which the solvent coordinates have been averaged over so that only solvent-mediated ion-ion forces need be treated, is relatively well developed. However the real forces at this level are even more poorly known than the forces at the Born-Oppenheim level referred to above. It is found that McMillan-Mayer level models can be brought into good agreement with solution thermodynamic data. The effort of comparing such models with appropriate non-thermodynamic experiments, such as diffraction experiments and NMR relaxation experiments, is still at an early stage.

Also attracting growing attention is the phase coexistence curve characteristic of ionic systems; it plays a role in some ionic solution phenomena, although examples in aqueous solutions are not known at this time. Other new features are the intense concentration dependence - at low concentration - of certain of the Harned coefficients that characterize mixed electrolyte solutions and the evidence for a solvent-separated state of the hydrophobic bond, the attractive force between hydrophobic ions, even those of zero charge, in water.

0-8412-0569-8/80/47-133-547$05.00/0

1. THERMODYNAMICS AND STRUCTURE

The thermodynamic theory of solutions is complete in the sense that the exact relations among thermodynamic coefficients are all known, the Gibbs-Helmholtz equation for example. However in practice it commonly is necessary to make predictions on the basis of incomplete data, therefore to make extrapolations and other approximations. Reliable approximations depend upon a knowledge of the solution structure.

The most widely used measure of structure in fluids is the pair correlation function (1-6) (or radial distribution function) $g_{ij}(r)$. It is defined so that

$$c_{i/j}(r) = c_i \; g_{ij}(r) \tag{1}$$

is the local concentration of particles of species i in a small volume at a distance r from the center of a particle of species j. Also $c_i = N_i/V$ is the bulk or stoichiometric concentration of species i. Examples are shown in Fig. 1. At small r, $g_{ij}(r)=0$ because the particles each occupy space from which other particles are excluded, basically because of the Pauli exclusion principle. Values of g_{ij} greater than unity reflect attraction between i and j, while values less than unity reflect repulsion. Indeed the potential $w_{ij}(r)$ of the force between i and j is given by the equation

$$g_{ij}(r) = e^{-w_{ij}(r)/k_B T} \tag{2}$$

where k_B is Boltzmann's constant and T the temperature.

If the functions g_{ij} for all of the species pairs in a fluid are known over a sufficient range of the state variables one can calculate the thermodynamic properties. (1-5) So forces determine structure through Eq. (2) and the thermodynamic properties are determined by the structure.

The potential w_{ij} is not generally the potential of the force acting between particles i and j in a vacuum, or even at infinite dilution in the solvent, in the case of a solution. Rather it is the potential of the force between particles i and j in the medium in which g_{ij} is measured. In this case the force is mediated by all of the other particles. An important class of problems in statistical mechanics deals with the calculation of g_{ij} or w_{ij} from models in which the forces in simpler situations are specified. Generally these calculations cannot be made exactly; the w_{ij} or g_{ij} merely are estimated on the basis of one of a number of approximation methods that have been developed for this purpose. (1-6) In this report I will describe some of the results of these approximation methods which are of interest here without going into the approximation methods themselves.

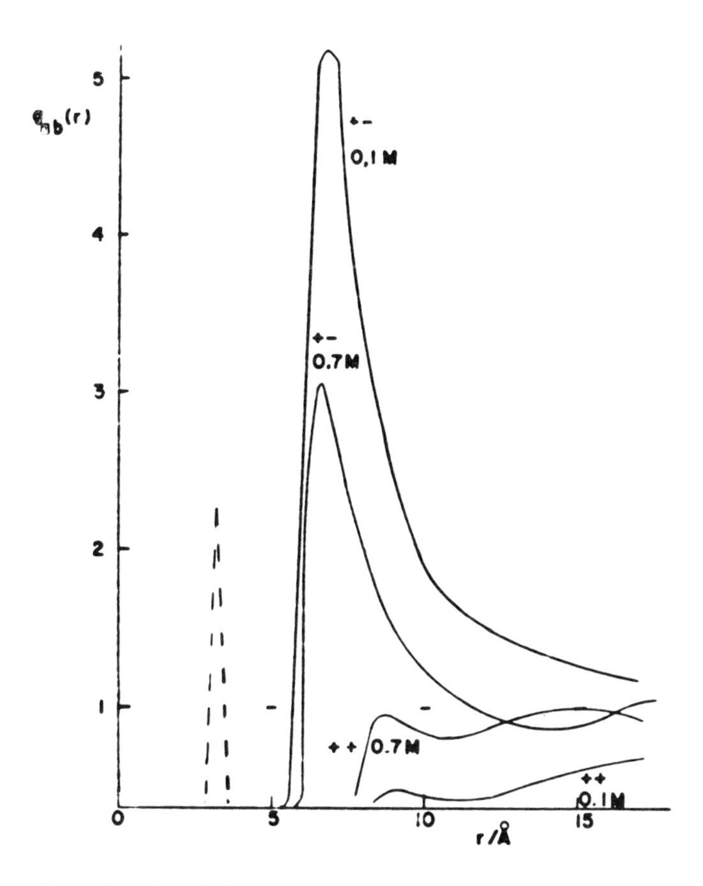

*Figure 1. Correlation functions in $NiCl_2(aq)$ at 0.1*M *and 0.7*M.

The solid curves below 7 Å are calculated accurately for a model (Section 4) that fits the osmotic coefficient data. The curves above 7 Å are merely schematic, showing in exaggerated form the oscillations that appear in g_{ab} at large r when the concentration is large, even for the models in Section 4. The dashed curve indicates the location and intensity of the peak in $g_{+-}(r)$ identified in aqueous $NiCl_2$ in neutron diffraction and EXAFS studies, as reviewed in Section 5.

2. HAMILTONIAN MODELS AT VARIOUS LEVELS

The word 'model' now is so often used to mean any set of approximations that it is convenient to use the term Hamiltonian model to mean a physical model. The model's Hamiltonian specifies the forces acting upon each particle in each possible configuration of the system, i.e. each set of locations of all of the particles. This may be done at several levels. (5, 8)

At the deepest level of interest here, the Schroedinger level, the particles are the nuclei and electrons making up the solution of interest. The Hamiltonian is well known. The laws of quantum mechanics must be used to calculate the wave functions which carry the information about the structure of the system.

Among the functions one can, at least in principle, calculate at the Schroedinger level is the Born-Oppenheimer (BO) potential surface, the potential of the forces among the nuclei assuming that at each nuclear configuration the time-independent Schroedinger equation is satisfied. We may think of this as the 'electron-averaged' potential. Such an N-body potential U_N often may be adequately represented as a sum of pair potentials

$$U_N(\underset{\sim}{r}_1, \underset{\sim}{r}_2, \ldots, \underset{\sim}{r}_N) = \sum_{\text{pairs}} u_{ij}(r_{ij}) \tag{3}$$

where the particle indices in the pair potentials may pertain to monatomic particles or to molecules; in the latter case the coordinates must in general include orientational and other internal coordinates as well as the center-to-center distance r_{ij}.

The program of calculating the BO-level potentials from Schroedinger level cannot often be carried through with the accuracy required for the intermolecular forces in solution theory. (9) Fortunately a great deal can be learned through the study of BO-level models in which the N-body potential is pairwise additive (as in Eq. (3)) and in which the pair potentials have very simple forms. (2,3,6) Thus for the hard sphere fluid we have, with σ=sphere diameter,

$$\begin{aligned} u(r) &= \infty \quad \text{if} \quad r<\sigma \\ &= 0 \quad \text{if} \quad r>\sigma \end{aligned} \tag{4}$$

while for the 6-12 fluid we have

$$u(r) = e_0[(\sigma/r)^{12} - 2(\sigma/r)^6] \tag{5}$$

The 6-12 potential is only qualitatively like the realistic potentials that can be derived by calculations at Schroedinger level for, say, Ar-Ar interactions. But it requires careful and detailed study to see how real simple fluids (i.e. one component fluids with monatomic particles) deviate from the behavior calculated from the 6-12 model. Moreover the principal structural features of simple fluids are already quite realistically given by the hard sphere fluid.

The lesson, that drastically simple Hamiltonian models are adequate to generate quite realistic fluid properties and hence to understand the structure of fluids, can be reinforced by many other examples. For the present Symposium the most important may be the Stillinger-Rahman series of studies of a BO-level

Hamiltonian model for water. (10,11)

In a McMillan-Mayer level model (MM-level) for a solution, the particles are the solute particles (i.e. the ions with positive, negative, or zero charge). The ion-ion potentials can, in principle, be generated by calculations in which one averages over the solvent coordinates in a BO-level model which sees the solvent particles. (4,5,12) Pairwise additivity (we use overbars for solvent-averaged potentials)

$$\bar{U}_N(r_1, r_2, \ldots, r_N) = \sum_{\text{pairs of ions}} \bar{u}_{ij}(r_{ij}) \tag{6}$$

is not so accurate or realistic as at BO-level, but is perhaps realistic enough for solutions in which the ionic concentrations do not much exceed 1M .

The simplest model for ionic solutions at the MM-level is the primitive model, (13,14)

$$\bar{u}_{ij} = u_{ij}^{HS} + e_i e_j/\varepsilon r \tag{7}$$

where u_{ij}^{HS} is the potential given in Eq. (4), with $\sigma \to \sigma_{ij}$, and where ε is the dielectric constant of the solvent medium. This is implicitly the model studied by Debye and Hückel, and in most later studies of ionic solution theory as well. Their well known result for the potential of average force,

$$w_{ij}(r) = u_{ij}^{HS}(r) + e_i e_j e^{-\kappa r}/\varepsilon r , \tag{8}$$

where κ^{-1} is the Debye shielding length, is not very accurate, even for this model, compared to some of the later results. Some of the advances are incorporated in the system of equations for the thermodynamic excess properties developed by K. S. Pitzer. (15)

In Section 4 we discuss some more refined MM-level models.

3. MM-LEVEL $\bar{u}_{+-}(r)$ FROM BO LEVEL

Until very recently there was no information about how an MM pair potential should look, based upon calculations from the deeper BO level. In the simplest BO level model for an ionic solution the solvent molecules are represented as hard spheres with centered point dipoles and the ions as hard spheres with centered charges. Now there are two sets of calculations, (16,17) by very different approximation methods, for this model where all of the spheres are 3Å in diameter, where the dipole moments are near 1 Debye, and where the ions are singly charged. The temperature is 25° and the solvent concentration is about 50M, corresponding to a liquid state. The dielectric constant of the model solvent is believed to be near 9.6.

As shown in Fig. 2 the primitive model with ε=9.6 is

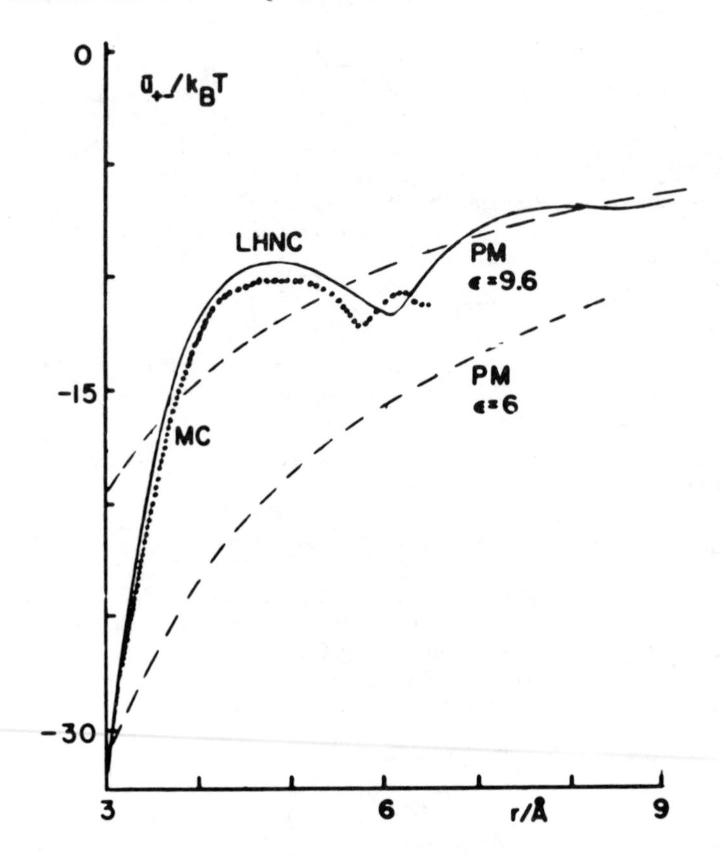

Figure 2. Solvent-averaged potential for charged hard-sphere ions in a dipolar hard-sphere solvent. MC approximation by Patey and Valleau (16) and LHNC approximation by Levesque, Weis, and Patey (17). Also shown are the primitive model functions for solvent dielectric constants 9.6 and 6.

only very close to the calculated $\bar{u}_{+-}(r)$ for $r>7Å$. At $r=3Å$, corresponding to a configuration in which the ions are in contact, the deviation of $\bar{u}_{+-}/k_BT$ from the primitive model is about 11 units. About 1 unit is expected on the basis of so-called liquid structure effects. Thus if the charges and dipole moments were made vanishingly small one would find $u_{+-}/k_BT \simeq -1$. The remaining 10 units is associated with the poor shielding of the ionic charges by the solvent dipoles when the ions are in contact. Indeed the contact value observed for $\bar{u}_{+-}/k_BT$ corresponds to a primitive model with $\varepsilon \simeq 6$ and the same ion sizes as the actual model, as shown in the figure.

The small- r behavior in Fig. 2 is quite surprising from the following point of view. For real 1-1 electrolytes in solvents with $\varepsilon \simeq 10$, e.g. tetrabutylammonium picrate in ethylene dichloride, the mass action constant for forming +- ion pairs, as estimated from the conductivity data, is rather close to what may be calculated from the primitive model, with a realistic ion size parameter, using the theory of Bjerrum (18). However the mass action constant for forming ion pairs estimated from the non-primitive model curve in Fig. 2 is larger than the primitive model result by about three orders of magnitude. This comparison suggests that the particular BO-level model treated in Fig. 2 is not very realistic. One does not yet know what additional features are required to improve the model.

Another method of calculation developed by Adelman and applied by him to a model for a 1-1 electrolyte in water gives much smaller deviations from the primitive model. (19)

We turn to $w_{ij}(r)$ in the case that both i and j are hydrophobic solute particles in water. The so-called hydrophobic bond (20) corresponds to a well in $w_{ij}(r)$ and a peak in the corresponding $g_{ij}(r)$. From a BO-level model for a solution of two Ne atoms in water, with the Bjerrum-like ST2 model for water (10), Geiger, Rahman, and Stillinger (11) find that their calculations suggest that $\bar{u}_{Ne,Ne}(r)$ has a minimum at an r value that corresponds to a solvent-separated hydrophobic bond. In a more detailed and specialized calculation along similar lines but for a model for Xe in ST2 water, Pangali, Rao, and Berne (21) find that $\bar{u}_{Xe,Xe}(r)$ has <u>two</u> wells, one near 4Å that corresponds to the usual idea of contact hydrophobic bonds and one near 7Å that corresponds very closely to the type of solvent-separated hydrophobic bond that one can get if the water near the hydrophobic species has the characteristic hydrogen bonded structure of water in the clathrate hydrates. Moreover the $\bar{u}_{Xe,Xe}(r)$ potential function is in excellent agreement with that calculated by Pratt and Chandler (22) by means of a very different approximation method and also a somewhat different model. Thus Pratt and Chandler are able to finesse the specification of the water-water pair potential in the BO-level model by introducing, at an appropriate point in the theory, the experimental (x-ray) distribution function (essentially $g_{OO}(r)$) for pure water.

The earlier rather complicated evidence for clathrate structures enforced by hydrophobic pairs (from EPR lineshape phenomena for paramagnetic hydrophobic solutes, (23)) and for two states of the hydrophobic bond (from thermodynamic excess functions (24,25)) is provided with a detailed background by these important theoretical developments.

4. REFINED MM-LEVEL MODELS

Extensive studies have been made of MM-level models with the potential (26)

$$\bar{u}_{ij}(r) = COR_{ij} + CAV_{ij} + e_i e_j/\varepsilon r + GUR_{ij} \qquad (9)$$

The COR term represents the short range repulsion between the ions, either by a power law $(\sigma_{ij}/r)^9$ or by an exponential function $e^{-r/\sigma_{ij}}$. Its parameters are calculated from the known sizes, e.g. crystal radii, of particles i and j . The CAV term, which varies like $1/r^4$, accounts for a particular dielectric effect, something like the one in the salting- out theory of Debye and McAulay. (27) It turns out to be numerically unimportant even for Setchenow coefficients. The Coulomb term is well known. The GUR term is intended to account for an effect pointed out some time ago by Gurney (28) and by H. S. Frank. (29) Thus if there is a region (the cosphere) around each ion within which the solvent is modified by the ion, then the force between two ions will have a contribution from the free energy changes when the cospheres overlap. This idea is quantified in the form (26)

$$GUR_{ij}(r) = A_{ij}V_{mu}(r)/V_w \tag{10}$$

where V_w is the molecular volume of the solvent and $V_{mu}(r)$ is the mutual volume of the cospheres of i and j when the distance between their centers is r . The amplitude of the effect, A_{ij} , the only parameter in Eq. (9) that is adjusted to fit the solution data, has the significance of the free energy change per molecule of water displaced from the overlapping cospheres. Most often the cospheres are assumed to include one molecular layer of solvent.

The osmotic coefficients calculated from Eq. (9) can be brought into good agreement with solution data up to about 1M for aqueous solutions of alkali (26) and alkaline earth halides, (30) tetraalkyl ammonium halides, (31) mixed electrolytes, where the Harned coefficients are measured, (32) and electrolyte-non electrolyte mixtures, where Setchenow coefficients are measured. (33) Equally good results have been found for excess enthalpies and volumes when these have been attempted.

Contributing to A_{ij} are, in addition to the solvent structural effects explicitly considered, contributions from dielectric saturation, from the liquid structure effects one has even in simple fluids, from solvent-mediated dispersion interactions of the ions, from charge-polarizability interactions of the ions, and so on. It is difficult to tell a-priori which effects are dominant or how big they are. However the collection of A_{ij} coefficients has characteristics that are consistent with the first named effect being dominant.

We find, from the models that fit the experimental data, that A_{ij}/k_BT is most often in the range from -0.3 to 0. It varies quite regularly with variation of species in a manner that is consistent with the conclusion that the important effects of ion size are explicitly accounted for in Eq. (9). (33,34) The most striking regularity is that if i and j are both species that are all or largely hydrophobic then A_{ij} tends to be nearly independent of species. This behavior makes it relatively easy to predict the thermodynamic excess functions of aqueous solutions of hydrophobic solutes with a certain accuracy, which doubtless could be improved by further work. (35)

One of the further refinements which seems desirable is to modify Eq. (9) so that it has wiggles (damped oscillatory behavior). Wiggles are expected in any realistic MM-level pair potential as a consequence of the molecular structure of the solvent; (2,3,10,11,21,22) they would be found even for two hard sphere solute particles in a hard-sphere liquid or for two $H_2{}^{18}O$ solute molecules in ordinary liquid H_2O, and are found in simulation studies of solutions based on BO-level models. In ionic solutions in a polar solvent another source of wiggles, evidenced in Fig. 2, may be associated with an oscillatory non-local dielectric function $\varepsilon(r)$. (36) These various studies may be used to guide the introduction of wiggles into Eq. (9) in a realistic way.

5. NON-THERMODYNAMIC DATA

Electrical conductivities are much easier to measure than thermodynamic coefficients and, of course, are affected by the solvent-averaged ion-ion forces. Using charged hard sphere models refined by incorporation of a Gurney-like term, Ebeling and coworkers (37,38) and Justice and Justice (39,40) have adjusted Gurney parameters for aqueous alkali halides and some other systems to fit the conductivity data. Quite remarkably the Gurney parameters adjusted to fit the conductivity data are very well correlated with the Gurney parameters adjusted to fit the thermodynamic excess functions. (Fig. 3) On this basis one could use conductivity data to find Gurney parameters in cases in which the thermodynamic data were incomplete and then use the results to predict thermodynamic excess functions which otherwise would be poorly known. A useful and significant correlation between the Gurney parameters and the solvation free energies also has been reported. (38)

Next we describe some work that is directed at the question, does the agreement with experiment of model potentials like the one in Eq. (9) imply that the models are right? It seems unlikely that the answer is affirmative because of the great variety of equations which, over the years, have been reported to give precision fits to the thermodynamic excess function data; apparently there is relatively little information in these data. More directly, we find that we can change some important aspects of the models within the scope of Eq. (9) and still fit the data for thermodynamic excess functions. Typical examples are given below.

For the rate constant k_{ab} of an activation controlled reaction of a solute particle of species a with one of species

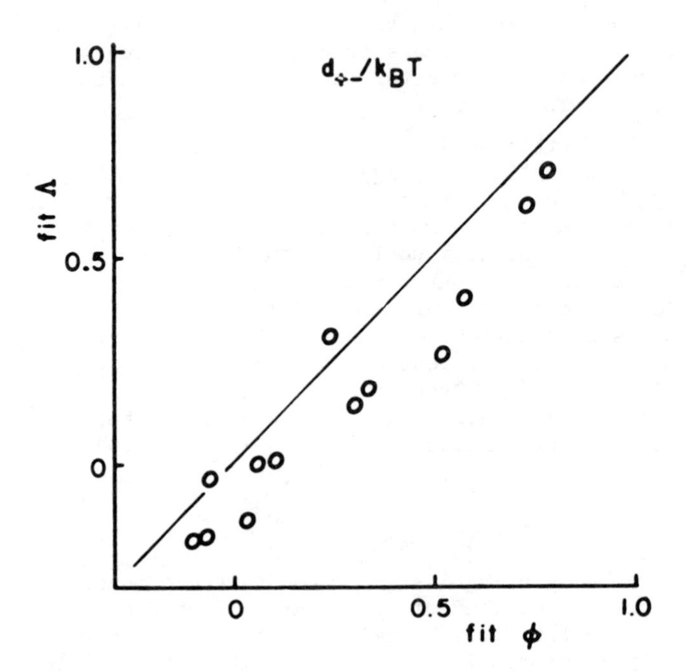

Figure 3. Gurney parameters for + − pairs determined by fitting data for osmotic coefficient φ and conductivity Λ (38). The line in the figure represents ideal correlation. The data are for alkali halides, except fluorides, in water. The parameter d_{+-} *is defined for a simpler MM-level model than Equation 9; in Ref. 38 it is reported that* $d_{+-}/k_BT = 0.75 + 3.6\ A_{+-}/k_BT$. *These correlations have been found by Justice and Justice as well (39, 40).*

b one has

$$k_{ab} = \int g_{ab}(r)\hat{k}_{ab}(r)\ d^3r \tag{11}$$

where $\hat{k}_{ab}(r)$ is the rate when the distance between the reactants is r . Comparing calculated and experimental k_{ab} is a much better way to test models than comparison of the thermodynamic excess functions because k_{ab} depends only on g_{ab} whereas the thermodynamic excess functions depend on all of the solute-solute g_{ij} , weighted by the concentration product $c_i c_j$. (34,41) While the theory for ordinary chemical kinetics is not sufficiently advanced so that one knows enough about $\hat{k}_{ab}(r)$ for this purpose, the situation is much better for spin relaxation induced by the dipolar spin-spin interaction, as in the process

$$^7Li^+(m) + Ni^{2+} = {}^7Li^+(m') + Ni^{2+} \quad ,$$

where $m \rightarrow m'$ represents a change in nuclear spin state of the ^7Li, in this case due to collision with the paramagnetic Ni^{2+}. The rate constant is determined by measuring T_1 of ^7Li as a function of Ni^{2+} concentration in aqueous solutions of LiCl and $NiCl_2$, with added $MgCl_2$ to keep the ionic strength fixed. The rate constant is calculated for models based on Eq. (9) for each one of the following cases.

A. Ni^{2+} and Li^+ both are rigidly hydrated so that the ionic radii used in Eq. (9) are respectively, 0.7+2.76 Å and 0.6+2.76Å.

B. Ni^{2+} is rigidly hydrated but Li^+ is not; its radius in Eq. (9) is only 0.6 Å.

C. Neither Ni^{2+} nor Li^+ is rigidly hydrated.

All three models may be fit to the thermodynamic data for aqueous solutions of $NiCl_2$, LiCl, and their mixtures by adjusting their Gurney parameters. Thus, it is difficult to tell from the thermodynamic data whether one model is more realistic than the others. However only model B is in satisfactory agreement with the spin relaxation data. (41)

At this time diffraction data for ion-ion distributions in aqueous solutions of moderate concentration are beginning to become available. In aqueous $NiCl_2$ solutions very refined neutron diffraction studies indicate that the Ni^{2+}-Cl^- pair correlation function has a peak near 3.1Å under conditions in which the Cl^- does not penetrate the $Ni(H_2O)_6^{2+}$ unit. (42) It is reported that EXAFS studies give the same result. (43) While the information is most welcome it is puzzling because a geometrical calculation indicates that the closest center to center distance for the Ni^{2+} and a Cl^- that does not penetrate the hydration shell is closer to 3.9Å. (7)

6. CORRESPONDING STATES FOR IONIC SYSTEMS

It has been proposed to define a reduced temperature T_r for a solution of a single electrolyte as the ratio of k_BT to the work required to separate a contact +- ion pair, and the reduced density ρ_r as the fraction of the space occupied by the ions. (44) The principal feature on the T_r,ρ_r corresponding states diagram is a coexistence curve for two phases, with an upper critical point as for the liquid-vapor equilibrium of a simple fluid, but with a markedly lower reduced temperature at the critical point than for a simple fluid (with the corresponding definition of the reduced temperature, i.e. the ratio of k_BT to the work required to separate a van der Waals pair.) In the case of a plasma, an ionic fluid without a solvent, the coexistence curve is for the liquid-vapor equilibrium, while for solutions it corresponds to two solution phases of different concentrations in equilibrium. Some non-aqueous solutions are known which do unmix to form two liquid phases of slightly different concentrations. While no examples in aqueous solution are known, the corresponding states principle for ionic fluids suggests that solutions of even 1-1 electrolytes in water might unmix above 300°C in some concentration range. It is difficult to be more precise because the coexistence curve apparently is sensitive to details in the pair potentials. (36)

7. MIXING COEFFICIENTS

In a symposium concerning industrial applications one expects solutions of single electrolytes to take second place in importance behind solutions of mixed electrolytes. The latter have scarcely been mentioned in this review. To redress the balance we recall here an interesting question regarding the mixing coefficient g_1 that appears in the expansion of the excess free energy of mixing two electrolytes with a common ion at constant molal ionic strength I.

$$\Delta_m G^{ex}(y,I) = y(1-y)I^2 WRT[g_0(I) + (1-2y)g_1(I) + \ldots] \quad (12)$$

Here y is the fraction of the ionic strength due to one of the electrolytes and W is the mass (kg) of solvent in the mixture. It is well known (4) that higher order limiting laws, depending upon only the charge type and ε , determine the behavior of g_0 at very low I. Recently it was found experimentally by Cassel and Wood (45) that in mixtures of different charge types at low I the coefficient g_1 (actually $\partial(g_1/T)/\partial(1/T)$) seemed to be diverging as $I \to 0$ and, moreover, that the species-dependence of the divergence indicated that the divergence was a limiting law phenomenon. Calculations by two approximation methods (46) showed that g_1 would indeed appear to diverge in the range studied by Cassel and Wood, and that the effect was indeed due to the long range Coulomb interactions, but the calculated and experimental coefficients disagreed in both sign and magnitude! On the other hand, more recent measurements in a different system by Khoo and coworkers (47) seem consistent with the theory, although the comparison has yet to be carefully developed.

8. ACKNOWLEDGEMENT

We are grateful for the support of the present work by the National Science Foundation.

LITERATURE CITED

1. Hill, T. L. "Statistical Mechanics," McGraw-Hill, New York, 1956.
2. Hansen, J. P. and McDonald, I. R., "Theory of Simple Liquids," Academic Press, N.Y., 1976.
3. Watts, R. O. and McGee, I. J. "Liquid State Physics," J. Wiley and Sons, N.Y., 1976.

4. Friedman, H. L., "Ionic Solution Theory," Interscience Publishers, N.Y., 1962.

5. Friedman, H. L. and Dale, W. D. T., in "Modern Theoretical Chemistry, Vol. V, Statistical Mechanics," Plenum Press, N.Y., 1977, Editor, Berne, Bruce J., pp. 85-135.

6. Andersen, H. C., Ann. Revs. Phys. Chem., 1975, 26, 145.

7. Friedman, H. L., Jour. Electrochem. Soc., 1977, 124, 421C.

8. Friedman, H. L., in "Protons and ions involved in fast dynamic phenomena," Elsevier, N.Y., 1978. Laszlo, P., editor.

9. Lie, G. C., Clementi, E., and Yoshimine, N., J. Chem. Phys., 1976, 64, 2314.

10. Rahman, A. and Stillinger, F. H., J. Chem. Phys., 1971, 55, 3336.

11. Geiger, A., Rahman, A., and Stillinger, F. H., J. Chem. Phys., 1979, 70, 263.

12. McMillan, W. G. and Mayer, J. E., J. Chem. Phys., 1945, 13, 276.

13. Friedman, H. L., J. Chem. Phys., 1960, 32, 1134.

14. Stell, G. R. and Høye, J., Faraday Discuss. Chem. Soc., 1977, 64, 16.

15. Pitzer, K. S., Accts. Chem. Res., 1977, 10, 371.

16. Patey, G. N. and Valleau, J. P., J. Chem. Phys., 1975, 63, 2334.

17. Levesque, D., Weis, J. J., and Patey, G. N., Physics Letters, 1978, 66A, 115.

18. Kraus, C. A., J. Phys. Chem., 1956, 60, 129.

19. Adelman, S. and Chen, J.-H., J. Chem. Phys., 1979, 70, 4291.

20. Kauzmann, W., Adv. Protein Chem., 1959, 14, 1.

21. Pangali, C., Rao, M., and Berne, B. J., J. Chem. Phys., 1979, 71, 2975.

22. Pratt, L. and Chandler, D., J. Chem. Phys., 1977, 67, 3683.

23. Jolicoeur, C. and Friedman, H. L., Ber. Bunsenges., 1971, 75, 248.

24. F. Franks in "Water, A Comprehensive Treatise," Plenum Press, N.Y.. 1974, Vol. 4, Ch. 1. Franks, F., editor.

25. Clarke, A. H., Franks, F., Pedley, M. D., and Reid, D. S., Faraday Trans. Chem. Soc. I, 1977, 73, 290.

26. Ramanathan, P. S. and Friedman, H. L., J. Chem. Phys., 1971, 54, 1086.

27. Debye, P. and McAulay, J., Physik. Z., 1925, 26, 22.

28. Gurney, R. W., "Ionic Processes in Solution," Dover, N.Y., 1953.

29. Frank, H. S., in "Chemical Physics of Ionic Solutions, edited by Conway, B. E. and Barradas, R. G., Wiley, N.Y., 1966, Z. Physik Chem. (Leipzig), 1965, 228, 364.

30. Friedman, H. L., Smitherman, A., and De Santis, R., J. Solution Chem., 1973, 2, 59.

31. Ramanathan, P. S., Krishnan, C. V., and Friedman, H. L., J. Solution Chem., 1972, 1, 237.

32. Friedman, H. L. and Ramanathan, P. S., J. Phys. Chem., 1970, 74, 3756.

33. Krishnan, C. V. and Friedman, H. L., J. Solution Chem., 1974, 3, 727.

34. Friedman, H. L., Krishnan, C. V., and Hwang, L. P., in "Structure of Water and Aqueous Solutions," Luck, W., editor, Verlag Chemies Gmbh, 1974, p. 169.

35. Rossky, P. J. and Friedman, H. L., J. Phys. Chem., 1979, submitted.

36. Dogonadze, R. R. and Kornyshev, A. A., J. Chem. Soc. Faraday Trans., 1974, 70, 2.

37. Bich, E., Ebeling, W., and Krienke, H., Z. Phys. Chem., Leipzig, 1976, 257, 549.

38. Wiechert, H., Krienke, H., Feistel, R., and Ebeling, W., Z. Phys. Chem. Leipzig, 1978, 259, 1057.

39. Justice, J.-C. and Justice, M.-C., Faraday Discuss. Chem. Soc., 1977, 64, 265.

40. Justice, M.-C. and Justice, J.-C., Pure and Applied Chem., 1979, 51, 1681.

41. Hirata, F., Holz, M., Hertz, H. G., and Friedman, H. L., to be submitted.

42. Soper, A. K., Neilson, G. W., Enderby, J. E. and Howe, R. A., J. Phys. C, 1977, 10, 1793.

43. Sandstrom, D. R., J. Chem. Phys., 1979, 71, 2381.

44. Friedman, H. L. and Larsen, B., J. Chem. Phys., 1978, 69, 92.

45. Wood, R. H. and Cassel, R. B., J. Phys. Chem., 1974, 78, 1924.

46. Friedman, H. L. and Krishnan, C. V., J. Phys. Chem., 1974, 78, 1927.

47. Khoo, K. H., Lim, T. K., and Chan, C. Y., Faraday Trans. Chem. Soc. I, 1978, 74, 2037.

RECEIVED January 31, 1980.

Activity Coefficients, Ionic Media, and Equilibrium Constants

R. M. PYTKOWICZ

Oregon State University, Corvallis, OR 97330

In this work the concepts of ionic medium, effective ionic strength and free versus total activity coefficients are examined. Then they are applied to the study of permissible and incorrect translations of equilibrium constants from one medium to another.

Ionic Media

Ionic media are solutions of background electrolytes which are concentrated enough so that the activity coefficients of the electrolytes of interest do not change during processes which are occurring. Typical ionic media are a 1 m $HClO_4$ or $NaClO_4$ solution and seawater.

Let us consider the dissolution-precipitation process in seawater in the following example. The normal concentrations of calcium and of carbonate in the near-surface oceanic waters are about $[Ca^{2+}] = 0.01$ and $[CO_3^{2-}] = 2 \times 10^{-4}$ M. The $CaCO_3$ in solution is metastable and roughly 200% saturated (1). Should precipitation occur due to an abundance of nuclei, $[CO_3^{2-}]$ will drop to 10^{-4} M but $[Ca^{2+}]$ will change by no more than 2%. Therefore, the ionic strength of the ionic medium seawater will remain essentially constant at 0.7 M. The major ion composition will also remain constant. We shall see later what the implications are for equilibrium constants.

It is important to realize that one must take into consideration ion association of the ionic media electrolytes because it affects the effective ionic strength (2). This in turn changes the activity coefficients of the ions under study. The effective ionic strength, I_e, of a 2-1 electrolyte CA_2 which associates is given by

$$I_e = 0.5\{4[C^{2+}]_F + [A^-]_F + [CA^+]\} \tag{1}$$

where the subscript F refers to free species. We shall see that I_e plays a key role in equilibria.

0-8412-0569-8/80/47-133-561$05.00/0

Electrolytes which often are considered to be completely dissociated but in effect are not are NaCl and $NaClO_4$ (3).

Ion-Pairing

In the preceeding section mention was made of ion association (ion-pairing) which, for the purposes of this paper, will refer to coulombic entities with or without cosphere overlap. Experimental support for ion-pairing has come from sound attenuation (4), Raman spectroscopy (5) and potentiometry (2, 3). Credibility has resulted from the model of Fuoss (6) applied by Kester and Pytkowicz (7).

Our method at present is not based upon theoretical models or departures from ideal behavior. It consists in the use of potentiometric determinations and literature values of activity coefficients, starting with HCl-$HClO_4$ electrolyte mixtures and with the assumption that $HClO_4$ is completely dissociated since the association constant pK = 7 is extremely small in this case. A comparison of experimental results with those calculated from the Fuoss (6) theory is presented in Table I. The theory is only valid approximately so that the order of magnitude agreement is fairly good, except in the cases of $MgCO_3^0$ and $CaCO_3^0$. Stoichiometric association constants K* are then obtained from the activity coefficients, expressions for K*, and from equations for the conservation of mass. The latter express the total concentration of a given ion as the sum of the concentrations of the free ion and of the ion-pairs. Values of K* and of the activity coefficients of free ions in ionic media depend only upon the effective ionic strength as is shown later.

Table I.
A comparison of stoichiometric association constants calculated from the Fuoss (6) model with Debye radii and from the measurements of Johnson and Pytkowicz (3).

Ion Pair	Calculated	Measured
$NaSO_4^-$	2.1	2.02
$MgSO_4^0$	4.6	10.2
$CaSO_4^0$	6.5	10.8
$NaHCO_3^0$	0.46	0.28
$MgHCO_3^+$	2.8	1.62
$CaHCO_3^+$	2.4	1.96
$NaCO_3^-$	2.0	4.25
$MgCO_3^0$	3.9	112.3 (?)
$CaCO_3^0$	4.9	279.6 (?)
$NaCl^0$	0.43	0.321
$MgCl^+$	2.5	1.91
$CaCl^+$	2.3	2.24

From the free activity coefficients and values of K* obtained in binary solutions it is possible then to calculate total (stoichiometric) activity coefficients in more complex solutions.

Our results lead to good predictions of activity coefficients in multicomponent systems from data measured in simple solutions. Also, they yield values similar to those of Pitzer and Kim (8) as is shown in Table II.

Table II.
A comparison of trace mean activity coefficients calculated by the methods of Johnson and Pytkowicz (3) and Pitzer and Kim (8).

System A/B	A (1)	A (2)	A (3)
$KCl/NaCl$	0.620	0.636	0.626
$KCl/CaCl_2$	0.639	0.628	0.662
$KCl/MgCl_2$	0.649	0.636	0.674
$NaCl/CaCl_2$	0.671	0.643	0.682
$NaCl/MgCl_2$	0.682	0.650	0.694
$CaCl_2/MgCl_2$	0.461	0.460	0.461

(1) Experimental
(2) Johnson and Pytkowicz
(3) Pitzer and Kim

For an artificial seawater consisting of Na^+, K^+, Mg^{2+}, Ca^{2+} and Cl^- (HCO_3^- and CO_3^{2-} missing) we obtain as examples that Na^+ is 85.1% and Ca^{2+} is 45.7% free.

Activity Coefficients

The ion-pair concept leads to two types of activity coefficients; $\gamma_{\pm T}$ for free plus dissociated species and $\gamma_{\pm F}$. These γ's are related by

$$a = \gamma_{\pm F}^2[F] = \gamma_{\pm T}^2[T] \tag{2}$$

where [F] and [T] are concentrations and a is the activity (2).

$\gamma_{\pm F}$ is a construct which includes effects such as those due to hydration, ion-cavity interactions, hard core, etc., but not to ion-pairing. Pytkowicz and Kester (2) and Johnson and Pytkowicz (3) have shown that $\gamma_{\pm F}$ depends only upon the I_e of the medium while $\gamma_{\pm T}$ is also a function of the composition.

It is important in equilibrium calculations to select the proper equation for activity coefficients when the authors do not actually measure them or if measured values in the medium of interest are not directly available. Much of what I say is known but some new concepts are entered into what follows.

Some measurements in single electrolyte solutions are available at concentrations of the order of 10^{-3} m although most of the data has been obtained in the range 0.1 - 6.0 m for cases when the solubility of the salt was not exceeded.

In single electrolyte solutions the Debye-Hückel limiting equation works up to 10^{-3} m. Its well known extended form used at m > 10^{-3} is

$$\log \gamma_{\pm T} = - \frac{|Z_C Z_A| A_D I^{0.5}}{1 + B_D a_D I^{0.5}} \tag{3}$$

however, it is entirely empirical because Frank and Thompson (9) showed that the ionic atmosphere is then coarse-grained. The sphere of radius $1/\kappa$ may contain only one or a fraction of an ion in contrast to the requirements of the theory. If equation (3) is used then a_D has to be calculated by curve-fitting to data for the electrolyte of interest. This fit to experimental data means that $\gamma_{\pm T}$ rather than $\gamma_{\pm F}$ is being determined.

In terms of empirical equations, the following one provides good fits from 10^{-3} to 1.0 m solutions according to our work. The Davies equation (10)

$$\log \gamma_{\pm} = - \frac{Z_C Z_A\ A_C I^{0.5}}{1 + B_C I^{0.5}} + C_C I + D_C I^{1.5} + E_C I^2 \tag{4}$$

in contrast to the above equations, is a predictive tool that works well up to I ≅ 0.1. Note should be made, however, to the effect that, if ion-pairing or complexing occurs, then the Davies equation only works if $\gamma_{\pm}$ is corrected for association. Thus, this equation yields $(\gamma_{\pm})_T = (\gamma_{\pm})_F$ when there is no association and $(\gamma_{\pm})_F$ when ion-pairing occurs.

Several approaches are available in the case of mixed electrolyte solutions. The Guntelberg equation can be used at very high dilutions to avoid the ambiguity in the meaning of a_D, the distance of closest approach, when several electrolytes are present. This equation is empirical and has fewer terms than the Debye-Huckel extended equation. I found it to yield poor agreement with experimental results even at m = 0.01 for NaCl at 25°C ($\gamma_{\pm\,calc}$ = 0.8985 and $\gamma_{\pm\,exp}$ = 0.9024). For the Davies equation for m = 0.20 one obtains $\gamma_{\pm\,calc}$ = 0.752 and $\gamma_{\pm\,exp}$ = 0.735 also for NaCl at 25°C.

An extended form of the Debye-Hückel equation is the hydration one of Robinson and Stokes (11). It contains two adjustable parameters, a_D and h, where h is related to the hydration number. It can be fitted to $\gamma_{\pm}$ for several electrolytes for concentrations in excess of 1 m. Their equation has the valuable feature of describing not only the salting-in but also the salting-out part of the $\gamma_{\pm}$ versus m curve. It should be noted, however, that the arbitrary allocation of non-hydration terms to the Debye-Hückel equation for the estimate of a_D at high concentrations weakens the theoretical validity of the hydration equation. Furthermore, it is an oversimplification to only consider the nonspecific coulombic and the hydration terms.

Note that, unless otherwise indicated, I am using I_e and $(\gamma_\pm)_T = (\gamma_\pm)_{exp}$ in this review. Furthermore, the simple equations presented so far do not include terms for the cosphere overlap, the hard core term, and ion-cavity interactions.

Next, predictive equations for activity coefficients in mixed electrolyte solutions, based upon results in simpler ones, will be mentioned. The work of Brønsted (12) and of Guggenheim (13) led to the specific interaction equation

$$\log \gamma_{\pm B} = - \frac{Z_C Z_A A_D I^{0.5}}{1 + I^{0.5}} + [2xB_{MX} + (B_{NX} + B_{MY})(1 - x)]m \quad (5)$$

$x/(1 - x)$ is the molal ratio of B to C where B is MX and C is NY. A similar condition applies to $\log \gamma_{\pm C}$. One can derive Harned's rule from this expression although it can also be done from the ion pairing model (2). The equations work well for I up to 0.1 m and permit the calculation of $\gamma_{\pm B}$ and $\gamma_{\pm C}$ in a mixture of the two electrolytes from data obtained in single electrolyte solutions. The method suffers, however, because the B coefficients are not allowed to vary with the ionic strength.

The statistical thermodynamic approach of Pitzer (14), involving specific interaction terms on the basis of the kinetic core effect, has provided coefficients which are a function of the ionic strength. The coefficients, as the stoichiometric association constants in our ion-pairing model, are obtained empirically in simple solutions and are then used to predict the activity coefficients in complex solutions. The Pitzer approach uses, however, a first term akin to the Debye-Hückel one to represent nonspecific effects at all concentrations. This weakens somewhat its theoretical foundation.

As most other methods, including our ion-pairing model which was described earlier, the Pitzer approach is empirical in practice. The interaction coefficients in this case are determined by curve fitting in single electrolyte solutions.

In the equations developed by Reilly and Wood (15) from the cluster integral model (16), $\gamma_\pm$ is calculated in complex solutions from excess properties of single salt solutions. Note that the cluster integral approach is based upon terms which represent the contributions of pair-wise ion interactions in various types of clusters to the potential interaction energy. Then, the partition function and the excess properties of the solution can be evaluated. The procedure is akin to the virial expansion in terms of clusters.

The more recent statistical thermodynamic work approaches the problem from the standpoint of the radial distribution function (17). In using this method, Ramanthan and Fridman (18), in contrast to the method of Pitzer, made the Gurney cosphere overlap, the ion-cavity repulsion, and repulsive core potential explicit. The curve fitting was done by means of the Gurney term.

The most fertile approaches so far, from the standpoint of predicting $\gamma_\pm$ in complex solutions from data in simple ones, have been that of Pitzer (14), and the ion-pair approach of Pytkowicz and Kester (2) and of Johnson and Pytkowicz (3). The lattice model of Pytkowicz and Johnson (19) is, at this time, an explanatory rather than a predictive tool.

The only three methods which do not require curve-fitting at present are the use of the Debye-Hückel limiting law and of the equations of Guntelberg and of Davies. Unfortunately these equations are of value only in very dilute and simple solutions.

Equilibrium Constants

Let us examine the following types of constants now in use and their relationships to the thermodynamic constants $K^{(t)}$:

a) Apparent dissociation constants which are of practical use in ionic media (20)

$$K_{HA}' = \frac{ka_H[A^-]_T}{[HA]} = K_{HA}^{(t)} \frac{(\gamma_{HA})_T}{(\gamma_A)_T} \tag{6}$$

b) Stoichiometric solubility products, such as

$$K_{sp}' = [Ca^{2+}]_T[CO_3^{2-}]_T = K_{sp}^{(t)}/(\gamma_{\pm CaCO_3})_T^2 \tag{7}$$

c) Free solubility product

$$K_{sp}'' = [Ca^{2+}]_F[CO_3^{2-}]_F = K_{sp}^{(t)}/(\gamma_{\pm CaCO_3})_F^2 \tag{8}$$

d) Stoichiometric association constant for ion-pairs

$$K^* = \frac{[CaCO_3^{\,o}]}{[Ca^{2+}]_F[CO_3^{2-}]_F} = K^{*(t)} \frac{(\gamma_{\pm CaCO_3})_F^2}{\gamma_{CaCO_3}} \tag{9}$$

e) Affinity constants for complexes

$$\beta = \frac{[PbCl_2]}{[Pb]_F[Cl^-]_F^2} = \beta^{(t)} \frac{(\gamma_{\pm PbCl_2})_F^3}{\gamma_{PbCl_2^{\,o}}} \tag{10}$$

$(\gamma_\pm)_F$ depends only upon the effective ionic strength of the medium. Thus, K_{sp}'', K^*, and β can be determined in one ionic medium such as $HClO_4$ and applied in another one such as seawater, provided that I_e is the same in the two solutions.

K_{HA}' and K_{SP}', on the other hand, must be determined and applied in the same medium because γ_T depends upon the composition of the solution. Note that ion-pairing models are required for both classes of constants to ascertain that I_e is indeed the same.

Literature Cited

1. Pytkowicz, R.M., J. Geol., 1965, 73,196-199.
2. Pytkowicz, R.M.; Kester, D.R., Am. J. Sci., 1969, 267, 217-229.
3. Johnson, K.; Pytkowicz, R.M., Am. J. Sci., 1978, 278, 1428-1447.
4. Fisher, F.H., Science, 1967, 157, 823.
5. Daly, F.P.; Brown, C.W.; Kester, D.R., J. Phys. Chem., 1972, 76, 3664-3668.
6. Fuoss, R.M., J. Am. Chem. Soc., 1958, 80, 5059-5061.
7. Kester, D.R.; Pytkowicz, R.M., Mar. Chem., 1975, 3, 365-374.
8. Pitzer, K.S.; Kim, J.J., Am. Chem. Soc. Jour., 1974, 96, 5701.
9. Frank, H.S.; Thompson, P.T., J. Chem. Phys., 1959, 31, 1086-1095.
10. Davies, C.W., "Ion Association", Butterworths, London, 1962.
11. Robinson, R.A.; Stokes, R.H., "Electrolyte Solutions", Butterworths, London, 1968.
12. Brønsted, J.N., J. Am. Chem. Soc., 1922, 44, 877-898.
13. Guggenheim, E.A., Philos. Mag., 1935, A, 588-643.
14. Pitzer, K.J., Phys. Chem., 1973, 77, 268-277.
15. Reilly, P.J.; Wood, R.H., J. Phys. Chem., 1969, 73, 4292-4297.
16. Friedman, H.L., "Ionic Solution Theory", Interscience, New York, 1962.
17. Vaslow, F., "Water and Aqueous Solutions", R.A. Horne, ed., Interscience, New York, 1972, pp. 465-518.
18. Ramanathan, P.S.; Friedman, H.L., J. Chem. Phys., 1971, 54, 1086-1099.
19. Pytkowicz, R.M.; Johnson, K., "Activity Coefficients in Electrolyte Solutions", R.M. Pytkowicz, ed., CRC Press, Coral Gables, Florida, In press.
20. Pytkowicz, R.M.; Ingle, S.E.; Mehrback, C., Limnol. Oceanogr., 1974, 19, 665-669.

RECEIVED February 11, 1980.

30

Flow Calorimetry of Aqueous Solutions at Temperatures up to 325°C

R. H. WOOD and D. SMITH–MAGOWAN

University of Delaware, Newark, DE 19711

1) Need

The fact that we are gathered together at a conference on "Thermodynamics on Aqueous Systems with Industrial Application" indicates the importance of thermodynamic data on aqueous solutions. In particular, there is a great need for data on high temperature aqueous systems. Because of the experimental difficulties, there are relatively few measurements on these systems and yet they are of very great industrial importance.

As the temperature of water approaches the critical temperature, 374°C, it becomes a low dielectric constant solvent with the dielectric constant approaching one tenth its value at 25°C. This is about the same dielectric constant as 1,1 dichloroethane at 25°C. Because of this large change in solvent properties there are corresponding very large changes in the properties of electrolytes dissolved in water. The few data available indicate[1-4] that the heat capacities of dilute aqueous solution become very large and negative as the temperature approaches the critical temperature and the data presented below will show that the relative apparent molal enthalpy, L_ϕ, becomes very large and positive at higher temperatures. These very large changes in thermodynamic properties make it very difficult to predict high temperature properties of solutions from low temperature measurements and increase the need for accurate measurements.

2) Information Obtained

Calorimetric measurements, when combined with the normally available room temperature thermodynamic properties, give values for free energy, enthalpy, heat capacity and even volume at high temperatures.

We have been actively developing two types of calorimeters which will operate at elevated temperatures and pressures. One type is a heat of mixing calorimeter to measure enthalpies of dilution in order to obtain differences in partial molal enthalpy

0-8412-0569-8/80/47-133-569$05.00/0

or heat contents:

$$\Delta H_D = -(L_{\phi \underline{m}_f} - L_{\phi \underline{m}_i})$$

This property simply considered is the first temperature derivative of the free energy or activity and can be used to obtain osmotic coefficients and activity coefficients by the relationships:

$$\phi(\underline{m},\mathfrak{T}) - \phi(\underline{m},T_2) = \{\underline{m}^{\frac{1}{2}}/2\nu R\}\int_{\mathfrak{T}}^{T_2} (\frac{\partial L_\phi}{\partial \underline{m}^{\frac{1}{2}}})d(\frac{1}{T})$$

and

$$\ell n\gamma(\underline{m},T) = [\phi(\underline{m},T)-1] + \int_0^m (\phi(\underline{m}',T)-1)d\ell n\underline{m}'$$

where $\mathfrak{T}$ is a reference temperature. We have also developed a heat capacity calorimeter for these extreme conditions.

The heat capacity is the second temperature derivative of the free energy and can be used to calculate the temperature dependence of equilibria by the relationship:

$$\frac{\Delta G(T,P)}{T} = \frac{\Delta G(\mathfrak{T},P)}{\mathfrak{T}} + \Delta H(\mathfrak{T},P)[\frac{1}{T} - \frac{1}{\mathfrak{T}}] + \int_{\mathfrak{T}}^{T}\int_{\mathfrak{T}}^{T'} (\Delta C_p\ dT'')\ d(\frac{1}{T'})$$

Similarly the temperature dependencies of the relative apparent molal heat content can be determined from the heat capacity by:

$$L_\phi(m,T_2) = L_\phi(m,\mathfrak{T}) + \int_{\mathfrak{T}}^{T_2}(C_{p\phi}(m,T) - C_p^\circ(0,T))dT$$

These calorimeters can be used to determine these thermal properties throughout a wide range of pressures. The pressure dependence can be used to calculate volumetric properties by means of the relationships:

$$(\frac{\partial H}{\partial P})_T = -T(\frac{\partial V}{\partial T})_P + V$$

$$(\frac{\partial C_p}{\partial P})_T = -T(\frac{\partial^2 V}{\partial T^2})_P$$

3) Advantages

In a flow calorimeter the thermodynamic properties are measured in a flowing stream contained in a small diameter tubing. Flow calorimetric techniques have been used for many years at room temperature because of their speed and convenience(5-9). For operations at high temperatures or with a high vapor pressure solvent, the advantages of using flow calorimetric techniques are overwhelming. In the first place there is no vapor space so there is no necessity for corrections for the change in the vapor space composition. Because of the small diameter tubing used, relatively thin walled tubing is strong enough to contain the high pressures and as a result the calorimeter can be constructed for very rapid thermal response. A third advantage is that experiments can be run consecutively without cooling and reloading the calorimeter. All that is necessary is to start pumping into the calorimeter from room temperature the fluids for the next measurement. In the following section we will show that these advantages of flow calorimetric techniques can be realized in practice for measurements on high temperature aqueous solutions by discussing the operation of several instruments that have been constructed in our laboratory.

4) Measurements of $\Delta_m H$

The first attempt in our laboratory to apply flow techniques to high temperature operation was the construction by Dr. E.E. Messikomer of a flow, heat-of-mixing calorimeter(10). Unfortunately, because the thermopiles used in this instrument did not work above 100°C the instrument was limited to this temperature. However, the results were encouraging because they showed that very rapid and accurate thermodynamic data could be obtained and that the operation of the calorimeter was as easy at 100°C as it was at room temperature.

Because of the encouraging results obtained with the first calorimeter, Dr. James Mayrath build a new version which was successfully operated up to 200°C(11). A schematic of this calorimeter is given in figure 1. Basically it is a heat-flow calorimeter in which the two liquids to be mixed are pumped at room temperature into a counter current heat exchanger, after which they are equilibrated with an aluminum calorimetric block. Next the two liquids are mixed, and the heat generated by the mixing process is extracted from the flowing stream by a series of thermopiles which can measure the heat extracted. Using several heat extractors guaranteed that most of the heat generated is measured by the thermopiles and the sum of the voltage on the thermopiles is then a measure of the rate of heat production.

A diagram of a typical run is given in figure 2 which shows the power generated by mixing of magnesium chloride with water at 200°C. In this calorimeter a heat of dilution takes 30 minutes and from an initial base line it takes about 15 minutes for the calorimeter to reach a steady state. The sensitivity of this calorimeter was equivalent to being able to detect a 2×10^{-4}K

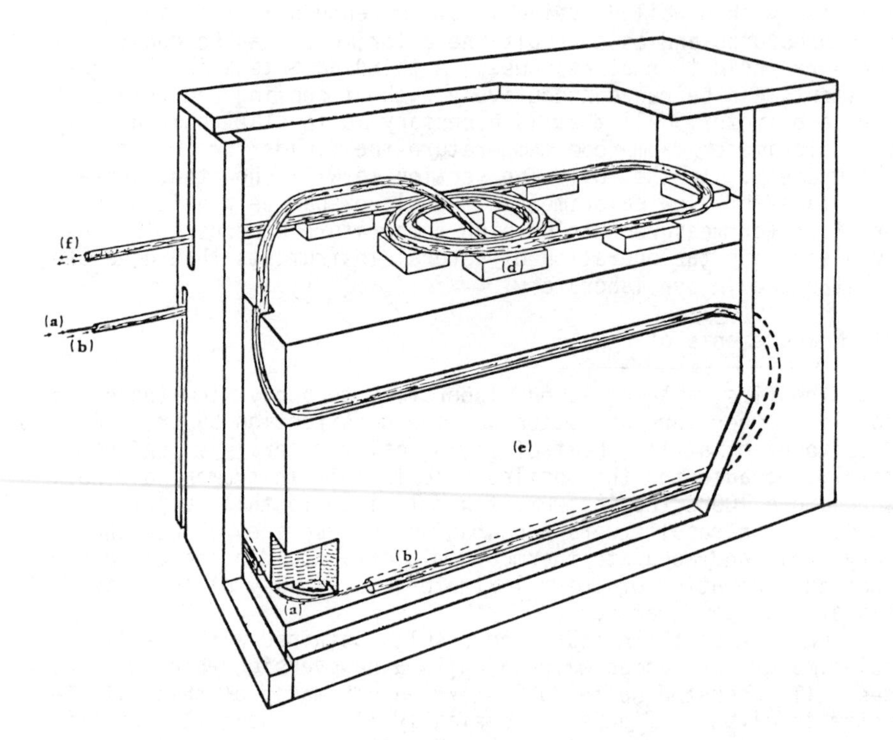

Figure 1. Flow heat of mixing calorimeter: (a and b) solutions to be mixed; (c) calorimetric block; (d) thermopiles for detecting heat flow; (e) exit for mixture

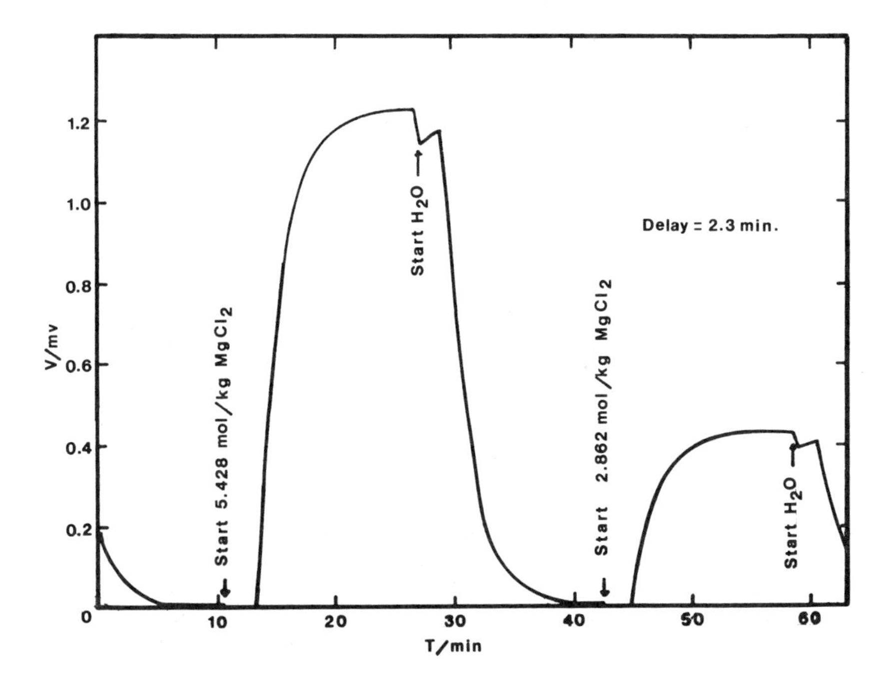

Figure 2. Results of an enthalpy of dilution run on aqueous $MgCl_2$ *at 473 K (thermopile voltage vs. time—maximum 1.23 ± 0.002; Q = 1.129 watts;* $\triangle T$ *= 12.8 K)*

temperature rise on mixing the calorimetric fluids. Some typical results from this calorimeter are shown in figures 3 and 4. It should be noted that the heat of dilution of magnesium chloride at 200°C is 40 kJ mol^{-1} and that this is close to the heat of reaction of H^+ with OH^- at room temperature. Thus, the heat effects in water at 200°C are extremely large and changing rapidly with temperature. This is a further reminder of the difficulty of predicting the properties of high temperature aqueous solutions from their properties at room temperatures.

5) Heat Capacity Measurements

The success of the mixing calorimeters further encouraged us to construct a flow, heat-capacity calorimeter using the basic design principals of Patrick Picker *et al*[8]. A schematic diagram of our heat capacity calorimeter is given in figure 5. The operation of the instrument is as follows: Water is pumped with a high pressure liquid chromatography pump through a 1.5 mm outside diameter hastalloy tube through a heat exchanger and onto a copper calorimetric block where it is equilibrated at the reference temperature. The fluid in the tube is then heated about 2°K and the resulting temperature rise detected by a thermistor. The stream then leaves the calorimetric block through the counter current heat exchanger and goes to a sample injection valve which allows a sample loop (containing 10 ml of the solution to be measured) to be interjected into the flow stream. The stream then goes through a second counter current heat exchanger and into an identical clorimetric unit with heater and thermistor to detect the temperature rise. After leaving the block through the counter current heat exchanger the solution exits the system through a back pressure regulator. In operation the pump and heaters are turned on with water flowing through both calorimetric units and a wheatstone bridge containing the two thermistors in opposite arms is balanced. After a steady state is reached the sample loop valve is opened and the sample solution flows through the second calorimetric unit. When the sample hits the heater and thermistor on the second unit there is a change in heat capacity and a consequent change in temperature. The heater on this unit is adjusted to rebalance the thermistor bridge and thus to keep the temperature rise exactly the same. The ratio of the power applied to the heater with water flowing and with the sample flowing is then the ratio of the volumetric heat capacities of the two solutions. The ratio of the mass flows through the calorimetric units in the two cases is just the ratio of the densities of water and of the solution to be measured at the temperature of the sample loop. This can be easily shown using the assumptions that 1) at constant composition, the mass flow is independent of temperature 2) at constant temperature, the volumetric flow is independent of the composition and 3) the heat and volume of mixing at the interface between the two solutions pro-

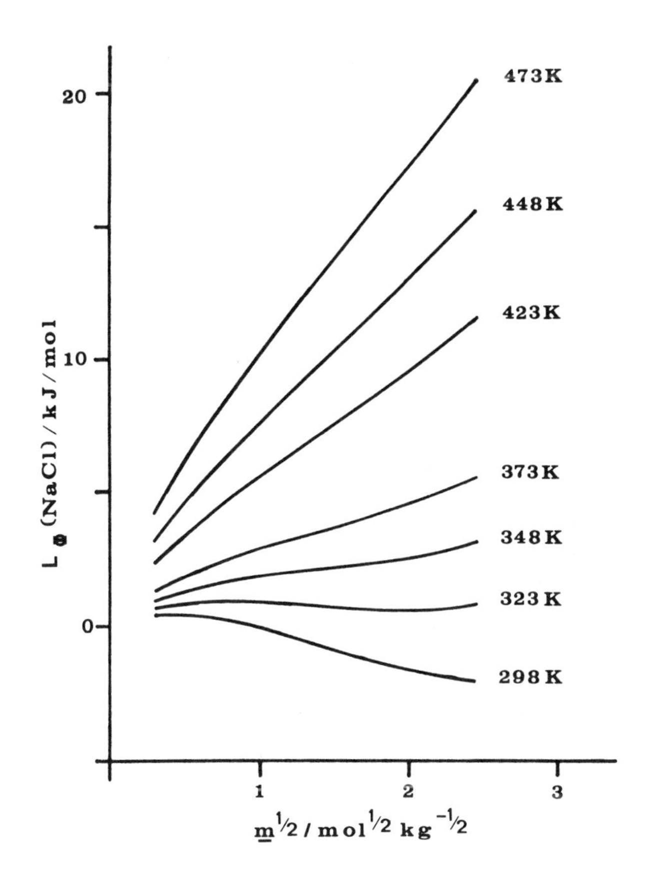

Figure 3. Apparent molal enthalpy of aqueous NaCl as a function of molality and temperature

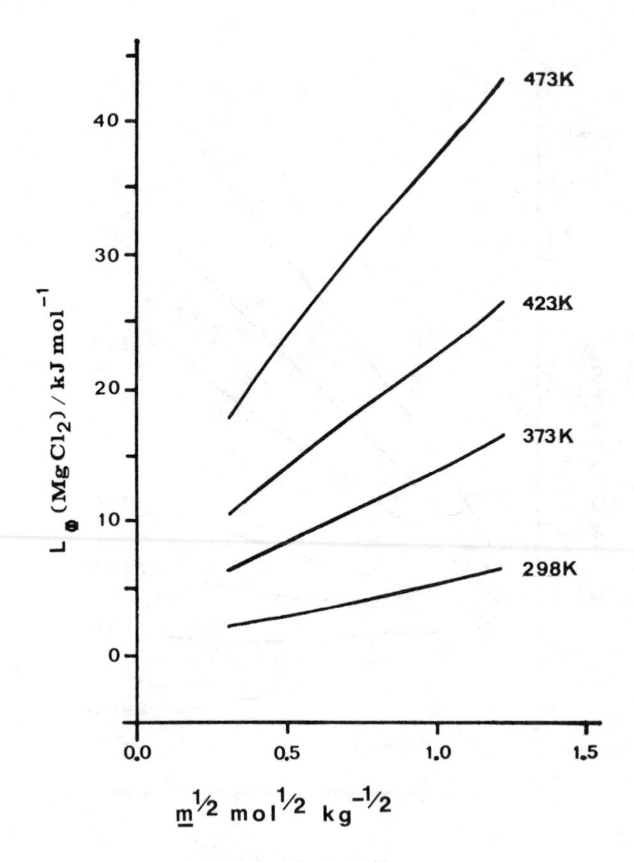

Figure 4. Apparent molal enthalpy of aqueous $MgCl_2$ as a function of molality and temperature

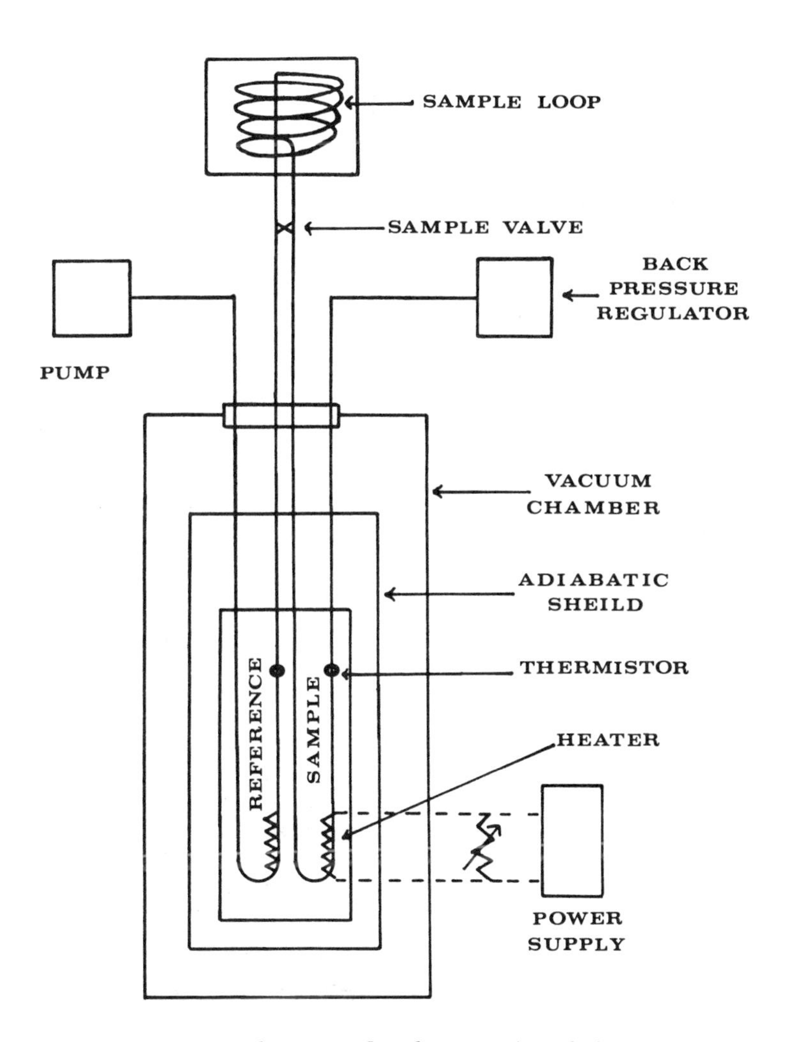

Figure 5. Schematic of flow, heat-capacity calorimeter

duces a negligible change in volume. Experience with the Picker instrument has shown that this latter assumption is quite accurate. The resulting equation for the heat capacities of the two solutions is

$$\frac{C_{p,2}}{C_{p,1}} = \frac{\dfrac{P_2}{\rho_1}}{\dfrac{P_1}{\rho_2}}$$

Where P_1 and P_2 are the powers in the heater with solutions 1 and 2 flowing and ρ_1 and ρ_2 are the densities of the solutions at the temperature of the sample loop.

The resulting instrument has a response time for changes of electrical power input of 50 seconds for 99% response, a sensitivity to a change in power of about 0.005% and a proven capability of operating with this sensitivity up to temperatures as high as 325°C.

The results of some measurements on sodium chloride solutions are given in figure 6. The difficulty of predicting properties of solutions at high temperatures is emphasized by these results. At low concentrations there is a very large change in heat capacity as temperature increases which is not present at higher concentrations. Indeed, the heat capacity differences at 25°C seems almost negligible compared with the changes found with changing temperature at low molality and with changing molality at high temperature.

It is difficult to compare these results with the previous results of other authors since we chose to make measurements along an isobar which was accessible at all temperatures. This allows convenient data reduction. Other authors have generally measured along the saturated water vapor pressure curve with consequent complications in data reduction. While at present, only this one isobar has been systematically investigated, a few measurements of the pressure dependence at 320K and 572K have been made. These results show that at all temperatures the pressure dependence of the heat capacity is appreciable and needs to be more fully evaluated, since the present body of volumetric data is not sufficiently precise to make these corrections accurately. In fact, our very limited results at different pressures suggest that the precision of this procedure may be sufficient to supplant volumetric determinations at high temperature and pressures.

6) Conclusion

These preliminary results show that the promise of flow calorimetric techniques for investigating the thermodynamic properties of high temperature aqueous solutions has been realized. Although there are many experimental difficulties in adapting

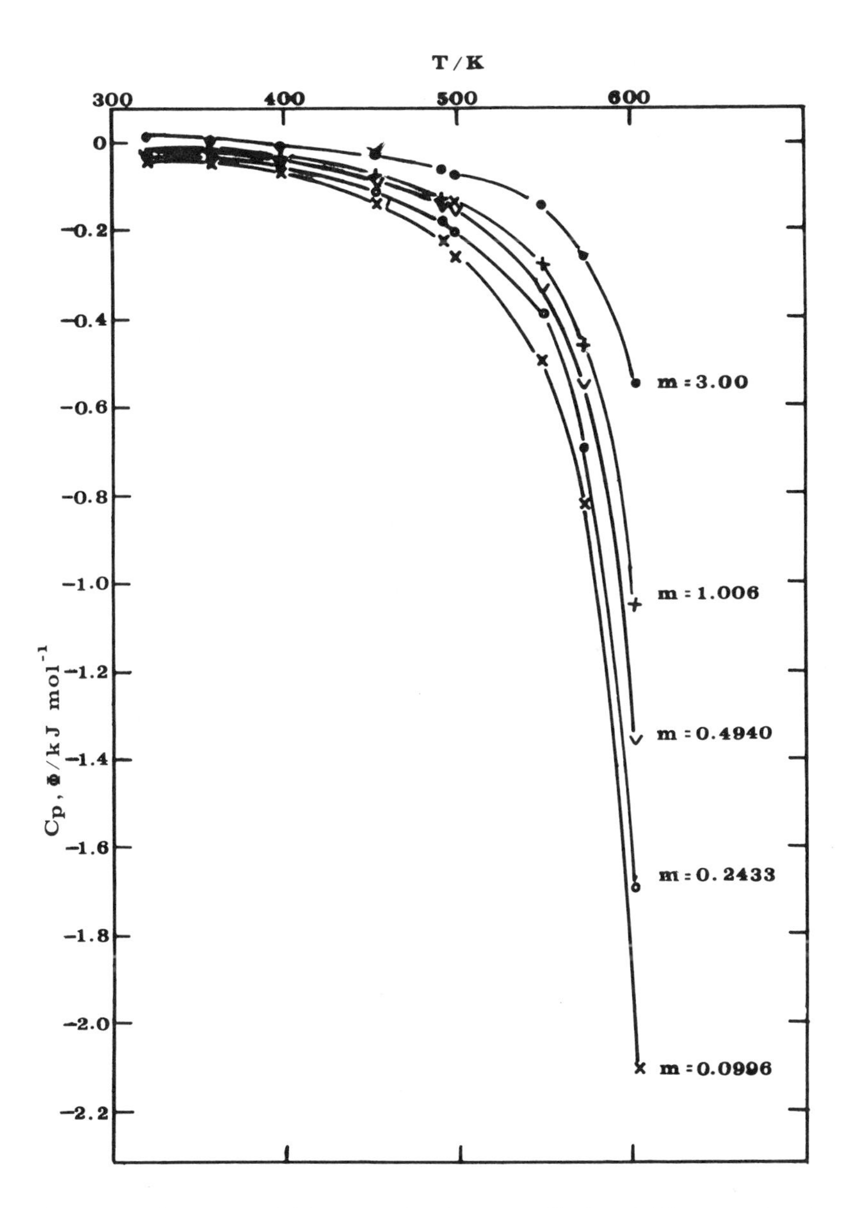

Figure 6. Apparent molal heat capacity of aqueous NaCl at 177 bars as a function of molality and temperature

these techniques to high temperature operation, these problems have now been solved and we have a rapid and sensitive measuring instrument. As a result we can now get down to the business of exploring aqueous solution chemistry at temperatures up to 325°C.

Literature Cited

1. a) Gardner, W.L.; Mitchell, R.E.; Cobble, J.W. J. Phys. Chem., 1969, 73, 2025.
   b) Cobble, J.W. and Murray, Jr., R.C. Discussions Faraday Soc., 1977, 64.
2. C-T. Liu and W.T. Lindsay. J. Soln. Chem., 1972, 1, 45.
3. Likke, S. and Bromley, L.A. J. Chem. Eng. Data, 1973, 18, 189.
4. Puchkov, L.V.; Styazhkin, P.S.; Fedorova, M.K. J. Applied Chem. (Russ.), 1976, 49, 1268.
5. Monk, P. and Wadsö, I. Acta Chem. Scand. 1968, 22, 1842.
6. Picker, P.; Jolicoeur, C.; Desnoyers, J.E. J. Chem. Thermodynamics, 1969, 1, 469.
7. Gill, S.J. and Chen, Y.-J. Rev. Sci. Instruments, 1972, 43, 774.
8. Picker, P.; Fortier, J.-L.; Philip, P.R; Desnoyers, J.E. J. Chem. Thermodyn., 1971, 3, 631.
9. Elliot, K.; Wormald, C.J. J. Chem. Thermodynamics, 1976, 8, 881.
10. Messikomer, E.E.; Wood, R.H. J. Chem. Thermodynamics, 1975, 7, 119.
11. Mayrath, J.E., "A Flow Microenthalpimetric Survey of Electrolyte Solution Thermodynamic Properties from 373K to 473K" Ph.D. Dissertation, University of Delaware, June, 1979.

RECEIVED January 30, 1980.

31

Review of the Experimental and Analytical Methods for the Determination of the Pressure-Volume-Temperature Properties of Electrolytes

FRANK J. MILLERO

Rosenstiel School of Marine and Atmospheric Science, University of Miami, Miami, FL 33149

The pressure-volume-temperature (PVT) properties of aqueous electrolyte and mixed electrolyte solutions are frequently needed to make practical engineering calculations. For example precise PVT properties of natural waters like seawater are required to determine the vertical stability, the circulation, and the mixing of waters in the oceans. Besides the practical interest, the PVT properties of aqueous electrolyte solutions can also yield information on the structure of solutions and the ionic interactions that occur in solution. The derived partial molal volumes of electrolytes yield information on ion-water and ion-ion interactions (1,2). The effect of pressure on chemical equilibria can also be derived from partial molal volume data (3).

I. Experimental Methods

The PVT properties of aqueous solutions can be determined by direct measurements or estimated using various models for the ionic interactions that occur in electrolyte solutions. In this paper a review will be made of the methods presently being used to determine the density and compressibility of electrolyte solutions. A brief review of high-pressure equations of state used to represent the experimental PVT properties will also be made. Simple additivity methods of estimating the density of mixed electrolyte solutions like seawater and geothermal brines will be presented. The predicted PVT properties for a number of mixed electrolyte solutions are found to be in good agreement with direct measurements.

A. One Atmosphere Densities. The densities or volume properties of solutions have been studied by a number of methods which are extensively reviewed elsewhere (4,5,6,7) of all of the methods, only the magnetic float (7-14), the hydrostatic balance (5,15-20), the vibrating flow densimeter (21,22), and dilatometric (23,24,25) methods give data with sufficient precision to study the densities of dilute solutions. For more concentrated

0-8412-0569-8/80/47-133-581$10.50/0

solutions (above 1 m) the classical pycnometer methods (4,5,6) can also be used to obtain reliable densities. I will not discuss all of these methods, but will briefly outline the three systems (hydrostatic balance, magnetic float and vibrating densimeters) that are presently being used to measure the densities of solutions. Since the PVT properties of water are well known (26-32) over a wide range of temperatures, most density measurements are made relative to water. By making relative density measurements ($\Delta\rho = \rho - \rho_o$), where ρ_o is the density of pure water, it is possible to make very precise measurements with relative ease.

1. Hydrostatic Balance. Many studies using a hydrostatic balance have been made of the densities of solids and liquids (20-25). By weighing solids of a known density (determined by independent measurements of mass and geometry) in a liquid, it is possible to determine the absolute density of the liquid. An analytical balance can be adapted to a high-precision liquid density apparatus by attaching a platinum wire or nylon string with a "sinker" to the balance arm (Figure 1). The sinker and wire are weighed in air and then in the solution. The surface of the solution is brought to the same mark on the wire each time so that it contributes the same (small) amount to the total volume (V). The difference between the weight in vacuum (W_V) and in water (W_o) is equal to the mass of displaced fluid. For relative density measurements the system is calibrated by weighing the sinker in a solution of known density (water). The volume of the sinker is determined from

$$V = (W_V - W_o)/\rho_o \tag{1}$$

where ρ_o is the density of water (28). Once the volume of the sinker is known, the density of an unknown solution can be determined from

$$\rho = (W_V - W)/V \tag{2}$$

where W is the weight in the solution of unknown density. The relative density ($\Delta\rho$) can be determined from

$$\Delta\rho = \rho - \rho_o = (W_o - W)/V = \left(\frac{W_o - W}{W_V - W}\right) \rho_o \tag{3}$$

The specific gravities (ρ/ρ_o) can be determined from the ratio of the weights in solution and water

$$\rho/\rho_o = (W_V - W)/(W_V - W_o) \tag{4}$$

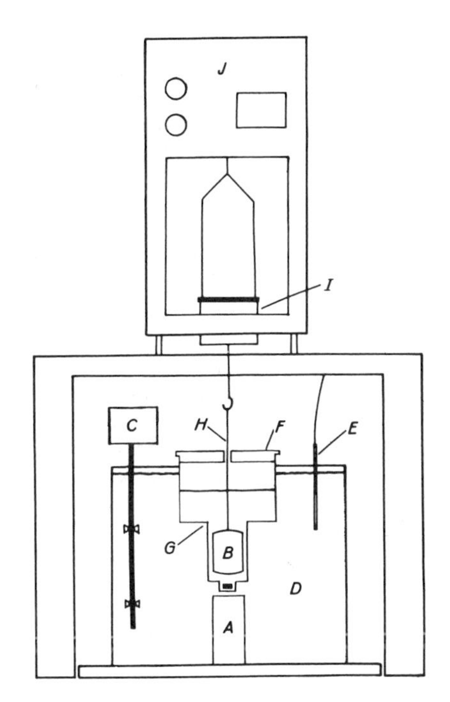

Figure 1. Sketch of the hydrostatic balance densitometer—(A) magnetic stirrer; (B) glass float; (C) stirring motor; (D) constant temperature bath; (E) nickel thermometer; (F) lucite plug; (G) sample container; (H) nylon wire; (I) suspension hook; (J) Mettler H20 balance (33)

A sketch of the hydrostatic balance that we have used (20,33,34) in my laboratory is shown in Figure 1. A 100g glass float (66 cm^3 in volume) is suspended by a nylon line (0.02 cm diameter) from a hook on the bottom pan of a Mettler (H-20) balance. The glass sample cell has a Lucite top with a hole in the center to accomodate the suspension line and is suspended in a constant temperature bath controlled to ± 0.001°C. A submersible magnetic stirrer agitates the solution inside the cell. A weight difference of $\pm 8 \times 10^{-5}$g gives a precision of ± 1 ppm in density. The accuracy of the system was checked by using NaCl solutions and found to be ± 2 ppm.

The largest uncertainty with using this method to measure densities arises from the surface tension effect between the liquid surface and the wire or string. To circumvent this problem many workers (4) have platinized the wire in the region at which it contacts the liquid surface. Another problem with this method is that it is difficult to control evaporation from the solution being studied. For precise work it is necessary to analyze an alloquot of solution after the density is determined.

2. Magnetic Float Densimeter. The floatation method of determining the density of a liquid is a modification of the hydrostatic method. In the floatation method, however, the float has no suspension thread or wire. Earlier workers adjusted the buoyancy of the float by mixing two liquids, adding Pt weights and by changing the temperature (4,7). In all of these early applications, the accuracy was limited by how accurately the coefficient of thermal expansion of the solution was known. Lamb and Lee (8) extended the accuracy and precision of the method by imposing a magnetic field on a float containing a soft iron core. They showed that this electromagnetic method of weighing was able to yield a precision of better than 1×10^{-7}g cm^{-3} in density. A number of workers (7,9-14) have modified this method by using a magnetic core in the float. A sketch of the magnetic float system which we have used for a number of years (7) is shown in Figure 2. The densimeter consists of the following major components: A) the solution container (110 cm^3); B) the magnetic float (32.2 cm^3); C) the pulldown solenoid; D) the main solenoid; and E) the support and leveling platform. The magnetic float is made of pyrex glass and contains a magnetic core. The pulldown solenoid C is used to bring the float into the field of the main solenoid (∿ 716 turns of Cu wire). The entire system is placed on the bottom of a constant temperature bath controlled to ± 0.001°C.

If a magnetic force from the main solenoid is used which is just sufficient to hold the float on the bottom of the solution container, the buoyancy forces give

$$\rho V = W + fi \tag{5}$$

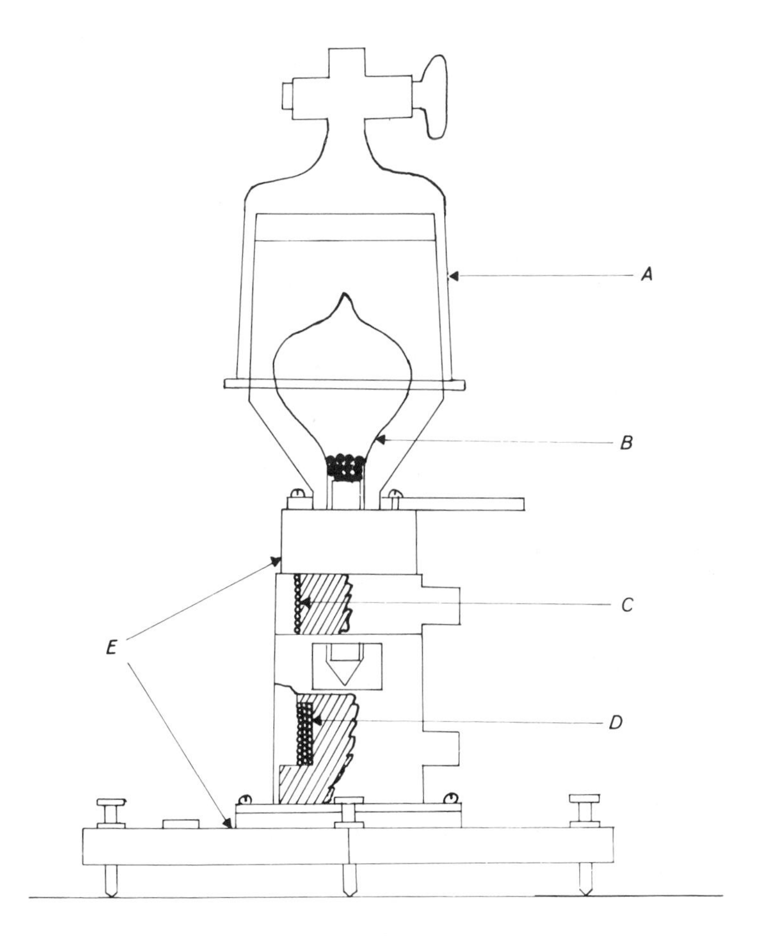

Review of Scientific Instruments

Figure 2. Sketch of the magnetic float densitometer: (A) solution container; (B) magnetic float; (C) pull down solenoid; (D) the main solenoid; (E) support and leveling platform (7)

where ρ is the density of the solution, *V* is the volume of the float, *W* is the weight of the float, *f* is an interaction constant and *i* is the current through the solenoid. The system is calibrated by measuring the current which is just sufficient to hold the float on the bottom in water when various Pt weights (w) are added to the float. The buoyancy equation is

$$\rho(V + w/\rho_{Pt}) = W + w + fi \qquad (6)$$

where ρ_{Pt} is the density of Pt. By rearranging this equation one obtains the linear equation

$$w(1 - \rho/\rho_{Pt}) = -fi + (\rho\ V - W) \qquad (7)$$

which can be used to determine f and V. The relative density of an electrolyte solution can be determined from

$$\Delta\rho = \rho - \rho_o = f\Delta i/(V + w/\rho_{Pt}) \qquad (8)$$

when the same weights are used ($\Delta i = i - i_o$). The precision of the system is ± 0.3 ppm and the accuracy in dilute solutions is ± 2 ppm. For more concentrated solutions the accuracy is limited by our present knowledge of the density of platinum (21.483 ± 0.002 g cm^{-3} at 25°C) (11). We have used this magnetic float system to study the volume properties of a number of electrolyte solutions (7,35-52).

By putting the system into a high pressure bomb (36) the magnetic system can be used to measure the densities of aqueous solutions over a wide pressure range with a precision of ± 11 ppm. We presently are developing a magnetic float system that can be used to measure the densities at low pressures to high temperatures (200°C).

3. Vibrating Flow Densimeter. One of the major advances made in making density measurements of solutions was the system developed by Kratky *et al*. (21) which measures the natural vibration frequency of a tube containing a liquid. The oscillating frequency (f) of the tube is related to its mass (m) by

$$f = 1/2\pi\ (k/m)^{1/2} \qquad (9)$$

where k is an instrument constant. Since the volume of the tube is constant, the density of a liquid contained in the tube is directly related to the period (τ) by

$$\rho = A + B\ \tau^2 \qquad (10)$$

The principle has been known for a long time, however, the modification made by Kratky *et al*. (21) and Picker *et al*. (22) have led to commercial instruments that can be used to measure densities routinely to a precision of ± 3 x 10^{-6}g cm^{-3}. The system described by Kratky et al. (21) (available commercially from Mettler) requires small volumes, has a wide operating temperature range, and a digital readout. Its major drawback is that it requires a long time to achieve thermal equilibrium. The flow densimeter developed by Picker et al. (22) requires only 5 to 7 cm^3 of solution and gives densities to a precision of ± 2 ppm in times of 5-10 minutes. A sketch of the Picker et al. (22) system is shown in Figure 3. A brass plate holds the stainless steel vibrating tube in place. Magnetic pickups are used to keep the tube vibrating at its natural frequency and sense the period of the vibration. A thermoregulated jacket surrounds the densimeter and the heat exchanger encompassing the liquid intake lines. The relative densities are determined by measuring the period in a solution and solvent

$$\Delta\rho = \rho - \rho_o = B(\tau^2 - \tau_o^2) \qquad (11)$$

The instrument constant B can be determined by measuring the τ in two fluids of known density. Air and water are used by most workers (22). In our laboratory we used seawater of known conductivity and pure water to calibrate our vibrating flow systems (53). The system gives accurate densities in dilute solutions, however, care must be taken when using the system in concentrated solutions or in solutions with large viscosities. The development of commercial flow densimeters has caused a rapid increase in the output of density measurements of solutions. Desnoyers, Jolicoeur and coworkers (54-69) have used this system to measure the densities of numerous electrolyte solutions. We have used the system to study the densities of electrolyte mixtures and natural waters (53,70-81). We routinely take our system to sea on oceanographic cruises (79) and find the system to perform very well on a rocking ship.

B. One Atmosphere Compressibilities. Most isothermal compressibility (β) measurements are made over extended pressure ranges and very few direct measurements have been made near 1 atm. It is difficult to obtain reliable values of β at 1 atm from high pressure PVT data due to the problems of extrapolating the compression [$k = (v^P - v^0)/v^0P$, where v is the specific volume] to zero applied pressure. If the compressions, volumes, or densities are fit to functions of P, the compressibility

$$\beta = \frac{1}{d}\left(\frac{\partial d}{\partial P}\right) = - \frac{1}{v}\left(\frac{\partial v}{\partial P}\right) \qquad (12)$$

at P = 0 is dependent upon the function used. Since there is no a priori method that can be used to obtain the correct pressure dependence, 1 atm compressibilities are more reliably determined by using sound speed measurements (29,82,83,84,85,86).

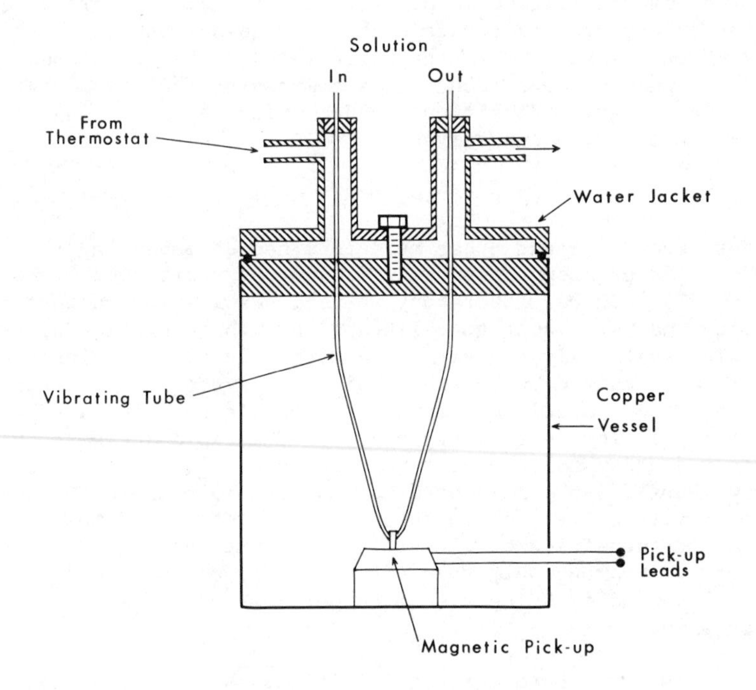

Figure 3. Sketch of the Sodev vibrating densitometer (22)

1. Piezometer Method. To measure the compressibility of liquids near 1 atm we have developed a piezometer system (87). A sketch of this system is shown in Figure 4. The piezometer, C, is a cylindrical glass vessel with a volume of ~ 450 cm^3. The piezometer contains the solution and ~ 330 gms of Hg. The top of the piezometer is fitted with a Taper joint for filling. A precision bore capillary, E, (2mm in diameter) is fitted to the bottom of the piezometer. The piezometer is suspended (B) in a brass or stainless steel pressure vessel, H. A glass boiler tube, J, encloses the upper portion of the capillary. The pressure vessel is filled with ethylene glycol which serves as a thermal and pressure medium. The entire apparatus is submerged in a constant temperature bath controlled to ± 0.001°C. The temperature inside the pressure vessel is monitored with a Hewlett-Packard quartz crystal thermometer (to determine when thermal equilibrium is reached after compression and decompression).

The compressibility of solutions are determined by measuring the volume change of Hg in the capillary when various pressures are applied to the system. The change in height of the meniscus (Δh) and the pressure change (ΔP) are dependent upon the volume of mercury (V_{Hg}), the volume of the solution (V_{Soln}), and the inside volume of the glass container to a reference mark (Ah). The total volume of the system is given by

$$V_T = V_{Soln} + V_{Hg} + Ah \tag{13}$$

where A is the cross sectional area of the capillary and h is the initial distance of the mercury meniscus below a reference mark. The differentiation of equation (13) with respect to pressure gives

$$\frac{\partial V_T}{\partial P} = -\frac{\partial V_{Soln}}{\partial P} - \frac{\partial V_{Hg}}{\partial P} + h\frac{\partial A}{\partial P} + A\frac{\partial h}{\partial P} \tag{14}$$

where the negative signs indicate a decrease in volume. Substituting $\partial A/\partial P = -(2/3)A\beta_g$, $\partial V_T/\partial P = -V_T\beta_g$, ($\beta_g$ is the compressibility of glass), $\beta_{Hg} = -(1/V_{Hg})(\partial V_{Hg}/\partial P)$, and $\beta_{Soln} = -1/V_{Soln}(\partial V_{Soln}/\partial P)$, we have

$$\beta_{Soln} = -\beta_{Hg}\frac{V_{Hg}}{V_{Soln}} + \beta_g\left(1 + \frac{V_{Hg}}{V_{Soln}} + \frac{A\Delta h}{3\,V_{Soln}}\right) + \frac{A\Delta h}{V_{Soln}\Delta P} \tag{15}$$

The system is calibrated by using known values of β for H_2O and Hg. If the same volume of Hg is used in the calibration and in the measurement of the unknown solution, the relative compressibilities are given by

$$\beta - \beta_o = \left(\frac{A\ \Delta h}{V\ \Delta P}\right)_{Soln} - \left(\frac{A\ \Delta h}{V\ \Delta P}\right)_{H_2O} \tag{16}$$

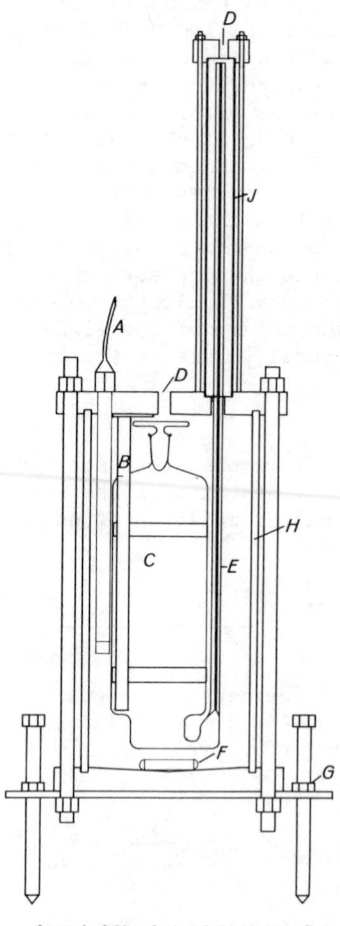

Figure 4. Isothermal compressibility apparatus (87)

This system has been shown to yield values of $\beta - \beta_0$ at 1 atm for D_2O, seawater and sea salts to an accuracy of ± 0.07 x 10^{-6} bar^{-1} (85,88,89,90). The greatest difficulty with using this system is that it takes a long time to reach thermal equilibrium.

2. Sound Speed Measurements. The most precise values of compressibility at 1 atm can be determined from speed of sound measurements. The adiabatic compressibility (β_S) is related to the speed of sound (u) by

$$\beta_S = 1/u^2\rho \tag{17}$$

The isothermal compressibilities (β) are related to the adiabatic values by the thermodynamic relationship

$$\beta = \beta_S + T\alpha^2/C_p\,\rho \tag{18}$$

where T is the absolute temperature, α is the thermal expansibility [$\alpha = 1/v(\partial v/\partial T)$] and C_p is the specific heat capacity at constant pressure of the solution. Since sound speeds can be measured to a precision of ± 0.02 m sec^{-1} (85), the adiabatic compressibilities can be determined to a precision of ± 0.002 x 10^{-6} bar^{-1}. The conversion of β_S to β can be made providing expansibility and heat capacity data are available (3). The differences between β and β_S for H_2O at 25°C is 0.473 x 10^{-6} bar^{-1} and increases to 0.810 x 10^{-6} bar^{-1} for a 1 m NaCl solution (3). Since specific heats are known to ± 0.001 j g^{-1} K^{-1} and expansibilities are known to ± 1.0 x 10^{-6} K^{-1}, values of $\beta - \beta_S$ can be estimated to ± 0.004 x 10^{-6} bar^{-1} for most electrolyte systems at 1 atm.

A sketch of the 1 atm sound velocimeter that we use in my laboratory (52,70,74,76,77,80,81,85,86,91,92) is shown in Figure 5. The system consists of a Nusonics "sing around" velocimeter (93) submerged in a constant temperature bath (A). The solution cell (C) has a volume of 80 cm^3 and is stirred by a submersible magnetic stirrer (B). A pulse from a transmitting transducer (D) travels through the solution, is reflected from the bottom (E) and is received by a receiving transducer (F). The received pulse is used to retrigger the first transducer to send a second pulse, repeating the process. The measured quantity is the pulse repetition frequency f. The sound speed is related to the measured frequency by (94)

$$1/f = \ell/u + \tau \tag{19}$$

where ℓ is the effective path length between the transducers and τ is the electronic delay time. The values of ℓ and τ are determined by calibrating the system with water (85) and using the sound velocities based on the measurements of Del Grosso and Mader (95). The relative sound speeds ($\Delta u = u - u_0$) can be determined from (94)

$$\Delta u = \ell(f - f_0)/[(1 - f\tau)(1 - f_0\tau)] \tag{20}$$

where the subscript zero refers to water. The advantage of using the velocimeter to measure relative sound speeds is that the values determined for ℓ and τ do not have a strong influence on Δu. For example, a 1% error in τ causes only a 0.03 m sec^{-1} error in Δu. The frequencies at a given temperature are reproducible to ± 0.5 Hz, which is equivalent to ± 0.02 m sec^{-1} in Δu. The precision of the relative sound speeds is ± 0.02 m sec^{-1} while the accuracies are thought to be ± 0.1 m sec^{-1} (85). We have used this system to determine the adiabatic compressibility of a number of aqueous solutions at 1 atm (52,70,74,76,77,80,81, 85,86,91,92).

As pointed out by Hayward (96) equation (17) cannot be used uncritically. In the presence of adsorption (which is known to occur for $MgSO_4$ solutions) (97-104) the velocity of sound (u_0) is shifted and becomes a complex function (105). The shift or dispersion, in velocity (Δu) caused by relaxation can be estimated from (105)

$$\Delta u = \frac{(\alpha_{MAX}/\pi)u_\infty(f/f_R)^2}{1 + (f/f_R)^2} \tag{21}$$

where α_{MAX} is the absorption per wavelength at the relaxation frequency, f is the frequency of the measurement (∿ 2 MHz), f_R is the relaxation frequency, and $u_\infty = u_0 + \Delta u$ is the sound velocity at infinite frequency. At concentrations of 0.5 m, equation (21) yields a value of Δu = 1.0 m sec^{-1} in $MgSO_4$ solutions (105). A shift of 1.0 m sec^{-1} in u will shift the adiabatic compressibility by 0.06 x 10^{-6} bar^{-1}. Since α has not been measured at high concentrations for many electrolytes, an exact correction cannot be made at present to all the published data measured at a fixed frequency.

C. High-Pressure PVT Properties. Three methods are presently being used to measure the high-pressure PVT properties of electrolyte solutions: volumetric methods, (26,32,106) high pressure magnetic float systems, (36,107,108) and high pressure speed of sound systems (109,110,111,112). I will not attempt to review all the modifications made to these systems.

1. High Pressure PVT Properties. The earlier volumetric methods (113,114) consisted of a piezometer, similar to the one shown in Figure 6, which is contained in a pressure bomb. The

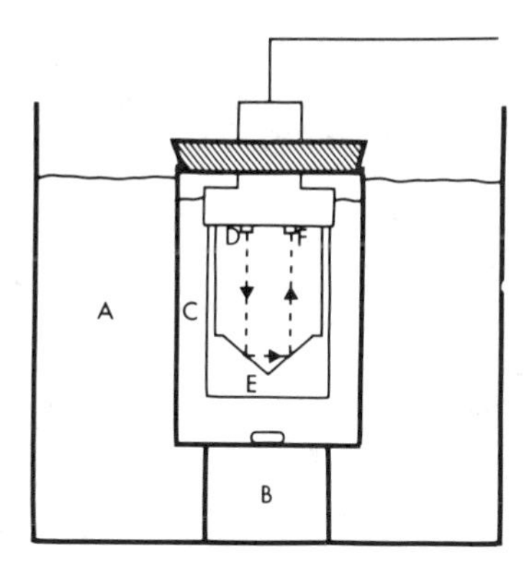

Figure 5. Sound vilocimeter for use at atmospheric pressure—(A) constant temperature bath; (B) magnetic stirrer; (C) solution cell; (D) transmitting transducer; (E) reflector; (F) receiving transducer (85)

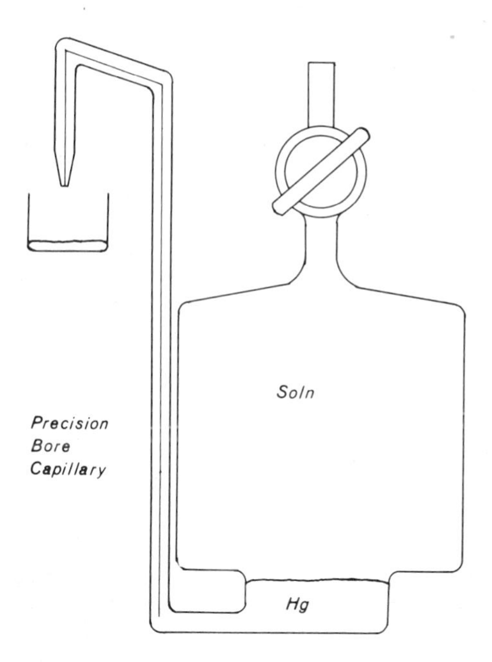

Figure 6. Weight dilatometer

Hg displaced after pressurization of the system is weighted after each measurement. The sensitivity of the method is limited to about 100 ppm by the discrete volume of the mercury drops and the small volume of the piezometer. More recent workers (106) have determined the changes in the volume of the system by using indirect methods to measure the height of Hg in the piezometer. Bradshaw and Schleicher (106) have designed a dilatometer system that can measure the expansion and compression of a solution under pressure. A sketch of their apparatus is shown in Figure 7. The volume change of the sample with temperature or pressure is obtained by measuring the change in height of mercury in the precision bore tubing section of the dilatometer. The tubing section is separated from the dilatometer flask by a tapered joint. The relative height of the mercury is determined by measuring the distance between a magnetic steel cylinder floating on the top of the mercury (floating core) and another cylinder firmly attached to the top of the mercury reservoir (ref. core). The dilatometer is mounted in a pressure vessel. The changes in the distance between the cores is measured by using a linear differential transformer with a micrometer head. The system was calibrated by making measurements on mercury. The precision of the system is ± 1 ppm/deg from 0 to 1000 bars applied pressure. The specific volumes of water determined with the system from 0 to 30°C are in excellent agreement with values determined from the sound derived equation of state of Chen, Fine and Millero (31) (see Figure 8).

Kell and Whalley (26,32) have described a volumometer method (Figure 9) to measure the volume of liquids. A calibrated piston (see Figure 9) is used to measure the change in volume of a solution with a change in temperature or pressure. The system is similar in principle to the apparatus described by Keyes (115). The pressure vessel of known volume is immersed in a thermostat. The pressure of the water is measured by means of pressure balances. The water is separated from the oil by mercury in a U tube or two legs of a W tube. The change in the volume of water in the pressure vessel due to a change in T or P is measured by the change of the position of the piston of the volumometer. The calibrated piston was advanced into the cylinder by means of a screw thread and was balanced by a similar piston. The pressure vessel and cylinder of the volumometer were jacketed by other pressure vessels to prevent deformation. The balancing pressure of the balancing water and the experimental water was measured by a three legged or W tube. The system was calibrated with water at 1 atm and mercury at high pressures. The system was used to measure the specific volume of water from 0 to 150°C to a precision of ± 10 ppm (26) and 150 to 350°C to a precision of 100 ppm (32). Fine and Millero (29) have shown that the high pressure volume measurements of Kell and Whalley from 0 to 100°C and 1 kbar disagreed from values derived from the sound speed measurements of Wilson (109) by ∿ 120 ppm. Kell and Whalley (30)

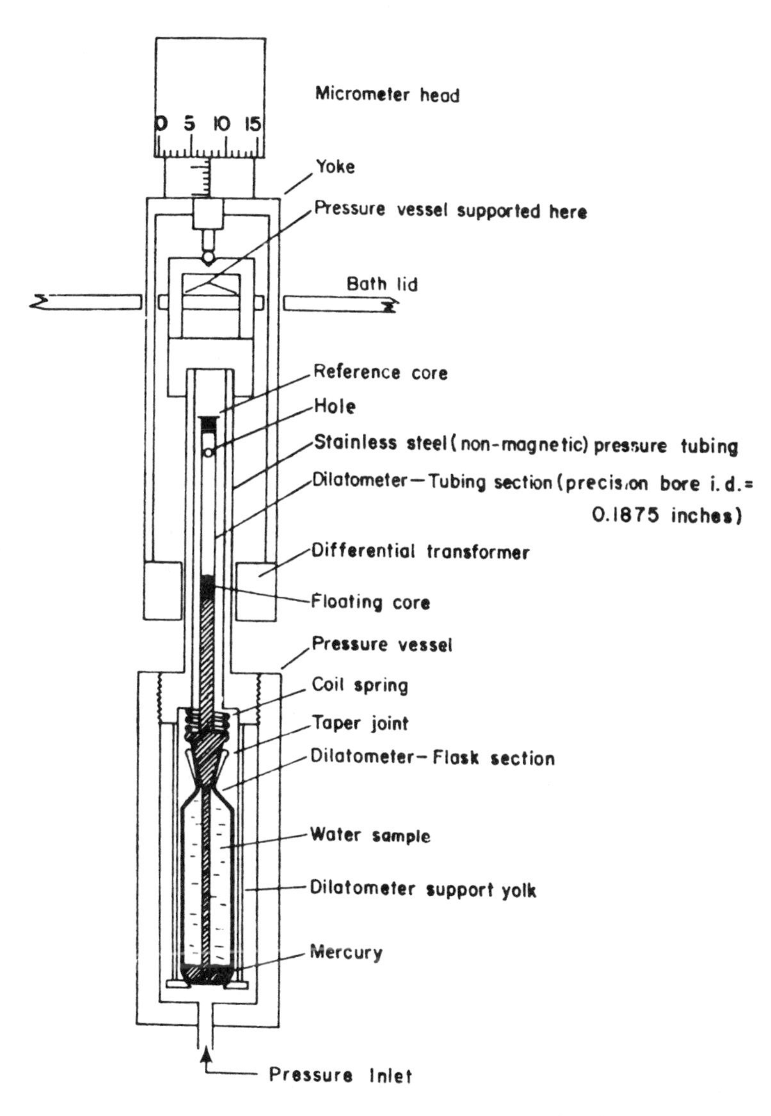

Deep-Sea Research

Figure 7. Dilatometer system (106)

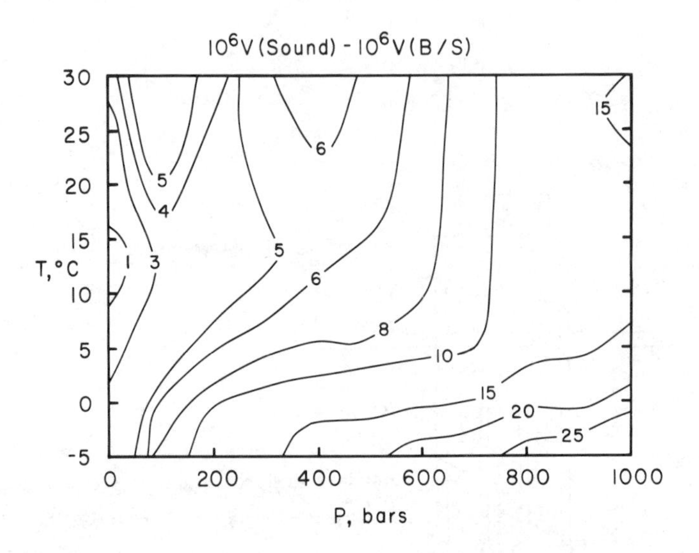

Figure 8. Comparisons of the specific volumes of water determined from sound speed measurements (31) and direct measurements (106)

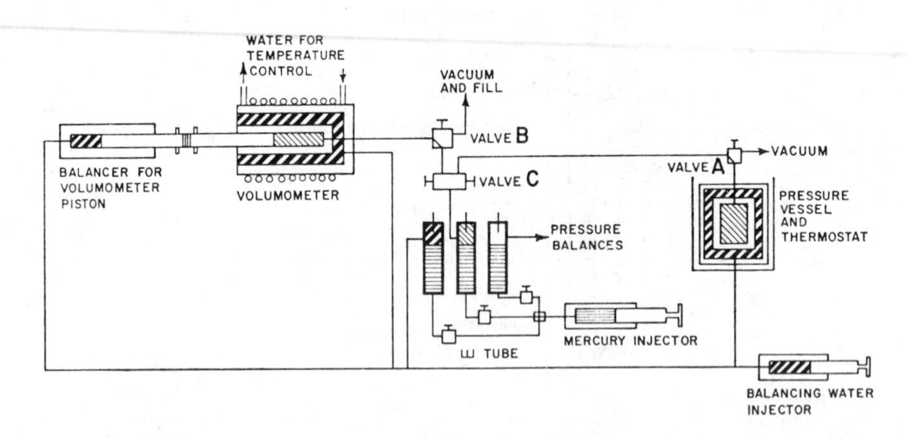

Figure 9. Diagram of the PVT apparatus (26) ((▨) balancing water; (▧) experimental water; (≡) mercury)

attributed this difference to an error in the compression of the pressure vessel. They used sound derived densities to redetermine the compression of the vessel and gave corrected values of v for water. Comparisons of the specific volumes of water at 1000 bar of various workers (29,30,106) with those calculated from the sound derived equation of state of Chen, Fine and Millero (31) are shown in Figure 10. The results are in good agreement and indicate that values of v for water are accurately known to ± 15 ppm from 0 to 100°C at 1000 bars. The differences at lower pressures are smaller.

The volumetric systems of Bradshaw and Schleicher (106) and Kell and Whalley (26,32) are the most precise methods of directly measuring the absolute densities or volumes at high pressures. These methods, however, are not ideally suitable for making systematic density studies of aqueous electrolyte solutions as a function of P and T because of the arduous nature of the experimental work and the loss of precision in very dilute solutions.

2. High Pressure Magnetic Float Methods. As discussed earlier the magnetic float method can be used to measure relative densities ($\Delta\rho$) with great precision. We have developed a high pressure magnetic float system (36) that is ideally suited to measure relative densities as a function of pressure. A sketch of the high pressure system is shown in Figure 11. The pressure bomb (B) has optical ports for observing the motion of the magnetic float and is made of berylium copper or stainless steel. The top (A) and Bottom (C) plugs were made of similar material. The bottom plug contains an insert plug (D) that supports the solenoid (G) which is 300 turns of No. 28 Cu wire. The window cones were machined from cast plexiglas rod and have a pitch of 30 degrees. The pressure is transmitted to the contents through port (H). The magnetic float is made of thick-wall (0.4 cm) pyrex glass tube and contains a permanent magnet. A high temperature Apiezon wax was used to hold the magnet in place.

The system is calibrated with water in the same manner as the 1 atm magnetic float system. The density of platinum as a function of pressure was estimated from the compressibility. Since the density of Pt decreases by only 0.009 g cm^{-3} from 0 to 1200 bars, the effect of pressure on the density of Pt can be neglected if weights below 1 g are used. Both f and V were found to be functions of pressure (36). Calibrations of the float at a given pressure indicate that the precision of the system is ± 10 ppm while the accuracies are about ± 20 ppm (116). We have used this system to measure the relative densities of D_2O (117), seawater (116,118), and some major sea salts (36, 119).

The major difficulties we have had with the high pressure magnetic system is that the floats easily break due to pressure shocks. This greatly disturbs one who has spent many hours calibrating the float in water at various pressures.

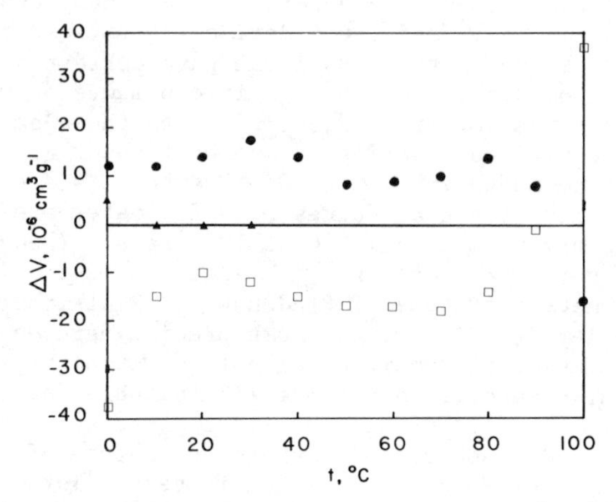

Journal of Chemical Physics

Figure 10. Comparisons of the specific volumes of water at 1000 bars calculated from the sound derived equation of state of Chen, Fine, and Millero (31) *and the data of various workers; (▲) Bradshaw and Schleicher; (●) Kell and Whalley; (□) Fine and Millero*

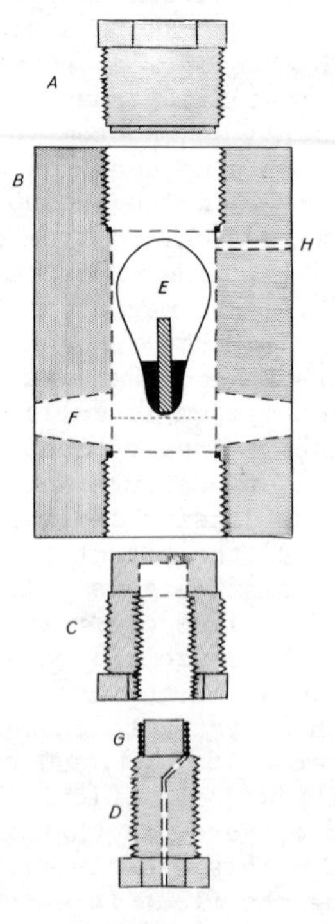

Journal of Solution Chemistry

Figure 11. High pressure magnetic float system (36)

3. High Pressure Sound Velocimeters. As discussed earlier, sound velocity measurements can yield precise compressibilities of solutions. Wilson (109) was the first to develop a high pressure sound velocimeter that could be used over a wide range of pressures and temperatures. He used the "sing around" system to measure the high pressure sound speeds of water (109), D_2O (120), and seawater (121) to a precision of ± 0.2 m sec^{-1} which is equivalent to ± 0.012 x 10^{-6} bar^{-1} in β_S. Barlow and Yazgan (111) have described a similar system used to measure speed of sound in water. Del Grosso and Mader (110) have developed a more elaborate inteferometer system and used it to measure the high pressure speed of sound in seawater solution to a precision of ± 0.05 m sec^{-1} (∿ ± 0.006 x 10^{-6} bar^{-1} in β_S). All of these systems are capable of making absolute measurements on pure water. Unfortunately, at present (112) the two sound studies made on pure water differ by as much as 1.22 m sec^{-1} at high temperatures and pressures (P = 800 bar and t = 93°C). Since the relative sound speeds of Wilson for D_2O and seawater are in good agreement with our independent studies (122,123), we feel that at present Wilson's results for water are the best available. We have (112) reanalyzed the sound speed measurements of Wilson for water and derived an equation for the speed of sound in water from 0 to 100°C and 0 to 1000 bars applied pressure (with σ = 0.08 m sec^{-1}). The 1 atm portion of the equation is based on the results of Del Grosso and Mader (95). We suggest that this equation be used to calibrate high pressure sound velocimeters until more reliable results become available.

We have designed a high pressure sound velocimeter (112,123, 124). A sketch of the system is shown in Figure 12. The velocimeter consists of a Nusonic high pressure single transducer "sing around" system suspended in a bomb. The high pressure bomb is made of stainless steel and has a volume of ∿ 500 cm^3. The bomb is submerged in a constant temperature bath controlled to ± 0.001°C. A Harwood Engineering controlled clearance piston gauge is used to set the pressures (± 0.1 bar at 1000 bars applied P).

The system was calibrated with water and the sound path (ℓ) was found to increase slightly as the pressure is increased (assuming τ is independent of pressure) (112). Since we are interested in measuring the relative speed of sound (Δu) in aqueous solutions, we use the velocimeter in a differential mode. The value of τ is not a function of pressure (or temperature), as the electronics are kept under ambient pressures. The values of Δu were determined from

$$\Delta u = u - u_o = \frac{(f - f_o)u_o}{f_o(1 - f\tau)} \quad (22)$$

where f and f_o are, respectively, the pulse repetition rates of the velocimeter in the solution and pure water. The length of

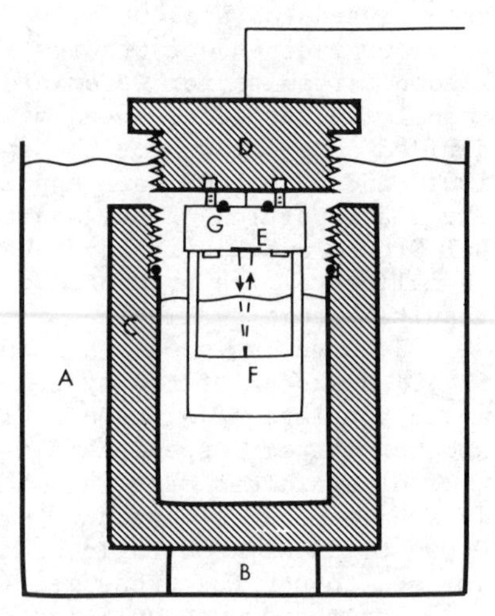

Figure 12. Sketch of the high pressure sound velocimeter: (A) constant temperature bath; (B)bomb stand; (C) pressure bomb; (D) plug; (E) transmitting transducer; (F) reflector; (G) O-ring (112)

the sound path ℓ is not needed to determine Δu at a given P and T. The electronic delay time is determined at 1 atm in water at various temperatures. The system has a precision of ± 0.19 m sec^{-1} from 0 to 50°C and 0 to 1000 bars. We have used this system to measure the speed of sound in D_2O (122), seawater (123), the major sea salts (91), and mixtures of the major sea salts (125).

To generate isothermal compressibilities from sound speeds, it is necessary to have reliable expansibility and heat capacity data (equation 18). We have developed an iterative method to convert high pressure sound speed to isothermal compressibilities (84). The effect of pressure on the volume of a solution $(\partial V/\partial P)_T$ at a constant pressure is given by

$$-(\partial v/\partial P)^P = v^P/u^P + T(\alpha^P v^P)^2/C_p^{\ P} \tag{23}$$

where the superscript P indicates a given pressure. The effect of pressure on the heat capacity can be determined from

$$C_p^{\ P} = C_p^{\ 0} - T_0\int^P (\partial^2 v/\partial T^2)dP \tag{24}$$

where $C_p^{\ 0}$ is the value at 1 atm (P = 0). Since the second term in equation (23) is small compared to v^P/u^P it can be treated as a perturbation term or adiabatic correction

$$-(\partial v/\partial P)^P = (v^2/u^2) + \Delta_S \tag{25}$$

To determine Δ_S it is necessary to use an equation of state to express the PVT properties. We have selected a second degree secant bulk modulus to represent the PVT properties of aqueous solutions (84)

$$K = Pv^0/(v^0 - v^P) = K^0 + AP + BP^2 \tag{26}$$

where K^0, A and B are functions of temperature and concentration. Differentiating this equation with respect to pressure yields

$$(\partial v/\partial P)^P = v^0(K^0 - BP^2)/(K^0 + AP + BP^2)^2 \tag{27}$$

The K^0 term is determined from the known 1 atm sound speeds (u^0), expansibilities (α^0), specific volumes (v^0), and heat capacities ($C_p^{\ 0}$)

$$K^0 = v^0/(\partial v/\partial P)^0 = v^0[(v^0/u^0)^2 + T(\alpha^0 v^0)^2/C_p^{\ 0}]^{-1} \tag{28}$$

The values of A and B are determined by an iterative computer technique using equations (23) and (27). As a first step, an arbitrary equation of state is used to generate $\partial^2V/\partial T^2$ and to calculate C_p^P, v^P and α^P. The equation of state for water (31) is used for electrolyte solutions. The values of C_p^P, v^P and α^P are combined with u^P to calculate $(\partial V/\partial P)^P$. The values of $(\partial V/\partial P)^P$ are fit to equation (27) by taking an arbitrary value of $B = B'$. The newly obtained equation of state is then used to generate C_p^P, v^P and α^P, which are again combined with the sound speeds to calculate $(\partial V/\partial P)^P$. The values are again fit to equation (27) where the B' term is replaced by the B term generated from the previous least squares fit. This process is repeated until the newly generated values of $(\partial V/\partial P)^P$ agree with the previous values to within $\pm$ 0.01 x 10^{-6} bar^{-1} at every concentration, temperature, and pressure. This method has been shown to yield values of v^P for water and seawater that agree with directly measured data to $\pm$ 10 x 10^{-6} g cm^{-3} from 0 to 1000 bars (31,126,127).

We prefer this method of generating PVT from sound speeds to the integration of the equation

$$\rho^P - \rho^0 = \int_0^P \frac{dp}{u^2} + T \int_0^P \left[\frac{(1/\rho)(\partial\rho/\partial T)^2}{C_p}\right] dP \qquad (29)$$

This equation can also be solved by an iterative method (128), however, the reliability of this method is in doubt (84) because it is the compressibility, not the specific volume or density that dominates (equation 23). This is also the reason why the precision of the sound speeds can only be compared directly with the precision of compressibility (not successive iterations of density).

The only difficulty with using this method is the lack of heat capacity data. With the wide spread use of the Picker *et al.* (129) heat capacity calorimeter one can usually find published heat capacities for most systems of interest (3) at 25°C. Since the C_p does not attribute much to the adiabatic correction, this is not a serious limitation.

II. High Pressure Equation of State of Electrolyte Solutions

Many attempts (130,131,132) have been made to derive an equation of state from molecular theory, but none of them have resulted in an adequate equation capable of expressing the results of direct measurements to within the experimental error. To meet this need, it has been necessary to employ empirical equations; the major justification for each is that it fits the experimental data. It is also desirable to choose an equation that gives reliable concentration, temperature, and pressure derivatives.

The most frequently used equations of state are summarized in Table I. Many of the equations used are based on the Tait (133) equation

$$k = \frac{v^0 - v^P}{v^0 P} = \frac{A}{\pi + P} \qquad (30)$$

or rearranged

$$v^P = v^0 \left(1 - \frac{AP}{\pi + P}\right) \qquad (31)$$

where k is the mean or secant compression, v^0 and v^P are the specific volumes at zero and P applied pressure. A is a constant and π is a temperature dependent parameter. Tait used this equation to fit the compression data of water and seawater solutions.

It was recently revealed by Hayward (134) that two generations of workers were misled into using what they came to call "the Tait equation". They did not realize that this well known equation was not Tait's original equation. The Tait equation was first misquoted by Tammann (135). The misquoted form of the Tait equation (which we shall refer to as the Tammann equation) is

$$\frac{1}{v^0}\left(\frac{\partial v}{\partial P}\right)^P = -\frac{C}{B + P} \qquad (32)$$

or integrated (B and C are independent of pressure)

$$v^P = v^0 - Cv^0 \, \ln \frac{B + P}{B} \qquad (33)$$

Comparisons with equation (30) reveal that Tammann replaced Tait's compression term, $(v^0 - v^P)/v^0 P$ by the corresponding differential coefficient $(\partial v/\partial P)^P/v^0$. In the low pressure range (below 500 bars) that concerns the majority of workers, $(v^0 - v^P)/v^0 P$ is almost equal to $(\partial v/\partial P)^P/v^0$, enabling one to understand the long history of misuse.

By rearranging the Tammann equation (32), we have

$$(\partial P/\partial v)^P = (\partial P/\partial v)^0 + a\,P \qquad (34)$$

where $(\partial P/\partial v)^0 = -B/Cv^0$ and $a = -1/Cv^0$. Differentiation of the Tait equation (30) with respect to pressure gives,

$$(\partial P/\partial v)^P = (\partial P/\partial v)^0 + a'P + b'P^2 \qquad (35)$$

TABLE I

STUDIES ON THE EQUATION OF STATE

Year	Investigator	Form
1888	Tait	$v^P = v^0 - v^0 AP/(\pi + P)$
1895	Tammann	$v^P = v^0 - Cv^0 \ln[(B + P)/B]$
1908	Ekman	$v^P = v^0 + AP + BP^2 + CP^3$
1909	Tumlirz	$v^P = B + D/(C + P)$
1935	Gibson	$v^P = v^0 - CX_1 \ln[(B + P)/B]$
1944	Murnaghan	$v^P = v^0 (1 + \eta\beta^0 P)^{-1/\eta}$
1965	Kell and Whalley	$v^P = v^0(1 + \sum_k \sum_\ell A_{k,\ell} t^k P^\ell)$
1967	Hayward	$v^P = v^0 - v^0 P/(K^0 + BP)$
1973	Millero and Coworkers	$v^P = v^0 - v^0 P/(K^0 + AP + BP^2)$
1976	Bradshaw and Schleicher	$v^P = [1 - \sum_{ij} A_{i,j} P^i S^i][v^0(0°C) + \int \frac{\partial v}{\partial T} dT]$

where $(\partial P/\partial v)^0 = -\pi/Av^0$, $a' = -2/Av^0$, and $b' = -1/A\pi v^0$. By comparing equations (34) and (35), it is clear that the Tammann equation contains one less pressure term than the Tait equation in bulk modulus form. [It should be pointed out that both equations have the same number of adjustable parameters, the term $(\partial P/\partial v)^0$, a' and b' are interrelated to each other.] The Tait equation (in bulk modulus form, equation 35) fits compression data about 100 times better than the Tammann equation (126). This is demonstrated in Figure 13 by examining $(\partial P/\partial v)^P$ determined from sound speed data (121). The maximum deviation of the fitted data is 1.068×10^2 bar g cm^{-3} for equation (34) and 1.1 bar g cm^{-3} for equation (35). These differences are equivalent to 0.2×10^{-6} bar^{-1} and 0.002×10^{-6} bar^{-1}, respectively, in compressibility and demonstrate that the Tait equation is superior to the Tammann equation.

The pressure dependence of $(\partial P/\partial v)^P$ can be tested by fitting $[(\partial P/\partial v)^P - (\partial P/\partial v)^0]/P$ vs. pressure. It is clearly shown in Figure 14 that the values of $[(\partial P/\partial v)^P - (\partial P/\partial v)^0]/P$ for pure water and 35 ‰ salinity seawater determined from sound speed data (112,123) increase almost linearly with pressure. This indicates that P^2 or even higher order terms are needed to represent $(\partial P/\partial v)$ over the pressure range of 0 to 1000 bars. In other words, the Tammann equation and the original Tait equation do not represent the PVT properties for pure or saline water within the accuracy of the data.

Tumlirz (136) presented an equation of state of the form

$$(v^P - B)\ (P + C) = D \tag{36}$$

where B, C, and D are temperature dependent parameters. Tumlirz's equation is just a rearrangement of the Tait equation. Replacing B, C, D in equation (36) by $v^0 - v^0A$, π and $v^0A\pi$, respectively, gives

$$(v^P - v^0 + v^0A)\ (P + \pi) = v^0A\pi \tag{37}$$

or

$$v^P = v^0 - \frac{v^0AP}{\pi + P} \tag{38}$$

which is the original Tait equation (30).

Gibson (137) extended the Tammann equation to solutions and published what is referred to as the Tait-Gibson equation

$$\left(\frac{\partial v}{\partial P}\right)^P = \frac{-CX_1}{B + P} \tag{39}$$

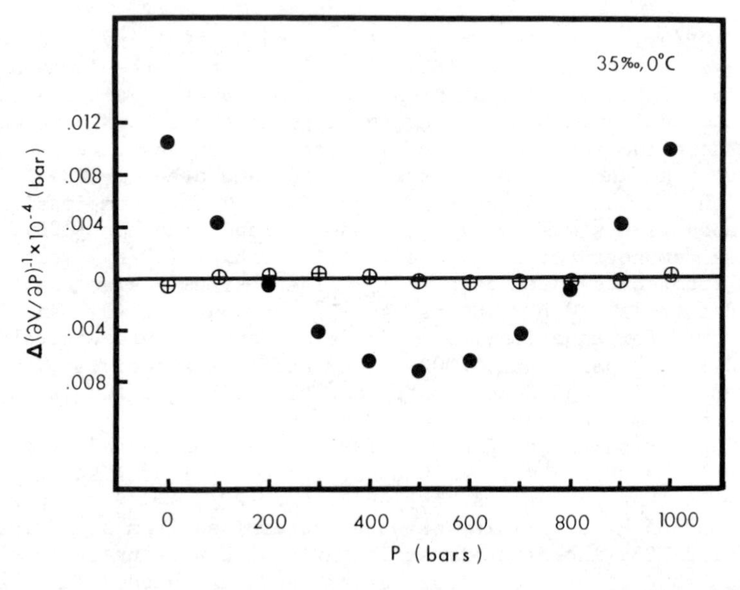

Journal of Marine Research

Figure 13. $(\partial P/\partial v)^P$ *determined from sound speed data (126) ((●) linear equation; (⊕) quadratic equation)*

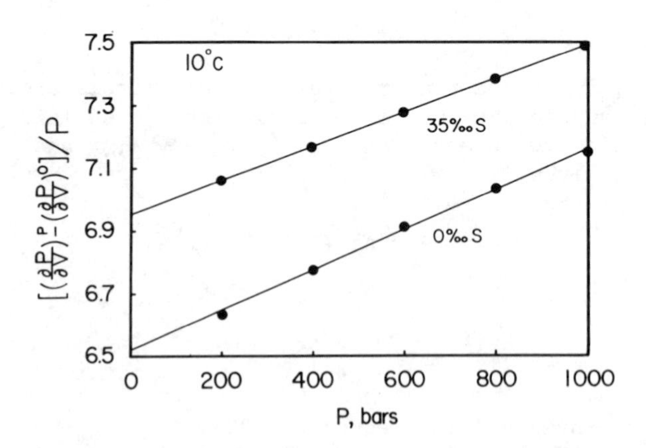

Figure 14. *Values of* $[(\partial P/\partial v)^P - (\partial P/\partial v)^0]/P$ *for pure water and 35‰ salinity seawater determined from sound speed data (112, 123).*

or integrated

$$v^P = v^0 - CX_1 \ln \frac{B + P}{B} \tag{40}$$

where X_1 is the weight fraction of water, C is a constant ($0.135/v^0$) and B is a function of temperature and concentration. For most of the systems examined, (114,137,138,139), C was found to be nearly independent of temperature and concentration.

Murnaghan (140,141) published an equation of the form

$$v^P = v^0 (1 + \eta\beta^0 P)^{-1/\eta} \tag{41}$$

in which β^0 is the isothermal compressibility at sea level pressure, η is $\partial(1/\beta)/\partial P$.

Kell and Whalley (26) fit the results of their pure water specific volume measurements to the polynomial

$$v = v^0(1 + \sum_k \sum_\ell A_{k,\ell} t^k P^\ell) \tag{42}$$

The standard deviation was reported to be 3 ppm at the lower pressures and 10 ppm at the higher pressures.

Li (142) published a variation of the Tait-Gibson equation for seawater from both compressibility and sound speed data

$$v^P = v^0 - (1 - 10^{-3}S)C \ln \frac{B + P}{B} \tag{43}$$

where S is the salinity in parts per thousand, C is a constant, and B is a temperature and salinity dependent parameter. Li examined the seawater results of Ekman (143) and Wilson (120) using this equation and found that the results could be fit to $\pm$ 10 x 10^{-6} cm^3 g^{-1} over the oceanic range of pressure, temperature, and salinity. He also noticed a difference at 1 atm between the compressibilities of Ekman and the sound derived data of Wilson. These differences were later shown by Lepple and Millero (88) to be due to errors in Ekman's work.

We attempted (126) to use Li's equation to represent Kell and Whalley's (26) as well as Fine and Millero's (29) pure water results and found that it could not fit the data to within the experimental error ($\pm$ 10 ppm).

Hayward (134) extensively investigated different forms of equations of state and suggested the reciprocal form of the Tait equation

$$\frac{1}{k} = \frac{v^0 P}{v^0 - v^P} = K = K^0 + BP \tag{44}$$

where $K^0 = \pi/A$ and $B = 1/A$ in equation (30). In equation (44) K is the secant bulk modulus, K^0 is the secant bulk modulus at an applied pressure P = 0, and equals the reciprocal of the isothermal compressibility, $1/\beta$, B is a function of concentration and temperature. Hayward concluded that the linear secant bulk modulus equation best represented the compression properties of liquids over the pressure range from zero to several hundred bars. With the addition of one or two extra terms, the equation can be extended to cover all liquids at pressures up to 12 Kb over a wide range of temperature.

Macdonald (144) analyzed several equations of state which had a variety of mathematical forms including the Tammann equation and the secant bulk modulus equation chosen by Hayward. (In his statistical analysis, Macdonald used the PVT data of Kell and Whalley (26) which has been shown to be in error (29). Thus, the conclusions of Macdonald may be questionable.) He disagreed with Hayward and selected the Murnaghan equation to be superior to either the Tammann equation or the linear secant modulus equation chosen by Hayward. If, however, the Tammann equation and the Murnaghan equation were both expanded to second order in pressure, then Macdonald found that the results obtained from both equations would agree. As shown earlier, the expansion of the Tammann equation to second order is equivalent to the bulk modulus form of the original Tait equation.

In work in our laboratory (29,31,84,116,118,126,127) we have used a second degree secant bulk modulus equation to fit direct measurements and sound derived specific volumes of solutions. The equation is given by

$$K = Pv^0/(v^0 - v^P) = K^0 + AP + BP^2 \qquad (45)$$

A plot of K for pure water as a function of pressure at different temperatures is shown in Figure 15. This figure clearly demonstrates that the B term is small and that A does not vary much with temperature. A similar behavior is found for seawater solutions at a given concentration (see Figure 16). Another interesting feature of K for electrolyte solutions is shown in Figure 17. The slopes (A) of K plotted vs. P are not strongly dependent upon concentration. These findings are similar to earlier work (114,137,138,139) of Gibson using the Tammann equation.

What this comparison indicates is that as a first approximation, the terms A and B in equation (45) are not strongly dependent upon temperature and concentration. The pressure dependence of K for electrolyte solutions can be thus estimated from the properties of pure water. Since $K^0 = 1/\beta^0$, the reciprocal of the 1 atm compressibility, it thus becomes possible to make reasonable estimates of v^P from 1 atm specific volume data (v^0) and compressibility data (β^0).

This can be demonstrated by considering the high pressure PVT properties of seawater. For seawater solutions the values of K^0, A and B at a given temperature are given by (127).

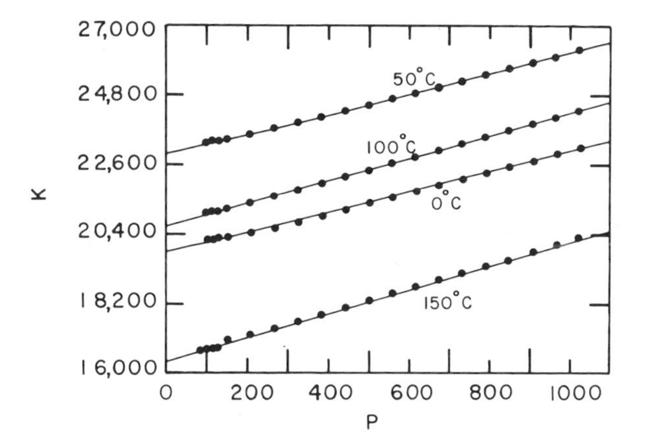

Figure 15. Plot of K for pure water as a function of pressure at different temperatures

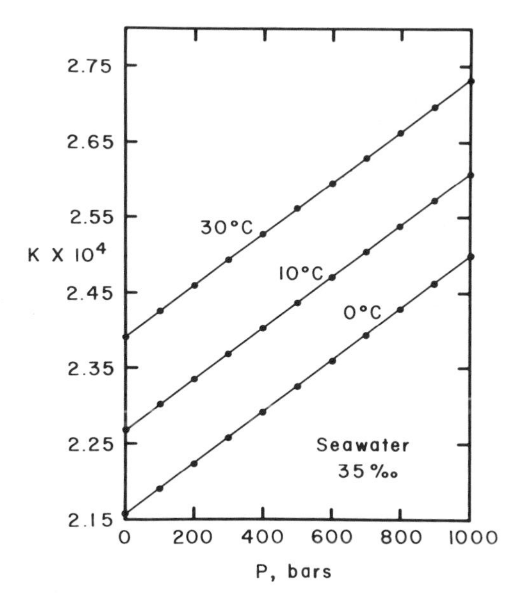

Figure 16. Seawater solutions at a given concentration

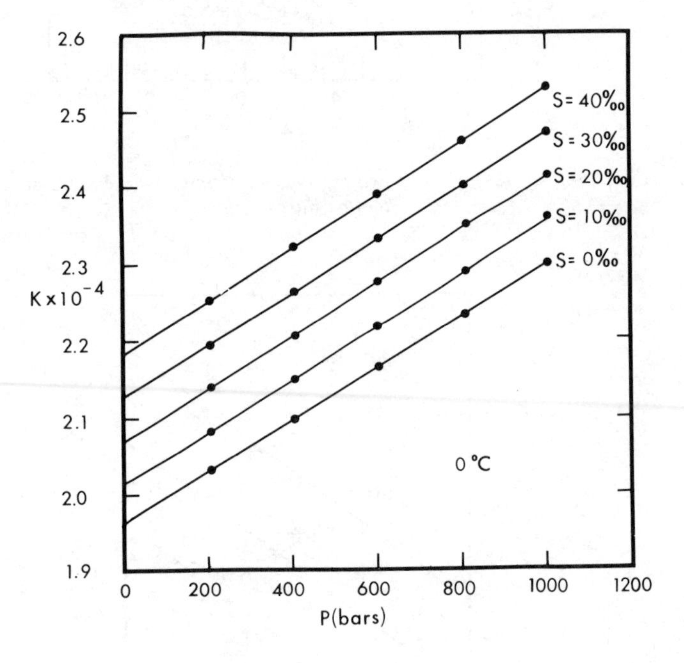

Figure 17. K for electrolyte solutions

$$K^0 = K_W^0 + a\,S + b\,S^{3/2} \quad (46)$$

$$A = A_W + c\,S + d\,S^{3/2} \quad (47)$$

$$B = B_W + e\,S \quad (48)$$

where S is the salinity (grams of salt per kg of solution) and the subscript W refers to pure water. For seawater of S = 35.0 the differences between V^P calculated with the full equation and with c, d, and e = 0 are shown in Table II. As is clearly demonstrated in this table, the specific volumes of seawater can be estimated to pressures of 1000 bars and from 0 to 40°C to ± 136 x 10^{-6} cm^3 g^{-1} using the equation of state for pure water (K_W^0, A_W and B_W) and the 1 atm values of K^0 for seawater (a and b) determined from sound speed measurements.

TABLE II

COMPARISONS OF THE MEASURED AND CALCULATED SPECIFIC VOLUMES OF 35 °/oo SALINITY SEAWATER USING PURE WATER HIGH PRESSURE

$\Delta v\ 10^6$ (Meas - Calc)[a]

P	0°C	10°C	20°C	30°C	40°C
0	0	0	0	0	0
100	3	3	2	1	1
200	10	9	7	4	2
300	20	17	14	10	5
400	34	29	23	16	9
500	50	43	34	24	15
600	67	58	46	34	19
700	85	74	61	44	26
800	102	91	76	54	32
900	119	109	92	67	39
1000	136	127	109	80	45

a) The calculated values of v^P were determined from $v^P = v^0(1 - P/K)$ where $K = K^0 + A_W P + B_W P^2$ ($K^0 = 1/\beta^0$ for seawater).

Unfortunately, at present, reliable PVT data for electrolytes at high pressures, temperatures and concentrations are not available to further test the applicability of these simple methods to natural waters. Reliable measurements of the speed of sound in aqueous electrolytes as a function of temperature, pressure and concentration should provide the data needed to test the postulation presented above. Since 1 atm measurements of v^0 and

β^0 are relatively easy to make over a wide temperature range, it would be possible to generate reasonable estimates of the high pressure PVT properties for electrolytes with the minimal experimental work.

III. Estimation of the PVT Properties of Mixed Electrolyte Solutions

In recent years a great deal of progress has been made in interpreting the physical-chemical properties of mixed electrolyte solutions (145,146). Simple additivity models have been developed (145) that can be used to examine how the physical-chemical properties are related to the ion-water and ion-ion interactions of the major constituents of natural waters. The models have been used to predict the density and compressibilities of seawater (48,86,88,118,147,148,149,150), river and estuarine waters (35,72,151,152), lakes (153), and brines (80) by using binary solution data for the ionic components. The predictions are made by using an additivity rule for the apparent molal properties first developed by Young (154) and modified by Wood and coworkers (155,156,157). The general form of Young's rule for the apparent molal volume of a mixed electrolyte solution is given by

$$\Phi_V = \sum_{MX} E_M E_X \phi_V(MX) \tag{49}$$

where E_i is the equivalent fraction of cation M and anion X, and $\phi_V(MX)$ is the apparent molal volume of MX at the ionic strength of the mixture. The value of ϕ_V for the mixture is related to the density by

$$\Phi_V = \frac{10^3(d_o - d)}{d_o N_T} + \frac{M_T}{d_o} \tag{50}$$

where N_T is the normality of the solution (equivalents per 1000 cm^3) and M_T is the mean equivalent weight ($M_T = \sum E_i M_i$, M_i is the equivalent weight of component i). The concentration dependence of ϕ_V is given by

$$\Phi_V = \Phi_V{}^o + S_V\, I_V{}^{1/2} + b_V\, I_V + \cdots\cdot \tag{51}$$

where $\Phi_V{}^o = \sum E_M E_X \phi_V{}^o\,(MX)$, $S_V = \sum E_M E_X S_V(MX)$, and $b_V = \sum E_M E_X b_V(MX)$. The first term is related to the ion-water interactions of the components of the mixture and the higher order terms are related to ion-ion interactions. Combining equations (50) and (51) one obtains

$$d = d_o + A_V I_V + B_V I_V^{3/2} + C_V I_V^2 + \ldots \quad (52)$$

where $A_V = 10^{-3}(M_T - d_o\Phi_V^o)k$, $B_V = -10^{-3} S_V d_o k$, and $C_V = -10^{-3} b_V d_o k$. The value of k is related to the ratio of equivalents and ionals in the mixture ($N_T = k\ I_V$). Since the $\phi_V^o(i)$ of the ionic components are additive at infinite dilution and the term S_V can be equated to the Debye-Hückel limiting law, it is possible to simplify the estimations of the terms A_V and B_V by using the ionic components of the solution rather than salts (145).

The use of equation (52) to estimate the density of a mixed electrolyte solution can be demonstrated by considering average seawater. The equivalent fraction of the cations and anions in seawater are shown in Figure 18. The solution mainly consists of the ions Na^+, Mg^{2+}, Cl^-, and SO_4^{2-}. The calculation of Φ_V^o and b_V for "sea salt" is given in Table III (151).

TABLE III

CALCULATION OF Φ_V^o AND b_V FOR SEA SALT AT 25°C

Ion	E_i	$\phi_V^o(i)$	$E_i\phi_V^o(i)$	b_V	$E_i b_V(i)$
Na^+	0.77268	-1.21	-0.935	1.078	0.833
Mg^{2+}	0.17573	-10.59	-1.861	-0.197	-0.035
Ca^{2+}	0.03390	-8.93	-0.303	0.242	0.008
K^+	0.01684	9.03	0.152	1.129	0.019
Sr^{2+}	0.00030	-9.08	-0.003	0.569	0.000_2
Cl^-	0.90078	17.83	16.061	-1.030	-0.928
SO_4^{2-}	0.09318	6.99	0.651	0.134	0.013
HCO_3^-	0.00318	24.28	0.077	-0.302	-0.010
Br^-	0.00139	24.71	0.034	-1.107	-0.001
CO_3^{2-}	0.00067	-1.89	-0.001	-0.780	-0.001
$B(OH)_4^-$	0.00014	21.84	0.003	3.630	0.001
F^-	0.00011	-1.16	-0.000_1	-0.538	-0.000_1
$B(OH)_3$	0.00054	39.22	0.021	1.300	0.001
			$\Phi_V^o = 13.896$		$b_V = -0.101$

Combining these values of Φ_V^o and b_V with $S_V = 2.150$, $M_T = 58.046$ and $k = 0.86906$, we have

$$(d - d_o)10^3 = 38.404\ I_V - 1.863\ I_V^{3/2} + 0.088\ I_V^2 \quad (53)$$

The values of $d - d_o$ calculated from equation (53) are compared to the measured values in Table IV. The agreement is quite reasonable considering the binary solution data at high concentrations are only reliable to ± 10 x 10^{-6} g cm^{-3}. Calculations at

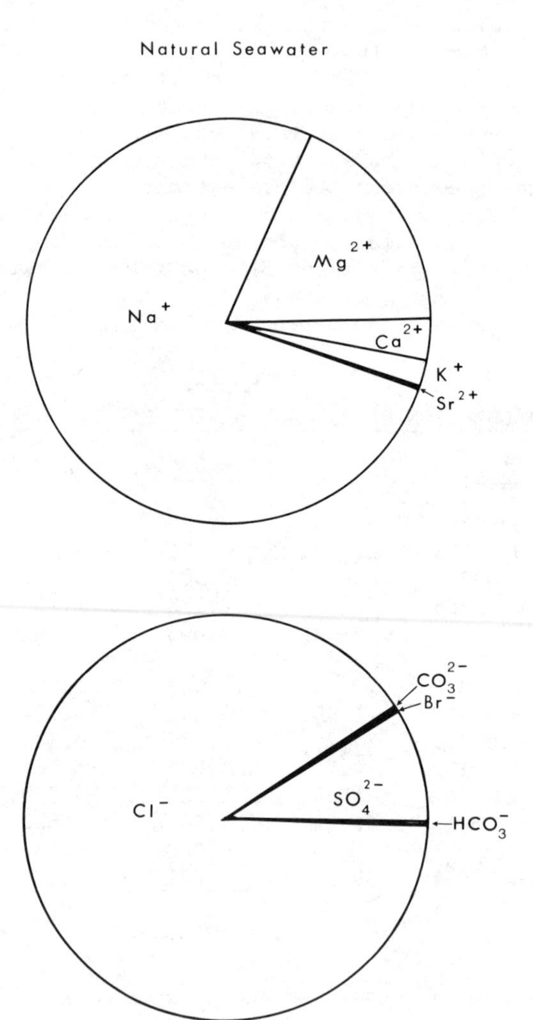

Figure 18. Equivalent fraction of cations and anions in seawater

the higher concentrations of natural brines ($I_V \cong 5.0$ usually agree to ± 250 x 10^{-6} g cm^{-3} with direct measurements (80) (see Table V). Predicted densities for saturated solutions to this accuracy are sufficient for most practical calculations. To make more reliable predictions, it is necessary to know the volume of mixing the common ion pairs of the major components (e.g., NaCl + $MgCl_2$, Na_2SO_4 + $MgSO_4$, $MgCl_2$ + $MgSO_4$ and Na_2SO_4 + NaCl).

TABLE IV
COMPARISON OF THE CALCULATED AND MEASURED DENSITIES OF SEAWATER SOLUTIONS AT 25°C

$(S(^o/oo)$	I_V	$(d - d_o)10^3$ Meas	Calc	Δ, ppm
5	0.09968	3.765	3.770	-5
10	0.20012	7.511	7.522	-11
15	0.30128	11.254	11.270	-16
20	0.40323	15.002	15.023	-21
25	0.50590	18.757	18.781	-24
30	0.60933	22.523	22.547	-24
35	0.71351	26.300	26.323	-23
				mean 18

TABLE V
COMPARISONS OF THE MEASURED AND CALCULATED RELATIVE DENSITIES OF ORCA BASIN BRINES AT 25°C[a]

Temp	$S(^o/oo)$	$10^3(d - d_o)$ Meas	Calc	Δ, ppm
15°C	250.85	194.76	194.51	250
25	250.85	191.56	191.52	40
35	250.85	189.37	189.45	80

a) From Ref. 80, I_V = 5.22

The compressibilities of mixed electrolyte solutions (48,86, 88,148) can also be estimated. The adiabatic apparent molal compressibility of a mixed electrolyte is given by

$$\Phi_{K(S)} = \sum E_M E_X \phi_{K(S)}(MX) \quad (54)$$

and is related to the adiabatic compressibility (β_S) by

$$\beta_S = \beta_{o(S)} + A_\beta I + B_\beta I^{3/2} + C_\beta I^2 \quad (55)$$

where $A_\beta = [\Phi^o_{K(S)} - \beta_{o(S)}\ \Phi_V{}^o]k$, $B_\beta = [S_K - \beta_{o(S)}\ S_V]k$, and $C_\beta = [B_K - \beta_{o(S)} b_V]k$. The speed of sound can be estimated from the predicted $\phi_{K(S)}$ using

$$u = \sqrt{1/\beta_{(S)} d} \tag{56}$$

Comparisons of the predicted and measured values of β_S and u for seawater solutions using the additivity method are shown in Table VI. The agreement is quite good and sufficient for most needs.

TABLE VI
COMPARISONS OF THE MEASURED AND CALCULATED VALUES OF β_S AND u FOR SEAWATER SOLUTIONS AT 25°C

	$(u - u_o)$, m sec^{-1}			$-[\beta_S - \beta_{o(S)}]$, 10^6 bar^{-1}		
S(°/oo)	Meas[a]	Calc[b]	Δ	Meas[a]	Calc[b]	Δ
5.024	5.50	5.49	0.01	0.495	0.495	0.000
10.057	10.96	10.91	0.05	0.980	0.978	0.002
15.142	16.44	16.38	0.06	1.461	1.458	0.004
20.000	21.60	21.61	-0.01	1.910	1.911	-0.001
24.943	26.91	26.92	-0.01	2.364	2.365	-0.001
30.031	32.41	32.41	0.00	2.827	2.827	0.000
35.003	37.72	37.78	-0.06	3.270	3.273	-0.003
35.005	37.72	37.78	-0.06	3.270	3.273	-0.003
40.025	43.18	43.21	-0.03	3.716	3.718	-0.002
			Mean ± 0.03			Mean ± 0.002

a) From ref. 85
b) From ref. 86

Once the density and compressibilities of mixed electrolyte solutions are known at 1 atm, values at high pressures can be made by using the secant bulk modulus equation of state. The major difficulty, at present, with using additivity methods to estimate the PVT properties of mixed electrolytes is the lack of experimental data for binary solutions over a wide range of concentrations and temperatures. Hopefully, in the near future we will be able to provide some of these data by measurements in our laboratory in Miami.

Acknowledgement

The author would like to acknowledge the support of the Office of Naval Research (N00014-75-C-0173) and the Oceanographic Section of the National Science Foundation (OCE77-28546) for this study.

Abstract

A brief review is made of the methods that are currently being used to determine the density (ρ) and compressibility (β) of electrolyte solutions as a function of pressure. The high pressure equations of state used to represent these properties are also discussed. The linear secant bulk modulus [$K = P\rho^P/(\rho^P - \rho^0)$] equation of state

$$K = K^0 + B\ P$$

is shown to give a reasonable representation of the high pressure PVT properties of water from 0 to 350°C, where $K^0 = 1/\beta^0$ the reciprocal of the 1 atm compressibility (applied pressure P = 0) and B is a constant. Since B is not strongly dependent upon temperature and concentration, the same equation can be used to estimate the high pressure PVT properties of electrolytes with reasonable accuracy, using 1 atm values of density and compressibility. Simple additivity methods of estimating the density and compressibilities of mixed electrolytes like seawater and geothermal brines is presented. The predicted PVT properties for a number of mixed electrolytes are in good agreement with direct measurements.

Literature Cited

1. Millero, F.J. Chem. Rev., 1971, 71, 147.
2. Millero, F.J., in "Water and Aqueous Solutions", R.A. Horne, Ed., Wiley and Sons: New York, 1972; p. 519.
3. Millero, F.J., in "Activity Coefficients in Aqueous Electrolyte Solutions", R.M. Pytkowicz, Ed., CRC Press: W. Palm Beach, Fla., 1979, in press.
4. Bauer, N.; Lewin, S.Z., in "Techniques of Organic Chemistry", A. Weissberger, Ed., Interscience: New York, 1959, p. 167.
5. Bowman, H.A.; Schoonover, R.M. J. Res. Nat. Bur. Stds., Sect. C, 1967, 71, 179.
6. Hidnert, P.; Peffer, E.L., "Densities of Solids and Liquids", NBS Circ. 487, U.S. Govt. Print. Off., Washington, D.C., 1950.
7. Millero, F.J. Rev. Sci. Instrum., 1967, 38, 1441.
8. Lamb, A.B.; Lee, R.E. J. Am. Chem. Soc., 1913, 35, 1666.
9. Geffcken, W.; Beckmann, Ch.; Kruis, A. Z. Phys. Chem. (Leipzig) Abt. B, 1933, 20, 398.
10. Hall, N.F.; Jones, T.O. J. Am. Chem. Soc., 1936, 58, 1915.
11. Ray, B.R.; MacInnes, D.A. Rev. Sci. Instrum., 1951, 22, 642.
12. Spedding, F.H.; Pikal, M.J.; Ayer, B.O. J. Phys. Chem., 1966, 70, 2440.
13. Franks, F.; Smith, H.T. Trans. Faraday Soc., 1967, 63, 2586.
14. Ubrich, D.V.; Kupke, D.W.; Beams, J.W. Proc. Nat. Acad. Sci. U.S., 1964, 52, 349.

15. Wirth, H.E. J. Am. Chem. Soc., 1937, 59, 2549.
16. Redlich, O.; Nielsen, L.E. J. Am. Chem. Soc., 1942, 64, 761.
17. Conway, B.E.; Verrall, R.E.; Desnoyers, J.E. Trans. Faraday Soc., 1966, 62, 2738.
18. Desnoyers, J.E.; Arel, M. Can. J. Chem., 1967, 45, 359.
19. Vaslow, F. J. Phys. Chem., 1966, 70, 2286.
20. Ward, G.K.; Millero, F.J. J. Soln. Chem., 1974, 3, 417.
21. Kratky, O.; Leopold, H.; Stabinger, H. Angew Phys., 1969, 2, 273.
22. Picker, P.; Tremblay, E.; Jolicoeur, C. J. Soln. Chem., 1974, 3, 377.
23. Geffcken, W.; Kruis, A.; Solana, L. Z. Phys. Chem., (Leipzig) Abt. B, 1933, 23, 175.
24. Hepler, L.G.; Stokes, J.M.; Stokes, R.H. Trans. Faraday Soc., 1965, 61, 20.
25. Wirth, H.E.; Bangert, F.K. J. Phys. Chem., 1972, 76, 3488.
26. Kell, G.S.; Whalley, E. Phil. Trans. R. Soc. Lond., 1965, 258, 565.
27. Kell, G.S. J. Chem. Eng. Data, 1967, 12, 66; ibid., 1970, 15, 119.
28. Kell, G.S. J. Chem. Eng. Data, 1975, 20, 97.
29. Fine, R.A.; Millero, F.J. J. Chem. Phys., 1973, 59, 5529.
30. Kell, G.S.; Whalley, E. J. Chem. Phys., 1975, 62, 3496.
31. Chen, C.T.; Fine, R.A.; Millero, F.J. J. Chem. Phys., 1977, 66, 2142.
32. Kell, G.S.; McLaurin, G.E.; Whalley, E. Phil. Trans. R. Soc. Lond., 1978, A360, 389.
33. Ward, G.K.; Millero, F.J. J. Soln. Chem., 1974, 3, 431.
34. Ward, G.K.; Millero, F.J. Geochim. Cosmochim. Acta, 1975, 39, 1595.
35. Millero, F.J.; Gonzalez, A.; Ward, G.K. J. Mar. Res., 1976, 34, 61.
36. Millero, F.J.; Knox, J.H.; Emmet, R.T. J. Soln. Chem., 1972, 1, 173.
37. Millero, F.J. J. Phys. Chem., 1967, 71, 4567.
38. Millero, F.J.; Drost-Hansen, W. J. Phys. Chem., 1968, 72, 1758.
39. Millero, F.J.; Drost-Hansen, W.; Korson, L. J. Phys. Chem., 1968, 72, 2251.
40. Millero, F.J. J. Phys. Chem., 1968, 72, 3209.
41. Millero, F.J.; Drost-Hansen, W. J. Chem. Eng. Data, 1968, 13, 330.
42. Millero, F.J. J. Phys. Chem., 1968, 72, 4589.
43. Millero, F.J. J. Phys. Chem., 1970, 74, 356.
44. Millero, F.J. J. Chem. Eng. Data, 1970, 15, 562.
45. Millero, F.J. J. Chem. Eng. Data, 1971, 16, 229.
46. Millero, F.J.; Hoff, E.V.; Kahn, L.A. J. Soln. Chem., 1972, 1, 309.
47. Millero, F.J. J. Soln. Chem., 1973, 2, 1.
48. Millero, F.J.; Lepple, F.K. Mar. Chem., 1973, 1, 89.

49. Millero, F.J.; Knox, J.H. J. Chem. Eng. Data, 1973, 18, 407.
50. Masterton, W.L.; Welles, H.; Knox, J.H.; Millero, F.J. J. Soln. Chem., 1974, 3, 91.
51. Millero, F.J.; Emmet, R.T. J. Mar. Res., 1976, 34, 15.
52. Chen, C.-T.; Millero, F.J. J. Soln. Chem., 1977, 6, 589.
53. Millero, F.J.; Lawson, D.; Gonzalez, A. J. Geophys. Res., 1976, 81, 1177.
54. Desnoyers, J.E.; de Visser, C.; Perron, G.; Picker, P. J. Soln. Chem., 1976, 5, 605.
55. Jolicoeur, C.; Philip, P.R.; Perron, G.; Leduc, P.-A.; Desnoyers, J.E. Can. J. Chem., 1972, 50, 3167.
56. Philip, P.R.; Desnoyers, J.E. J. Soln. Chem., 1972, 1, 353.
57. Leduc, P.-A., Desnoyers, J.E. Can. J. Chem., 1973, 51, 2993.
58. Desnoyers, J.E.; Page, R.; Perron, G.; Fortier, J.-L.; Leduc, P.-A.; Platford, R.F. Can. J. Chem., 1973, 51, 2129.
59. Millero, F.J.; Perron, G.; Desnoyers, J.E. J. Geophys. Res., 1973, 78, 4499.
60. Perron, G.; Desnoyers, J.E.; Millero, F.J. Can. J. Chem., 1974, 52, 3738.
61. Fortier, J.-L.; Leduc, P.-A.; Desnoyers, J.E. J. Soln. Chem. Chem., 1974, 3, 523.
62. Leduc, P.-A.; Fortier, J.-L.; Desnoyers, J.E. J. Phys. Chem., 1974, 78, 1217.
63. Jolicoeur, C.; Boileau, J. J. Soln. Chem., 1974, 3, 889.
64. Perron, G.; Desnoyers, J.E.; Millero, F.J. Can. J. Chem., 1975, 53, 1134.
65. Perron, G.; Fortier, J.-L.; Desnoyers, J.E. J. Chem. Thermodyn., 1975, 7, 1177.
66. Jolicoeur, C.; Boileau, J.; Bazinet, S.; Picker, P. Can. J. Chem., 1975, 53, 716.
67. Desrosiers, N.; Desnoyers, J.E. Can. J. Chem., 1975, 53, 3206.
68. Perron, G.; Desrosiers, N.; Desnoyers, J.E. Can. J. Chem., 1976, 54, 2163.
69. Wen, W.-Y.; Lo Surdo, A.; Jolicoeur, C.; Boileau, J. J. Phys. Chem., 1976, 80, 466.
70. Millero, F.J.; Ward, G.K.; Chetirkin, P.V. J. Biol. Chem., 1976, 251, 4001.
71. Millero, F.J.; Gonzalez, A.; Brewer, P.G.; Bradshaw, A. Earth Planet. Sci. Lett., 1976, 32, 468.
72. Millero, F.J.; Kremling, K. Deep-Sea Res., 1976, 23, 1129.
73. Millero, F.J.; Chetirkin, P.V.; Culkin, F. Deep-Sea Res., 1977, 24, 315.
74. Millero, F.J.; Gombar, F.; Oster, J. J. Soln. Chem., 1977, 6, 269.
75. Millero, F.J.; Laferriere, A.; Chetirkin, P.V. J. Phys. Chem., 1977, 81, 1737.
76. Millero, F.J.; Lo Surdo, A.; Shin, C. J. Phys. Chem., 1978, 82, 784.

77. Lo Surdo, A.; Shin, C.; Millero, F.J. J. Chem. Eng. Data, 1978, 23, 197.
78. Millero, F.J.; Forsht, D.; Means, D.; Gieskes, J.; Kenyon, K. J. Geophys. Res., 1978, 83, 2359.
79. Millero, F.J.; Means, D.; Miller, C.M. Deep-Sea Res., 1978, 25, 563.
80. Millero, F.J.; Lo Surdo, A.; Chetirkin, P.V.; Guinasso, N.L. Limnol. Ocanogr., 1979, 24, 218.
81. Lo Surdo, A.; Bernstrom, K.; Jonsson, C.-A.; Millero, F.J. J. Phys. Chem., 1979, 83, 1255.
82. Mathieson, J.G.; Conway, B.E. J. Soln. Chem., 1974, 3, 455.
83. Sakurai, M.; Nakajima, T.; Komatsu, T.; Nakagawa, T. Chem. Lett. (Japan), 1975, 971.
84. Wang, D.P.; Millero, F.J. J. Geophys. Res., 1973, 78, 7122.
85. Millero, F.J.; Kubinski, T. J. Acoust. Soc. Am., 1975, 57, 312.
86. Millero, F.J.; Ward, G.K.; Chetirkin, P.V. J. Acoust. Soc. Am., 1977, 61, 1492.
87. Millero, F.J.; Curry, R.W.; Drost-Hansen, W. J. Chem. Eng. Data, 1969, 14, 422.
88. Lepple, F.K.; Millero, F.J. Deep-Sea Res., 1971, 18, 1233.
89. Millero, F.J., Lepple, F.K. J. Chem. Phys., 1971, 54, 946.
90. Millero, F.J.; Ward, G.K.; Lepple, F.K.; Hoff, E.V. J. Phys. Chem., 1974, 78, 1636.
91. Chen, C.-T.; Chen, L.-S.; Millero, F.J. J. Acoust. Soc. Am., 1978, 63, 1795.
92. Lo Surdo, A.; Millero, F.J. J. Soln. Chem., 1979, in press.
93. Greenspan, M.; Tschiegg, C.E. Rev. Sci. Instrum., 1957, 28, 897.
94. Garsey, R.; Roe, F.J.; Mahoney, R.; Litovitz, T.A. J. Chem. Phys., 1969, 50, 5222.
95. Del Grosso, V.A.; Mader, C.W. J. Acoust. Soc. Am., 1972, 52, 1442.
96. Hayward, A.T.J. Nature, 1969, 221, 1047.
97. Leonard, R.; Combs, P.; Stridmore, L. J. Acoust. Soc. Am., 1949, 21, 63.
98. Liebermann, L. J. Acoust. Soc. Am., 1948, 20, 868.
99. Bies, D. J. Chem. Phys., 1955, 23, 428.
100. Wilson, O.; Leonard, R. J. Acoust. Soc. Am., 1954, 26, 223.
101. Kurtze, G.; Tamm, K. Acustica, 1953, 3, 33.
102. Smith, W.C.; Barrett, R.E.; Beyer, R.T. J. Acoust. Soc. Am., 1951, 23, 71.
103. Tamm, K.; Krutze, G.; Kaiser, R. Acustica, 1954, 4, 380.
104. Atkinson, G.; Petrucci, S. J. Phys. Chem., 1966, 70, 3122.
105. Atkinson, G., personal communication, 1976.
106. Bradshaw, A.; Schleicher, K.E. Deep-Sea Res., 1970, 17, 691, ibid., 1976, 23, 583.
107. Fahey, P.F.; Kupke, D.W.; Beams, J.W. Proc. Nat. Acad. Sci. U.S., 1969, 63, 548.

108. Haynes, W.H.; Stewart, J.W. Rev. Sci. Instrum., 1971, 42, 1142.
109. Wilson, W.D. J. Acoust. Soc. Am., 1959, 31, 1067.
110. Del Grosso, V.A.; Mader, C.W. J. Acoust. Soc. Am., 1972, 52, 961.
111. Barlow, A.J.; Yazgan, E. Br. J. Appl. Phys., 1966, 17, 807; ibid., 1967, 18, 645.
112. Chen, C.-T.; Millero, F.J. J. Acoust. Soc. Am., 1976, 60, 1270.
113. Adams, L.H. J. Am. Chem. Soc., 1931, 53, 3769.
114. Gibson, R.E. J. Am. Chem. Soc., 1937, 59, 1521.
115. Keyes, F.G. Proc. Am. Acad. Arts Sci., U.S., 1933, 68, 505.
116. Chen, C.-T.; Millero, F.J. Deep-Sea Res., 1976, 23, 595.
117. Emmet, R.T.; Millero, F.J. J. Chem. Eng. Data, 1975, 20, 351.
118. Emmet, R.T.; Millero, F.J. J. Geophys. Res., 1974, 79, 3463.
119. Chen, C.-T.; Emmet, R.T.; Millero, F.J. J. Chem. Eng. Data, 1977, 22, 201.
120. Wilson, W.D. J. Acoust. Soc. Am., 1961, 33, 314.
121. Wilson, W.D. J. Acoust. Soc. Am., 1960, 32, 1357.
122. Chen, C.-T.; Millero, F.J. J. Acoust. Soc. Am., 1977, 62, 553.
123. Chen, C.-T.; Millero, F.J. J. Acoust. Soc. Am., 1977, 62, 1129.
124. Chen, C.-T. Ph.D. Dissertation, University of Miami, Miami, Florida, 1977.
125. Chen, C.-T.; Millero, F.J., in preparation.
126. Fine, R.A.; Wang, D.-P.; Millero, F.J. J. Mar. Res., 1974, 32, 433.
127. Chen, C.-T.; Millero, F.J. J. Mar. Res., 1978, 36, 657.
128. Crease, J. Deep-Sea Res., 1962, 9, 209.
129. Picker, P.; Leduc, P.-A.; Philip, P.R.; Desnoyers, J.E. J. Chem. Thermodyn., 1971, 3, 631.
130. Ben-Naim, A., in "Water - A Comprehensive Treatise", Vol. 1, Chapt. 11, F. Franks, Ed., Plenum Press: New York, 1972.
131. Weres, O.; Rice, S.A. Am. Chem. Soc., 1972, 94, 8983.
132. Gibbs, J.H.; Cohen, C.; Fleming, P.D. III; Porosoff, H. J. Soln. Chem., 1973, 2, 277.
133. Tait, P.G. "Voyage of H.M.S. Challenger HMSO", II, London, 1888; p. 1.
134. Hayward, A.T.J. Br. J. Appl. Phys., 1967, 18, 965.
135. Tammann, G. Zeit. Phys. Chem., 1895, 17, 620.
136. Tumlirz, O. Math. Naturwiss., Kl., 1909, 118A, 203.
137. Gibson, R.E. J. Am. Chem. Soc., 1935, 57, 284.
138. Gibson, R.E.; Loeffler, O.H. J. Phys. Chem., 1939, 43, 207.
139. Gibson, R.E.; Loeffler, O.H. J. Am. Chem. Soc., 1941, 63, 898.
140. Murnaghan, F.D. Proc. Nat. Acad. Sci. U.S., 1944, 30, 244.
141. Murnaghan, F.D. Proc. Symp. Appl. Math., 1949, 1, 158.

142. Li, Y.-H. J. Geophys. Res., 1967, 72, 2665.
143. Ekman, V.W. Pubs. Circonst. Cons. Perm. Int. Explor. Mer., 1908, 43, 1.
144. Macdonald, J.R. Rev. Mod. Phys., 1969, 41, 316.
145. Millero, F.J., in "The Sea", Vol. 5, Chapter 1, E.D. Goldberg, Ed., Wiley: New York, 1974; pp. 3-80.
146. Millero, F.J. Ann. Rev. Earth Planet. Sci., 1974, 2, 101.
147. Millero, F.J. Deep-Sea Res., 1973, 20, 101.
148. Millero, F.J. J. Soln. Chem., 1973, 2, 1.
149. Millero, F.J., in "Structure of Water and Aqueous Solutions Solutions", W. Luck, Ed., Verlag Chemie, Weinheim, Bergstr., 1974; pp. 513-522.
150. Millero, F.J. Nav. Res. Rev., 1974, 27, 40.
151. Millero, F.J., in "Marine Chemistry in the Coastal Environment", Chapter 2, T.M. Chruch, Ed., ACS Symp. Ser. 18, Washington, D.C., 1975; p. 25.
152. Millero, F.J. Thalassia Jugoslavica, 1978, 14, 1.
153. Millero, F.J. Earth Planet. Sci. Lett., 1980, in press.
154. Young, T.F. Rec. Chem. Progr., 1951, 12, 81.
155. Wood, R.H.; Anderson, H.L. J. Phys. Chem., 1966, 70, 1877.
156. Reilly, P.J.; Wood, R.H. J. Phys. Chem., 1969, 73, 4292.
157. Wood, R.H.; Reilly, P.J. Ann. Rev. Phys. Chem., 1970, 21, 287.

RECEIVED January 31, 1980.

HYDROMETALLURGY, OCEANOGRAPHY, AND GEOLOGY

32

Application of Thermodynamics in Hydrometallurgy

A State-of-the-Art Review

HERBERT E. BARNER
Kennecott Copper Corp., Lexington, MA 02173

ROGER N. KUST
Exxon Research and Engineering Co., Florham Park, NJ 07932

In the past several years interest in hydrometallurgical processing of ores has grown significantly. This growth can be attributed to several factors, among which are the necessity for environmental protection and pollution control, the increase in cost and decrease in availability of oil and natural gas, the increased complexity of ores processed, and the increased exploitation of non-sulfide ores. Whether or not hydrometallurgical processing will provide satisfactory answers to many of the problems facing the minerals industry today is yet to be demonstrated. However, there have been several instances in recent years where hydrometallurgical processing, totally or in part, has proven to be advantageous.

Hydrometallurgical processes involve the treating of a raw ore or concentrate with an aqueous solution of a chemical reagent in a reactor. The desired metal values are leached from the ore or concentrate, and the residue, after washing, is rejected as tailings. The leach liquor, containing one or more metal values in solution is then processed to the metals. The leaching reactor can be a conventional stirred tank or autoclave, or, as in the case of dump leaching, can be a large pile of rejected ore. In in-situ type leaching, the reactor is the ore-body itself, into which a suitable lixiviant is pumped.

Hydro processes operate at lower temperatures than pyro processes, usually 50-250°C, and as a result the rates at which reactions occur are frequently several orders of magnitude slower. Consequently, in the development of such processes, kinetic studies become important. However, application of thermodynamics can still give valuable insight into the nature of various processes and should be used to determine process limitations.

One important way in which thermodynamics is utilized in hydrometallurgical process development is as a guide to experimental planning and process selection and evaluation. A high degree of accuracy is not needed at this stage of process development. Estimates which are within 15 to 20% of the true value are frequently adequate for making feasibility calculations.

0-8412-0569-8/80/47-133-625$05.00/0

There are numerous areas which can be treated by thermodynamic analysis. Some of them are:

1. Calculation of the solubility of simple and complex salts and gases, including estimation of "metal loading" in leach liquors
2. Estimation of vapor pressures of volatile components
3. Determination of the extent of reaction or position of equilibrium under various conditions of temperature, pressure and concentrations.
4. Estimation of extent of complex ion and ion pair formation.
5. Calculation of distribution coefficients for ion exchange processes.

Earlier Reviews

A number of earlier reviews on this subject have been published. MacDonald (1) has addressed the electrochemistry of metals in aqueous systems at high temperature, including the extrapolation of low temperature data and the activity coefficients of ions, water and dissolved gases. Two contributions from the Warren Spring Laboratory in England (2, 3) have reviewed pressure hydrolysis at high temperature and the extrapolation of potential-pH (Pourbaix) diagrams to high temperatures. Kwok and Robbins (4) have also considered the thermodynamics of high-temperature solutions with emphasis on separation of metal values from leach solutions; the $CuSO_4$ - H_2SO_4-H_2O was treated in some detail.

Peters (5) has reviewed the leaching of copper, nickel, zinc, lead and molybdenum concentrates in terms of the thermodynamic stability of the sulfide minerals of these metals. Process developments associated with the most favorable decomposition paths are considered.

From a geochemistry point of view, Helgeson (6, 7, 8) has presented the properties of water at high temperature and pressure, Debye-Huckel parameters as a function of temperature and pressure, and the partial molal properties of electrolytes. Helgeson (9) has in addition published a comprehensive monograph on the thermodynamic aspects of hydrothermal ore-forming solutions. Barnes (10) has also presented a review of hydrothermal geochemistry with emphasis on thermodynamic interpretation and experimental measurements at high temperature and pressure.

Species in Solution

Before any thermodynamic estimates can be made, there must be an assessment of the major solution, solid phase, and gaseous phase species present in a given system. In some instances physico-chemical measurements must be made to identify the species to be considered, but frequently chemical intuition and judgement flavored with experience are sufficient to provide a basis for initial estimating. Generally, for metal species in solution one needs to know how metal ions are hydrated, the number of ligands associated with the metal, the charges on the complex ions present, and whether or not the metal species are primarily mononuclear. For many metals such as copper, zinc, nickel, silver, etc. there is reasonable agreement as to what type of complexes are formed with the relatively simple ligands of interest to hydrometallurgy, e.g., hydroxide, chloride, sulfate, nitrate, cyanide, thiocyanate. For a few metals, such as molybdenum and tungsten, the species formed are extremely complex in nature and there is little agreement on their constitution, particularly in weak acid or neutral solutions.

A recent report on the recovery of metal values from geothermal brine indicated the presence of lead, silver, copper, and iron in the cationic form. Because the brine contained 155 g/l of chloride the metals must be present as chloroanions. The presence of chloroanions rather than aquated cations suggests a different type of processing and would certainly alter the results of thermodynamic estimates.

Equilibrium Constants and Free Energy of Formation

It is necessary to consider a number of equilibrium reactions in an analysis of a hydrometallurgical process. These include complexing reactions that occur in solution as well as solubility reactions that define equilibria for the dissolution and precipitation of solid phases. As an example, in analyzing the precipitation of iron compounds from sulfuric acid leach solutions, McAndrew, et al. (11) consider up to 32 hydroxyl and sulfate complexing reactions and 13 precipitation reactions. Within a restricted pH range only a few of these equilibria are relevant and need to be considered. Nevertheless, equilibrium constants for the relevant reactions must be known. Furthermore, since most processes operate at elevated temperatures, it is essential that these parameters be known over a range of temperatures. The availability of this information is discussed below.

Data at $25^{\circ}C$. Free energy of formation, equilibrium constants and related data at $25^{\circ}C$ exist for a wide range of minerals, other solids, gases and aqueous species, including ions and complexes (see later discussion on data sources). Availability of data at this reference temperature is usually not a major stumbling block.

On the other hand, new solution species are being identified. For example, some polynuclear species and some ion pair complexes are now recognized as being more significant in aqueous solutions than previously thought. There is therefore a need to develop, extend and up-date the data on new species which come to be recognized as significant.

Temperature Effects. The most reliable source of information on the temperature dependence of a specific equilibrium constant is experimental measurement. Occasionally, sufficient experimental data are available. In the more usual situation, however, only sporadic high-temperature data, if any at all, are available. It is then necessary to use some form of extrapolation procedure to extend the $25^{\circ}C$ data to higher temperatures.

A number of approaches are available to extrapolate low-temperature equilibrium constants. Various aspects of this problem have been discussed and some comparisons to experimental data have been made. See, for example, Criss and Cobble (12), Helgeson (13), MacDonald (1), and Manning and Melling (3). We are not, however, aware of any recent comprehensive evaluation, error analysis or overall assessment. We summarize below what we believe to be the present situation.

The equation that relates the Gibbs free energy of a chemical reaction at temperature T to that at the reference temperature 298°C is given by

$$\Delta G^o_T = \Delta G^o_{298} - \Delta S^o_{298}\,\Delta T + \int_{298}^{T} \Delta C^o_p\, dT - T\int_{298}^{T} \Delta C^o_p\, d(\ln T) \qquad (1)$$

The various approaches for estimating ΔG^o_T (and hence the equilibrium constant at elevated temperature) correspond to the assumptions made in estimating the heat capacity term, ΔC^o_p.

1. $\Delta C^o_p = 0$. With this assumption, the last two terms in Equation (1) vanish. It can be shown that the resulting equation is equivalent to the integrated form of the van't Hoff equation where the heat of reaction term is assumed to be constant,

$$\Delta G^o_T = \Delta G^o_{298} + T\Delta H^o_{298}\left(\frac{1}{T} - \frac{1}{298}\right) \qquad (2)$$

For aqueous systems, this assumption can lead to an error of several orders of magnitude at 250°C-300°C (3, 13, 14), as illustrated in Figure 1 for the dissociation of HSO_4^-. Clearly, the use of this assumption in aqueous systems should be avoided.

2. $\Delta C^o_p(T) = \Delta C^o_p(298°K)$. Since C^o_p values at 298°K are available for most ions, this assumption permits estimation of ΔG^o_T for many reactions of interest. This assumption has not been widely tested, but is certainly a substantial improvement over assuming, $\Delta C^o_p = 0$ (see Figures 1 and 2).

3. ΔC^o_p proportional to absolute temperature. Harned and Owens (15) have observed that

$$\Delta C^o_p = 2cT \qquad (3)$$

This leads to the expression

$$\Delta G^o_T = \Delta G^o_{298} - \Delta S^o_{298}(T-298) + c(T-298)^2 \qquad (4)$$

where the constant c can be evaluated from Eq. (3) at 298°K. Khodakovskiy (16) used this assumption to extrapolate dissociation constants for acids, Henry's law constants for gases and solubility products for salts. In a variation of this technique Khodakovskiy (17) has extrapolated partial molal heat capacities of ions and neutral molecules by using $\bar{C}_p = bT$.

4. ΔC^o_p estimated from "correspondence principle". This approach, proposed by Criss and Cobble (12), appears to be the most widely used one. It rests on what has been called the entropy correspondence principle which states that the partial molal entropy of an ion at temperature T is related to its partial molal entropy at 298°K by a relation

$$\bar{S}^o_T = a_T + b_T\,\bar{S}^o_{298}(\text{absolute}) \qquad (5)$$

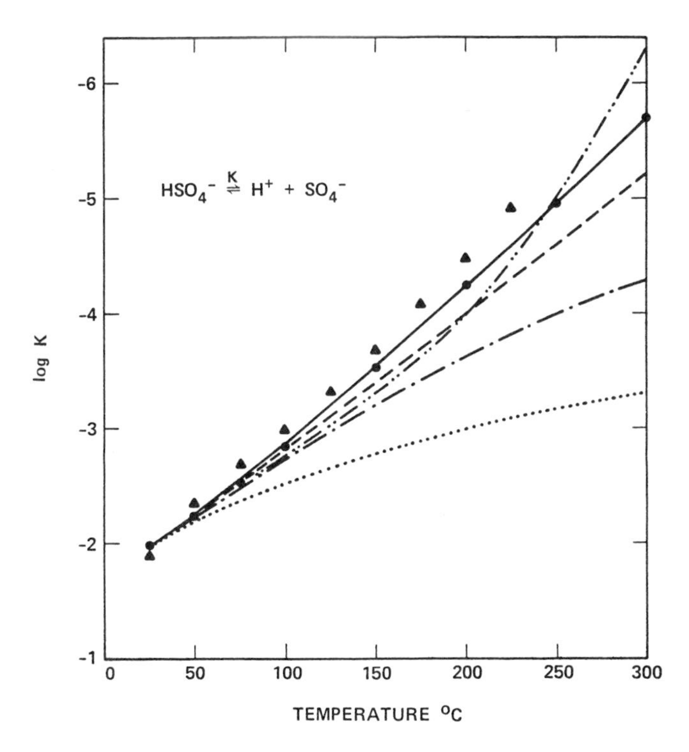

Figure 1. Extrapolation of HSO_4^- dissociation constants ((●) (62); (▲) (63); (· · ·) $\Delta Cp = 0$; (– – –) $\Delta Cp = \Delta Cp$ at 25°C; (– · –) $\Delta Cp \propto$ abs. temp.; (——) ΔCp from Correspondence Principle (Criss & Cobble); (– · · –) Helgeson)

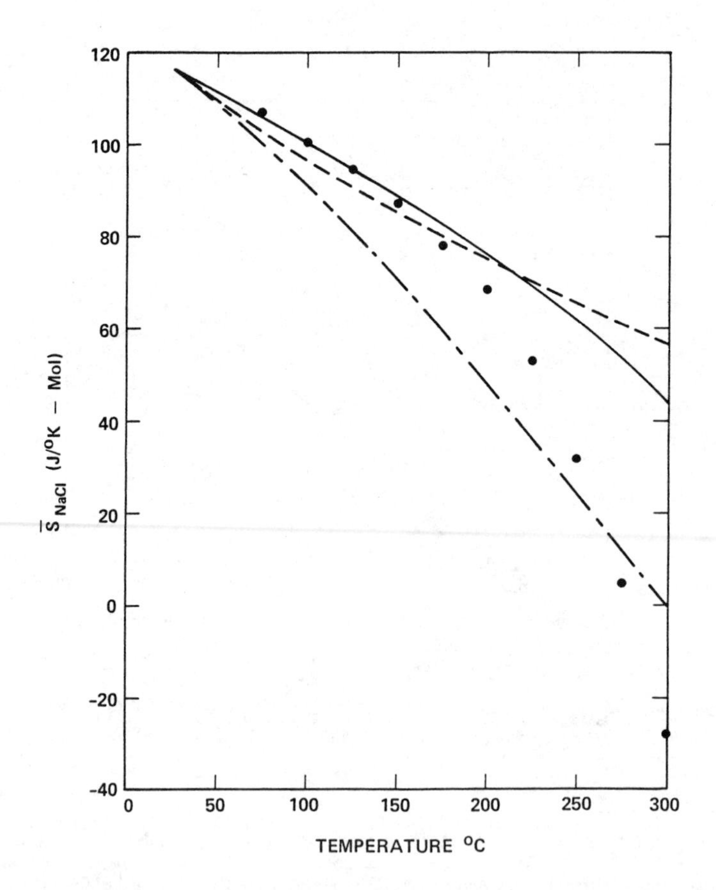

Figure 2. Standard partial molal entropy for NaCl in aqueous solution (experimental data based on Liu and Lindsay, taken from Ref. 1) (– – –) Cp° = constant = Cp° (298°K); (– · –) Cp° ∝ abs. temp.; (——) Cp° from Correspondence Principle (Criss & Cobble); (●) experimental (Liu & Lindsay)

The coefficients a_T and b_T are dependent on temperature and the type of ion (cations, OH^- and simple anions, oxyanions and acid oxyanions) and have been established by Criss and Cobble by analyzing available experimental data. $\bar{S}^o_{298}$(absolute) is the 298°K entropy on an absolute scale corresponding to the partial molal entropy of H^+(aq) as -5.0 cal deg^{-1} mole^{-1} at 298°K. The absolute and conventional partial molal entropy for any ion are related by

$$\bar{S}^o_{298}\text{ (absolute)} = \bar{S}^o_{298}\text{ (conventional)} - 5.0\, z \tag{6}$$

where z is the ionic charge.

Criss and Cobble calculate the average heat capacity of an ion between 298°K and temperature T in terms of its partial molal entropy:

$$\left.\bar{C}^o_p\right]^T_{298} = (\bar{S}^o_T - \bar{S}^o_{298})/\ln (T/298) \tag{7}$$

and hence, the free energy of reaction from Eq. (1) becomes

$$\Delta G^o_T = \Delta G^o_{298} - \Delta S^o_{298}\, \Delta T + \left.\Delta C^o_p\right]^T_{298} \left[(T-298) - T \ln T/298\right] \tag{8}$$

Having evaluated the average heat capacity of ions by Eq. (7),

$$\left.\Delta C^o_p\right]^T_{298}$$

for the chemical reaction is obtained by merely adding the average heat capacities of products minus reactants. (The average heat capacities of solids and gases are calculated from generally well-known C_p data.) Thus, ΔG^o_T can be estimated from Eq. (8).

The Criss and Cobble procedure has been widely used and is generally accepted to yield reliable results up to the 150°C - 200°C range. The parameters at higher temperatures were obtained by Criss and Cobble by extrapolation, so the predictions here may be expected to be less reliable. The comparison in Figure 1 suggests that the dissociation of HSO_4^- is predicted reasonably well up to 300°C. On the other hand, Figure 2 indicates that the entropy of sodium chloride electrolyte is not predicted well above 200°C. Tremaine and Goldman (18) have also reported significant discrepancies for this method in predicting the free energy functions of Ca^{2+}, Sr^{2+} and Ba^{2+} above 200°C.

While the above comparisons are hardly conclusive, they do suggest caution in applying the correspondence principle as it presently exists to temperatures above 200°C.

5. Statistical Thermodynamic Approach. Helgeson (13) has described the dissociation for complexes (such as for $NH_4OH \rightleftharpoons NH_4^+ + OH^-$) in terms of two functions - an electrostatic temperature function and a non-electrostatic (dielectric) temperature function. The following equation has been suggested to obtain ΔG_T for the dissociation of complexes in solution:

$$-\Delta G_T^o = \Delta S_r^o \left[T_r - \frac{\theta}{\omega} \left\{ 1 - \exp \left[\exp(b+aT) - c + (T - T_r)/\theta \right] \right\} \right] + \Delta H_{T_r}^o \tag{9}$$

Values for the coefficients a, b, c, θ and ω are given by Helgeson; Tr is a reference temperature, usually taken to be 298°K. Eq. (9) appears to be reliable up to about 200°C, but breaks down at higher temperatures. Note that for this approach there is no need for C_p^o data, even at the 298°K reference temperature; the method is particularly useful for complexes where such data are often not available.

The (admittedly limited) comparisons shown in Figures 1 and 2 suggest that either the assumption of a constant ΔC_p or the correspondence principle leads to reasonable estimates of the free energy of formation at temperatures up to 150°C to 200°C. Beyond this range, the extrapolations are in doubt. It appears that an overall assessment of the correspondence principle with emphasis on temperatures of 200°C and above, and with refinement of the Criss and Cobble parameters would be very desirable.

Activity Coefficients

In applying thermodynamics to the aqueous systems important in hydrometallurgy, one quickly meets the problem of activity coefficients. Most hydrometallurgical systems can be classed as strong electrolytes with ionic strength values ranging from 0.1m to over 6m. Values of thermodynamic parameters can vary several orders of magnitude as the ionic strength increases from infinite dilution to 6m.

Various attempts have been made to increase the valid range of the Debye-Huckel equation to regions of high ionic strength by the use of empirically fitted parameters. Several of these equations are listed in table I.

None of these extensions has been really satisfactory and they are not very useful at high ionic strength. The Davies equation (19) differs from the others in providing an additional term which alters the response of the activity coefficient to changes in ionic strength, particularly at higher values. The authors have had some success with this type of equation by replacing the .2 factor in the second term with a variable. The variable can be determined by experiment at a particular set of conditions.

In the last decade, however, several significant advances have been made in the treatment of activity coefficients. Meissner and colleagues (20, 21, 22, 23) were reasonably successful in correlating a large amount of experimental activity coefficient data in terms of a reduced activity coefficient. They developed a series of curves of reduced activity coefficient vs ionic strength. If information is available which can locate the system of interest on one curve, then values of the activity coefficient at other ionic strengths can be obtained by following the curve. Further, a procedure was devised, to allow selection of a reference curve when no experimental reference value is available.

Table I

Equations for Estimating Individual Ion Activity Coefficient

Equation		Range, m
$\log \gamma = -AZ^2 I^{\frac{1}{2}}$	Debye-Huckel	$I < .005$
$\log \gamma = \frac{-AZ^2 I^{\frac{1}{2}}}{1 + BaI^{\frac{1}{2}}}$	Extended Debye-Huckel	$I < 0.1$
$\log \gamma = \frac{-AZ^2 I^{\frac{1}{2}}}{1 + I^{\frac{1}{2}}}$	Guntelberg	$I < 0.1$
$\log \gamma = \frac{-AZ^2 I^{\frac{1}{2}}}{1 + I^{\frac{1}{2}}} + 0.2I$	Davies	$I < 0.5$

A one-parameter equation was developed by Bromley (24), again by correlation of published experimental data, which allows estimation of activity coefficients with an error of less than 10% up to ionic strength of 6m. Values of the parameters for over 170 salts of various charge types were calculated. A method was also given for calculating individual ion activity. Cognet and Renon (25) have used this formulation in determining the effect of chloride and sulfate ion on the liquid ion exchange behavior of copper (II).

An important series of papers by Professor Pitzer and colleagues (26, 27, 28, 29), beginning in 1973, has laid the ground work for what appears to be the most comprehensive and theoretically founded treatment to date. This treatment is based on the ion interaction model using the Debye-Huckel ion distribution and establishes the concept that the effect of short range forces, that is the second virial coefficient, should also depend on the ionic strength. Interaction parameters for a large number of electrolytes have been determined.

An important application of Pitzer's work is that of Whitfield (30) who developed a model for sea water. Single ion activity coefficients for many trace metals in sea water are tabulated over the ionic strength range of 0.2m to 3.0m.

It is interesting to note that the system developed by Bromley may be described as a simplification of the Pitzer system.

Volatile weak electrolytes, such as ammonia, carbon dioxide, and hydrohydrogen sulfide are of great interest in hydrometallurgy. The vapor-liquid equilibria of these electrolytes have been treated recently by Edwards, et al. (31, 32). In this treatment, Pitzer's equations for activity coefficient were used up to 170°C and for electrolyte concentrations of 10-20 molal.

From the foregoing discussion we conclude that some sophisticated tools are now available by which the activity coefficient in hydrometallurgical systems can be addressed. What is lacking is the actual application of these tools by the industry. The next step in establishing the accuracy of the available approaches lies in providing a broader data base for complex multicomponent systems which can be used for parameter refinement. The lack of data is most serious in the weak electrolyte area, but even familiar systems such as those encountered in sulfuric acid leaching need attention.

Sources of Data

Sources of thermodynamic data which the authors have found to be valuable in hydrometallurgical applications have been assembled in Table II.

With respect to parameters for activity coefficient equations, the literature referenced earlier under this topic should be consulted.

Computer Programs

Very few generalized computer-based techniques for calculating chemical equilibria in electrolyte systems have been reported. Crerar (47) describes a method for calculating multicomponent equilibria based on equilibrium constants and activity coefficients estimated from the Debye Huckel equation. It is not clear, however, if this technique has been applied in general to the solubility of minerals and solids. A second generalized approach has been developed by OIL Systems, Inc. (48). It also operates on specified equilibrium constants and incorporates activity coefficient corrections for ions, non-electrolytes and water. This technique has been applied to a variety of electrolyte equilibrium problems including vapor-liquid equilibria and solubility of solids.

Several computer-based techniques have been developed for more specific applications. Truesdell (45) describes a computer program for calculating equilibrium distributions in natural water systems, given concentrations and pH. Edwards, et al. (31, 32) have developed computer programs for treating volatile weak electrolytes such as ammonia, carbon dioxide and hydrogen sulfide systems; however, in their present state these programs (presumably) do not accommodate metallic species in solutions. A number of computer programs for generating potential -pH diagrams have been described (49, 50, 51). Helgeson (52) describes a method often used within geosciences whereby compositional changes, mass transfers and the order of appearance of stable and metastable phases are determined in tracing reaction paths from an initial set of conditions to a final state of equilibrium.

Generalized computer techniques for calculating vapor-liquid equilibria and solubility relationships in electrolyte systems are not readily available in the metallurgical industry.

Table II

Useful Data Compilations for Hydrometallurgical Applications

NBS Technical Notes Series 270 (33)	Comprehensive source of 25°C data; compounds and aqueous species.
Latimer (34)	Somewhat dated, but still of interest for 25°C data on some compounds.
Garrels and Christ (35)	25°C data; compounds and aqueous species.
Sillen and Martell (36)	Valuable collection of experimental data on stability constants.
Smith and Martell (37)	Extensive compilation of stability constants.
Milazzo and Coroli (38)	Compilation of standard electrode potentials.
JANAF (39)	Comprehensive source of data for compounds.
Barner and Scheuerman (40)	Properties of ions and complexes extrapolated to high temperatures by methods of Criss and Cobble, and Helgeson.
Barin and Knacke (41)	Thermochemical data for inorganic compounds.
Kelley (42)	High temperature heat-content, heat-capacity and entrophy data for inorganic compounds.
Robie and Waldbaum (43)	Thermochemical data for minerals and inorganic compounds.
Naumov, Ryzhenko and Khodakovsky (17)	Thermochemical data for inorganic compounds; Henry's constants; dissociation constants; ionic and neutral species in aqueous solution.
Helgeson (44)	Equilibrium constants for hydrothermal systems; heat capacities of ions at elevated temperatures.
Truesdell (45)	Equilibrium constants for reactions important in natural water systems.
Barnes, Helgeson and Ellis (46)	Ionization constants in aqueous solutions.

Table III

Some Published Applications

Bianchi and Longhi (53)	Potential -pH diagram (25°C) for copper in sea water; activity coefficient corrections included.
Biernat and Robins (54)	Potential -pH diagrams for $Fe-H_2O$ and $Fe-S-H_2O$ systems; temp. up to 300°C.
Brook (55)	Potential -pH diagrams for Al, Cu, Cr, Pb, Ni, Sn and $Ti-H_2O$ systems; temp. up to 150°C.
Linkson, Nobbs and Lake (56)	Potential -pH diagrams relating to reduction of Cu (II) from acid leach solutions; temp. up to 100°C.
Ferreira (57)	Potential -pH diagrams for $S-H_2O$, $Cu-S-H_2O$ and $Fe-S-H_2O$ systems; temp. to 150°C.
Cognet and Renon (25)	Thermodynamic interpretation of Cu(II) extraction by LIX 64 organic reagent.
Manning (2)	Precipitation of metal oxides and hydroxides by hydrolysis; includes interpretation in terms of potential -pH diagrams.
Kwok and Robbins (4)	Precipitation of metals by heating solutions to high temperatures; separations involving Cu, Fe, Ni, Zn, Cd and Cr.
McAndrew, Wang and Brown (11)	Precipitation of iron compounds from sulfuric and leach solutions; temp. up to 140°C.
Meddings and Mackiw (58)	Thermodynamic feasibility and equilibria in gaseous reduction of dissolved metal species.
Vega and Funk (59)	Prediction of solid-liquid equilibria in multicomponent salt systems based on parameters estimated from binary and ternary systems.
Bratt and Gorden (60)	Purification of solutions for electrolytic production of zinc.
Osseo-asare Fuerstenau (61)	Potential -pH and other stability diagrams for the Cu-, Ni-, and $Co-NH_3-H_2O$ systems at 25°C.

Some Published Applications

In many of the published applications of thermodynamics in hydrometallurgy, activity coefficients have been either omitted or crudely estimated. No doubt, this has been due in part to the difficulties in estimating ionic activity coefficients at high ionic strengths. However, with the recent surge of developments, some of the more current studies have addressed the activity coefficient problem more realistically. Representative published applications are presented in Table III.

A large number of publications deal with the construction and interpretation of potential -pH (Pourbaix) diagrams, and some of these have been included in Table III. Most of these studies avoid the question of activity coefficients because the stability fields are calculated for arbitrarily specified activities of the species in solution.

Other references in Table III discuss applications in precipitation of metal compounds, gaseous reduction of metals from solution, equilibrium of copper in solvent extraction, electrolyte purification and solid-liquid equilibria in concentrated salt solutions. The papers by Cognet and Renon (25) and Vega and Funk (59) stand out as recent studies in which rational approaches have been used for estimating ionic activity coefficients. In general, however, few of the studies are based on the more recent developments in ionic activity coefficients.

Conclusions

Most hydrometallurgical systems operate in the $50^{o}C$ to $250^{o}C$ temperature range and can be classified as strong electrolytes with ionic strengths ranging from 0.1m to 6m or higher. Furthermore, experimental data are seldom available in the regions of interest. Consequently, the successful use of thermodynamics requires that extrapolations be made in temperature, and that estimates be made of ionic activity coefficients.

Our review of the literature and assessment of the state of affairs in practice leads us to the following conclusions:

1. Reasonably reliable procedures exist for extrapolating low-temperature free energy data to temperatures in the $150^{o}C$ -$200^{o}C$ range. Beyond this range extrapolations are in doubt. An overall assessment and refinement of the Criss and Cobble correspondence principle with emphasis on temperatures of $200^{o}C$ and above would be valuable.
2. Recent advances in the treatment of activity coefficients now allow realistic predictions of ionic activity to be made. However, actual application of these newer tools by the industry is lacking.
3. Greater emphasis on generalized computer techniques for calculating solubility and vapor-liquid equilibria would be helpful in encouraging greater application of thermodynamics in practice.

Abstract

The application of thermodynamics in hydrometallurgy is surveyed. The types of systems and problems encountered are reviewed and the ranges of conditions (temperature, pressure, ionic strength) typically approached are considered. The difficulties encountered in making thermodynamic estimates in industrial applications are discussed, with particular reference to the assessment of species and temperature effects, and the estimation of activity coefficients.

Literature Cited

1. MacDonald, D.D. in "Modern Aspects of Electrochemistry"; Bockris, J.O.; Conway, B.E., Ed. Plenum Press: New York, 1975; 11, 141.

2. Manning, G.D. "Pressure Hydrolysis-A Survey", Warren Spring Laboratory LR 134(ME), 1970.

3. Manning, G.D.; Melling, J. "Potential (Eh)-pH Diagrams at Elevated Temperatures - A Survey", Warren Spring Laboratory LR 128(ME), 1971.

4. Kwok, O.J.; Robins, R.G. in "International Symposium on Hydrometallurgy", AIME, 1973, 1033.

5. Peters, E. Metall. Trans. B, 1976, 7B, 505.

6. Helgeson, H.C.; Kirkham, D.H. Amer. J. Sci., 1974, 274, 1089.

7. Helgeson, H.C.; Kirkham, D.H. Amer. J. Sci., 1974, 274, 1199.

8. Helgeson, H.C.; Kirkham, D.H. Amer. J. Sci., 1976, 276, 97.

9. Helgeson, H.C. "Complexing and Hydrothermal Ore Deposition", Pergamon Press: New York, 1964.

10. Barnes, H.L. "Conference on High Temperature-High Pressure Electrochemistry in Aqueous Solutions"; NACE, 1976, 14-23.

11. McAndrew, R.T.; Wang, S.S.; Brown, W.R. CIM Bull., Jan. 1975, 101.

12. Criss, C.M.; Cobble, J.W., J. Am. Chem. Soc., 1964, 86, 5390.

13. Helgeson, H.C. J. Phy. Chem., 1967, 71, 3121.

14. Lewis, D. Arkiv for Kemi, 1970, 32, 385.

15. Harned, H.S.; Owens, B.B. "The Physical Chemistry of Electrolyte Solutions", Reinhold Publishing Corp.: New York, 1958.

16. Khodakovskiy, I.L.; Ryzhenko, B.N.; Naumov, G.B., Geochem. Interl., 1968, 6, 1200.

17. Naumov, G.B.; Ryzhenko, B.N.; Khodokovsky, I.L. "Handbook of Thermodynamic Data"; Atomizdat: Moscow, 1971. (Available in English translation through NTIS, U.S. Dept. of Commerce, PB-226, 722).

18. Tremaine, P.R.; Goldman, S. J. Phys. Chem., 1978, 82, 2317.

19. Davies, C.W. J. Chem. Soc., 1938, 2093.

20. Meissner, H.P.; Tester, J.W. Ind. Eng. Chem. Proc. Des. Develop., 1972, 11, 128.

21. Meissner, H.P.; Kusik, C.L. AIChE Journal,, 1972, 18, 294.

22. Meissner, H.P.; Peppas, N.A. AIChE Journal, 1973, 19, 806.

23. Meissner, H.P.; Kusik, C.L.; Tester, J.W. AIChE Journal, 1972, 18, 661.

24. Bromley, L.A. AIChE Journal, 1973, 19, 313.

25. Cognet, M.C.; Renon, H. Hydrometallurgy, 1976/1977, 2, 305.

26. Pitzer, K.S. J. Phys. Chem., 1973, 77, 268.

27. Pitzer, K.S.; Mayorga, G. J. Phys. Chem., 1973, 77, 2300.

28. Pitzer, K.S.; Mayorga, G. J. Solution Chem., 1974, 3, 539.

29. Pitzer, K.S.; Kim, J.J. J. Am. Chem. Soc., 1974, 96, 5701.

30. Whitfield, M. Geochimica et Cosmochimica Acta., 1975, 39, 1545.

31. Edwards, T.J.; Newman, J.; Prausnitz, J.M. AIChE Journal, 1975, 21, 248.

32. Edwards, T.J.; Maurer, G.; Newman, J.; Prausnitz, J.M. AIChE Journal, 1978, 24, 966.

33. "Selected Values of Chemical Thermodynamic Properties", National Bureau of Standards, Technical Notes 270-3: 1968; 270-4: 1969, 270-5: 1971; 270-6: 1971; 270-7: 1973.

34. Latimer, W.M. "The Oxidation States of the Elements and Their Potentials in Aqueous Solutions", Prentice Hall: Englewood Cliffs, N.J., 1952.

35. Garrels, R.M.; Christ, C.L. "Solution Minerals and Equilibria", Harper and Row: New York, 1965.

36. Sillen, L.G.; Martell, A.E. "Stability Constants of Metal-Ion Complexes", The Chemical Society, London: 1964.

37. Smith, R.M.; Martell "Critical Stability Constants", Plenum Press: 1976.

38. Milazzo, G.; Caroli, S. "Tables of Standard Electrode Potentials", J. Wiley & Sons: New York, 1978.

39. JANAF Thermochemical Tables, Dow Chemical Co.: Michigan, 1965.

40. Barner, H.E.; Scheuerman, R.V. "Handbook of Thermochemical Data for Compounds and Aqueous Species"; J. Wiley & Sons: New York, 1978.

41. Barin, I.; Knacke, O. "Thermochemical Properties of Inorganic Substances", Springer Verlag: Berlin, 1973.

42. Kelley, K.K. "Contributions to the Data on Theoretical Metallurgy", Bureau of Mines, U.S. Dept. of the Interior XIII. Bulletin 584, High Temperature Heat Content, Heat Capacity and Entropy Data for the Elements and Inorganic Compounds, 1960.

43. Robie, R.A.; Waldbaum, D.R. "Thermodynamic Properties of Minerals and Related Substances at 298.15^{o}K and One Atmosphere Pressure and at Higher Temperatures", U.S. Dept. of the Interior, Geological Survey Bulletin 1259: 1968.

44. Helgeson, H.C. Amer. J. Sci., 1969, 267, 729.

45. Truesdell, A.H.; Jones, B.F. J. Research U.S. Geol. Survey, 1974, 2, 233.

46. Barnes, H.L.; Helgeson, H.C.; Ellis, A.J. in "Handbook of Physical Constants"; S.P. Clark, Jr., Ed.; Geol. Soc. Amer. Mem. 97, 401-412.

47. Crerar, D.A. Geochimica et Cosmochimica Acta., 1975, 39, 1375.

48. Zemaitis, J.F., Industrial Research, Nov. 1975, 70.

49. Froning, M.H.;Shanley, M.E.; Verink, E.D. Corros. Sci., 1976, 16, 371.

50. Brook, P.A. Corros. Sci., 1974, 11, 389.

51. Duby, P. "The Thermodynamic Properties of Aqueous Inorganic Copper Systems", International Copper Research Association: New York, 1977.

52. Helgeson, H.C. Mineral/Soc. Amer. Spec. Paper, 1970, 3, 155.

53. Bianchi, G.; Longhi, P. Corr. Sci., 1973, 13, 853.

54. Biernat, R.J.; Robbins, R.G. Electrochimica Acta, 1972, 17, 1261.

55. Brook, P.A. Corros. Sci., 1972, 12, 297.

56. Linkson, P.B.; Nobbs, D.M.; Lake, I.A. "Advances in Extractive Metallurgy", IMM, 1977, 111.

57. Ferreira, R.C.H. in "Leaching and Reduction in Hydrometallurgy", The Institution of Mining and Metallurgy: London, 1975, 67.

58. Meddings, B.; Mackiw, V.N. Unit Processes in Hydrometallurgy M.E. Wadsworth; F.T. Davis, Ed., AIME: 1963, 345.

59. Vega, R.; Funk, E.W. Desalination, 1974, 15, 225.

60. Bratt, G.C.; Gordon in "Research in Chemical and Extractive Metallurgy"; Woodcock, Jenkins, Willis, Ed., 197.

61. Osseo-Asare, K.; Fuerstenau, D.W. AIChE Symp. Series, 1978, 74, No. 173, 1.

62. Marshall, W.L.; Jones, E.V. J. Phys. Chem., 1966, 70, 4028.

63. Lietzke, M.H.; Stonghton, R.W.; Young, T.F. J. Phys. Chem., 1961, 65, 2247.

RECEIVED January 31, 1980.

33

Equilibrium and Kinetic Problems in Mixed Electrolyte Solutions

RICARDO M. PYTKOWICZ and M. R. COLE

Oregon State University, Corvallis, OR 97331

Mixed aqueous electrolyte solutions such as body fluids, rivers, lakes, oceans and, at times, laboratory and industrial fluids present important problems which are not found in single electrolyte solutions. New perceptions and results are being obtained in complex media and some examples will be covered in this paper.

The behavior of carbonates will be used to illustrate heterogeneous processes, with emphasis upon the formation of inorganic surface coatings and solid solutions. This is a vital topic in the study of solid-solution interactions since it is coatings rather than bulk phases which are sensed by liquid solutions. Homogeneous reactions will be studied in terms of the competition of coulombic ion pairs with true complexes for anions. An extended form of the phase rule will be used.

Phase Rule

The phase rule is often used in the form $f = c - p + 2$ to ascertain the number of degrees of freedom of a system even when the concentration units in the aqueous phase are molal (m) or molar. This is not correct because the phase rule is derived in terms of mole fractions (X). Thus, an additional quantity, the total number of moles, is required to convert X into m. This is illustrated by equations below which we shall find useful later on.

For the system CO_2-H_2O with two phases, vapor and aqueous solution, if we assume for simplicity that pH_2O and the activity coefficients for all the components are known, the equations are

$$P = pCO_2 + \sum_i p_i \qquad (1)$$

$$[CO_2] = k_{CO_2}\, pCO_2 \qquad (2)$$

$$K_h' = [H_2CO_3]/[CO_2][H_2O] \qquad (3)$$

0-8412-0569-8/80/47-133-643$05.00/0

$$K_1' = a_H[HCO_3^-]/[H_2CO_3] \quad (4)$$

$$K_2' = a_H[CO_3^{2-}]/[HCO_3^-] \quad (5)$$

$$K_w' = [H^+][OH^-]/[H_2O] \quad (6)$$

$$[H^+] = [HCO_3^-] + 2[CO_3^{2-}] + [OH^-] \quad (7)$$

$$TCO_2 + TH_2O = [CO_2] + [H_2CO_3] + [HCO_3^-] + + [H_2O] + [H^+] + [OH^-] \quad (8)$$

Thus, there are eight equations in nine unknowns.

The phase rule tells us that $f = 2 - 2 + 2 = 2$ so that, if P and T are fixed then the system is at equilibrium and is specified. This is true in terms of mole fractions in solution but, for molalities, $f_m = f + 1 = 3$ and $c_m = f_m - 2$. The term c_m is the number of compositional variables which must be specified. Its value is $3 - 2 = 1$ in accord with the analysis of the above system of equations, that is, in accord with a system of eight equations in nine unknowns. In the absence of a vapor phase, $f = 3$ and $f_m = 4$ with $c_m = 2$.

Solid-Solution Interactions

At this point I shall focus upon the effect of the formation of $Ca_xMg_{1-x}CO_3$ solid solutions, either in bulk phases or in surface coatings, upon the kinetic and the equilibrium behavior of carbonates. My thoughts have resulted in part from the extensive literature in this field, with special reference to the papers by Plummer and Mackenzie (1) and by Wollast and Reinhard-Derie (2).

Consider an aqueous solution containing Ca^{2+}, Mg^{2+} and Cl^- ions to which enough Na_2CO_3 is added to induce supersaturation of $CaCO_3$. It has been well established that, for $(Mg^{2+})/(Ca^{2+}) < 4$, magnesian calcites of formula $Ca_xMg_{1-x}CO_3$ precipitate while for ratios above four aragonite settles down (3, 4). Seawater falls in the latter category.

The reason for the aragonite precipitation is not its intrinsic stability. The solubilities of calcium carbonates increase from pure calcite to magnesian calcites containing up to about 10 mole percent $MgCO_3$, to aragonite, and then on to high magnesian calcites. Thus, aragonite is less stable than low-magnesian calcites (5). I suspect that aragonite comes down because, at high values of $[Mg^{2+}]$, the calcite nuclei are quite soluble due to their high magnesium contents and the formation of critical nuclei is improbable. This type of rate control of the polymorph formed is one aspect of typical solid-mixed electrolyte solution systems.

Let us focus upon the low $[Mg^{2+}]/[Ca^{2+}]$ case as it yields interesting solids and is relevant to many freshwater systems. A steady state is eventually reached when a solid is exposed to a solution and this state may be the result of rate factors or of thermodynamic ones. We shall see the thermodynamic case which occurs is not the conventional one alone.

It is known that calcites formed in the presence of Mg^{2+} ions turn out to be magnesian calcites with $0.70 \leq x < 1$ (1, 6). The calcites may be bulk precipitates as, for example, in marine cements or, in the case of seeded runs, may form coatings of a different composition from that of the bulk phase. Under special circumstances dolomite may result (6).

Different rates of precipitation can cause different amounts of Mg^{2+} to be incorporated into the solid. One may expect that a high rate, achieved by a larger initial addition of Na_2CO_3, should cause a larger uptake of $MgCO_3$ and an increase in the solubility of calcite. This would occur due to collisions of Mg^{2+} with the growing calcite crystals without as much of a chance for equilibration as in slower runs. This is indeed shown to be the case in Table I in which the runs were seeded with calcite so that we are observing the effect of coatings and to which different amounts of Na_2CO_3 were added. The principle is the same for bulk phases formed in unseeded runs, namely, that metastable solids can be formed and may persist for long times. The particular importance of surface coatings is that they, rather than the internal bulk phases, govern the interactions with aqueous solutions. The increase in the final pH results from an enhanced solubility of the magnesium calcites as the pH increases with a higher $[CO_3^{2-}]$ of the magnesium calcites. The increase solubility results from a higher magnesium content.

$$K'^{(Ca)}_{sp} = [Ca^{2+}][CO_3^{2-}] \quad (9)$$

$$K'^{(Mg)}_{sp} = [Mg^{2+}][CO_3^{2-}] \quad (10)$$

$$X_{CaCO_3} + K_{MgCO_3} = 1 \quad (11)$$

where the K_{sp} values pertain to the presence of a solid solution. These are three equations in four unknowns so that there is an additional degree of compositional freedom. In terms of the phase rule the system with the components $CaCO_3$-$MgCO_3$-CO_2-H_2O and with

Table I
Seeded runs at several initial supersaturations

(Ca^{2+}) (molal)	(Mg^{2+}) (molal)	Na_2CO_3 added (m moles/kg-SW)	Calcite added (g/kg-SW)	pH_f
0.01	0.01	1.1	0.82	7.79
0.01	0.01	2.3	0.82	7.83
0.01	0.01	3.5	0.82	7.90

one solid, one liquid, and one vapor phase has f = 3, f_m = 4, and c_m = 2. This agrees with what equations (1) through (11) tell us. The variables to be fixed may be, for example, a_H and $x = X_{CaCO_3}$. This tells us that, for each solid composition, there is only one equilibrium aqueous solution of a given composition $y = [Ca^{2+}]/\{[Ca^{2+}] + [Mg^{2+}]\}$. This is shown in Fig. 1 by $X_{f1}Y_{f1}$ and $X_{f2}Y_{f2}$. X_i is the initial supersaturated solution while X_{f1} and X_{f2} are the aqueous solutions in equilibrium with the solids Y_{f1} and Y_{f2}. We shall see next why the initial aqueous composition X_i is shown to yield two (actually many) equilibrium systems.

Conventional thermodynamics shows that for each X_{fi} there is one and only one Y_{fi} but does not tell us anything about the relationship between initial solids and aqueous solutions and the final equilibrium system. This is where the approach of Wollast and Reinhard-Derie (2), upon which I have elaborated somewhat, comes in.

These authors (2) presented their argument for dissolution. In the case of precipitation, which I treat, it starts with

$$D_i = P + D \tag{12}$$

where D_i is the amount of initial solutes and P and D are the amounts of the precipitate and of D_i later on. The ratio solid/solution implies the solid surface/total solutes ratio. Let x be the mole fraction of $CaCO_3$ in the solid and y be $m_{Ca}/(m_{Ca} + m_{Mg})$. Then, for calcium

$$y_i D_i = xP + yD \tag{13}$$

Through the manipulation of Eqns. (12) and (13) one arrives at

$$y = \frac{y_i D_i}{D} - \frac{D_i - D}{D} x \tag{14}$$

This is the condition for conservation of mass shown in Fig. 2. Note that I set $x = X_{Ca}$ while Wollast and Reinhard-Derie use the symbol $x = X_{Mg}$. The equilibrium condition is obtained in a straightforward manner from the ratio of $K_{sp}^{(Ca)} = \gamma_{\pm CaCO_3}^2 K_{sp}^{'(Ca)}/\lambda_{CaCO_3} X_{CaCO_3}$ and of a similar expression for $K_{sp}^{(Mg)}$.

This figure shows that the equilibrium values x_f, y_f depend upon the solid/water ratio. There is no thermodynamic inconsistency in this because, in terms of Fig. 1 all that Fig. 2 implies is that X_i for two solid/water ratios will yield two values of X_f, each with its thermodynamically corresponding Y_f (see Figure 1).

Thus, larger solid/water ratios such as are encountered in pore waters of sediments lead to smaller $MgCO_3$ contents in the equilibrium magnesian calcites although in either case the magnesium content of the solid increases. Wollast and Reinhard-Derie presented data to support the theory from the standpoint of dissolution and some of our results for the precipitation case

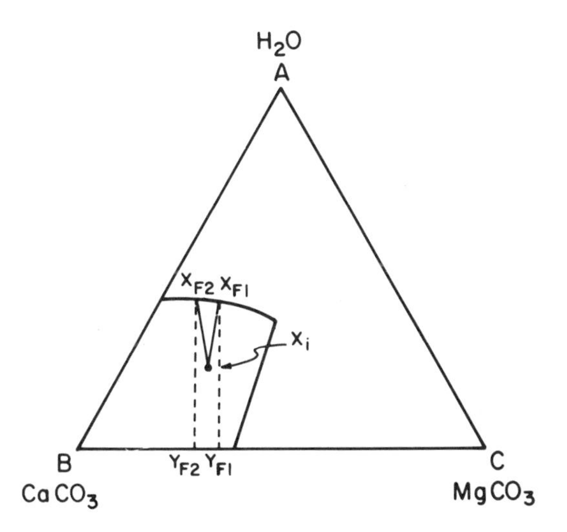

Figure 1. Triangular solid solution–aqueous solution equilibria

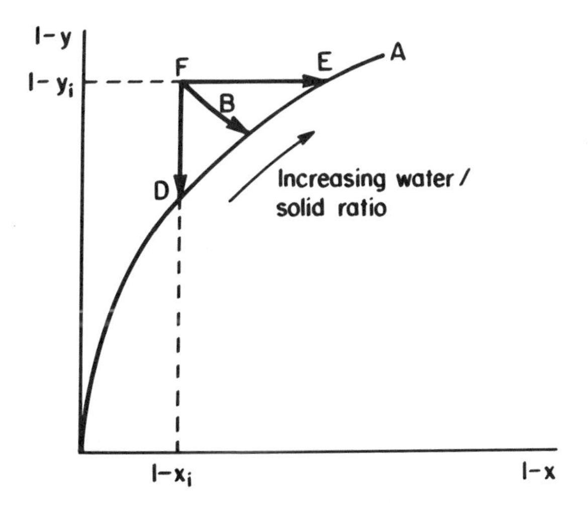

Figure 2. Equilibrium Curve A and conservation of mass Curve B. The equilibrium point D corresponds to a very large solid/water ratio and the reverse is true for E.

are shown in Table II. A kinetic control would lead to a faster precipitation and, consequently, to a higher $MgCO_3$ content and solubility for a larger addition of calcite as more surface for nucleation would be available.

Table II
The equilibrium pH in the multistate equilibrium as a function of the solid/water ratio (Cole and Pytkowicz, in preparation) for artificial seawaters with modified (Mg^{2+}). The last two columns for calcite refer to successive additions of solid.

(Ca^{2+}) (molal)	(Mg^{2+}) (molal)	Na_2CO_3 added (mol/kg SW)	Calcite added (g/kg SW)	Steady State pH	Calcite added	Steady state pH
0.01	0.01	3.27	0.62	----	0.62	7.88
0.01	0.01	1.61	0.20	7.95	0.20	7.89
0.01	0.01	1.31	0.07	8.01	0.07	8.01

We observe the opposite trend, expected from the multiphase hypothesis as a higher solid/solution ratio leads to a smaller mole fraction of $MgCO_3$ and, consequently, to a lower solubility (and pH) of the solution. In the first run the high Na_2CO_3 added is offset by the large amount of calcite added. In the dissolution process the magnesium content of the solid decreases and does so to a greater extent for smaller solid/solution ratios.

For Table III the solubility products of $CaCO_3$ and $MgCO_3$ were placed in the form

$$\lambda = \frac{\gamma_{\pm}^{2} K_{sp}'}{X\, K_{sp}} \qquad (15)$$

λ was calculated for known values of $\gamma_{\pm}$ and K_{sp}' in seawater and from assumed values of X. Then y_f was calculated from the equilibrium equation and compared to $y_f = 0.01/(0.01 + 0.05) = 0.167$ for seawater. It can be seen that the most stable calcite in seawater must contain roughly less than one mole percent $MgCO_3$.

The time behaviors for the various types of processes are shown in Figure 3. The three upper curves correspond to a kinetic control while the three lower ones result from thermodynamic control. It should be noted that at intermediate rates of precipitation the two types of mechanisms may act at the same time in terms of the difference between the ion product and K_{sp}'.

The important conclusion is that complex controlling processes can occur in solubility phenomena in mixed electrolyte solutions. This is especially true of surface coatings formed kinetically or by multistate thermodynamics and which prevent the aqueous solution from interaction with internal bulk phases. One should remember of course that, when the degree of supersaturation is large enough for bulk precipitation to occur, the kinetic and multiphase thermodynamic processes studied above will apply to the actual bulk phases.

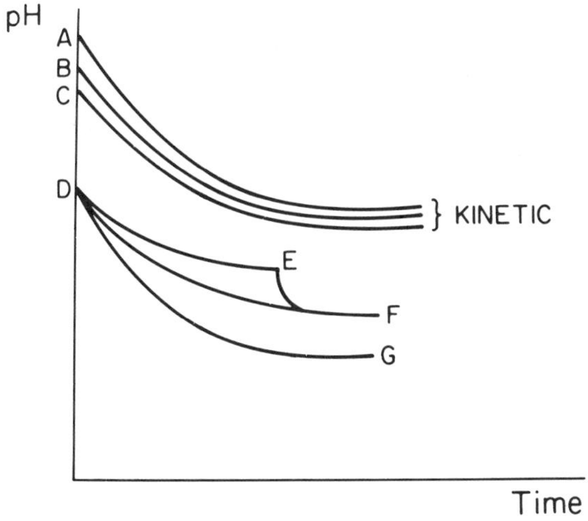

Figure 3. Controls of precipitation (idealized curves) A, B, C result from the addition of different amounts of Na_2CO_3. At D and E 1g of calcite was added per kg of seawater for Curve DEF, at D 2g were added for curve DF, and 3g for DG.

Table III
Calcite mole fraction X, solid state activity coefficients λ, and Y_F, the solute mole fractions of calcium at equilibrium in seawater.

X	λ_{CaCO_3}	λ_{MgCO_3}	y_f
0.99	1.13	166	0.174
0.98	1.14	83.0	0.178
0.97	1.15	55.3	0.179
0.88	1.27	13.8	0.218

Ion Pair-Complex Competition

The concepts of ionic media, free and total activity coefficients, properties of equilibrium constants, and effective ionic strength to be used here were examined in an earlier paper in this volume.

This section is based upon the papers of Johnson and Pytkowicz (7) and of Sipos *et al*. (8). The concepts used are the formation of coulombic ion-pairs between the major ions of seawater ($NaCl^o$, $NaHCO_3{}^o$, $NaCO_3{}^-$, $NaSO_4{}^-$, and similar pairs for Ca^{2+} and Mg^{2+}) and the formation of true complexes such as $PbCl^+$, $PbCl_2{}^o$, $CdCl^+$, $PbOH^+$, $PbCO_3{}^o$, etc. The coexistence of these two types of entities implies competition, e.g., for Cl^- ions, one has $NaCl^o$ and $PbCl^+$, and yields trace metal speciation quite different from those obtained in the absence of ion pairs. In Table IV are shown the fractions of the major ions which are free and ion-paired.

Table IV
Seawater Speciation (7).

	% Free	M-Cl	$M-SO_4$	$M-HCO_3$	$M-CO_3$
Na^+	84.0	12.2	3.8	0.0	0.0
K^+	78.5	17.2	4.3	0.0	0.0
Mg^{2+}	50.9	39.1	9.7	0.2	0.0
Ca^{2+}	45.8	43.8	10.0	0.3	0.1
etc.					

$$a = \gamma_T(T) = \gamma_F(F)$$

$$100\% \text{ change in } CO_3{}^{2-} \rightarrow 4\% \text{ change in } Ca^{2+}$$

$$(\gamma_{CO_3})_T = (\gamma_{CO_3})_F/[1 + K^*_{NaCO_3}(Na^+)_F + \ldots]$$

$$K_{sp}' = K_{sp}/(\gamma_{Ca})_T(\gamma_{CO_3})_T$$

$$K^* = K^*(t)\,(\gamma_{Ca})_F\,(\gamma_{CO_3})_F/CaCO_3{}^o$$

The solution of the competition problem consists essentially in solving a system of equations containing the stoichiometric stability constants for the ion-pairs and the complexes plus expressions for the conservation of mass of the type

$$[Na^+]_T = [Na^+]_F + [NaCl^o] + [NaHCO_3^o] + [NaCO_3^-] + [NaSO_4^-] \quad (16)$$

and

$$[Pb^{2+}]_T = [Pb^{2+}]_F + [PbCl^+] + [PbCl_2^o] + [PbOH^+] + + [PbOH_2^o] + [PbCO_3^o] \quad (17)$$

Originally, the stoichiometric stability constants β for the lead and the cadmium complexes with chloride had been determined in $NaCl$-$NaClO_4$ solutions and it had been assumed that the NaCl was completely dissociated. The nominal ionic strength was one molal. The constants were later corrected by replacing the actual free chlorides for the total chlorides in the calculation of

$$\beta_1^{(c)} = \frac{(MCl^-)}{(M^{2+})_F\ (Cl^-)_F} = \beta_1 \frac{(Cl^-)_T}{(Cl^-)_F} \quad (18)$$

and

$$\beta_2^{(c)} = \frac{(MCl_2^o)}{(M^{2+})_F\ (Cl^-)_F^2} \quad (19)$$

where (c) indicates the corrected values. This was done at the effective ionic strength of the seawater of interest which was roughly 0.7.

Uncorrected and corrected results for the trace metal speciation of lead are presented in Table V. It can be seen that the competition between true complexes and coulombic ion-pairs modifies considerably the speciation of lead. The fractions do not add up to 100% because species such as $PbCl_2^o$, $PbCl_3^-$, etc., were not entered into the table.

A glossary of symbols can be found in Table VI.

Table V
Lead speciation in seawater corrected and uncorrected for Cl^- ion-pairs.

	Uncorrected	Corrected
$[Pb^{2+}]$	1.94%	0.57%
$[PbCl^+]$	4.48%	1.57%
$[PbOH^+]$	61.55%	35.89%
$[PbCO_3^o]$	22.53%	44.33%

Table VI
Symbols.

c	number of components
D	amount of solute during the course of an experiment
D_i	amount of initial solutes
f	number of degrees of freedom
$K'_{sp}(Ca)$	stoichiometric solubility product of $CaCO_3$
$K'_{sp}(Mg)$	stoichiometric solubility product of $MgCO_3$
m	molality
$[Na^+]_F$	free ion concentration of Na^+
$[Na^+]_T$	total concentration of Na^+
P	amount of solid during the course of an experiment; also the pressure
p	number of phases
pCO_2	partial pressure of CO_2
pH_f	final pH
T	temperature
X	mole fraction
x	mole fraction of component such as $CaCO_3$ in the solid
y	solute mole fraction of the element in the aqueous solution
$\gamma_\pm$	mean molal activity coefficient

Literature Cited

1. Plummer, N.L.; Mackenzie, F.T., Am. J. Sci., 274, 61-83, 1974.
2. Wollast, R.; Reinhard-Derie, D., "The Fate of Fossil Fuel CO_2 in the Ocean", N. Andersen and A. Malahoff, eds., pp. 479-493, Plenum, New York, 1977.
3. Kitano, Y.; Hood, D.W., J. Oceanogr. Soc. Japan, 18, 141-145, 1962.
4. Möller, P.; Rajagopalan, G., Z. Phys. Chem., Neve Folge, 94, 297-314, 1975.
5. Chave, K.E.; Deffeyes, K.S.; Weyl, P.K.; Garrels, R.M.; Thompson, M.E., Science, 137, 33-34, 1962.
6. Berner, R.A., "Principles of Chemical Sedimentology", Mc-Graw Hill, New Jersey, 1971.
7. Johnson, K.; Pytkowicz, R.M., Am. J. Sci., 278, 1428-1447, 1978.
8. Sipos, L.; Raspos, B.; Nürnberg, H.W.; Pytkowicz, R.M., Mar. Chem., In press.

RECEIVED February 14, 1980.

34

Thermodynamics of High-Temperature Aqueous Systems

What the Electricity Generating Industry Needs to Know

D. J. TURNER

Central Electricity Research Laboratories, Kelvin Avenue, Leatherhead, England

In the conversion of fossil and nuclear energy to electricity, the value of high temperature solution phase thermodynamics in improving plant reliability has been far less obvious than that of classical thermodynamics in predicting Carnot cycle efficiency. Experimental studies under conditions appropriate to modern boiler plant are difficult and with little pressure from designers for such studies this area of thermodynamic study has been seriously neglected until the last decade or two.

A recent editorial in the journal "Corrosion" (1) referred to the lack of thermodynamic data in high temperature water as "appalling". In the author's opinion this is no exaggeration, since it is unlikely that methods of treating boiler water or of predicting the long term corrosion behaviour of boiler plant will ever be much better than empirical until a much better understanding of the solution phase chemistry is available. In the present circumstances it is hardly surprising that laboratory corrosion tests have frequently provided an inadequate basis for designing more reliable steam generators (2).

Water of various degrees of purity is the normal heat transfer fluid employed and a number of important problems with modern boiler water circuits are markedly influenced by solution composition. Most problems arise where solutions can concentrate and the compositions of such solutions can only be obtained by calculation from thermodynamic data. This paper concentrates on the kind of aqueous phase data which are currently most needed. Many of the needs overlap with those of geochemical interest. However, since Barnes (3) has recently reviewed the latter field, specifically geochemical needs will not be discussed. "High temperature" in this paper is generally taken to mean within about 100°C of the critical point of water (374°C), though some important problems which occur at lower temperatures are also considered.

The solvent properties of water change considerably between 25°C and the critical point with the result that qualitative conclusions based on room temperature experience can be totally

0-8412-0569-8/80/47-133-653$06.75/0

misleading when applied to boiler conditions. Though there are at present too few reliable high temperature data to allow many quantitative predictions, current knowledge is sufficient to allow quite important qualitative conclusions to be drawn.

An attempt is made here to define the main types of data needed. Available data and estimation procedures are considered as are some important experimental and more fundamental problems which complicate certain types of study. Finally, some examples are given of attempts to apply thermodynamic arguments to a variety of power station problems. First, however, an indication is given of circuit conditions and the more economically important problems in which high temperature aqueous solutions play a significant role.

1.Power Station Water Circuits

Fig. 1 represents a not quite typical high pressure boiler water circuit operating with the water in the drum at about 350°C and at a pressure of about 165 bar. Most modern plant (except for water reactors which operate nearer 300°C) employs roughly these conditions though with supercritical units (pressure above the critical pressure) the fluid can be considered liquid-like well above the critical temperature. Since there is no phase change in a supercritical unit, no drum is provided and the boiler is a "once-through" unit. Sub-critical once-through boilers are also used (for example in Advanced Gas Reactors, AGR) and here the water is simply allowed to evaporate to dryness in the boiler tubes. To drive one 500 MW turbine typically requires boiling 1.5 million kg of water an hour; the boiler capacity would be about 0.5 million kg.

The heat required can be supplied by burning fossil fuel, by gas heated in a reactor core (e.g. AGR and Magnox), by liquid metal (e.g. fast breeder reactors) or by water heated in a reactor (Pressurized Water Reactor, PWR). The steam generator of a PWR is constructed quite differently from the other types in that the heating fluid (primary circuit water) rather than the boiling fluid is inside the tubing. In a Boiling Water Reactor (BWR), the boiler itself is in the reactor core with the fuel cans inside the boiler tubes. In all units except the water reactors, the steam is considerably superheated before passing to the turbine and condenser which it will leave as liquid water under a vacuum at typically 30°C and 40 mbar.

Few high pressure drum boilers are equipped with the condensate polishing plant illustrated in Fig. 1 but once-through boilers (having no drum to allow the accumulation of impurities in the water phase) normally are. The water is pumped from the condenser back through feed-heaters and into the boiler again.

The materials of construction vary considerably with different boiler designs. Conventional drum boilers generally use carbon steel boiler tubes and stainless steel superheater tubes.

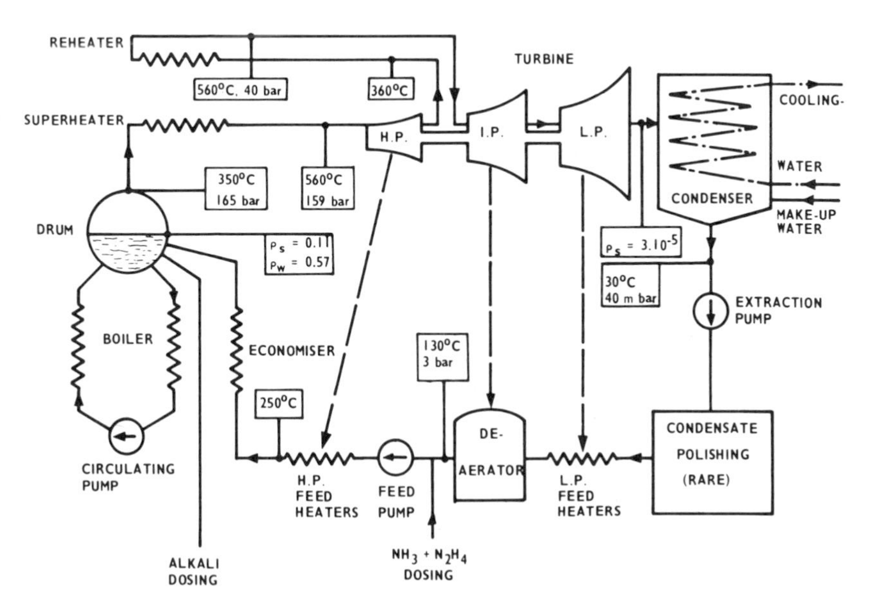

Figure 1. Simplified water circuit for a high-pressure boiler ((——) main circuit; (— —) bled steam lines; (— · —) cooling water)

Nuclear stations tend to employ ferritic chrome steels, stainless steels and nickel based alloys in their boilers. Turbines contain a variety of steels while condensers are usually of brass or, increasingly, of titanium. Low pressure feed heaters have traditionally been made of brass also, but increasingly steels are used.

Chemical treatment is primarily aimed at minimizing the corrosion rate of the circuit construction materials. The exact methods chosen will depend on the design of the circuit as well as on the nature of these materials. The usual approach is to maintain low oxygen levels and keep the water slightly alkaline. In a drum boiler a variety of solid alkalis can be used but the most common ones are dilute caustic soda or mixtures of Na_3PO_4 and Na_2HPO_4. With once-through boilers ammonia is normally used instead and it is sometimes also used in drum boilers and PWR secondary circuits. Hydrazine is frequently dosed into the feed train to control O_2 levels. However, an alternative to the low oxygen alkaline solution regime is to rely on the protective haematite film which is formed in very high purity water in the presence of controlled quantities of oxygen. This approach is favoured in some once-through plant (particularly in Germany) following the work of Freier (4,5) and a similar chemistry is used in most BWR plant where the continuous radiolytic production of oxygen makes it impractical to maintain low oxygen levels. A somewhat more detailed summary of methods of boiler-water treatment, including the application of chelating agents on-load, has been given recently (6).

The most serious current problems where knowledge of solution compositions are required are of three main types: boiler integrity, turbine integrity and out-of-core radioactivity. The financial costs of the first two problems arise mainly when failures in modern, efficient plant require their replacement by plant which is considerably more expensive to run. The third problem manifests itself in the need to share any extensive maintenance work in highly radioactive areas among hundreds of men so that none exceeds his permitted radiation dose.

Plate 1 illustrates what can be the consequences of a stress corrosion failure in a turbine and Plate 2 the consequences of two forms of boiler tube corrosion: tube thinning and hydrogen embrittlement. The economic consequences of such problems and certain other areas where information on high temperature solutions is needed have been discussed elsewhere (6).

Illustrations of what can and cannot be done (on the basis of currently available thermodynamic data) towards understanding and solving a variety of water circuit problems are briefly discussed in Section 3.

2. Thermodynamic Considerations

Types of Equilibria. In all the problems discussed it is the

Plate 1. Consequences of turbine disc failure (Hinkley)

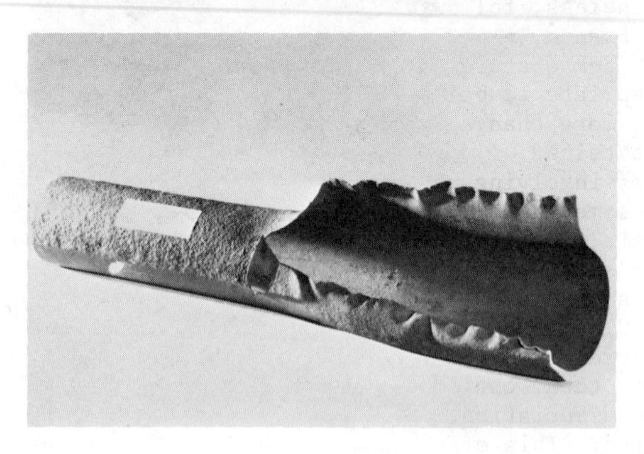

Plate 2. Corrosion damage of boiler tubing

free energy properties which are of concern, though enthalpies, entropies and (molar) volumes may, of course, be used to obtain them. With strong electrolytes, concern is primarily with how free energies change with concentration and it will be seen that knowledge in this field is far more advanced and in most respects less troublesome than with standard free energies of reaction where much larger uncertainties can be introduced. Accordingly most attention will be devoted to free energies and equilibrium constants. Clearly hydrolysis constants, solubilities, steam-water partition coefficients and chelation equilibria are involved in the problems described as are redox equilibria and the formation of ordinary (monodentate) complexes and ion pairs.

Difficulties at High Temperature. A number of difficulties combine to explain the singular lack of certain kinds of thermodynamic data in high temperature solution.

Experimentally these start with the need to contain such a good solvent and corrosive liquid as water simultaneously at high temperature and pressure. Few metals (for containment), insulating materials (for electro-chemical measurements) or windows (for optical measurements) are inert in more than a small range of solutions. Furthermore many reactions which can be ignored for kinetic reasons at 25°C are likely to proceed fast at 300°C. Thus it is doubtful whether one could study the chemistry of FeIII in the presence of hydrogen or of FeII using a standard perchlorate supporting electrolyte.

Nevertheless, following the pioneering work of Prof. E.U. Franck, experimental data of various kinds have been obtained at temperatures and pressures considerably higher than those appropriate to boiler water. Without this lead it is doubtful whether more than a fraction of the data now available would have been obtained.

Systems involving aqueous transition metal ions are inevitably somewhat complicated because of their relatively high charge, their tendency to form coordination complexes and their tendency to exist in more than one oxidation state. At high temperatures, the dielectric constant of water is very much lower than at 25°C so that the ability of water to stabilize highly charged species is greatly reduced and reactions which lead to a reduction in total charge are correspondingly favoured. These include ion association, complex formation with anions, hydrolysis and reduction. This effect makes it difficult or impossible to obtain meaningful reaction free energies from techniques which require pH buffers or supporting electrolytes.

A slightly different aspect of the same problem makes it exceptionally difficult to estimate certain types of equilibrium constant at high temperatures from data at low temperatures. The well known relationships between equilibrium constant, K, ΔG^{o}, ΔH^{o} and ΔS^{o} may conveniently be written

$$\ln K = -\frac{\Delta H^o}{RT} + \frac{\Delta S^o}{R}$$

which, if ΔH^o and ΔS^o are reasonably independent of temperature predicts a linear plot of ln K vs. 1/T. With gases and solids, ΔC_p^o is usually relatively small and such plots approximate well to straight lines over wide temperature ranges. Even when there is some curvature, ΔC_p^o can often be estimated within reasonably well defined limits.

Log K for any endothermic reaction (positive ΔH^o) which produces ions will inevitably go through a maximum as a function of reciprocal temperature because, above some temperature, the decreasing solvating power of water must make the reaction exothermic. The ionic dissociation of water is an example as seen in Fig. 2. An alternative way of saying this, of course, is that the reaction has a large negative ΔC_p^o. Estimation is difficult because, even where it is known at 25°C, there is no reason to believe that ΔC_p^o is independent of temperature. $\bar{C}_p^o$ for NaCl decreases from -92 J K^{-1} mol^{-1} at 25°C to -836 J K^{-1} mol^{-1} at 300°C (7), and Criss (8) has suggested why such large negative values could be expected at high temperature.

An even more serious problem can arise when dissolved species expected to predominate at high temperatures are undetectable at 25°C or are only present at concentrations which are too low for them to be adequately characterized thermodynamically. Examples are certain transition metal chloro-complexes (9,10) and mixed complexes of such metals with hydroxide and another ligand (11,12). Thus it seems that chloride complexing so alters the aqueous chemistry of copper and gold that supposedly inert gold components in autoclaves are reversibly oxidized by CuII (10) and it is likely that mixed oxine and hydroxy complexes of FeII contribute considerably to the gross under-estimation (by a factor of up to 10^8) of magnetite solubility in oxine (12,14).

These sort of problems make it difficult to obtain reliable high temperature data on the aqueous chemistry of transition metal ions. Unfortunately the necessary timescales for even the simpler experimental studies are frequently too long for a Ph.D. student to make reasonable progress in 3 years from scratch or for industrial researchers to make much reportable progress before the patience of those supporting the work is exhausted. Results can be reported far more rapidly from, for example, corrosion experiments and since corrosion theories are in general of so little predictive value, each relevant alloy/electrolyte combination needs its own study. In such circumstances it is hardly surprising that thermodynamic studies have been (with a few notable exceptions) relatively poorly supported, while corrosion data continue to be amassed without any reliable thermodynamic framework within which to understand them.

Thermodynamic Data Available. Excellent reviews of available results up to 1973 were presented at a recent conference by most of the major contributors to the field (15-20). That of Franck (15) covers various properties of water and solutions up to 1000°C and 100 kbar. The self-dissociation constant of water, K_W is reported over this range but due to the large compressibility of water near the critical point, it is not easy to extrapolate K_W to the lower densities appropriate to SVP (saturated vapour pressure) conditions above 300°C.

Marshall's extensive review (16) concentrates mainly on conductance and solubility studies of simple (non-transition metal) electrolytes and the application of extended Debye-Huckel equations in describing the ionic strength dependence of equilibrium constants. The conductance studies covered conditions to 4 kbar and 800°C while the solubility studies were mostly at SVP up to 350°C. In the latter studies above 300°C deviations from Debye-Huckel behaviour were found. This is not surprising since the Debye-Huckel theory treats the solvent as incompressible and, as seen in Fig. 3, water rapidly becomes more compressible above 300°C. Until a theory which accounts for electrostriction in a compressible fluid becomes available, extrapolation to infinite dilution at temperatures much above 300°C must be considered untrustworthy. Since water becomes infinitely compressible at the critical point, the standard entropy of an ion becomes infinitely negative, so that the concept of a standard ionic free energy becomes meaningless.

The work described by Marshall (16), together with the vapour pressure studies on 1:1 and 1:2 electrolytes up to 300°C reported by Lindsay and Liu (17) and recent theoretical work by Silvester and Pitzer (21) and by Helgeson and Kirkham (22) provide a good understanding of the behaviour of simple electrolytes over wide ranges of temperature and concentration. However, as just seen, the behaviour under SVP conditions above 300°C becomes decreasingly well defined towards the critical point.

The review of Martynova (18) covers solubilities of a variety of salts and oxides up to 10 kbar and 700°C and also available steam-water distribution coefficients. That of Lietzke (19) reviews measurements of standard electrode potentials and ionic activity coefficients using Harned cells up to 175-200°C. The review of Mesmer, Sweeton, Hitch and Baes (20) covers a range of protolytic dissociation reactions up to 300°C at SVP. Apart from the work on Fe_3O_4 solubility by Sweeton and Baes (23), the only references to hydrolysis and complexing reactions by transition metals above 100°C were to aluminium hydrolysis (20) and nickel hydrolysis (24) both to 150°C. Nikolaeva (24) was one of several at the conference who discussed the problems arising when hydrolysis and complexing occur simultaneously. There appear to be no experimental studies of solution phase redox equilibria above 100°C.

In view of the findings of Lindsay and Liu (17) that at 300°C $MgCl_2$ behaves like a 1:1 electrolyte, $MgCl^+ + Cl^-$, it seems

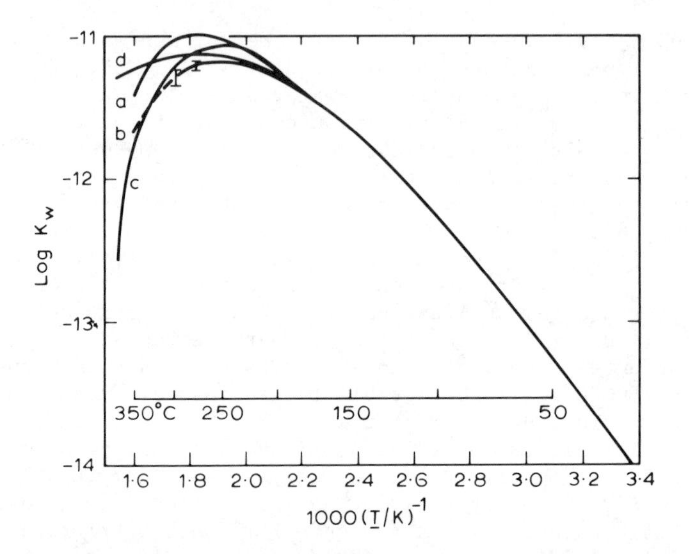

Figure 2. Temperature dependence of molal ion product of water: (a) Fisher and Barnes (36); (b) Sweeton et al. (35); (c) Sirota and Shviraev (37); (d) Correspondence Principle estimate of Lewis (91)

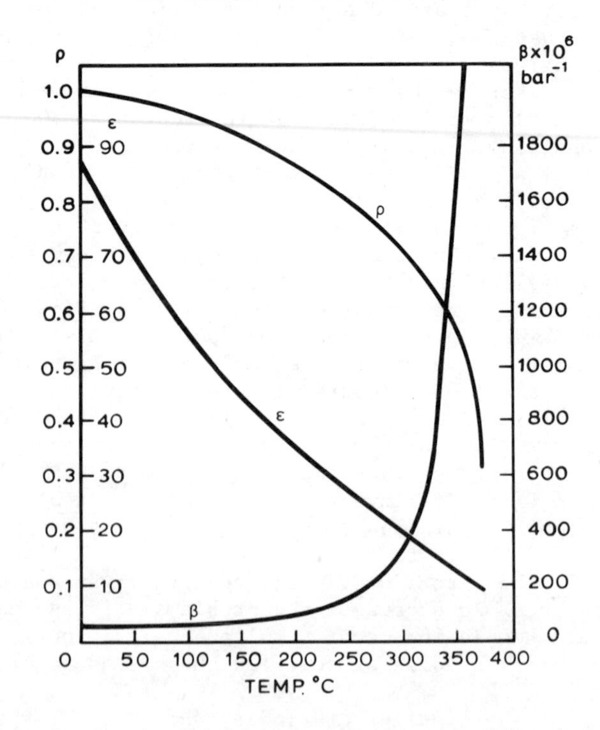

Figure 3. Density, dielectric constant, and compressibility of water

unlikely that the readily polarized Fe^{2+} ion remains uncomplexed at 300°C in Cl^--containing media. Sweeton and Baes (23) interpreted the solubility of Fe_3O_4 in HCl without invoking Cl^--complexing. Thus it seems unlikely that reliable thermodynamic data on the aqueous Fe^{2+} ion can be obtained from their data at 300°C without independent evidence concerning the extent of Cl^--complexing. The interpretation with regard to anionic species seems to be unambiguous, however.

Exactly the same problem arises with the recent studies of NiO solubility by Tremaine and Leblanc (25) and again the thermodynamic data on the aqueous anionic species at 300°C are likely to be more reliable than on the Ni^{2+} ion. There is good spectroscopic evidence for complex formation in chlorides of nickel (II), (26) cobalt (II) (27), and copper (II) (28) at 300°C and above. Most of the work was done at rather high Cl^- concentrations but qualitatively the effects of dielectric constant and concentration are as expected. A noteworthy feature (which estimation procedures will have to allow for) is the change from 6 to 4 coordination at the lower pressures (150-300 bar) and the higher Cl^- concentrations. This change appears to take place with only 2 or 3 Cl^- ions coordinated to the metal (at least in the case of Ni(II)).

Since 1973, progress has been made in all the fields reviewed and a more complete review of Franck's work has appeared (29). For the present purposes it is necessary to concentrate on the two areas which are least well studied: simple electrolytes between 300°C and the critical point and the chemistry of aqueous transition metal cations. A number of studies which do not fall into these categories must, however, be mentioned because of their direct relevance to boiler water chemistry. These are studies of sodium phosphate phase equilibria (30), ammonia dissociation (31) and isopiestic studies of calcium and magnesium chlorides (32). Two studies of the self-dissociation of D_2O have also appeared (33,34).

The K_W results of Sweeton, Mesmer and Baes (35) plotted in Fig. 2 were reported in 1974 and although they only extend to 300°C they may well be more accurate above this temperature than the experimental results of Fisher and Barnes (36), since, as mentioned, earlier, the Debye-Huckel theory may not give reliable extrapolations to infinite dilution at temperatures where water is highly compressible. While their work (35) involves extrapolation to infinite dilution as well as to higher temperatures it is very encouraging to note that their ΔC_p^o at 300°C (-960 J K^{-1} mol^{-1}) is of the magnitude expected on the basis of the NaCl studies referred to in Section 2. The conductance results of Sirota and Shviriaev (37) above 300°C also seem more consistent with the results of Sweeton, Mesmer and Baes (35), than with those of Fisher and Barnes (36). Marshall and Franck's recent representation of data up to 1000°C and 10,000 bars (38) predicts high temperature SVP results somewhat lower than those of Sirota and Shviriaev (37).

Marshall has extended his high temperature solubility studies (39,40,41) and has begun some work on liquid-vapour critical temperatures of solutions (42,43) which should prove valuable. Some of Marshall's higher temperature results (>300°C) have been questioned (44) and there do seem to be unexplained differences between studies in stainless steel and titanium vessels (45).

The problem of measuring the thermodynamic properties of aqueous transition metal ions above 100°C has also received some attention with studies on Fe^{3+} complexing with Cl^- (46), Br^- (47) and SO_4^{2-} (48) up to 150°C and the formation of anionic hydroxy complexes of Pb^{2+} up to 300°C (49).

Preliminary heats of solution of $CoCl_2$ and $CuCl_2$ have been measured up to 300°C by Cobble and Murray (50). Hydrolysis was suppressed by HCl addition so that when the work is completed and when the extent of Cl^- complexing (and Cu^{2+} reduction) can be allowed for the data will prove extremely valuable. Preliminary concentration cell studies on the Cl^- complexing of Cd^{2+} and Ni^{2+} up to 170°C (51) support the conclusions given earlier that such complexing with first row transition metal ions is likely to be significant by 300°C.

The solubility of Fe_3O_4 has been studied in the temperature range 500-600°C by Chou and Eugster (52) using the HCl fugacity buffer developed earlier (53). Under the conditions used both HCl and the soluble iron species $FeCl_2$ are completely associated. Clearly the derived thermodynamic data are also of potential value but more work on Cl^--complexing is needed before they can throw light on the aqueous chemistry of Fe^{2+} under high pressure boiler conditions.

Many of the 25°C oxidation potential estimates of Latimer (54) were obtained simply from a knowledge of what reactions proceed and what do not. Hence preparative and decomposition experiments in simple autoclaves are also of considerable value provided that full experimental details are published. Swaddle's group has performed a number of such studies on the transition metals from which boiler water circuits are made (55,56,57) and also on species of more direct relevance to laboratory studies (58,59,60). Quite trivial unexpected observations in autoclave studies can be used to place limits on equilibrium constants. In complex systems, unique interpretations will usually be impossible but the observations may still prove useful if they can be supplemented by estimated data (10, 61).

Estimation Procedures. There are basically two ways which have been developed to deal with the fact that heat capacity terms are large in reactions involving ions. One is based on empirical relationships (the entropy correspondence principle) between ionic entropies at different temperatures which Criss and Cobble (62) developed and checked to 200°C. Lewis (63) has checked a number of its predictions against available experimental evidence and has found the method reasonably satisfactory for several systems up to 250°C. He (64) and others, for example Macdonald and colleagues (65,66) have used the method to estimate equilibrium constants to 300°C and above.

It cannot be claimed, however, that the method has been extensively checked above 250°C and it appears that it inevitably must become unreliable over 300°C. The average heat capacities of ions calculated by Criss and Cobble (67) show no sign that $\bar{C}_p^o$ values of ions are becoming increasingly negative with temperature, although, as was seen in Section 2, the effect is becoming considerable by 300°C. Nevertheless many equilibrium constants could probably be obtained to within an order of magnitude at 300°C if a reliable estimate could be made of the thermodynamic properties of any uncharged species or ion pairs. Unfortunately, there is, as yet, no reliable method of characterising such species if (as will frequently be the case) they are only stable at high temperatures. With the self-dissociation of water this is not, of course, a problem and, as can be seen in Fig. 2, for K_W the entropy correspondence method only begins to manifest its underestimation of the magnitude of ΔC_p^o above 300°C.

It is not a problem either for the protonation constant of S^{2-} (i.e. the reciprocal of the second dissociation constant of H_2S) some estimates of which are shown in Fig. 4. Neither Cobble's estimate (68), using the correspondence principle (curve a) nor Pohl's (69) extrapolation (curve b) using an empirical equation due to Harned and Embree (70) is showing any indication of the expected minimum in K. The extrapolation used by Khodakovskii et al (71) (curve c) is based on the more frequently used expression of Harned and Robinson (72) and a different selection of low temperature data. While their result looks more reasonable it is difficult to have much confidence in any of the results even up to 200°C. The apparent failure of the correspondence principle may arise as much from the choice of low temperature data as a failure of the relationship itself.

The disagreement between the calculated standard free energies of formation of aqueous Fe^{2+} and those deduced by Sweeton and Baes (23) has been commented on by the author (9) and by Tremaine, Van Massow and Shierman (73). In view of the problem at 300°C with Cl^--complexing (discussed earlier) it seems unlikely to the author that the thermodynamics of dissolution of magnetite in acid solution are quite as well characterized as is suggested by the calculations of Tremaine et al (73).

The second method of estimation which has so far been developed is based on consideration of those ΔC_p^o values which were available in 1967 when Helgeson developed it (74). This method essentially separates electrostatic and non-electrostatic contributions to ΔS^o and ΔC_p^o and Helgeson has compared the predictions of a number of different assumptions concerning ΔC_p^o with published high temperature equilibrium constants. He concluded that the best one to make is that ΔC_p^o is proportional, at each temperature, to the electrostatic contribution, $\Delta C_{p,e}^o$. This

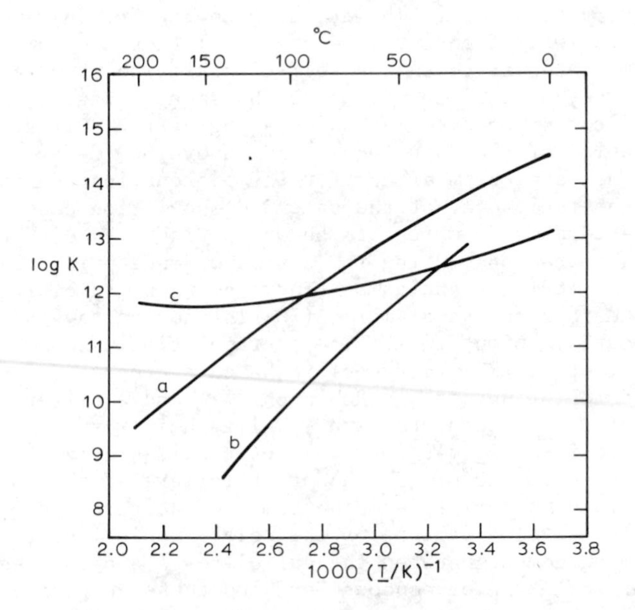

Figure 4. Protonation constant of S^{2-}: (a) Cobble (68); (b) Pohl (69); Khodakovskii, Zhogina, and Ryzhenko (71)

assumption, combined with a simple electrostatic model, leads to an explicit relationship between log K and temperature which includes in it the temperature dependence of the dielectric constant of water:

$$\ln K = \frac{\Delta S^{o}(298)}{RT}\{298 - \frac{\theta}{\omega}(1 - \exp\left[\exp(b + aT) - C + (T - 298)/\theta\right])\} - \frac{\Delta H^{o}(298)}{RT} \quad \ldots (1)$$

Here C(= exp (b + 298a)), ω(= 1 + aCθ), b and θ are constants in an expression for the temperature dependence of ε, the dielectric constant of water:

$$\varepsilon = \varepsilon_{o} \exp\left[- \exp(b + aT) - T/\theta\right] \quad \ldots (2)$$

Equation (1) was a simplification derived from the following more complete, but less readily usable equation:

$$\ln K = \frac{S_{e}^{o}(298)}{RT}\{298 - T + \frac{\theta}{\omega}(1 - \exp\left[\exp(b + aT) - C + (T - 298)/\theta\right])\}$$

$$- \frac{\Delta H^{o}(298)}{RT} + \frac{\Delta S^{o}(298)}{R} + \frac{\alpha}{R}\left[\ln(T/298) - 1 + 298/T\right] + \frac{\beta(T - 298)^{2}}{2RT} \quad \ldots (3)$$

The first term represents how the electrostatic contributions differ from those at 25°C, ΔS_{e}^{o} (298) being the electrostatic contribution to the reaction's standard entropy at 25°C. The last two terms derive from the assumption that the non-electrostatic part of ΔC_{p}^{o}, $\Delta C_{p,n}^{o}$, can be represented by

$$\Delta C_{p,n}^{o} = \alpha + \beta T \quad \ldots (4)$$

Following Helgeson (74) a term λT^{2} (in (4)) is ignored. On his model the contribution of each ion to ΔS_{e}^{o} is given by

$$\bar{S}_{e}^{o} = -A\left[\exp\{\exp(b + aT)\} + T/\theta\right]\left[1/\theta + a\exp(b + aT)\right]/\varepsilon_{o} \quad \ldots (5)$$

where $A = (Ze)^{2}N/2r$, Ze is the charge on the ion, r its radius and N, Avogadro's number. ε_{o} is defined in equation (2) and takes the value 305.7. The electrostatic model is crude and the choice of r to be employed is somewhat arbitrary, but Helgeson's model to a certain extent allows for this by taking up uncertainties in the electrostatic contribution in α and β. This was a quite intentional feature of the model because it is believed that much of the unreliability of hydration models arises from non-electrostatic contributions. Helgeson used equation (3) to curve fit all sufficiently reliable experimental data and from this obtained best fit values of α, β and ΔS^o_e for a number of equilibria.

The author has recently attempted to use this method to estimate the equilibrium constant, K_h, of the reaction

$$3I_2 + 3H_2O \quad = \quad 5I^- + IO_3 + 6H^+$$

up to 300°C from experimental data at 25°C and 60°C (75). The reaction is, of course, a severe test as it produces 12 ions from 3 molecules of neutral solute. Equation (1) is totally unsatisfactory since it fails to predict the expected maximum in the equilibrium constant as seen in Fig. 5. Thus an attempt was made to use a calculated ΔS^o_e (via equation (5)) and see if a better estimate was possible using what Helgeson's results (74) suggested were reasonable values of α and β. Two assumptions were tried, one was that $\Delta C^o_{p,n}$ is independent of temperature - i.e. $\beta = 0$ - and the other made use of the observation that in Helgeson's Table 2 there is an approximate relationship between α and β, $\alpha = -313$ (±48). These estimates are also shown in Fig. 5 and it is clear from the divergence of results above 100°C that the method is too sensitive to the values of α and β to be of use at least in this case.

The author believes (75) the correspondence principle method (as used by Lewis (64) based on $\bar{C}^o_p$ $(I_2) = 65$, although uncertain due to lack of appropriate data on I_2, provides the best estimate. Almost certainly free energy approaches like Helgeson's can be improved by better ionic hydration models. To this end a number have been qualitatively compared (76) and checked against experimental data on NaCl (77,78). More extensive calculations based on one (fixed hydration) model have also been presented (79) and found to predict ionic free energies better than the correspondence principle between 150 and 275°C. At higher temperatures, however, the model is less satisfactory. Much more work is needed in this area since, if such methods are to prove reliable in the difficult region between 300°C and the critical point, the hydration models must be as free as possible from empirical fitting parameters.

3. Applications

There follow some examples of attempts to apply thermodynamic arguments to a number of plant problems. Attention is directed as much to what can and cannot be done with currently available data as to the practical significance of the results. Areas where work is particularly needed are stressed.

Generation of Corrosive Environments. The materials of construction of a water circuit are, of course, selected to be

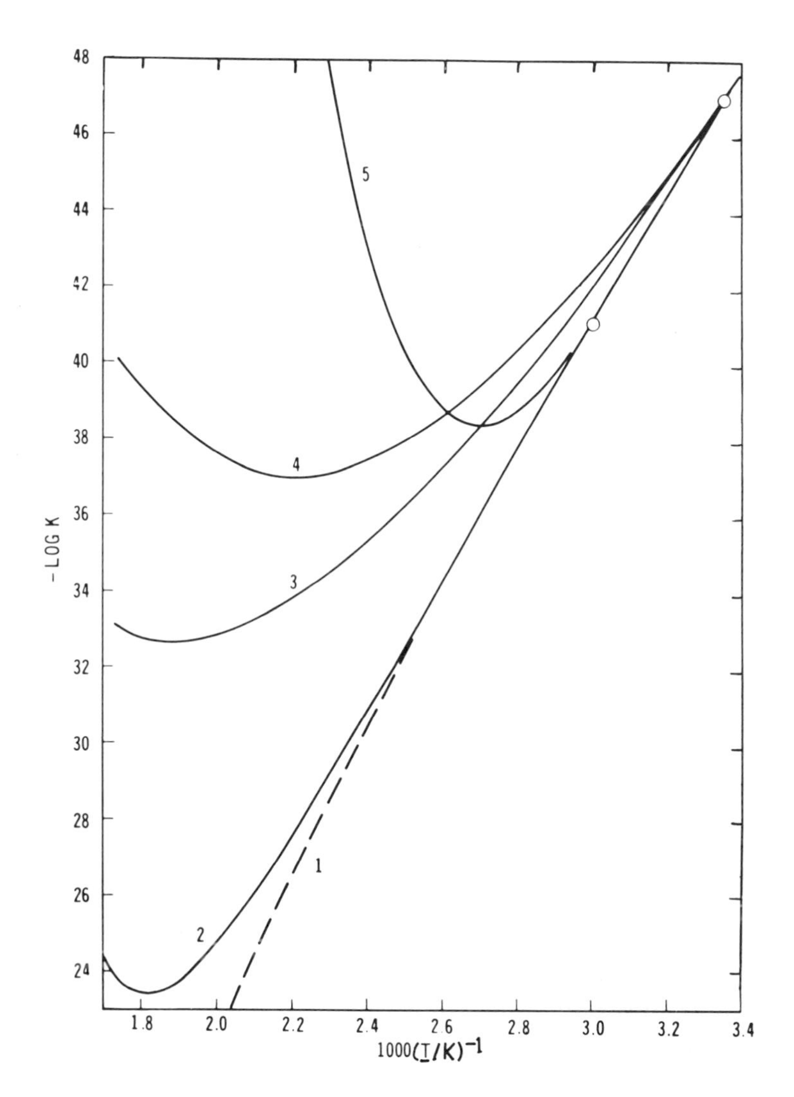

Figure 5. Predictions of K_h: (1) Helgeson Equation 1; (2) Equation 3, $\beta = 0$; (3) entropy correspondence $\overline{C}_p°(I_2(aq)) = 0$; (4) entropy correspondence $\overline{C}_p°(I_2(aq)) = 65$; (5) Equation 3, $\alpha = -313\beta$

corrosion resistant in whatever solutions they are expected to see. Most serious problems arise because impurities in or additives to the water are able to concentrate, sometimes by many orders of magnitude. This can occur where water is boiling on thin porous oxide layers under high heat fluxes and in regions of turbines where the steam is nominally dry but of such a pressure that concentrated solutions could be in equilibrium with it. Simulations of the former occurrence have shown concentration factors of 10^5 to be possible while in theory the latter situations could result in even higher concentrations. In PWR steam generators, electrolytes can concentrate between the tubes and their support plates. The accelerated production of corrosion product leads to "denting" by crushing of the supported tubes.

If the impurity leaking (via the condensers) to a boiler is sea-water, the chemistry is fairly simple and it is easy to predict roughly under what conditions the concentrating liquid will go acid due to Mg^{2+} hydrolysis (80,81). However, without more reliable data on this hydrolysis reaction the best that can be hoped for in estimating the pH is ±0.5 pH unit at 300°C (80). The much more complex situation which arises when the condenser leak allows in river or lake water can be dealt with formally (82,83) but the uncertainties in the data are usually too large to yield reliable pH estimates.

We have been able, however, on occasions to use a very simple model to help understand specific plant problems where river water analyses were available and on one occasion to show that at different times the boiler water had (as corrosion evidence suggested) alternated between acidic and alkaline conditions. The model assumes that by 350°C any normally dissociated multi-charged ions will be sufficiently unstable that they will undergo whatever appropriate hydrolysis reactions can reduce their charge to unity. Whether the water goes acid or alkaline then simply depends on whether the total (equivalent) concentration of multiply charged cations exceeds or is smaller than the concentration of multiply charged anions.

When a 60 MW turbine at Hinkley A power station disintegrated in 1969 from stress corrosion cracking of a low pressure turbine disc (consequences shown in Plate 1) it was considered that NaOH solutions were most probably involved (84) and it was soon found that if NaOH were the sole electrolyte present its maximum concentration (based on vapour pressure depression) was sufficient to have caused the cracking. However, it was also found that in mixtures it was only the free NaOH which led to rapid stress corrosion cracking. Considerations of acid gas solubility and solution thermodynamics showed that at the CO_2 and acetate levels present it was most unlikely that free NaOH was present in sufficient quantity to be responsible for the Hinkley failure (85).

Ammonium acetate and ammonium carbonate had also been found to induce stress corrosion cracking of the appropriate steels but similar thermodynamic arguments showed that neither electrolyte could in practice approach the levels required to cause cracking (86,85). Subsequent attempts to improve the estimates (87) confirmed these findings. The effect of acid gases like CO_2 and acetic acid on reducing the free NaOH concentrations in turbines makes one wonder if it may be possible to purify boiler water too much for the good of the turbine. If this is so, then some of the observations recently made by Bussart, Curran and Gould (88) on the effect of water chemistry on modern large turbines may not, after all, be "paradoxical".

It was eventually concluded that the most likely chemical culprit in the Hinkley turbines was molybdate which can be formed from the MoS_2 lubricant under the conditions used during turbine assembly (89) or leached from the steel under stagnant conditions in sufficient quantity to induce stress corrosion (90).

Understanding Corrosion Processes. The application of Pourbaix diagrams to corrosion problems is well known and will not be considered here. Much effort has gone into producing such diagrams for high temperature use. A recent paper (91) lists 15 references to the subject. The diagrams are particularly useful in interpreting corrosion or electrochemical studies conducted at controlled potential. However, with few exceptions (92) little attention has been given to the role of solution phase additives and impurities in influencing the composition of the corrosion film, although quite subtle compositional differences across a corrosion film have been discussed in terms of redox potential (93,94). Since all chemical reactions will be much faster above 300°C than at 25°C it seems likely that redox buffering by solution components should be more predictable thermodynamically (once the data become available) at the higher temperatures.

A feature of corrosion studies which has been stressed recently (2) is the complete failure of laboratory tests on their own to predict how reliable operation of some nuclear steam generators can be maintained. At least a part of this problem is likely to arise from different redox and/or pH conditions imposed by the solution in autoclave tests and in plant conditions and many low level contaminants could be involved. In view of what has been said earlier concerning the role of Mo(VI) in stagnant water it is clear that some data, at least on the thermodynamics of aqueous Mo species, should be sought at high temperatures. Some molybdate appears to be able to enter solution through the vapour at 250°C (61), so the contamination problem is not necessarily solved by the use of liners. Presumably other species capable of influencing redox potentials and pH can also contaminate solutions through the vapour.

It seems to the author that until some means is available for estimating the pH and redox potential of solutions both in autoclave studies and under specific local plant conditions there will always be doubt about the predictive value of many corrosion studies carried out in autoclaves.

Out of Core Radioactivity. In water reactors the main source of out of core activity arises from the corrosion products of the out of core circuit which are transported into the core and, after neutron irradiation, subsequently transported back. Since the core of a PWR will inevitably be hotter than the steam generator, there has been interest, particularly at Atomic Energy of Canada Ltd. (A.E.C.L.), in the effect of temperature on the solubility of the various corrosion products present and how this influences their transportation round the circuit (95). The solubilities could, even with perfect data, only be tied down to a range of levels for most elements because of the presence of radiolytically produced H_2 and O_2 at levels which are not at equilibrium. The system is further complicated by the presence of mixed spinels as well as pure oxides amongst the corrosion products. Despite these complications, however, a combination of detailed sampling and thermodynamic rationalization (based mainly on estimated data (65, 66) is resulting in a greatly improved understanding of the processes involved (96). There is little doubt that the experimental programme being pursued by A.E.C.L. will lead to better understanding of the behaviour of corrosion products in all types of plant.

Thermodynamic arguments have also been used in support of work on decontaminating the circuits of BWRs (11). It was shown that conventional citric acid cleaning solutions could not dissolve either Fe_2O_3 or important Co-containing spinels unless quite strong reducing agents were present and it was also shown that the risk of electrodeposition ^{60}Co on steel surfaces during the decontamination is greatly reduced under strongly reducing conditions. There was some evidence from results on decontamination of the Winfrith SGHWR (Steam Generating Heavy Water Reactor) that ^{60}Co may have been electrodeposited either during early decontaminations or on load; the latter seems possible thermodynamically though unlikely kinetically under normal (radiation-free) conditions. The high temperature thermodynamic data were, however, considered insufficiently reliable to be certain of the significance of some of the plant observations. It is, however, clear that electrodeposition of ^{60}Co on load (as well as during decontamination) is unlikely under strongly reducing conditions such as those which are nominally maintained in PWR primary circuits.

The release of radioactive iodines from BWR circuits, first into the steam phase and then into the turbine hall, has also been considered thermodynamically (75). A re-analysis of some experimental data of Styrikovich et al (97), suggested that iodates were not, as had been tentatively proposed, likely to be present. Styrikovich's prediction of HIO as a principal species under BWR conditions was confirmed, but it was concluded that his experiments had not measured its steam/water partition coefficient. In view of the meagre experimental evidence, however, more work on this system is desirable.

New Methods of Chemical Treatment. As described elsewhere (98) considerable success has been achieved in treating low pressure boilers with polyamino-carboxylic acid type chelating agents. Some years ago the author suggested that weaker, but more thermally stable, chelating agents such as oxine (8 hydroxy quinoline) might find application in maintaining once-through boilers free of debris. The main doubt (98,99) was whether such bidentate chelating agents are strong enough (thermodynamically) to be of any value and early estimates of magnetite solubilities up to 200°C (14) were not encouraging. Apart from some recent work of Alexander (100) there is still hardly any data on metal ion chelation above 100°C. However it has been possible to estimate roughly some relevant high temperature stability constants (12) and crudely correct them on the basis of measured iron levels dissolved from Fe_3O_4 by oxine (101). On this basis the chances of using oxine successfully in a once-through boiler look good, catechol may be effective and dicarboxylic acids may be usable in an adaption of a Russian method of treating super-critical boilers (102) to suit sub-critical once-through boilers.

Estimation in this field is, at the moment, inevitably grossly approximate because of the lack of high temperature data and the likelihood (discussed in Section 2) of forming mixed hydroxy-chelate complexes at high temperatures. Experimental work in this area is particularly needed.

4. Concluding Remarks

There are many additional ways in which thermodynamic arguments could, in principle, be used both for pure prediction and for rationalizing plant findings. It is also necessary to quantify most of the rather qualitative conclusions discussed in Section 3. The least predictable systems involve the behaviour of transition metal ions at high temperatures and until a good deal more work is done to disentangle their complexing and hydrolysis reactions it is unlikely that much progress can be made. The author very much hopes that the complexity of the transition metal systems will not inhibit work in this area.

The loss to a user of thermodynamics in my field would have been considerable if Sweeton and Baes had been put off beginning or reporting their (23) Fe_3O_4 solubility work by the fear that on its own, it might not yield a complete answer to the understanding of Fe_3O_4 solubility. In this particular case the decision was probably not difficult as there may be no problem below 250-300°C. The question which worries me is how people are to be encouraged to study systems which are obviously more complex, experimentally difficult and unlikely, on their own, to yield reliable thermo-dynamic reaction free energies, (e.g. Fe_3O_4 solubility at higher temperatures or studies on mixed complexes).

Similarly, if Styrikovich et al (97), had worried about iodates and not given what the present author believes is an incorrect interpretation of the results, the data might still be unavailable. While a more detailed tabulation of the raw data would have been better, the data themselves are valuable in providing the only available experimental work on the behaviour of iodine in high temperature water. It is to be hoped that, increasingly, Journals will provide facilities for authors to tabulate their raw data. The Journal of Chemical Thermodynamics is to be congratulated on the amount of data they printed in the paper of Sweeton and Baes (23). If the formation constant of $FeCl^+$ and related constants can eventually be measured or estimated, and if a reanalysis proves necessary, the data are all there to use.

In view of the difficulties discussed in Section 2 it seems that many of the more important equilibria of relevance to power station operation will not be directly measurable. It is certain, therefore, that great emphasis will have to be placed on methods of estimating high temperature data. It also seems clear that, if these are to be checked up to 350°C, a variety of experimental techniques may well prove necessary to sort out usable thermodynamic data from experiments which, on their own, cannot give them. Alternatively, if estimation procedures can be developed which are substantially free from empirical fitting parameters, they may not require extensive checking.

This review was carried out at the Central Electricity Research Laboratories and is published by permission of the Central Electricity Generating Board.

5.Literature Cited

1. Staehle, R.W., Corrosion, 1977, 33, 1

2. Faster, D., Kernenergie, 1979, 22, 118

3. Barnes, H.L., "High Temperature High Pressure Electrochemistry in Aqueous Solutions", N.A.C.E.-4, Houston, Texas, 1976, p.14

4. Freier, R.K., VGB-Speisewassertagung, 1969, p.11

5. Freier, R.K., VGB-Speisewassertagung, 1970, p.8

6. Turner, D.J., paper presented at symposium "Industrial Uses of Thermochemical Data", 1979, sponsored by NPL and Chemical Society, University of Surrey, Oct.

7. Liu, C. and Lindsay, W.T. Jr., J. Solution Chem., 1972, 1, 45

8. Choi, Y-S. and Criss, C.M., Faraday Disc. of Chem. Soc., 1977, 64, 204

9. Turner, D.J., C.E.R.L. Note, 1976, RD/L/N 165/76

10. Turner, D.J., C.E.R.L. Note, 1979, draft "Some Unexpected Reactions Between Metals and Aqueous Solutions at High Temperature; Part I. Gold and Platinum Dissolution"

11. Turner, D.J., C.E.R.L. Note, 1976, RD/L/N 213/76

12. Turner, D.J., C.E.R.L. Note, 1978, RD/L/N 11/78

13. Passell, T.O., E.P.R.I. Jnl., 1977, 2, 42

14. Bawden, R.J., C.E.R.L. Note, 1976, RD/L/N 72/76

15. Franck, E.U., "High Temperature Pressure Electrochemistry in Aqueous Solutions", N.A.C.E.-4, Houston, Texas, 1976, p.109

16. Marshall, W.L., ibid, p.117

17. Lindsay, W.T. and Liu, C., ibid, p.139

18. Martynova, O.I., ibid, p.131

19. Lietzke, M.H., ibid, p.317

20. Mesmer, R.E., Sweeton, F.H., Hitch, B.F. and Baes, Jr.,C.F., ibid, p.365

21. Silvester, L.F. and Pitzer, K.S., J. Phys. Chem., 1977, 81, 1822

22. Helgeson, H.C. and Kirkham, D.H., Amer. J. Sci., 1976, 276, 97

23. Sweeton, F.H. and Baes, C.F., Jr., J. Chem. Thermodynamics, 1970, 2, 479

24. Nikolaeva, N.M., "High Temperature High Pressure Electrochemistry in Aqueous Solutions", N.A.C.E.-4, Houston, Texas, 1976, p.146

25. Tremaine, P.R. and Leblanc, J.C., "The Solubility of Nickel Oxide and Hydrolysis of Ni^{2+} in Water to 573 K". In press

26. Lüdemann, H.D. and Franck, E.U., Ber Bunsenges. physik. Chem., 1968, 72, 514

27. Lüdemann, H.D. and Franck, E.U., Ber Bunsenges. physik. Chem., 1967, 71, 455

28. Scholz, B., Lüdemann, H.D. and Franck, E.U., Ber. Bunsenges. physik. Chem., 1972, 76, 406

29. Franck, E.U., "Phase Equilibria and Fluid Properties in the Chemical Industry", ed. T.S. Storvik and S.I. Sandler, A.C.S. Symposium, Series No. 60, Chap. 5, p.99

30. Broadbent, D., Lewis, G.G. and Wetton, E.A.M., J. Chem. Soc. Dalton Trans., 1977, 464

31. Hitch, B.F. and Mesmer, R.E., J. Soln. Chem., 1976, 5, 667

32. Holmes, H.F., Baes, Jr., C.F. and Mesmer, R.E., J. Chem. Thermodynamics, 1978, 10, 983

33. Shoesmith, D.W. and Lee, W., Can. J. Chem., 1976, 54, 3553

34. Mesmer, R.E. and Herting, D.L., J. Soln. Chem., 1978, 7, 901

35. Sweeton, F.H., Mesmer, R.E. and Baes Jr., C.F., J. Soln. Chem., 1974, 3, 191

36. Fisher, J.R. and Barnes, H.L., J. Phys. Chem., 1972, 76, 90

37. Sirota, A.M. and Shviriaev, Yu.V., "High Temperature High Pressure Electrochemistry in Aqueous Solutions", N.A.C.E.-4, Houston, Texas, 1976, p.169

38. Marshall, W.L. and Franck, E.U., Paper presented at International Association for the Properties of Steam, Sept. 1979, Munich, West Germany

39. Marshall, W.L., J. Inorg. Nucl. Chem., 1975, 37, 2155

40. Marshall, W.L. and Slusher, R., J. Inorg. Nucl. Chem., 1975, 37, 2165

41. Marshall, W.L. and Slusher, R., J. Inorg. Nucl. Chem., 1975, 37, 2171

42. Marshall, W.L. and Jones, E.V., J. Inorg. Nucl. Chem., 1974, 36, 2313

43. Marshall, W.L. and Jones, E.V., J. Inorg. Nucl. Chem., 1974 36, 2319

44. Templeton, C.C., J. Chem. Thermodynamics, 1976, 8, 99

45. Marshall, W.L., ibid, p.100

46. Nikolaeva, N.M. and Tsvelodub, L.D., Zh. Neorg. Khim., 1977, 22, 380

47. Nikolaeva, N.M., Zh. Neorg. Khim., 1977, 22, 2447

48. Nikolaeva, N.M. and Tsveladub, L.D., Zh. Neorg. Khim., 1975, 20, 3033

49. Tugarinov, I.A., Ganeev, I.G. and Khodakovskii, I.L., Geokhimya, 1975, 9, 1345

50. Cobble, J.W. and Murray, Jr., R.C., Faraday Disc. of Chem. Soc., 1977, 64, 144

51. Bawden, R.J. and Turner, D.J., unpublished results

52. Chou, I.-M. and Eugster, H.P., Amer. J. Sci., 1977, 277, 1296

53. Frantz, J.D. and Eugster, H.P., Amer. J. Sci., 1973, 273, 268

54. Latimer, W.L., "Oxidation Potentials", Prentice Hall Inc., New Jersey, 1952

55. Swaddle, T.W., Lipton, J.H., Guastalla, G. and Bayliss, P., Can. J. Chem., 1971, 49, 2433

56. Kong, P-C., Swaddle, T.W. and Bayliss, P., Can. J. Chem., 1971, 49, 2442

57. Gainsford, A.R., Sisley, M.J., Swaddle, T.W. and Bayliss, P., Can. J. Chem., 1975, 53, 12

58. Henderson, M.P., Miasek, V.I. and Swaddle, T.W., Can. J. Chem., 1971, 49, 317

59. Newton, A.M. and Swaddle, T.W., Can. J. Chem., 1974, 2751

60. Fabes, L. and Swaddle, T.W., Can. J. Chem., 1975, 53, 3053

61. Turner, D.J., C.E.R.L. Note, 1979, draft "Some Unexpected Reactions Between Metals and Aqueous Solutions at High Temperature. Part II. Steam Volatilization of Components from Stainless Steel"

62. Criss, C.M. and Cobble, J.W., J. Amer. Chem. Soc., 1964, 86, 5385

63. Lewis, D., Arkiv. Kemi., 1971, 32, 385

64. Lewis, D., Aktieb. Atomenergi Report, 1971, AE-436

65. Macdonald, D.D., Shierman, G.R. and Butler, P., Atomic Energy of Canada Ltd. Reports, 1972, AECL-4136, 4137, 4138, 4139

66. Macdonald, D.D. and Rummery, T.E., Atomic Energy of Canada Ltd. Report, 1973, AECL-4140

67. Criss, C.M. and Cobble, J.W., J. Amer. Chem. Soc., 1964, 86, 5390

68. Cobble, J.W., J. Amer. Chem. Soc., 1964, 86, 5394

69. Pohl, H.A., J. Chem. and Eng. Data, 1962, 7, 295

70. Harned, H.S. and Embree, N.D., J. Amer. Chem. Soc., 1934, 56, 1050

71. Khodakovskii, I.L., Zhogina, V.V. and Ryzenko, B.N., Geokhimiya, 1965, 827

72. Harned, H.S. and Robinson, R.A., Trans. Faraday Soc., 1940, 36, 973

73. Tremaine, P.R., Von Massow, R. and Shierman, G.R., Thermochimica Acta., 1977, 19, 287

74. Helgeson, H.C., J. Phys. Chem., 1967, 71, 3121

75. Turner, D.J., "Water Chemistry of Nuclear Reactor Systems", British Nuclear Energy Society, London, 1978, p.489

76. Turner, D.J., C.E.R.L. Note, 1969, RD/L/N 25/69

77. Bawden, R.J., C.E.R.L. Note, 1977, RD/L/N 85/77

78. Turner, D.J., Faraday Disc. of Chem. Soc., 1977, 64, 231

79. Tremaine, P.R. and Goldman, S., J. Phys. Chem., 1978, 82, 2317

80. Turner, D.J., "High Temperature High Pressure Electrochemistry in Aqueous Solutions", N.A.C.E.-4, Houston, Texas, 1976, p.188

81. Bawden, R.J., C.E.R.L. Note, 1973, RD/L/N 20/73

82. Bawden, R.J., C.E.R.L. Note, 1977, RD/L/N 87/77

83. Bawden, R.J., C.E.R.L. Note, 1979, RD/L/N 100/79

84. Hearn, B. and D. de G. Jones, C.E.R.L. Report, 1971, RD/L/R 1699

85. Bawden, R.J. and Turner, D.J., C.E.R.L. Note, 1973, RD/L/N 213/73

86. Turner, D.J., C.E.R.L. Note, 1971, RD/L/N 269/71

87. Turner, D.J., C.E.R.L. Note, 1974, RD/L/N 238/74

88. Bussert, B.W., Curran, R.M. and Gould, G.C., Trans. ASME, 1978, Paper No. 78-JPGC-Pwr-9

89. Turner, D.J., C.E.R.L. Note, 1974, RD/L/N 204/74

90. Newman, J.F., C.E.R.L. Note, 1974, RD/L/N 215/74

91. Lewis, D., Chem. Scripta, 1974, 6, 49

92. Lewis, D., Aktieb. Atomenergi Report, 1977, AE-514

93. Bignold, G.J., Garnsey, R. and Mann, G.M.W., Corros. Sci., 1972, 12, 1325

94. Turner, D.J., C.E.R.L. Note, 1979, draft "Thermodynamics and the Nature of Oxides Formed on Iron-Chromium Alloys"

95. Tomlinson, M., "High Temperature High Pressure Electrochemistry in Aqueous Solutions", N.A.C.E.-4, Houston, Texas, 1976, p.221

96. Montford, B. and Rummery, T.E., Atomic Energy of Canada Ltd. Report, 1975, AECL-4444

97. Styrikovich, M.A., Martynova, O.I., Katkovskaya, K.Ya., Dubrovskii, I.Ya. and Smirnova, I.N., Atom Energya, 1964, 17, 45 (transl. p.735)

98. Turner, D.J., J. Appl. Chem., 1972, 22, 983

99. Turner, D.J., "High Temperature High Pressure Electrochemistry in Aqueous Solutions", N.A.C.E.-4, Houston, Texas, 1976, p.256

100. Alexander, R.D., "Dissociation Constants in Water at High Temperatures: 1, 10-Phemanthroline, Related Bases and their Complexes with Iron (II)", Ph.D. Thesis, University of Surrey, 1976

101. Osborne, O., Wilson, J.S., Fried Jr., A.R. and Pryor Jr. W.M., Dow Chemical Co. Report, 1973, Final Report on ASME Contract C-9-2-D

102. Margulova, T.Kh., Yalova, R.Ya., Bulovko, A.Yu. and Krol', A.Ya., Thermal Eng., 1972, 19, 114

RECEIVED January 31, 1980.

35

The Computation of Pourbaix Diagrams

T. I. BARRY

Chemical Standards Division, National Physical Laboratory,
Teddington, Middlesex, TW11 OLW, UK

Pourbaix diagrams illustrate graphically the dominant solution or precipitate species of a component or components as a function of pH and oxidation potential (1). They are particularly useful for defining the conditions for selective precipitation or solution in hydrometallurgical extraction (2) and for passivation of metals. However, they are tedious to produce manually, especially when a number of components are present. The purpose of this paper is to demonstrate the principles of automatic computation for simple and complex systems and to illustrate these by reference to the copper and sulphur systems both separately and combined. The same methods are applied to the delineation of the conditions under which various chloride complexes of copper will predominate as a function of chloride activity rather than pH.

Principles

Figure 1 shows a Pourbaix diagram for sulphur compounds in water. The point of the diagram is to indicate which compound of sulphur has the highest activity at combinations of pH and oxidation potential. It shows for example that in oxidising, alkaline solutions the sulphate ion dominates, whereas in reducing, acid solutions aqueous hydrogen sulphide is the major sulphur compound. In acid conditions of intermediate oxidation potential a wedge-shaped region is found which defines the circumstances under which precipitation of sulphur can occur.

An odd feature of the diagrams, as described here, is that unless the solution compounds have equal and constant activity coefficients, the concentrations of these compounds in their respective zones are not equal and could in principle be very different. To remove this anomaly the activity coefficients could readily be incorporated provided they were independent of pH and pE, the variables of the system.

0-8412-0569-8/80/47-133-681$05.00/0

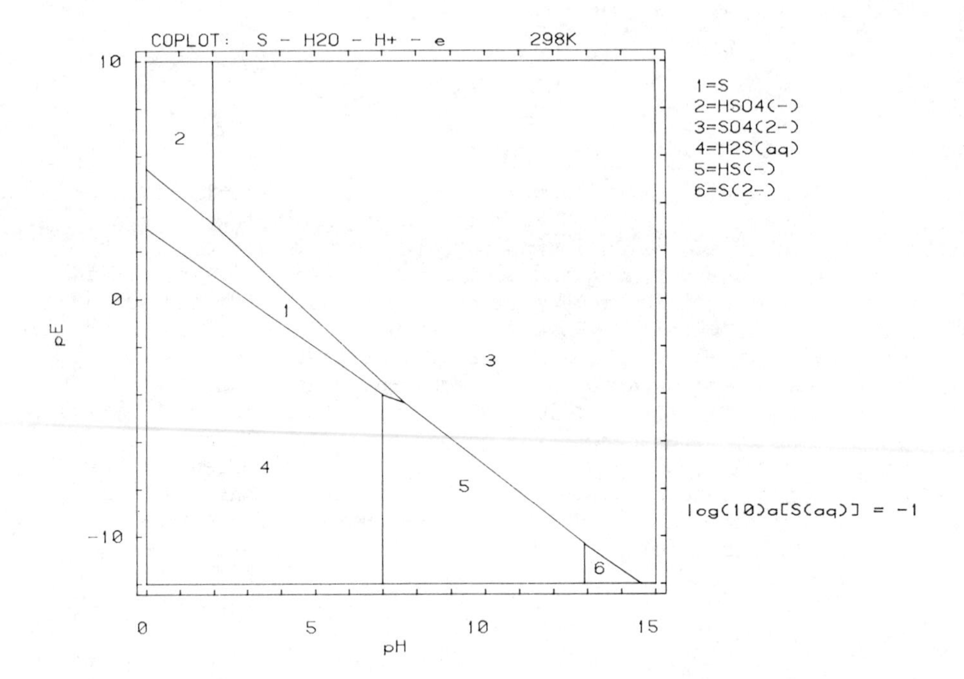

Figure 1. A pH–pE diagram for the S–H_2O–H^+–e system at 298.15 K. The activities of the predominant sulphur compounds in solution are 10^{-1}.

The diagrams calculated by the methods described differ to some extent from many available in the literature in that they omit the predominant solution compound in areas where a condensed compound also forms. This greatly simplifies the diagrams without impairing their usefulness for most purposes.

The system defined by figure 1 has four components, $S - H_2O - H^+ - e$. It is important to note that these fall into two classes. Sulphur will be classified as a type 1 component to signify that we are interested in what compounds it forms. Condensed type 1 compounds, have an activity of 1 if present and > 1 if absent. The activities of the predominant type 1 compounds in solution can be fixed at any desired value. In figure 1 they are set at 10^{-1}. Thus the sulphate ion has an activity of 10^{-1} in its area and HS^- has an activity of 10^{-1} in its area. On the line between these areas HS^- and the sulphate ion coexist with equal activities and the position of the coexistence line between these two compounds can be obtained by solution of the equation for the equilibrium

$$SO_4^{2-} = HS^- + 4H_2O - 9H^+ - 8e.$$

The remainder of the components are of type 2. Their activities are either fixed, as for H_2O, or they form the independent variables of the system, H^+ and e. The activities of H^+ and the notional activity of the electron are expressed in logarithmic form as pH and pE where pE is related to the oxidation potential, E_h, by the relation

$$E_h = \frac{RT \ln 10}{F} pE \qquad 1$$

There are 6 sulphur compounds to be considered and therefore 15 possible coexistence lines of which only 9 represent stable equilibria. Moreover, even these 9 lines are valid (i.e. correspond to stable coexistence) along only part of their possible extents. The problem then is to determine the equations for the lines and the range over which they are valid, and to provide a method for plotting them.

Methods for a simple case

The principles of the method become much clearer if applied to a particular rather than the general case.

1 The first step is to list the compounds of the type 1 component together with their Gibbs energies of formation at the chosen temperature, or the function $[\Delta H^o(f, 298) + G^o(T) - H^o(298)]$ which is much easier to calculate in a data-bank. The two functions must not be used for different substances in the same calculation.

The data used in this paper are taken from the monograph by Duby (3) on aqueous systems of copper.

Table I

No.	Compound Formula	$\Delta_f G^o(298)$ /kJ mole^{-1}
1	S	0
2	HSO_4^-	-756.01
3	SO_4^{2-}	-744.63
4	$H_2S(aq)$	-27.87
5	HS^-	12.05
6	S^{2-}	85.77

This is not a complete list of type 1 compounds in the system but it is sufficient to demonstrate the principles. In this table, the formulae of the compounds are just as essential items of data as the Gibbs energies, since, as will be seen, they are used to generate the stoichiometry numbers in the equations for reactions.

2 The next stage is to choose a reference compound of the type 1 component. Unless it is very far from stable in the system, it is advantageous to choose the element, rhombic sulphur in this case.

3 Equations are then written for the conversion of one mole of the reference compound to all other compounds of the type 1 component using the type 2 components to balance the equations.

$$1,2 \quad S = HSO_4^- - 4H_2O + 7H^+ + 6e$$
$$1,3 \quad S = SO_4^{2-} - 4H_2O + 8H^+ + 6e$$
$$1,4 \quad S = H_2S(aq) - 2H^+ - 2e$$
$$1,5 \quad S = HS^- - H^+ - 2e$$
$$1,6 \quad S = S^{2-} - 2e$$

The numbers to the left of the equation identify the compounds of sulphur participating in each equation. Note that any other equation can be generated by subtraction of pairs of these equations. Thus equation 3,5 is generated by subtracting equation 1,3 from equation 1,5. The order is important.

$$3,5 \quad SO_4^{2-} = HS^- + 4H_2O - 9H^+ - 8e$$

4 The standard Gibbs energy changes for the reactions of the reference compound are then calculated. They are listed in column 2 of table 2.

Non-standard Gibbs energy changes can be calculated from the following equation.

$$\Delta G = \Delta G^o + RT \sum_{g=1}^{m+1} \nu_g \ln a_g \qquad 2$$

where g defines the position of the compounds in the equation and m is the number of components. For the system $S - H_2O - H^+ - e$, there are 4 components and, therefore, 5 compounds in each equation. The stoichiometry number for compound g in the equation is given the symbol ν_g, which may be zero. The stoichiometry number for the reference compound is always -1, as it always lies on the left of the equation and is always assigned 1 mole.

For a given diagram the activity of the sulphur solution species can be set to any desired value. The activity of elemental sulphur and water are both unity. The logarithms of the activities of the hydrogen ion and the electron are for convenience assigned the symbols X and Y respectively. Thus X = -pH and Y = -pE. For the particular case of reaction 1,3 the following equation is obtained for $\Delta G/R'T$ where $R' = R \ln 10$

$$-\frac{\Delta G}{R'T} = (-\frac{\Delta G^o}{R'T} - \log a_2) - 8X - 6Y \qquad 3$$

where a_2 is the predetermined activity of the sulphur compounds in solution. Log a_1 = 0 and is therefore omitted.

In the general case it is convenient to assign the symbol ψ to the bracketed term where

$$\psi = -\frac{\Delta G^o}{R'T} - \sum_{g}^{m-1} \nu_g \log a_g \qquad 4$$

and where the summation now omits the variables of the diagram to be calculated.

When it is remembered that $-\Delta G^o/R'T = \log K$ it can be seen that ψ has the nature of log K adjusted for the activities of the fixed components.

Equation 2 can now be rewritten

$$\chi = \psi - \alpha X - \beta Y \qquad 5$$

where α and β are stoichiometry coefficients for the independently variable components and where

$$\chi = -\Delta G/R'T \qquad 6$$

At equilibrium $\chi = 0$ and lines of coexistence between the compounds of the type 1 component have the algebraic form

$$\psi - \alpha X - \beta Y = 0 \quad \text{(equilibrium condition)} \qquad 7$$

Stage 4 is completed by making a table of values of ψ, α and β for the chemical equations 1,2 to 1,6. If the maximum activity of the solution compounds of sulphur is set at 10^{-1}, the following table is obtained:

Table II

Parameters for the $S-H_2O$ system

Equation Number	ΔG^o/kJ mole^{-1}	log K	ψ	α	β
1,2	192.707	-33.763	-32.763	7	6
1,3	204.087	-35.757	-34.757	8	6
1,4	-27.87	4.882	5.882	-2	-2
1,5	12.050	-2.111	-1.111	-1	-2
1,6	85.772	-15.027	-14.027	0	-2

The last three columns of table 2 form the basic set of input data for further calculation. Values of ψ, α and β for any other reaction can readily be obtained by subtraction as follows

$$\psi(ij) = \psi(1j) - \psi(1i)$$
$$\alpha(ij) = \alpha(1j) - \alpha(1i) \text{ etc} \qquad 8$$

Thus only the data for reactions of the reference compound need to occupy space in the computer core.

5 Key features of Pourbaix diagrams are the points of intersection between the coexistence lines. In a simple diagram, three compounds of a dependent component can coexist at these points. Thus, if compounds i, j and k coexist at a point, 3 coexistence lines must radiate from the point, ij, jk and ik. The coordinates of potential triple intersection points can be determined by simultaneous solution of pairs of equations. For example the coordinates of the equilibrium point between sulphur, SO_4^{2-} and HS^- are determined by solution of the equations

$$1,3 \qquad -34.757 - 8X - 6Y = 0$$
$$1,5 \qquad -1.111 + X + 2Y = 0$$
$$X(1,3,5) = -7.618 \text{ , pH} = 7.618$$
$$Y(1,3,5) = 4.3645 \text{, pE} = -4.3645$$

6 The next step is to determine whether the calculated points are valid or not. The simplest test is to determine which reactions for formation of the sulphur compounds have the highest values of χ, i.e. the lowest (most negative) ΔG. If a potential intersection point, X(ijk), Y(ijk) is valid then the values of $\chi(1,i)$, $\chi(1,j)$ and $\chi(1,k)$ should fulfil three criteria. They should be higher than any other value of $\chi(1,g)$, greater than or equal to zero and equal to each other except for rounding off errors.

Referring to table 2 it is found that for X = -7.618, Y = 4.3645

$$\chi(1,2) = -5.624$$

$$\chi(1,3) = 0$$

$$\chi(1,4) = -0.625$$

$$\chi(1,5) = 0$$

$$\chi(1,6) = -5.298$$

Note that $\chi(1,1)$ need not be calculated as it is always zero. The results show the triple intersection is valid because $\chi(1,1)$, $\chi(1,3)$ and $\chi(1,5)$ are higher than the remaining values of χ. Thus coexistence lines 1,3, 1,5 and 3,5 do indeed radiate from pH = 7.618, pE = -4.3645. Valid intersection points are stored together with values of i, j and k.

7 For the purpose of computer calculation potential coexistence lines fall into a number of classes.

- a they are nowhere valid (other compounds are always more stable).
- b They are valid only outside the boundaries of the diagram.
- c They make two valid intersections with the diagram boundaries but make no intersections within it.
- d They make one valid intersection with the diagram boundary and one valid intersection within the diagram.
- e They make two valid intersections within the diagram.

Intersections with the diagram boundary must therefore be calculated for each potential line and tested for validity in an analogous way to the triple intersections. Once two valid intersections of either kind have been found for a line it need not be reconsidered.

For simplicity intersections between pairs of lines ij and jk can be taken in numerical order, evaluating X(ijk) and Y(ijk) with $i < j < k$ and $k \le n$, where n is the number of compounds. The maximum number of intersection points that need to be examined is $\binom{n}{3}$, i.e. n(n-1)(n-2)/6. For 40 compounds it is 9880. This is not a very large number for a computer to handle because the actual calculations are very trivial. Nevertheless, methods are available for improving the programme efficiency as briefly described under program comparisons.

It should explicitly be stated at this point that it is not possible, except by accidental coincidence, for an intersection between two lines i,j and k,l to be valid if either i or j is not equal to either k or l. Thus such intersections need never be considered.

The coordinates of the two valid intersection points for each real line can be stored in an array. When all intersections have been calculated, this array constitutes the output file to the plotter, which should be programmed to draw straight lines between the coordinate pairs corresponding to a given line. It is useful to note that the lines themselves do not need to be identified in the output to the plotter. All that is necessary is to have a method for labelling the areas they enclose.

The following method can be used to label the areas in which the compounds predominate. The X and Y coordinates respectively are summed for the corners of the areas occupied on the diagram by each compound in turn, including the diagram corners if the compound is stable there. The average values of X and Y are used to centre the label. A flag can be assigned to any compound that is nowhere predominant, so that its label can be omitted from the diagram.

The methods described above were used to produce figure 1 for the $S - H_2O$ system and figure 2 for the $Cu - H_2O$ system. In common with practical experience, figure 2 shows there is no combination of pH and oxidation potential for which the cuprous ion is the dominant compound of copper.

Methods for more complex diagrams

Up to this point the paper has been concerned only with simple diagrams for which there is only one type 1 component. If more than one type 1 component is present each area of the diagram corresponds in general to as many compounds as there are type 1 components. Moreover, if the type 1 components form compounds between each other, in this case Cu_2S and CuS, it is necessary to fix from the start the proportions of the type 1 components. Logically, the proportions should correspond with the values set for the solution concentrations. Thus, if the concentration of sulphur in solution greatly exceeds that of copper, each area will correspond to one copper compound, which may also be a compound of sulphur, and one sulphur compound which is not also a compound of copper.

Ideally a computer system for calculation of Pourbaix diagrams would comprise a database for a wide range of compounds coupled to a computer program that would undertake the following tasks.

1 Upon input of the system expressed as components the computer would retrieve data for compounds of the type 1 components with each other and with the type 2 components. Data are given in table 4.

If variables other than pH and pE are to be used as coordinates of the diagram, they must be specified, together with any other type 2 components, so that data can be retrieved.

2 The type 1 compounds would be sorted into lists. For example, if the proportion of sulphur substantially exceeded that of copper, the lists would comprise

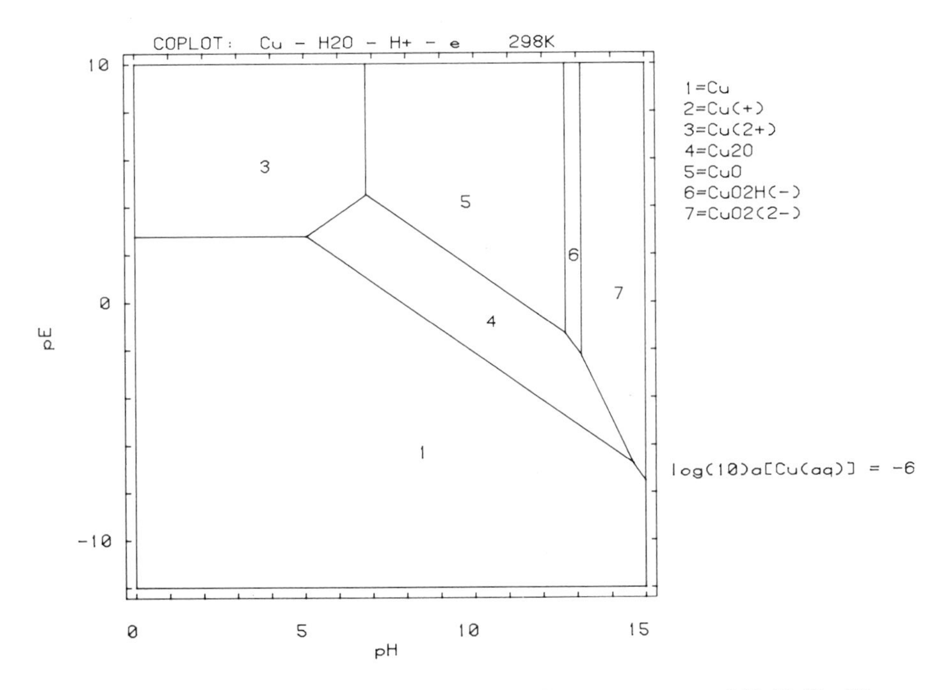

Figure 2. A pH–pE diagram for the Cu–H_2O–H^+–e system at 298.15 K. The activities of the predominant copper compounds in solution are 10^{-6}.

Copper Compounds				Sulphur Compounds	
1	Cu	6	CuO_2H^-	1	S
2	Cu^+	7	CuO_2^{2-}	2	HSO_4^-
3	Cu^{2+}	8	Cu_2S	3	SO_4^{2-}
4	Cu_2O	9	CuS	4	H_2S
5	CuO			5	HS^-
				6	S^{2-}

If, on the other hand the diagram were required for the case where the proportion of copper exceeded that of sulphur, Cu_2S and CuS would occur in the sulphur rather than the copper list.

3 Equations would be written in matrix form for formation of all possible combinations of type 1 compounds selected one at a time from each list, using as reference the elements or compounds in proportions adequate to ensure that the dominant type 1 component is always present independently. In the chosen example sulphur is the dominant component and an S:Cu ratio of 2:1 is adequate for this purpose. Equations can readily be calculated by matrix algebra for all possible reactions of the reference compounds. Eg

$$(CuS + S) = 0.5Cu_2S + 1.5SO_4^{2-} - 6H_2O + 12H^+ + 9e$$

$$(CuS + S) = 0.5Cu_2S + 1.5HS^- - 1.5H^+ - 3e$$

Note that (CuS + S) is used as the reference state rather than (Cu + 2S) because it is the combination of lower Gibbs energy. The combination (CuS + S) is treated as a single compound in the programs used at NPL with a stoichiometry number of -1 as it is on the left of the equation.

The computer would store the stoichiometry numbers in an array. Because the number of combinations may in some cases be large, it may be desirable to eliminate incompatible combinations such as $H_2S(aq) + CuO_2^{2-}$ either manually or automatically on the basis of previously calculated diagrams for the individual systems.

4 For each equation values of ψ, α and β calculated. If the same values of α and β are found for more than one equation, only the equation corresponding to the highest value of ψ is selected for reasons given in the following section.

5 The intersection points between coexistence lines and of coexistence lines with the boundaries of the diagram are calculated in the same way as already described for simple diagrams. However, in systems of more than one type 1 component quadruple intersections can occur at which for example HSO_4^-, SO_4^{2-}, Cu^{2+} and Cu_2S coexist. In such a case four values of χ will be found to be equal at the intersection point showing that it is the

intersection point of four lines.

The resultant diagram is shown in figure 3. The correspondence with figure 1 can readily be seen but the stability of the copper sulphides results in great differences from figure 2. The sharp changes in direction of the boundaries of the regions where the sulphides of copper predominate show how marked are the effects of change of ionic state of the sulphur. The sulphides are stable only at intermediate oxidation potentials.

If copper rather than sulphur had been made the dominant type 1 component the $Cu - H_2O$ diagram would form part of the diagram and it would be overlain by a region in which Cu_2S was also present. Neither CuS nor S would appear on the diagram.

Variations of Pourbaix diagrams

The chief limitation of Pourbaix diagrams is that they show only the dominant compound in any particular area. They do not show the presence of other compounds, which may be of comparable concentration nor the fact that the activities of solution species are continuous functions.

A particular problem is found for systems in which a complexing agent such as Cl^- or NH_3 is present. In such cases a number of different complexes may be found for the same valence state of the cation. Because the valence state is constant these must be represented on a Pourbaix diagram only by the dominant complex. Even excluding partly hydrolysed complexes (by selecting acid conditions), there are at least 11 ionic forms of Cu^+ and Cu^{2+}

Cu^+	$CuCl_3^{2-}$	$CuCl_2$ aqueous
CuCl solid	Cu^{2+}	$CuCl_3^-$
CuCl aqueous	$CuCl^+$	$CuCl_4^{2-}$
$CuCl_2^-$	$CuCl_2$ solid	

It would be possible to consider chloride as a type 1 component in this system so that the diagram could reveal areas in which Cl_2(gas), Cl^-, ClO_3^- and ClO_4^- predominated. However, to do so here would obscure the question of how to deal with the various chloride complexes of cuprous and cupric ions, which is the principle concern of this section.

The equations for the formation of all chloride complexes of Cu^{2+} differ only in the number of Cl^- ions necessary to balance the equations. The stoichiometry numbers, α for H^+ and β for the electron are all zero and two respectively, as shown in table 3. A method for selecting the dominant complex must be included in the program if the program is not to fail by division by zero. The method is simply to choose the equation for which ψ is highest.

In table 3 data for the cupric complexes are presented. $\psi(0)$ has been evaluated for a chloride activity of 10^0 and a copper ion activity of 10^{-6}, and $\psi(-1)$ for the case where the chloride

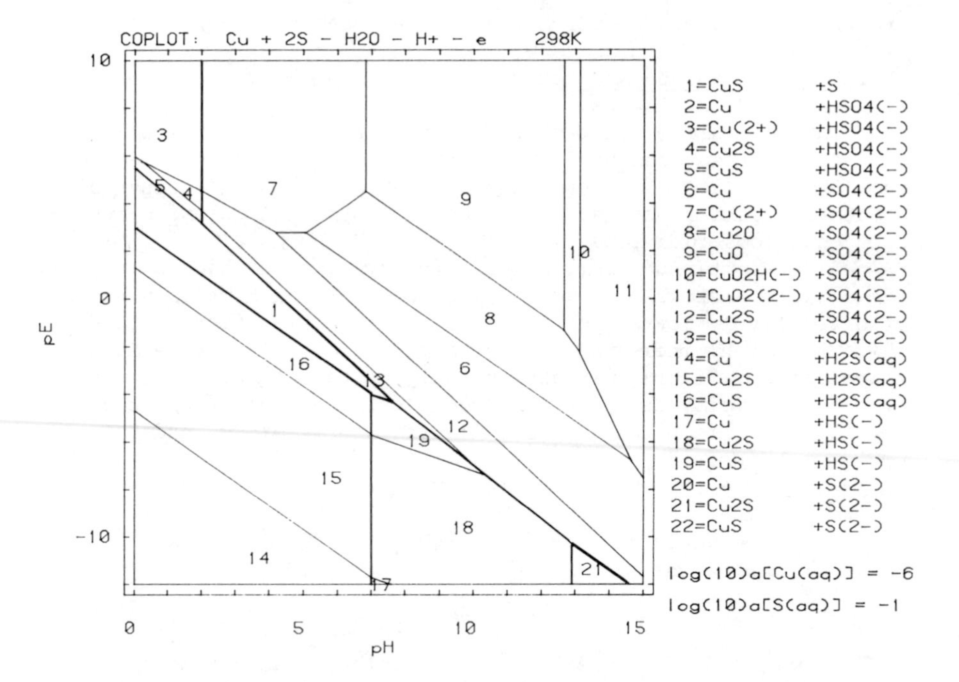

Figure 3. A pH–pE diagram for the Cu–S–H_2O–H^+–e system at 298.15 K where the sulfur is present in excess of copper. The activities of the predominant sulfur and copper solution species are 10^{-1} and 10^{-6}, respectively.

activity has been changed to 10^{-1}.

Table III

Parameters for chloride complexes of the cupric ion

Equation	$\log_{10}K$	$\psi(0)$	$\psi(-1)$	α	β	α'
$Cu = Cu^{2+} + 2e$	-11.51	-5.51	-5.51	0	2	0
$Cu = CuCl^{+} - Cl^{-} + 2e$	-11.05	-5.05	-6.05	0	2	-1
$Cu = CuCl_2(c) - 2Cl^{-} + 2e$	-15.57	-15.57	-17.57	0	2	-2
$Cu = CuCl_2(aq) - 2Cl^{-} + 2e$	-11.32	-5.32	-7.32	0	2	-2
$Cu = CuCl_3^{-} - 3Cl^{-} + 2e$	-13.72	-7.72	-10.72	0	2	-3
$Cu = CuCl_4^{2-} - 4Cl^{-} + 2e$	-16.05	-10.05	-14.05	0	2	-4

The highest value of $\psi(0)$ (chloride activity 10^0) is found for the formation of $CuCl^+$, indicating that this compound will dominate all other cupric species under these conditions. However, at a chloride activity of 10^{-1}, only ten times lower, the uncomplexed ion, Cu^{2+}, becomes dominant.

Figure 4 shows a Pourbaix diagram for this case. Even at this relatively low activity the presence of chloride has a profound effect by increasing the stability of cuprous by complex formation, cf figure 2. However, the effect of changing the chloride activity cannot readily be predicted from figure 4 alone. It would be possible to draw families of diagrams on the same plot, each individual diagram corresponding to a particular chloride activity, but diagrams of this kind are very confusing, particularly as in the present case, when the dominant species in certain zones would change from one individual diagram to another. An advantage to the user of a computer-based system of diagram calculation from a databank is that diagrams need be produced only for systems and conditions of direct relevance. There is no need to pack the diagrams with confusing information.

Fortunately a simple solution is available to the problem of representing all the chloride complexes. The algebra of Pourbaix diagrams remains unchanged if log $a(Cl^-)$ replaces pH as one of the variables. Figure 5 shows a diagram for the $Cu-H_2O-Cl^--H^+-e$ system under conditions of low pH and low solute activity, where the only stable compounds of copper are copper itself, Cu^+, Cu^{2+} or their complexes with Cl^-. The new stoichiometry coefficients are listed under α' in table 3. $CuCl_2(c)$ is eliminated because it has a lower value of ψ than $CuCl_2(aq)$ but the same α' and β.

Figure 5 shows that at very high chloride activities the cuprous complex, $CuCl_3^{2-}$, becomes very dominant, being oxidised to cupric complexes only above pE = 10. This information is of value because methods of stabilising particular valence states such as Cu^+ offer means of separating metals that would be difficult to separate in their normal valence states.

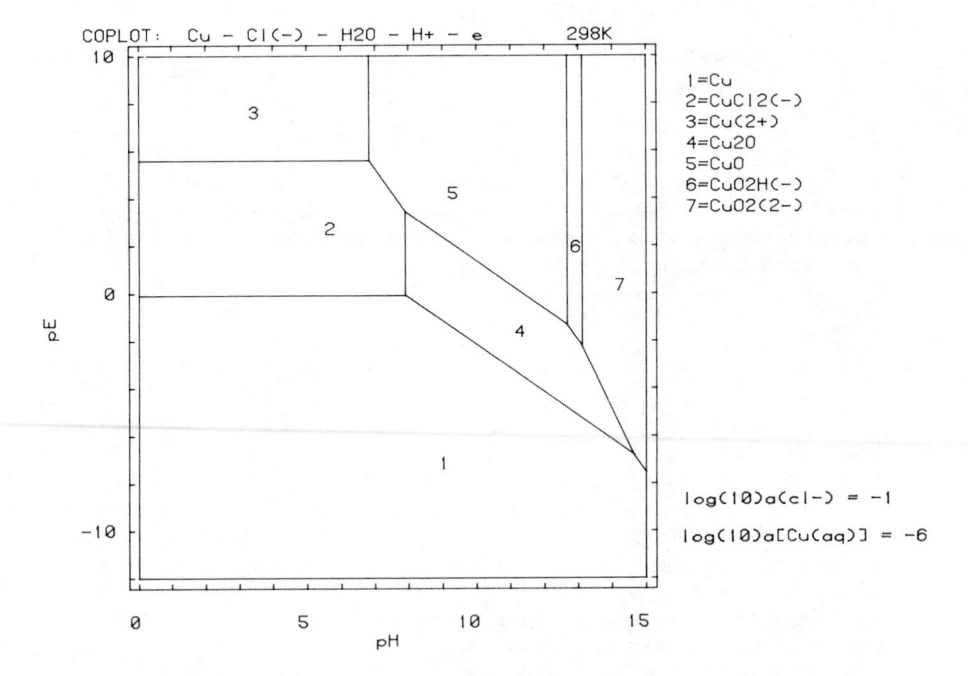

Figure 4. A pH–pE diagram for the Cu–Cl^-–H_2O–H^+–e system at 298.15 K. The activity of Cl^- is 10^{-1}; the activities of the predominant copper solution species are 10^{-6}. The diagram can represent only the most stable chloride complexes of cuprous and cupric and not the proportions of the various complexes.

Table IV

Data used in this paper are taken almost exclusively from the review by Duby(3) in order that diagrams should be consistent with his. ΔG^{o} for CuCl(aq) was estimated from data given by Smith and Martell(6).

Compound	$\Delta_f G^o$(298K)/kJ mole^{-1}	Compound	$\Delta_f G^o$(298K)/kJ mole^{-1}
H^+	0	CuO	-127.90
e	0	$HCuO_2^-$	-258.57
H_2O	-237.18	CuO_2^{2-}	-183.68
S	0	Cu_2S	-87.44
HSO_4^-	-756.01	CuS	-53.14
SO_4^{2-}	-744.63	CuCl(c)	-119.66
H_2S	-27.87	CuCl(a)	-96.65
HS^-	12.05	$CuCl_2^-$	-240.16
S^{2-}	85.77	$CuCl_3^{2-}$	-376.56
Cl^-	-131.26	$CuCl^+$	-68.20
Cu	0	$CuCl_2$(c)	-173.64
Cu^+	50.63	$CuCl_2$(a)	-179.90
Cu^{2+}	65.69	$CuCl_3^-$	-315.47
Cu_2O	-147.90	$CuCl_4^{2-}$	-433.46

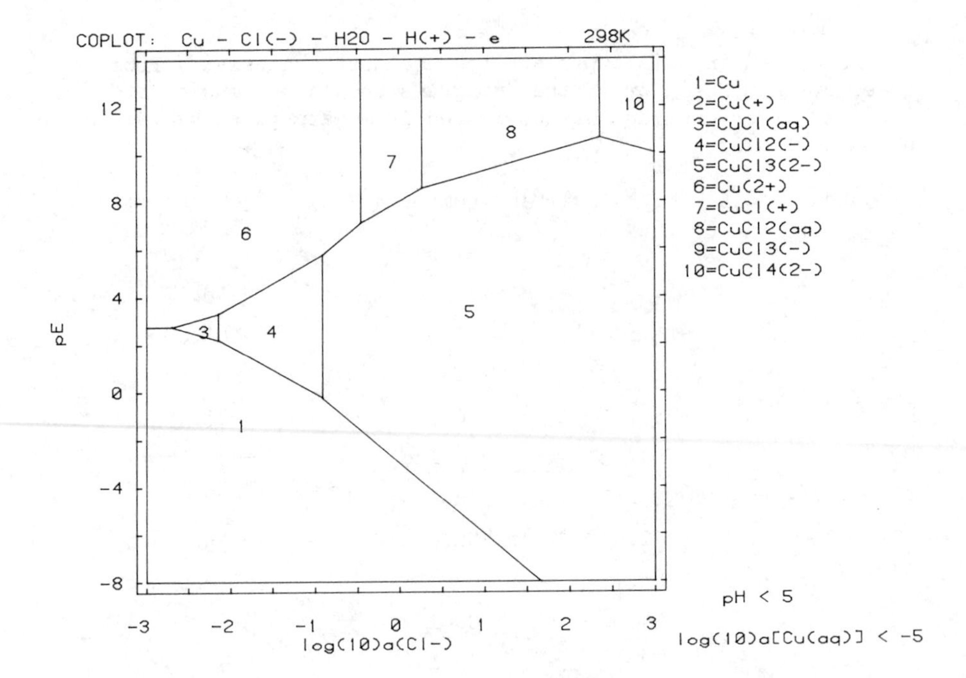

Figure 5. A predominance area diagram for the $Cu–Cl^{-1}–H_2O–H^+–e$ system at 298.15 K analogous to a pH–pE diagram but in which $log(10)a(Cl^-)$ replaces pH as a variable. The pH is less than 5. The activity of the predominant copper solution species are 10^{-6}.

Even this diagram does not give a clear impression of the relative proportions of the various copper compounds present in solution. However, provided no polynuclear species are present, it is a relatively simple matter to use the values of χ to evaluate these proportions and to plot them as a function of a single variable. Figure 6 shows a diagram of this kind using the same data as figure 5 calculated for pE = 10 and variable chloride activity under the assumption that all compounds have the same activity coefficient. It would not be difficult to allow for different values of activity coefficients if these were known.

Program comparisons

The actual program used at NPL was written by N.P. Barry on the basis of the methods described previously. It is written in FORTRAN and has been implemented on IBM 370 and UNIVAC 1100 computers operated by computer bureaux. Vector algebra is employed. The reason why the graphs have double boundaries is that the calculation can be performed for boundaries of any convex polygon of up to 30 sides. This permits calculations to be restricted to the stability range of particular components, for example, that of water or chloride.

Because the program is not coupled to a data base, the input at present comprises values of ψ, α and β for reactions of the reference compound(s). The output comprises the number of lines in the diagram including those of the boundaries, the coordinates of the ends of each line, the number of compounds and the coordinates of the label for each compound in number order. Compounds that are not present are assigned the coordinates 0,0.

The plotting program is written on a Tektronix 4051 graphics micro-computer which is coupled to a 4662 plotter.

A much simpler method of producing Pourbaix diagrams is to divide the coordinate system up into as many points as can conveniently be plotted by a printer. At each point the function χ is then evaluated for each equation 1,i and the value of i determined that gives the highest value of χ. A symbol corresponding to i can then be printed. The method has the virtues that this simple procedure completes the operation and a plotter is not required. This is the basis of the method described by Duby.(3) Its disadvantages are that it is rather slow, it has a poor resolution and the result is not visually satisfying.

A number of other methods have been described as reviewed by Linkson, Phillips and Rowles.(4) Some of these use the point by point method briefly described above and others, using a convex-polygon method, search for the boundaries of one predominance region at a time. An advantage of this approach is that the number of intersections that need to be considered can be reduced substantially as follows. The boundaries of any convex polygon necessarily lie in a sequence of progressively changing slope. Thus testing for intersections with lines in order of their slope

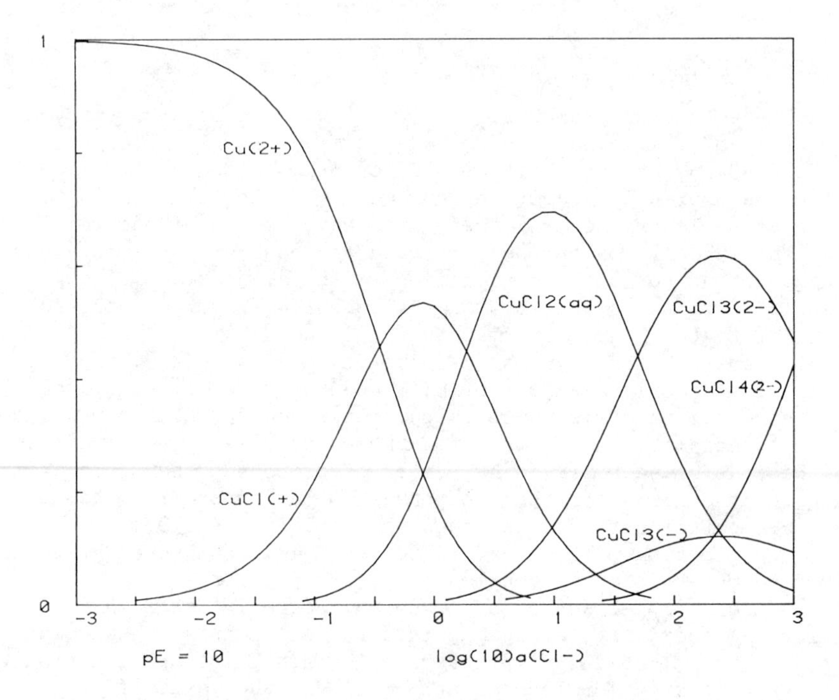

Figure 6. A diagram showing the proportions of various chloride complexes of copper calculated as a function of chloride activity for pH < 5, pE = 10, and copper ion activities less than about 10^{-6}. Note that the cuprous complex $CuCl_3^{2-}$ is dominant at very high chloride activities. The calculation is based on a number of assumptions (see text) that are unlikely to be entirely valid: chloride activities much greater than 100 cannot be achieved readily.

greatly improves the rate of finding valid intersections. Perhaps the most significant development is that of Turnbull(5) who has linked a computational program to a database. Moreover, up to 5 type 1 compounds can be considered.

In considering the value of different approaches to automatic drawing of Pourbaix diagrams the efficiency of the computer program is only one factor. The simplicity and adequacy of input and output operations are equally or more important. The value of a coupled database is high because it greatly reduces input time and the probability of error. It should not be necessary to call for compounds by formula name for data retrieval since it is clearly possible to call for all the compounds in a given system and to have a means of deleting or adding compounds. This is the method used in the operation of the Gibbs energy minimisation calculations on condensed and gaseous substances using the NPL system MTDATA.

Literature Cited

1 Pourbaix, M. "Atlas of Electrochemical Equilibria in Aqueous Solutions"; Pergammon: New York, 1966.
2 Burkin, A.R. Proc. Conf. "Industrial Use of Thermochemical Data", Barry, T.I., Ed. To be published by The Chemical Society: London.
3 Duby, P., "The Thermodynamic Properties of Aqueous Inorganic Copper Compounds"; Monograph IV, International Copper Research Association Inc., 1977.
4 Linkson, P.B.; Phillips, B.E.; Rowles, C.D, Minerals Science Engineering, 1979, 11, 65.
5 Turnbull, A.G, Extraction Metallurgy Symposium, University of New South Wales: Sydney, 1977.
6 Smith, R.M.; Martell, A.E. "Critical Stability Constants," Vol 4: "Inorganic Complexes," Plenum Press: New York, 1976.

RECEIVED January 31, 1980.

36

Applications and Needs of Thermodynamics in Electrometallurgy

T. J. O'KEEFE and R. A. NARASAGOUDAR

Graduate Center for Materials Research, University of Missouri—Rolla, Rolla, MO 65401

I. Introduction.

Historically, there has always been a concerted effort to provide a firm thermodynamic base for the development of processes involving metal extraction and refining. There can be no argument that equilibrium principles can and have been applied to advantage in aqueous processing. Certain inherent characteristics in the thermodynamic approach have prevented its use as extensively as it might be used in certain areas of electrometallurgy. The initial science of electrochemistry was almost exclusively thermodynamic in nature, and explanations of cell behavior were sought using the familiar equation of Nernst:

$$E = E^{o} + \frac{RT}{nF} \ell n \frac{a_{\text{oxidized species}}}{a_{\text{reduced species}}} \tag{1}$$

where E and E^{o} are the equilibrium and the standard potentials respectively, F is the Faraday constant, R the gas constant, T the absolute temperature, n the moles of electrons and a is the activity. The E or E^{o} could be used in determining the spontaneity of a reaction, depending on the state of reactants and products, by the expression:

$$\Delta G = -nFE \tag{2}$$

The use of these expressions is effectual only in cases where there is no extensive deviation in the system behavior due to charge transfer overpotential or other kinetic effects.(1) The calculated threshold or thermodynamic energy requirement (2) (ΔG in the previous equation) is often much lower than actually encountered, but is still useful in estimating an approximate or theoretical minimum energy required for electrolysis. Part of the difficulty in applying thermodynamics to many systems of industrial interest may reside in an inability to properly define the activities or nature of the various species involved in the

0-8412-0569-8/80/47-133-701$05.00/0

reactions. Many electrochemical reactions involving metals, such as cathodic reduction, cementation, gaseous reduction from solution or corrosion, are very susceptible to changes caused by the presence of trace concentrations of certain chemicals in solution. Their effects can be helpful or harmful, depending on the particulars, but the conditions resulting from the presence of the trace impurities are most often difficult to control and predict.

One of the objectives of this paper will be to show some specific examples of these effects in electrolysis and illustrate the substantial need for a better understanding of the thermodynamics of the solution chemistry involved in electrodics. Some of these needs are more obvious and have been indicated previously (3) and include such items as ΔG^o, K_{so} and Cp data on the systems of interest. However, much more extensive information is necessary on adsorption phenomena, complex ion formation and the equilibrium concentrations of these influential species. This need has always existed but it is even more important now if the current challenges being imposed by energy and materials shortages and environmental control are to be met.

II. Extraction Processes.

A. Flowsheet Design. The basic process flowsheet for the extraction and refining of metals when electrolysis is involved is similar in many respects. The specifics may change to suit the physical needs and chemistry of the particular metal system in question, but the objectives of the unit operations and unit processes are comparable in general terms. As an example, the processing scheme for the production of electrolytic zinc is shown in Figure 1. Zinc is a good example because it incorporates many of the items that must be addressed by the extractive metallurgist if a pure metal is to be produced efficiently. The very active standard potential of zinc forces considerable care to be taken during processing to insure that troublesome impurities, ones that can reduce current efficiency, are removed from the system. The final quality of the metal is also important, for many applications require special high grade metal of 99.99% purity. Time does not permit an extensive evaluation of the roast, leach, purification and electrolysis steps individually. Briefly, the objective is to convert the ZnS to an oxide which can be leached in sulfuric acid electrolyte. Impurities such as Cu, Ni, Co, Sb and As are then removed by cementation using zinc powder prior to electrolysis. Throughout the procedure a close scrutiny is made of the types and concentrations of various chemical species present. A typical chemical analysis range required for satisfactory electrolysis is given in Table I. The very low contents for certain trace metals, such as nickel, cobalt, copper, antimony, and germanium are notable, and if these levels aren't attained then decreased current efficiency can result. Elements such as lead and cadmium must be minimized to insure proper metal purity.

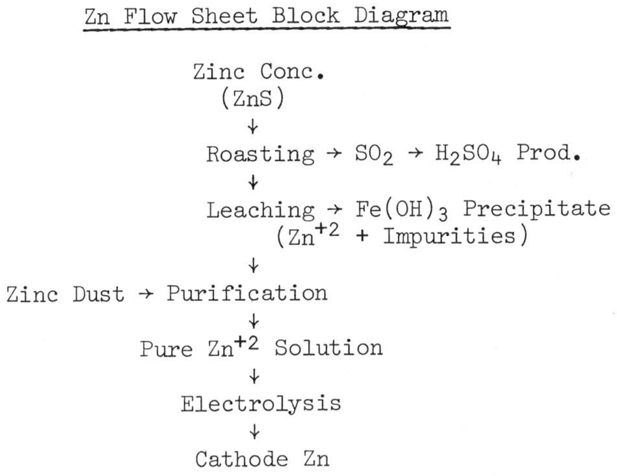

Zn Flow Sheet Block Diagram

Zinc Conc.
(ZnS)
↓
Roasting → SO_2 → H_2SO_4 Prod.
↓
Leaching → $Fe(OH)_3$ Precipitate
(Zn^{+2} + Impurities)
↓
Zinc Dust → Purification
↓
Pure Zn^{+2} Solution
↓
Electrolysis
↓
Cathode Zn

Figure 1. Zinc flow sheet block diagram

Table I. Analysis of Neutral Purified Solution for Zn Electrolysis.(4) (mg/ℓ)

Element	Solution Prepared in Laboratory	Actual Plant Solution
Zn	130,000	170,000
Mn	<0.1	3,500
Cd	1.3	1.7
Sb	0.01	-
As	<0.04	-
Sb+As	-	0.04
Ge	0.002	0.004
Co	<0.1	0.1
Ni	<0.05	<0.05
Fe	0.4	5.9
Cu	0.08	0.3
F^-	0.1	1.0
Cl^-	1.0	370

The processing of copper, tin, manganese, nickel, etc., follows a similar pattern, but the emphasis required at the various stages of processing may change with the nature of the metal. For example, current efficiency in manganese electrolysis is even more sensitive to impurities than zinc. Simultaneous hydrogen evaluation from impurities in the deposit is not a problem with copper because of its nobility. However, the final purity required for copper intensifies the need to eliminate undesirable impurities that may co-deposit and contaminate the cathode or adversely affect electrocrystallization or deposit growth by causing a rough, irregular surface.

B. Recycling. Besides the production of metal from ore, it will become increasingly common in the future to have waste residue as the starting feed. The technical difficulties encountered in treating waste residues are quite formidable. The material is often in a finely divided state and difficult to handle. Such a waste becomes a likely candidate for a combined hydro-electro processing scheme, but there are other complicating factors. The chemical composition of wastes is often quite varied, unlike most ore concentrates obtained by flotation and other mineral dressing operations. Undesirable trace impurities are often more abundant and less readily eliminated than those found in the ore. If the production of zinc using such materials were examined, the problems that might result due to a variety of entrained impurities from multiple sources are evident. Thus, electrolytic metal production promises to become even more complicated than it is at present, and an improved level of understanding of the thermodynamic behavior of these undesirable elements would be very useful.

C. Selected Applications. Some brief examples of the types of uses that can be made of thermodynamics in electrometallurgy are given. Pourbaix diagrams, which have received rather extensive use, are excluded since they are discussed in greater detail in other papers that are a part of this symposium.

One of the steps in the leaching and purification of zinc electrolyte is the oxidation of ferrous to ferric iron, with the subsequent precipitation of $Fe(OH)_3$. Calculations (5) may be performed to determine the influence of pH and oxygen pressure on the final equilibrium ratio obtained for these two ions.

The overall reaction is:

$$4Fe^{+2} + O_2 + 4H^+ = 4Fe^{+3} + 2H_2O$$

with

$$K = \frac{(a_{Fe^{+3}})^4}{(a_{Fe^{+2}})^4 (a_{H^+})^4 \, p_{O_2}} = 10^{31} \qquad (3)$$

Substitution into this expression reveals that at constant pH, the Fe^{+2}/Fe^{+3} ratio goes from 1.0 (arbitrary reference) at 0.3 atmospheres O_2 to 0.6 at 2 atmospheres. Similarly, at constant p_{O_2} this ratio changes from 1.0 at pH 5 to 6.2 at pH 5.8. (For example, the activity ratio decreases from 2.42×10^{-8} to 1.50×10^{-8} at $a_{H+} = 1$ when the p_{O_2} is increased from 0.3 to 2 atmospheres. At $p_{O_2} = 1$, the ratio increases from 1.78×10^{-3} at pH 5 to 1.12×10^{-2} at pH 5.8).

The elimination of impurities by cementation with zinc dust can be estimated by examining an expression of the type:

$$Ni^{+2} + Zn = Ni + Zn^{+2}$$

from which the equilibrium concentration of Ni^{+2} is found to be about 10^{-17} using a reaction potential E° of 0.51 volts. Using estimated activity coefficients, this value becomes approximately 5×10^{-13}. Thus, the indication is that nickel may be successfully removed from aqueous solution using zinc dust and this is borne out in commercial practice. The situation is somewhat complicated in actual practice, as activators, such as copper and arsenic or antimony are commonly needed to effect a removal to the desired level. As mentioned previously, it is in areas such as these that additional thermodynamic analyses are needed if the role of the activators is to be properly understood.

III. Trace Metals and Additives in Electrodeposition.

A. Role of Metal Ions. The previous discussion alluded to the somewhat involved behavior encountered at an electrode when even limited amounts of certain elements are present in solution. Certainly, attention must be paid to the composition of an electrolyte when parts per million or even parts per billion can have a major impact on the processing. Table II lists data that show the influence of certain metal impurities on zinc electrowinning current efficiency. The impurity sites offer localized regions where, because of a more noble potential and lower overpotential, hydrogen ions are selectively reduced rather than zinc. The behavior of impurities is also difficult to predict because their influence on deposit growth or current efficiency is time dependent. A critical time often exists for various concentrations of an impurity beyond which a drastic drop in current efficiency occurs, an example of which is shown in Figure 2. The exact cause of this effect of plating time is not known, but it is thought to be related to some condition on the cathode surface which causes a relative increase in impurity concentration. Once initiated, the local area is subjected to a pitting type of corrosion, a condition which forces an autocatalytic dissolution of the deposited zinc. The condition can become so catastrophic that there is complete dissolution of the zinc previously deposited, which corresponds to a negative current efficiency. There are practical, industrial implications to this phenomenon since the time of deposition and other processing parameters are dictated by the quality of the electrolyte. In certain instances, it may be desired to plate for 48 hours but reduced efficiency with time forces the

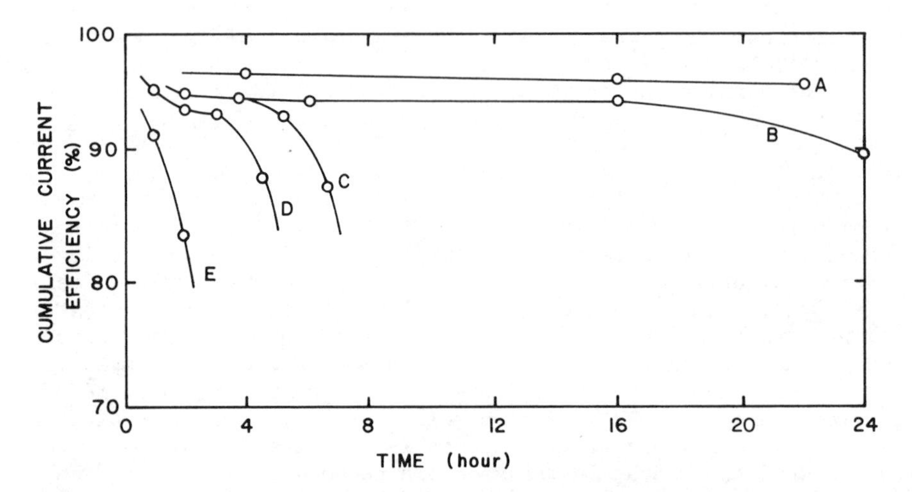

Figure 2. Effect of nickel concentration on the incubation period in zinc electrowinning (7) at 40°C and 75 A/ft² ((A) 0.0 mg/L; (B) 1 mg/L; (C) 1.75 mg/L; (D) 2.5 mg/L; and (E) mg/L

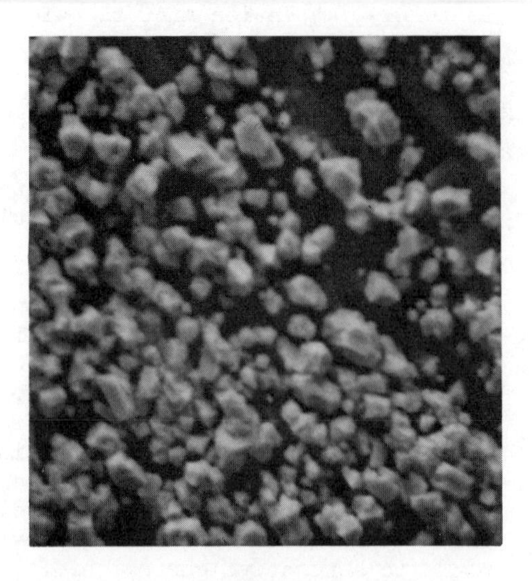

Figure 3. Scanning electron micrograph for the deposit obtained from acidified zinc sulfate electrolyte (9) (0.77M Zn^{++}, 1M H_2SO_4), × 880

cathodes to be pulled and stripped after only 24 hours.

In copper electrolysis, there are only a very limited number of impurities that will cause a current efficiency decrease, and these occur particularly in electrowinning. One troublesome impurity is iron, as the ferric ion is preferentially reduced at the cathode to ferrous ion. The ferrous ion is subsequently reoxidized at the anode. The actual current efficiency obtained is dependent on the iron concentration. (8) For example, an electrolyte containing 10 gpl iron gives a 77% current efficiency, but with 1 gpl, a 90% value can be obtained.

Cathode deposit morphology can also be affected by the presence of trace elements in solution. An SEM micrograph of a zinc deposit from a pure or unadulterated electrolyte and time duration of two minutes at 40 mA/cm^2 is shown in Figure 3. When 40 parts per billion of antimony was added to this solution, and a similar electrodeposition cycle was performed, an approximate ten-fold increase in crystallite size was obtained (see Figure 4).

The influence of antimony at a level of 300 ppm in copper electrolysis is also significant. The morphologies of deposits made from a pure acid-copper sulfate electrolyte and from an identical solution to which the antimony was added are shown in Figures 5 and 6. There are many other combinations of impurities and electrolytes which exhibit this changing surface appearance and deposit orientation besides those selected as examples. Anion effects are also not uncommon, with the halogens often causing the more notable changes.

B. Role of Protein Additives. Protein additives have been used extensively in the electrolytic production of metals. There are few, if any, commercial operations that do not rely almost exclusively on the use of addition agents to assist in process efficiency and cathode growth control. The animal glues, synthetic protein colloids, and gelatin are the most common reagents employed. However, there are a multitude of other organic and inorganic additives employed. The presence of low concentrations of active organics in an electrolyte can have a major influence on the polarization behavior and deposit growth. The influence of an organic additive on zinc deposition can be seen by comparing the structure in Figure 3 with that in Figure 7, which was made with 80 parts per million glue in the electrolyte. The fibrous appearance is due to the basal plane orientation being perpendicular to the substrate.

The deposit orientation is altered when additives are present as seen from Table III, and the degree of change is a function of both concentration and type of glue added. There are a variety of leveling agents besides glue used in copper electrolysis. A few of the more prominently mentioned include thiourea, Separan, gum arabic, jaguar C, Avitone, lignon sulphonate, casein and goulac being examples. Chlorine is added (usually as HCl) to assist in controlling the silver in the cathode, but it also can influence the deposition process. It would be very desirable to be able to

Figure 4. Scanning electron micrograph for the deposit obtained from acidified zinc sulfate electrolyte (9) (0.77M Zn^{++}, 1M H_2SO_4) containing 40 ppb Sb, × 920

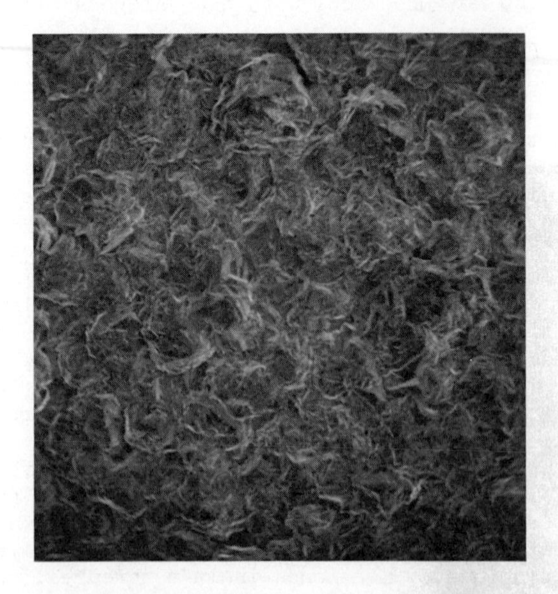

Figure 5. Copper deposit with no additives (10) × 80

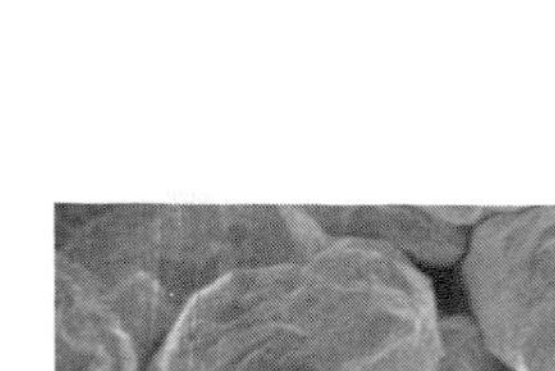

Figure 6. Copper deposit with 300 ppm Sb (10) × 80

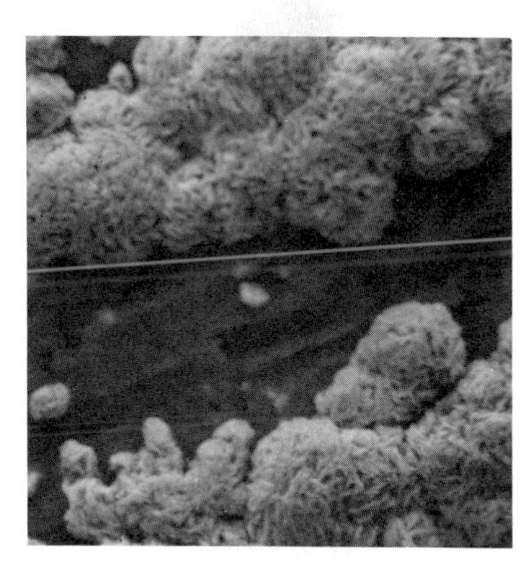

Figure 7. Scanning electron micrograph for the deposit obtained from acidified zinc sulfate electrolyte (9) (0.77M Zn^{++}, 1M H_2SO_4) containing 80 ppm glue, × 880

Table II. Effect of Impurities on Current Efficiency for Zn Electrowinning. (6)

Element	Concentration (mg/ℓ)	% Change in C.E. Relative to Pure Zn Solution
Ag	10	-79
Au	4	-83
As	1	-28
Cd	12	+4
Co	10	-15
Cu	15	-36
Ge	1	-56
In	5	-54
Ni	10	-16
Pb	10	+5
Pt	2	-69
Sb	2	-77
Sn	10	-32

Table III. Effect of Additives on the Orientation of Cu Deposit. (11)

Addition (mg/ℓ)	(111)	(200)	(220)	(311)	(222)	(400)	(331)	(420)
Blank	0	0	100	0	0	0	0	0
TCP-EZ3, 2.4	1	1	100	9	0	0	2	0
TCP-EZ3, 4.8	45	25	100	54	3	1	16	34
TCP-EZ3, 9.6	100	100	64	50	7	3	19	29
TCP-EZ3, 19.2	47	100	2	6	2	5	4	2
TCP-69, 2.4	4	6	100	4	0	0	8	2
TCP-69, 4.8	3	3	100	4	0	3	9	5
TCP-69, 9.6	5	0	100	16	0	0	7	0
TCP-69, 19.2	2	2	100	14	0	0	5	1
TCP-69, 38.4	100	2	4	5	0	0	5	0

determine the thermodynamic contribution to the orientation changes. Unfortunately, available techniques and data do not allow such evaluations to be made, but it is a goal worth attaining.

C. Chemical Interactions in Solution. The rather dramatic effect of trace metal ions and additives on electrodeposition was reviewed briefly in the previous sections for copper and zinc. The examples given were typical of the response to low concentrations of numerous chemical species observed for many electrolytic systems. The problem of deposit control is difficult enough because of the strong influence of these minor constituents, but it is complicated even further due to interactions among the various ions and additives in the electrolyte.

Zinc morphologies varied substantially when either antimony or glue was in solution. When both are present, an interaction occurs which tends to eliminate the effects of both. This is revealed not only in the structure, but in the current efficiency as well. The detrimental effect of antimony is shown to be eliminated by adding the proper amount of glue (see Figure 8).

In the past, a number of factors were responsible for the difficulty in evaluating the point when the proper ratio was attained. First, there is no fast, reliable means for chemically analyzing the amount of organics in these strong electrolytes. Secondly, even though the antimony concentration can be determined, it is found that other ions (e.g. fluorine) can cause the "effective" concentration of the antimony to vary from the actual physical amount present. This can be demonstrated by measuring the degree of initial polarization on cyclic voltammograms from solutions containing antimony and glue. Figure 9 illustrates both the strong influence of the single additives and the interaction between them that appears to reduce the effect of each. Undesirable synergistic effects can also be obtained, an example of which is given in Figure 10 for the combination of cobalt and antimony.

As might now be expected, certain similar occurrences are present in copper electrolysis. Table IV lists the concentration of antimony analyzed in the copper cathode when various levels of the additives were in the electrolyte. The data indicate a critical concentration of antimony in solution is necessary (approximately 300 ppm) before appreciable amounts are taken into the cathode. Glue does not alter the results for 600 ppm antimony but chloride ion reduces the effective antimony to 200 ppm or less. To further complicate matters, when 30 ppm chloride ion is present and titanium cathode blanks are used for the copper deposition, an incomplete or lacy structure is obtained. If 30 ppm glue is, in turn, added to this solution, the effect of chloride ion is counteracted and complete coverage is again obtained.

IV. Summary and Conclusions.

A very limited view of some of the technical problems that might be encountered in electrometallurgy has been given.

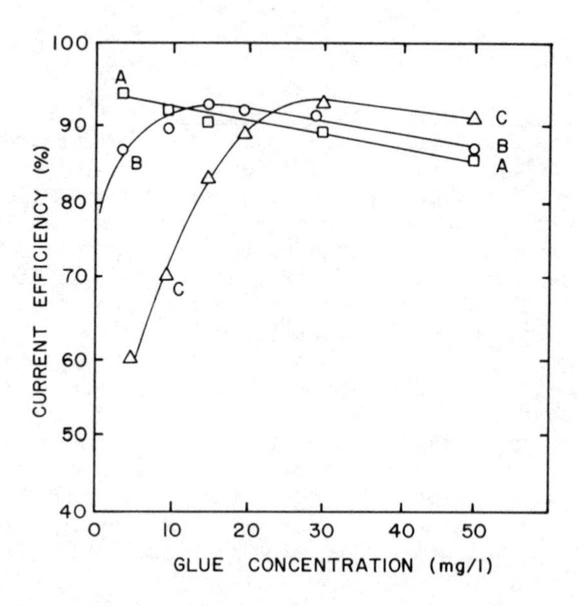

Figure 8. Effect of glue addition on current efficiency (12) of zinc electrowinning for antimony-containing solutions ((A) 0.0 mg/L Sb; (B) 0.04 mg/L Sb; and (C) 0.08 mg/L Sb)

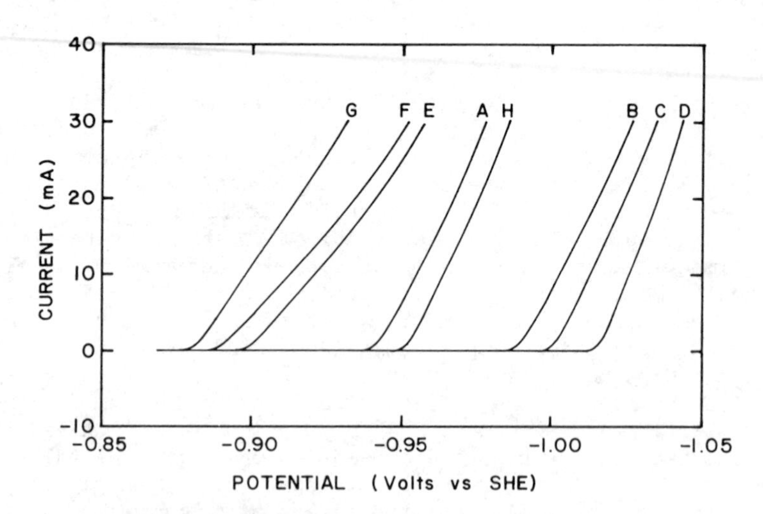

Figure 9. Transient polarization curves for acidified zinc sulfate electrolyte (9) (0.77M Zn^{++}, 1M H_2SO_4): (A) no additions; (B) 10 ppm glue; (C) 20 ppm glue; (D) 40 ppm glue; (E) 10 ppb Sb; (F) 20 ppb Sb; (G) 40 ppb Sb; (H) 40 ppb Sb + 20 ppm glue. Area of aluminum cathode = 1.18 cm^2.

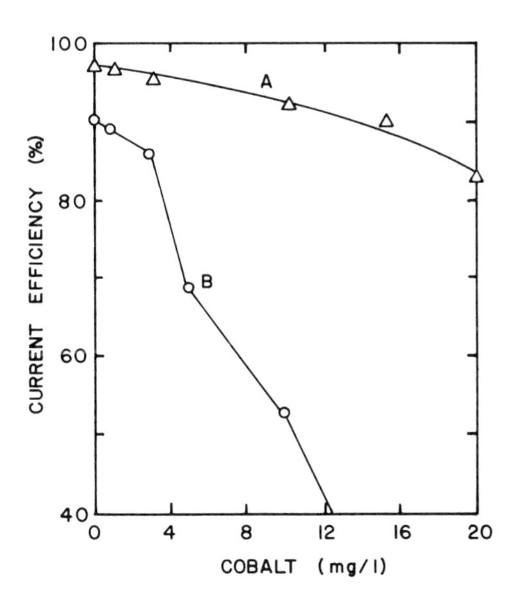

Figure 10. Current efficiency for zinc electrowinning vs. cobalt concentration for prepared electrolytes (4) containing 65 g/L Zn and 100 g/L H_2SO_4; T = 45°C, current density = 40 mA/cm²; (A) 0.0 ppb Sb; (B) 28 ppb Sb

Table IV. Atomic Absorption Analysis of Antimony in the Copper Deposit. (9)

Addition (ppm)	% Antimony in Deposit
Sb 60	0.0023
Sb 100	0.0033
Sb 200	0.0058
Sb 300	0.12
Sb 600	0.30
Sb 600, glue 7·5	0.26
Sb 600, glue 15	0.37
Sb 600, glue 30	0.35
Sb 600, glue 60	0.31
Sb 600, Cl^- 1·5	0.0027
Sb 600, Cl^- 3·5	0.0038
Sb 600, Cl^- 7·5	0.0023
Sb 600, Cl^- 15	0.0035
Sb 600, Cl^- 30	0.0027

Hopefully, it has been shown that the existing ability to truly control the electrodeposition of metals is limited indeed and that an extensive amount of work is necessary in this area. A better understanding of the actual mechanisms of reduction will probably come through a kinetic evaluation. However, it may be equally important to provide a better understanding of the thermodynamics and solution chemistry of these complex systems even before polarization occurs. The need to gain more insight into the interaction among chemical species in solution, particularly between trace impurities and organics, can't be overlooked. Identification of the species as they exist in the double layer and the determination of their thermodynamic properties would be of assistance in gaining insight into their influence on electrode processes. Some sort of thermodynamic interaction coefficient for trace metals and organics in various high concentration aqueous electrolytes is badly needed. This factor may or may not be a strict activity coefficient similar to that encountered in high temperature pyromettalurgy. At this stage, so little is understood about these parameters in industrial operations that it is even difficult to identify the specific problem. The consequences, however, are more easily determined. With the ever increasing pressure to improve processing efficiencies in this era of predicted shortages, it is obvious the technical problems will magnify proportionally. If this challenge is to be met, then a much higher level of understanding of our electrolytic systems must evolve. The role of thermodynamics in attaining this goal could be very substantial, and undoubtedly a greater emphasis must be placed on a more complete characterization of electrolyte systems if there is to be any meaningful level of success attained.

Literature Cited.

1. Bockris, J. O'M.; Drazic, D.; "Electrochem Science," Barnes and Noble, 1972, p. 79.
2. Ettel, V. A.; CIM Bulletin, 1977, 70 (783), 179-187.
3. Somasundaran, P.; Fuerstenau, D. W.; "Research Needs in Mineral Processing," NSF Workshop ENG-75-09322, Columbia University, August 1975.
4. Fosnacht, D. R.; Ph.D. Dissertation, University of Missouri-Rolla, 1978.
5. Bratt, G. C.: Gordon, A. E.; "Research in Chem. and Ext. Met.," Australian Inst. of Mining and Metallurgy, 197-210, 1967.
6. Bratt, G. C.; The Australian Institute of Mining and Metallurgy Conference at Tasmania, 1977.
7. Fukubayashi, H.; Ph.D. Dissertation, University of Missouri-Rolla, 1972.
8. Biswas, A. K.; Davenport, W. G.; "Extractive Metallurgy of Copper," Pergamon Int. Library, 20, p. 329, 1976.

9. Lamping, B. A.; O'Keefe, T. J.; Met. Trans., 1976, 7B, pp. 551-558.
10. O'Keefe, T. J.: Hurst, L. R.; J. Applied Electrochem., 1978, 8, pp. 109-119.
11. Lorenz, W. P.; O'Keefe, T. J.; Sonnino, C. B., Surface Technology, 1978, 6, pp. 179-190.
12. Robinson, D. J.; O'Keefe, T. J.; J. Applied Electrochem, 1976, 6, pp. 1-7.

RECEIVED January 31, 1980.

37

Activity Coefficients at High Concentrations in Multicomponent Salt Systems

EDWARD W. FUNK[1]

Corporate Research—Science Laboratories, Exxon Research and Engineering Company, Linden, NJ 07036

Phase behavior in concentrated aqueous electrolyte systems is of interest for a variety of applications such as separation processes for complex salts, hydrometallurgical extraction of metals, interpretation of geological data and development of high energy density batteries. Our interest in developing simple thermodynamic correlations for concentrated salt systems was motivated by the need to interpret the complex solid-liquid equilibria which occur in the extraction of sodium nitrate from complex salt mixtures which occur in Northern Chile (Chilean saltpeter). However, we believe the thermodynamic approach can also be applied to other areas of technological interest.

Understanding of phase behavior in concentrated salts systems requires liquid-phase activity coefficients for the electrolytes and for water in the multicomponent system. Although there is a large number of experimental data (1,2,3) for ternary aqueous electrolyte systems, few equations are available to correlate the activity coefficients of these systems in the concentrated region. The most successful present techniques are those discussed by Meissner and co-workers (4,5) and Bromley (6)

Our approach is different from previous methods in two basic aspects. First, we define our standard state as the saturated solution and, second, we define our activity coefficients in a way similar to that commonly used for nonelectrolytes. This approach allows a simple thermodynamic treatment of the concentrated region although the approach is not appropriate, either practically or theoretically for the highly dilute region.

To illustrate the use of our thermodynamic treatment for concentrated systems, we have selected the system HCl-$NaCl$-H_2O where there are complete and precise thermodynamic data for the ternary system and the constituent binaries from low molality

[1] Work done at Universidad Técnica del Estado, Santiago, Chile.

0-8412-0569-8/80/47-133-717$05.75/0

up to highly concentrated solutions. Figure 1 shows the mean ionic activity coefficients of HCl and NaCl as functions of the total molality at 25°C; this figure shows that the activity coefficients of the electrolytes show a large change with composition in the concentrated region.

We present results describing the solid-liquid and the vapor-liquid equilibria in the NaCl-HCl-H_2O system. In the first part, purely empirical relations are used to describe the activity coefficients and the second part includes use of a semi-empirical model (7) to describe the compositional dependence of the activity coefficients.

The final section of the paper discusses areas of applications of this thermodynamic techniques to systems of practical interest and also limitations of our approach.

Thermodynamics of Dilute Solutions

The activity coefficients in dilute aqueous solutions (molalities less than 0.2) can be described using the Brönsted-Guggenheim theory (8). The equations for the activity coefficients are derived by first defining the excess Gibbs energy as the difference between the Gibbs energy of the solution and that when each ion is in its standard state of infinite dilution. For convenience, this expression is divided into two parts; the first leads to the Debye-Hückel limiting law, and the second is the correction to the Debye-Hückel theory.

The first part of the excess Gibbs energy is not written explicitly, since it is not possible to form a simple, intuitive expression that gives, after appropriate differentiation, the Debye-Hückel limiting law. However, the second part of the excess Gibbs energy is very similar to the Wohl expansion used for non-electrolyte solutions (9). Considering a ternary aqueous solution of ions j, k and a common ion i, the excess Gibbs energy (g^E) is:

$$g^E = g^E \text{ (Debye-Hückel)} + 2A_{ij}x_i x_j + 2A_{ik}x_i x_k \qquad (1)$$

where ion j belongs to electrolyte 2 and k to electrolyte 3. x_i is the ionic mole fraction of ion i, and the parameters A_{ij}, A_{ik} characterize the interactions between pairs of ions. As suggested by the Brönsted rule, no terms appear for the interaction between ions of like sign.

Equation 1 is differentiated to calculate the ionic activity coefficients using the same technique as for non-electrolytes. Combining these ionic activity coefficients to form the mean ionic activity coefficients, γ_i, we obtain the Brönsted-Guggenheim equations for two 1-1 electrolytes with a common ion:

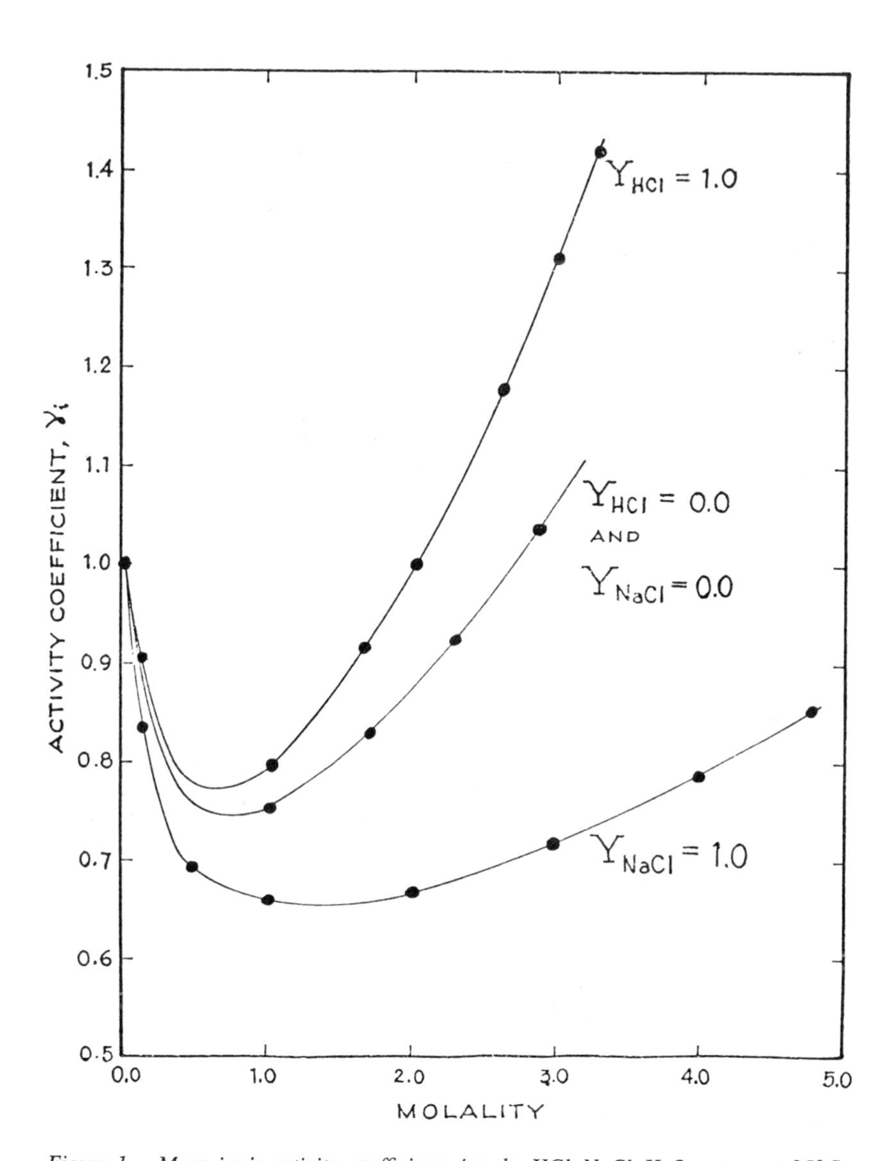

Figure 1. Mean ionic activity coefficients for the HCl–NaCl–H_2O system at 25°C

$$\ln\gamma_2 = -cI^{1/2}/(1 + I^{1/2}) + 2A_{ij}m_2 + (A_{ij} + A_{ik})m_3 \qquad (2)$$

$$\ln\gamma_3 = -cI^{1/2}/(1 + I^{1/2}) + 2A_{ik}m_3 + (A_{ij} + A_{ik})m_2 \qquad (3)$$

where c is the Debye-Hückel constant, $m_2 m_3$ are the molalities of the electrolytes, and I is the ionic strength. The interaction parameters of Equations 2 and 3 can be readily calculated using only data for the binary solutions. However, without further interactions terms, the Brönsted-Guggenheim equations are limited to molalities below 0.2.

Harned's Rule

Activity coefficients in concentrated solutions are often described using Harned's rule (1). This rule states that for a ternary solution at constant total molality the logarithm of the activity coefficient of each electrolyte is proportional to the molality of the other electrolyte. The expressions for the activity coefficients are written:

$$\log \gamma_2 = \log \gamma_{2(0)} - \alpha_{23}m_3 \qquad (4)$$

$$\log \gamma_3 = \log \gamma_{3(0)} - \alpha_{32}m_2 \qquad (5)$$

where subscript 2 refers to HCl and 3 to NaCl. The activity coefficient $\gamma_{2(0)}$ is calculated at the total molality of the solution and assuming the absence of 3; there is a similar definition for $\gamma_{3(0)}$. The parameters α_{23} and α_{32} characterize the interactions occuring between electrolytes 2 and 3. Harned's rule correlates the experimental activity coefficients for most ternary aqueous electrolytes solutions.

For dilute solutions, Equations 4 and 5 reduce to the Brönsted-Guggenheim equations, and the parameters α_{23} and α_{32} can be expressed in terms of the interaction parameters of the Brönsted-Guggenheim theory. For concentrated solutions, Harned's rule is a simple empirical extension of the Brönsted-Guggenheim theory. Thus, it is surprising how well the rule describes activity coefficients in highly concentrated solutions.

Figure 2 presents the parameters of Equations 4 and 5 as functions of temperature and total molality. The experimental data compiled and discussed by Harned and Owen (1) were used to calculate the experimental parameters shown in Figure 2. The parameter α_{23} changes significantly with temperature and molality, however becomes independent of molality at high molalities. On the other hand, α_{32} varies in a complicated manner with temperature and molality. The results of Figure 2 show that the parameters of Harned's rule cannot be reliably

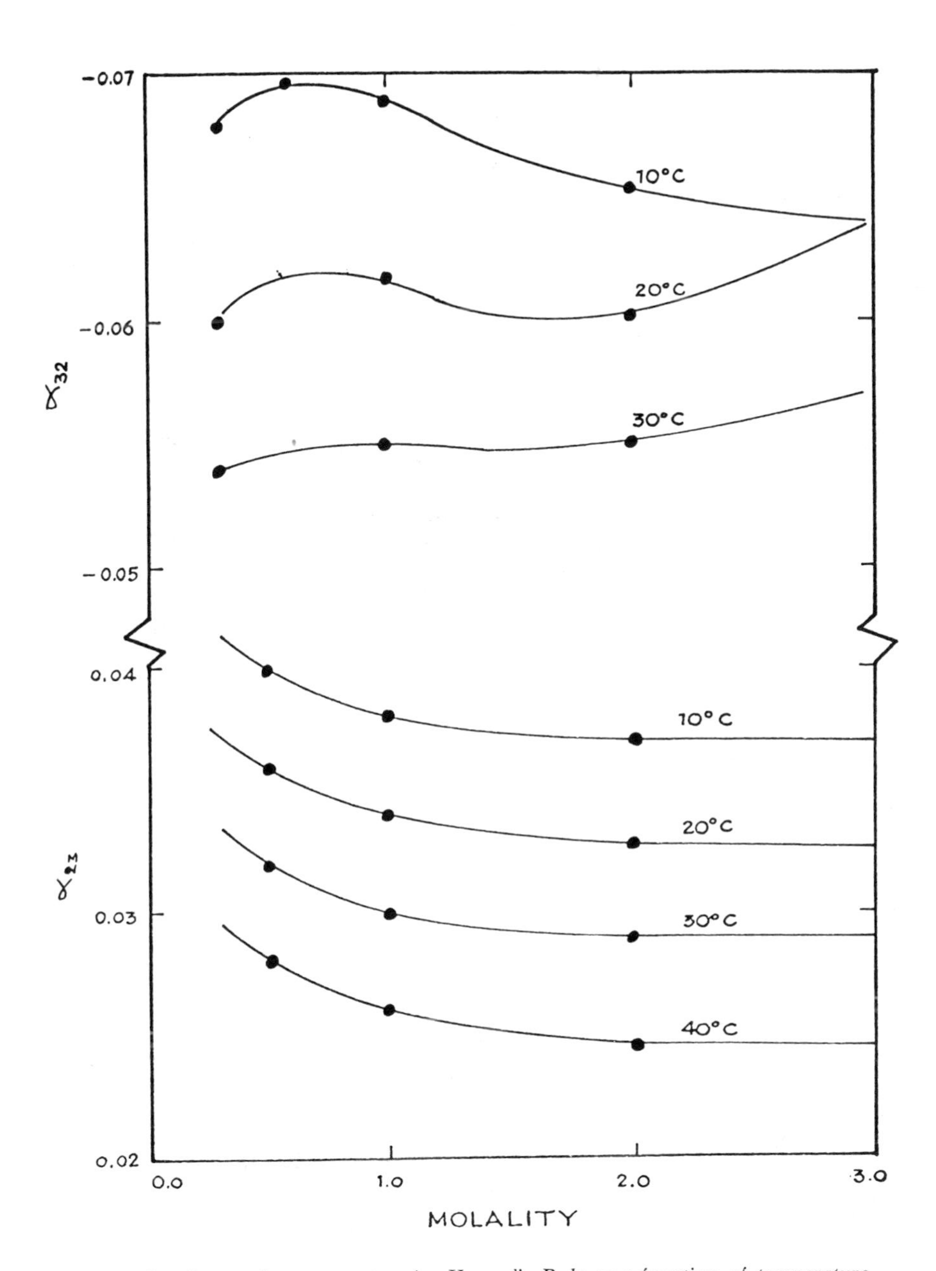

Figure 2. Interaction parameters for Harned's Rule as a function of temperature and total molality

extrapolated to unstudied conditions of temperature and total solution molality.

The success of Harned's rule for ternary solutions is largely fortuitous, and the rule has no theoretical basis to expect that it would be useful for solutions containing more than two electrolytes. Furthermore, for high concentrations of several electrolytes, activity coefficients such as $\gamma_{3(0)}$ are hypothetical. There are, unfortunately, few experimental data available to test Harned's rule for concentrated solutions of three or more electrolytes.

Thermodynamics of Concentrated Solutions

The Brönsted-Guggenheim equations provide a highly satisfactory description of the activity coefficients in dilute solutions; however, their empirical extension to concentrated solutions (Harned's rule) introduces several serious problems.

For concentrated solutions, the activity coefficient of an electrolyte is conveniently defined as though it were a non-electrolyte. This is a practical definition for the description of phase equilibria involving electrolytes. This new activity coefficient Γ_i can be related to the mean ionic activity coefficient by equating expressions for the liquid-phase fugacity written in terms of each of the activity coefficients. For any 1-1 electrolyte, the relation is:

$$x_i \Gamma_i f_i^\circ = m_i^2 \gamma_i^2 H_i \tag{6}$$

where the activity coefficient of electrolyte i alone in solution, Γ_i°, is normalized such that:

$$\Gamma_i^\circ \to 1.0 \quad \text{as} \quad x_i \to x_i^* \tag{7}$$

The standard state for the mean ionic activity coefficient is Henry's constant H_i, f_i° is the standard-state fugacity for the activity coefficient Γ_i, and x_i is the mole fraction of electrolyte i calculated as though the electrolytes did not dissociate in solution. The activity coefficient Γ_i° is normalized such that it becomes unity at some mole fraction x_i^*. For NaCl, x_3^* is conveniently taken as the saturation point. Thus Γ_3° is unity at 25°C for the saturation molality of 6.05. The activity coefficient of HCl is normalized to be unity at an HCl molality of 10.0 for all temperatures. These standard states have been chosen to be close to conditions of interest in phase equilibria.

Figure 3 shows the activity coefficient of HCl in aqueous solution, Γ_2°, as a function of liquid-phase composition for 10, 25 and 50°C. Experimental activity-coefficient data given by Harned and Owen (1) were used in conjunction with

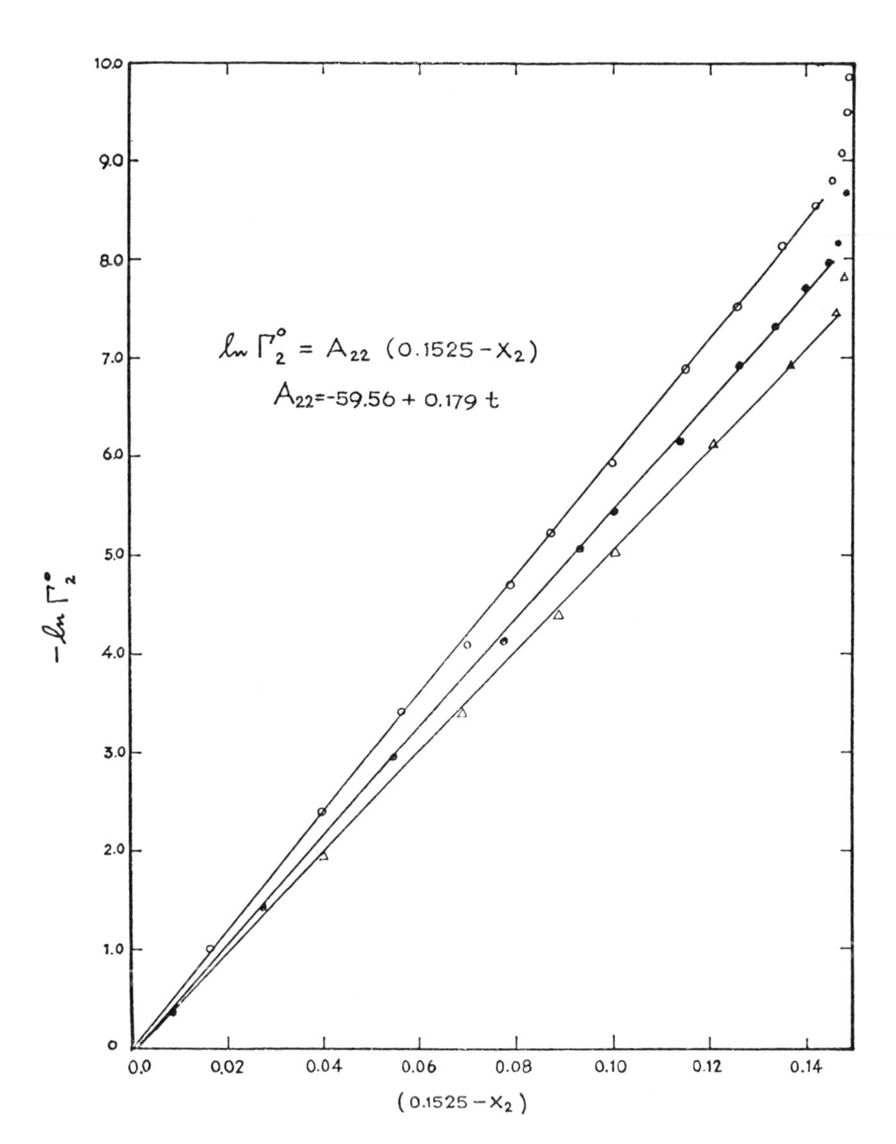

Figure 3. Activity coefficient of HCl as a function of temperature and composition (experimental points: (○) 10°C; (●) 25°C; (△) 50°C)

Equation 6 to calculate experimental values of Γ_2°; the ratio of standard states appearing in Equation 6, f_i°/H_i, was calculated using experimental data for the standard-state composition and the normalization condition, Equation 7. The activity coefficient shown in Figure 3 changes smoothly with composition and does not have a minimum such as the mean ionic activity coefficient of HCl shown in Figure 1. Figure 4 shows the activity coefficient of NaCl, Γ_3°, as a function of $x_3^*-x_3$ for various temperatures from 0 to 50°C. Again the mean ionic activity coefficient data compiled by Harned and Owen were transformed using Equation 6 to obtain the Γ_3° activity coefficients shown in Figure 4. For both HCl and NaCl the activity coefficient Γ_i° changes rapidly with liquid-phase mole fraction only in dilute solutions. These concentrations are not of present interest, since they are already well described by the Brönsted-Guggenheim equations.

The activity coefficients of HCl(2) and NaCl(3) at molalities above 0.2 in their respective binary solutions can be calculated by:

$$\ln \Gamma_2^\circ = (-59.56 + 0.179t)\ [0.1525 - x_2] \qquad (8)$$

$$\ln \Gamma_3^\circ = -28.05\ [x_3^* - x_3] - 0.0436\ e^{50(0.0594 - x_3)} \qquad (9)$$

from 0 to 50°C and where the change of the saturation mole fraction, x_3^*, with temperature accounts for the temperature variation of the activity coefficient of NaCl. The data of Seidell (3) were used to express x_3^* as:

$$x_3^* = 0.0990 + 0.00028\ (t/10) \qquad (10)$$

from 0 to 50°C where t is the temperature in °C.

For ternary aqueous solutions of HCl and NaCl, the following semiempirical equations are proposed to describe the activity coefficients of the electrolytes:

$$\ln \Gamma_2 = \ln \Gamma_2^\circ + A_{23}x_3 \qquad (11)$$

$$\ln \Gamma_3 = \ln \Gamma_3^\circ + A_{32}x_2 \qquad (12)$$

where A_{23} and A_{32} are parameters used to characterize the interactions between the two different electrolytes and, like the parameters in Harned's rule, must be calculated from data for the ternary mixture. The activity coefficient Γ_i° is calculated at the molality of electrolyte i, not at the total molality. Since Γ_i° is calculated using Equation 8 or 9, Equations 11 and 12 are limited to the calculation of Γ_i at electrolyte molalities, m_i, above 0.2.

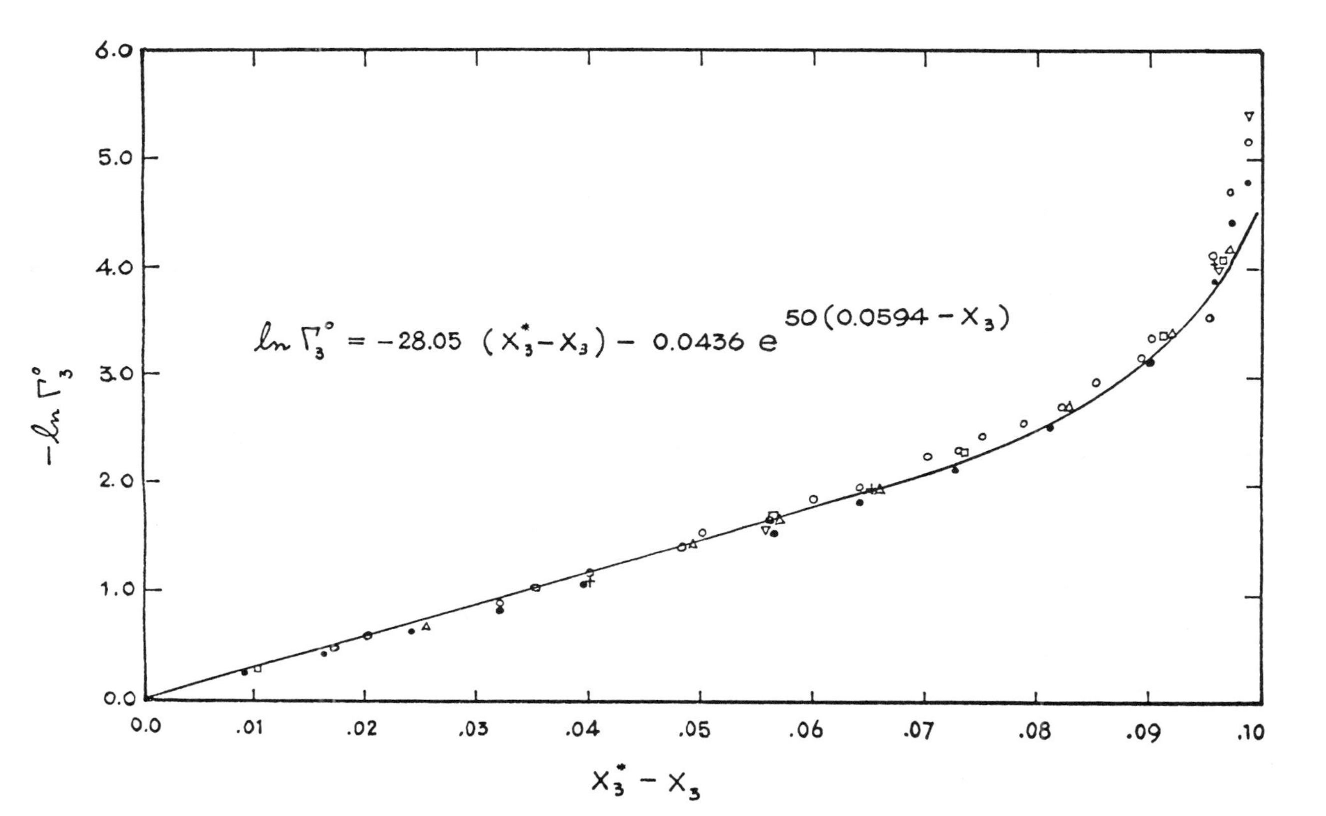

Figure 4. Activity coefficient of NaCl as a function of temperature and composition (experimental points: (●) 0°C; (○) 10°C; (▽) 20°C; (□) 30°C; (+) 40°C; (△) 50°C

The Γ°_i terms of Equations 11 and 12 give the contribution to the excess Gibbs energy due to the electrolytes not being at their respective standard-state concentrations. The terms containing the interaction parameters give the contribution to the excess Gibbs energy due to interactions between electrolytes 2 and 3. This second part of the excess Gibbs energy has the same form used in the Brönsted-Guggenheim theory, Harned's rule, and non-electrolyte solutions. Equations 11 and 12 can be easily extended to solutions containing more than two electrolytes.

Equations 11 and 12 were fit to the experimental activity coefficients of HCl and NaCl as described by Harned's rule. Figure 2 was used to calculate the parameters for Harned's rule from 10 to 40°C and for total molalities 0.2 to approximately 6. The interaction parameter A_{23} is independent of the total molality; the parameter A_{32} decreases with total molality but appears to reach a constant value at high molalities. Both parameters are weak functions of temperature and can be expressed by:

$$A_{23} = 48.50 - 0.70\ (t/10) \tag{13}$$

$$A_{32} = 68.80 - 1.33\ (t/10) - 3.20\ m. \tag{14}$$

from 0 to 50°C and in the total molality range 0.2 to approximately 10 m. Activity coefficients of HCl calculated using Equations 11 and 13 are within $\pm$ 1% of the experimental values. For NaCl, deviations between experimental and calculated activity coefficients are less than $\pm$ 3%. Estimates of the activity coefficients at total molalities above 10 are probably not reliable.

Activity Coefficient of Water in Concentrated Solutions

Equations 8 and 9 can be used with the Gibbs-Duhem equation to calculate $\Gamma^\circ_{1(i)}$, the activity coefficient of water, for each of the binary systems. The Gibbs-Duhem equation for a binary aqueous electrolyte solution is written:

$$\ln \Gamma^\circ_{1(i)} = \int_{x_i = x_i^*}^{x_i} \left(-\frac{x_i}{x_1}\right) d \ln \Gamma^\circ_i \tag{15}$$

where the activity coefficient of water is normalized such that it becomes unity when the activity coefficient of the electrolyte is unity. Substitution of Equation 8 into Equation 15 gives:

$$\ln \Gamma^\circ_{1(2)} = (-59.56 + 0.179t)\ \left[(x_2^* - x_2) + \ln \frac{(1-x_2^*)}{(1-x_2)}\right] \tag{16}$$

for the activity coefficient of water in the $HCl-H_2O$ system, where $x_2^*=0.1525$. Figure 5 shows a comparison between predicted and experimental activity coefficients of water in the $HCl-H_2O$ system at 10, 25 and 50°C; Equation 16 predicts accurate values of $\Gamma_{1(2)}^{\circ}$ except in very dilute solutions. For such solutions, the Brönsted-Guggenheim theory can be used to calculate the activity coefficient of water.

The activity coefficient of water in the $NaCl-H_2O$ system can be well described by substitution of only the first term of Equation 9 into Equation 15. The resulting expression for the activity coefficient of water is:

$$\ln \Gamma_{1(3)}^{\circ} = -28.05 \left[(x_3^* - x_3) + \ln \frac{(1-x_3^*)}{(1-x_3)}\right] \quad (17)$$

The second term of Equation 9 is only important for the calculation of the activity coefficient of NaCl at low concentrations, and makes little contribution to the integral in Equation 15. Equation 17 predicts the activity coefficients of water within 1% of the experimental values for molalities above 0.2.

For the ternary solution, the Gibbs-Duhem equation can be easily integrated to calculate the activity coefficient of water when the expressions for the activity coefficients of the electrolytes are written at constant molality. For Harned's rule, integration of the Gibbs-Duhem equation gives the activity of water as:

$$\frac{-55.51}{Y_3 m^2} \log \frac{a_w(Y_3)}{a_w(Y_3=0)} = Y_3 (\alpha_{23} + \alpha_{32}) - 2\,\alpha_{23} \quad (18)$$

for a constant total molality of m. The fraction of total electrolyte in solution as i is expressed by Y_i; $a_w(Y_3)$ is the activity of water for a given value of Y_3. The standard state for water is pure water at the temperature of the system. The Brönsted-Guggenheim equations can also be substituted into the Gibbs-Duhem equation to calculate the activity of water in dilute solutions. Harned and Robinson (8) give the result and a detailed discussion of the Brönsted-Guggenheim equations.

Equations 11 and 12 are not written for constant molality, and can not be easily used with the Gibbs-Duhem equation to obtain an analytical expression for the activity of water in the ternary solution. However, it is possible to propose a separate equation for the activity coefficient of water that is consistent with the proposed model of concentrated solutions.

The activity coefficient of water in the ternary solution $\Gamma_{1(2,3)}$, is estimated by:

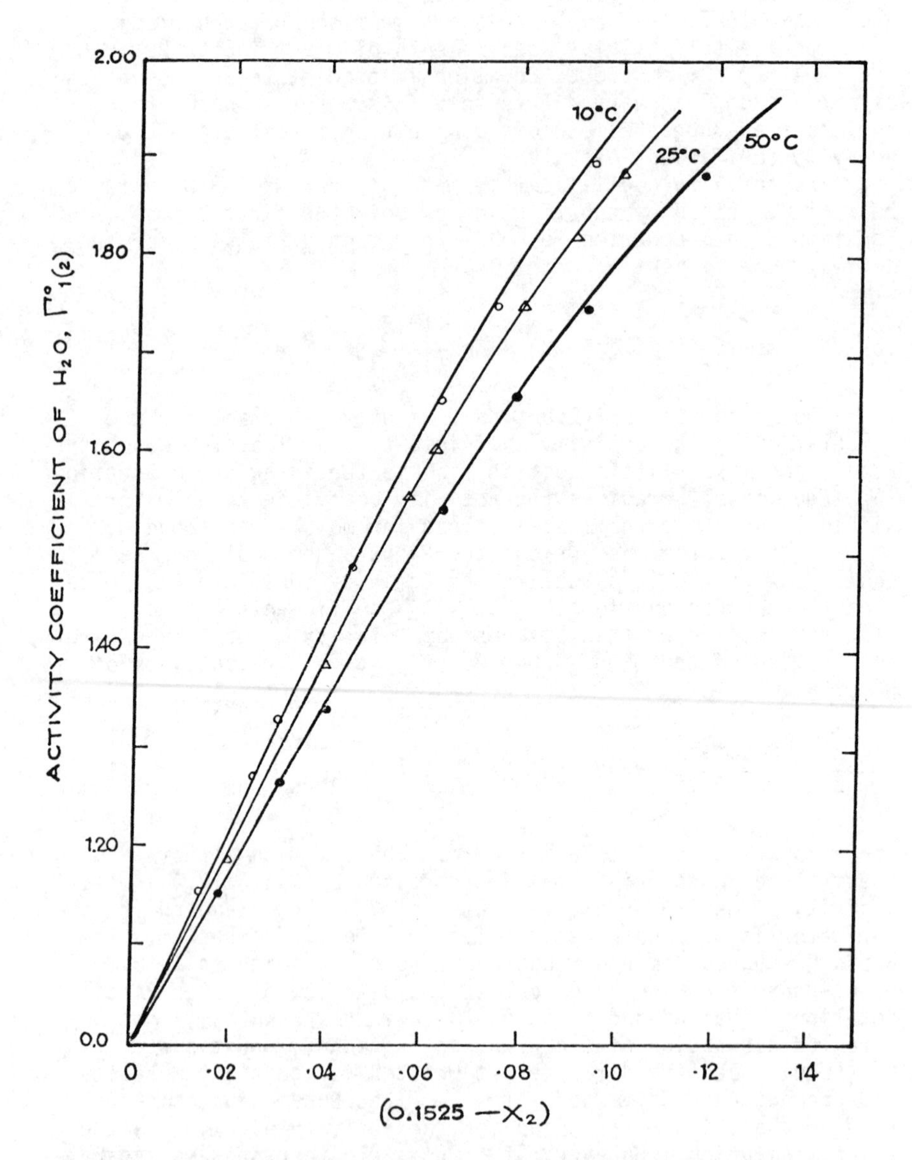

Figure 5. Activity coefficients of water in the HCl–H_2O system at 10° (○), 25° (△), and 50°C (●)

$$\ln \Gamma_{1(2,3)} = Y_2 \ln \Gamma^{\circ}_{1(2)} + Y_3 \ln \Gamma^{\circ}_{1(3)} \quad (19)$$

The standard-state fugacity of water in the ternary solution, $f^{\circ}_{1(2,3)}$, is expressed similarly by:

$$\ln f^{\circ}_{1(2,3)} = Y_2 \ln f^{\circ}_{1(2)} + Y_3 \ln f^{\circ}_{1(3)} \quad (20)$$

where $f^{\circ}_{1(i)}$ is the standard-state fugacity of water in the binary solution of electrolyte i. The activity coefficient $\Gamma^{\circ}_{1(i)}$ is calculated using the mole fraction of i in the ternary solution and either Equation 16 or 17. The standard-state fugacity of water in the ternary solution changes with Y_i since the standard-state fugacity of water is different in each binary system. Equation 20 is similar to the expression derived by O'Connell and Prausnitz (10) for the composition dependence of Henry's constant in a mixed solvent. Equation 19 estimates the activity coefficient of water in the ternary solution using only data for the binary mixtures; therefore, it cannot be expected to give very precise results.

Vapor-liquid equilibrium data for the two binary systems (11) were used to calculate the standard-state fugacities required in Equations 6 and 20. In the temperature range 0-50°C, there fugacities can be expressed by:

$$\ln f^{\circ}_2 = 1.332 + 0.781\ (t/10) \quad (21)$$

$$\ln f^{\circ}_{1(2)} = 0.885 + 0.610\ (t/10) \quad (22)$$

$$\ln f^{\circ}_{1(3)} = 1.485 + 0.591\ (t/10) \quad (23)$$

where the fugacity is in mm. Hg.

Figure 6 compares experimental and calculated activity coefficients of water in the ternary system at 25°C and a total molality of 3.0. Equation 18 was used to express the experimental activity coefficients. Agreement between experimental and calculated values is surprisingly good considering that Equation 19 contains no ternary parameters. The activity coefficient of water in the HCl-$NaCl$-H_2O system is not a strong function of composition, and Equation 19 provides an adequate description of the activity coefficients.

Vapor-Liquid Equilibria

Equations 11 and 19 express the necessary liquid-phase activity coefficients for the calculation of vapor-liquid equilibria in the HCl-$NaCl$-H_2O system. Equation 11 is very convenient

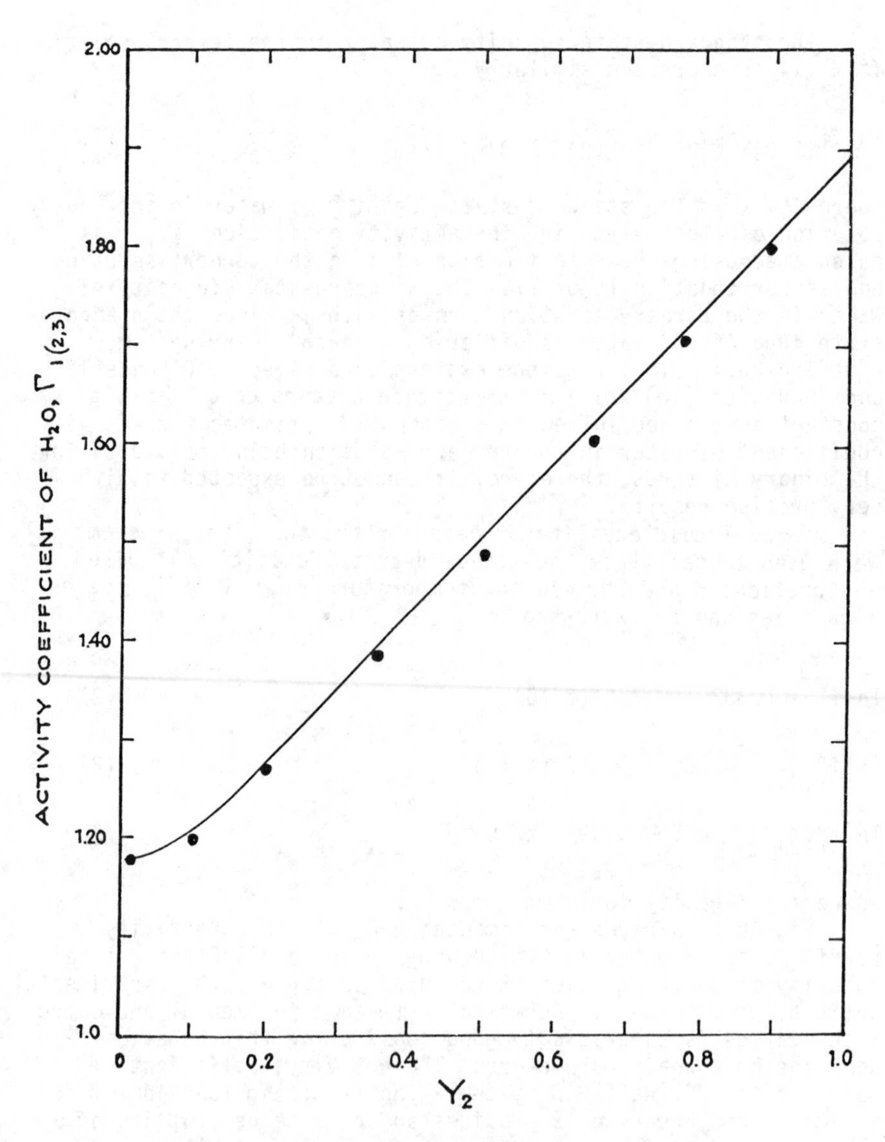

Figure 6. Activity coefficients of water in the HCl–NaCl–H_2O system at 25°C and total molality of 3.0 ((●) experimental)

for vapor-liquid equilibrium calculations, since HCl is treated as a non-electrolyte in both liquid and vapor phases. This avoids the customary equality (12) between the fugacity of HCl vapor and the mean ionic fugacity of HCl.

At moderate pressures, the virial equation of state, truncated after the second virial coefficient, can be used to describe the vapor phase. As suggested by Hirschfelder, et. al. (13) the temperature dependence of the virial coefficients is expressed:

$$B_{11} = 49.85\ [1.0 - 0.328\ e^{1288/T}] \qquad (24)$$

$$B_{22} = 77.43\ [1.0 - 0.704\ e^{400/T}] \qquad (25)$$

where T is the temperature in °K. The correlation of Pitzer (14) was used to calculate the second virial coefficients of HCl, and the experimental data of O'Connell and Prausnitz (15) were used to calculate B_{11} for water. The cross virial coefficients are estimated by:

$$B_{12} = 63.69\ [1.0 - 0.516\ e^{735/T}] \qquad (26)$$

where simple mixing rules were used for the three parameters in Equations 24 and 25.

Figure 7 shows the predicted vapor-phase mole fractions of HCl at 25°C as a function of the liquid-phase molality of HCl for a constant NaCl molality of 3. Also included are predicted vapor-phase mole fractions of HCl when the interaction parameter A_{23} is taken as zero. There are unfortunately no experimental vapor-liquid equilibrium data available for the HCl-NaCl-H_2O system; however, considering the excellent description of the liquid-phase activity coefficients and the low total pressures, it is expected that predicted mole fractions would be within 2-3% of the experimental values.

Solid-Liquid Equilibria

The solid-liquid equilibrium for NaCl is rigorously expressed by:

$$\frac{f_3^S}{f_3^\circ} = x_3 \Gamma_3 \qquad (27)$$

where f_3^S is the fugacity of solid sodium chloride. The ratio of fugacities in Equation 27 is the solubility product of NaCl and can be determined using solubility data. At saturation in the NaCl-H_2O system, Γ_3° is unity and the saturation mole frac-

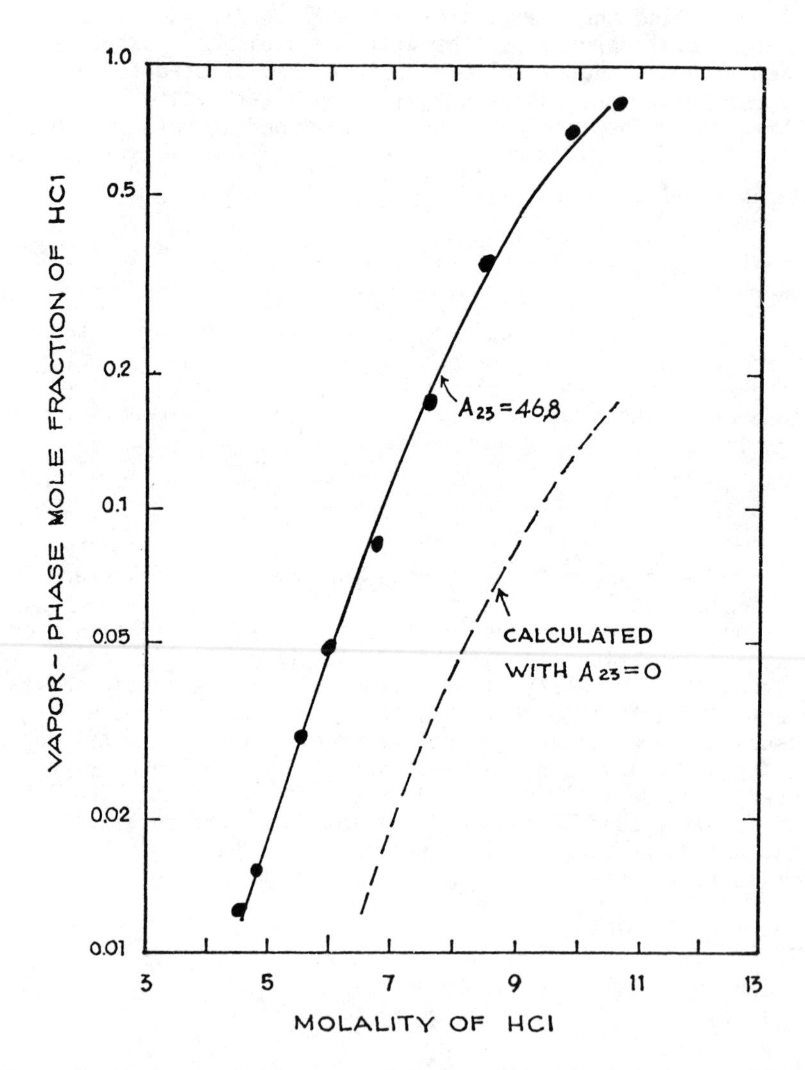

Figure 7. Predicted vapor–liquid equilibria in the HCl–NaCl–H_2O system at 25°C for NaCl molality of 3.0

tion, x_3^*, is equal to the solubility product. Therefore, Equation 10 can be used to express the solubility product of NaCl as a function of temperature.

The selected normalization of the NaCl activity coefficient has two particular advantages for solid-liquid equilibria. First, the solubility product is calculated directly from available solubility data; no activity-coefficient data are required. Second, the activity coefficient of NaCl has a clear interpretation; it provides a quantitative measure of how HCl changes the solubility of NaCl from its standard-state value of x_3^*

Solid-liquid equilibrium data (16) for the HCl-NaCl-H_2O system at 25°C were used with Equation 27 to calculate experimental activity coefficients of NaCl. Table 1 shows a comparison between the experimental activity coefficients and those calculated using Equation 12. The agreement between experimental and calculated activity coefficients is very good, and Equation 12 should be useful for predictions of solid-liquid equilibria at other temperatures.

Equation 27 is similar to the solid-liquid equilibrium relation used for non-electrolytes. As in the case of the vapor-liquid equilibrium relation for HCl, the solid-liquid equilibrium expression for NaCl is simple since the electrolyte is treated thermodynamically the same in both phases.

Extension of Technique by Vera and Co-Workers

Vera and co-workers (7,17,18) have extended the thermodynamic correlation and made two additions. First, they have developed a semi-empirical expression for the excess Gibbs energy in place of the simple empirical equations originally used (Equations 8 and 9). Also, while they use a standard state of the electrolyte of a saturated solution, they change the standard state of water back to the conventional one of pure water.

The expression for the excess Gibbs energy suggested by Correa and Vera is:

$$g_{1i}^E/RT = Ax \ln x + Bx^{1/2} + Kx + cx^{3/2} + Dx^2 + \ldots \quad (28)$$

to describe a binary system of electrolyte i in aqueous solution and where A, B, C, D, . . . are adjustable parameters. K is not an independent parameter and is determined by the normalization condition for the activity coefficient. Application of standard thermodynamics leads to the following expression for the activity coefficient of water:

$$\ln \Gamma_1 = ax + bx^{1/2} + cx^{3/2} + dx^2 + \ldots \quad (29)$$

TABLE I

CALCULATED ACTIVITY COEFFICIENTS OF NaCl AND EXPERIMENTAL VALUES OBTAINED FROM SOLID-LIQUID EQUILIBRIUM DATA

m_2	m_3	Γ_3(Exp.)	Γ_3(Cal.)
0.00	6.126	1.00	1.00
1.00	5.096	1.20	1.24
2.00	4.054	1.51	1.53
3.00	3.100	1.97	2.07
5.00	1.884	3.27	3.27
6.00	1.020	6.09	5.83
7.50	0.496	12.73	12.10
8.50	0.306	20.69	20.10

where A=-a, B-2b, C=-2c, and D=-d. Use of the Gibbs-Duhem equation and the normalization condition leads to the expression for the activity coefficient of the electrolyte:

$$\ln \Gamma_i = (K-a) - a \ln x + bx^{-1/2} + \alpha_1 x + \alpha_2 x^{1/2} + cx^{3/2} + dx^2 \, .. \quad (30)$$

where α_1=a-2d and α_2=b-3c, and where the parameter K is given by:

$$K = a + a \ln x^* - bx^{*-1/2} - x^* - x^{*-1/2} - cx^{*3/2} - dx^{*2} \; . \; . \quad (31)$$

These equations do not reduce to the Debye-Hückel model for dilute solutions and are thus only justified for the treatment of very concentrated solutions.

This model has been applied by Vera and Vega (17) to the NaCl-HCl-H_2O system. Table 2 presents their fit to the vapor pressures of water and the activity coefficients in the NaCl-H_2O system. As can be seen in Table 2, the agreement between the model and the experimental data is very good down to 0.2 molality. In a similar way, it was also found possible to obtain an excellent fit to the experimental data for the HCl-H_2O system.

Figure 8 presents the solid-liquid equilibria in the NaCl-HCl-H_2O system at 30°C; the results of the Vera approach are in very good agreement with the experimental data.

Vera (18) has also examined how his model can be used to predict the activity of water in supersaturated salt solutions. Again, the agreement of the model with experimental data is surprisingly good given the simple nature of the model.

Conclusions

We have presented a thermodynamic technique which is useful for the correlation of thermodynamic data of aqueous electrolyte systems in the concentrated region. The approach was illustrated using the ternary system of HCl-NaCl-H_2O. The correlation gives a good description of solid-liquid and vapor-liquid equilibria; the two ternary parameters required to calculate the activity coefficients of the electrolytes are simple functions of the temperature and the total molality.

This correlation is limited to relatively concentrated solutions (above 0.2 molal for the HCl-NaCl-H_2O system) and has no theoretical significance in dilute solutions since the treatment does not explicitly take into account ionic dissociation of the electrolytes. Furthermore, the correlation is also limited to systems with a common ion.

The proposed correlation is, however, a simple technique for phase-equilibrium calculations in concentrated solutions. The phase-equilibrium relation for volatile electrolytes, such as HCl, has the advantage that the electrolyte in aqueous solution

TABLE II

COMPARISON OF EXPERIMENTAL AND CALCULATED PROPERTIES FOR THE SYSTEM $NaCl-H_2O$- TECHNIQUE OF CORREA AND VERA (7)

	P/P°		$\gamma_\pm$	
Molality	Exp.	Calc.	Exp.	Calc.
0.2	0.9934	0.9935	0.735	0.720
0.4	0.9868	0.9867	0.693	0.694
0.6	0.9802	0.9801	0.673	0.675
1.0	0.9669	0.9667	0.657	0.659
1.2	0.9601	0.9599	0.654	0.657
1.4	0.9532	0.9531	0.655	0.656
1.8	0.9389	0.9389	0.662	0.662
2.0	0.9316	0.9316	0.668	0.668
2.4	0.9166	0.9168	0.683	0.682
2.6	0.9089	0.9091	0.692	0.692
2.8	0.9011	0.9012	0.702	0.702
3.0	0.8932	0.8931	0.714	0.714
3.2	0.8851	0.8850	0.726	0.727
3.6	0.8686	0.8684	0.753	0.754
4.0	0.8515	0.8513	0.783	0.785
4.2	0.8428	0.8426	0.800	0.801
4.4	0.8339	0.8338	0.817	0.819
4.6	0.8250	0.8249	0.835	0.837
4.8	0.8160	0.8157	0.854	0.856
5.0	0.8068	0.8068	0.874	0.875
5.2	0.7976	0.7973	0.895	0.897
5.4	0.7883	0.7881	0.916	0.918
5.6	0.7788	0.7790	0.939	0.939
5.8	0.7693	0.7696	0.962	0.962
6.0	0.7598	0.7600	0.986	0.986
6.17			1.006	1.006

Canadian Journal of Chemical Engineering

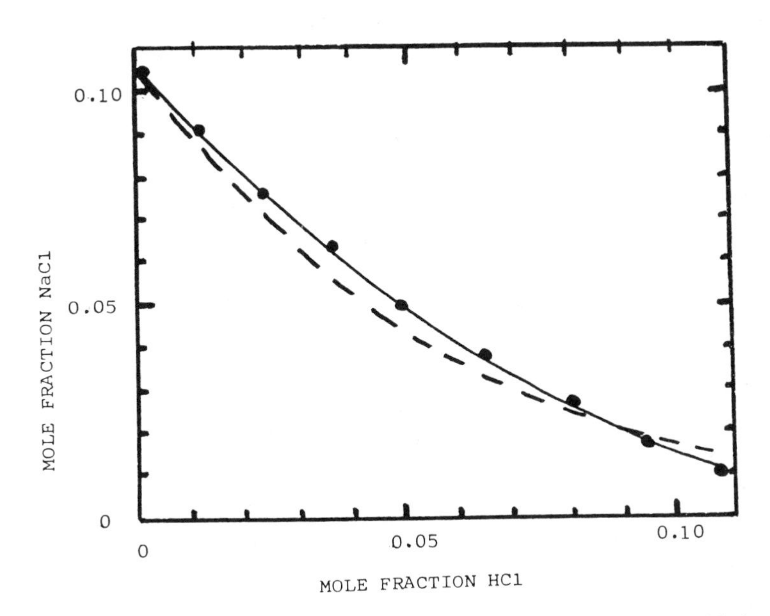

Canadian Journal of Chemical Engineering

Figure 8. Solid–liquid equilibria for the HCl–NaCl system at 30°C—correlation of Vega and Vera (17) ((●) experimental; (– – –) calculated)

is treated as a non-electrolyte, consistent with its state in the vapor phase. Similarly, in the description of solid-liquid equilibria, sodium chloride is considered as the same thermodynamic component in both solid and liquid phase.

The correlation has also been applied to a wide variety of ternary and quarternary systems by Vega and Funk (19). Again, the correlation is very good for the concentrated region and was effective in describing complex equilibria occuring in the quaternary system of $NaCl-NaNO_3-Na_2SO_4-H_2O$.

The work of Vera and co-workers has led to a semi-empirical expression for the excess Gibbs energy which is consistent with our choice of the saturated solution as the standard state for the electrolyte. Vera has, however, shown that pure water is a more convenient standard state for H_2O in place of the saturated solution used by Vega and Funk (19). This is particularly convenient for ternary and higher systems since it avoids the complication of having a composition-dependent standard state.

The correlation presented in this paper can be very simply applied to phase-equilibrium calculations for concentrated electrolyte systems, however, care must be taken to remember that it is basically a correlational approach and not a molecular model for aqueous electrolyte solutions.

Abstract

Semi-empirical equations are proposed to describe activity coefficients at high concentrations in multicomponent salt system. The approach is limited to systems containing a common ion. The proposed equations highlight the region of high concentration and thus complement the various thermodynamic treatments developed for dilute aqueous solutions. The proposed equations for the activity coefficients of the electrolytes contain two parameters determined using activity coefficient data for the ternary system. The activity coefficient of water is well described using only data of the binary subsystems. The activity coefficients in the proposed equations have been defined similarly to those of non-electrolytes. However, they are easily related to the familiar mean ionic activity coefficients. These new equations have potential use in processing of salt systems, hydrometallurgy and interpretation of geological data.

Literature Cited

1. Harned, H. S., Owen, B. B., "Physical Chemistry of Electrolyte Solutions", 3rd Edition, Reinhold, New York, New York, 1958.

2. Robinson, R. A., Stokes, R. H., "Electrolyte Solutions", 2nd Edition, Butterworths, London, 1970.

3. Seidell, S., Linde, W. F., "Solubilities", Vol. 2, ACS, Washington, D.C., 1965.

4. Meissner, H. P., Kusik, C. L., AIChE Journal, 1972, 18, 294.

5. Meissner, H. P., Kusik, C. L., Tester, J. W., ibid., 1972, 661.

6. Bromley, L. A., AIChE Journal, 1973, 19, 313.

7. Correa, H. A., Vera, J. H., Can. J. Chem. Eng., 1975, 53, 204.

8. Harned, H. S., Robinson, R. A., "Multicomponent Electrolyte Solutions", Pergamon, London, 1968.

9. Prausnitz, J. M., "Molecular Thermodynamics of Fluid-Phase Equilibria", Prentice-Hall, Englewood Cliffs, New Jersey, 1969.

10. O'Connell, J. P., Prausnitz, J. M., Ind. Eng. Chem. Fund., 1964, 3, 347.

11. Perry, R. H., Chilton, C. H., "Chemical Engineers' Handbook", 5th Edition, McGraw Hill, New York, 1973.

12. Hala, E., Proc. Int. Symp. Distill., 1969, 3, 8.

13. Hirschfelder, J. O., McClure, F. T., Weeks, I. F., J. Chem. Phys., 1942, 10, 201.

14. Lewis, G. N., Randall, M., "Thermodynamics", Revised at Pitzer, K. S., Brewer, L., 2nd Edition, McGraw-Hill, New York, 1961.

15. O'Connell, J. P., Ind. Eng. Chem. Fund., 1964, 3, 347.

16. Åkerlöf, G., Turck, H. E., J. Am. Chem. Soc., 1934, 56, 1875.

17. Vega, R., Vera, J. H., Can. J. Chem. Eng., 1976, 54, 245.

18. Vera, J. H., Can. J. Chem. Eng., 1977, 55, 484.

19. Vega, R., Funk, E. W., Desalination, 1974, 15, 225.

RECEIVED January 31, 1980.

Modeling the Chemical Equilibria in Solid-Liquid Reactions

Application to Leaching of Ores

KNONA C. LIDDELL[1] and RENATO G. BAUTISTA

Ames Laboratory USDOE and Department of Chemical Engineering,
Iowa State University, Ames, IA 50011

Classical thermodynamic approaches have limited utility when applied to many heterogeneous reaction systems. The fluid phase in such a system may contain a large number of chemical species reacting relatively rapidly with one another. The solid-fluid reaction may be considerably slower and it often happens that a heterogeneous reaction does not reach equilibrium. After a lengthy observation period, the measured concentrations for many systems give a mass action quotient much smaller than the equilibrium constant for the heterogeneous reaction. This is true for the dissolution of several copper sulfides by Fe^{3+} or O_2 in leach dumps, for example. Under some conditions, it is also true for a gas-liquid heterogeneous reaction such as

$$4Fe^{2+} + O_2 + 4H^+ \rightarrow 4Fe^{3+} + 2H_2O. \quad (1)$$

If the discrepancy in the rates of the fluid phase and heterogeneous reactions is great enough, one is justified in considering that the species in the fluid adjust rapidly enough to the changes in fluid phase analytical concentrations from reaction of the solid that chemical equilibrium is maintained <u>within</u> the fluid even though the fluid-solid reaction itself is far from equilibrium. As far as the fluid phase is concerned, reaction of the solid causes perturbations in the analytical (or total) concentrations but these perturbations are applied gradually in infinitesimally small steps; the equilibrium concentrations within the fluid adjust smoothly and remain at equilibrium as the heterogeneous process proceeds.

Following the changes in fluid phase concentrations accompanying heterogeneous reaction is potentially of great value in hydrometallurgy and other applied areas. Knowing how the fluid composition changed during a mineral processing operation, for example, would be useful in optimizing the process.

Equilibrium concentrations in a solution can be calculated by

[1] Current address: Department of Chemical Engineering, Montana State University, Bozeman, Montana 59717

0-8412-0569-8/80/47-133-741$05.00/0

solving a set of simultaneous nonlinear algebraic equations consisting of mass action and material balance expressions. (If ions are present, a charge balance may be written as well.) The analytical (total) concentrations of the solution's components must be known and appear in the material balance equations.

The analytical concentrations change during the heterogeneous reaction. These changes are governed by the stoichiometry of the heterogeneous reaction and are proportional to one another. For example, chalcopyrite is dissolved in acidic ferric chloride solution according to

$$CuFeS_2 + 4FeCl_3 \rightarrow CuCl_2 + 5FeCl_2 + 2S^0. \quad (2)$$

The analytical concentrations needed for a description of such a solution are those of $FeCl_3$, $CuCl_2$, $FeCl_2$ and HCl. For a constant solution volume, the analytical concentration of $FeCl_3$ drops while those of $CuCl_2$ and $FeCl_2$ increase in the ratio -4:1:5. The analytical concentration of HCl is unchanged because H^+ is not a reactant. The total number of negative ions (or Cl^- concentration) remains the same while the identities of the positively charged ions change during reaction.

It is possible to calculate the equilibrium solution concentrations during a heterogeneous reaction by changing the analytical concentrations to take account of the ongoing transfer of reactant out of and product into the solution but it is usually impractical to do so. Large amounts of computer time would be required for such a computation unless the system of equations is a small one and the changes in analytical concentrations are fairly large.

The method presented here provides a more convenient means of obtaining this type of information.

Development of the Partial Equilibrium Model

In a solid-fluid reaction system, the fluid phase may have a chemistry of its own, reactions that go on quite apart from the heterogeneous reaction. This is particularly true of aqueous fluid phases, which can have acid-base, complexation, oxidation-reduction and less common types of reactions. With rapid reversible reactions in the solution and an irreversible heterogeneous reaction, the whole system may be said to be in "partial equilibrium". Systems of this kind have been treated in detail in the geochemical literature (1) but to our knowledge a partial equilibrium model has not previously been applied to problems of interest in engineering or metallurgy.

Physically, the solid and the fluid are linked by the mass transfer between them. The equilibrium concentrations in the solution are continually changing as the analytical concentrations change; the adjustments are constrained to be such that the mass action expressions and balance equations are always satisfied.

The derivation of the equations of the model is discussed in more detail elsewhere (1,2) but is summarized here.

Assume for the moment that there is a single irreversible reaction. Let ξ be a progress variable describing the extent of reaction; $0 \leq \xi \leq 1$. As ξ increases in infinitesimal increments $d\xi$, the solution's analytical concentrations are perturbed and the activities of the species in solution change. Thus for each solute species s_j, a_j is a function of ξ.

Taking a basis of 1 kg of solution, the chemical equation for the irreversible reaction can be written

$$\sum_i \bar{N}_i S_i + \sum_j \bar{n}_j \bar{s}_j = 0 \tag{3}$$

where $\bar{N}_i$ is the number of moles of solid S_i formed and $\bar{n}_j$ is the number of moles of solution species s_j formed. Now consider how the solution is affected: changes in the activity a and the molality m, of a solution phase species are related by

$$da_j = d(\gamma_j m_j) \tag{4}$$

while the concentration change is related to the heterogeneous reaction by

$$\bar{n}_j d\xi = dm_j . \tag{5}$$

This assumes that the solution phase activity coefficients do not change during an incremental change in the extent of the heterogeneous reaction.

For each solution reaction, there is a mass action expression

$$K = \prod_l a_l^{\nu_l} . \tag{6}$$

Because the activities change as the heterogeneous reaction progresses, this may be differentiated with respect to ξ, giving

$$0 = \sum_l \nu_l \frac{d \ln a_l}{d_\xi} . \tag{7}$$

Substituting from Equations (4) and (5) and neglecting the dependence of ln a_l on m_l, and activity changes for any but the solute species

$$0 = \sum_j \frac{\nu_j \bar{n}_j}{m_j} \tag{8}$$

Equation (8) is the basic equation of the partial equilibrium model. There is one equation of this form for every fluid phase reaction that is at equilibrium. The unknowns to be determined are the $\bar{n}_j$. Once an increment $d\xi$ of suitable size is selected, the $\bar{n}_j$ give the molality changes through Equation (5).

Two additional points about Equation (8) need to be discussed here. Equation (8) contains m_j in the denominator. Thus the solution concentrations must be known before the first increment $d\xi$ is taken and none of them can be zero. In practice this means that the set of nonlinear equations (mass action and balance equations) describing the fluid phase in its initial unperturbed equilibrium state must be solved once. Further, Equation (8) does not completely describe a heterogeneous system at partial equilibrium.

Consider the following simple example. Suppose there are just two reactions occurring. In Equation (9), solid S dissociates to give the soluble species M_1 and M_2; product M_1 dimerizes in Equation (10). The second reaction is reversible but the first is not.

$$S \rightarrow M_1 + M_2 \tag{9}$$

$$M_1 + M_1 \rightleftharpoons D \; . \tag{10}$$

One would like to know the changes in the amounts of the four species for an increment $d\bar{\xi}$ of reaction (9). There are four unknowns, $\bar{N}_S$, $\bar{n}_{M_1}$, $\bar{n}_{M_2}$ and $\bar{n}_D$. Applying Equation (8) gives

$$0 = \frac{\bar{n}_D}{m_D} - \frac{2\bar{n}_{M_1}}{m_{M_1}} \tag{11}$$

Two material balances can be written

$$0 = \bar{N}_s + \bar{n}_{M_1} + 2\bar{n}_D \tag{12}$$

$$0 = \bar{N}_s + \bar{n}_{M_2} \; . \tag{13}$$

With only three equations, the problem is not as yet determinate. Two possible approaches can be taken. One is to introduce kinetic information on the rate of dissociation, using for example

$$\frac{-1}{\nu} \frac{d\bar{N}_s}{dt} = r_S \tag{14}$$

where r_S is a function of concentration, temperature, etc., and t is time. Equations (11-14) would give the number of moles of each substance formed directly as a function of time but may be rather difficult to solve. In order to make use of Equation (5) to obtain concentration changes, moreover, a relationship between ξ and t is required. The other possibility is to remove a variable. Dividing each variable by $-\bar{N}_S$ gives

$$n_{M_1} = \frac{\bar{n}_{M_1}}{-\bar{N}_S} = \frac{\text{moles } M_1 \text{ formed}}{\text{moles S reacting}} \tag{15}$$

$$n_{M_2} = \frac{\bar{n}_{M_2}}{-\bar{N}_S} = \frac{\text{moles } M_2 \text{ formed}}{\text{moles S reacting}} \tag{16}$$

$$n_D = \frac{\bar{n}_D}{-\bar{N}_S} = \frac{\text{moles D formed}}{\text{moles S reacting}} \tag{17}$$

The sign convention implicit in Equation (3) is preserved; n_j is positive for a product and negative for a reactant. Then Equations (11-13) become

$$0 = \frac{n_D}{m_D} - \frac{2n_{M_1}}{m_{M_1}} \tag{18}$$

$$0 = -1 + n_{M_1} + 2n_D \tag{19}$$

$$0 = -1 + n_{M_2} \tag{20}$$

Once m_D and m_{M_1} are known, this system is easy to solve.

Now suppose that under some conditions formation of a trimer may be important. There is an additional reaction to take into account

$$D + M_1 \rightleftharpoons T \tag{21}$$

and an additional unknown $\bar{n}_T$ (or n_T). However, there is also an additional equation that can be written (and an extra term to be added to Equation (19)).

$$0 = \frac{\bar{n}_T}{m_T} - \frac{\bar{n}_D}{m_D} - \frac{\bar{n}_{M_1}}{m_{M_1}} \tag{22}$$

Considering more species that are involved in equilibrium reactions makes the number of unknowns larger, therefore, but an equation is added to the system along with each new unknown.

That is not the case for species that participate only in irreversible reactions. Consider another structural form of the solid, S', with a different reaction rate.

$$S' \rightarrow M_1 + M_2 \tag{23}$$

The unknown $\bar{N}_S$ appears in the material balance equations but there are no additional equations to be appended to the set.

Each irreversible reaction causes a disparity between the number of unknowns and the number of equations, forcing either (1) scaling of the variables, (2) use of kinetic data or (3) both.

Setting up the model equations for various situations is discussed in more detail below but we digress here to outline the steps involved in applying the partial equilibrium model. They are as follows:

(1) Decide what reactions need to be considered and which of them are irreversible and which are reversible.
(2) List the unknowns.
(3) Write an equation of the form of Equation (8) for each reversible reaction. To obtain the initial fluid phase concentration m_j, it is necessary to know the analytical concentrations in the solution at the start of the heterogeneous reaction as well as equilibrium constants for the reversible reactions.
(4) Write as many independent mass balances among the unknowns as possible. If there are ions in the fluid phase, a charge balance expressing electroneutrality can be written if desired.
(5) Check the equations written in step (4) to be sure they are all independent.
(6) Reduce the number of equations by scaling them and/or select kinetic data. If kinetic data are to be used, it is advisable to change the variables, replacing t by ξ and C_i by $\bar{n}_i$ or $\bar{N}_i$.

Application of the Partial Equilibrium Model

We now consider the various fine points in the use of the model and then discuss several applications.

One difference between conducting and nonconducting media is that in the former case a charge balance may take the place of one of the material balances. For the same number of solution species, however, the number of independent balance equations is the same in the two cases.

Material balances can be written for moieties which are conserved during the reaction, such as the atoms of a particular element or the total charge, or for reactant or product species if the stoichiometry is unambiguous. Oxidation-reduction reactions may be particularly troublesome. In the following situation, for example, one cannot write a material balance relating protons to water molecules. Consider the oxidation of O_2 to H_2O and the equilibrium dissociation of H_2O.

$$\ldots + O_2 + 4H^+ \longrightarrow \ldots + 2H_2O \qquad (24)$$

$$2H^+ + 2OH^- \rightleftharpoons 2H_2O \qquad (25)$$

The ratio of H^+ to H_2O is 2:1 in Equation (24) and 1:1 in Equation (25); it is not possible to write an equation that describes the overall changes in the amounts of these two species.

Once the chemical reactions of the system have been identified, the number of unknowns chosen affects the size of the set of equations but has no influence on the number of equations that need to be used. To a certain extent, the unknowns to be solved for can be selected arbitrarily; a variable may be excluded from the list of unknowns as long as (1) it is not itself of physical interest and (2) it appears only in the irreversible reactions. Omitting such a variable does not affect the imbalance between the number of unknowns and the total number of equilibrium and balance equations.

Suppose the reactions of the system are the following:

$$S + R^+ \longrightarrow P_1^{\,+} + P_2 \qquad (26)$$

$$P_1^{\,+} \rightleftharpoons (P_1A)^0 \qquad (27)$$

Assume for the moment that it is the concentration changes for P_1^+ and $(P_1A)^0$ as solid S reacts that are sought. Material balance equations can be written for P_1^+ and $(P_1A)^0$ as well as a charge balance but trial and error shows that it is impossible to do so without introducing $\bar{n}_{R^+}$ and $\bar{n}_{A^-}$. P_2 need not appear, however. A set of three independent balances is:

$$\bar{n}_{P_1^+} = -\bar{N}_S - \bar{n}_{(P_1A)^0} \qquad (28)$$

$$\bar{n}_{(P_1A)^0} = -\bar{n}_{A^-} \qquad (29)$$

$$\bar{n}_{R^+} + \bar{n}_{P_1^+} = \bar{n}_{A^-} \tag{30}$$

(Equation (30) assumes that A^- was the only anion present before reaction began.) The equation for the equilibrium reaction is

$$0 = \frac{\bar{n}_{(P_1A)^0}}{m_{(P_1A)^0}} - \frac{\bar{n}_{P_1^+}}{m_{P_1^+}} - \frac{\bar{n}_{A^-}}{m_{A^-}} \tag{31}$$

There are four equations and five unknowns. If P_2 is of physical interest, the unknown $\bar{n}_{P_2}$ can be included; an independent material balance for it is easily written:

$$\bar{n}_{P_2} = -\bar{N}_S \tag{32}$$

There are now six unknowns and five equations. The imbalance between the number of unknowns and the number of equations is unchanged. If there is no reason to include the extra variable, it is best to leave it out, thereby simplifying the computation as much as possible.

This example also shows that the amount of kinetic information needed to complete the set of equations (one equation in this case) depends on the number of independent irreversible reactions (one) and not on the number of chemical species involved only in the irreversible reactions (two) or on the number of these species appearing as unknowns.

With an aqueous fluid phase of high ionic strength, the problem of obtaining activity coefficients may be circumvented simply by using apparent equilibrium constants expressed in terms of concentrations. This procedure is recommended for hydrometallurgical systems in which complexation reactions are important, e.g., in ammonia, chloride, or sulfate solutions.

Sometimes the equilibrium reaction for the formation of a species in solution may be written in more than one way. For example, the hydrolysis reactions of ferric ion have this characteristic; the formation of $FeOH^{2+}$ may be written as

$$Fe^{3+} + H_2O \rightleftharpoons FeOH^{2+} + H^+ \tag{33}$$

or

$$Fe^{3+} + OH^- \rightleftharpoons FeOH^{2+} \tag{34}$$

Quite apart from the question of whether it is desirable to take H_2O as a variable, it may be best in a strongly acid solution to work from Equation (33) rather than Equation (34). This is because, in such a solution, the hydroxide ion concentration is orders of magnitude smaller than the concentrations of the other species and the accuracy with which the model equations can be solved becomes an important consideration.

If kinetic data are to be used, it is necessary to transform the variables to conform with those of the partial equilibrium model. The units used in the model equations for $\bar{N}_i$ and $\bar{n}_j$ are moles formed/kg of solution. Thus the mass of solution in the reacting system from which the kinetic data comes must be known. Frequently, one will know the volume and have to approximate the density. A relation between ξ and t is also needed. For this, the mass of solid originally present must be known. The amount of solid reacing, $-\Delta\bar{N}_S$, for a time interval Δt can be obtained from rate curves or calculated from an integrated rate equation. The fraction of the original mass reacting in the time interval gives an approximate value of ξ, e.g.,

$$\frac{-\Delta\bar{N}_S}{\bar{N}_{S,0}} = \xi_{av} \tag{35}$$

where ξ_{av} is interpreted as the extent of reaction averaged over the time interval Δt and $\bar{N}_{S,0}$ is the mass of the solid present before the heterogeneous reaction. It is then possible to plot $\bar{N}_S$ versus ξ_{av} and to relate variables by an algebraic equation. It is this algebraic equation that is used to complete the set of equations of the partial equilibrium model. Once the complete set is solved, the concentration changes can be given as functions of time.

This technique can be generalized if two or more independent heterogeneous reactions occur. In this situation, rate data must be available for each reaction under comparable conditions.

If a single solid reacts irreversibly in separate reactions, $\Delta\bar{N}_S$ is the sum of two or more contributions. Ideally, the kinetic data will include overall reaction rates under conditions when all the irreversible reactions occur simultaneously as well as the individual rates of the different reactions. Then the fraction of the total irreversible process due to each of the reactions can be determined. Like $\Delta\bar{N}_S$, ξ_{av} is a sum of terms. For the same initial mass of solid $\bar{N}_{S,0}$, $\Delta\bar{N}_{S,i}$ and ξ_i are obtained from individual rate curves. Unless each $\Delta\bar{N}_{S,i}$ depends on its ξ_i in the same way, the relative importance of the various heterogeneous reactions changes with the total extent of reaction. In any event, it is necessary to express each $\Delta\bar{N}_{S,i}$ as a function of ξ_{av}.

We now discuss in detail setting up the partial equilibrium model for a particular case. The dissolution of chalcopyrite, $CuFeS_2$, has been studied extensively in the laboratory (3,4,5) and we have been interested in it because of its importance in dump leaching. Under dump leaching conditions, two dissolution reactions have been identified for this mineral (3,4,5):

$$CuFeS_2 + 4Fe^{3+} \rightarrow Cu^{2+} + 5Fe^{2+} + 2S^0 \tag{36}$$

$$CuFeS_2 + O_2 + 4H^+ \rightarrow Cu^{2+} + Fe^{2+} + 2S^0 + 2H_2O \tag{37}$$

The equilibrium constants for both reactions are very large: 6.76×10^{20} for Equation (36) and 8.62×10^{54} for Equation (37) (2). Concentration data from dump leaching operations indicates that neither reaction is close to equilibrium. Hence these are irreversible reactions. The leach solution is a sulfate medium and the solution chemistry involved a number of species. Sulfate and bisulfate ions are in equilibrium, Fe^{3+} undergoes a series of reactions producing hydroxo complexes and Cu^{2+}, Fe^{3+} and Fe^{2+} form a number of complex ions with sulfate and bisulfate ions.

The solid sulfur product need not be chosen as an unknown. Near room temperature, only a small percentage of it is oxidized to soluble sulfur-containing anions(4). It can be assumed, therefore, that none of the sulfur atoms originally present in the solid chalcopyrite enter the solution. The sulfur product is not recovered in the leaching process and does not affect the solution chemistry.

To a first approximation, the concentration and activity of the water molecules do not change during dissolution so $\bar{n}_{H_2O}$ can be neglected.

The unknown parameters of interest are $\bar{N}_{CuFeS_2}$, $\bar{n}_{O_2}$, $\bar{n}_{H^+}$, the various Fe(III), Fe(II) and Cu(II) species, $\bar{n}_{SO_4^{2-}}$ and $\bar{n}_{HSO_4^-}$.

To obtain the initial equilibrium concentrations of the various ions, the solution is taken to contain $Fe_2(SO_4)_3$, $FeSO_4$, H_2SO_4 and a small amount of $CuSO_4$. Leach liquor is recycled after the recovery step so traces of $CuSO_4$ are always present. Analytical concentrations of these substances and the equilibrium constants for each equilibrium reaction must be known. Mass balances for Fe(III), Fe(II), Cu(II) and SO_4^{2-} and a charge balance supplement the mass action equations. This nonlinear set of equations can be solved by the well-known Newton-Raphson method (6).

In writing balance equations for the partial equilibrium model, two quantities are absolutely conserved. These are the total number of sulfate moieties and the net charge in solution. The resulting equations are:

$$\begin{aligned} 0 = {} & \bar{n}_{SO_4^{2-}} + \bar{n}_{HSO_4^-} + \bar{n}_{FeSO_4^+} + 2\bar{n}_{Fe(SO_4)_2^-} \\ & + \bar{n}_{FeHSO_4^{2+}} + \bar{n}_{FeSO_4^0} + \bar{n}_{FeHSO_4^+} + \bar{n}_{CuSO_4^0} \end{aligned} \tag{38}$$

$$0 = \bar{n}_{H^+} - 2\bar{n}_{SO_4{}^{2-}} - \bar{n}_{HSO_4{}^-} + 3\bar{n}_{Fe^{3+}} + 2\bar{n}_{FeOH^{2+}} + \bar{n}_{Fe(OH)_2{}^+} + 4\bar{n}_{Fe_2(OH)_2{}^{4+}} + \bar{n}_{FeSO_2{}^+} - \bar{n}_{Fe(SO_4)_2{}^-} + 2\bar{n}_{FeHSO_4{}^{2+}} + 2\bar{n}_{Fe^{2+}} + \bar{n}_{FeHSO_4{}^+} + 2\bar{n}_{Cu^{2+}} - \bar{n}_{OH^-} \quad (39)$$

Material balances can also be written for Fe(III), Fe(II) and Cu(II) from the chemical equations for the two dissolution reactions. Expressing other quantities in terms of the amount of reactant consumed gives

$$\bar{N}_{CuFeS_2} = 1/4\bar{n}_{Fe(III)} + \bar{n}_{O_2} \quad (40)$$

$$\bar{n}_{Fe(II)} = -5/4\bar{n}_{Fe(III)} - \bar{n}_{O_2} \quad (41)$$

$$\bar{n}_{Cu(II)} = \bar{N}_{CuFeS_2} \quad (42)$$

where

$$\bar{n}_{Fe(III)} = \bar{n}_{Fe^{3+}} + \bar{n}_{FeOH^{2+}} + \dots + \bar{n}_{Fe(SO_4)_2{}^-} + \bar{n}_{FeHSO_4{}^{2+}} \quad (43)$$

$$\bar{n}_{Fe(II)} = \bar{n}_{Fe^{2+}} + \bar{n}_{FeSO_4{}^0} + \bar{n}_{FeHSO_2{}^+} \quad (44)$$

$$\bar{n}_{Cu(II)} = \bar{n}_{Cu^{2+}} + \bar{n}_{CuSO_4{}^0} \quad (45)$$

These five balance equations are all independent.

Among the solution phase species H^+, $SO_4{}^{2-}$, $HSO_4{}^-$, Fe^{3+}, $FeOH^{2+}$, $Fe(OH)_2{}^+$, $Fe_2(OH)_2{}^{4+}$, $FeSO_4{}^+$, $Fe(SO_4)_2{}^-$, $FeHSO_4{}^{2+}$, Fe^{2+}, $FeSO_4{}^0$, $FeHSO_4{}^{2+}$, Cu^{2+}, and $CuSO_4{}^0$, independent mass action expressions can be written for the formation of $HSO_4{}^-$, $FeOH^{2+}$, $Fe(OH)_2{}^+$, $Fe_2(OH)_2{}^{4+}$, $FeSO_4{}^+$, $Fe(SO_4)_2{}^-$, $FeHSO_4{}^{2+}$, $FeSO_4{}^0$, $FeHSO_4{}^+$, and $CuSO_4{}^0$, giving an additonal 10 equations.

There are a total of 15 equations and 17 unknowns.

Each unknown $\bar{n}_i$ can be divided by $-\bar{N}_{CuFeS_2}$ to reduce the number of unknowns by one but clearly the use of kinetic information cannot be avoided.

The only kinetic data that permits direct comparison of the rates of Reactions (36) and (37) is that of Baur, Gibbs and Wadsworth (3). After a very brief initial period of rapid reaction, Reactions (36) and (37) are followed. Extrapolation of the data of Baur, Gibbs and Wadsworth to long times and transformation of variables indicate that after the initial rapid reaction, chalcopyrite and oxygen react in a constant ratio, i.e., the fraction of the total copper dissolved due to O_2 is a constant. A constant value can then be assigned for n_{O_2}; fortuitously, the completed set of algebraic partial equilibrium equations is linear in this case.

Summary and Conclusion

We have solved the equations of the partial equilibrium model for a number of different initial analytical concentrations and choices of n_{O_2}. The results of the greatest immediate practical importance concern the effects of the initial analytical $Fe_2(SO_4)_3$ concentration and the fraction of copper dissolved by O_2 and H^+. It was assumed that the dissolved oxygen concentration did not limit the reaction, i.e., oxygen entered the solution as fast as it was used up in the reaction so that a constant dissolved oxygen level was maintained. The model equations were solved for successive reaction increments until either the total Fe(III) concentration had dropped to 10^{-4} m or the solubility product for precipitation of amorphous $Fe(OH)_3$ was exceeded. Precipitation of Fe^{3+} is a major problem in dump leaching operations and should be prevented if possible. Typical results are shown in Figure 1. At the same value of $|n_{O_2}|$, the final copper recovery is increased by increasing the initial $Fe_2(SO_4)_3$ concentration. For a given $Fe_2(SO_4)_3$ concentration, increasing the fraction dissolved by O_2 and H^+ causes an increase in the final Cu(II) concentration but the curve eventually drops off sharply as $Fe(OH)_3$ precipitates. Moreover, the decline in copper recovery occurs at smaller values of $|n_{O_2}|$ for higher ferric sulfate concentrations.

The complete results of the calculations applying the partial equilibrium model to the ambient temperature dissolution of chalcopyrite by ferric ion and by oxygen and acid are being reported in another publication (7). The concentration changes occurring during the chalcopyrite dissolution in the dump can be predicted by the model. This in effect indicates how the influent leaching solution concentration should be changed in order to maximize copper dissolution and prevent the precipitation of trivalent iron. This can best be attained by maintaining a high Fe(III) concentration and by high acid concentrations. A very low concentration of dissolved oxygen is

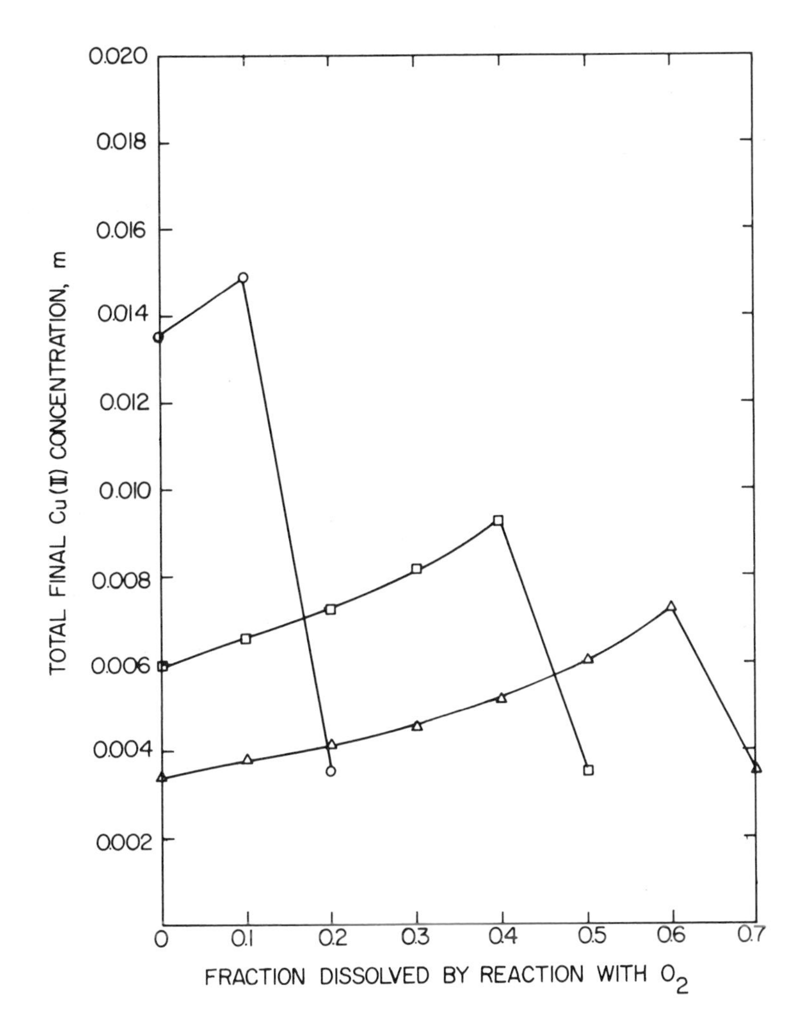

Figure 1. Final Cu(II) concentration as a function of the fraction of chalcopyrite dissolved by O_2 and H^+ for various initial $Fe_2(SO_4)_3$ concentrations: $C_{Fe_2(SO_4)_3}$: (○) *0.025*m; (□) *0.010*m; (△) *0.005*m; C_{FeSO_4}—*0.2*m; C_{CuSO_4}—*0.001*m; $C_{H_2SO_4}$*0.01*m

indicated since high levels of dissolved oxygen have a deleterious effect on the acid concentrations. This in turn affects the precipitation of ferric hydroxide.

Abstract

In many solid-liquid systems of practical interest, the heterogeneous reactions do not reach equilibrium. Reactions among dissolved species, however, generally are faster. Hence it is often possible to regard the heterogeneous reactions in such a system as producing changes in the concentrations of some of the liquid phase species, thereby causing the equilibria in the solution to shift. Classifying the reactions in such a system as reversible or irreversible makes possible the development of a mathematical model of the changes in solution concentration accompanying a heterogeneous process. The derivation and use of the model equations are discussed. The information that would be needed to calculate concentrations in an equilibruim liquid phase system must be supplemented by kinetic information in order to make the heterogeneous problem determinate. The information needed includes: (1) the stoichiometry of each reaction, reversible or irreversible, taking place; (2) analytical concentrations in the solution phase at the start of the heterogeneous reactions; (3) equilibrium constants for the reversible reactions, which may be either true thermodynamic or apparent constants; and (4) kinetic data in the heterogeneous reactions. A set of equations is obtained that includes a linear equation for each reversible liquid phase reaction, material balances and, in the case of an aqueous liquid phase, a charge balance; following a change of variable, kinetic equations complete the set. The equations are algebraic in form and in favorable cases may all be linear. This type of model may readily be applied to various hydrometallurgical systems. Mineral leaching is discussed in detail.

Terms and Solutions

a_j activity of solute species

K equilibrium constant

m_j molality of solution phase species s_j; moles/kg solution

n_j number of moles of solution phase species s_j formed per mole of solid reacting

$\bar{n}_j$ number of moles of solution phase species s_j formed per kg of solution

$\bar{N}_i$ number of moles of solid S_i formed per kg of solution

γ_j activity coefficient of solution phase species s_j

ν_1 stoichiometric coefficient of species 1 in equilibrium constant

ξ progress variable for extent of reaction; fraction reacted

Acknowledgment

This research was supported in part through the financial support of Monsanto, Union Carbide and Phillips Petroleum Corporations in the form of graduate fellowships, the National Science Foundation Energy Traineeship, and Graduate Assistantship from the Engineering Research Institute and the Ames Laboratory, USDOE to one of the authors (K.C.L.). This work was also supported by the U.S. Department of Energy, contract No. W-7405-Eng-82, Division of Basic Energy Sciences, Chemical Sciences Program.

Literature Cited

1 Helgeson, H. C., Geochim. Cesmochim. Acta, 1968, 32, 853.
2 Liddell, K. C., A Mathematical Model of the Chemistry of the Dump Leaching of Chalcopyrite, 1979, Ph.D. Thesis, Iowa State University, Ames, Iowa.
3 Baur, J. P.; Gibbs, H. L.; Wadsworth, M. E., Initial-Stage Sulfuric Acid Leaching Kinetics of Chalcopyrite Using Radiochemical Techniques, 1974, United States Bureau of Mines Report of Investigations 7823, Washington, D.C.
4 Dutrizac, J. E., Met. Trans. B, 1978, 9B, 431.
5 Munoz, P. B.; Miller, J. D.; Wadsworth, M. E., Met. Trans. B, 1979, 10B, 149.
6 Carnahan, B.; Luther, H. A.; Wilkes, J. O., "Applied Numerical Methods", 1969, John Wiley and Sons, New York, NY.
7 Liddell, K. C.; Bautista, R. G., Met. Trans. B, 1980.

RECEIVED January 30, 1980.

INDEX

B

C

D

F

K

L

M

N